WE WANT TO HEAR FROM YOU!!

By sharing your opinions about this book, you will help us ensure that you are getting the most value for your textbook dollars. After you've used the book for awhile, please fill out this form, fold, tape and drop it in the mail.

Course Title:_____ Text Title & Author:_____

1. Are you a major in this subject? ❑ yes ❑ no ❑ undecided
   Were you required to purchase this text? ❑ yes ❑ no

2. Did you purchase this book: ❑ for yourself? ❑ for yourself and at least one other student?
   Was a copy available when you needed one? ❑ yes ❑ no

3. Was a study guide available for purchase? ❑ yes ❑ no ❑ don't know
   If yes, did you purchase it? ❑ yes ❑ no   Might you purchase it in the future? ❑ yes ❑ no

4. Were any other supplements to the text available (for example, software, a workbook, etc.)? ❑ yes ❑ no
   If yes, what? _____
   Did you purchase any other supplement? ❑ yes ❑ no

5. How far along in the course are you? ❑ only starting ❑ less than midway ❑ more than midway ❑ completed

6. How much have you used this text? ❑ only skimmed it ❑ read/studied a few chapters
   ❑ read/studied most chapters ❑ read/studied entire text

7. Have you read the introductory material (such as the preface)? ❑ yes ❑ no
   Do you feel you know how to effectively use this book? ❑ yes ❑ no

8. Even if you've only skimmed the text, please rate your perception of it in terms of the following:

a) Value as a reference	❑ highly valuable	❑ somewhat valuable	❑ not valuable
b) Readability	❑ consistently clear	❑ sometimes unclear	❑ generally unclear
c) Illustrations/photos	❑ very effective	❑ somewhat effective	❑ ineffective
d) Design/use of color	❑ very effective	❑ somewhat effective	❑ ineffective
e) Study help in the text	❑ very effective	❑ somewhat effective	❑ ineffective
f) Level	❑ too difficult	❑ appropriate	❑ too easy /not challenging
g) Problems	❑ too difficult	❑ appropriate	❑ too easy / not challenging
h) OVERALL PERCEPTION:	❑ better than average	❑ average	❑ less than average

9. Do you find the examples in the text relevant to you? ❑ yes ❑ no
   Note any that you find particularly relevant: _____

10. By looking at the text, do you think it treats the subject as interestingly as possible? ❑ yes ❑ no ❑ hard to tell

11. What do you like most about this book? _____
    What *don't* you like about this book?_____

12. At the end of the semester, what do you intend to do with this text?
    ❑ keep for future reference ❑ sell back to bookstore or other students ❑ unsure

(continued on back)

Name _____ School _____

May we quote you?   ❏ Yes   ❏ No

Student Comments _____

_____

_____

_____

_____

_____

_____

_____

Tear card at perforation, fold in half, tape and mail.

 **WILEY**
Publishers Since 1807

THANK YOU FOR YOUR HELP!

Geography

REALMS, REGIONS, AND CONCEPTS

Eighth Edition

SILVER ANNIVERSARY

Geography

REALMS, REGIONS, AND CONCEPTS

Eighth Edition

H. J. de Blij
University of South Florida, St. Petersburg

Peter O. Muller
University of Miami

Chapter Opener Maps which appear in this text are from *Goode's World Atlas*, © Rand McNally, R.L. 93-S-115, and are used with permission.

JOHN WILEY & SONS, INC.

New York ■ Chichester ■ Brisbane ■ Toronto ■ Singapore

EXECUTIVE EDITOR: Christopher Rogers
ACQUISITIONS EDITOR: Nanette Kauffman
MARKETING MANAGER: Catherine Faduska
DEVELOPMENTAL DIRECTOR: Johnna Barto
DESIGN DIRECTOR: Madelyn Lesure/Ann Marie Renzi
PRODUCTION EDITOR: Jeanine Furino
PHOTO EDITOR: Mary Ann Price
ILLUSTRATION EDITOR: Edward Starr
ASSISTANT MANUFACTURING MANAGER: Mark Cirillo
SENIOR PHOTO EDITOR: Mary Ann Price
PHOTO ASSISTANT: Kim Khatchatourian
PHOTO EDITOR: Alexandra Truitt
PRODUCTION MANAGER: Linda Muriello
TEXT DESIGN: Nancy Field
COVER DESIGN: Edward A. Butler
FRONT COVER GLOBE MAP OF THE REALMS: Deasy GeoGraphics Laboratory, The Pennsylvania State University. Ocean data copyright © 1995 The Living Earth, Inc.
BACK COVER PHOTO: Grant V. Faint/The Image Bank

This book was set in Times Roman by University Graphics and printed and bound by Von Hoffman Press. The cover was printed by Phoenix Color.

Library of Congress Cataloging in Publication Data:

de Blij, H.J.
    Geography : realms, regions, and concepts / H.J. de Blij, Peter O.
Muller.—8th ed.
        p.   cm.
    At head of title: Silver anniversary.
    Includes bibliographical references.
    ISBN 0-471-11946-6 (cloth : alk. paper)
    1. Geography.   I. Muller, Peter O.   II. Title.
    G128.D42   1997   96-18955
    910—dc20   CIP

Printed in the United States of America

10 9 8 7 6 5 4 3 2 1

To the memory of
Harry J. Schaleman, Jr.
(1928–1995)

Professor of Geography
University of South Florida

Cherished friend, generous colleague,
and global ambassador for our discipline.

PREFACE

Over more than a quarter of a century, *Geography: Realms, Regions, and Concepts* has reported (and sometimes anticipated) trends in the discipline of Geography and developments in the world at large. In seven preceding editions, *Regions*, as the book is generally called, has explained the modern world's great geographic realms and their physical and human contents, and has introduced geography itself, the discipline that links the study of human societies and natural environments through a fascinating, spatial approach. From old ideas to new, from environmental determinism to expansion diffusion, and from historic developments to current ones, from decolonization to devolution, *Regions* has provided geographic perspective on our transforming world.

The book before you, therefore, is an information highway to geographic literacy. The first edition appeared in 1971, at a time when school geography in the United States (not in Canada) was a subject in decline. It was a precursor of a dangerous isolationism in America, and geographers foresaw the looming cost of geographic illiteracy. Sure enough, the media during the 1980s began to report that polls, public surveys, tests, and other instruments were recording a lack of geographic knowledge at a time when our world was changing ever faster and becoming more competitive by the day. Various institutions, including the National Geographic Society, banks, airline companies, and a consortium of scholarly organizations mobilized to combat an educational dilemma that had resulted substantially from a neglect of the very topics this book is about.

Before we can usefully discuss such commonplace topics as our ''shrinking world,'' our ''global village,'' and our ''distant linkages,'' we should know what the parts are, the components that do the shrinking and linking. This is not just an academic exercise. You will find that much of what you encounter in this book is of immediate, practical value to you—as a citizen, a consumer, a traveller, a voter, a jobseeker. North America is a geographic realm with global interests and involvements. Those interests and involvements require countless, often instantaneous decisions. Such decisions must be based on the best possible knowledge of the world beyond our continent. That knowledge can be gained by studying the layout of our world, its environments, societies, resources, policies, traditions, and other properties—in short, its regional geography.

Realms and Concepts

This book is organized into thirteen chapters. The Introduction discusses the world as a whole, outlining the physical stage on which the human drama is being played out, providing environmental information, demographic data, political background, and economic indices. The remaining twelve chapters all focus on one of the world's major geographic realms.

Geographic concepts and ideas are placed in their regional settings in all 13 chapters. Most of these approximately 150 concepts are primarily geographical, but others are ideas about which, we believe, students of geography should have some knowledge. Although such concepts are listed on the opening page of every chapter, we have not, of course, enumerated every geographic notion used in that chapter. Many colleagues, we suspect, will want to make their own realm-concept associations, and as readers will readily perceive, the book's organization is quite flexible. It is possible, for example, to focus almost exclusively on substantive regional material, or, alternately, to concentrate mainly on conceptual issues.

The Eighth Edition

Ever since this book was conceived, it has been a challenge to keep abreast of the rapid changes affecting the world.

The decolonization of Africa and other areas, the growing economic power of the Pacific Rim, the devolution of the Soviet Union, the transition in South Africa, the collapse of Yugoslavia—these are just a few of the more dramatic changes that have occurred over the period of the life of this book. The unification effort in Europe, the resurrection of Christian churches in Russia and of Islamic forces in Central Asia, the pandemic of AIDS, and the gradual weakening of many nation-states are among other processes transforming the world's geography.

The Eighth Edition incorporates the current stages of long-range developments as well as the impacts of recent events. Throughout the first seven editions of *Regions*, geographic realms were classified according to their economic geographies into developed and underdeveloped groups. The justification for this dichotomy, however, weakened over time. Certain realms not only contained ''developed'' as well as ''underdeveloped'' countries; stages of development *within* individual countries varied so greatly that generalizing about them became impractical. We therefore abandoned this particular concept and shifted our focus to the enormous regional and social disparities that are arising in countries where some provinces are developing rapidly while others stagnate. True, some countries remain almost entirely devoid of modernizing development, as we note in the Introduction. But not whole realms—any more.

Another world-transforming process is urbanization. Many of our readers will have noted a small news report in the media recently, to the effect that the upward curve of urbanizing humanity is about to pass the 50 percent mark. For the first time in human history, a majority of people live in cities and towns. We take note of this momentous milestone by introducing a new feature to this edition of *Regions*: ''Among the Realm's Great Cities''— 34 brief commentaries on the geographies of many of the world's leading cities, accompanied by newly drawn maps. From burgeoning Mexico City to mushrooming Shanghai the engines of change, and the crucibles of transculturation, are the great cities.

Regional transformations sometimes require new nomenclature. In the Sixth Edition of *Regions*, we reintroduced the name Turkestan to denote the region of Muslim-influenced countries in Central Asia, newly independent following the collapse of the Soviet Union. In the Seventh Edition, we acknowledged the emergence of a functional region extending from East Asia to eastern Australia, and called it Australasia. In the Eighth Edition, we take note of the diminishing contrasts and increasing similarities among three Pacific Rim economic powerhouses—Japan, South Korea, and Taiwan—and identify a region of the East Asian realm we call the Jakota Triangle. Among other new regional names to be found in this new edition are Near Abroad (Russia), Republic of the Pampas (Brazil), and Gauteng (South Africa).

The elimination of the developed-underdeveloped duality allowed us to restructure *Regions* more rationally in the context of current world developments. We have repositioned Australia to follow Southeast (and East) Asia, confirming Australia's ongoing reorientation toward the Pacific Rim and away from its Anglo-European heritage. We have expanded our coverage of the Pacific Realm because of the significance of developments there.

Several chapters have been substantially restructured and revised; all have been shortened. The changing map of unifying Europe receives due attention, but the regional discussion of European states has been condensed by more than one-third. The evolving Russian Federation, its role in the Transcaucasian Transition Zone, and its problems in the Near Abroad also are the subject of revision. The chapter on North America gives greater emphasis to Canada, particularly the devolutionary forces centered in Quebec. New cartography and text marks the revision of the Middle America chapter, especially concerning Mexico as well as the Central American republics. Recognizing the growing economic integration of politically democratizing South America, we have recast that chapter to focus more effectively on giant Brazil and its social, economic, and devolutionary problems, on intra-realm linkages, and on Pacific Rim developments in Chile. The complex chapter on North Africa/Southwest Asia contains new cartography on fossil fuels and Israel and the West Bank. The chapter on Subsaharan Africa reintegrates South Africa into its regional framework and presents that state's new political geography. In South Asia, a new focus is on important developments occurring in India's Maharashtra State, where a new industrial era has arrived but not without serious sociopolitical consequences. As noted earlier, a new conceptualization of the East Asian realm incorporates the Jakota Triangle. In the Southeast Asia chapter, greater attention is given to Pacific Rim-related developments, notably in Thailand, Malaysia, and Vietnam. Brief but tightly-focused chapters on the Australian and Pacific realms also benefit from significant revision.

Data Sources

As with every new edition of this book, all quantitative information was updated to the year of publication and checked rigorously. In addition to the major revisions described above, hundreds of other modifications were made, many in response to readers' and reviewers' comments. Some readers found our habit of reporting urban population data within the text disruptive, so we now tabulate these at the beginning of the ''Regions of the Realm'' section of each chapter. The stream of new spellings of geographic

[1]Discussion of Pacific Rim areas is integrated into the appropriate realms but can be easily located on the pages with red bars in the outside margins.

names continues, and we pride ourselves on being a reliable source for current and correct usage (the government of the former Soviet Republic of Kazakhstan, for example, insists on its preference for Kazakstan, now the internationally approved form). While we have made major structural changes to this edition, we have not abandoned our attention to geographic detail. The population figures used in the text are our projections for 1997 (unless otherwise indicated) and are consistent with the national demographic data displayed in Appendix A. The chief source that we used as a basis for developing our projections was the 1995 *World Population Data Sheet* published by the Population Reference Bureau, Inc. The urban population figures—which entail a far greater problem in reliability and comparability—are mainly drawn from the most recent (1995) database published by the United Nations' Population Division. For cities of less than 750,000, we developed our own estimates from a variety of other sources. At any rate, the urban population figures used here are estimates for 1997 and they represent *metropolitan-area totals* unless otherwise specified.

Cartography

This Eighth Edition continues the innovation of the Seventh, when atlas-style maps from the most recently available edition of *Goode's World Atlas* were used as opening maps for each chapter. In the current edition, two maps were drawn in the Rand McNally style to serve as special openers: the map of North Africa/Southwest Asia and the map of the Pacific Realm; the South Asia map was substantially expanded from its *Goode's* source.

Readers of earlier editions of *Regions* will note that our tradition of updating, enhancing, and improving our own thematic cartography continues. Some 150 new or exhaustively revised maps appear in this edition, including those on Italy (regions and populations), Mexico (both the internal States and the maquiladora zone), South Africa (old and new provinces), the Jakota Triangle (emerging region on the Pacific Rim), and China (ethnolinguistic areas).

Users of this book should note that the spelling of some names on these thematic maps does not always match that on the *Goode's World Atlas* maps. This is not unusual; you will even find inconsistencies among various atlases. Almost invariably, we have followed the very latest standards set by the United States Board of Geographic Names.

Photography

The map undoubtedly is geography's closest ally, but there are times when photography is not far behind. Whether from space, from an aircraft, from the tallest building in town, or on the ground, a photograph can, indeed, be worth a thousand words. When geographers perform field re-

search in some area of the world, they are likely to maintain a written record that correlates with the photographic one. Sometimes the combination produces a useful illustration, and in this book many of the authors' photographs, taken in the field, are reproduced with accompanying "field notes." Other photographs, selected to illustrate particular points in the text, have routine captions.

Pedagogy

Over all its editions, this book has been structured to help students learn important geographic concepts as well as appreciate the constantly changing world.

Continuing Special Features

Atlas Map   As in the Seventh Edition, a comprehensive map of the region opens each chapter. The maps are reproduced from the 19th revised edition (1994) of *Goode's World Atlas* (the maps for Chapters 6 and 12 have been created in the Atlas style).

Key Ideas & Concepts   Each chapter begins with a chronological list of the boldfaced key geographic ideas and concepts discussed in the chapter and serves as a review aid for students.

List of Regions   Also on the chapter opening page, a list of the regions within the particular realm provides a preview and helps to organize the chapter. For ease of identification, the diamond symbol that denotes the regions list here also appears beside each region heading in the chapter.

Major Geographic Qualities   Near the beginning of each realm chapter, we list, in boxed format, the major geographic qualities that best summarize that portion of the Earth's surface.

Focus on a Systematic Field   Also located near the opening of each of the twelve realm chapters is a Focus on a Systematic Field essay that covers a major topical subfield of human or physical geography. Each of these overviews was carefully selected so that its contents tie in to regional-geographic material subsequently developed in the chapter. (The entire program is depicted in Figure I-13.)

Sidebar (Blue) Boxes   Special topical and sometimes controversial issues are highlighted in blue boxed sections. These boxes allow us to include interesting and current topics without interrupting the flow of material within chapters.

Pronunciation Guide   Our readers have lauded the inclusion of Pronunciation Guides at the end of each chapter as a convenient reference. In choosing words for inclusion (largely place names), we decided not to list words that were pronounced the way they were written unless we thought mispronunciation was likely. Although we strive for authenticity throughout, we aimed for Americanized rather than native-language pronunciations. For many place names, our initial guide was the current edition of *Webster's New Geographical Dictionary*. In choosing the phonetic presentation method, we kept things as simple as possible by avoiding a formalized symbol system that would have required constant decoding. Accordingly, we employed a syllabic phonetic-spelling system with stress syllables capitalized. (For example, we pronounce our surnames duh-BLAY and MULL-uh.) The most frequently used vowel sounds would translate as follows: *ah* as in father, *oh* as in tone, *au* as in out, and *uh* as in banana.

Appendices and References   At the end of the book, we have included four appendices: (A) Population Data on the World's Countries, (B) a guide to Map Reading and Interpretation, (C) an overview of Career Opportunities in Geography, and (D) Video Links. In addition, lengthy lists of *References and Further Readings* are provided for each chapter. Those lists are followed by the Gazetteer, our *Geographical Index* of map names, and by the main index.

New Special Features

Two-part Chapter Organization   To help the reader to logically organize the material within chapters, we have broken most of the regional chapters, into two distinct parts: first, ''Defining the Realm'' includes the general physiographic, historical, and human-geographic background common to the realm, and the second section, ''Regions of the Realm,'' presents each of the distinctive regions within the realm (denoted by ❖ symbol).

Among the Realm's Great Cities   We have added this new feature which reflects the growing process and influence of urbanization worldwide. More than thirty brief profiles of the world's leading cities are presented, each accompanied by newly drawn maps.

Urban Population Tables   Early in the ''Regions'' section of each chapter, we have included a table reporting the most up-to-date urban population data (based on 1997 estimates drawn from the sources listed above). Readers should find this format less disruptive than citing the population when the city is mentioned in the text.

From the Field Notes   In the Eighth Edition we have introduced a new feature that has proven effective in some of our other textbooks. Many of the photographs in this book were taken by the senior author while doing field-

work. The extensive captions, *From the Field Notes*, provide valuable insights into how a geographer observes and interprets information in the field.

Video Link Icons   One of the most distinctive ancillaries available with this book are the video series. These videos are from different sources: *Geography On Location Video Series with H.J. de Blij* (John Wiley & Sons) and *Good Morning America* (ABC). A video icon and number appear in the margins beside the text discussions of pertinent locations; the specific videos and sources are included in a number list at the end of the book for easy reference in Appendix D.

Ancillaries

A broad spectrum of print and electronic ancillaries are available to accompany this edition of *Regions*. Additional information regarding these ancillaries can be obtained by contacting John Wiley & Sons.

- **Student Study Guide.** Prepared by text co-author Peter Muller and Elizabeth Muller Hames, University of Miami, this top-selling Study Guide contains learning objectives, self-test questions, practice exams, term paper pointers, 48 outline maps, and map exercises, all specifically designed to help students get the most out of their world regional geography course.
- **Instructor's Manual.** Prepared by Laurie Molina, Florida State University, this effective teaching resource provides outlines, descriptions, and key terms to help professors bring the concepts presented in *Regions* to life in the classroom. This new edition of the Instructor's Manual also connects each chapter to the wide variety of media products available to complement the text.
- **Test Bank.** Prepared by longstanding test bank author Ira Sheskin, University of Miami, this expanded test bank contains over 1,500 test items including multiple choice, fill-in, matching, and essay questions, all coordinated with the new edition of *Regions*.
- **Computerized Test Bank.** The test bank is available in computerized form for either IBM or Macintosh.
- **Overhead Transparencies.** 100 full-color maps are available as overhead transparencies.
- **Slide Set.** All 100 full-color maps available as transparencies are also available as full-color 35 mm slides.
- **Geography On Location Video Series with H.J. de Blij.** These three videotapes, comprised of 31 five-minute segments narrated by H.J. de Blij, contain exclusive, up-to-date footage of geographically significant locations throughout the world. The videotapes

not only delve into the physical, economic, historical, social, and political issues relevant to these places, but also provide insights from a uniquely geographic point of view. To find out more about the Geography On Location Videotapes, contact your Wiley representative.

- **Good Morning America Videotapes.** These seven videotapes, sold separately or as a set, are a collection of H.J. de Blij's appearances as Geography Editor on this popular ABC morning news show. Each tape contains 10–15 segments discussing geographically significant issues and world events. To find out more about the Good Morning America Videotapes, contact your Wiley representative.

- **Geography CD-ROM.** This innovative CD-ROM provides a rich, multi-media environment which allows for flexible use of an extensive bank of unique videos, animations, photos, and maps, including thematic map overlays. The videos included on the CD-ROM have been selected from the Geography On Location Videotapes and specifically tailored to the CD-ROM. The program features an intuitive, easy-to-learn interface, and its case approach focuses on four fascinating regions plus an introductory chapter on regional geography. For more information regarding the CD-ROM, contact John Wiley & Sons.

- *Regions* Eighth Edition is also featured as the companion text to **The Power of Place—World Regional Geography**, a PBS television course and video resource produced in collaboration with the Annenberg/CPB Project. *The Power of Place* is a series of twenty-six half-hour video programs organized around the *Regions* text. Each program contains documentary-style case studies that focus on one of eleven geographic realms. Videocassettes can be purchased individually or as a thirteen-tape set. A *Study Guide* and *Faculty Guide* are also available to supplement the programs. For information regarding using *The Power of Place* as a television course, contact the PBS Adult Learning Service at 1-800-257-2578. To purchase videocassettes for institutional or classroom use, contact The Annenberg/CPB Multimedia Collection at 1-800-LEARNER.

Acknowledgments

In the twenty-five years since the publication of the First Edition of *Geography: Realms, Regions, and Concepts*, we have been fortunate to receive advice and assistance from literally hundreds of people. One of the rewards associated with the publication of a book of this kind is the steady stream of correspondence and other feedback it generates. Geographers, economists, political scientists, education specialists, and others have written us, often with fascinat-

ing enclosures. We make it a point to respond personally to every such letter, and our editors have communicated with many of our correspondents as well. Moreover, we have considered every suggestion made—and many who wrote or transmitted their reactions through other channels will see their recommendations in print in this edition.

Student Response

A good part of the correspondence we receive comes from student readers. On this occasion, we would like to extend our deep appreciation to the more than 600,000 students around the world who have studied from the first seven editions of our text. In particular, we thank the students from 150 different colleges across the country who took the time to send us their opinions of the Seventh Edition. A special recognition goes to the 26 students in Professor David Kromm's Winter 1994–1995 class at Kansas State University who shared with us their honest evaluations of the textbook. In planning the Eighth Edition, we have made a special effort to respond to what these students have told us.

- Students told us they found the maps and graphics attractive and functional. *We have enhanced the map program with exhaustive updating and have added 34 new maps of the world's major cities.*

- Several students also commented that the chapters were long and sometimes wordy. *We have divided the chapters into two parts: the first part covers the general physiography, history, and human geography; the second part is devoted to the various regions within the realm. To add interest for today's student, the authors have added a new feature highlighting the Great Cities of each realm from the point of view of a contemporary visitor. We have also enlivened the layout with a fresh new design.*

- Generally students have told us that they found the pedagogical devices quite useful. *We have kept the study aids the students cited as effective: lists of key ideas and concepts, major geographic qualities, and pronunciation guides.*

Faculty Feedback

Faculty members from a large number of North American colleges and universities continue to supply us with vital feedback and much-appreciated advice. Our publishers commissioned a number of reviews, and we are most grateful to the following professors for showing us where the written text could be strengthened and made more precise:

H. Gardiner Barnum, University of Vermont

Olwyn and Brian Blouet, College of William and Mary (Virginia)

Douglas E. Heath, Northampton Community College (Pennsylvania)

Mohammed S. Kamiar, Florida Community College of Jacksonville

Philip L. Keating, University of Miami (Florida)

David Kromm, Kansas State University

Charles Lieble, Valdosta State University (Georgia)

Ian MacLachlan, University of Lethbridge (Alberta)

Tom L. Martinson, Auburn University (Alabama)

Curtis Thomson, University of Idaho

Helen J. Turner, University of Memphis (Tenn.)

In addition, several faculty colleagues from around the world assisted us with earlier editions, and their contributions continue to grace the pages of this book. Among them are:

James P. Allen, California State University, Northridge

Stephen S. Birdsall, University of North Carolina

J. Douglas Eyre, University of North Carolina

Fang Yong-ming, Shanghai, China

Edward J. Fernald, Florida State University

Ray Henkel, Arizona State University

Gil Latz, Portland State University (Oregon)

Melinda S. Meade, University of North Carolina

Henry N. Michael, Temple University (Pennsylvania)

Clifton W. Pannell, University of Georgia

J. R. Victor Prescott, University of Melbourne (Victoria)

Canute Vander Meer, University of Vermont

We also received input from a much wider circle of academic geographers. The list that follows is merely representative of a group of colleagues across North America to whom we are grateful for taking the time to share their thoughts and opinions with us:

Mel Aamodt, California State University-Stanislaus

R. Gabrys Alexson, University of Wisconsin-Superior

Nigel Allan, University of California-Davis

John L. Allen, University of Connecticut

Jerry R. Aschermann, Missouri Western State College

Joseph M. Ashley, Montana State University

Theodore P. Aufdemberge, Concordia College (Michigan)

Edward Babin, University of South Carolina-Spartanburg

Marvin W. Baker, University of Oklahoma

Thomas F. Baucom, Jacksonville State University (Alabama)

Gouri Banerjee, Boston University (Massachusetts)

J. Henry Barton, Thiel College (Pennsylvania)

Steven Bass, Paradise Valley Community College (Arizona)

Klaus J. Bayr, University of New Hampshire-Manchester

James Bell, Linn Benton Community College (Oregon)

William H. Berentsen, University of Connecticut

Riva Berleant-Schiller, University of Connecticut

Thomas Bitner, University of Wisconsin

Warren Bland, California State University-Northridge

Davis Blevins, Huntington College (Alabama)

S. Bo Jung, Bellevue College (Nebraska)

Martha Bonte, Clinton Community College (Idaho)

George R. Botjer, University of Tampa (Florida)

Kathleen Braden, Seattle Pacific University (Washington)

R. Lynn Bradley, Belleville Area College (Illinois)

Ken Brehob, Elmhurst, Illinois

James A. Brey, University of Wisconsin-Fox Valley

Robert Brinson, Santa Fe Community College (Florida)

Reuben H. Brooks, Tennessee State University

Larry Brown, Ohio State University

Lawrence A. Brown, Troy State-Dothan (Alabama)

Robert N. Brown, Delta State University (Mississippi)

Randall L. Buchman, Defiance College (Ohio)

DiAnn Casteel, Tusculum College (Tennessee)

John E. Coffman, University of Houston (Texas)

Dwayne Cole, Grand Rapids Baptist College (Michigan)

Barbara Connelly, Westchester Community College (New York)

Willis M. Conover, University of Scranton (Pennsylvania)

Omar Conrad, Maple Woods Community College (Missouri)

Barbara Cragg, Aquinas College (Michigan)

Georges G. Cravins, University of Wisconsin

John A. Cross, University of Wisconsin-Oshkosh

William Curran, South Suburban (Illinois)

Armando Da Silva, Towson State University (Maryland)

David D. Daniels, Central Missouri State University

Rudolph L. Daniels, Morningside College (Iowa)

Satish K. Davgun, Bemidji State University (Minnesota)

James Davis, Illinois College

James L. Davis, Western Kentucky University

Keith Debbage, University of North Carolina-Greensboro

Dennis K. Dedrick, Georgetown College (Kentucky)

Stanford Demars, Rhode Island College

Thomas Dimicelli, William Paterson College (New Jersey)

D.F. Doeppers, University of Wisconsin

Ann Doolen, Lincoln College (Illinois)

Steven Driever, University of Missouri-Kansas City

Keith A. Ducote, Cabrillo Community College (California)

Walter N. Duffet, University of Arizona

Christina Dunphy, Champlain College (Vermont)

Anthony Dzik, Shawnee State University (Kansas)

Dennis Edgell, Firelands BGSU (Ohio)

James H. Edmonson, Union University (Tennessee)

M.H. Edney, State University of New York-Binghamton

Harold M. Elliott, Weber State University (Utah)

James Elsnes, Western State College

Dino Fiabane, Community College of Philadelphia (Pennsylvania)

G.A. Finchum, Milligan College (Tennessee)

Ira Fogel, Foothill College (California)

Robert G. Foote, Wayne State College (Nebraska)

G.S. Freedom, McNeese State University (Louisiana)

Owen Furuseth, University of North Carolina-Charlotte

Richard Fusch, Ohio Wesleyan

Gary Gaile, University of Colorado-Boulder

Evelyn Gallegos, Eastern Michigan University & Schoolcraft College

Jerry Gerlach, Winona State University (Minnesota)

Lorne E. Glaim, Pacific Union College (California)

Sharleen Gonzalez, Baker College (Michigan)

Daniel B. Good, Georgia Southern University

Gary C. Goodwin, Suffolk Community College (New York)

S. Gopal, Boston University (Massachusetts)

Robert Gould, Morehead State University (Kentucky)

Gordon Grant, Texas A&M University

Donald Green, Baylor University (Texas)

Gary M. Green, University of North Alabama

Mark Greer, Laramie County Community College (Wyoming)

Stanley C. Green, Laredo State University (Texas)

W. Gregory Hager, Northwestern Connecticut Community College

Ruth F. Hale, University of Wisconsin-River Falls

John W. Hall, Louisiana State University-Shreveport

Peter L. Halvorson, University of Connecticut

Mervin Hanson, Willmar Community College (Minnesota)

Robert J. Hartig, Fort Valley State College (Georgia)

James G. Heidt, University of Wisconsin Center-Sheboygan

Catherine Helgeland, University of Wisconsin-Manitowoc

Norma Hendrix, East Arkansas Community College

James Hertzler, Goshen College (Indiana)

John Hickey, Inver Hills Community College (Minnesota)

Thomas Higgins, San Jacinto College (Texas)

Eugene Hill, Westminster College (Missouri)

Louise Hill, University of South Carolina-Spartanburg

Miriam Helen Hill, Indiana University Southeast

Suzy Hill, University of South Carolina-Spartanburg

Robert Hilt, Pittsburg State University (Kansas)

Priscilla Holland, University of North Alabama

Robert K. Holz, University of Texas-Austin

R. Hostetler, Fresno City College (California)

Lloyd E. Hudman, Brigham Young University (Utah)

Janis W. Humble, University of Kentucky

William Imperatore, Appalachian State University (North Carolina)

Richard Jackson, Brigham Young University (Utah)

Mary Jacob, Mount Holyoke College (Massachusetts)

Gregory Jeane, Samford University (Alabama)

Scott Jeffrey, Catonville Community College (Maryland)

Jerzy Jemiolo, Ball State University (Indiana)

Sharon Johnson, Marymount College (New York)

David Johnson, University of Southwestern Louisiana

Jeffrey Jones, University of Kentucky

Marcus E. Jones, Claflin College (South Carolina)

Matti E. Kaups, University of Minnesota-Duluth

Colleen Keen, Gustavus Adolphus College (Minnesota)

Gordon F. Kells, Mott Community College

Susanne Kibler-Hacker, Unity College (Maine)

J.W. King, University of Utah

John C. Kinworthy, Concordia College (Nebraska)

Albert Kitchen, Paine College

Ted Klimasewski, Jacksonville State University (Alabama)

Robert D. Klingensmith, Ohio State University-Newark

Lawrence M. Knopp, Jr., University of Minnesota-Duluth

Terrill J. Kramer, University of Nevada

Arthur Krim, Salve Regina College (Rhode Island)

Elroy Lang, El Camino Community College (California)

Christopher Lant, Southern Illinois University-Carbondale

A.J. Larson, University of Illinois-Chicago

Larry League, Dickinson State University (North Dakota)

David R. Lee, Florida Atlantic University

Joe Leeper, Humboldt State University (California)

Yechiel M. Lehavy, Atlantic Community College (New Jersey)

John C. Lewis, Northeast Louisiana University

Caedmon S. Liburd, University of Alaska-Anchorage

T. Ligibel, Eastern Michigan University

Z.L. Lipchinsky, Berea College (Kentucky)

Allan L. Lippert, Manatee Community College (Florida)

John H. Litcher, Wake Forest University (North Carolina)

Li Liu, Stephen F. Austin State University (Texas)

William R. Livingston, Baker College (Michigan)

Cynthia Longstreet, Ohio State University

Tom Love, Linfield College (Oregon)

K.J. Lowrey, Miami University (Ohio)

Robin R. Lyons, University of Hawaii-Leeward Community College

Susan M. Macey, Southwest Texas State University

Christiane Mainzer, Oxnard College (California)

Harley I. Manner, University of Guam

James T. Markley, Lord Fairfax Community College (Virginia)

Sister Mary Lenore Martin, Saint Mary College (Kansas)

Kent Mathewson, Louisiana State University

Dick Mayer, Maui Community College (Hawaii)

Dean R. Mayhew, Maine Maritime Academy

J.P. McFadden, Orange Coast College (California)

Bernard McGonigle, Community College of Philadelphia (Pennsylvania)

Paul D. Meartz, Mayville State University (North Dakota)

Inez Miyares, Arizona State University

Bob Monahan, Western Washington University

Keith Montgomery, University of Wisconsin-Marathon

John Morton, Benedict College (South Carolina)

Anne Mosher, Louisiana State University

Robert R. Myers, West Georgia College

Yaser M. Najjar, Framingham State College (Massachusetts)

Jeffrey W. Neff, Western Carolina University

David Nemeth, University of Toledo (Ohio)

Raymond O'Brien, Bucks County Community College (Pennsylvania)

John Odland, Indiana University

Patrick O'Sullivan, Florida State University

Bimal K. Paul, Kansas State University

James Penn, Southeastern Louisiana University

Paul Phillips, Fort Hays State University (Kansas)

Michael Phoenix, William Paterson College (New Jersey)

Jerry Pitzl, Macalester College (Minnesota)

Billie E. Pool, Holmes Community College (Mississippi)

Vinton M. Prince, Wilmington College (North Carolina)

Rhonda Reagan, Blinn College (Texas)

Danny I. Reams, Southeast Community College (Nebraska)

Jim Reck, Golden West College (California)

Roger Reede, Southwest State University (Minnesota)

John Ressler, Central Washington University

John B. Richards, Southern Oregon State College

David C. Richardson, Evangel College (Missouri)

Susan Roberts, University of Vermont

Wolf Roder, University of Cincinnati

James Rogers, University of Central Oklahoma

James C. Rose, Tompkins/Cortland Community College (New York)

Thomas E. Ross, Pembroke State University (North Carolina)

Thomas A. Rumney, State University of New York-Plattsburgh

George H. Russell, University of Connecticut

Rajagopal Ryali, Auburn University at Montgomery (Alabama)

Perry Ryan, Mott Community College

Adena Schutzberg, Middlesex Community College (Massachusetts)

Sidney R. Sherter, Long Island University (New York)

Nanda Shrestha, University of Wisconsin-Whitewater

William R. Siddall, Kansas State University

David Silva, Bee County College (Texas)

Morris Simon, Stillman College (Alabama)

Kenn E. Sinclair, Holyoke Community College (Massachusetts)

Robert Sinclair, Wayne State University (Michigan)

Everett G. Smith, Jr., University of Oregon

Richard V. Smith, Miami University (Ohio)

Carolyn D. Spatta, California State University-Hayward

M.R. Sponberg, Laredo Junior College (Texas)

Donald L. Stahl, Towson State University (Maryland)

Elaine Steinberg, Central Florida Community College

D.J. Stephenson, Ohio University Eastern

Herschel Stern, Mira Costa College (California)

Reed F. Stewart, Bridgewater State College (Massachusetts)

Noel L. Stirrat, College of Lake County (Illinois)

George Stoops, Mankato State University (Minnesota)

Joseph P. Stoltman, Western Michigan University

P. Suckling, University of Northern Iowa

T.L. Tarlos, Orange Coast College (California)

Michael Thede, North Iowa Area Community College

Derrick J. Thom, Utah State University

Curtis Thomson, University of Idaho

S. Toops, Miami University (Ohio)

Roger T. Trindell, Mansfield University of Pennsylvania

Dan Turbeville, East Oregon State College

Norman Tyler, Eastern Michigan University

George Van Otten, Northern Arizona University

C.S. Verma, Weber State College (Utah)

Graham T. Walker, Metropolitan State College of Denver (Colorado)

Deborah Wallin, Skagit Valley College (Washington)

Mike Walters, Henderson Community College (Kentucky)

J.L. Watkins, Midwestern State University (Texas)

P. Gary White, Western Carolina University (North Carolina)

W.R. White, Western Oregon University

Gary Whitton, Fairbanks, Alaska

Gene C. Wilken, Colorado State University

Stephen A. Williams, Methodist College

P. Williams, Baldwin-Wallace College

Morton D. Winsberg, Florida State University

Roger Winsor, Appalachian State University (North Carolina)

William A. Withington, University of Kentucky

A. Wolf, Appalachian State University

Joseph Wood, George Mason University (Virginia)

Richard Wood, Seminole Junior College (Florida)

Sophia Hinshalwood, Montclair State College (New Jersey)

George I. Woodall, Winthrop College (North Carolina)

Stephen E. Wright, James Madison University (Virginia)

Leon Yacher, Southern Connecticut State University

Donald J. Zeigler, Old Dominion University (Virginia)

In assembling the current revision, we are indebted to the following people for advising us on a number of matters:

James D. Fitzsimmons, U.S. Bureau of the Census (D.C.)

Gary A. Fuller, University of Hawaii

Chip Hoehler, Lake Wales, Florida

Richard C. Jones, University of Texas at San Antonio

Ian MacLachlan, University of Lethbridge (Alberta)

Gabriel Marcella, U.S. Army War College (Pennsylvania)

Pai Yung-feng, New York City

John D. Stephens, H. M. Gousha Company (Texas)
George E. Stuart, National Geographic Society (D.C.)

We also record our appreciation to those geographers who ensured the quality of this book's ancillary products: Ira M. Sheskin (University of Miami) prepared the *Test Bank* and manipulated a large body of demographic data to derive the tabular display in Appendix A; Laurie Molina (Florida State University/Florida Geographic Alliance) prepared the *Instructor's Manual*; and Elizabeth Muller Hames (newly-minted M.A., University of Miami) co-authored the *Study Guide* and prepared the Geographical Index. At the University of Miami's Department of Geography, we are most thankful for the support we continue to receive from faculty colleagues Tom Boswell, Phil Keating, Bin Li, Jan Nijman, and Ira Sheskin. Moreover, the enthusiastic office staff tirelessly performs an array of critical supporting tasks, all supervised by superb senior secretary and office manager, Hilde Al-Mashat.

Personal Appreciation

At John Wiley & Sons we are privileged to work with a team of outstanding textbook publishing professionals. As authors we are most acutely aware of these talents, on a daily basis, during the critical production stage, and no performances could exceed those of Senior Production Editor Jeanine Furino (our quarterback), Illustration Editor Edward Starr, and Photo Assistant Kim Khatchatourian. Others who played a leading role in this process were Design Director Madelyn Lesure, Photo Editor Alexandra Truitt, Developmental Director Johnna Barto, Copy Editor Betty Pessagno, Production Manager Linda Muriello, Illustration Manager Ishaya Monokoff, Don Larson of Mapping Specialists, Inc., and Photo Editor Mary Ann Price. The Geography Editor's office smoothly guided this project and kept us moving forward during three administrative tenures, and we especially thank former Editor Frank Lyman, Interim Editor (also Executive Editor) Chris Rogers, and newly-arrived, current Editor Nanette Kauffman. Others at Wiley who gave us support and encouragement were Senior Marketing Manager Cathy Faduska, Photo Research Manager Stella Kupferberg, Media Editor Fadia Stermasi, Supplements Editor Francine Banner as well as Bethany Balch, Felicia Reid, Barbara Bredenko, Eric Stano, and Ann Berlin.

Finally, and most of all, we thank our wives, Bonnie and Nancy, for seeing us through the hectic schedule of our fifth collaboration on this volume in the past 12 years.

H.J. de Blij
Boca Grande, Florida

Peter O. Muller
Coral Gables, Florida

May 10, 1996

A GEOGRAPHY STUDENT'S BEGINNING

College is not an easy thing, that much I know from my own personal experience. Suffering three disappointing false starts after graduating from high school in 1982, I returned for one last shot in the spring of 1990. My previous failures were, looking back with the advantage of hindsight, due largely to a lack of focus and an absence of conviction. My obvious talents were within the fine arts—specifically drawing, photography, and graphic design. However, the academic study of the arts continually left me frustrated. Art was no longer fun. What began as a hobby suddenly became a burden. Thus, the spring semester of 1990 was much like the previous disappointments.

In the fall, I decided on a new approach. I enrolled in what would be the first of many courses under the grand heading of geography. It was something I had always been interested in, but never really knew where to begin. After all, geography's importance was, and still is, sadly misunderstood throughout many college campuses. Nonetheless, as registration closed at Golden West College, I found myself enrolled in two art classes and Geography 100—World Geography.

Three weeks into the semester I dropped the two art classes. A week after that, I formally declared my new major as Geography. The course was challenging and engaging. It was a fresh new discipline that literally opened a window of understanding and appreciation to a much larger world than I had ever known. Don Schmitz, my instructor, was teaching his first class ever. He breathed life into the subject and presented his own knowledge with a conviction that seems too rare in academia these days. Likewise, the textbook he chose for the course was written in much the same way.

As a student, I must come clean and admit that I have never read a textbook and actually enjoyed the experience. However, as a professional geographer, I would personally like to thank Dr. de Blij and Dr. Muller for their vision and foresight in *Geography: Regions and Concepts*. For me, it is not a simplistic text about geography. In fact, the most unlikely catalyst to my own success as a student of geography is contained between the covers of that book, but doesn't even exist in the body of the text.

In "Appendix C" I learned about what geography was, where it had been, and where it was going. I read discussions about the many sub-disciplines within geography and their uniquely related career paths. I read about a professional organization called the Association of American Geographers and became a student member by semester's end. Most importantly, I uncovered a column that mentioned the importance of internships. Specifically, the appendix provided general information on the National Geographic Society's Geography Intern Program. Suddenly, in the flash of a moment, I was focused, and I had conviction.

Nearly five years and 115 semester hours later at 8:00 a.m. on August 28, 1995, I was not only a recent graduate of California State University-Fullerton, but I was also on the Washington, D.C. Metro on my way to the headquarters of the National Geographic Society. My far reaching goal became realized as I was about to begin my first day as a "geography intern"—one of seven students selected from all reaches of the United States for the 16-week session.

Annually, some 175 applications are submitted to the Geography Intern Program. Of those, only 16 to 24 are selected to participate. Interns are given practical experience within various divisions of the Society during three separate sessions. For me, the experience was more valuable than I could have ever expected. As a cartographer, I was honored and privileged to work alongside some of the finest professionals within the Society's Cartographic Division. The division and the Society as a whole have always held the highest standards when it comes to the accuracy and quality of their products and it has set the benchmark for the future of my career. Ironically, upon completing my internship I immediately went to work as a freelance cartographer for the Society's Book Division and I am now a permanent resident of the D.C. area. As my friend and one-time professor Dr. Robert Arnold wrote to me, ''Life has many unforeseen turns . . .''

Suggesting that the appendix of an introductory geography text focused my ambition may seem a bit absurd. However, while a simple, good textbook supplements the instruction of a good professor, and vice versa, a great textbook goes beyond the typical formula. A great textbook attempts to reach students outside of the classroom experience. It interacts, through a synchronous combination of practical and informative chapters and thoughtful appendices, with those who are genuinely curious about the subject and where it might lead them into the future. On page 615 of my copy of *Geography: Regions and Concepts, Fifth Edition*, exists an inviting doorway ideally marked for the introductory student of geography. I confidently stepped through it and have since resisted looking back to see the plume of dust I left behind.

Louis Spirito

Louis Spirito continues to work as a freelance cartographer for the National Geographic Society. He is planning to begin graduate studies in Cartography at George Mason University in the near future.

Brief Contents

CONTENTS

CHAPTER 2

Russia's Fractious Federation   109

DEFINING THE REALM   113

REPUBLICS OF THE RUSSIAN REALM   129

REGIONS OF THE RUSSIAN REALM   140

CHAPTER 3

North America:
The Postindustrial Transformation   155

DEFINING THE REALM   156

CHAPTER 4

Middle America:
Collision of Cultures   199

CHAPTER 5

South America:
Continent of Contrasts   233

CHAPTER 6

North Africa/Southwest Asia: The Energy of Islam   272

CHAPTER 7

Subsaharan Africa: Realm of Reversals   327

CHAPTER 8

South Asia:
Resurgent Regionalism   371

CHAPTER 9

East Asia:
Realm of Titans   413

World Regional Geography: Physical and Human Foundations

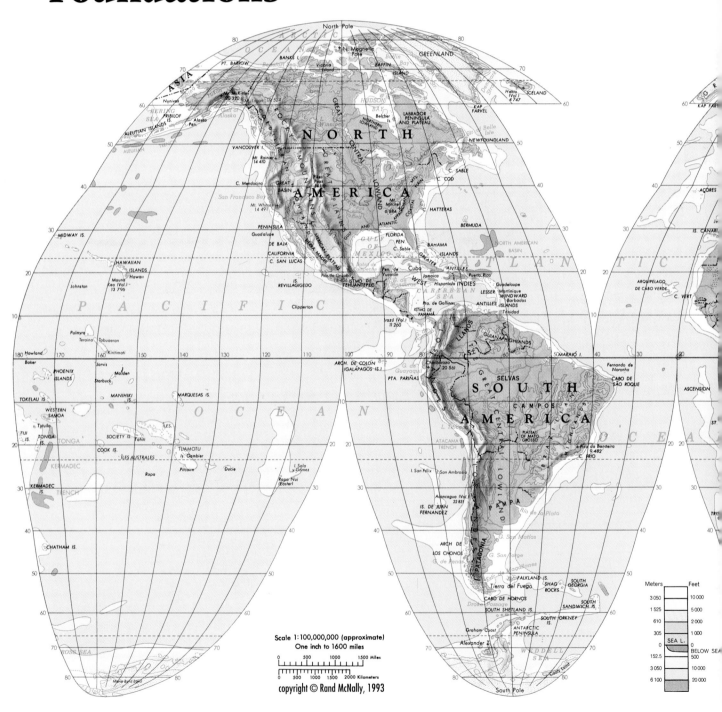

Scale 1:100,000,000 (approximate)
One inch to 1600 miles

copyright © Rand McNally, 1993

IDEAS & CONCEPTS

Geographic realm	Formal region	Cultural landscape	Globalization
Spatial perspective	Spatial system	Population density	Iconography
Transition zone	Hinterland	Urbanization	Regional geography
Regional concept	Functional region	Population growth	Systematic fields
Area	Scale	State	
Boundaries	Physiography	Development	
Location	Culture	Core area	

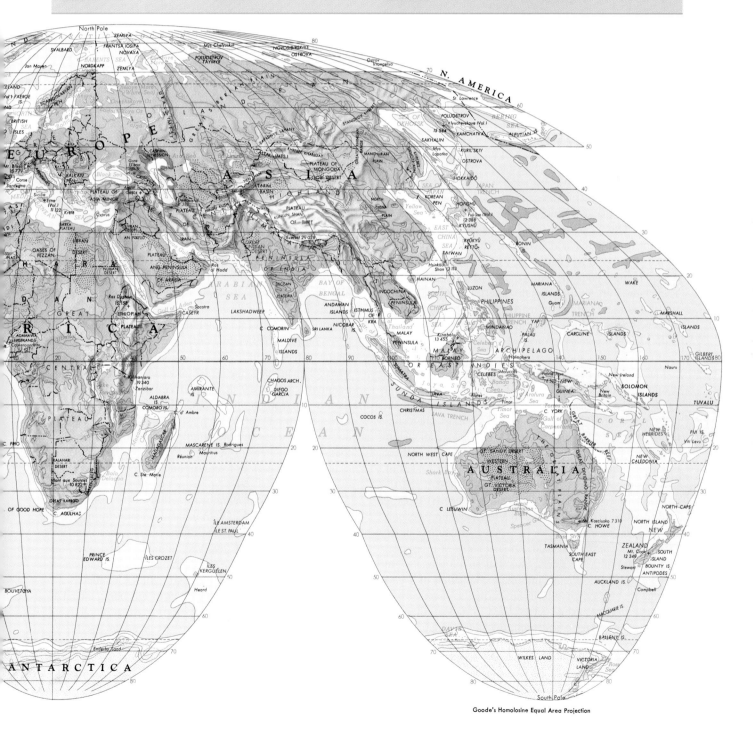

Goode's Homolosine Equal Area Projection

What a time this is to be studying geography! The world is undergoing an historic transformation, one of those momentous political, economic, and social upheavals about which we have read in history books. It is happening today, and as in the past, new alliances are forming, old unions are fracturing, novel ideas are spreading, older notions are fading. We hear from our political leaders about the emergence of a "New World Order," but such pronouncements are premature. A future world order is indeed in the offing, but it is too early to tell what it will be like. Of this we can be sure: the new world map will look quite different from the old. Our task is to understand the ongoing changes, to make sense of the new directions our world is taking. Geography is our most powerful ally in this mission.

GEOGRAPHIC PERSPECTIVES

In this book we take a penetrating look at the geographic framework of the contemporary world, the grand design that is the product of thousands of years of human achievement and failure, movement and stagnation, revolution and stability, interaction and isolation. Ours is an interconnected world of travel and trade, tourism and television, a global village—but the village still has neighborhoods. Their names are Europe, South America, Southeast Asia, and others familiar to all of us. We call such global neighborhoods **geographic realms**, and when we subject these realms to geographic scrutiny, we find that they each have their own identity and distinctiveness.

Geographers study the locations and distributions of features on the Earth's surface. These features may be the landmarks of human occupation or the properties of the natural environment, or both: one of the most interesting themes in geography has to do with the interrelationships of natural environments and human societies. Geographers, in their research, investigate the reasons for, or causes of, these distributions. Their approach to the human and the natural world is guided by a **spatial perspective**. Just as historians focus on time and chronology, geographers concentrate on space and place. The spatial structure of cities, the layout of farms and fields, the networks of transportation, the system of rivers, the pattern of climate—all this and much more goes into the examination of a geographic realm. As you will discover, the language of geography is full of spatial terms: area, distance, direction, clustering, proximity, accessibility, isolation, and many others.

In this book we use the geographic perspective and geography's spatial terminology to investigate the world's great geographic realms. We will find that each of these realms possesses a special combination of cultural, organizational, and environmental properties. These characteristic qualities are imprinted on the landscape, giving each realm its own traditional attributes and social settings. As we come to understand the human and natural makeup of those geographic realms, we learn not only *where* they are located (always a crucial question in geography, and often the answer is not as simple as it may appear), but also *why they are located where they are*, how they are constituted, and what their future is likely to be in our changing world.

It is not enough, however, to study the world from a geographic viewpoint without learning something about the discipline of geography itself. Beginning in this introductory chapter, we introduce the fundamental ideas and concepts that make modern geography what it is. Not only will these geographic ideas enhance your awareness of the many dimensions of our complex, multicultural, interconnected world, but also many of them will remain useful to you long after you have closed this book. Welcome to geography . . . realms, regions, *and* concepts!

REALMS AND REGIONS

Geographers, like other scholars, seek to establish order from the countless pieces of information (data) with which they are confronted. Biologists have established a system of classification, or *taxonomy*, to categorize millions of plants and animals into a hierarchical system consisting of seven ranks. In descending order, we humans belong to the animal *kingdom*, the *phylum* (division) named chordata, the *class* of mammals, the *order* of primates, the *family* of hominids, the *genus* designated *Homo*, and the *species* known as *Homo sapiens*. Geologists classify the Earth's rocks into three major (and many subsidiary) categories and then fit these into a complicated geologic time scale that spans hundreds of millions of years. Historians define eras, ages, and periods to conceptualize the sequence of the events they study.

Geography, too, employs systems of classification. When geographers deal with urban problems, for instance, they employ a classification scheme based on the sizes and functions of the places involved. Some of the terms in this classification are familiar words in our everyday language: megalopolis, metropolis, city, town, village, hamlet.

In regional geography, the focus of this book, the challenge is rather different. We, too, need a hierarchical framework to accommodate the areas of the world we study, from the largest to the smallest. But our classification scheme is horizontal, not vertical. It is *spatial*. Our equivalent of the biologists' overarching kingdoms (of plants, animals) is the Earth's natural partitioning into landmasses and oceans. The next level is the division of the inhabited landmasses into geographic realms based, as noted earlier, on human as well as physical (natural) properties.

Realms and Their Criteria

In any classification system, criteria are the key. Not all animals are mammals, and fewer still are primates: the criteria for inclusion in that biological order are more specific and restrictive. A dolphin may look and appear to behave like a fish, but dolphins anatomically and functionally qualify as belonging to the order of mammals.

Geographic realms are based on sets of spatial criteria. First, they are the largest units into which the inhabited world can be divided. The criteria on which such a broad regionalization is based include both physical (that is, natural) and human (or social) yardsticks. South America is a geographic realm, for example, because it is physically a continent and culturally dominated by a certain set of social norms. The realm called South Asia, on the other hand, lies on a Eurasian landmass shared by several other geographic realms; high mountains, wide deserts, and dense forests combine with a distinctive social fabric to create this well-defined realm centered on the multicultural state of India.

Second, geographic realms are the result of the interaction of human societies and natural environments, a *functional* interaction that is revealed by farms, mines, fishing ports, transport routes, dams, bridges, villages, and countless other features on the landscape. Under this criterion, Antarctica is a continent but not a geographic realm.

Third, geographic realms must represent the most comprehensive and encompassing definition of the great clusters of humankind in the world today. China lies at the heart of such a cluster, as does India. Africa constitutes a geographic realm from the southern margin of the Sahara (an Arabic word for "desert") to the Cape of Good Hope, and from its Atlantic to its Indian Ocean shores.

Figure I-1 displays the 12 world geographic realms based on these criteria. As we will show in more detail later, not only waters but also deserts and mountains mark the borders of these realms; we will also discuss the position of these boundaries as each realm is examined. For the moment, keep in mind the following:

■ *Where geographic realms meet,* **transition zones,** *not sharp boundaries, mark their contacts.*

We need only remind ourselves of the border zone between the geographic realm in which most of us live, North America, and the adjacent realm of Middle America. The line on Figure I-1 coincides with the boundary between Mexico and the United States, crosses the Gulf of Mexico, and then separates Florida from Cuba and the Bahamas. But Hispanic influences are strong in North America far to the north of this boundary, and U.S. economic influence is strong to the south of it. The line, therefore, represents an ever-changing zone of regional interaction. Again, there are many ties between South Florida and the Bahamas, but the Bahamas more strongly resemble a Caribbean society than a North American one.

In Africa, the transition zone from Subsaharan to North Africa is so wide and well defined that we have put it on the world map; elsewhere, transition zones tend to be nar-

Cities are the engines of change, the crucibles of cultural transformation. A large sector of Miami (Florida) has become almost totally Hispanicized as a result of the immigration of hundreds of thousands of Cubans and other Spanish-speaking people from Middle America. The peaceful absorption of this population is one of the unsung triumphs of American civilization; in the process, part of Miami became known as "Little Havana," its cultural landscape dominated by Cuban motifs. The tattered U.S. flag and the parking signs (still in English) symbolize the transition of metropolitan Miami, whose Hispanic population was 4 percent in 1950 and exceeded 50 percent in 1997.

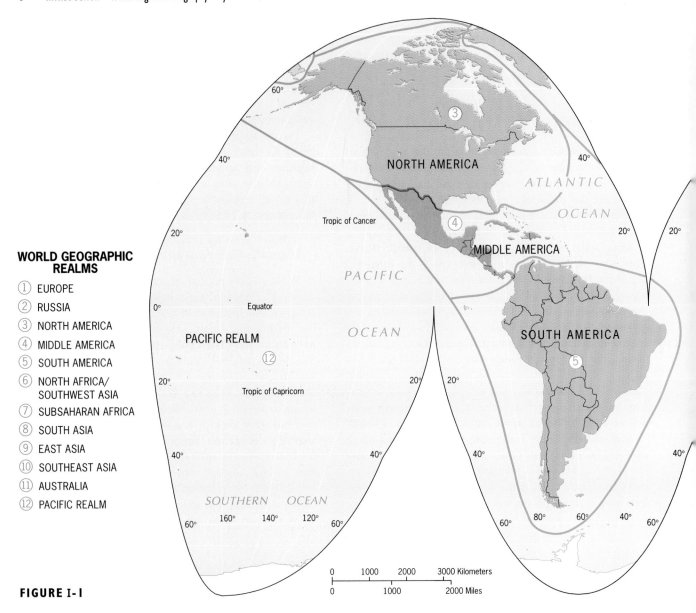

WORLD GEOGRAPHIC REALMS

① EUROPE
② RUSSIA
③ NORTH AMERICA
④ MIDDLE AMERICA
⑤ SOUTH AMERICA
⑥ NORTH AFRICA/ SOUTHWEST ASIA
⑦ SUBSAHARAN AFRICA
⑧ SOUTH ASIA
⑨ EAST ASIA
⑩ SOUTHEAST ASIA
⑪ AUSTRALIA
⑫ PACIFIC REALM

FIGURE I-1

rower and less easily represented. In the late 1990s, such countries as Belarus (between Europe and Russia) and Kazakstan (between Russia and Muslim Southwest Asia) lie in inter-realm transition zones. Remember, over much (though not all) of their length, borders between realms are zones of regional change.

■ *Geographic realms change over time.*

Had we drawn Figure I-1 before Columbus' time, the map would have looked quite different: Amerindian states and peoples would have determined the boundaries in the Americas; Australia and New Guinea would have constituted one realm, and New Zealand would have been part of the Pacific realm. The colonization, Europeanization, and Westernization of the world changed that map dramatically. During the four decades following the end of

World War II there was relatively little change, but the years since 1985 have witnessed far-reaching realignments that are still going on. Over six editions of this book, between 1971 and 1991, we made only minor adjustments to a realms framework that was essentially stable. Now, in the late 1990s, we continue to make major revisions, underscoring the transformation of our world as the twenty-first century rapidly approaches.

As we try to envisage what a New World Order will look like on the map, note that the 12 geographic realms can be divided into two groups: (1) those dominated by one major political entity (North America/United States, Middle America/Mexico, South America/Brazil, South Asia/India, and East Asia/China as well as Russia and Australia), and (2) those in which there are many countries but no dominant state (Europe, North Africa/Southwest Asia,

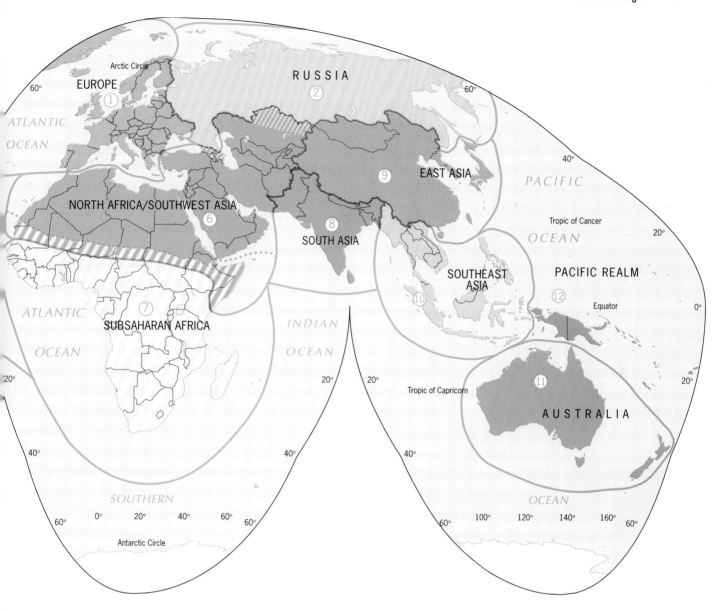

Subsaharan Africa, Southeast Asia, and the Pacific realm). For several decades we have known a world dominated by two major powers, the United States and the former Soviet Union. Will a multipolar world arise from the rubble of that bipolar world? That is a question to be addressed in the pages that follow.

Regions and Their Criteria

The spatial division of the world into geographic realms establishes a broad global framework, but for our purposes a more refined level of spatial classification is needed. This brings us to an important organizing concept in geography: the **regional concept**. To continue the analogy with biological taxonomy, we now go from phylum to order. To

establish regions within geographic realms, we need more specific criteria.

Let us use the North American realm to demonstrate the regional idea. When we refer to a part of the United States or Canada (e.g., the South, the Midwest, or the Prairie Provinces), we employ a regional concept—not scientifically, but as a matter of everyday communication. We reveal our *perception* of local or distant space as well as our mental image of the region to which we make reference.

But what exactly is the Midwest? If you were asked to draw this region on the North American map, how would you do it? Regions are easy to imagine and describe, but they can be very difficult to outline on a map. One way to go about defining the Midwest is to use the borders of states: certain states are part of this region, others not. Another way is to use agriculture as the chief criterion: where corn and soybeans occupy a certain percentage of the farmland, there lies the Midwest. Each method results in a different delimitation; a Midwest based on states is different from a Midwest based on farm production. And therein lies an important principle: regions are scientific devices that allow us to make spatial generalizations, and they are based on artificial criteria we establish for the purpose of constructing them. If you were doing research on the geography behind politics, then a Midwest region based on state boundaries would make sense. If you were studying agricultural distributions, then you would need a different definition of the Midwest.

Given these different dimensions of the same region, are there properties all regions have in common? Let us identify several. To begin with, all regions have **area**. This would seem to be so obvious as not to need elaboration, but there is more to this idea than first meets the eye. Regions may be intellectual constructs, but they are not abstractions: they exist in the real world, and they occupy space on the Earth's surface.

It follows that regions have **boundaries**. Occasionally, nature itself draws sharp dividing lines, for example, along the crest of a mountain range or the margin of a forest. More often, regional boundaries are not self-evident and must be determined on the basis of criteria established for that purpose. Take, for instance, the notion of a Corn Belt in the agricultural heartland of the United States. When you travel into that region, you see more and more fields of corn until this crop dominates the farmlands. But where exactly is the limit of this region we call the Corn Belt? That depends on specific criteria. First we would establish a unit area, say, one-quarter square mile. Next we would create a grid consisting of quarter-square-mile units and superimpose it on an area larger than the corn-growing zone. Then we would decide that in order to be part of the

Corn Belt, more than 50 percent of the farmland in any unit area must be devoted to growing corn. In addition, we might decide that a unit area is part of the Corn Belt even if it contains less than 50 percent corn, as long as it is completely surrounded by spatial units that do qualify. Result: a Corn Belt region boundary that is both visible and quantitatively defined. Of course, not all regional boundaries can be so specifically delimited, and neighboring regions, like adjacent realms, sometimes display transitional borderlands.

All regions also have **location**. Often the name of a region contains a locational clue, as in Amazon Basin or Indochina (a region of Southeast Asia lying between India and China). Geographers refer to the *absolute location* of a place or region by reference to the Earth's grid system; this provides the latitudinal and longitudinal extent of the region with respect to grid coordinates. A far more practical measure is a region's *relative location*, that is, its location with reference to other regions. Again, the names of some regions reveal aspects of their relative locations, as in *Eastern* Europe and *Equatorial* Africa.

Many regions are marked by a certain *homogeneity*, or sameness. Homogeneity may lie in a region's human (cultural) properties or its physical (natural) characteristics, or both. Siberia, a vast region of northeastern Russia, is marked by a sparse human population residing in widely scattered, small settlements of similar form; frigid climates; extensive areas of permafrost (permanently frozen subsoil); and cold-adapted vegetation. This dominant uniformity makes it one of Russia's natural and cultural regions, extending from the Ural Mountains in the west to the shores of the Pacific Ocean in the east. When regions display a measurable and often visible internal homogeneity, they are referred to as **formal regions**. But not all formal regions are visibly uniform. For example, a region may be

Interconnections in the spatial system: Garden City, Iowa. Roads and grain elevators link the town to its surroundings. The spatial homogeneity of the region is quite evident from the farmscape beyond.

delimited within which one particular language is spoken by, say, 90 percent of the people or more. This cannot be seen in the landscape, but the region is a reality and its boundaries can be drawn quite accurately based on this criterion. It, too, is a formal region.

Other regions are marked *not* by their internal sameness, but by their functional integration, that is, the way they work. These regions are defined as **spatial systems**, and are formed by the areal extent of the activities that define them. Take the case of a large city with its surrounding zone of suburbs, urban-fringe countryside, satellite towns, and farms. The city supplies goods and services to this encircling zone, and it buys farm products and other commodities there. The city is the heart, the *core* of this region, and its surrounding zone of interaction is called its **hinterland**. But eventually the city's influence wanes on the outer periphery of that hinterland, and there lies the boundary of the **functional region** of which it is the focus. A functional region, therefore, is forged by a structured, urban-centered system of interaction. It has a core and also a periphery. As we shall see later, core-periphery contrasts in some parts of the world are becoming so strong that they endanger the stability of countries.

To a greater or lesser degree, all human-geographic regions are *interconnected*, linked to other regions in various ways. We know that the borders of geographic realms sometimes take on the character of transition zones, and so it is with neighboring regions. Trade, migration, education, television, computer linkages, and other interactions ensure a blurring of regional boundaries. These are among the links in the fast-growing interdependence among the world's peoples, and they help to reduce the differences that still divide us. Understanding these differences is an important step toward lessening them further.

REGIONS AT SCALE

In this book we examine the geography of the world at the level of detail of the region, using regional concepts as we proceed. But all regions are parts of realms, which is why this book is called *Geography: Realms, Regions, and Concepts*.

As we have just noted, regions come in many sizes. Some are huge, for example, the Russian formal region of Siberia. Others are comparatively small, as in the case of the functional region constituted by Boston and its hinterland. Some geographers have tried to establish a terminology to distinguish between larger and smaller regions; occasionally, you will see a reference to a *subregion* within a region. But no generally accepted nomenclature has ever emerged, so we use the term *region* for bounded spaces large and small. At the level of detail in our maps, we must focus on the larger regions, referring occasionally to

smaller regions "nested" within the larger ones. And this brings us to the geographic concept of scale.

The map is the geographer's strongest ally. It does for geography what taxonomic (and other) classification systems do for the other sciences. Maps display enormous quantities of information; they suggest interrelationships; they answer questions; they lay out spatial problems for researchers to investigate (see Appendix B, Map Reading and Interpretation). Many maps, moreover, are simply fascinating. No personal library is complete without a good atlas.

Maps represent the surface of the Earth (and other features of the planet, present and past) at various levels of generalization. These levels of generalization can be gauged by looking at the map's **scale**, that is, the ratio of the distance between two places on a map and the actual distance between those two places on the Earth's surface.

Consider the four maps in Figure I-2. On the first map (upper left), most of the North American realm is shown, but very little spatial information can be provided, although the political boundary between Canada and the United States is evident. On the second map (upper right), eastern and central Canada are depicted in sufficient detail to permit display of the provinces, several cities, and some physical features (Manitoba's major lakes) not shown on the first map. The third map (lower left) shows the main surface communications of the province of Quebec and immediate surroundings, the relative location of Montreal, and the St. Lawrence and Hudson/James Bay drainage systems. The fourth map (lower right) reveals the metropolitan layout of Montreal and its adjacent hinterland in considerable detail.

Each of the four maps has a scale designation, which can be shown as a bar graph (in miles and kilometers [km] in this case) and as a fraction—1:103,000,000 on the first map. The fraction is a ratio indicating that one unit of distance on the map (one inch or one centimeter) represents 103 million of the same units on the ground. The smaller the fraction (in other words, the larger the number in the denominator), the smaller the scale of the map. Clearly, this representative fraction (RF) on the first map (1:103,000,000) is the smallest, and that of the fourth map (1:1,000,000) is the largest. Comparing maps number 1 and number 3, we find that on the linear scale, number 3 has a representative fraction that is more than four times larger than that of number 1. When it comes to areal representation, however, 1:24,000,000 is more than 16 times larger than 1:103,000,000 because the linear difference prevails in both dimensions (the length and the breadth of the map).

In a book that surveys the major regions of each world realm, it is necessary to operate at relatively small scales. When studying smaller regions in greater detail, our ability to specify criteria and to "filter" the factors we employ increases as we work at larger scales. That method will often come into play when urban areas are the topic of

EFFECT OF SCALE

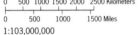

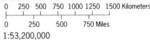

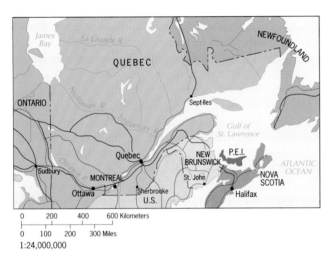

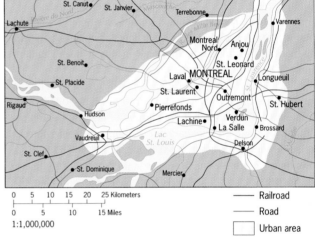

FIGURE I-2

concern (as suggested by the map of metropolitan Montreal in Fig. I-2). But most of the time our view will be more macroscopic and general: a small-scale perspective on the world's geographic realms and regions.

THE NATURAL STAGE

Although the world-scale regions we will define and explore in this book are the products of human activity, we will have frequent occasion to refer to the physical setting against which the human drama is being played out. For all our industrial capacity and technological progress, the natural environment still plays a prominent role in human fortunes.

Significantly, we have learned that environments have changed in the past, sometimes quite rapidly. Archeologists have excavated cities that long ago rose to power in splen-

dor amid plenty of water, farm produce, and raw materials such as wood and fiber. Today their ruins lie buried under desert sand. In Roman times, farmlands near North Africa's Mediterranean coast yielded large amounts of produce, amply watered by rainfall and irrigation systems. Today the Roman aqueducts lie in ruins, and fields are abandoned. And now we worry about the future: about *desertification*, the drying up of areas in the path of expanding deserts; about global warming, the prospect of warmer temperatures than have prevailed over past decades; and about associated climatic shifts. Will polar ice-sheets melt and cause sea levels to rise? Will croplands stop yielding their produce? Which regions of the world will be most severely affected by environmental change?

These are important questions. We must not study geographic realms and regions without reference to the physical stage they occupy. *Physical geography*—the spatial study of the earth's natural phenomena and their systems, processes, and structures—provides us with the necessary

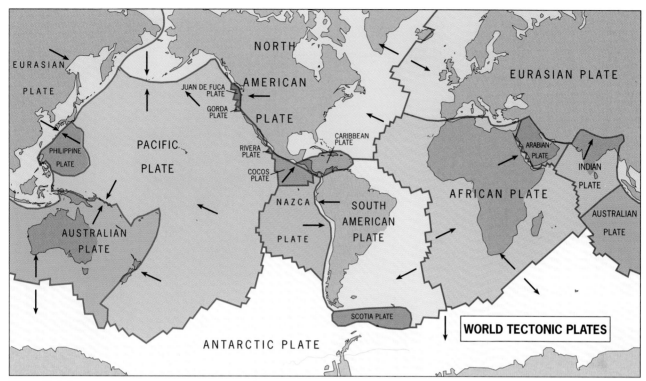

FIGURE I-3

information. It was a physical geographer, the climatologist Alfred Wegener, who nearly a century ago used spatial analysis to propose a then-revolutionary hypothesis: the hypothesis of *continental drift*. Wegener studied the outlines of the continents and, marshalling a vast array of evidence, suggested that the landmasses had at one time in the Earth's history been united. Since then, they had drifted apart, he argued, but the opposite sides of the Atlantic Ocean still fit together like pieces of a jigsaw puzzle, especially Africa and South America. What was revolutionary about this idea was that the seemingly stable, immovable continents were in fact mobile.

Continental drift, Wegener argued in a book published in 1915, explained the great variety of natural landscapes on the continents. For example, as South America drifted westward, away from Africa and into the Pacific, its leading edge crumpled up like the folds of an accordion, forming the Andes Mountains. This, too, is why North America's greatest mountains lie in the west, while those of Australia (which moved eastward into the Pacific) lie in the eastern margin of that landmass.

Wegener did not live to see his hypothesis accepted; many prominent geologists quickly derided it as nonsense. But as is so often the case, a geographer's spatial theory pointed the way for the research of others. Eventually, geologists did discover that the thin crust of the Earth consists of a set of *tectonic plates*, great slabs of solid material that not only make up ocean floors but also carry continents.

And (to no geographer's surprise), those plates (Fig. I-3) are indeed mobile. They move slowly, a few centimeters (an inch or two) per year, propelled by heat-driven convection cells in the molten rock deep below the crust. A couple of inches per year is not much, but the Earth's history is measured in millions of years. So the continental landmasses travel thousands of miles, opening and closing oceans and changing the world map of land and water bodies.

As the great tectonic plates collide, the one that consists of lighter (less heavy) rock rides up over the weightier one in a gigantic collision that creates just what Wegener said it does: great mountain chains. This collision is accompanied by numerous earthquakes and volcanic eruptions. In Figure I-3, note that the North American and South American Plates are moving toward the Pacific, as is the Eurasian Plate. Thus the Pacific Ocean is surrounded by a nearly continuous plate-collision zone, which is called the Pacific Ring of Fire. To those realms and regions that contain parts of this zone of crustal instability, this is no mere abstraction. Japan, for example, lies near the colliding edges of three plates. Disastrous earthquakes are a fact of life there, and volcanoes threaten many communities. On the American side of the Pacific, too, earthquakes and volcanism create a high-risk zone all along the western (leading-edge) margins of the continents.

The physical landscapes of the continents reveal the effects of millions of years of movement. Today, a giant

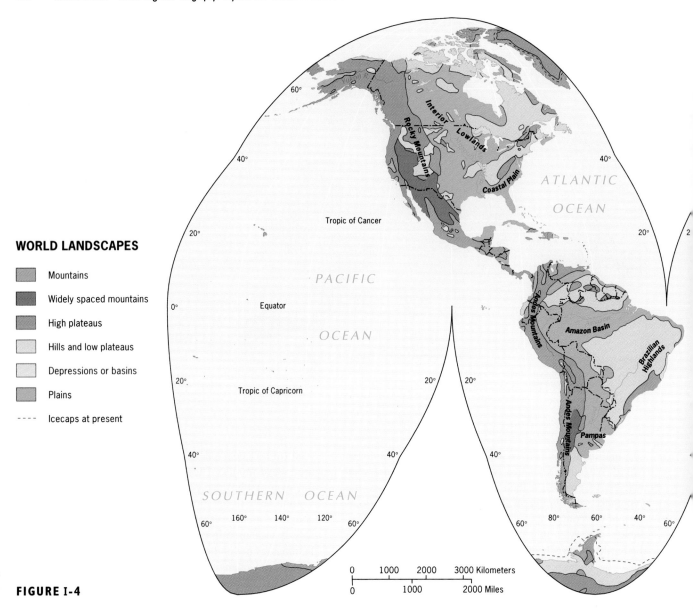

WORLD LANDSCAPES

- ▢ Mountains
- ▢ Widely spaced mountains
- ▢ High plateaus
- ▢ Hills and low plateaus
- ▢ Depressions or basins
- ▢ Plains
- ---- Icecaps at present

FIGURE I-4

mountain range is forming where the Indian Plate is pushing into the heart of Asia; the Himalayas, capped by Mount Everest (presently the highest mountain on the planet), are the manifestations of this collision. But note that the Himalayas are only one link in a mountainous plate-collision chain that extends southeastward from the Mediterranean to mainland Southeast Asia and beyond (Fig. I-4). Life in and near such zones is always affected by risk.

As research continues, the map of the Earth's tectonic plates is still being drawn, so Figure I-3 will not be the last word. But it does help to explain why Africa, once the center of Wegener's postulated supercontinent (which he named *Pangaea*, meaning ''all-earth'') does not have the lengthy, linear mountains all other continents possess. Africa, as we shall see in Chapter 7, is often—and appropriately—called the ''plateau continent.'' It is one of the Earth's oldest pieces of real estate, geologically speaking.

As we investigate each geographic realm, we will take note of the landscapes that form its physical base (Fig. I-4). All the continents contain old geologic cores known as *shields* that often create vast expanses of plains (low-lying flatlands) or plateaus (higher-elevation flatlands). Between these low-relief expanses and the high mountain ranges (Andes, Rockies, Alps, etc.) lie hills of many different sizes. There is no generalizing about the Earth's landscapes, and we will have to examine them one realm and region at a time.

Glaciations

To complicate matters, even as the continental landmasses move, the Earth periodically undergoes *glaciation*, a time during which global temperatures drop and environments

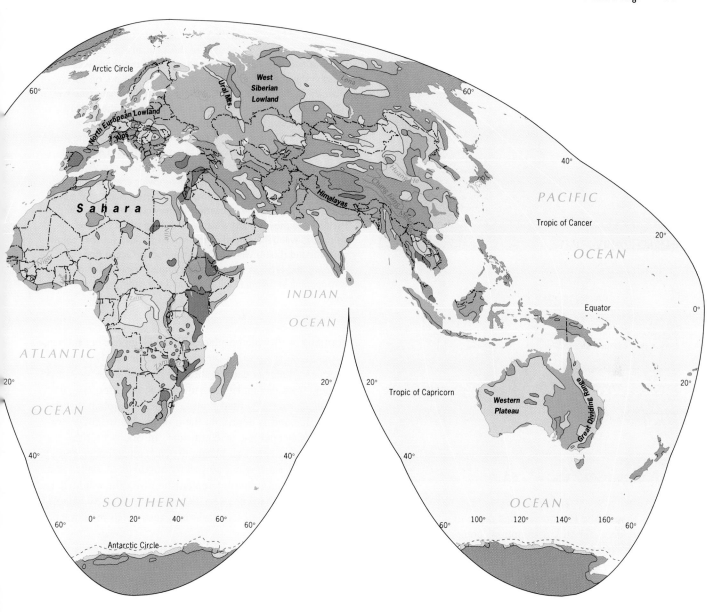

change. A glaciation actually is not just one cooling phase but many. Over millions of years, the temperature swings back and forth, from warmer to colder and back to warmer again; and each time it gets colder, drastic changes occur. Continental-scale, glacial icesheets expand. Sea levels drop. Climates change. Plant life shifts equatorward. Species of animals and plants die out, but others, better adapted, survive and succeed.

Over the past 20 million years, the Earth has been cooling, and during the most recent 5 to 6 million years we have been in the grip of a glaciation—the Late Cenozoic glaciation, as geologists have named it. (They used to call it the *Pleistocene* glaciation until they recently discovered that it started well before the Pleistocene epoch, our epoch of emergence, began.) We humans are the products of that glaciation. Our distant ancestors were among species that managed to adapt when, in their East African homeland,

the climate cooled, trees died out, drought prevailed, and animals long hunted disappeared. In the several million years that followed, the climate warmed and cooled again many times—as many as two dozen times, perhaps more. We seem repeatedly to have survived by adaptation during cold glacial times, then thrived and expanded during the next *interglaciation* (the global warm-up that occurs between two glaciations).

About 15,000 years ago, the icesheets were back for the last (or rather, most recent) time, and the map of the Northern Hemisphere looked as in Figure I-5. Since then, the glaciers have again withdrawn, but this has been no ordinary interglaciation. In the geologic eye-blink of the present interglaciation, the entire modern history of human civilization is playing itself out—from rock shelters to megacities, from stone tools to space shuttles, from human millions to billions. Geologists call ours the Holocene ep-

From the Field Notes
"The land of New Zealand's South Island dropped off precipitously under the waters of the Tasman Sea here, along the southwestern coast. As we approached the shore we crossed the contact zone between the Pacific Plate, on which most of the South Island lies, and the Australian Plate (see Fig. I-3). Subduction along this zone creates some of the world's most spectacular scenery as the crust is pushed precipitously downward, to be recycled in the mantle below . . . On the flank of Mauna Kea, Hawaii, basaltic lava pours forth and new land is created even as it is lost by subduction elsewhere. A soft, steady rumble accompanied this slow-moving eruption that consumed trees and (elsewhere) even houses, reached the sea, and changed the island's coastline in previous phases. As the Pacific Plate moves northwestward, the Big Island's active volcanoes will be deprived of their hot-spot lava sources below, a new island (Loihi) will rise from the Pacific Ocean floor to the east, and erosion will begin to reduce Hawaii's landmass as is happening in Oahu, Maui, Kauai, and the other islands to the west."

och, as though the Pleistocene has come to an end. There is no evidence for this assumption: compared to earlier global glaciations, the present one has just begun.

These days we worry about global warming, and we are right in doing so if only to reduce the pollutants we are spewing into the atmosphere. But over the long term, the greater threat may come from a return of the glaciers, preceded (as in the past) by a time of worldwide weather extremes. In recent weather perturbations, some geographers see a pattern similar to those that presaged earlier cooling and glacial expansion. Imagine it: a world with 6 billion people facing a major reduction in its habitable living space. That would make global warming a mere inconvenience by comparison.

Water—Essence of Life

Water is the key to life on Earth. Without water, our planet would be as barren as the Moon or Mars. It would seem that we should have no shortage of this vital commodity: more than 70 percent of the Earth's surface is covered by the water of the oceans, seas, lakes, and streams. In fact, water shortages occur in many areas of our planet, even in

normally well-watered places. Steadily increasing human populations are putting unprecedented demands on water supplies. Water is carried through modern aqueducts and tunnels from distant rivers and dams to mushrooming metropolises. In the United States, farmers complain that the cities' demand is depriving them of needed irrigation water, as is happening in California. In South Florida, the urban area centered on Miami depends on an underground reservoir of fresh water tapped by hundreds of wells—but when this *aquifer* is not replenished by rain water, it is invaded by salt water from the adjacent ocean. So Miami, like many other cities, occasionally restricts its inhabitants' water use.

Such problems seem minor compared to what is happening elsewhere. Southern Africa, in the early 1990s, was in the grip of the worst drought in its history; in Zimbabwe, a plateau country in this region's interior, almost all the livestock perished, crops failed, and starvation threatened. Death-dealing droughts have struck West Africa, Northeast Brazil, Ethiopia, Central Asia, and other areas during the past 20 years alone.

No place in the world is immune to water shortages. Conflicts loom over actual and proposed dam construction. Wars have been fought over water sources. As human numbers continue to grow and climatic fluctuations increase, the water crisis (for such it already is, for hundreds of millions of us) will become far more significant than oil shortages of the recent past.

Ocean water as such is of little use to humanity: it is made useful through a mechanism that brings moisture from the oceans to the landmasses. This mechanism is the *hydrologic cycle*, which functions as a circulation system. Water evaporates into the air from the salt-water ocean surface, leaving the salt behind; the moisture-laden air mass then drifts over land, where (by various atmospheric

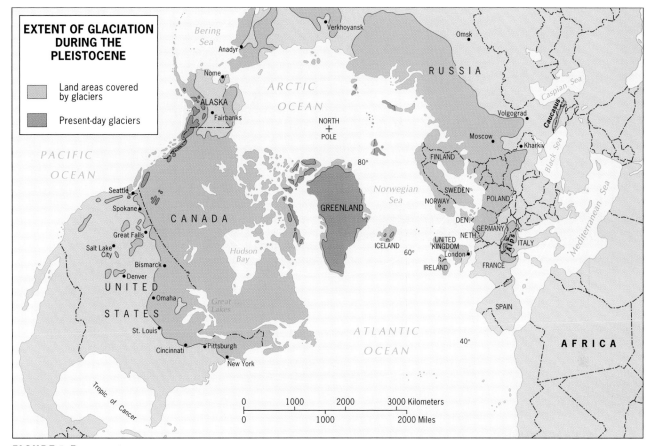

FIGURE I-5

From the Field Notes

"Fly across the western United States toward California, and the cultivated Central Valley looms like a giant oasis in the desert. Water turns the barren countryside into a carpet of green, its limits marked by boundaries as sharp as any on the Earth's surface. The irrigation channel seen in the center of this photograph sustains the crops in the fields; should the wells and mountain snows that feed this ribbon of life ever fail, the desert will be back.... In China's Loess Plateau, water is a precious commodity carried long distances to farm fields in a system of elaborate aqueducts that resemble those the Romans built 2000 years ago in North Africa and the Middle East. From Egypt to Ecuador and from California to China, our planetary surface carries the marks of humanity's quest for water—a quest that will put countries on a collision course as human numbers multiply and regional water sources falter."

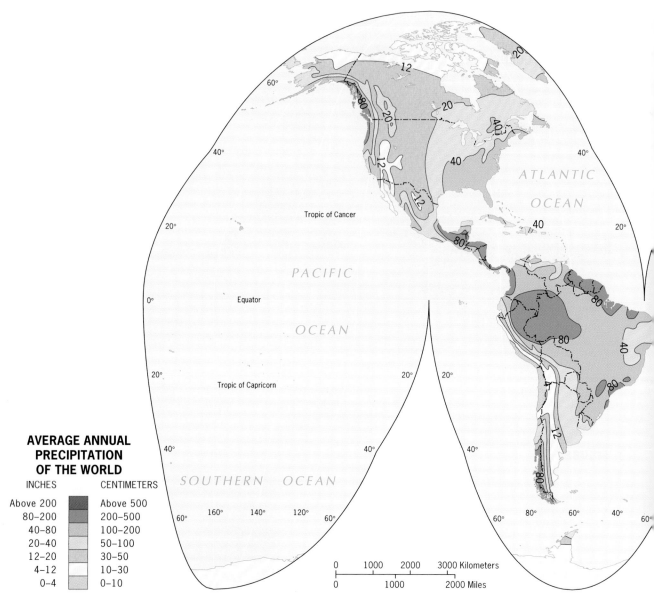

FIGURE I-6

processes) condensation occurs and fresh-water precipitation falls. This precipitation replenishes underground reservoirs, moistens soils, sustains plants and animals, and fills streams. Much of it returns to the oceans as runoff via rivers, and the endless cycle repeats itself.

Precipitation Distribution

The hydrologic cycle is neither constant nor infallible. It can and does fail, imperiling life where and when this happens. It also can be enhanced under certain unusual circumstances—for example, during 1992, when a perturbation known as the El Niño-Southern Oscillation (ENSO), a tropical Pacific climate anomaly, sent a series of moisture-laden air masses toward California and Texas. In

Southern California, then in the grip of a five-year drought, torrential rains brought mudslides as well as relief. In Texas, where the rains lasted longer, they caused devastating floods that drowned thousands of livestock.

Remember this when you examine the global precipitation map (Fig. I-6): it shows the *average* amount of precipitation received by all areas of the world during the comparatively brief time period in which measurements have been taken. Compare it to Figure I-1, and you will gain an impression of the overall water availability in each of the world's major geographic realms.

Figure I-6 reveals an equatorial zone of heavy rainfall, where annual totals often exceed 80 inches (200 centimeters [cm]), extending from mainland Middle America through the Amazon Basin, across smaller areas of West

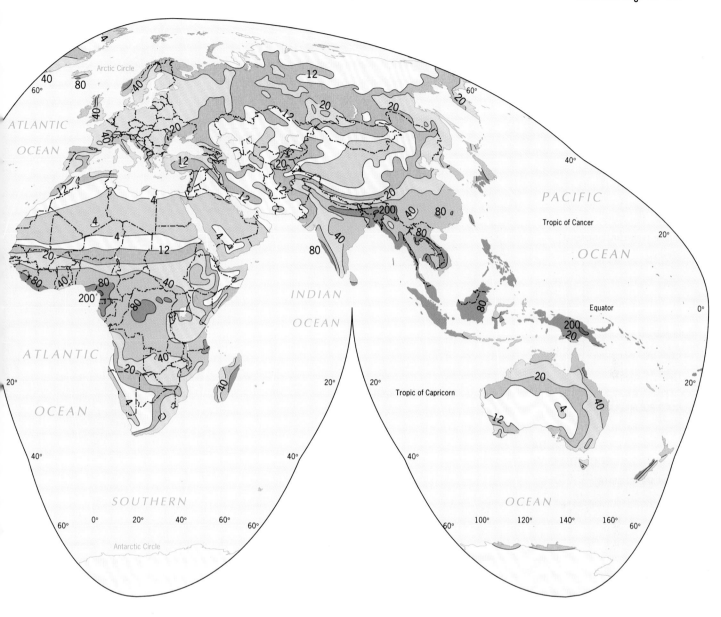

and Equatorial Africa, and into South and Southeast Asia. This low-latitude zone of high precipitation gives way to dry conditions in both poleward directions. In equator-straddling Africa, for example, the arid Sahara lies to the north of the tropical humid zone and the Kalahari Desert lies to the south. Interior Asia and central Australia also are very dry, as is southwestern North America.

The general pattern of equatorial wetness and adjacent dryness is broken along the coasts of all the continents, and it is possible to discern a certain consistency in this spatial distribution of precipitation. Eastern coasts of continents and islands in tropical as well as mid-latitude locations receive comparatively heavy rainfall, as in the southeastern United States, eastern Brazil, eastern Australia, and southeastern China. Furthermore, a narrow zone of higher pre-

cipitation exists at higher latitudes on the western margins of the continents, including the Pacific Northwest coast of the United States and Canada, the southern coast of Chile, the southern tip of Africa, the southwestern corner of Australia, and most important, the western exposure of the great Eurasian landmass—Europe.

The distribution of world precipitation, as reflected in Figure I-6, results from an intricate interaction of global systems of atmospheric and oceanic circulation as well as heat and moisture transfer. Although the analysis of these systems is a part of the subject of physical geography, we should remind ourselves that even a slight change in one of them can have a major effect on a region's habitability. Again, it is important to remember that Figure I-6 displays the average annual precipitation on the continents, but no

WORLD CLIMATES
After Köppen–Geiger

A HUMID EQUATORIAL CLIMATE

Af	No dry season
Am	Short dry season
Aw	Dry winter

B DRY CLIMATE

| BS | Semiarid |
| BW | Arid |

h=hot
k=cold

C HUMID TEMPERATE CLIMATE

Cf	No dry season
Cw	Dry winter
Cs	Dry summer

a=hot summer
b=cool summer
c=short, cool summer
d=very cold winter

D HUMID COLD CLIMATE

| Df | No dry season |
| Dw | Dry winter |

E COLD POLAR CLIMATE

| E | Tundra and ice |

H HIGHLAND CLIMATE

| H | Unclassified highlands |

0   1000   2000   3000 Kilometers
0   1000   2000 Miles

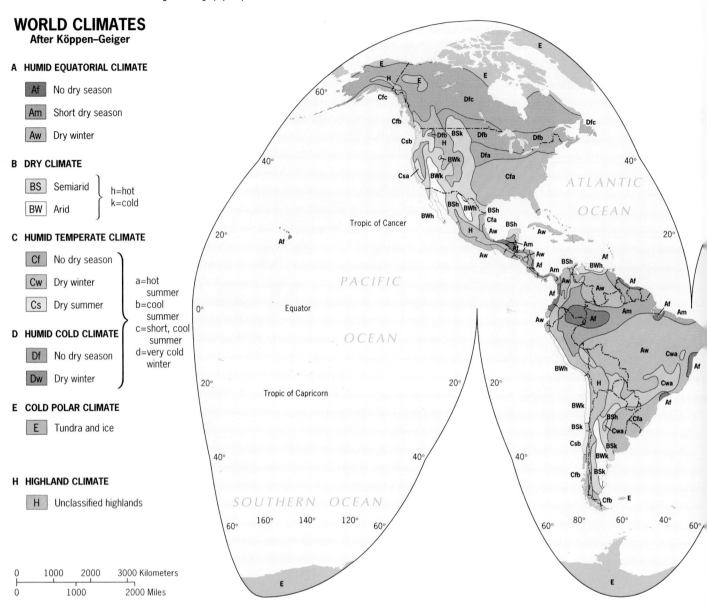

FIGURE I-7

place on Earth is guaranteed to receive, in any given year, precisely its average rainfall. In general, the variability of precipitation increases as the recorded average total decreases. In other words, rainfall is least dependable just where reliability is needed most—in the drier portions of the inhabited world.

Precipitation and Human Habitation

Even heavy year-round precipitation is no guarantee that an area can sustain large, dense populations. In the equatorial latitudes, heavy precipitation combined with high temperatures leads to the faster destruction of fallen leaves and branches by bacteria and fungi. This severely inhibits the formation of humus, the dark-colored upper soil layer consisting of nutrient-rich decaying organic matter, which is vital to soil fertility. The drenched soil is also subjected to the removal of its best nutrients through *leaching*—the dissolving and downward transport of nutrients by percolating water—so that only oxides of iron and aluminum remain at the surface to give the tropical soil its characteristic reddish color. Such tropical soils (called oxisols) nurture the dense rainforest, but they cannot support crops without massive fertilization. The rainforest thrives on its own decaying vegetative matter, but when the land is cleared, the leached soil proves to be quite infertile. Not surprisingly, the Amazon and Zaïre (Congo) basins are not among the world's most populous regions.

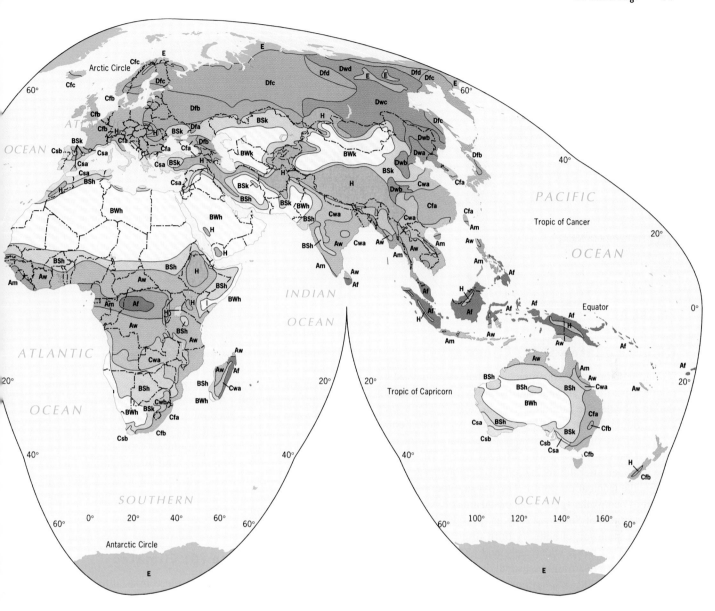

Climatic Regions

It is not difficult to discern the significance of precipitation distribution on the map of world climates (Fig. I-7). Delimiting climatic regions, however, has always presented problems for geographers. In the first place, climatic records are still scarce, short-term, or otherwise inadequate in many parts of the world. Second, weather and climate tend to change gradually from place to place, but the transitions must be presented as authoritative-looking lines on the map. In addition, there is always room for argument concerning the criteria to be used and how these criteria should be weighed. Vegetation is a response to prevailing climatic conditions. Should boundary lines between climate re-

gions, therefore, be based on vegetative changes observed in the landscape no matter what precipitation and temperature records show? The debate continues. World vegetation is mapped in Figure I-8; compare it to Figure I-7 and draw your own conclusions!

Figure I-7 displays a regionalization system devised by Wladimir Köppen and modified by Rudolf Geiger. This scheme, which has the advantage of comparative simplicity, is represented by a set of letter symbols. The first (capital) letter is the critical one: the *A* climates are humid and tropical; the *B* climates are dominated by dryness; the *C* climates are humid and comparatively mild; the *D* climates reflect increasing coldness; and the *E* climates mark the frigid polar and near-polar areas.

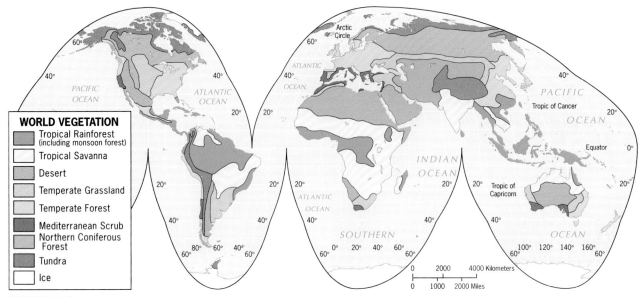

FIGURE I-8

Humid Equatorial (*A*) Climates

The humid equatorial, or tropical, climates are marked by high temperatures all year and by heavy precipitation. In the *Af* subtype, the rainfall arrives in substantial amounts every month; but in the *Am* areas, there is a sudden enormous increase due to the arrival of the annual wet *monsoon* (the Arabic word for "season" [see p. 375]). The *Af* subtype is named after the vegetation association that develops there—the tropical rainforest. The *Am* subtype, prevailing in part of peninsular India, in a coastal area of West Africa, and in sections of Southeast Asia, is appropriately referred to as the monsoon climate. A third tropical climate, the savanna (*Aw*), has a wider daily and annual temperature range and a more strongly seasonal distribution of rainfall. As Figure I-6 indicates, savanna rainfall totals tend to be lower than those in the rainforest zone, and the associated seasonality is often expressed in a "double maximum." This means that each year produces two periods of increased rainfall separated by pronounced dry spells. In many savanna zones, inhabitants refer to the "long rains" and the "short rains" to identify those seasons; a persistent problem in these regions is the unpredictability of the rain's arrival. Savanna soils are not among the most fertile, and when the rains fail the specter of hunger arises on these humid grasslands. Savanna regions are far more densely peopled than rainforest areas, and millions of residents of the savanna subsist on what they manage to cultivate. Rainfall variability under the savanna regime is their principal environmental problem.

Dry (*B*) Climates

Dry climates occur in lower as well as higher latitudes. The difference between the *BW* (true desert) and the moister

BS (semiarid steppe) varies but may be taken to lie at about 10 inches (25 cm) of annual precipitation. Parts of the central Sahara in North Africa receive less than 4 inches (10 cm) of rainfall. A pervasive characteristic of the world's arid areas is an enormous daily temperature range, especially in subtropical deserts. In the Sahara, there are recorded instances of a maximum daytime shade temperature of over 120°F (49°C) followed by a nighttime low of 48°F (9°C). Soils in these arid areas tend to be thin and poorly developed; soil scientists have an appropriate name for them—aridisols.

Humid Temperate (*C*) Climates

As the map shows, almost all these mid-latitude climate areas lie just beyond the Tropics of Cancer and Capricorn (23½° North and South latitude, respectively). This is the prevailing climate in the southeastern United States from Kentucky to central Florida, on North America's west coast, in most of Europe and the Mediterranean, in southern Brazil and northern Argentina, in coastal South Africa and Australia, and in eastern China and southern Japan. None of these areas suffers climatic extremes or severity, but the winters can be fairly cold, especially away from temperature-moderating water bodies. These areas lie about midway between the winterless equatorial climates and the summerless polar zones. Some fertile and productive soils have developed under this regime, as we will note in our discussion of the North American and European realms.

The humid temperate climates range from quite moist, as along the densely forested coasts of Oregon, Washington, and British Columbia, to relatively dry, as in the so-called Mediterranean (dry-summer) areas that include not

From the Field Notes

"A mountain forest near the equator in Kenya, East Africa displayed many of the characteristics of an equatorial rainforest: tall trees with a nearly continuous canopy, great species diversity, much wood and leaf litter. But we noted differences as well: more undergrowth, fewer trees in the middle range of height. Cooler temperatures caused by elevation changed the forest's composition, but the same elevation ensured ample moisture to sustain it . . . The shorelines of the Strait of Magellan, looking toward the Tierra del Fuego side, made us think of those luxuriant African forests. Here the rocks were bare, the slopes snow-dusted, the clouds white with ice rather than dark gray with rain, the water frigid. This scene represented the *E* climates (Fig. I-7), a touch of Antarctica at the southern tip of South America."

only coastal southern Europe and northwestern Africa but also the southwestern tips of Australia and Africa, central Chile, and Southern California. In these Mediterranean environments, the scrubby, moisture-preserving vegetation creates a natural landscape very different from that of richly green Western Europe.

Humid Cold (*D*) Climates

The humid cold (or "snow") climates may be called the continental climates, for they seem to develop in the interior of large landmasses, as in the heart of Eurasia or North America. No equivalent land areas at similar latitudes exist in the Southern Hemisphere; consequently, no *D* climates occur there at all.

Great annual temperature ranges mark these humid continental climates, and very cold winters and relatively cool summers are the rule. In a *Dfa* climate, for instance, the warmest summer month (July) may average as high as 70°F (21°C), but the coldest month (January) might average only 12°F (−11°C). Total precipitation, a substantial part of which comes as snow, is not very high, ranging from over 30 inches (75 cm) to a steppe-like 10 inches (25 cm). Compensating for this paucity of precipitation are cool temperatures that inhibit the loss of moisture via evaporation and evapotranspiration (moisture loss to the atmosphere from soils and plants).

Some of the world's most productive soils lie in areas under humid cold climates, including the U.S. Midwest, parts of southern Russia and Ukraine, and Northeast China. The period of winter dormancy (when all water is frozen) and the accumulation of plant debris during the fall combine to balance the soil-forming and enriching processes. The soil differentiates into well-defined, nutrient-rich layers, and a substantial store of organic humus accumulates. Even where the annual precipitation is light, this environment sustains extensive coniferous forests.

Cold Polar (*E*) and Highland (*H*) Climates

Cold polar (*E*) climates are differentiated into true icecap conditions, where permanent ice and snow keep vegetation from gaining a foothold, and the tundra, where up to four months of the year may have average temperatures above freezing. Like rainforest, savanna, and steppe, the term *tundra* is vegetative as well as climatic, and the boundary between the *D* and *E* climates in Figure I-7 corresponds quite closely to that between the northern coniferous forests and the tundra (Fig. I-8).

Finally, the *H* climates—unclassified highlands mapped in gray (Fig. I-7)—resemble the *E* climates in a number of ways. High elevations and the complex topography of major mountain systems often produce near-Arctic climates above the tree line, even in the lowest latitudes (such as the Andes in western equatorial South America).

Later, when we investigate the reasons behind the location of the world's great clusters of human population, we should remember that water has been important to human life throughout history. The valley and delta of the lower Nile River in northeastern Africa, the basin of the Ganges River in India and Bangladesh, and the plain of the Huang He (Yellow River) in China all contain recent alluvial (river-deposited) soils, and through their nearly legendary fertility they sustain many millions of people. Today, more than 95 percent of Egypt's 65 million people live within 12 miles (20 kilometers) of the Nile waterway. Hundreds of millions of Indians and Chinese depend directly on alluvial soils in the river basins of the Ganges and Huang He, where crops range from corn to cotton, wheat to jute, rice to soybeans. These are only the most prominent examples of such alluvium-based agglomerations. In Pakistan, Bangladesh, Vietnam, and many other countries, the alluvial soils of river valleys provide in abundance what the generally infertile upland soils do not.

These riverine population clusters also provide a link with humanity's past. In the fertile valleys of Southwest Asia's rivers, the art and science of irrigation may first have been learned. And two of the world's oldest continuous cultures—Egypt and China—still retain heartlands near their ancient geographic hearths of thousands of years ago.

REGIONS AND CULTURES

Whenever we explore a geographic realm or region, it is important to assess the physical stage that forms its base—its total physical geography, or **physiography**. Still, the realms and regions to be discussed in the chapters that follow are defined primarily by human-geographic criteria, and one of those criteria, culture, is especially significant. Therefore, we should carefully consider the culture concept in a regional context.

Anthropologists, when they define the term **culture**, tend to concentrate on abstractions: learning, knowledge and its transmission, and behavior. More than a half century ago, Ralph Linton defined culture as ''the sum total of the knowledge, attitudes, and habitual behavior patterns shared and transmitted by the members of a society.'' Marvin Harris, in 1971, wrote that culture is ''the learned patterns of thought and behavior characteristic of a population or society.'' Literally hundreds of definitions of the culture concept exist. As with the regional concept in geography, we are reminded that such definitions are arbitrary and designed for particular purposes.

Geographers are most interested in the imprints of culture and its associated patterns of behavior on the landscape. Thus we will examine how members of a society perceive and exploit their available resources, the way they maximize the opportunities and adapt to the limitations of their natural environment, and the way they organize the portion of the Earth that is theirs. Human works remain etched on the Earth's surface for a very long time: Egyptian pyramids, Chinese walls, and Roman roads and bridges still mark countrysides millennia later. Over time, regions take on certain dominant qualities that collectively constitute a regional character, a personality, a distinct atmosphere. This is one of the crucial criteria in our division of the human world into major geographic realms and regions.

Cultural Landscapes

As we noted, geographers are particularly concerned with the impress of culture on the Earth's physical surface. Culture is expressed in many ways as it gives visible character to a region. Often a single scene in a photograph or picture can reveal, in general terms, in which part of the world it was made. The architecture, forms of transportation, goods being carried, and clothing of the people (all these are part of culture too) reveal enough to permit a guess as to which region of the world is represented. This is so because the people of any culture are active agents of change; when they occupy their portion of the Earth's available space, they transform the land by building structures on it, creating lines of transport and communication, parceling out the fields, and tilling the soil (among countless other activities).

The composite of human imprints on the Earth's surface is called the **cultural landscape**, a term that came into general use in geography during the 1920s. Carl Ortwin Sauer (for several decades a professor of geography at the University of California, Berkeley) developed a school of cultural geography that focused on the concept of cultural landscape. In a paper written in 1927 and titled ''Recent Developments in Cultural Geography,'' Sauer proposed his most straightforward definition of the cultural landscape: *the forms superimposed on the physical landscape by the activities of man.* He stressed that such forms result from the operation of cultural processes—causal forces that shape cultural patterns—that unfold over a long period of time and involve the cumulative influences of successive occupants.

Sometimes these successive groups are not of the same culture. Settlements built by European colonizers over a century ago are now occupied by Africans; minarets of Islam still rise above the buildings of certain Eastern European cities, recalling an earlier period of dominance by the Muslim Ottoman Empire. In 1929, Derwent Whittlesey introduced the term *sequent occupance* to categorize these successive stages in the evolution of a region's cultural landscape.

The durability of the concept of cultural landscape is underscored by its redefinition in 1984 by the eminent scholar John Brinckerhoff Jackson. His statement closely parallels Sauer's: *a composition of man-made or man-modified spaces to serve as infrastructure or background for our collective existence.* Thus the cultural landscape consists of buildings and roads and fields—and more. But it also possesses an intangible quality, an atmosphere or flavor, a sense of place that is often easy to perceive yet difficult to define. The smells and sights and sounds of a traditional African market are unmistakable, but try re-

cording those qualities on maps or in some other objective way for comparative study!

More concrete properties are easier to observe and record. Take, for instance, the urban ''townscape''—a prominent element of the overall cultural landscape—and compare a major U.S. city with, say, one in Japan. Visual representations of these two metropolitan scenes would reveal the differences quickly, of course, but so would maps of the two urban places. The American city with its rectangular layout of the *central business district* (*CBD*) and its widely dispersed, now heavily urbanized suburbs contrasts sharply with the clustered, space-conserving Japanese metropolis. Using a rural example, we see that the spatially lavish subdivision and ownership patterns of American farmland look unmistakably different from the traditional African countryside, with its irregular, often tiny patches of land surrounding a village. Still, the totality of a cultural landscape can never be captured on a photograph or map because the personality of a region involves far more than its prevailing spatial organization. One must also include its visual appearance, its noises and odors, the shared experiences of its inhabitants, and even their pace of life.

From the Field Notes

"The Atlantic-coast city of Bergen, Norway displayed the Norse cultural landscape more comprehensively, it seemed, than any other Norwegian city, even Oslo. The high-relief site of Bergen creates great vistas, but also long shadows; windows are large to let in maximum light. Red-tiled roofs are pitched steeply to enhance runoff and inhibit snow accumulation; streets are narrow and houses clustered, conserving warmth . . . The coastal village of Mengkabong on the Borneo coast of the South China Sea represents a cultural landscape seen all along the island's shores, a stilt village of the Bajau, a fishing people. Houses and canoes are built of wood as they have been for centuries. But we could see some evidence of modernization: windows filling wall openings, water piped in from a nearby well."

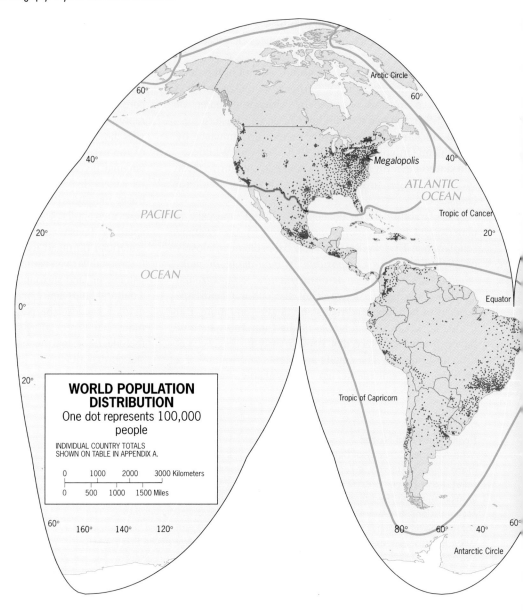

FIGURE I-9

Culture and Ethnicity

Language, religion, and other cultural traditions often are durable and persistent. Culture is not necessarily based in ethnicity, however, and as we study the world's human-geographic regions, we should be aware that peoples of different ethnic stocks can achieve a common cultural landscape, while people of the same ethnic background can be divided along cultural lines.

The recent events in the former Eastern European country of Yugoslavia are a good case in point. As the old Yugoslavia broke up in the early 1990s, its component parts were first affected and then engulfed by what was often described as "ethnic" conflict. When the crisis reached Bosnia-Herzegovina in the heart of Yugoslavia,

three groups fought a bitter civil war. These groups were identified as Bosnians, Serbs, and Muslims—but in fact they were all ethnic Slavs. (Yugoslavia means "Land of the South Slavs.") What distanced them from one another, and what kept the conflict going, was cultural tradition, not ethnicity. The Bosnians and the Serbs had developed different communities in different parts of the former Yugoslavia. The Muslims were descendants of Slavs who, a century ago or more, had been converted by the Turks (who once ruled here) to Islam. All three of these groups feared domination by the others, and so South Slav turned against South Slav in what was, in truth, a culture-based conflict.

Even though the post–Cold War world is very young, it has already witnessed numerous intraregional conflicts;

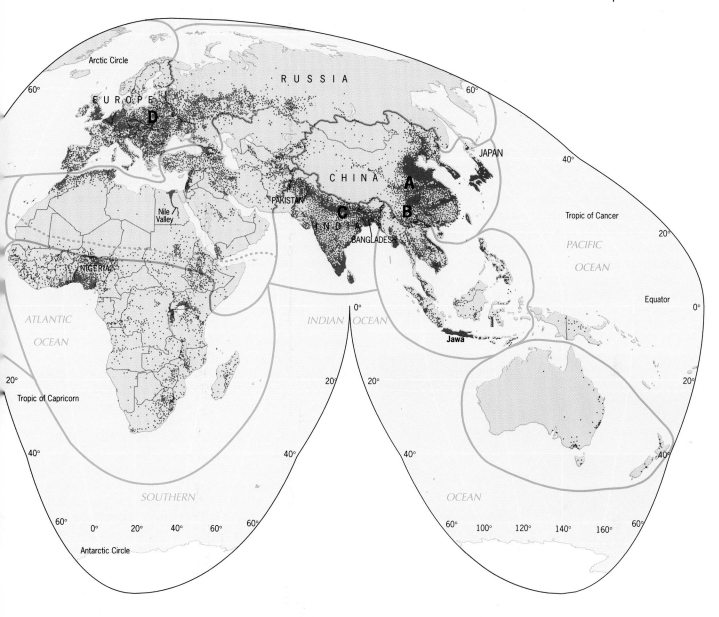

in later chapters, we will explain some of these in geographic context. Not all of them have ethnicity at their roots. Culture is a great unifier; it can also be a powerful divider.

REALMS OF POPULATION

Earlier we noted that population numbers by themselves do not define geographic realms or regions. Population distributions, and the functioning society that gives them common ground, are more significant criteria. That is why it is possible to identify one geographic realm with barely more than 20 million people (Australia) and another with 1.5 billion inhabitants (the East Asian realm). Neither population numbers nor territorial size alone can delimit a geographic realm.

Nevertheless, the map of world population distribution (Fig. I-9) suggests the relative location of several of the world's geographic realms, based on the strong clustering in certain areas. Before we examine these clusters in some detail, let us remember that the Earth's human population is fast approaching 6 billion—6 thousand million people confined to the landmasses that constitute less than 30 percent of our planet's surface, much of which is arid desert, rugged mountain terrain, or frigid tundra. (Remember that

Fig. I-9 is a mid-1990s still picture of an ever-changing scene; the explosive growth of humankind continues.) After thousands of years of relatively slow growth, world population during the past two centuries has been expanding at an increasing rate. It took about 17 centuries following the time of the birth of Christ for the world to add 250 million people; now that same number is being added about every *two and a half years*!

Demographic (population-related) issues will arise repeatedly as we survey the world's most crowded realms in later chapters. For the moment, we will confine ourselves to the overall global situation. For that purpose it is interesting to compare the map of world population distribution to the demographic data arrayed in Appendix A (pp. A-1– A-15). This table provides information about the total populations in each of the geographic realms as well as their individual countries. Also, if you compare Figure I-9 to the maps of terrain (Fig. I-4), precipitation (Fig. I-6), and climate (Fig. I-7), you will note that some of the world's

largest population concentrations lie in the fertile basins of major rivers (China's Huang He and Chang Jiang, India's Ganges). We live in a modern world, but old ways of life still prescribe the location of hundreds of millions on this Earth.

Major Population Clusters

The world's greatest population cluster, **East Asia**, lies centered on China and includes the Pacific-facing Asian coastal zone from the Korean Peninsula to Vietnam. The map indicates that the number of people per unit area—the **population density**—tends to decline from the coastal zone toward the interior. Note, however, the ribbon-like extensions marked *A* and *B* (Fig. I-9). These, as the map of world landscapes (Fig. I-4) confirms, are populations concentrated in the valleys of China's major rivers, the Huang He and the Chang Jiang (the Yellow and Yangzi [or Long], respectively). Here in East Asia, the great majority of the people are farmers, not city dwellers. True, there are great cities in China (such as Beijing and Shanghai), but the total population of these urban centers is far outnumbered by the farmers—those who live and work on the land and whose crops of rice and wheat feed not only themselves but also the people in those cities.

The **South Asia** population cluster lies centered on India and includes the populous neighboring countries of Bangladesh and Pakistan. This huge agglomeration of humanity focuses on the broad plain of the lower Ganges River (*C* in Fig. I-9). This cluster is nearly as large as that of East Asia and at present growth rates will overtake East Asia early in the next century. As in East Asia, the overwhelming majority of the people are farmers, but here in South Asia the pressure on the land is even greater, and farming is not as efficient. As we note in Chapter 8, the population issue looms very large in this realm.

From the Field Notes

"We watched as bulldozers attacked the tropical forest inland from the East Malaysian city of Kota Kinabalu on the northwestern coast of the island of Borneo, knocking down the trees, hauling them to a clearing, and stacking them for burning. A supervisor explained that this would be the site of a new suburb of "KK," complete with shopping center, school, and medical clinic. 'Our population is growing fast,' he said, 'and soon we'll go even deeper in.' Borneo is not yet a densely peopled island, but its natural environments, habitats, and wildlife are under increasing threat . . . In the port of Kota Kinabalu we saw another side of deforestation in Sabah. Stacks of sawn lumber from the tropical forest lay ready for export (the ship in the background was bound for Japan); trucks arrived continuously, loaded with wood. When countries have few economic options, they sell their natural heritage to high bidders in the wealthier world."

The third-ranking population cluster, **Europe**, also lies on the world's biggest landmass but at the opposite end from China. The European cluster, including western Russia, counts over 700 million inhabitants, which puts it in a class with the two larger Eurasian concentrations—but there the similarity ends. Here, the key to the linear, east-west orientation of the axis of population (*D* in Fig. I-9) is not a fertile river basin but a zone of raw materials for industry. Europe is among the world's most highly urbanized and industrialized realms, its human agglomeration sustained by forges and factories rather than paddies and pastures.

The three world population concentrations just discussed (East Asia, South Asia, and Europe) account for more than 3.3 billion of the world's 5.9 billion people. Nowhere else on the globe is there any population cluster with a total of even half of any of these. Look at the dimensions of the landmasses on Figure I-9 and consider that the populations of South America, Subsaharan Africa, and Australia together total less than that of India alone. In fact, the next-ranking cluster is **Eastern North America**, comprising the east-central United States and southeastern Canada; however, it is only about one-quarter the size of the smallest of the Eurasian concentrations. As Figure I-9 clearly shows, this region does not possess the large, contiguous, high-density zones of Europe or East and South Asia. Like the European cluster, much of the population of this region is concentrated in several major metropolitan centers; the rural areas remain relatively sparsely settled. The heart of the North American cluster lies in the urban complex that lines the U.S. northeastern seaboard from Boston to Washington, D.C., and includes New York, Philadelphia, and Baltimore—the great multimetropolitan agglomeration urban geographers refer to as *Megalopolis.*

Urbanization and Population Growth

The degree to which people in the world's geographic realms and regions are concentrated in cities, therefore, is a defining property of those regions. This process of **urbanization** goes on in all the world's realms, and in 1997, 43 percent of humankind (2.5 billion) resided in cities and towns. Urbanization, however, is occurring faster in some realms than in others. South America, for instance, has nearly 75 percent of its population agglomerated in urban areas; Subsaharan Africa is only 27 percent urbanized. And *within* these realms, some regions are more urbanized than others.

Another relevant criterion in the assessment of geographic realms and regions is their rate of **population growth**. Of the four major clusters on the global population map (Fig. I-9), the two largest also are growing faster compared to the third and fourth (Europe and North America, respectively). Size discrepancies among these concentrations, therefore, are increasing. The issue of population growth will come up repeatedly as we discuss the countries and societies in fast- and slow-growing regions, because economic progress and demographic conditions are closely intertwined.

REALMS, REGIONS, AND STATES

Whenever geographers are confronted by the need to define (and justify!) boundaries, they look for precedents—for example, an existing grid that can be put to use or data from a previous effort that may prove helpful. A case in point is the Township-and-Range system of land division in the United States west of the Appalachian Mountains. When you fly over this area, you will see a persistent rectangular pattern on the ground, the product of the Ordinance of 1785, when it was decided to survey and delimit this then-remote area before it would be opened to settlement. The country beyond the Appalachians was divided into squares of 6 by 6 miles, so that each square (or "township") contained 36 square miles. Next, each of these square miles (640 acres) was subdivided into four 160-acre squares, which was to be the smallest parcel of land a settler could purchase (later, smaller parcels were also allowed). Today, the Township-and-Range system is etched into the entire cultural landscape, having controlled the location of towns, roads, farms, and fields (see photo p. 6). For geographers, the system is a great help: when it comes to delimiting a Corn Belt, a Wheat Belt, or some other economic region, this ready-made grid is invaluable.

On a global scale we have at our disposal only one existing framework, and it is not a regular grid: the international boundary system that marks the territorial limits of the world's 180-plus countries (Fig. I-10). As the map shows, countries range in size from the largest, Russia (6.6 million square miles [17.1 million sq km]), to entities so tiny they cannot be shown. Irregular as this framework may be, it nonetheless helps us in our effort to define regions within the world's great geographic realms. Although we will often refer to the political entities shown in Figure I-10 as countries, the appropriate geographic term for them is *states.*

The **state** has been developing for thousands of years, ever since agricultural surpluses made possible the growth of large and powerful cities that could command hinterlands and control peoples far beyond their walls. But the modern state is a relatively recent phenomenon. Just a little over a century ago, there still were open, unclaimed, unbounded areas that sometimes served as buffer zones be-

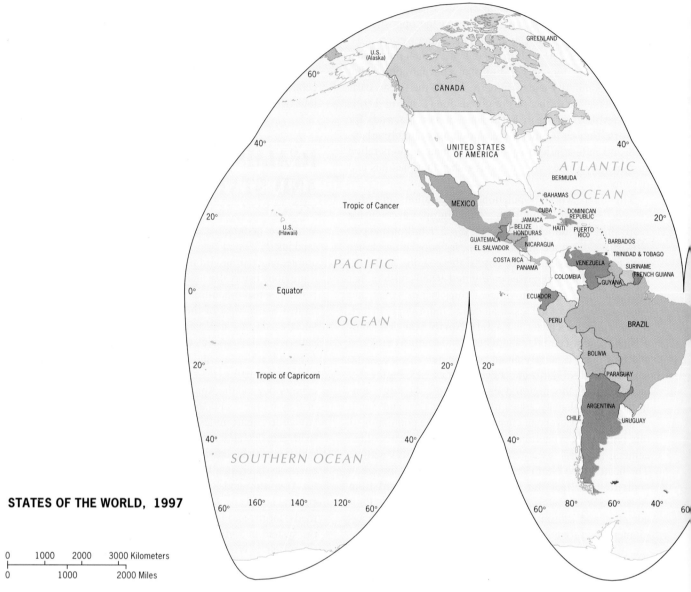

STATES OF THE WORLD, 1997

FIGURE I-10

tween rival states. The boundary framework we see on the world political map today substantially came about during the nineteenth century, and the independence of dozens of former colonies (which made them states as well) occurred during the twentieth century. Today, the *European state model*—a clearly and legally defined territory inhabited by a population governed from a capital city by a representative government—is gaining ground in the aftermath of collapsed colonial and communist empires.

As Figures I-1 and I-10 suggest, geographic realms mostly are assemblages of states, and the borders between realms frequently coincide with the boundaries between countries—for example, between North America and Middle America along the U.S.–Mexico boundary. But a realm boundary can also cut *across* a state, as in the case of the one between Subsaharan Africa and the Muslim-dominated

realm of North Africa/Southwest Asia. Here the boundary takes on the properties of a wide transition zone, but it still divides states such as Chad and Sudan. The transformation of the margins of the former Soviet Union, too, is creating similar cross-country transitions. Newly independent states such as Belarus (between Europe and Russia) and Kazakstan (between Russia and Muslim Southwest Asia) lie in zones of regional change.

Most often, however, geographic realms consist of groups of states whose boundaries also mark the limits of the realms. Look at the case of Southeast Asia, for instance. The northern border of this geographic realm coincides with the political boundary that separates China (a realm practically unto itself) from Vietnam, Laos, and Myanmar (Burma). Its western border is defined by the boundary between Myanmar and Bangladesh (which is part of the

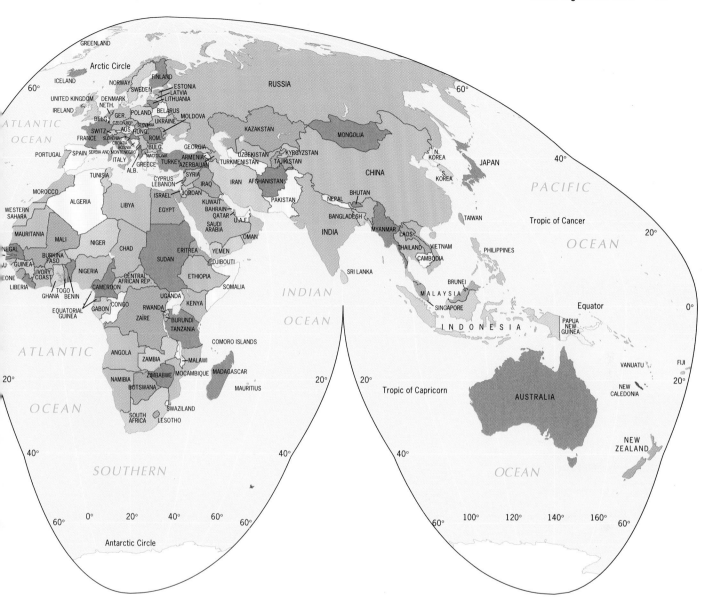

South Asian realm). Here, the state boundary framework helps delimit geographic realms.

The world boundary framework is even more useful in the delimitation of regions *within* geographic realms. We shall discuss regional divisions every time we introduce a geographic realm, but an example is appropriate here. In the Middle America realm, we recognize four regions. Two of these lie on the mainland: Mexico, the giant of the realm, and Central America, the seven comparatively small countries located between Mexico and the Panama-Colombia border (which is also the boundary with the South American realm). Central America often is misdefined in news reports; the correct regional definition is based on the politico-geographical framework.

To our earlier criteria of physiography, population distribution, and cultural geography, therefore, we now add political geography as a determinant of world-scale geographic regions. In doing so we should be aware that the global boundary framework continues to change and that boundaries are created (as in 1993 between the Czech Republic and Slovakia) as well as eliminated (e.g., between former West and East Germany in 1990). But the overall system, much of it resulting from colonial and imperial expansionism, has turned out to be quite durable, despite the predictions of some geographers that the "boundaries of imperialism" would be replaced by newly negotiated ones in the postcolonial period.

Toward the end of this book, we will take note of a recent, ominous development in boundary-making: the extension of boundaries onto and into the oceans and seas. It is a process that has been consuming the last of the Earth's open frontiers, with consequences that are yet uncertain.

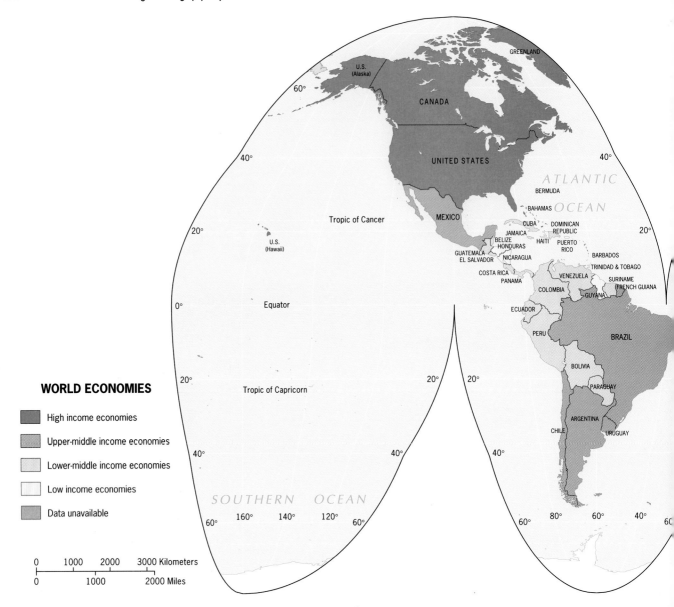

FIGURE I-II

PATTERNS OF DEVELOPMENT

Finally, we turn to economic geography for criteria that allow us to place countries, regions, and realms in regional groupings based on their level of **development**. The field of economic geography focuses on spatial aspects of the ways people make a living and thus deals with patterns of production, distribution, and consumption of goods and services. As with all else in this world, these patterns exhibit considerable variation. Individual states report their imports and exports, farm and factory production, and many other economic data to the United Nations and other

international agencies. From such information it is possible to determine the comparative economic well-being of the world's countries.

The World Bank, one of the agencies that monitor economic conditions, classifies countries into four groups: (1) high-income, (2) upper-middle-income, (3) lower-middle-income, and (4) low-income countries. These groupings display regional clustering when mapped (Fig. I-11). The higher income economies are concentrated in Europe, North America, and along the western Pacific Rim, most notably Japan and Australia. The low-income countries dominate in Africa and parts of Asia.

As Figure I-11 shows, several geographic-realm boundaries can be discerned on this economic map, including that between North America and Middle America, between

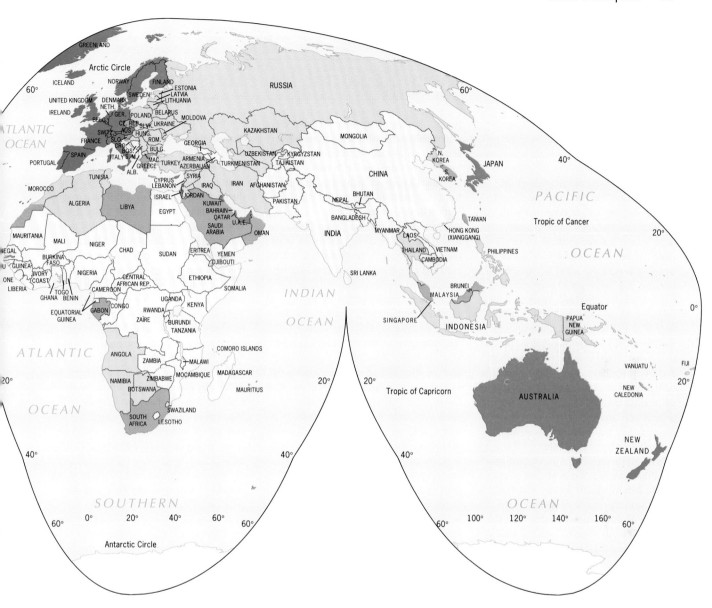

Australia and Southeast Asia, and between Russia and East Asia. Within the great geographic realms, you can see some distinct regional boundaries: between Western and Eastern Europe, between Brazil and Andean South America (the region to the west), and between the oil-rich Arabian Peninsula and the countries of the Middle East to its northwest.

Figure I-11 reveals the level of economic development of the world's countries—based, we should remind ourselves, on the political framework as the reporting grid. Until very recently, economists and economic geographers divided the world, on this basis, into developed countries (DCs) and underdeveloped countries (UDCs). But that distinction has lost much of its relevance, for reasons Figure I-11 cannot show. Today, many countries, whatever their level of income, display cores of development (where they often resemble the richest societies) as well as peripheries of poverty and underdevelopment (see box titled "Core-Periphery Regional Relationships"). Some comparatively wealthy countries, for example, Spain and Saudi Arabia, still include areas of severe underdevelopment. Other countries, mapped as low-income economies, have given rise to bustling, skyscrapered cities whose streets are jammed with luxury automobiles and whose shops are crammed with opulent goods. As the rich cores get richer and the poor peripheries get poorer, the averages reported as "national" economic statistics lose much of their geographic meaning. Perhaps an index of *regional disparity* within each economy would give us a truer picture.

It is therefore no longer appropriate to divide the world

■ ■

Core-Periphery Regional Relationships

Ours is a world of uneven development. The advantages lie with the developed countries (DCs), and economic indicators show that the gap between the DCs and many underdeveloped countries (UDCs) is growing. How has this situation become so extreme? And what perpetuates it?

Regional contrasts in development have been attributed to many factors. Climate has been blamed for it (as a conditioner of human capacity), as has cultural heritage (e.g., resistance to innovation), colonial exploitation, and, more recently, neocolonialism. Obviously, the distribution of accessible natural resources and the territorial location of countries play a major part. Opportunities for interaction and exchange have always been greater in certain areas than in others. Isolation from the mainstreams of change has been a disadvantage. These factors, to a greater or lesser degree depending on location, certainly have played their roles.

One of the hallmarks of national spatial organization is the evolution of **core areas**, foci of human activity that function as the leading regions of control and change. Most commonly today, we tend to associate this regional concept with a country's heartland—its largest population cluster, most productive and influential region, the area possessing the greatest centrality and accessibility that usually contains a powerful capital city.

This is nothing new. Anthropologists have unearthed evidence of the earliest states, proving that city-dominated cores formed their foci. These were small urban

centers, but later they were recast as city-states, culture hearths, seats of empires, and sources of modern technological revolutions. Consider this: such cores would not have developed without contributions from their surrounding areas. One of the earliest developments in ancient cities was the creation of organized armed forces to help rulers secure taxes and tribute from the people living in the countryside. Thus core and periphery (margin) became functionally tied together: the core had requirements, and the periphery gave.

Have modern times ended this core-periphery relationship? The answer not only is negative—the situation has actually worsened. You can see core (*have*) and periphery (*have-not*) relationships in every country in the world. In many African countries (Kenya, Senegal, and Zimbabwe, to name just three examples in different regions), the former colonial capital now lies at the focus of the national core area. Its tall buildings and government centers symbolize the concentration of privilege, power, and control. Here are the trappings of development, the paved roads and streetlights, hospitals and schools, corporate headquarters and businesses, industries and markets. Surrounding the capital are farms producing high-priced goods, especially perishables, for the city dwellers who can afford to buy them. But a little farther out into the countryside, life swiftly changes. In the periphery, the core area is distant, even ominous—always a concern. What will be paid for crops on the markets in the core? What will the

cost be of goods made there and needed here, in the countryside? How high will taxes be? This core-periphery arrangement is a system that keeps regional development contrasts high. The core may give the impression of being a thriving metropolis and may remind one of a city in a developed country. But the periphery is the land where underdevelopment reigns.

Looking at the world as a whole from this perspective, we see that core-periphery systems now encompass not just individual countries but the entire globe. Today, the whole of Western Europe functions as a core area; North America forms a second core area; and Japan is a third. Therefore, what was said about that powerful city-region in an otherwise underdeveloped country applies, at the macroscale, to these global core areas as well. Whether in Europe, North America, or Japan, power, money, influence, and decision-making organizations are concentrated here.

Imagine that you, as a geographer, are asked to locate a factory (say, a textile plant) to sell products in core-area markets. Salaries and wages in the core area are high, so you look for an alternative in the periphery, where wages are much lower. Raw materials also can be bought locally. So you advise the corporation to establish its factory in a certain country in the periphery, thereby initiating a relationship between core-based business and periphery-based producers. Now, however, the corporate management expresses concern over political stability in that peri-

■ ■ ■

geographically into developed and underdeveloped realms. East Asia is dominated by low-income economies, but it includes Japan, one of the world's richest states. Europe is one of the world's highest-income realms, but not in Albania, Moldova, or Ukraine.

Not long ago, countries also were categorized as being part of a First World (capitalist), Second World (socialist/communist), Third World (underdeveloped), Fourth World (severely underdeveloped), or even Fifth World (poorest).

This classification, too, has lost most of its significance in the fast-changing world of the 1990s. The most meaningful distinction we can presently make is based on the notion of *advantage*: there are countries that have it and many others that do not. Such advantage takes many forms: geographic location (to be on the coast is better than to be landlocked!), raw materials, governmental system, political stability, productive skills, and much more. As the gap between the advantaged and the disadvantaged states widens phery country. Soon you are talking to government representatives about guarantees, and the economic connection also becomes a political one.

Is all this necessarily disadvantageous to the countries in the periphery, the mainly underdeveloped countries, the countries of the disadvantaged world? After all, the core-area corporations and businesses make investments, stimulate economies, and put people to work in the countries of the periphery. The problem is that a dependency relationship develops that soon tends to diminish any such advantage. The profits earned return mostly to the core area and benefit the periphery only minimally. When you see that ultramodern high-rise hotel towering above the sandy beaches of a Caribbean island and watch its staff at work, you observe both sides of it. True, an investment was made here and jobs were created. But the hotel's profits go back to the multinational corporation's bank account—in New York, Tokyo, or London.

Another problem has to do with the sociopolitical effects. That textile factory we mentioned needs a manager and other administrative personnel. These people tend to be (or become) part of the *elite*, the "upper class" in that periphery country, a kind of cadre of representatives of the world core area in the disadvantaged realms. Because they represent core-area interests, these people have divided loyalties: what is best for core-area enterprises is not always in the best interest of their home country. Such examples of core-periphery interactions abound, and the en-

tire world is now interconnected to a far greater degree than many of us realize. Superimposed on the traditional cultural landscapes of the disadvantaged countries is a growing global network of interaction that reaches into even the smallest village store in the remotest part of a UDC. The numerous geographic realms and regions we will come to recognize in this book are, above all, enmeshed in a worldwide economic-geographic system—a system tightly oriented to the world's core areas.

Thus the countries of the periphery find themselves locked within a global economic system over which they have no control. Even countries that find themselves possessing commodities in high demand in the core (such as the oil of the OPEC states) have difficulty converting their temporary advantage into longer-term parity with core-area powers. Countries that depend heavily for their income on the export of raw materials such as strategic minerals (or single crops such as bananas or sugar) are at the mercy of those who do the buying. But there are other reasons for the depressing condition of many periphery countries. Like the contemporary city, the world core area beckons constantly, siphoning off the skilled and professional people from the periphery. How many doctors, engineers, and teachers from India are working in England and the United States? And how badly India now needs such trained people! Every loss is magnified—but oppressive political circumstances in periphery countries contribute as "push" factors in this "brain drain."

Another aspect of the continuing underdevelopment of periphery countries involves traditional cultural attitudes and values. The introduction of core-area tentacles into outlying regions where traditional culture is strong may lead to a reaction, perhaps a resurgence of fundamentalism. This, in turn, restricts further economic change. The invasion of modern ways can be unsettling. Do not forget that the long traditions of societies in the periphery have generated large bureaucracies whose interests are threatened by the kind of change development brings with it. Occasionally, the tangible presence of the external (core) installation such as an office building, airline agency, or retail establishment is the target of violent attack. Such incidents symbolize ideological opposition to the intrusion they represent.

The countries of the periphery, therefore, confront severe problems of many kinds. They are pawns in a global economic game whose rules they cannot touch, let alone change. Their internal problems are intensified by the aggressive involvement of core-area interests. Their resource use (such as allocation of soils to producing local food versus export crops) is strongly affected by foreign interference. They suffer far more than core-area countries do from environmental degradation, overpopulation, and mismanagement. They possess inherited disadvantages that have grown, not lessened, over time. The widening gaps that result between enriching cores and persistently impoverished peripheries clearly are a threat to the future of the world.

(and it *is* widening), the stability of the political world is at risk, and with it the advantages of the "haves" over the "have nots."

As we proceed with our investigation of the world's geographic realms, the concept of development will be examined in regional context and from several viewpoints. To a considerable degree, Figure I-11 is a continuing reflection of events that began long before the Industrial Revolution: Europe, even by the middle of the eighteenth cen-

tury, had laid the foundations for its colonial expansion. The Industrial Revolution then magnified Europe's demands for raw materials, and its manufactures increased the efficiency of its imperial control. Western countries thereby gained an enormous head start, while colonial dependencies remained suppliers of resources and consumers of the products of the Western industries. Thus was born a system of international exchange and capital flow that changed very little when the colonial period came to an end. Developing countries, well aware of their predicament, accused the developed world of perpetuating its long-term advantage through *neocolonialism*—the entrenchment of the old system under a new guise.

Symptoms of Underdevelopment

Although it is no longer appropriate to divide the world into developed and underdeveloped geographic realms, make no mistake: there still *are* regions within realms, and countries within regions, that suffer from underdevelopment. As we will discover during our global journey, underdevelopment takes many forms and displays many

symptoms; it also has numerous causes. To get an idea of some of the symptoms, see Appendix A and compare, say, Bangladesh or Moçambique with Japan or Canada. Population growth rates in the UDCs are higher, life expectancies shorter, urbanization rates lower, incomes smaller. As we will find, the poorer countries suffer from high infant and child mortality rates, poor overall health and sanitation, inefficient farming, unsatisfactory diets, hugely overcrowded cities, and many other ills. Many UDCs are trapped in a global economic system in which the export of unfinished raw materials is their only source of outside income.

No doubt the world economic system works to the disadvantage of the UDCs, but sadly it is not the only obstacle that the less advantaged countries face. Political instability, corruptible leaderships and elites, misdirected priorities,

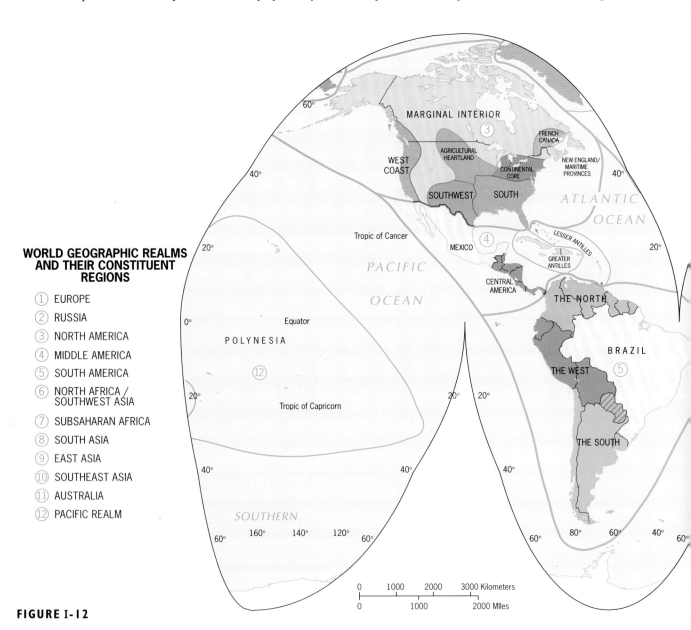

WORLD GEOGRAPHIC REALMS AND THEIR CONSTITUENT REGIONS

① EUROPE
② RUSSIA
③ NORTH AMERICA
④ MIDDLE AMERICA
⑤ SOUTH AMERICA
⑥ NORTH AFRICA / SOUTHWEST ASIA
⑦ SUBSAHARAN AFRICA
⑧ SOUTH ASIA
⑨ EAST ASIA
⑩ SOUTHEAST ASIA
⑪ AUSTRALIA
⑫ PACIFIC REALM

FIGURE I-12

misuses of aid, and traditionalism are among the conditions that commonly inhibit development. External interference by interests representing powerful DCs has also had negative impacts on the economic as well as political progress of many UDCs, especially during the period of the Cold War, when domestic strife was magnified into major conflict when the United States and the now-defunct Soviet Union took opposite sides. Angola, Ethiopia, Afghanistan, Vietnam, Nicaragua, and other UDCs suffered incalculably as a result of such superpower involvement.

National development, therefore, remains a powerful factor in the delimitation of regions within geographic realms. Economic development is in part a legacy of earlier conditions in our changing world, affected by colonialism, superpower conflict, and other circumstances. Yet it is only one criterion in our geographic regionalization.

THE REGIONAL FRAMEWORK

At the beginning of this Introduction, we outlined a map of the great geographic realms of the world (Fig. I-1). We then addressed the task of dividing these realms into regions, and we enumerated several criteria to be used for that purpose. The result is Figure I-12. This global framework is much more than a regionalization of cultural landscapes: it also reflects criteria of physical, political, economic, urban, and historical geography. It is a spatial synthesis of human geography in all its manifestations, not cultural geography alone.

Twelve geographic realms form the fundamental structure of our global survey. Of course, we will examine the bases for the delimitation of each one of them, but our

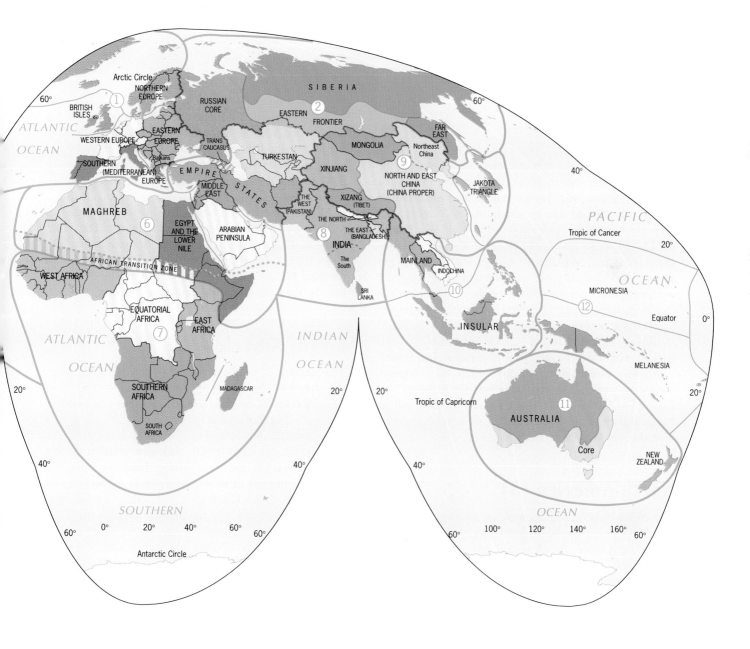

focus will be on the regions they contain. Follow along on Figure I-12 as we prepare for the chapters ahead. Here are the geographic realms *and* their regions.

Europe (1)

Europe merits recognition as a world realm despite the smallness of its area on the western, peninsular margin of the Eurasian landmass. Certainly Europe's size is no measure of its global significance. European influences, innovations, empires, and revolutions have transformed the world far beyond its borders. Time and again—despite internal wars, the loss of colonial empires, and the impact of external competition—Europe has proved to contain the human and natural resources needed to overcome disaster.

Europe for more than half a century has been engaged in an historic program of unification, the European Union. Yet Europe has not lost its regional identities. It has five regions: Western Europe; the British Isles; Northern Europe; Southern (Mediterranean) Europe; and Eastern Europe.

Russia (2)

Russia, the heart of the former Soviet Empire, remains the largest territorial state in the world, even after the loss of its 14 colonies in 1991. The Russian domain was achieved by the tsars, forged into an empire by the Soviets, and ruined by communist mismanagement. World War II dealt the Soviet Union a devastating blow, but it galvanized Russian nationalism, and after the war the state rose to a position of world superpower with a mighty military and a pioneering space program. However, the Soviet Union was a colonial empire, and as Europe's colonial domains fell apart in the postwar era, Russia's increasingly began to show signs of breakdown as well. On December 25, 1991, the Union of Soviet Socialist Republics passed into history.

Russia's vast realm can be divided into many regions, but for our purposes we recognize four: the Russian Core, centered on Moscow; the Eastern Frontier; Siberia; and the Far East. Each of these, given their dimensions and internal diversity, has subregions (see Chapter 2).

North America (3)

The North American geographic realm consists of two countries, the United States and Canada. This postindustrial realm is characterized by high levels of urbanization, sophisticated technology, unmatched mobility, and massive consumption of the planet's resources and commodities. This is also a realm of pluralistic societies troubled by the reality that minorities tend to remain separate from the dominant culture. In the late 1990s, a strong separatist movement seeks to make Quebec a sovereign state.

North America is divided into eight regions. The Continental Core is the urban-industrial heartland of the realm; today it is completing its transformation into a postindustrial complex whose functions are increasingly shared with the South, the Southwest, and the West Coast. The world's most productive grain and livestock farming dominates the Agricultural Heartland. Independence-minded French Canada and the New England/Maritime Provinces region are struggling to avoid being left behind as the economic spotlight continues to shift away from northeastern North America. The vast, resource-rich Marginal Interior awaits development in the twenty-first century. Note that many regions straddle the U.S.–Canada border—signifying economic interaction that will intensify as the new North American Free Trade zone takes full effect in the near future.

Middle America (4)

Middle and South America together are often called Latin America because of the Spanish and Portuguese imprint that resulted from the European expansion and the destruction of Amerindian cultures. This ''Latin'' imprint is far more pervasive in South America than in Middle America, however, and it represents one geographic basis for distinguishing between Middle and South American realms.

Middle America consists of four regions. Mexico, the realm's giant, by itself constitutes a region. The seven states of Central America form the second region. The larger islands of the Caribbean (the Greater Antilles) create the third Middle American region, and the arc of smaller islands (the Lesser Antilles) constitutes the fourth.

South America (5)

The continent of South America also is a geographic realm, one shared by Portuguese-influenced Brazil and a former Spanish colonial domain now divided into nine separate countries. The Roman Catholic religion continues to be a major cultural influence, and systems of land ownership, tenancy, taxation, and tribute also were transferred from the Old World to the New. Today, South America still carries the impress of its source region in the architecture of its cities, in the visual arts, in its music, and in other ways. But the realm is also forging new linkages, with trade expansion leading the way.

South America is divided into four regions. These are the giant of this realm, Brazil; the Caribbean-facing countries of the North; the Amerindian-influenced, mountainous West; and the countries of the ''Southern Cone,'' where South America is economically most developed.

From the Field Notes

"In Middle and South America, conservation of nature's legacy and history's heritage has been impeded by rapid population growth, often uncontrolled frontier penetration, external economic intervention, and lack of money to pay for it. Costa Rica is world renowned for its conservation efforts. The Carara Reserve is one of dozens of parks and preserves that protect the country's varied ecosystems. We walked in silence through the dense forest, and could not help comparing this experience to that in interior Brazil, where vast tracts of forest lay in smoldering ashes. But even Costa Rica is not immune from the forces of destruction. Walking along the road toward the bridge where we had left our car, we saw a huge tract where the forest had been cut down and cattle grazed. 'Your hamburgers,' said our Costa Rican colleague when I asked him about it . . . Not only nature, but also history has suffered neglect. One of South America's greatest architectural legacies lies in the old Brazilian city of Salvador in Bahia State. When we first saw it nearly 15 years ago, cathedrals, churches, villas, and ornate public buildings were decaying and collapsing after two centuries of dereliction. Then the United Nations declared Salvador's old town a World Heritage site, and funding was made available to begin restoration. Our most recent visit, in 1995, proved the Brazilians as enthusiastic as anyone for historic preservation; all it took was the resources to make it possible."

North Africa/Southwest Asia (6)

This vast and sprawling geographic realm is known by several names, none of which is completely satisfactory: the Islamic realm, the Arab World, and the dry world. (Some geographers use the term *Naswasia*, and *Afrasia* is occasionally seen as well.) The Islamic (Muslim) religion is this realm's overriding cultural feature, often vividly expressed in the cultural landscape. Aridity dominates the natural environment, so that people are clustered where there is water—often in widely separated settlements. Such isolation perpetuates cultural discreteness, and we can distinguish regional contrasts within this realm quite clearly.

There are seven regions, three in North Africa and four in Southwest Asia. Egypt forms a region by itself, and to the west lies the North Africa dominated by the Maghreb. Southward across all of Africa lies the transition zone referred to earlier. In Southwest Asia the pivotal region is often called (for want of a more appropriate term) the Middle East, a crescent of five countries lying between the Mediterranean and the Persian Gulf. Southward lies the Arabian Peninsula, and to the north stretches a group of non-Arab (Empire) states in a region from Turkey to Afghanistan. East of the Caspian Sea, anchoring the realm's northeast, lies the group of five Muslim countries (Turkestan) now freed from Soviet communist rule, where Islam is reviving.

Subsaharan Africa (7)

Between the southern margins of the Sahara and the southernmost cape of South Africa lies the geographic realm of Subsaharan Africa. Its border with the realm to the north is a broad transition zone with several countries straddling it (see Fig. I-12). Elsewhere, however, coastlines mark the limits of this realm, the cradle of humanity, a culturally very distinctive part of the world. European colonialism left its imprints, notably in the political boundary framework and in the export-oriented transport system. But African traditional cultures persisted; many hundreds of languages are spoken here, and numerous community religions endure despite Christian and Islamic penetrations. Africa also is the least developed and least urbanized of the major realms. It decidedly remains a realm of farmers.

The realm consists of four regions: West Africa, East Africa, Equatorial Africa, and Southern Africa. The last of these is dominated by South Africa, where a momentous political and social transformation is under way.

South Asia (8)

The familiar triangular shape of India outlines a subcontinent in itself—a clearly discernible physical realm bounded by mountain ranges and seas, and inhabited by a population that constitutes one of the greatest human concentrations on Earth. The scene of ancient civilizations, India became the cornerstone of the vast British colonial empire. From the colonial period and its often violent aftermath emerged six states: India, Pakistan, Bangladesh, Sri Lanka, Nepal, and Bhutan.

This is a realm of many languages and great religions and consequently of deep cultural divisions. Although India is dominantly a Hindu society, its Muslim minority is larger than the entire population of any single country in the North Africa/Southwest Asia realm. The South Asian realm consists of five regions: India's Ganges Plain-dominated heartland at its center, Pakistan to the west, the mountainous north comprised of Nepal and Bhutan, Bangladesh in the east, and India's Dravidian South and island Sri Lanka to the south.

East Asia (9)

China lies at the heart of a vast East Asian sphere that encompasses the world's largest population agglomeration. Territorially, this realm extends from the Russian border not far from Siberia to the tropical coasts of the South China Sea, and from the westernmost Pacific islands to the deserts and rugged highlands of inner Asia. Demographically, China's 1.3 billion people dominate this realm; historically, they are heirs to one of the world's great ancient culture hearths.

China is East Asia's demographic giant; Japan dominates this realm economically. But Japan's supremacy is eroding as other *economic tigers* are emerging; and Japanese investments and trade links are increasingly enmeshing Japan in the East Asian sphere. Japan anchors a still-evolving, still-disjointed new region, the western *Pacific*

PACIFIC RIM: EMERGING REGION

Regions are subject to change. Even geographic realms can change in short order: when the Soviet Union collapsed in 1991, a functional realm disintegrated and the outlying pieces were welded onto neighboring realms. The Baltic States rejoined Europe. The countries of former Soviet Central Asia turned their gaze from Moscow to Mecca. The Russian realm emerged from the wreckage of the U.S.S.R.

Today we witness the evolution of a region born of a string of economic miracles on the shores of the Pacific Ocean. It is a still-discontinuous region, but its foci lie on the Pacific's western margins: in Japan, South Korea, Taiwan, coastal China, Thailand, Malaysia. The regional term *Pacific Rim* has come into use to describe this dramatic development, which in a few decades has redrawn the map of the Pacific periphery—not only in East and Southeast Asia but also in Australia, in Canada and the United States, and even in South America's Chile. It is a superb example of a functional region: economic activity in the form of capital flows, raw material movements, and trade links are generating urbanization, industrialization, and labor migration. In the process, cultural landscapes from Jakarta to Santiago are being transformed.

New regions sometimes require new geographic terminology. Just as the landmass containing the "continents" of Europe and Asia is called *Eurasia*, the emerging Pacific Rim stretches from northern Asia into Australia and is aptly named the Pacific Rim of *Austrasia*. As a regional appellation, Austrasia refers to the regional realignment of western Pacific coastlands from Japan's Hokkaido to New Zealand.

Rim, that extends from southeasternmost Russia in the north to New Zealand in the southeast (see box titled "Pacific Rim: Emerging Region"). Because Pacific Rim developments currently affect several realms and regions facing the Pacific Ocean, the reader is alerted to this momentous phenomenon by the red-edged pages in each of the chapters in which they are discussed.

The East Asian geographic realm displays five regions. China Proper, that is, eastern and northeastern China, forms the core, with North Korea a (probably temporary) component. The great mountain ranges and plateaus of Xizang (Tibet) lie in the southwest. The vast deserts of Xinjiang lie in the west. Mongolia occupies the north. And in the east lies the Jakota Triangle, the economically advanced region anchored by Japan but also including South Korea and Taiwan.

Southeast Asia (10)

Southeast Asia's corner of the world is a particularly varied mosaic of ethnic and linguistic groups. This realm has been the scene of countless contests for power and primacy and

From the Field Notes
"Wherever we went in Vietnam, the need for infrastructural improvements was evident. Roads need widening and surfacing. Bridges must be built or improved. Electrical power networks await modernization. Water supply is undependable and polluted. Communications systems are in poor repair and often chaotic, as this jumble of wires, typical of many a street in Saigon (Ho Chi Minh City) shows. Telephone service, indispensable to businesses, remains inadequate . . . In Shanghai, much of the old city is not being repaired or improved; it is simply being swept away by bulldozers, and modern high-rises mushroom where traditional neighborhoods lay. The wholesale transformation of Shanghai reflects the price being paid by many Pacific Rim cities and the communities they encompass. Economic development can exact a high cost on those in its path."

has been called the Eastern Europe of Asia. During the colonial period, the term *Indochina* came into use to denote the eastern portion of mainland Southeast Asia. The term is a good one because it reflects the major sources of cultural influence that have affected the entire realm. Ethnic affinities are with China, but cultural imprints (Hindu, Buddhist, even Muslim) came from or via India.

Spatially, the realm's discontinuities are quite obvious: it consists of a peninsular mainland (where populations tend to be clustered in river basins) and thousands of islands forming the archipelagoes of Indonesia and the Philippines. The two regions, therefore, are based on this mainland–island distinction (Fig. I-12).

But remote Australia, too, is changing. Pacific Rim developments affect its regional geography today.

The regions of this realm are defined by physical as well as human geography: in Australia, a highly urbanized, two-part functional core and a vast, desert-dominated interior; and in New Zealand, two main islands that exhibit considerable physical and cultural contrast.

Australia (11)

Australia, with its distant neighbor, New Zealand, constitutes a geographic realm by virtue of its continental dimensions, insular separation, and dominantly Western cultural heritage. Territorially larger than Europe, economically far ahead of its Southeast Asian and Pacific neighbors, and demographically distinct from all the world's realms, Australia merits separate identity in Figure I-12.

Pacific Realm (12)

Between Asia and Australia to the west and the Americas to the east lies the immense Pacific Ocean, larger than all the landmasses on Earth combined. In this great ocean lie tens of thousands of islands large and small. This fragmented, culturally complex area constitutes the Pacific geographic realm.

The Pacific realm has been divided into three regions as a matter of long-term custom. The most populous region, anchored by New Guinea, is named Melanesia (*melas* means black, in reference to the very dark skin, hair, and eyes of the majority of the people). Immediately to the north lies Micronesia (*micro* means small, in reference to the smallness of many of this region's islands). And to the east lies Polynesia (*poly* means many), the huge central Pacific region of numerous islands extending from the Hawaiian archipelago (island chain) southeastward to Easter Island and as far southwestward as New Zealand.

In Australia and New Zealand, two realms meet: Australia's aboriginal peoples are Melanesians, and New Zealand's Maori population is of Polynesian ancestry.

LINKAGES

We now have before us the comprehensive framework we need to enlarge our scale and deepen our perspective. Be-

fore we begin, one alert: do not let the regional boundaries in Fig. I-12 mislead you into assuming that the realms and regions of the world exist in isolation from one another. We noted that regions change, and change results from interaction. The Pacific Rim phenomenon is a current example of such change and interaction, and it is happening in a realm whose largest country, China, tried to isolate itself from the outside world for several decades after mid-century. We will see that international interaction is moving us toward **globalization**, the gradual reduction of regional contrasts resulting from increasing cultural, economic, and political exchanges. In an ideal world, regional differences would be eliminated altogether. But in the real world of the closing twentieth century, realms and regions retain their **iconographies**, their unifying symbols and habits of culture and tradition. To understand these iconographies is to be able to accommodate them, a critical capacity for navigating our multicultural world. Our interconnected world still has its regional neighborhoods, and geography is the vehicle of choice to visit them.

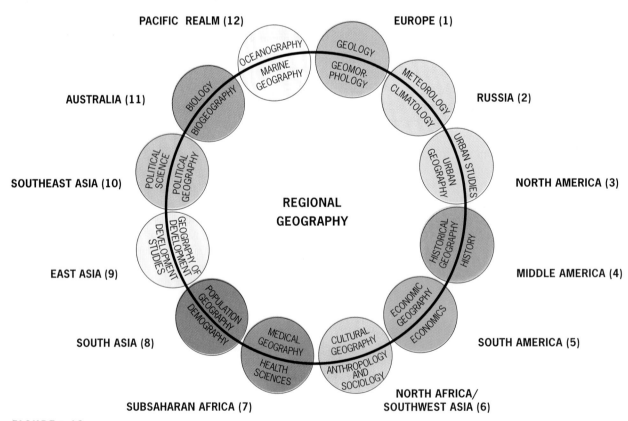

THE RELATIONSHIP BETWEEN REGIONAL AND SYSTEMATIC GEOGRAPHY

FIGURE I-13

TWO SIDES OF GEOGRAPHY

As this introductory chapter demonstrates, our world regional survey is no mere description of places and areas. We have combined the study of realms and regions with a look at geography's ideas and concepts—the notions, generalizations, and basic theories that make the discipline what it is. We continue this method in the chapters ahead so that we will become better acquainted with the world *and* with geography.

By now you are aware that geography is a wide-ranging, multifaceted discipline. Geography often is described as being a social science, but that is only half the story: in fact, geography straddles the divide between the social and the physical (natural) sciences. Many of the ideas and concepts you will encounter have to do with the many interactions between human societies and natural environments.

Regional geography allows us to view the world in an all-encompassing way. As we have seen, regional geography borrows information from many sources to create an overall image of our divided world. Those sources are not random. They form the topical, or **systematic fields** of geography. Figure I-13 shows that regional geography is nurtured by a variety of geographic fields, each of which in turn has a link to disciplines outside geography. Consider the discipline of anthropology: its discoveries are crucial to cultural geography, and cultural geography in turn contributes to regional geography.

These systematic fields of geography are so named because their approach is global, not regional. Take the study of cities, urban geography. Urbanization is a worldwide process, and urban geographers can identify certain human activities that all cities of the world exhibit in one form or another. But cities also display regional properties. The model Japanese city is quite distinct from, say, the African city. Regional geography, therefore, borrows from the systematic field of urban geography, but it injects the regional perspective as well.

In this book we tell you about a dozen of these systematic fields, but there are more. These discussions will strengthen your understanding not only of the geographic realm under study, but also of a systematic field that is

WHAT DO GEOGRAPHERS DO?

■ ■ ■

A systematic spatial perspective and an interest in regional study are the unifying themes and enthusiasms of geography. Geography's practitioners include physical geographers, whose principal interests are the study of geomorphology (land surfaces), research on climate and weather, vegetation and soils, and the management of water and other natural resources. There also are geographers who concentrate their research and teaching on the ecological interrelationships between the physical and human worlds; they study the impact of humankind on our globe's natural environments and the influences of the environment (including such artificial contents as air and water pollution) on human individuals and societies.

Other geographers are regional specialists, who often concentrate their work for governments, planning agencies, and multinational corporations on a particular region of the world. Still other geographers—who now constitute the largest group of practitioners—are devoted to certain topical or systematic subfields such as urban geography, economic geography, cultural geography, and many others (see Fig. I-13); they perform numerous tasks associated with the identification and resolution (through policy-making and planning) of spatial problems in their specialized areas. And, as in the past, there remain many geographers who combine their fascination for spatial questions with technical know-how.

Computerized cartography, geographic information systems, remote sensing, and even environmental engineering are among specializations listed by the 10,000-plus professional geographers of North America. In Appendix C, you will find considerable information on the discipline, how one trains to become a geographer, and the many exciting career options that are open to the young professional.

■ ■ ■

particularly relevant to it. They also give us an opportunity to further develop the theme of *practical* geography (see box titled "What Do Geographers Do?" for an overview of the geography profession and Appendix C for a discussion of career opportunities). As the book unfolds, we will build a foundation of systematic as well as regional knowledge.

■ PRONUNCIATION GUIDE ■

Afghanistan (aff-GHANN-uh-stan)
Afrasia (aff-FRAY-zhuh)
Alluvial (uh-LOO-vee-ull)
Angola (ang-GOH-luh)
Antilles (an-TILL-eeze)
Archipelago (ark-uh-PELL-uh-goh)
Aridisol (uh-RIDDY-soll)
Aquifer (ACK-kwuh-fer)
Austrasia (aw-STRAY-zhuh)
Bahia (bah-EE-wah)
Bajau (bah-YAU)
Bangladesh (bang-gluh-DESH)
Beijing (bay-ZHING)
Belarus (bella-ROOSE)
Bhutan (boo-TAHN)
Borneo (BOAR-nee-oh)
Bosnia-Herzegovina (BOZ-nee-uh
    hert-suh-goh-VEE-nuh)
Brahmaputra (brahm-uh-POOH-truh)
Buddhist (BOOD-ist)
Carara (kuh-RAHR-uh)
Caribbean (kuh-RIB-ee-un/karra-BEE-
    un)
Caspian (KASS-spee-un)
Cenozoic (senno-ZOH-ick)
Chang Jiang (chahng-jee-AHNG)
Chile (CHILLI/CHEE-lay)
Coniferous (kuh-NIFF-uh-russ)
Cyprus (SYE-pruss)
Czech (CHECK)
Dhaka (DAHK-uh)
Dravidian (druh-VIDDY-un)
El Niño (ell-NEEN-yoh)
Ethiopia (eeth-ee-OH-pea-uh)
Fenneman (FENN-uh-munn)
Fungi (FUN-jye)
Ganges (GAN-jeeze)
Geiger (GHYE-gherr)
Han (HAHN)
Hierarchical (hire-ARK-uh-kull)
Himalayas (him-AHL-yuzz/
    himma-LAY-uzz)
Hindu (HIN-doo)
Ho Chi Minh (hoh-chee-MINN)
Hokkaido (hoh-KYE-doh)
Holistic (hoh-LISS-tick)

Holocene (HOLLO-seen)
Homogeneity (hoh-moh-juh-NAY-uh-
    tee)
Homo sapiens (hoh-moh SAY-pea-enz)
Huang He (HWAHNG-HUH)
Humus (HYOO-muss)
Indonesia (indo-NEE-zhuh)
Islam (iss-LAHM)
Jakarta (juh-KAHR-tuh)
Jakota (juh-KOH-duh)
Jawa (JAH-vuh)
Kalahari (kalla-HAH-ree)
Kauai (KAU-eye)
Kazakstan (kuzz-uck-STAHN)
Kenya (KEN-yuh)
Kikuyu (kee-KOO-yoo)
Köppen (KER-pun)
Kota Kinabalu (KOH-tuh
    kin-uh-buh-LOO)
Laos (LAUSS)
Loess (LERSS)
Loihi (loh-EE-hee)
Magellan (muh-JELL-un)
Maghreb (mahg-GRAHB)
Malaysia (muh-LAY-zhuh)
Manitoba (manna-TOH-buh)
Maori (MAH-aw-ree/MAU-ree)
Maui (MAU-ee)
Mauna Kea (mau-nuh-KAY-uh)
Megalopolis (meh-guh-LOPP-uh-liss)
Melanesia (mella-NEE-zhuh)
Mengkabong (MENG-kah-bong)
Micronesia (mye-kroh-NEE-zhuh)
Maçambique (moh-sum-BEAK)
Moldova (moal-DOH-vuh)
Mongolia (mung-GOH-lee-uh)
Montreal (mun-tree-AWL)
Moscow (MOSS-kau)
Muslim (MUZZ-lim)
Myanmar (mee-ahn-MAH)
Naswasia (nass-SWAY-zhuh)
Nei Mongol (nay-MUNG-goal)
Nepal (nuh-PAHL)
New Guinea (noo-GHINNY)
Nicaragua (nick-uh-RAH-gwuh)
Niger (nee-ZHAIR)

Nigeria (nye-JEERY-uh)
Oahu (uh-WAH-hoo)
Oxisol (OXY-soll)
Pakistan (PAH-kih-stahn)
Pangaea (pan-GAY-uh)
Paraguay (PAHRA-gwye)
Philippines (FILL-uh-peenz)
Physiography (fizzy-OGG-ruh-fee)
Pleistocene (PLY-stoh-seen)
Polynesia (polla-NEE-zhuh)
Quebec (kwuh-BECK)
Sabah (SAHB-ah)
Sahara (suh-HARRA)
Saigon (sye-GAHN)
Salvador (SULL-vuh-dor)
Santiago (sahn-tee-AH-goh)
Sauer (SOUR)
Senegal (sen-ih-GAWL)
Shanghai (shang-HYE)
Siberia (sye-BEERY-uh)
Sisal (SYE-sull)
Slav (SLAHV)
Slovakia (sloh-VAH-kee-uh)
Spatial (SPAY-shull)
Sri Lanka (sree-LAHNG-kuh)
Steppe (STEP)
Sudan (soo-DAN)
Taiwan (tye-WAHN)
Thailand (TYE-land)
Tibet (tuh-BETT)
Tierra del Fuego (tee-ERRA dale
    FWAY-goh)
Tokyo (TOH-kee-oh)
Turkestan (TERK-uh-stahn)
Ukraine (yoo-CRANE)
Ural (YOOR-ull)
Uruguay (OO-rah-gwye)
Vietnam (vee-et-NAHM)
Wegener (VAY-ghenner)
Yugoslavia (yoo-goh-SLAH-vee-uh)
Xinjiang (shin-jee-AHNG)
Xizang (sheedz-AHNG)
Yangzi (YANG-dzee)
Zaïre (zah-EAR)
Zimbabwe (zim-BAHB-way)

Resilient Europe: Confronting New Challenges

IDEAS & CONCEPTS

Relative location
Geomorphology
Infrastructure
Areal functional
    specialization
Model
Von Thünen's Isolated State
Industrial location
Nation-state
Centripetal forces
Centrifugal forces
Spatial interaction principles
    Complementarity
    Transferability
    Intervening opportunity

Primate city
Metropolis
Central business district
    (CBD)
Devolution
Supranationalism
Site
Situation
Conurbation
Shatter belt
Balkanization
Irredentism

REGIONS

Western Europe
The British Isles
Northern (Nordic) Europe

Mediterranean Europe
Eastern Europe

In so many ways, Europe has for centuries been the heart of the world. European empires spanned the globe and transformed societies far and near. European capitals were the focal points of trade networks that controlled distant resources. Millions of Europeans migrated from their homelands to the New World as well as to newly settled parts of the Old, creating new societies from North

America to Australia. In agriculture, in industry, and in political organization, Europe went through revolutions—and then exported those revolutions across much of the globe, serving to consolidate the European advantage. Yet during the twentieth century, Europe twice plunged the world into war. In the aftermath of World War II (1939–

1945), Europe's weakened powers lost the colonial possessions that for so long had provided wealth and influence, and the continent was divided until 1990 by an ideological Iron Curtain. Resilient Western Europe's recovery and Eastern Europe's rejection of communism have been the dominant events of the past half-century.

DEFINING THE EUROPEAN REALM

These internal regional groupings constitute major elements of Europe's human geography, whose overall framework is an intricate mosaic of 38 countries. *Western Europe* includes the large countries of Germany and France; the so-called Low Countries of Belgium, the Netherlands, and Luxembourg; and the Alpine countries of Switzerland, Austria, and tiny Liechtenstein. *Eastern Europe* consists of the former communist satellites of Poland, the Czech Republic, Slovakia, Hungary, Romania, Bulgaria, Albania, and the remnants of former Yugoslavia—Serbia-Montenegro, Slovenia, Croatia, Bosnia, and Mace-

donia; to these 12 countries, we must now also add the five republics of the former Soviet Union that joined Eastern Europe upon independence in 1991—Latvia, Lithuania, Belarus, Moldova, and Ukraine. Three additional regional groupings constitute the remainder of Europe: the offshore *British Isles* (the United Kingdom and Ireland); Norden or *Northern Europe* (Denmark, Norway, Sweden, Finland, Estonia, and the island country of Iceland); and *Mediterranean* or *Southern Europe* (Spain, Portugal, Italy, Greece, and the island microstate of Malta). This long list of states notwithstanding, when we add everything together, Europe

RELATIVE LOCATION: EUROPE IN THE LAND HEMISPHERE

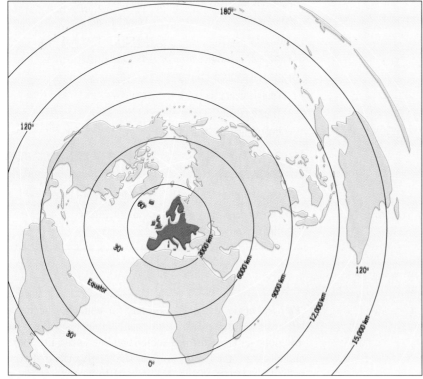

Azimuthal equidistant projection centered on Hamburg, Germany

FIGURE 1-1

is still a world geographic realm of quite modest proportions on the peninsular margin of western Eurasia (see Fig. I-12). Yet, despite its comparatively small size, for more than 2,000 years the European realm has been a leading focus of human achievement, a hearth of innovation and invention.

Europe's human resources have been matched by its large and varied raw-material base; whenever the opportunity or the need arose, the realm's physical geography proved to contain what was required. In fact, for so limited an area (slightly less than two-thirds the size of the United States), Europe's internal natural diversity is probably unequaled. From the warm shores of the Mediterranean to the frigid Scandinavian Arctic, from the flat coastlands of the North Sea to the soaring peaks of the Alps, and from the moist woodlands and moors of the Atlantic fringe to the semiarid prairies north of the Black Sea, Europe presents an almost infinite range of natural environments. The insular and peninsular west contrasts strongly against a more interior, continental east. A raw-material-laden backbone extends across the middle of Europe from England eastward to Ukraine, yielding coal, iron ore, and many other valuable minerals. This diversity is not confined to the physical makeup of the continent. The European realm contains peoples of many different cultural-linguistic stocks, including not only Latins, Germanics, and Slavs but also numerous minorities such as Finns, Hungarians, and various Celtic-speaking groups (as Fig. 1-6 shows).

This diversity of physical and human content alone, of course, is not Europe's greatest asset. If differences in habitat and culture automatically led to rapid human progress, Europe would have had many more competitors for its world position than it did. But in Europe, there were virtually unmatched advantages of scale and geographic proximity.

Globally, Europe's **relative location**—at the heart of the *land hemisphere*—is one of maximum efficiency for contact with the rest of the world (Fig. 1-1). Regionally, Europe is also far more than a mere western extremity of the Eurasian landmass. Almost nowhere is Europe far from that essential ingredient of European development—the sea—and the water interdigitates with the land here as it does nowhere else on Earth. Southern and Western Europe consist almost entirely of peninsulas and islands, from Greece, Italy, and the Iberian Peninsula (Spain and Portugal) to the British Isles, Denmark, and the Scandinavian Peninsula (Norway and Sweden). Southern Europe faces the Mediterranean, and Western Europe virtually surrounds the North Sea as it looks out over the Atlantic Ocean. Beyond the Mediterranean lies Africa and across the Atlantic are the Americas. Europe has long been a place of contact between peoples and cultures, of circulation of goods and ideas. The hundreds of miles of navigable waterways; the easily traversed bays, straits, and channels between numerous islands and peninsulas and the mainland; and the highly accessible Mediterranean, North, and Baltic seas all provided the routeways for these exchanges. Later, even the oceans became avenues of long-distance spatial interaction.

This historic advantage of moderate distances applies on the mainland as well. Europe's Alps may form a transcontinental divider, but what they separate still lies in close juxtaposition. (Alpine passes have for centuries provided several corridors for overland contact.) Consider Rome and Paris: the distance between these long-time control points of Mediterranean and northwestern Europe is less than that between New York and Chicago. No place in Europe is very far from anyplace else on the continent, although nearby places are often sharply different from each other in terms of economy and outlook. Short distances and large differences make for much interaction—and that has been the hallmark of the geography of Europe for over a millennium.

THE MAJOR GEOGRAPHIC QUALITIES OF EUROPE

1. The European realm consists of the western extremity of the Eurasian landmass, a locale of maximum efficiency for contact with the rest of the world.

2. Europe's lingering and resurgent world influence results largely from advantages accrued over centuries of global political and economic domination.

3. The European natural environment displays a wide range of topographic, climatic, vegetative, and soil conditions and is endowed with many industrial resources.

4. Europe is marked by strong internal regional differentiation (cultural as well as physical), exhibits a high degree of functional specialization, and provides multiple exchange opportunities.

5. European economies are dominated by manufacturing, and the level of productivity has been high; levels of development generally decline from west to east.

6. Europe's nation-states emerged from durable power cores that formed the headquarters of world colonial empires. A number of those states are now bedeviled by internal separatist movements.

7. Europe's population is generally well off, highly urbanized, well educated, enjoys long life expectancies, and constitutes one of the world's three largest population clusters.

8. Europe has made important progress toward international economic integration. The push toward still stronger and broader coordination continues.

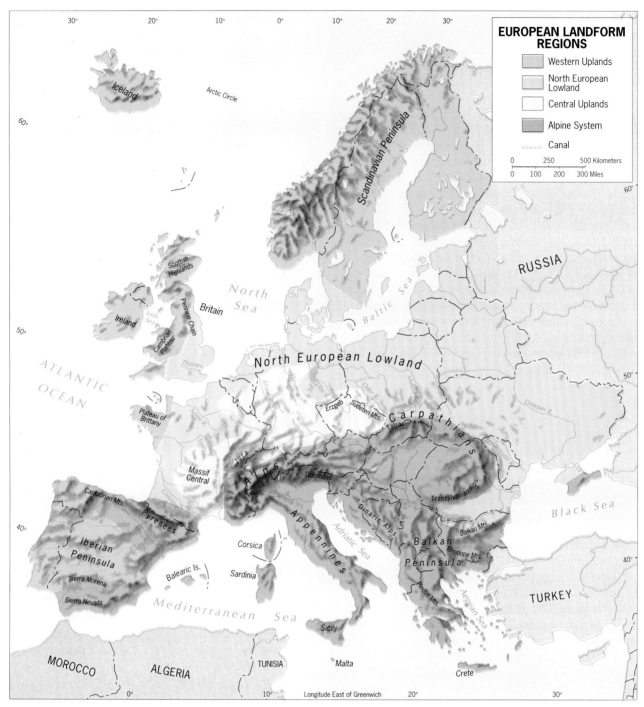

FIGURE 1-2

LANDSCAPES AND OPPORTUNITIES

Europe may be small in area, but its physical landscapes are varied and complex. It would be easy to identify a large number of physiographic regions, but in doing so we might lose sight of the broader regional pattern—a pattern that has much to do with the way the European human drama unfolded. Accordingly, Europe's landscapes can be grouped regionally into four units: the Central Uplands, the Alpine Mountains in the south, the Western Uplands, and the great North European Lowland (Fig. 1-2).

The very heart of Europe is occupied by an area of hills and small plateaus, with forest-clad slopes and fertile val-

■ ■ ■ ■ ■ ■ ■ ■ ■ ■ ■ ■ **FOCUS ON A SYSTEMATIC FIELD** ■ ■ ■ ■ ■ ■ ■ ■ ■ ■ ■

Geomorphology

One of geography's binding interests lies in the relationships between human societies and the natural environments that influence them. By definition, the natural environment means more than climate and weather: it also comprises soils and vegetation, streams and slopes, terrain and relief. When archeologists study ancient societies, they are as interested in learning about the environmental circumstances as they are in the communities' artifacts. Their research team is likely, therefore, to include a scientist who will investigate the **geomorphology** of the setting.

Geomorphologists study the evolution of landscape. As with all geographic pursuits, this can be done at various scales. A geomorphologist working with a team of archeologists at a particular site in East Africa, Southwest Asia, or Middle America unravels the environmental history of that particular place, and his or her maps will be large-scale. Other geomorphologists investigate the evolution of entire mountain ranges or drainage basins, working, therefore, at a smaller scale. Still others try to understand the development of whole continental landscapes over the course of tectonic-plate events, creating models that suggest how the global physical stage has changed over tens of millions of years.

Geo (Earth) + *morphology* (form, structure) links geography and geology in a very productive way. To comprehend how the Earth's surface has been sculpted by nature's forces requires much knowledge of geology. To understand how nature's work and humanity's endeavors interrelate demands thorough grounding in geography. In practice, geomorphologists focus primarily on nature's processes, often without any direct linkage to human implications. But to understand human history, geographers could not do without what geomorphologists discover.

The Earth's still-thin crust is continuously deformed by *tectonic forces* that bend and break it. Entire continental landmasses ride on *plates* that pull apart along mid-oceanic ridges amid fiery submarine volcanism and collide along *subduction zones* marked by linear mountain ranges

and rocked by earthquakes. Entire regions are deformed by tilting, warping, and doming. Tectonic forces affect different crustal rocks in different ways: *igneous* rocks such as granite and basalt, solidified from the molten state, tend to be hard and resistant, but *sedimentary* rocks such as sandstone and shale are more easily crumpled or folded into lengthy ridges. The *metamorphic* rocks, altered by heat and pressure (granite becomes gneiss, sandstone converts into quartzite, shale hardens into slate), often show the shattering effect of tectonic action.

Even before a phase of tectonic activity has run its course, the rocky surface is attacked by *weathering* (disintegration caused by temperature change and chemical action) and *erosion*, the removal of weathered rock and its further breakdown by streams, glaciers, wind, and coastal wave action. Again, some rocks are more susceptible to these processes than others. Hard crystalline rocks (igneous and metamorphic) tend to withstand them longer than softer sedimentary strata. Igneous rock masses often stand out high above the landscape—for example, in those spectacular domes that tower over the setting of Rio de Janeiro, Brazil; sedimentary layers form many of the deep valleys in mountain ranges exhibiting complex geology.

The several agents of erosion (streams, glaciers, wind, waves) all create distinctive *landscapes*. We are familiar with these: the river-sculpted scenery of the Appalachians, the dramatic mountains of glaciated Alaska, the windswept deserts of the Southwest, the cliffs and beaches of California's shoreline. Behind such first impressions, however, lies a complex geomorphic story. Geomorphologists often focus on a single *landform* (say, a volcano), a single stream and its valley, a glacial deposit to unravel the forces and factors that contributed to the overall landscape's evolution.

With all this in mind, let us look closely at the map of Europe's landscape regions (Fig. 1-2). Probably the most famous landscape region in all of Europe is the mountainous Alps of Switzerland, a

historic barrier between northwest and southeast. As the map's legend indicates, the Alps of Switzerland and Austria are part of a greater *Alpine System* that also includes the rugged Pyrenees, the historic border between Spain and France, Italy's Appennines, Romania's Carpathians, and the mountain ranges of the Balkan Peninsula.

The Alpine System does not, however, represent all of Europe's high-relief topography. Note the discontinuous region mapped as the Western Uplands in Figure 1-2, which includes most of the Scandinavian Peninsula, the western British Isles, and the western part of the Iberian Peninsula. This zone of mountains and rugged plateaus is much older than the Alpine System, and much of it has been worn down by erosion. Its geomorphic history goes back to the time when the continental landmasses were assembled into one supercontinent (see pp. 331-332), and the Atlantic Ocean had not yet opened up. The mountains of Scotland and Norway were continuations of North America's Appalachians, separated by the divergence of the North American and the Eurasian tectonic plates over the past 200 million years.

Between these two high-relief zones of Europe lie the North European Lowland and the plateau-like Central Uplands. Note that these zones contain the very heart of Europe's human geography: the Paris Basin, London and southeastern England, and most of Germany. Fertile soils, natural waterways augmented by artificial canals, good harbors, and comparatively uninterrupted surface routes enhanced circulation, interaction, and economic progress. In the Introduction we discussed the core-periphery concept; here is a good example. In physical terms, Europe's core is generally defined by the regions mapped in green and yellow in Figure 1-2, the North European Lowland and the Central Uplands. The periphery, with a few exceptions in northern Italy and northeastern Spain, coincides with the Western Uplands and the Alpine System. Human and physical geographies often are strongly intertwined.

■ ■ ■

The Alpine System forms Europe's mountain backbone, extending from the Pyrenees on the French–Spanish border to the Carpathians and Balkan Mountains in the east. A formidable cultural as well as physiographic divide, the Alps were glaciated during the Pleistocene and remain snowcapped today, dominating the magnificent scenery of eastern France (shown here in Savoy on the Swiss border), Switzerland, and part of Austria. Increasing road freight traffic, resulting from the economic interaction between Italy and Western Europe, and expanding tourism attracted by these spectacular landscapes, are causing environmental degradation in the Alps.

leys. These *Central Uplands* also contain the majority of Europe's productive coalfields. When the region emerged from its long medieval quiescence and stirred with the stimuli of the Industrial Revolution, towns on the Uplands' flanks grew into cities, and farmlands gave way to mines and factories.

The Central Uplands are flanked on the south by the much higher Alpine Mountains (and to the west, north, and east by the North European Lowland). The *Alpine Mountains* include not only the famous Alps themselves but also other ranges that belong to this great mountain system. The Pyrenees between Spain and France (one of Europe's few true physical barriers), Italy's Appennines, former Yugoslavia's Dinaric ranges, and the Carpathians of Eastern Europe are all part of this Alpine system, which extends even into North Africa (as the Atlas Mountains) and eastward

into Turkey and beyond. Although the Alps are rugged and imposing, they have not prevented spatial interaction: traders have operated through their mountain passes for many centuries.

Europe's western margins are also quite rugged, but the *Western Uplands* of Scandinavia, Scotland, Ireland, France's Brittany, Portugal, and Spain are not part of the Alpine system (maximum elevations are markedly lower than those in the Alps). This western arc of highlands represents older geologic mountain building, contrasting sharply with the relatively young, still active, earthquake-prone Alpine Mountains. Scandinavia's uplands form part of an ancient geologic shield underlain by old crystalline rocks now bearing the marks of the Pleistocene glaciation. Spain's central plateau, or *Meseta*, is also supported by comparatively old rocks, now mostly worn down to a table-land.

The last of Europe's landform regions is also its most densely populated. The *North European Lowland*—also known as the Great European Plain—extends in a gigantic arc from southern France through the Low Countries across northern Germany and then eastward through Poland, from where it fans out deeply into southwestern Russia. Note, too, that southeastern England, Denmark, and the southern tip of Sweden also belong to this region (Fig. 1-2). Most of the North European Lowland lies below 500 feet (150 m) in elevation, and local relief rarely exceeds 100 feet (30 m). Beyond this single topographic factor there is much to differentiate it internally. In France, this region includes the basins of three major rivers—the Garonne, Loire, and Seine. In the Netherlands, for a good part it is made up of land reclaimed from the sea, enclosed by dikes and lying below sea level. In southeastern England, the higher areas of the Netherlands, northern Germany and Denmark, southern Sweden, and farther eastward, it bears the marks of the Pleistocene glaciation that withdrew only a few thousand years ago (see Fig. I-5). Each of these particular areas affords its own opportunities as soils and climates vary, giving rise to some of the world's most productive (and prestigious) agricultural pursuits.

The North European Lowland has also been one of Europe's major avenues of human contact. Entire peoples have migrated across it; armies have repeatedly marched through it. As settlement took place, agricultural diversity became the norm, and land use came to be dominated by intensive farming organized around a myriad of villages from which farmers commuted to their nearby fields. Today, centuries later, it is still not possible to speak of a ''dairy belt'' or ''wheat belt'' in Europe like those in North America; even where one particular crop dominates the farming scene, different crops are likely to grow just a few fields away.

Finally, Europe's great lowland possesses yet another crucial advantage: its multitude of navigable rivers, emerging from higher adjacent areas and wending their way to

EUROPE: THE EASTERN BOUNDARY

■ ■ ■

The European realm is bounded on the west, north, and south by Atlantic, Arctic, and Mediterranean waters, respectively. Europe's eastern boundary, however, has always been a matter for debate. Some scholars place this boundary at the Ural Mountains (deep inside Russia), thereby recognizing a "European" Russia and, presumably, an "Asian" one as well. Others argue that because there is a continuous transition from west to east (which continues into Russia), there is no point in trying to define any boundary.

Still, the boundary used in this chapter—marking Europe's eastern boundary as the border with Russia—has geographic justifications. Eastern Europe shares with Western Europe its fragmentation into several states possessing distinct cultural geographies, a political condition that sets it apart from the Russian giant to the east. Historical and cultural contrasts also mark large segments of this boundary (see Fig. 1-6). The former Soviet Union's 1945–1990 hegemony in Eastern Europe—a dominance lasting well back into pre-1917, tsarist times for several western republics of the U.S.S.R. that proclaimed independence in 1991—did not wipe out certain differences in ideology and nationalist loyalties between the two regions. Eastern Europe retains its discrete nationalisms, latent and actual conflicts, and pressures; these qualities are most vividly demonstrated by the upheavals that have swept across the region since 1989. Despite its own internal disunity, Russia remains Eurasia's territorial giant: several times as large as Europe, historically consolidated by force, and a potentially destabilizing influence on the emerging order of post-communist Europe.

■ ■ ■

the sea. In addition to the three rivers of France already mentioned, the Rhine-Meuse (Maas) river system serves one of Europe's most productive industrial and agricultural areas, reaching the sea via the Netherlands; the Weser, Elbe, and Oder cross northern Germany; and the Vistula traverses Poland. In southeastern Europe, the Danube rivals the Rhine in regularity of flow and navigability. Thus, north of the Alpine Mountain system, Europe's major rivers create a radial pattern outward from the continent's interior highlands. In this way, the natural waterways as well as the land surface of the North European Lowland favor traffic and trade. Over many centuries, the Europeans have improved the situation still further by connecting navigable stretches of rivers with artificial canals (Fig. 1-2). These waterways, and the roads and railroads that followed later, combined to bring tens of thousands of localities into contact with one another. Thus new techniques and innovations could spread rapidly, and trading connections and activity intensified continuously.

HERITAGE OF ORDER

Modern Europe was peopled in the wake of the Pleistocene's most recent glacial retreat—a gradual withdrawal that caused cold tundra to turn into deciduous forest and ice-filled valleys into grassy vales. On Mediterranean shores, Europe witnessed the rise of its first great civilizations—on the islands and peninsulas of Greece, and later in Italy. Greece lay exposed to the influences radiating from the advanced civilizations of Mesopotamia and the Nile valley (see map p. 279), and the intervening eastern Mediterranean was crisscrossed by maritime trade routes.

Ancient Greece

As the ancient Greeks forged their city-states and intercity leagues, they made impressive intellectual achievements as well (which peaked during the fourth century B.C.). Their political science and philosophy have influenced politics and government ever since, and great accomplishments were also recorded in such fields as architecture, sculpture, literature, and education. Because of the fragmentation of Greece's habitat, there was local experimentation and suc-

From the Field Notes

"The ancient Greeks certainly chose their building sites effectively. One can see the Acropolis from all parts of the city, its Parthenon towering over the townscape. (*Acro* in Greek means 'high point,' and *polis*, of course, means *city*). Several buildings can be seen rising on the Acropolis, among which the Parthenon is the incomparable masterpiece of Phidias, architect-engineer of ancient Athens. Work began in 447 B.C. The great columns are 34 feet high; they swell gently in the middle and lean very slightly inwards. Thus the Parthenon is designed as the base of a great pyramid with an apex that would be 3,000 feet high. This temple to Athena was adorned with beautiful sculptures, almost all of which have been taken away. But its magnificent lines have inspired architects ever after, from Roman times to the twentieth century, and undoubtedly beyond."

cess, followed by active exchanges of ideas and innovations. But internal discord was always present as well; in the end it ensured Greece's decline. By 147 B.C., the Romans had defeated the last sovereign Greek intercity league. Nevertheless, what the ancient Greeks had accomplished was not undone: they had transformed the eastern Mediterranean into one of the cultural cores of the world, and Greek culture became a major component of Roman civilization.

The Roman Empire

The Roman successors to ancient Greece made their own essential contributions. The Greeks never achieved politico-territorial organization on the scale accomplished by Imperial Rome. Much additional progress was made in such spheres as land and sea communications, military organization, law, and governmental administration. During its greatest expansion (in the second century A.D.), the Roman Empire extended from Britain to the Persian Gulf and from the Black Sea to Egypt. Given the variety of cultures that had been brought under Roman control and the resulting exchange of ideas and innovations, there were many opportunities for regional interaction—particularly in southern and western Europe. Areas that had hitherto supported only subsistence modes of life were drawn into the greater economic framework of the state, and suddenly there were distant markets for products that had never found even local markets before. Foodstuffs and raw materials now flowed into Rome from most of the Mediterranean Basin. With a population of perhaps a quarter-million, the city itself was the greatest single marketplace of the Empire and the first metropolitan-scale urban center in Europe.

That urban tradition came to characterize Roman culture throughout the Empire, and many cities and towns founded by the Romans continue to prosper to this day. Roman urban centers were also connected by an unparalleled network of highway and water routes, facilities that all formed part of an **infrastructure** needed to support economic growth and development. (Today, a modern state's infrastructure would include railroads, airports, energy-distribution systems, telecommunications networks, and the like.) More than anything else, however, the Roman Empire left Europe a legacy of ideas—concepts that long lay dormant but eventually played their part when Europe again discovered the path of progress. In political and military organization, effective administration, and long-term stability, the Empire was centuries ahead of its time. Moreover, never was a larger part of Europe unified by acquiescence than it was under the Romans, and at no time did Europe come closer to obtaining a *lingua franca* (common language) than it did during the age of Rome.

Finally, Europe's transformation under Roman rule heavily involved the geographic principle of **areal functional specialization**. Before the Romans brought order and connectivity to their vast domain, much of Europe was inhabited by tribal peoples whose livelihoods were on a subsistence level. Many of these groups lived in virtual isolation, traded little, and fought over territory when encroachment occurred. Peoples under Rome's sway, however, were brought into Roman economic as well as political spheres, and farmlands, irrigation systems, mines, and workshops appeared. Thus Roman-dominated areas began to take on a characteristic that has marked Europe ever since: *particular peoples and particular places concentrated on the production of particular goods.* Parts of North Africa became granaries for urbanizing (European) Rome; Elba, a Mediterranean island, produced iron ore; the Cartagena area of southeastern Spain mined and exported silver and lead. Many other locales in the Roman Empire specialized in the production of particular farm commodities, manufactured goods, or minerals. The Romans knew how to exploit their natural resources; at the same time, they also learned to use the diversified productive talents of their subjects.

DECLINE AND REBIRTH

The eventual breakdown and collapse of the Empire in the fifth century A.D. could not undo what the Romans had forged in the spreading of their language, in the dissemination of Christianity (in some ways the sole strand of permanence through the ensuing Dark Ages), and in education, the arts, and countless other spheres. But ancient Rome's decline was attended by a momentous stirring of Europe's peoples as Germanic and Slavic populations moved to their present positions on the European stage. The Anglo-Saxons invaded Britain from Danish shores, the Franks moved into France, the Allemanni traversed the North European Lowland and settled in Germany. Capitalizing on the disintegration of Roman power, numerous kings, dukes, barons, and counts established themselves as local rulers. Europe was in turmoil, and its weakness invited invasion from North Africa and Southwest Asia. In Iberia, the Arab-Berber Moors conquered a large area; in Eastern Europe, the Ottoman Turks extended their Islamic empire. The townscapes of southern Spain and the Balkans still carry the cultural imprints of these Muslim invasions.

After nearly a thousand years of feudal fragmentation during the Dark and Middle Ages, modern Europe began its emergence in the second half of the fifteenth century. At home, monarchies strengthened at the expense of feudal lords and landed aristocracies and, in the process, forged the beginnings of nation-states. Abroad, Western Europe's developing states were on the threshold of discovery—the discovery of continents and riches across the oceans. Eu-

The ancient Romans permanently influenced cultural landscapes throughout their empire, building towns, roads, aqueducts, and other structures. This spectacular Roman aqueduct (here crossing the Gard River) has stood the test of nearly 2,000 years near Avignon in the Provence region of southeastern France.

rope's emerging powers were fired by a new national consciousness and pride, and there was renewed interest in Greek and Roman achievements in science and government. Appropriately, this period is referred to as Europe's *Renaissance*.

The new age of progress and rising prosperity was centered in Western Europe, whose countries lay open to the new pathways to wealth—the oceans. Now the highly competitive monarchies of Western Europe engaged in economic nationalism that operated in the form of *mercantilism*. The objectives of this policy were the accumulation of as large a quantity of gold and silver as possible, and the use of foreign trade and colonial acquisition to achieve that end. Mercantilism was promoted and sustained by the state; precious metals could be obtained either by the conquest of peoples in possession of them or indirectly by achieving a favorable balance of international trade. Thus there was stimulus not only to seek new territories where such metals might lie, but also to produce goods at home that could be sold profitably abroad. The matrix of modern states in Western Europe was beginning to take shape, and the spiral had been entered that was to lead to great empires and a period of world domination.

THE REVOLUTIONS OF MODERNIZING EUROPE

Strife and dislocation punctuated Europe's march to world domination. Much of what was achieved during the Renaissance was destroyed again as powerful monarchies struggled for primacy; religious conflicts dealt death and misery; and the beginnings of parliamentary government fell under new tyrannies. Nevertheless, revolutions in several spheres were in the making. Economic developments in Western Europe ultimately proved to be the undoing of absolute monarchs and their privileged, land-owning nobilities. The city-based merchant was gaining wealth and prestige, and the traditional measure of affluence—land—began to lose its status in these changing times. The merchants and businesspeople of Europe were soon able to demand political recognition on grounds the nobles could not match. Urban industries were thriving; Europe's population, more or less stable at about 100 million since the mid-sixteenth century, was on the increase.

The Agrarian Revolution

This transformation was heightened by an ongoing *agrarian revolution*—the significant metamorphosis of European farming that preceded the Industrial Revolution and helped make possible a sustained population increase during the seventeenth and eighteenth centuries. The Netherlands, Belgium, and northern Italy (soon joined by England and France) paved the way with their successes in commerce and manufacturing. The stimulus provided by expanding urbanization and markets led to the more efficient organization of land ownership and agriculture. At the same time, methods of soil preparation, crop rotation, cultivation, harvesting, and livestock feeding improved; more effective farm equipment and storage and distribution systems were also developed. In the growing cities and towns, farm products fetched higher prices. New crops were introduced, especially from the Americas; the potato now became a European staple. More and more of Europe's farmers were drawn from subsistence into profit-driven market economies. Later, the manufactured products of the Industrial Revolution further stimulated the transformation of the realm's agriculture.

As new forces and processes began to reshape the economic geography of Europe, certain scholars tried to interpret the new spatial patterns they produced. In 1826, the economist Johann Heinrich von Thünen (1783–1850) fashioned one of the world's first geographical *models* (see box titled "Models in Geography"). For four decades von Thünen, who owned a large farming estate in northeastern Germany, studied the effects of distance and transportation costs on the location of productive activity. Eventually, he published a classic work entitled *The Isolated State*, and his methods in many ways constitute the foundations of modern location theory.

Von Thünen's Isolated State model was so named because he wanted to establish, for purposes of theoretical analysis, a self-contained country devoid of outside influences that would disturb the internal workings of the economy. Thus he created a sort of regional laboratory within which he could identify the factors that influence the locational distribution of farms around a single urban center. To do this, he made a number of limiting assumptions. First, he stipulated that the soil and climate would be uniform throughout the region. Second, no river valleys or mountains would interrupt a completely flat land surface. Third, there would be a single centrally positioned city in the Isolated State, and the latter would be surrounded by an empty, unoccupied wilderness. Fourth, the farmers in the Isolated State would transport their own products to market by oxcart, directly overland and straight to the central city. This, of course, is the same as assuming a system of radially converging roads of equal and constant quality; with such a system, transport costs would be directly proportional to distance.

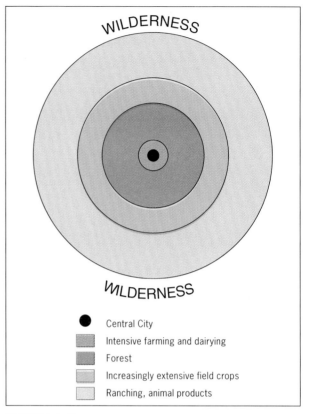

FIGURE 1-3

Von Thünen integrated these assumptions with what he had learned from the actual data collected while running his estate, and he now asked himself: What would be the ideal spatial arrangement of agricultural activities within the Isolated State? He concluded that farm products would be raised in a series of concentric zones outward from the central market city. Nearest to the city would be grown those crops that perished easily and/or yielded the highest returns (such as vegetables), because this readily accessible farmland was in great demand and therefore quite expensive; dairying would also be carried on in this innermost zone. Farther away would be potatoes and grains. Eventually, since transport costs to the city increased with distance, there would come a line beyond which it would be uneconomical to produce crops. There the wilderness would begin.

Von Thünen's model incorporated four zones or rings of agricultural land use surrounding the market center (Fig. 1-3). The first and innermost belt would be a zone of intensive farming and dairying. The second zone, according to von Thünen, would be an area of forest used for firewood and timber (still important as building material in his time). Next, there would be a third ring of increasingly extensive field crops. The fourth and outermost zone would be occupied by ranching and animal products beyond which would begin the wilderness that isolated the region from the rest of the world.

MODELS IN GEOGRAPHY
■ ■ ■

A current, widely used approach to generalization in both human and physical regional geography is the development of conceptual **models**. Peter Haggett, in his book *Locational Analysis in Human Geography*, offers an especially lucid definition: *in model-building we create an idealized representation of reality in order to demonstrate its most important properties.*

The use of models by geographers is necessitated by the complexity of reality: to understand how things work, we must first filter out the main spatial processes and their responses from the myriad details with which they are embedded in a highly complicated world. Models therefore provide a simplified picture of reality in order to convey, if not the entire truth, then at least a useful and essential part of it. The theory-based derivation of the von Thünen model (Fig. 1-3) and its *empirical*, or real-world, application to contemporary Europe (Fig. 1-4) offer a classical demonstration of this geographic method.

■ ■ ■

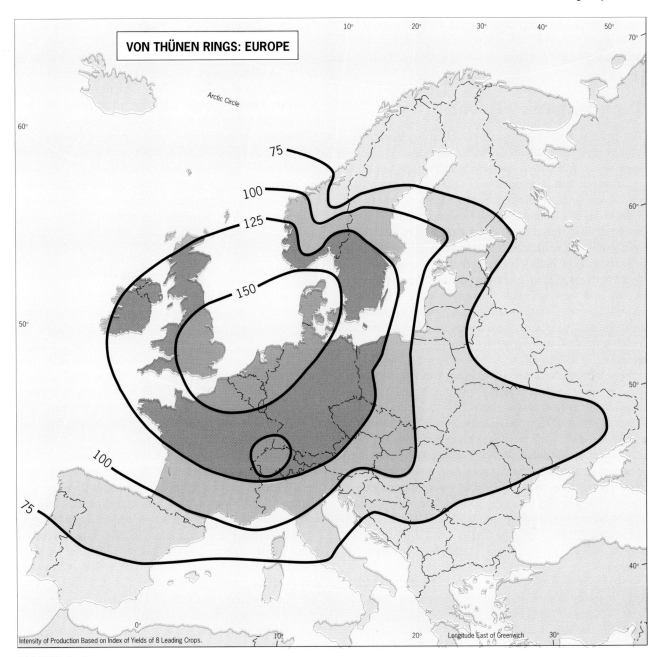

VON THÜNEN RINGS: EUROPE

Arctic Circle

75

100

125

150

100

75

Intensity of Production Based on Index of Yields of 8 Leading Crops.

Longitude East of Greenwich

FIGURE 1-4

Von Thünen knew, of course, that the real Europe (or world) did not present idealized situations exactly as he had postulated them. Transport routes serve certain areas more efficiently than others. Physical barriers can impede the most modern of surface communications. External economic influences invade every area. But von Thünen wanted to eliminate these disruptive conditions in order to discern the fundamental processes that shaped the spatial layout of the agricultural economy. Later, the distorting factors could be introduced one by one and their influence measured. First, however, he developed his model in theoretical isolation, basing it on total regional uniformity.

It is a great tribute to von Thünen that his work still commands the attention of geographers. The economic-geographic landscape of Europe has changed enormously since his time, but geographers still compare present-day patterns of economic activity to the Thünian model. Such a comparison was made by Samuel van Valkenburg and Colbert Held, whose map of Europe's agricultural intensity reveals a striking ring-like concentricity (Fig. 1-4). The overriding spatial change since von Thünen's time is the improvement in transportation technology, which permitted the Isolated State to expand from *micro-* to *macroscale*. Thus the model is no longer centered on a single city but rather on the vast urbanized area lining the southern coasts of the North Sea—which now commands a continent-wide Thünian agricultural system.

The Industrial Revolution

Alongside its advances in agriculture, Europe also had developed significant industries before the *Industrial Revolution* began. In Flanders (western Belgium) and England, specialization had been achieved in the manufacturing of woolen and linen textiles; in eastern Germany's Saxony, iron ore was mined and smelted. European manufacturers produced a wide range of goods for local markets, but their quality was often surpassed by textiles and other wares from India and China. This gave European entrepreneurs the incentive to refine and mass-produce their products. Because raw materials could be shipped home in virtually unlimited quantity, if they could find ways to mass-produce these commodities into finished goods, they could bury the Asian industries under growing volumes and declining prices.

Now the search for better machinery was on, especially improved spinning and weaving equipment. In the 1780s, James Watt and others succeeded in devising a steam-driven engine, and soon this invention was adapted for various uses. About the same time, it was realized that coal (converted into carbon-rich coke) was a vastly superior substitute for charcoal in smelting iron. These momentous innovations had a rapid effect. The power loom revolutionized the weaving industry. Iron smelters, long dependent on Europe's dwindling forests for fuel, could now be concentrated near coalfields. Engines could move locomotives as well as power looms. Ocean shipping entered a new age.

England had an enormous advantage, for the Industrial Revolution occurred when British influence reigned worldwide and the significant innovations were achieved in Britain itself. The British controlled the flow of raw materials, they held a monopoly over products that were in global demand, and they alone possessed the skills necessary to make the machines that manufactured the products. Soon the fruits of the Industrial Revolution were being exported, and the modern industrial spatial organization of Europe began to take shape (Fig. 1-5). In Britain, manufacturing regions, densely populated and heavily urbanized, developed near coalfields in the English Midlands, at Newcastle to the northeast, in southern Wales, and along Scotland's Clyde River in the Glasgow area.

In mainland Europe, a belt of major coalfields extends from west to east, roughly along the southern margins of the North European Lowland, due eastward from southern England across northern France and Belgium, Germany (the Ruhr), western Bohemia in the Czech Republic, Silesia in southern Poland, and the Donets Basin in eastern Ukraine. Iron ore is found in a broadly similar belt, and the industrial map of Europe reflects the resulting concentrations of economic activity (Fig. 1-5). Another set of manufacturing regions emerged in and near the growing urban centers of Europe, as the same map demonstrates. London—already Europe's leading urban focus and Britain's richest domestic market—was typical of these developments. Many local industries were established here, taking advantage of the large supply of labor, the ready availability of capital, and the proximity of so great a number of potential buyers. Although the Industrial Revolution thrust other places into prominence, London did not lose its primacy: industries in and around the British capital multiplied.

Industrial and Urban Intensification

It is not surprising that the industrialization of Europe—or, rather, its industrial *intensification* following the Industrial Revolution—also became a focus of geographic research. What influences affected industrial location? How was Europe's industrialization channeled? Again, the first important studies were conducted by German scholars, mostly during the second half of the nineteenth century. Much of this work was incorporated in a volume by Alfred Weber (1868–1958), published in 1909 and entitled *Concerning the Location of Industries*. Like von Thünen, Weber began with a set of limiting assumptions in order to minimize the complexities of the real Europe. But unlike von Thünen, Weber dealt with activities that take place at particular *points* rather than across large areas. Manufacturing plants, mines, and markets are located at specific places, and so Weber created a model region marked by sets of points where these activities would occur. He eliminated labor mobility and varying wage rates, and thereby could calculate the ''pulls'' exerted on each point in his theoretical region.

In the process, Weber discerned various factors that affect **industrial location**. He recognized what he called ''general'' factors that would affect all industries—dominated by transport costs for raw materials and finished products—and ''special'' factors (such as perishability of foods). He also differentiated between ''regional'' factors (transport and labor costs) and ''local'' factors. The local factors, Weber argued, involve *agglomerative* (concentrating) and *deglomerative* (deconcentrating) forces. Take the case of London discussed above: industries located there, in large part, because of the advantages of locating together. The availability of specialized equipment, a technologically sophisticated labor force, and a large-scale market made London (as well as Paris and other big cities not positioned on rich deposits of natural resources) an attractive site for many manufacturing plants that could benefit from agglomeration. On the other hand, such concentration may over time create strong disadvantages—

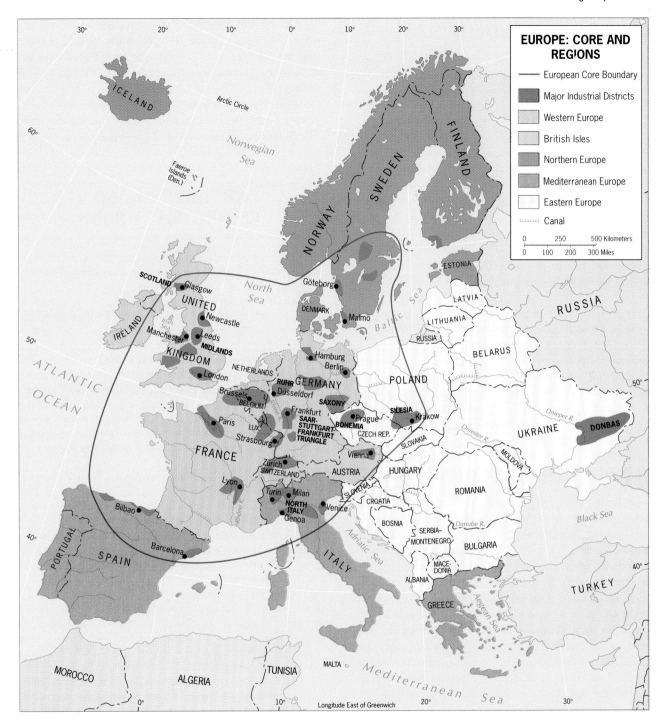

FIGURE 1-5

chiefly competition for space, rising land prices, congestion, and environmental pollution. Eventually an industry might move away and deglomerative forces would set in.

Europe's industrialization also speeded the growth of many of its cities and towns. In Britain in the year 1800, only about 9 percent of the population lived in urban areas; but by 1900, some 62 percent resided in cities and towns (today that proportion surpasses 90 percent). All this was happening while the total population skyrocketed as well.

As industrial modernization came to Belgium and Germany, to France and the Netherlands, and to other parts of Western Europe, the entire urban pattern changed. The nature of this process—the growth and strengthening of towns and cities—and related questions also became topics of geographic study, just as agriculture and industrialization had. (Principles of urban geography are discussed in the ''Focus on a Systematic Field'' at the beginning of Chapter 3.)

Political Revolutions

Europe had had long experience with experiments in democratic government, but the *political revolution* that swept the realm after 1780 brought transformation on an unprecedented scale. Overshadowing these events was the French Revolution (1789–1795), but France's—and Europe's—political catharsis lasted into the twentieth century as the rising tide of nationalism eventually affected every monarchy on the continent.

In France, the popular revolution had plunged the country into years of chaos and destruction. Only when Napoleon took control in 1799 was stability restored. Napoleon personified the new French republic, and he reorganized France so completely that he laid the foundations of the modern nation-state (a concept treated in the following section). He also built an empire that extended from German Prussia to Spain and from the Netherlands to Italy. Although his armies were ultimately repulsed, Napoleon had forever changed the political spatial structure of Europe. His French forces had been joined by nationalist revolutionaries from all over the continent; one monarchy after another had been toppled. Even after his final defeat at Waterloo in 1815, there were popular uprisings in Spain, Portugal, Italy, and Greece. By then Europe had had its first real taste of democracy and nationalist power—and it would not revert to its old ways.

The Rise of the Nation-State

As Europe went through its periods of rebirth and revolutionary change, the realm's politico-geographical map was transformed. Smaller entities were absorbed into larger units, conflicts resolved (by force as well as negotiation), boundaries defined, and internal divisions reorganized. European nation-states were in the making.

But what is a nation-state and what is not? The question centers in part on the definition of the term *nation*. That definition usually involves measures of homogeneity: a nation should comprise a group of tightly knit people who speak a single language, have a common history, share the same ethnic background, and are united by common political institutions. Accepted definitions of the term suggest that many states are not nation-states because their populations are divided in one or more important ways. But cultural homogeneity may not be as important as a more intangible "national spirit" or emotional commitment to the state and what it stands for. One of Europe's oldest states, Switzerland, has a population that is divided along linguistic, religious, and historical lines—but Switzerland has proved to be a most durable nation-state nonetheless, marking its seven hundredth anniversary in 1991.

A **nation-state**, therefore, may be defined as a political unit comprising a clearly delineated territory and inhabited by a substantial population, sufficiently well organized to possess a certain measure of power, the people considering themselves to be a nation, with certain emotional and other ties that are expressed in their most tangible form in the state's legal institutions, political system, and ideological strength. Supporting this structure, of course, is a government that constantly works to ensure that the forces unifying the state prevail over those that would drive it apart (see box titled ''Centripetal and Centrifugal Forces'').

CENTRIPETAL AND CENTRIFUGAL FORCES

■ ■ ■

Political geographers use the terms *centripetal* and *centrifugal* to identify forces within a state that tend, respectively, to bind that political system together and to pull it apart.

Centripetal forces tend to tie the state together, to unify and strengthen it. A real or perceived external threat can be a powerful centripetal force. More important and lasting, however, is a sense of commitment to the governmental system, a recognition that it constitutes the best option. This commitment is sometimes focused on the strong charismatic qualities of one individual—a leader who personifies the state, who captures the population's imagination. (The origin of the word *charisma* lies in the Greek expression that means "divine gift.") At times such charismatic qualities can submerge nearly everything else. Juan Peron's lasting popularity in Argentina is a case in point. Jomo Kenyatta in Kenya, Josip Broz Tito in Yugoslavia, Charles De Gaulle in France, Mao Zedong in China, and Jawaharlal Nehru in India all possessed similar charisma and individually played dominant roles (extending well beyond their lifetimes) in binding their states.

Centrifugal forces are disunifying or divisive. They can cause deteriorating internal relationships. Religious conflict, racial strife, linguistic cleavages, and contrasting regional outlooks are among the major centrifugal forces. During the 1960s, the Vietnam (Indochina) War became a major centrifugal force in the United States, the aftermath of which is still felt three decades later. Newly independent countries often find tribalism a leading centrifugal force, sometimes strong enough to threaten the very survival of the whole state system (as the Biafra conflict of the 1960s did in Nigeria).

The degree of strength and cohesion of the state depends on a surplus of centripetal forces over divisive centrifugal forces. It is difficult to measure such intangible qualities, but some attempts have been made in this direction—for example, by determining attitudes among minorities and by evaluating the strength of regionalism as expressed in political campaigns and voter preferences. When the centrifugal forces become excessively strong and cannot be checked, even by external imposition, the state breaks up (as the Soviet Union and Yugoslavia did in 1991, and Czechoslovakia in 1993); or it undergoes revolutionary internal change that makes it, in effect, a new entity, as Iran became after the ouster of the shah and as Cuba became after the victory of Castro's forces.

■ ■ ■

This definition of the nation-state essentially identifies the European model that emerged in the course of the realm's long period of evolution and revolutionary change. France is often cited as the best example among Europe's nation-states, but Poland, Hungary, Sweden, the United Kingdom, Spain, and newly reunified Germany are also among countries that satisfy the terms of the definition to a great extent. Belgium and Moldova are examples of European states that cannot at present be designated as nation-states. The former Yugoslavia and Czechoslovakia never were nation-states; their actual and latent centrifugal (disunifying) forces were too strong to allow long-term stability.

THE EUROPEAN REALM TODAY

The nation-states of Europe are among the world's oldest, and the colonial empires of European powers were among the most durable. Despite many disruptive wars and revolutions, European nations survived, and from their long-term stability they forged a confidence—based on strong individual identities—that is a hallmark of European culture. Although Europe quite clearly constitutes a geographic realm, it exhibits little geographic homogeneity.

It is sometimes postulated that Europe may be viewed as a regional unit because its peoples share Indo-European languages (Fig. 1-6), Christian religious traditions (see Fig. 6-1), and common European (Caucasian) racial ancestry. Yet these human cultural and physical traits extend well beyond Europe and, in any case, are not strong unifying elements within the European mosaic. As Figure 1-6 shows, Hungarians and Finns are among European groups who do not speak Indo-European languages; indeed, for so small a realm, Europe is a veritable Tower of Babel. As for common religious traditions, Europe has a history of intense and destructive conflict over issues involving religion; shared Christian principles, for instance, have done little to bring cohesion to Northern Ireland. And again, Europe's purported common racial ancestry masks strong differences in observable physical characteristics between Spaniard and Swede, Scot and Sicilian.

In terms of economic development, centuries of European exploitation of overseas domains amassed fortunes at home and established a global influence that survived two world wars and continued when the colonial era ended. Europe in 1945 entered a postwar era of reconstruction, realignment, and resurgence. Reconstruction was aided by the billions of dollars made available by the United States through the Marshall Plan (1948–1952). Realignment came in the form of the Iron Curtain separating the Soviet-dominated East from Western Europe; it also involved the

emergence of several international "blocs" consisting of states seeking to promote multinational cooperation. The resurgence of Europe continued despite the loss of colonial empires, recurrent political crises, and other obstacles; once again, Europe's momentum had carried the day. Contemporary Europe emerged from the recently completed postwar era with many strengths. Among its most important geographic properties are those discussed in the following sections.

Intensifying Spatial Interaction

Greater international cooperation is a logical outcome of the peaceful postwar era because Europe's environments and resources—probably to a larger degree than in any world area of similar size—have long presented outstanding opportunities for human contact and interaction. Conceptually, **spatial interaction** is best organized around a triad of principles developed by the American geographer Edward Ullman (1912–1976): (1) complementarity, (2) transferability, and (3) intervening opportunity.

Complementarity occurs when one area has a surplus of a commodity demanded by a second area. The mere existence of a resource in a locality is no guarantee that trade will develop; that resource must specifically be needed elsewhere. Thus complementarity arises from regional variations in both the supply and demand of human and natural resources. **Transferability** refers to the ease with which a commodity may be transported between two places. Sheer distance, in terms of both the cost and time of movement, may be the major obstacle to the transferability of a good. Therefore, even though complementarity may exist between a pair of areas, the problems of economically overcoming the distance separating them may be so great that trade cannot begin. The third interaction principle, **intervening opportunity**, holds that potential trade between two places, even if they satisfy the necessary conditions of complementarity and transferability, will develop only in the absence of a closer, intervening source of supply.

European economic geography has always been stimulated by internal complementarities. A specific example involves Italy, which ranks foremost among Mediterranean countries in economic development but lacks adequate coal supplies. For a long time its industries have depended on coal imports from the rich deposits of Western Europe. At the same time, Italian farmers grow crops that cannot be raised in the cooler climate north of the Alps. Citrus fruits, olives, grapes, and early vegetables are in high demand in Western Europe's markets. Hence, Italy imports northwestern Europe's coal, and northwestern Europe imports Italian fruits and wines. This case of double complemen-

FIGURE 1-6

tarity is not counteracted by transferability restrictions: the physical barrier of the Alps has long been breached by rail and highway routes, and the two-way trade flow is most attractive to shippers because freight-carrying vehicles will not have to return home empty. Moreover, because northwestern Europe and Italy are the closest sources of coal

The French have long been leaders in high-speed train technology. This TGV is passing through Montpellier on a new line linking Marseille and the Spanish border.

and fruits, respectively, there are no intervening opportunities to disrupt spatial interaction, and a thriving mutual exchange of such commodities is generated.

Historically, there have been countless examples of this kind throughout Europe, all dependent on efficient transportation linkages. A good circulation system has characterized Europe since Roman times, and the steady improvement of transport technology spawned trading relationships involving ever more numerous and distant places. Today's excellent network of railroads, highways, and air routes is fully integrated because of the efforts of the postwar planners of a united Europe. In the 1990s, two major projects have been completed that will significantly enhance northwestern Europe's already splendid passenger- and freight-rail operations. The long-awaited tunnel beneath the English Channel opened in 1994, for the first time directly connecting Britain with the continent and halving the surface travel time (to three hours) between London and Paris. And by 1998, a new high-speed rail network will be in place to link all the major cities of Western Europe, modeled on France's *TGV (train à grande vitesse)* that connects Paris with most other French cities at average speeds of 185 miles (300 km) per hour.

Assets of Urban Tradition

Europe ranks among the world's most highly urbanized realms. No less than 72 percent of the European population today lives in towns and cities. Within the realm, that urban proportion rises to 84 and 81 percent, respectively, in the regions of Northern and Western Europe. In certain countries, it rises even higher: Belgium, 97 percent; the United Kingdom, 92 percent; the Netherlands, 89 percent; and Germany, 85 percent.

Large cities are the crucibles of their nations' cultures. In his 1939 study of the pivotal role of great cities in the development of national cultures, American geographer Mark Jefferson postulated the law of the **primate city**, which stated that "a country's leading city is always disproportionately large and exceptionally expressive of national capacity and feeling." Although rather imprecise, this "law" can readily be demonstrated using European examples. Certainly Paris personifies France in countless ways, and there is nothing in England to rival London. In both of these primate cities, the culture and history of a nation and empire are indelibly etched in the urban landscape. Similarly, Vienna is a microcosm of Austria, Warsaw is the heart of Poland, Stockholm is Sweden, and Athens is Greece. Today, each of these (together with the other primate cities of Europe) sits atop a hierarchy of urban centers that has captured the lion's share of national population growth since World War II.

At the intraurban scale, the appearance of the cityscape in Europe contrasts markedly with the urban scene in the United States. Lawrence Sommers has underscored several reasons for this dissimilarity:

Age is a principal factor, but ethnic and environmental differences also play major roles in the appearance of the European city. Politics, war, fire, religion, culture, and economics also have played a role. Land is expensive due to its scarcity, and capital for private enterprise development has been insufficient, so government-built housing is quite common. Prices for real estate and rent have been government controlled in many countries. Planning and zoning codes as well as the development of utilities are determined by government policies [that strongly] control urban land development.

The internal spatial structure of the European **metropolis** (i.e., the central city and its suburban ring) is typified by the urban pattern of the London region (Fig. 1-7). The metropolitan area remains focused on the large city at its center, especially the downtown **central business district (CBD)**, which is the oldest part of the urban agglomeration and contains the region's largest concentration of business, government, and shopping facilities as well as its wealthiest and most prestigious residences. Wide residential sectors radiate outward from the CBD across the rest of the central city, each one home to a particular income group and thereby perpetuating spatially the strong class system of European society. Beyond the central city lies a sizeable suburban ring, but residential densities here are much higher than in the United States because the European tradition is one of setting aside recreational spaces (in "greenbelts") and living in apartments rather than detached single-family houses. There is also a far greater reliance on public transportation, which further concentrates the suburban development pattern. That has allowed many non-

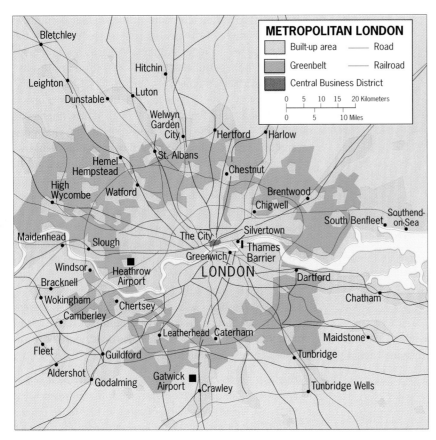

METROPOLITAN LONDON

☐ Built-up area   ——— Road

☐ Greenbelt   ——— Railroad

■ Central Business District

0   5   10 15   20 Kilometers

0   5   10 Miles

FIGURE 1-7

residential activities to suburbanize as well, and today ultramodern outlying business centers increasingly compete with the CBD in many parts of urban Europe (see the box titled "Paris's *La Défense*").

Political and Economic Transformation

Today, in the aftermath of the collapse of communism in both Eastern Europe and the former Soviet Union in the early 1990s, the European realm has entered a new era. As the Soviet Empire unraveled, it fragmented into 15 newly independent states along the lines of the former republics of the U.S.S.R. Five of these fledgling countries—Latvia, Lithuania, Moldova, Ukraine, and Belarus (formerly Belorussia)—are included in an expanded, redefined Eastern Europe that now stretches as far eastward as the Russian border; a sixth former Soviet republic, Estonia, is historically and culturally part of Northern Europe. This latest geographic realignment has enlarged the European realm's territorial dimensions by 21 percent, the number of its countries to 38, and its total population by 16 percent since 1990.

At the same time, a single new Europe has taken the place of the two Europes that until only a few years ago threateningly faced each other across the Iron Curtain. While this political transformation has markedly lessened the danger of an East-West military confrontation, it has done little to reduce the realm's internal cultural divisions. Indeed, the swift revival of nationalisms and old hostilities in Eastern Europe—which almost immediately led to brutal civil warfare in disintegrating Yugoslavia—are a reminder that many difficult challenges face a restructuring Europe in its quest for international cooperation. These problems notwithstanding, the Europeans have long recognized the disadvantages of functioning within their highly fragmented mosaic and have already made considerable progress toward unification. For more than half a century, Western Europe has been forging the economic linkages that are the cornerstone of today's *European Union*. This 15-member multinational organization, which incorporates almost all of the realm's core area (Fig. 1-5), offers the best hope for integrating the new Europe as it develops and implements plans to expand its operations. We now examine in some detail the interplay of forces involving the challenge of political disintegration and the opportunity for greater economic integration.

Devolution

In our earlier discussion of nation-states, we noted that centrifugal forces in Yugoslavia and Czechoslovakia recently

PARIS'S *LA DÉFENSE*

■ ■ ■

Mushrooming "suburban downtowns" are transforming the U.S. metropolitan landscape, and these "edge cities" are now beginning to appear in Western Europe too. The best known and most important suburban business complex in Europe is *La Défense*, located just west of Paris, whose growth has been encouraged over the past two decades in order to relieve congestion in the French capital's CBD. By the mid-1990s, it had become evident that the planners had carried out their mission only too well: this burgeoning activity center has not only achieved (unintended) primacy in its own metropolis but has also become continental Europe's largest business district, whose total annual business transactions exceed the gross national product of the Netherlands! Among other things, *La Défense* is now home to Europe's largest concentration of computer-age office space, its biggest shopping mall, more than 150 corporate headquarters facilities, the European head offices of dozens of prestigious multinational corporations, a massive hotel–convention–restaurant–entertainment complex, and a workforce of about 150,000 people (more than half of them in management positions).

Perhaps the most striking aspect of *La Défense* is that it is so much more than just another large-scale suburban downtown. Its centerpiece, the enormous Grande Arche (completed in the early 1980s), anchors the west end of Paris's "Historic Axis" that now stretches over 5 miles (8 km) from the Louvre up the Champs Elysées to the Arc de Triomphe and then outward to *La Défense*. This architectural jewel (surrounded by the more than 50 unique sculptures that line its vast esplanade) has already become one of the most famous Parisian monuments, attracting perhaps 30 million visitors a year. Despite those numbers, the regional planners of *La Défense* have designed a new transport network that is proving capable of handling the huge traffic flow. Besides five expressway connections and behemoth parking garages for automobiles, five separate rail systems converge here—regular Metro (subway) lines, an express Metro line, a commuter railroad line, a long-distance passenger-rail line, and a new TGV line (see p. 59). This represents some of the most advanced transportation planning in the world; clearly, the thriving success of *La Défense* will have important lessons to teach the city builders of the twenty-first century.

■ ■ ■

Europe's leading corporate complex, and in many ways its most modern and architecturally most innovative, is *La Défense*, west of the old city. Stand on top of the Arc de Triomphe, and at the end of the avenue your gaze is concentrated on the open interior of the huge Grande Arche seen in the foreground of this photograph. Steel-and-glass towers of various shapes and sizes surround this remarkable structure at the end of the esplanade; *La Défense* has itself become a tourist attraction to rival the treasures of the old city on the Seine.

reached a level that triggered the political disintegration of both countries. In other parts of the realm, too, similar tensions are now heightening and could lead to the future breakup of many additional states. The term **devolution** has come into use to describe the process whereby regions or peoples within a state demand and gain political strength and sometimes autonomy at the expense of the center—through negotiation or active rebellion. Most states exhibit internal regionalism, but the process of devolution is set into motion when a key centripetal force—the nationally accepted idea of what a country stands for—erodes to the point that a regional secession movement is launched. This slide toward separatism is most likely to occur in countries whose governments have had problems from the beginning in trying to forge a viable nation-state. Indeed, both Yugoslavia and Czechoslovakia were quickly patched together by powerful outsiders (the victorious Allies of World War I at the 1919 Versailles Peace Conference) from the Eastern European ethnic crazy-quilt left behind by the defeated Austro-Hungarian Empire.

The devolution concept itself was first used to summarize the course of political events in the United Kingdom, where the resurrection of regional separatism is something of a geographic irony. This state is dominated, in terms of population as well as political and economic power, by England, the historic core area of the British Isles. The country's three other politico-geographical entities—Scotland, Wales, and Northern Ireland—were acquired over several centuries and attached to England. Time, however, has failed to submerge regionalism in the United Kingdom, notwithstanding the development introduced by the Industrial Revolution and its aftermath, and the period of empire and comparative wealth. During the 1960s and 1970s, London was forced to confront a virtual civil war in Northern Ireland as well as a rising tide of separatism in Scotland and Wales. Scottish nationalism proved to be a particularly

D

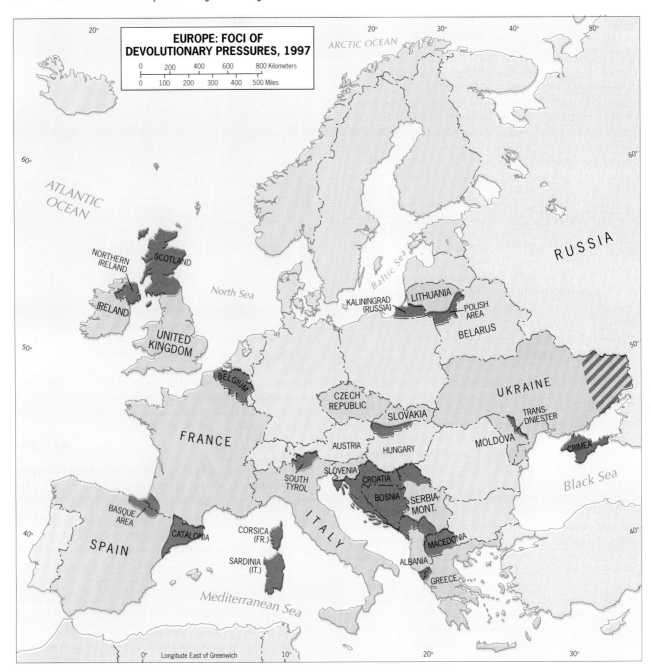

FIGURE 1-8

potent force in British politics because Scotland was then becoming the center of the country's vital new energy industries based on the discovery of major oil and gas deposits beneath the adjacent North Sea. The central government responded by allowing the nationalists to vote in favor of their own assemblies with (initially) limited legislative and executive powers. Although this proposal lost on a technicality (even though a majority of Scots *did* vote in favor of it), the outcome had the effect of perpetuating the separatist movement, and its goals remain under formal discussion in London. Thus the devolution process con-

cerning Scotland's place in the United Kingdom is far from over—with opinion polls indicating that support for Scottish independence has been rising steadily since the early 1990s.

The emergence and strength of regionalism in the United Kingdom, one of Europe's most durable states, underscores the potential impact of devolution elsewhere on the continent. Another European state attempting to adjust to strong regional forces is Spain, which has signed autonomy agreements with the leaders of the northern Basque region and the northeastern province of Catalonia. These agreements have given both areas their own parliaments, recognized their languages as official and equal to Spanish, and transferred powers of local taxation and education. Other signs of devolutionary tensions on the map of Europe can be seen in Figure 1-8. France has had to contend with a drive for secession on its Mediterranean island of Corsica. A similar movement demands autonomy for the neighboring island of Sardinia, which is part of Italy; the Italians also face another separatist drive in their far north in South Tyrol, where the Alps of Austria, Switzerland, and Italy meet. Besides the already-collapsed states of Yugoslavia and Czechoslovakia, the map also shows that devolutionary pressures affect Belgium—where a deep ethnic division persists between the country's Flemish and French speakers—as well as several of the newly independent states that emerged along the western rim of Russia following the dissolution of the Soviet Union in 1991.

Regionalism in Europe today is further heightened by the emergence of subnational regions as new hubs of economic power and influence. This is particularly true of city-regions anchored by booming commercial centers, such as France's Lyon, Italy's Milan, Spain's Barcelona, and Germany's Stuttgart. Most noteworthy is that these cities have developed direct linkages and relationships with one another, thereby bypassing the capital cities and central governments of their respective nation-states. Moreover, this new interregional web is being built with the encouragement of the European Union, with an eye not only toward forging a more open and flexible Europe but also toward building a future in which this expanding network of regions may supplant the old framework of nation-states.

European Unification

Geographic contradictions abound in today's world. One of the more fascinating is that while the fragmenting forces of devolution surface at the subnational scale in Europe, opposing forces at the international level continue to build toward stronger ties among the realm's states. Thus integration and disintegration are proceeding simultaneously. The year 1994 marked the fiftieth anniversary of the *Benelux* agreement, which established an economic union that linked the three small Western European countries of *Bel*gium, the *Net*herlands, and *Lux*embourg. More importantly, it simultaneously ignited a still-ongoing experiment in international cooperation and association that is culminating in the formation of a multicountry European Union which could, some hope and believe, eventually lead to a United States of Europe.

European countries are not alone in trying to forge a mutually beneficial union, nor is this the first time that the Europeans have tried to achieve such a goal. But this particular European experiment of the late twentieth century is the greatest accomplishment of its kind to date. Political geographers define the voluntary association of three or

more countries as **supranationalism**. Such an association may involve economic, cultural, and political spheres. The key factor involves sovereignty: Are the participant countries willing to give up some sovereignty for the betterment of their union?

In the case of Europe, that is indeed the situation. It all began in 1944, just before the end of World War II, when the Benelux organization was formed. Then, soon after the war, a farsighted U.S. Secretary of State, George Marshall, proposed a massive infusion of American aid to the struggling countries of postwar Western Europe. The Marshall Plan (1948–1952) not only boosted national economies: it

---

SUPRANATIONALISM IN EUROPE
■-■-■

1944 Benelux Agreement signed.

1947 Marshall Plan proposed.

1948 Organization for European Economic Cooperation (OEEC) established.

1949 Council of Europe created.

1951 European Coal and Steel Community (ECSC) Agreement signed (effective 1952).

1957 Treaty of Rome signed, establishing European Economic Community (EEC) (effective 1958), also known as the Common Market and "The Six."
European Atomic Energy Community (EURATOM) Treaty signed (effective 1958).

1959 European Free Trade Association (EFTA) Treaty signed (effective 1960).

1965 EEC-ECSC-EURATOM Merger Treaty signed (effective 1967).

1968 All customs duties removed for intra-EEC trade; common external tariff established.

1973 United Kingdom, Denmark, and Ireland admitted as members of EEC, creating "The Nine."

1979 First general elections for a European Parliament held; new 410-member legislature meets in Strasbourg.

1981 Greece admitted as member of EC (EEC has been shortened), creating "The Ten."

1986 Spain and Portugal admitted as members of EC, creating "The Twelve."
Single European Act ratified, targeting a functioning European Union in the 1990s.

1990 Charter of Paris signed by 34 members of the Conference on Security and Cooperation in Europe (CSCE).
Former East Germany, as part of newly reunified Germany, incorporated into EC.

1991 Maastricht meeting charts European Union (EU) course for the 1990s.

1993 Single European Market goes into effect.
Modified European Union Treaty ratified, transforming EC into EU.

1995 Austria, Finland, and Sweden admitted into EU, creating "The Fifteen."

■-■-■

FIGURE 1-9

also led directly to the need for a multinational European economic-administrative structure. This, Europeans realized, was no time for national boundaries and troublesome tariffs to interfere with the flow of raw materials and finished products. The Europeans also worried that future misunderstandings might lead to renewed conflict, so they established the Council of Europe, a deliberative body that met in Strasbourg, France (a city that is often mentioned as a possible capital for a future united Europe).

These two initiatives in the economic and political spheres produced one success after another, although traditional European rivalries did not disappear. In 1957,

six Western European countries—France (then-West) Germany, Italy, and the Benelux group—signed the far-reaching Treaty of Rome. This agreement established, as of 1958, the so-called *Common Market*—also known as the *European Economic Community* (or *EEC* for short), and later simply the *European Community* (*EC*). Not to be outdone, the British, who did not join the Common Market, in 1959 led six other countries to create the *European Free Trade Association* (*EFTA*). But this group of seven (the United Kingdom, Sweden, Norway, Denmark, Switzerland, Austria, and Portugal) was no match for the EEC in terms of productive capacity, markets, or raw materials.

NATO IN THE NINETIES

■ ■ ■

NATO was established at the height of the Cold War in 1950 by 10 Western European countries (as well as the United States and Canada) in order to provide for their collective defense against the menacing Soviet Union. The Soviets responded by similarly aligning their satellite countries to the east of the Iron Curtain—under a treaty known as the Warsaw Pact—and the two military blocs then tensely confronted each other for the next four decades. The recent collapse of the U.S.S.R. and the Warsaw Pact has greatly diminished the threat of invasion from the east but complicated the issues surrounding European security that had been rather simple before 1991.

As the 1990s unfolded, it became clear that new supranational military arrangements were needed to accommodate the changed balance of power. The search for new answers has revolved around four questions. What would be the new military role of the United States in Europe now that it no longer needed to station large numbers of troops there to deter a Soviet attack? Is post-communist Russia now a friend or foe, and how should this still-powerful superpower remnant be treated? How could the now mostly democratic, former Warsaw Pact countries be absorbed into NATO's structure? And how should Europe's internal defense capabilities now be expanded to make up for the reduced presence of U.S. forces and weapons?

These issues continue to be intensely debated today, but by 1996 it was possible to discern the broad outlines of NATO's transformation. A commitment has been made to extend the alliance eastward over the next few years. New member-states can only be admitted if they satisfy certain basic conditions: they must be governed democratically, military forces must be under civilian control, market economies must prevail, and domestic human rights must be fully protected. (The Eastern European countries likeliest to be among the first entrants are the Czech Republic, Poland, Hungary, and the Baltic states.) Russia itself will probably never join,

but a special security relationship linked to NATO's expansion will be worked out. An intermediate step toward these goals was taken in 1994 with the Partnership for Peace agreements that were concluded between NATO and all of the ex-communist countries except former Yugoslavia (Fig. 1-10). These agreements were designed to avoid provoking Russia (which also became a Partner for Peace), and represent a sort of halfway house of military cooperation with NATO that is open to any European country. If an attack on a Partner should occur, however, NATO is only obligated to "consult" and not come to its immediate aid militarily.

Within NATO, the Europeans have made it clear that they want the United States to continue to play a leading role—as underscored in the late-1995 deployment of troops to enforce the fragile peace agreement that terminated the hostilities in Bosnia. But the United States and its allies have also come to an understanding that Europe will need to become more active in its own defense through the strengthening of NATO's "European pillar." This effort is specifically aimed at military operations in which the United States does not need or want to participate. The coordinating agency is the Western European Union (WEU), a long-neglected security organization established during Marshall Plan days. Although it consists of only 10 full members (the EU countries that also belong to NATO, except Denmark), 17 additional states are affiliated through either observer or associate status to give WEU the breadth it needs to accomplish its new tasks (Fig. 1-10).

After decades in the background, important issues concerning European self-defense are increasingly moving to the forefront. As in so many other spheres, this will compel France, the United Kingdom, Germany, and other countries to confront their different views and forge new avenues of cooperation. Thus still-evolving NATO is becoming ever more deeply involved in the European unification movement.

■ ■ ■

Soon EFTA lost its relevance as some of its members (including the British) began to join the expansion-minded EC (see box titled ''Supranationalism in Europe''). Now the Community began to grow, first to 9 members, then to 10 and 12, and, by 1995, to 15 (Fig. 1-9). Today, this group of 15 countries forms the vanguard of the *European Union*.

In the political sphere, too, progress was made that many had believed impossible. The Council of Europe *did* evolve into an embryonic European Parliament, and in 1979 the first Europe-wide elections were held to send 410 representatives—now elected by the people and their parties rather than appointed by governments—to Strasbourg. The European Parliament began to reflect more accurately the political mainstream of what at the time was still Europe west of the Iron Curtain. Elections now are held at

regular five-year intervals. Those of 1984 and 1989 enlarged the Parliament to 518 members; the 1994 election was especially significant because it was the first after the consolidation of the European Union.

While all these events were occurring, supranational cooperation in other spheres also took place. EURATOM (European Atomic Energy Community) was founded to develop the peaceful uses of nuclear energy, in which France is now a world leader. ESRO (European Space Research Organization) coordinates Europe's successful space programs. And NATO (North Atlantic Treaty Organization) forms the umbrella for the military activities of member countries (see box titled ''NATO in the Nineties'').

But the spotlight today is on ''The Fifteen,'' the countries that form the European Union. (To remember which

G

FIGURE 1-10

countries these are, use the *4–5–6* method: the *four* giants—Germany, France, Italy, and the United Kingdom; the *five* inner countries [all neighbors of Germany]—Belgium, the Netherlands, Luxembourg, Denmark, and Austria; and the *six* outer countries—Ireland, Sweden, Finland, Greece, Spain, and Portugal.) In 1986, the (then) 12 member-states (Austria, Finland, and Sweden did not join until 1995) signed the momentous Single European Act, promising to work toward the achievement and implementation of some 279 specific goals by the end of 1992.

In late 1991, The Twelve met at Maastricht in the Netherlands to confront the inevitable difficulties these goals raised. The British, especially, had reservations about giving more power to the European Parliament; they also continued to have doubts about eventual political union (which is one of the 279 goals). The prospect of a common currency (to be called the ''Euro'') scheduled to be introduced at the end of 1999, further worried several of The Twelve: imagine Britain without the pound, France without the franc, Germany without its mark! Nonetheless, a central European banking system with a single European currency was approved through the Economic and Monetary Union (EMU).

These are just a few of the still-unresolved items before the members of the emerging European Union; common foreign and defense policies, policies on immigrant workers' rights, and many other matters also remain to be worked out. In the meantime, the Charter of Paris was signed by The Twelve as well as 22 other European states in 1990. This charter acknowledged and approved Germany's reunification, ratified Europe's present boundaries, and established the first greater-European mechanism for settling disputes among member countries. In fact, the Charter itself may be regarded as less important than the list of signatories, now including Eastern European countries.

During the mid-1990s, the still-evolving supranational process proved to be one of difficulty in the aftermath of the Treaty on European Union concluded at Maastricht. For all its recent convergence, the Union still (and for a very long time to come) will be a patchwork of states with ethnic traditions and histories of conflict and competition. Economic success and growing well-being tend to submerge such differences, but these divisive forces are bound to resurface when the EU faces troubled economic or social times. That is precisely what happened during the early 1990s as Europe's economic recession deepened and resurgent nationalism prompted many member-states to re-examine the concessions they made at Maastricht. Their leaders quickly assembled and explained that the intention

From the Field Notes
"No place in Europe displays the flag of the European Union as liberally as does the City of Maastricht, where the European Union Treaty of 1991 was signed. The flag, seen here at City Hall, shows twelve yellow stars (representing the signatories to the Treaty) against a blue background. Although three additional states joined the EU in 1995, and others may join later, the flag will remain as is, and will not, as the U.S. flag does, change to reflect changing times."

at Maastricht was to respect the separate national identity and diversity of each country. Their words were reassuring perhaps, but they signaled an unmistakable shift away from the path leading toward a "United States of Europe." These misgivings aside, the modified Union Treaty was finally ratified by each of The Twelve in 1993 and the EU was immediately launched. The Single European Market—with more than 95 percent of its 279 goals attained—also began operating in 1993, allowing the unrestricted flow of goods, services, and capital among The Twelve (who became The Fifteen in 1995 with the admission of Austria, Finland, and Sweden).

If the European Union succeeds, Europe will become an even more formidable presence in the economic and political geography of the world than it has been in the past. The Fifteen contain just under 375 million people,

constitute one of the world's richest markets, and produce about 40 percent of all global exports. Not surprisingly, other European states want to join the EU. As long as these potential members are politically stable and economically strong, no insurmountable difficulties loom. But as Figure 1-9 shows, almost all of those remaining countries are located in turbulent Eastern Europe, with several bedeviled by political problems that could powerfully strain the organization. Moreover, under the rules of the EU, the richer countries must subsidize the poorer ones; thus the entry of impoverished Eastern European states (most of whom also lag in privatizing their economies) would greatly add to the burden that this rule imposes on the wealthier Western and Northern European members. An even more difficult challenge would involve Turkey. Many Western European countries would like to see Turkey join The Fifteen, thereby widening the EU's reach into the Muslim world. Turkey has indicated its interest in joining, but a referendum in Greece shows that more than 90 percent of voters there are opposed to Turkey's admission. Other EU members have expressed concern over Turkey's human rights record, particularly its treatment of the Kurdish minority (see Chapter 6, p. 302), which would not meet the standards set by the Union.

In the late 1990s, these concerns have combined to slow the expansion of the EU until at least the turn of the century. The dream of closer European unity still exists, but the record of the past 40 years shows that progress toward

supranational goals tends to be cyclical. The recurring challenge is to modify the EU's structure, originally designed for six countries, because each enlargement of the organization brings bigger problems of balancing among increasingly diverse member-states in population size, wealth, and cultural background. Visions of a political United States of Europe notwithstanding, the engine of European unification is fueled by economics. When economic times are good, supranationalism thrives. And until recently, Europe (especially the old EEC countries) prospered. When times became more difficult, so did the path toward further integration.

REGIONS OF THE EUROPEAN REALM

The European geographic realm may not be large territorially, but it nevertheless incorporates much environmental, cultural, and economic diversity. Europe extends from Atlantic shores to Russian borders, and from Arctic cold to Mediterranean warmth. We can better understand Europe's spatial variation by grouping its 38 countries into regions based on their proximity, environmental similarities, historic associations, cultural commonalities, social parallels, and economic linkages. On such bases, we identify five regions of Europe: (1) Western Europe; (2) the British Isles; (3) Northern (Nordic) Europe; (4) Mediterranean Europe; and (5) Eastern Europe.

As the following discussion shows, European regions (like their counterparts all over the world) are subject to change. Western, Northern, and Eastern Europe were spatially modified by the collapse of the Soviet Union in 1991. Changing core-periphery relationships also continue to alter the geographic framework of this complex realm, notably through the expansion of the European Union.

❖ WESTERN EUROPE

Western Europe is the heart of the realm, the hub of its economic power, the focus of its unifying drive. This is the Europe of industry and commerce, of great cities and bustling interaction, of functional specialization and areal interdependence. This is Europe at its most dynamic, where countries founded empires while forging democratic governments and where devastating war and astounding recovery have molded cultural landscapes.

Western Europe, therefore, is a functional rather than a formal region. It is defined by its productivity, circulation, and interaction in numerous spheres, by ways and stan-

dards of living that set it apart from other regions in all directions. As such, it is a multicultural region with a history of conflict. Indeed, the region we recognize today, which arose from the destruction of World War II, could be shattered again by political developments. It has happened before.

Western Europe is dominated by two of Europe's giants, Germany and France, flanked by the three Benelux states (Belgium, the Netherlands, and Luxembourg) in the northwest and by the three Alpine states (Switzerland, Austria, and the microstate of Liechtenstein) in the east. With a

Major Cities of the Realm

City	Population* in millions
Amsterdam, Netherlands	1.1
Athens, Greece	3.8
Barcelona, Spain	2.8
Berlin, Germany	3.3
Brussels, Belgium	1.1
Frankfurt, Germany	3.6
London, UK	7.3
Lyon, France	1.3
Madrid, Spain	4.1
Milan, Italy	4.3
Paris, France	9.5
Prague, Czech Rep.	1.2
Rome, Italy	2.9
Stuttgart, Germany	2.6
Vienna, Austria	2.1
Warsaw, Poland	2.4

*Based on 1997 estimates

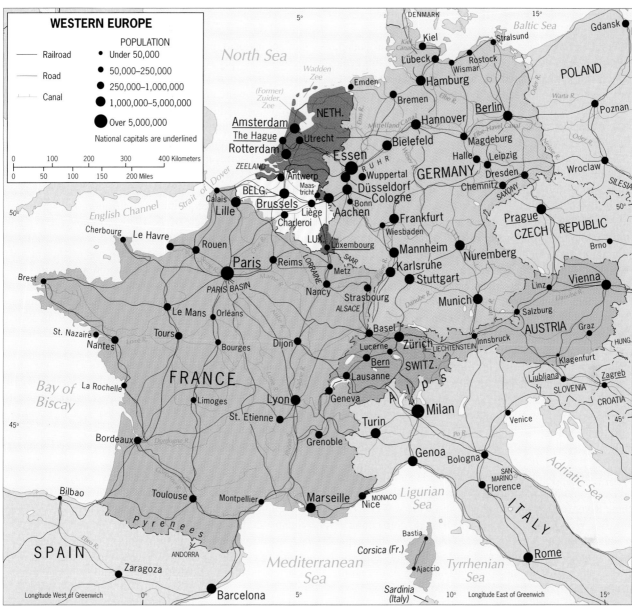

FIGURE 1-11

combined population of more than 180 million representing eight of the world's richest economies, Western Europe is a powerful force on the international stage (Fig. 1-11).

Dominant Germany

Germany—for nearly half a century a country divided into a larger, more prosperous, capitalist West and a smaller, poorer, communist East—was reunited in 1990 during the collapse of communism. Reunification erased the sealed border between West and East and toppled the infamous Berlin Wall in the divided capital; but after the celebrations

were over, the Germans faced a formidable economic and social challenge. That so much progress has been made toward the reintegration of the former East into the economic sphere of the West in less than a decade is testimony to the strength of the national economy.

Before World War II, the young German Empire had expanded into what is today western Poland and incorporated a wide area along the Baltic Sea as far east as coastal Lithuania. At that time, Germany possessed three major industrial complexes: the *Ruhr* near the Netherlands border in the west, the *Saxony* area along the Czech border in the center, and *Silesia* in the east (Fig. 1-11). Germany's enormous industrial machine supported armed forces capable of plunging the world into war; domestic raw materials,

B

ranging from coal and fuels to iron and alloys, helped make this possible.

When Germany was dismembered after the war ended in 1945, and its eastern boundaries were redrawn, this economic geography changed drastically. Silesia became part of the newly delimited state of Poland. Saxony lay in communist East Germany. The new boundary framework left West Germany with the Ruhr, but also with a much-diminished resource base.

Nevertheless, West Germany soon developed Europe's most successful economy, aided by the Marshall Plan, by political stability, and by the enlightened policies of the powers that had defeated it. In just 15 years, between 1949 and 1964, West Germany's gross national product (GNP) tripled while industrial output rose 60 percent. Millions of emigrants from communist East Germany and German-speaking refugees from Eastern Europe were absorbed by West Germany's market-driven economy; unemployment was virtually nonexistent, and hundreds of thousands of Turkish and other foreign workers came to take jobs unfilled or unwanted by Germans.

Geography had much to do with West Germany's "economic miracle." As Figure 1-11 shows, Germany borders every one of the other Western European countries except Liechtenstein, and its surface transport systems (railroads, highways, navigable rivers, artificial waterways) are second to none in the region—indeed, in the realm. Seemingly somewhat landlocked by all these neighbors, Germany in fact benefits from the services and efficiency of one of the world's largest ports, Rotterdam (Netherlands). The Ruhr industrial complex, as the map shows, is linked by the Rhine River (one of the world's busiest waterways) to this port. But West Germany never was a single-core country. In addition to the cluster of cities marking the Ruhr, major urban centers anchor industrial and agricultural zones in several other parts of the country: Hamburg in the north (itself a major port), Frankfurt in the center, Stuttgart and Munich in the south.

The powerful West German economy stood among the world's largest producers of iron, steel, chemicals, motor vehicles, machinery, textiles, and a host of other goods, in addition to a wide range of farm products. Unencumbered by major military expenditures, the West German economy consumed prodigious quantities of imported raw materials and sold its output in markets all over the world. As the giant of the original Common Market, and under the security umbrella of NATO, West Germany translated its geographic advantages into prosperity.

But all growing economies go through periods of slowdown, and West Germany was no exception. In the 1970s, the outlines of Germany's future problems could be discerned: an aging population putting strain on the country's capacity to provide social services; a growing number of unemployed requiring help from the state; a rising anti-

foreigner sentiment marked by murderous attacks on Turkish and African guest workers; and the revival of extremism in national politics. When the price of petroleum increased steeply and the world was plunged into the "oil crisis" of the mid-1970s, West Germany suffered severely, and its economic miracle had come to an end. In the Ruhr, aging heavy industries lost ground on world markets, and diversification and modernization were slow in coming. Suddenly, the future of this huge complex was in doubt.

When reunification with East Germany occurred in 1990, West Germany did not appear to be in the best position to absorb the accompanying stresses. The 1980s had seen considerable economic revitalization, but there was no return to the miracle of the 1950s. Moreover, East Germany, with outdated manufacturing plants, deteriorated infrastructures, polluted environments, and inefficient, collectivized farming, was in worse condition than its communist-period statistics had suggested. West Germans were forced to pay a higher sales tax as well as an income tax surcharge to help meet the huge cost of reunification. When an economic recession struck in the early 1990s, the future of the new Germany appeared clouded.

By the late 1990s, however, another of Germany's impressive recoveries was under way, and the integration of East with West was progressing rapidly. The flood of westward-migrating *Ossies* (East Germans) abated, equalization programs proceeded, and the new Germany's dominant position in Western Europe as well as the European Union was sustained.

The political geography of Germany today is an extension of the federal framework with which West Germany was endowed after the Second World War (Fig. 1-12). The country is divided into 16 States, or *Länder*, each representing a traditional cultural or political node in the experience of the German nation. Such subnational administrative restructuring reflects the EU's desire to forge a more open Europe, dominated less by self-serving nation-states and more by the direct interactions of their key component regions. We will see evidence of this also in France and Spain.

Six of Germany's States lie in the former East Germany. The leading State by many measures is Brandenburg, at the heart of which lies the historic city of Berlin, once-and-future capital of the country (Fig. 1-12). Before 1990, the smaller city of Bonn served as West Germany's headquarters, but Berlin is the nation's primate city and the functions of government will be transferred when a major construction project in Berlin, now in progress, is completed. South of Brandenburg lies Saxony, the old German industrial complex again part of the unified state. Resource-rich but burdened by outdated and inefficient factories and high unemployment, Saxony has encountered difficulties in the new era. The old cities of Leipzig and Dresden reveal in their cultural landscapes the aftermath of war, the bleak-

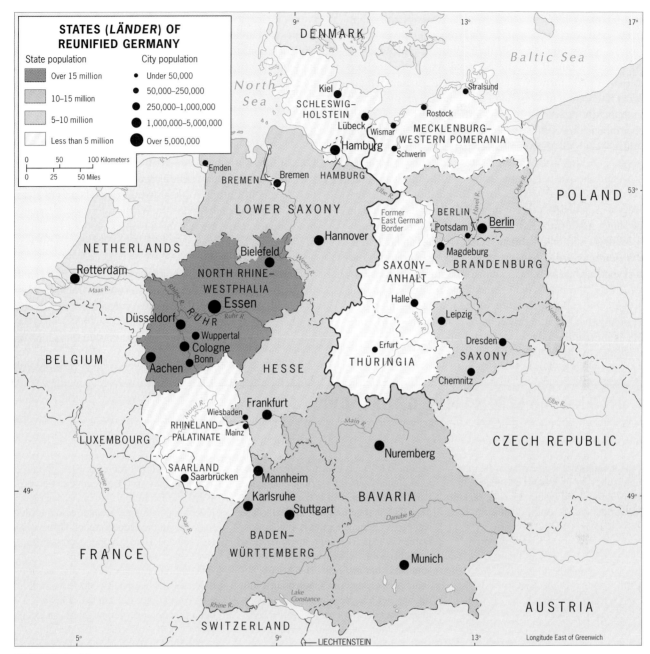

FIGURE 1-12

ness of communism, and the dislocation arising from the transition to capitalism.

Among the 10 *Länder* of former West Germany, the leading State in many respects is North Rhine-Westphalia, home to the Ruhr manufacturing complex as well as the long-time West German capital, Bonn. As Figure 1-12 shows, North Rhine-Westphalia virtually coincides with the hinterland of the great port of Rotterdam and contains six major cities in addition to the capital. Immediately to the southeast lies Hesse, with its major Rhine-Main industrial zone centered on Frankfurt, Germany's financial head-

quarters and air-transport hub. South of Hesse lie Baden-Württemberg, centered on Stuttgart, and Bavaria, anchored by Munich; here, in Germany's "Sunbelt," has emerged a booming high-technology heartland.

As the map shows, the important port cities of Hamburg and Bremen have their own status as *Länder*, populous North Sea gateways to the river-and-canal system that links them to most of northern Germany. Both ports benefit from the expansion of their hinterlands as a result of the elimination of the boundary between East and West Germany.

The extension of West Germany's political system into

East Germany has proceeded with far less difficulty than many observers anticipated, and representatives from Eastern States now sit with their Western colleagues in the Federal Council and the National Assembly, the two houses of Parliament. While the economic, social, and political integration of West and East continue, problems will undoubtedly arise, but major hurdles have been cleared. The dimensions of those problems are seen in the changed vital statistics of the reunited country: as a result of the merger, all kinds of measures showed declines, from life expectancies and per capita incomes to health conditions and urbanization levels. Before reunification, West Germany was about 92 percent urbanized; after its union with the much more rural East, the index stood at 85 percent.

With 82 million inhabitants, Germany has about 23 million more residents than any other European country; one in seven Europeans is a German. Some Europeans are concerned by Germany's enlargement and ultimate economic strengthening; the domination of the European Union by a Germany that twice in this century plunged the world into war is among the factors that fuel opposition to the unifi-

cation concept in some quarters. Right-wing extremist activities and attacks on foreigners in Germany seem to justify such doubts, but the counterargument is that countries participating in multinational associations are less likely to take unilateral actions. On this matter, proponents of the European Union agree.

France

One of those proponents is the other leading Western European country, France. The French and the Germans have been rivals in Europe for centuries. France is an old state, by most measures the oldest in Western Europe. Germany is a young country, created in 1871 after a loose association of German-speaking states had fought a successful war against . . . the French!

Territorially, France is much larger than Germany, and the map suggests that France has a superior relative location, with coastlines on the Mediterranean Sea, the Atlantic Ocean, and, at Calais, even a window on the North Sea. But France does not have any good natural harbors, and its

---

■ AMONG THE REALM'S GREAT CITIES . . .

Paris

If the greatness of a city were to be measured solely by its number of inhabitants, Paris (9.5 million) would not even rank in the world's top 20. But if greatness is

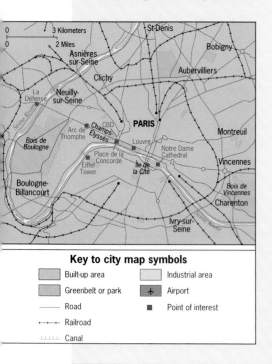

measured by a city's historic heritage, cultural content, and international influence, Paris has no peer. Old Paris, near the Île de la Cité that housed the original village and carries the eight-century-old Notre Dame Cathedral, contains an unparalleled assemblage of architectural and artistic landmarks old and new. The Arc de Triomphe, erected by Napoleon in 1806 (though not completed until 1836), commemorates the emperor's victories and stands as a monument to French neoclassical architecture, overlooking one of the world's most famous streets, the Champs Elysées, which leads to the grandest of city squares, the Place de la Concorde, and on to the magnificent palace-turned-museum, the Louvre. Even the Eiffel Tower, built for the 1889 International Exposition over the objections of locals who regarded it as ugly and unsafe, became a treasure. From its beautiful Seine River bridges to its palaces and parks, Paris embodies French culture and tradition. It is perhaps the ultimate primate city in all the world.

As the capital of a globe-girdling empire, Paris was the hearth from which ra-

diated the cultural forces of Francophone assimilation, transforming much of North, West, and Equatorial Africa, Madagascar, Indochina, and many smaller colonies into societies on the French model. Distant cities such as Dakar, Abidjan, Brazzaville, and Saigon acquired a Parisian atmosphere. France, meanwhile, spent heavily to keep Paris, especially Old Paris, well maintained—not just a relict of history, but a functioning, vibrant center, an example to which other cities should aspire.

Today, Old Paris is ringed by a new and very different Paris. Stand on top of the Arc de Triomphe and look in the other direction from the Champs Elysées, and the tree-lined avenue gives way to *La Défense,* an ultramodern high-rise complex that is one of Europe's leading business districts (see p. 61). But from atop the Eiffel Tower you can see as far as 50 miles (80 km), and discern a Paris visitors rarely experience: grimy, aging industrial quarters, and poor, crowded neighborhoods where discontent and unemployment fester—and where Muslim immigrants cluster in a world apart from the splendor of the old city.

rivers and inland waterways are not navigable far inland by oceangoing ships. France has no equivalent, internally or externally, to Rotterdam.

The map of Western Europe (Fig. 1-11) reveals a significant demographic contrast between France and Germany. France has one dominant city, Paris, at the heart of the Paris Basin, France's core area. No other city in France comes close to Paris in terms of population or centrality: Paris has 9.5 million residents, whereas its closest rival, Lyon, has only 1.3 million. Germany has no city to match Paris, but a number of cities with populations between 1 and 5 million lie dispersed throughout the country. And as Appendix A shows, Germany is substantially more highly urbanized overall than France.

Why should Paris, without major raw materials nearby, have grown so large? Whenever geographers investigate the evolution of a city, they focus on two important locational qualities: its **site** (the physical attributes of the place it occupies) and its **situation** (its location relative to surrounding areas of productive capacity, other cities and towns, barriers to access and movement, and other aspects of the greater regional framework in which it lies).

The site of the original settlement at Paris lay on an island in the Seine River, a defensible place at a point where the river was often crossed. This island, the *Île de la Cité*, was a Roman outpost 2,000 years ago; for centuries its security ensured continuity. Eventually the island became overcrowded and the city expanded along the banks of the river (Fig. 1-13).

Soon the settlement's advantageous situation stimulated its growth and prosperity. Its fertile agricultural hinterland thrived and, as an enlarging market, Paris's focality increased steadily. The Seine River is joined near Paris by several navigable tributaries (the Oise, Marne, and Yonne). When canals extended these waterways even farther, Paris was linked to the Loire Valley, the Rhône-Saône Basin, the Lorraine (an industrial area), and the northern border with Belgium. When Napoleon reorganized France and built a radial system of roads—followed later by railroads—that focused on Paris from all parts of the country, the city's primacy was assured (Fig. 1-13, inset map). The only disadvantage in Paris's situation lies in its seaward access: oceangoing ships can sail up the Seine River only as far as Rouen. Between the coastal port of Le Havre and Rouen, the meandering Seine is lined by rural villages interspersed with docking facilities.

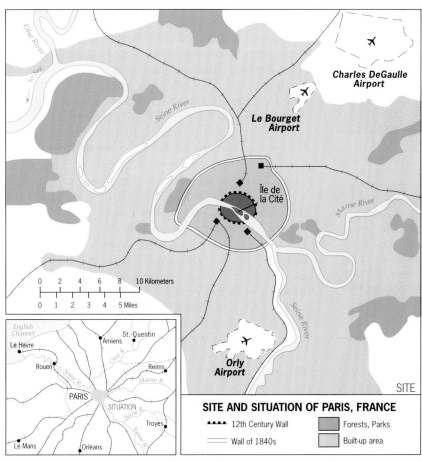

SITE AND SITUATION OF PARIS, FRANCE

▪▪▪▪ 12th Century Wall

═══ Wall of 1840s

Forests, Parks

Built-up area

FIGURE 1-13

From the Field Notes

"Unlike England's Thames River (photo p. 80), the Seine has a meandering course in an often steep-sided valley; large ships can sail upstream only as far as Rouen, not to the French capital of Paris. At the river's mouth, the port of Le Havre is no great natural harbor, so that the lower Seine is lined with smaller riverside ports handling such commodities as oil, coal, and (as in the photo at the top) scrap iron ... Interspersed are the waterfront villages for which Normandy is known, historic settlements that attract a steady stream of tourists during the summer."

As Paris evolved into one of Europe's greatest cities, France's industrial development was less spectacular. Various explanations have been given for this, ranging from the lesser availability of high-quality coal (and therefore its higher cost) and the lower efficiency of transport systems here to the notion that French manufacturers preferred to specialize in such products as high-quality textiles and precision equipment. Meanwhile, French agriculture re-

mained Europe's most productive. From the Paris Basin to the Mediterranean coast and from the banks of the Rhine to the shores of the Bay of Biscay, France has a wide range of soils and climates, and its farms produce an almost limitless diversity of products. French wines, notably those of the Bordeaux, Burgundy, and Champagne districts, and French cheeses reach markets around the globe.

Today, France's economic geography is dominated by new high-tech industries, especially transportation and telecommunications. France is a leading producer of high-speed trains, aircraft, fiber-optic communications systems, and various space-related technologies. It also is the world leader in the development of a nuclear power grid, which today supplies more than 75 percent of the country's electricity and thereby reduces its dependence on imported oil.

When Napoleon reorganized France, he broke up the country's large traditional subregions and established more than 80 small *départements* (additions and subdivisions later increased this number to 96). Each *département* had representation in Paris, but the power was concentrated in the capital, not in the individual *départements*. France became a highly centralized state, and remained so for nearly two centuries (see inset map, Fig. 1-14).

Today, France is decentralizing. A new subnational framework of 22 historically significant provinces, groupings of *départements* called *regions* (Fig. 1-14), has been established to accommodate the devolutionary forces felt throughout Europe and, indeed, throughout the world. These regions, though still represented in the Paris government, have substantial autonomy in such fields as taxation, borrowing, and development spending. The cities that anchor them benefit because they are the seats of governing regional councils that can act to attract investment not only within France but also from abroad. Lyon, France's second city and headquarters of the region named Rhône-Alpes, has become a focus for growth industries and multinational firms. The region is evolving into an economic juggernaut, a driving force in the European economy. As such, Lyon and its hinterland have become one of those subnational economic powerhouses now called the *motors of Europe* (others being Italy's Lombardy, Spain's Catalonia, Germany's Baden-Württemberg) that function as self-standing entities with their own international business connections to places as far away as China and Chile.

Until after mid-century, France lay at the center of a global colonial empire that extended from Algeria to Madagascar and from West Africa to the South Pacific. That empire was lost (except for a few remnants), and today France's cities are home to nearly 3 million immigrants from the former colonies (almost 5 percent of the total population of 59 million). Political and social crises in the former colonies tend to spill over into France; in the 1990s, a struggle between Islamic extremists and the government in Algeria led to acts of terrorism in Paris and other cities. The French have long maintained policies of active in-

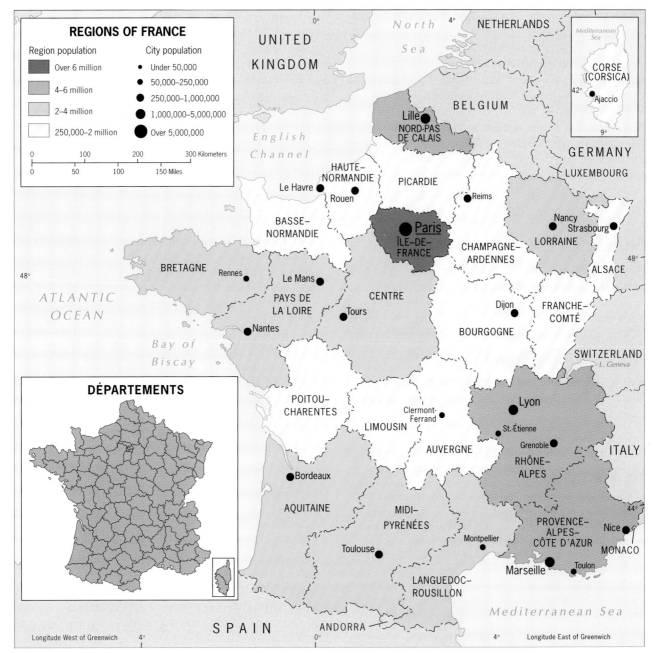

REGIONS OF FRANCE

Region population
- ■ Over 6 million
- ■ 4–6 million
- ■ 2–4 million
- □ 250,000–2 million

City population
- • Under 50,000
- • 50,000–250,000
- ● 250,000–1,000,000
- ● 1,000,000–5,000,000
- ● Over 5,000,000

DÉPARTEMENTS

FIGURE 1-14

volvement in their former empire, including armed support for threatened governments, and they show no signs of giving up the last of their holdings in South America (French Guiana), the Caribbean, the Atlantic, or the Pacific. France's South Pacific dependencies cost Paris billions of dollars annually, but despite opposition and even insurrection, the French Tricolor continues to fly over such far-flung capitals as Papeete (Tahiti), Saint-Denis (Réunion), and Nouméa (New Caledonia).

This global gulag of colonial relics reflects France's ambivalence regarding its changing status in this changing world. Still harboring imperial (and superpower) notions, the French defy international opinion and explode nuclear bombs in an atoll in French Polynesia even as they take the lead in the international community's search for a solution to the conflict in former Yugoslavia. A French president visits (formerly French-colonial) Vietnam to secure closer ties even as Paris seeks to loosen its bonds with the states of its former Subsaharan African empire. French leaders want to make their country a cornerstone of the European Union; French public opinion waxes and wanes on the matter. Still powerful in the country's affairs, French farmers besiege the capital when European Union policies hurt their interests.

France's regional reorganization, coupled with advances in internal and international surface and air transportation, reinforces one of the world's richest and most diversified economies. When the first-ever direct link between France and Britain, the Channel Tunnel, opened in 1994, another of France's regions—Nord-Pas de Calais—stood to gain, transforming its failing Rustbelt economy into one based on transport and trade. A pillar of Western Europe, France remains a powerful force in the European Union.

Benelux

Three political entities are crowded into the northwestern corner of Western Europe: Belgium, the Netherlands, and tiny Luxembourg, collectively referred to by their first syllables (*Be-Ne-Lux*). These states are also frequently called the Low Countries; this label is very appropriate because most of the land is extremely flat and lies near (and in the Netherlands even below) sea level. Only toward the southeast, in Luxembourg and eastern Belgium's Ardennes, is there a hill-and-plateau landscape with elevations in excess

From the Field Notes

"Hundreds of bicycles are parked under the overhang of an office building in Utrecht, Netherlands. We watched the commuters arrive in the morning, filling the rows of bike stalls, then leave after work. From executives in business attire to office workers in jeans, all came to work by bicycle. Imagine the size of the parking lot required if these people had driven to work; imagine also the amount of gasoline saved, and pollution avoided. The continued use of the bicycle in the automobile era also has had an effect in limiting the sprawl of Dutch (and other European) cities. In the often cold and rainy weather of the Netherlands, the shorter the bike trip the better, so there is less attraction to living in a distant suburb."

of 1,000 feet (300 m). As with Germany and France, there are major differences between Belgium and the Netherlands. Indeed, there are such contrasts that the two latter countries find themselves in a position of complementarity.

Belgium is marked by two industrial corridors. One is the coal-based east-west axis through Charleroi and Liège, where there are heavy industries. The second corridor of lighter and more varied manufacturing extends north from Charleroi through Brussels to the major port of Antwerp. The diversified industrial products of these areas include metals, chemicals, furniture, and specialties such as pianos, soaps, and cutlery. The Netherlands, in contrast, has a large agricultural base (along with its vitally important transport functions); it can export dairy products, meats, vegetables, and other foods. Hence, there was mutual advantage in the Benelux economic union of the 1940s: it facilitated the reciprocal flow of needed imports to both countries, and it doubled the domestic market.

The Benelux countries are among the most densely populated on Earth, and space is truly at a premium. Some 26 million people inhabit an area about the size of Maine (home to 1.2 million). For centuries the Dutch have been expanding their living space—not at the expense of their neighbors, but by wresting it from the sea. The greatest project so far is the draining of almost the entire Zuider Zee (Sea) begun in 1932 and scheduled for completion after the turn of the century. In the southwest the islands of Zeeland are being connected by dikes and the water is being pumped out, creating additional *polders* (reclaimed lands).

Three cities—Amsterdam, Rotterdam, and The Hague—anchor the Netherlands' triangular core area. Amsterdam, the constitutional capital, remains very much the focus of the Netherlands, with a bustling commercial center, a busy port, and a variety of light manufactures. Rotterdam, Europe's busiest port, is the shipping gateway to Western Europe, commanding the entries to both the Rhine and Meuse (Maas) rivers. Its modern development mirrors that of Germany's Ruhr industrial region and the adjacent Rhine valley; thus ongoing manufacturing decline in the Rhine hinterland as well as increased competition from other European ports is eroding Rotterdam's historic situational advantage. The third city in the triangle, The Hague, is the seat of the Dutch government and the home of the United Nations' World Court.

Collectively, these three cities of the triangular core have spawned a **conurbation**, the term geographers use to define the huge multimetropolitan complexes formed by the coalescence of two or more major urban areas. This particular coalescence, known as *Randstad*, has created a ring-shaped urban complex that surrounds a still-rural center (the literal translation of *rand* is edge or margin). A more precise labeling of the conurbation, however, would be Randstad-Holland, because *Holland* (meaning "hollow country") refers specifically to the Dutch heartland that faces the North Sea in these lowest-lying western provinces of the Netherlands.

Belgium, the Netherlands, and Luxembourg, with their limited space and modest resource base, have turned to managing international trade as a productive economic activity. Brussels, Belgium's capital, also has acquired a significant political role, serving (with Strasbourg) as one of the co-capitals of the European Union, headquarters of NATO, and administrative center for numerous international economic organizations. Hundreds of multinational corporations have their offices here, enhancing Brussels' role as a financial center and commercial-industrial complex.

Belgium, however, remains a three-region amalgam with an uncertain political future. Devolutionary forces from time to time threaten to sever Dutch- (Flemish-) speaking Flanders (with 55 percent of the population) from French-speaking Wallonia (33 percent). In the early 1990s, such a prospect led to a politico-geographical reorganization of the country, establishing three coequal regions (the third: Brussels in a quasi-federal arrangement). Flemish nationalism, however, remains a potent force.

The Alpine States

Switzerland and Austria share a landlocked situation and the mountainous topography of the Alps—and little else. Austria is a member of the European Union; Switzerland is not. Austria is a monolingual society; Switzerland is multilingual. Austria has a large primate city; multicultural Switzerland does not. Austria has a substantial range of domestic raw materials (including oil); Switzerland does not. Austria is twice as large as Switzerland and has a larger population than its neighbor to the west; Austria also has much more cultivable land than Switzerland.

And yet, as Appendix A indicates, Switzerland, not Austria, is the leading state in Western Europe's Alpine subregion in terms of per capita GNP and other indices. The world geomorphic map (Fig. I-4) suggests that mountainous countries face certain limits on development; mountainous terrain, and its frequent corollary, *landlocked location*, tend to inhibit the dissemination of ideas and innovations, obstruct circulation, constrain agriculture, and divide cultures. Tibet, Afghanistan, and Ethiopia seem to prove the point. That is why Switzerland is such an important lesson in human geography. The Swiss, through their skills and abilities, have overcome a seemingly restrictive environment and have converted it into an asset that keeps them in the forefront of prosperous European countries.

At various times in their history, the Swiss employed their Alpine environments to act as middlemen in interregional trade, to use the water cascading from their mountains to build hydroelectric power plants to support highly specialized industries, and to attract millions of tourists. Meanwhile, Swiss farmers perfected ways to maximize the productivity of mountain pastures and valley soils. True, the Swiss speak German in the north, French in the west, Italian in the southeast, and even a bit of Romansch in remote highlands (Fig. 1-6); but such cultural heterogeneity has long been overcome by economic success. Neutrality, security, and stability made Switzerland a haven during war, a world banking giant, a global magnet for money.

The financial center of Zürich (Switzerland's largest city, located in the German-speaking sector), the international city of Geneva (in the far southwest and thus in the French sector), and the capital of Bern (near the French-German linguistic divide) each has its own precision industries, some of historic vintage. Swiss quality is an international standard, whether in watches, instruments, or specialty foods.

Switzerland is therefore a Western European country, part and parcel of the European core—but even the prosperous Swiss confront problems in this changing realm. Cross traffic between Western and Mediterranean Europe, coupled with a relentlessly growing stream of tourists, is damaging Alpine environments. Switzerland's traditional neutrality and aloofness, to the point that it remains the only Western European country not yet a member of the EU, is causing concern for a future in which the Swiss could lose their competitive edge in various fields.

Austria, by contrast, joined the European Union in 1995, following a referendum that revealed considerable opposition to the move despite the majority that favored it. German-speaking, overwhelmingly Catholic, and a remnant of the Austro-Hungarian Empire, Austria has a historical geography that is far more reminiscent of unstable Eastern Europe than Switzerland's. Even Austria's physical geography seems to demand that the country look eastward: it is at its widest, lowest, and most productive in the east, where the eastward-flowing Danube links it to Hungary, its old ally in the anti-Muslim campaigns of past centuries.

Vienna, by far the Alpine subregion's largest city, also lies on the country's eastern perimeter. One of the world's most expressive primate cities, Vienna is an outdoor museum of monumental art and architecture, replete with palaces, castles, mansions, public gardens, statues, magnificent churches, and grand public buildings. Many of these structures commemorate victories over the Ottoman Turks, here in this outpost of Western culture during a time when much of Eastern Europe was under Islamic rule. Today, Vienna represents an outpost of another sort: the easternmost city of Western Europe, situated on the doorstep of transforming Eastern Europe much as Singapore sits at the threshold of Southeast Asia's economies. Corporations doing business in the uncertain commercial climates of Eastern Europe are establishing headquarters in Vienna, using the city and its excellent infrastructure and comparatively uncorrupt institutions as a base from which to reach into ex-communist neighbors.

As the eastern branch of the Germanic culture that dominates the heart of Western Europe, Austria is part and parcel of this region. Its cultural landscapes are those of the West, its economy is closely tied to the region (more than 40 percent of exports and imports are with Germany),

and its political system is representative of the region's democracies. When Austria joined the EU, it met the Maastricht criteria without difficulty. But joining the EU made Austria's eastern boundary something more than an ordinary interregional border: here now lies the great divide between Europe's "ins" and the "outs" that desire to join but are not yet eligible. Becoming part of the EU changed Austria's relative location in ways we cannot yet foretell.

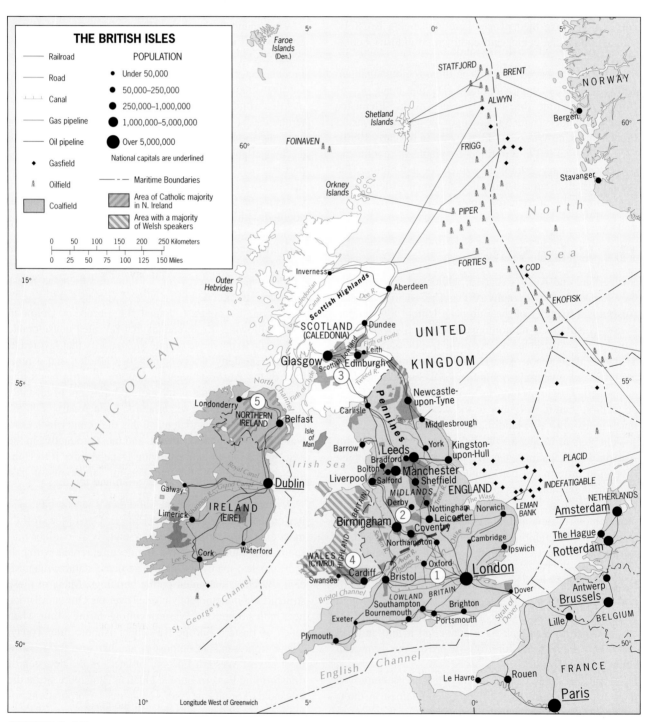

FIGURE 1-15

❖ THE BRITISH ISLES

Off the coast of mainland Western Europe lie two major islands, surrounded by a constellation of tiny ones, that constitute the British Isles, a discrete region of the European realm (Fig. 1-15). The larger of the two major islands, which also lies nearest to the mainland (a mere 21 miles, or 34 km, at the closest point), is the island called *Britain*; its smaller neighbor to the west is *Ireland*.

Considerable name confusion attaches to these islands as well as to the countries on them. They still are called the British Isles, although British dominance over most of Ireland ended three generations ago. The nation-state that occupies Britain and a small corner of northeastern Ireland is officially called the United Kingdom of Great Britain and Northern Ireland—United Kingdom for short and UK by abbreviation. But this country often is referred to simply as Britain, and its people are known as the British. The nation-state of Ireland officially is the Republic of Ireland (*Eire* in Irish Gaelic), but it does not include all of the island of Ireland.

How convenient it would be if physical and political geography coincided! Unfortunately, this is not the case. During the long British occupation of Ireland, which is overwhelmingly Catholic, Protestants from northern Britain settled in large numbers in northeastern Ireland. When British domination ended in 1921, the Irish were set free—except in that corner in the north, where London kept control to protect the Protestant settlers there. That is why the country to this day is officially known as the United Kingdom of Great Britain *and Northern Ireland*.

Northern Ireland (Fig. 1-15) was home not only to Protestants from Britain, but also to a substantial population of Irish Catholics who found themselves on the wrong side of the border when Ireland was liberated. Ever since, there has been conflict that intermittently has engulfed Northern Ireland and spilled over into Britain and even into Western Europe. It has been (and remains) one of Europe's most costly struggles.

Although all of Britain lies in the United Kingdom, political divisions exist here as well. *England* is the largest of these units, the center of power from which the rest of the region was originally brought under unified control. The English conquered *Wales* in the Middle Ages, and *Scotland*'s link to England, cemented when a Scottish king ascended the English throne, was ratified by the Act of Union of 1707. Thus England, Wales, Scotland, and Northern Ireland became the *United Kingdom*.

The British Isles form a distinct region of Europe for several reasons. Britain's insularity provided centuries of security from turbulent Europe, protecting the evolving British nation as it achieved a system of parliamentary government that had no peer in the Western world. Having united the Welsh, Scots, and Irish, the British set out to

forge what was to become the world's largest colonial empire. An era of mercantilism and domestic manufacturing (the latter based on water power from streams flowing off the Pennines, Britain's mountain backbone) foreshadowed the momentous Industrial Revolution, which transformed Britain—and much of the world. British cities became synonyms for specialized products as the smokestacks of factories rose like forests over the urban scene. London on the Thames River anchored an English core area that mushroomed into the headquarters of a global political, financial, and cultural empire. As recently as the Second World War (1939–1945), the narrow English Channel ensured the United Kingdom's impregnability against German attack, giving the British time to organize their war machine. When the United Kingdom emerged from that conflict as a leading power among the victorious Allies, it seemed that its role in the postwar era was assured.

Two unexpected developments changed that prospect: the collapse of colonialism and the resurgence of mainland Europe. In a few short decades, an empire of centuries was lost and the United Kingdom became, if not just another European country, then certainly a lesser power in the world. And in a few short years mainland Europe's economies, fueled by the funds of the Marshall Plan, rose to challenge Britain's. When the supranationalism movement gathered momentum on the mainland, the United Kingdom, still enmeshed in Commonwealth ties to its former colonies, at first remained aloof. As we noted earlier, the British tried to counter the old EEC by taking a leading role in EFTA, but eventually (in the early 1970s) the United Kingdom opted to join what has become the European Union. It was a marriage of necessity rather than enthusiasm: unlike most of the peoples in other EU countries, the British public never had the chance to vote on EU participation in a referendum. Polls indicate that such a referendum might lead to rejection, but the government, aware of the economic cost of exclusion, took the necessary step.

Ever since joining the EU, London has sought to restrain EU moves toward tighter integration and tougher regulations. The British view of the EU is that the organization should impose a minimum of rules; federalism is not an option. In that respect, Britain's historic, insular standoffishness remains unchanged.

The United Kingdom

The United Kingdom, with an area about the size of Oregon and a population of nearly 60 million (Oregon: 3.3 million) by European standards is a large country. Only Germany has more people, and only 9 of Europe's 38 countries are larger territorially. As we will see, the UK has considerable spatial variation. On the basis of a combination of physiographic, economic, political, and cultural cri-

From the field Notes

"Sail up the Thames River toward London, and after passing Tilbury on the port side you see flatlands that seem to lie barely above water level. Then something comes into view that does not appear on many maps of the area: a row of aluminum domes standing in the water from one bank to the other.... Pass between them, and you see one of Europe's defenses against rising water: the Thames Barrier. Accumulating sediments are causing the lower Thames area to subside, creating an ever-greater danger of floods. The Thames Barrier erects a set of barrages, like giant doors, when the need arises, protecting the urban and industrial area upstream against flooding."

teria, the United Kingdom is divided into five subregions: (1) Affluent Southern England centered on the London area, (2) Stagnant Northern England at the heart of Britain, (3) Individualistic Scotland in the far north, (4) Intractable Wales to England's west, and (5) Embattled Northern Ireland across the North Channel (Fig. 1-15).

Between the border with Scotland and a line approximately from the Bristol Channel to The Wash lies the heart of Britain, the hearth of the Industrial Revolution. Even here, obsolescence has overtaken what was once ultramodern. The cities of the Midlands (Manchester, Sheffield, Leeds), once synonymous with industrial power, now form an English Rustbelt plagued by social as well as economic problems. In virtually every economic indicator, the Stagnant North lags behind England as a whole, and far behind the comparatively prosperous South. The Midlands cities, of course, are not unique in this respect: the new era of high technology in the world's industrialized countries has overtaken many great manufacturing centers of the past, leaving them with unemployment problems, racial tensions, deteriorating neighborhoods, and shrinking economies. The old port city of Liverpool, once the thriving gateway to an overseas empire, has seen its docks all but abandoned. Liverpool's hopes now rest on a thread: the Channel Tunnel and the link it provides to burgeoning Western Europe.

As industrial decline afflicts the Stagnant North, an opposite trend marks the South. In Affluent Southern England, high-technology as well as service industries dominated by financial, banking, engineering, communications, and energy-related activities are booming. As Figure 1-15 suggests, this subregion consists essentially of metropolitan London (mapped in Fig. 1-7) and its hinterland, and it is home to more than one-third of the United Kingdom's nearly 60 million inhabitants. This is Britain at its most modern, most Europe-oriented, best served by internal communications systems and external connections.

At the opposite (northern) end of the island of Britain lies Scotland, a subregion marked by Scandinavian landscapes and strong cultural individualism. Scotland is no small outpost: it is twice as large as the Netherlands and has more than 5 million inhabitants (about as many as Denmark). Most of the population is concentrated in the Scottish Lowland anchored by Glasgow in the west and Edinburgh in the east, where coalfields and iron ores enabled the Scots to participate in Britain's Industrial Revolution, creating (among other manufacturing) a world-renowned shipbuilding industry. When decline and obsolescence overtook the area, an alternative emerged: as Figure 1-15 shows, the rock layers beneath the North Sea off Scotland's east coast contain major reserves of oil. Edinburgh's port of Leith, once a fishing port and then a shipbuilding center,

A

■ AMONG THE REALM'S GREAT CITIES...

London

Sail westward up the meandering Thames River toward the heart of London, and be prepared to be disappointed. London does not overpower or overwhelm with spectacular skylines or beckoning beauty. It is, rather, an amalgam of towns—Chelsea, Chiswick, Dulwich, Hampstead, Islington—each with its own social character and urban landscape. Some of these towns come into view from the same river the Romans sailed 2,000 years ago: Silvertown and its waterfront urban renewal; Greenwich with its famed Observatory; Thamesmead, the model modern commuter community. Others somehow retain their identity in the vast metropolis that remains (Fig. 1-7)—in many ways, Europe's most civilized and cosmopolitan city. And each contributes to the whole in its own way: every part of London, it seems, has its memories of empire, its memorials to heroes, its monuments to wartime courage.

Along the banks of the Thames, London displays the heritage of state and empire: The Tower and the Tower Bridge, the Houses of Parliament (officially

known as the Palace of Westminster), the Royal Festival Hall. Step ashore, and you find London to be a memorable mix of the historic and (often architecturally ugly) modern, of the obsolete and the efficient, of the poor and the prosperous. Public transportation, by world standards, is excellent; traffic, however, often is chaotic, gridlocked by narrow streets. Recreational and cultural amenities are second to none. London seems to stand with one foot in the twenty-first century and the other in the nineteenth. Its airports are ultra-modern. But when the Eurostar TGV train from Paris emerges from the Channel Tunnel it has to slow down for the final hour in its three-hour run, because the rails into London cannot carry it at its normal high speed.

London remains one of the world's most livable cities for its size (7.3 million) in large measure because of farsighted urban planning that created and maintained, around the central city, a so-called Greenbelt set aside for recreation, farming, and other nonresidential, noncommercial uses (Fig. 1-7). Although London's growth

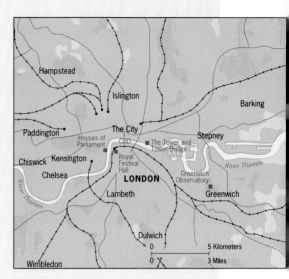

eroded this Greenbelt, in places leaving only "green wedges," the design did preserve crucial open space in and around the city, channeling suburbanization toward a zone at least 25 miles (40 km) from the center. The map reveals the design's continuing impact.

---

now became a major focus for North Sea oil drilling and pipeline equipment. Farther north, Inverness specialized in the manufacture and repair of oil platforms. Meanwhile, some high-tech development, notably in the hinterland of the old industrial city of Glasgow (in a corridor popularly called Silicon Glen), is compensating somewhat for the decline of heavy manufacturing. But none of this has been enough to halt a rising sense among the Scots that, in the United Kingdom and in the greater European Union sphere, they are economically as well as politically disadvantaged. Scottish nationalism has waxed and waned, but polls indicate that a majority of Scots favor stronger home rule and want a Scottish Parliament to sit in Edinburgh (where a facility is ready and waiting). As we noted in our earlier discussion of Europe's devolutionary forces, support for independence is rising again in the late 1990s.

Wales, one-fourth the size of Scotland and with just 2.9 million inhabitants, contains some of Britain's most rugged and remote landscapes, where ancient Celtic peoples found refuge from invaders. Their languages still survive after centuries of domination by the English. Wales, too, was engulfed by the Industrial Revolution based on its high-

quality coal reserves, but when the better deposits were exhausted and production costs mounted, competition from manufacturing centers in England proved too strong. Soon, Wales's countryside was ravaged by strip mines, its underemployed people were dislocated, and its cities became riddled with slums. As a result, many Welsh citizens left their historic homeland in search of opportunities elsewhere. Cardiff, the Welsh capital, at one time was the world's largest coal-exporting port. Today it carries the marks of decline and stagnation. So it is no surprise that Welsh politics, like those of Scotland, are roiled by economic as well as cultural issues. Welsh nationalism has not yet achieved as strong a voice as Scotland's, but devolutionary forces are at work here too.

Northern Ireland remains London's most serious domestic political problem. With a population of 1.7 million occupying the northeastern one-sixth of the island of Ireland, here we observe the troubled legacy of British colonial rule. About two-thirds of the people in Northern Ireland trace their ancestry to Scotland or England and adhere to Protestant religious beliefs; some 34 percent are Roman Catholics, who share their Catholicism with virtually the entire population of the Irish Republic on the other side of the border. Although Figure 1-15 suggests that there are majority areas of Protestants and Catholics in Northern Ireland, no clear separation exists; mostly they live in clusters throughout the territory, including walled-off neighborhoods in the major cities of Belfast and Londonderry. Partition is no solution to a conflict that has raged for three decades at a cost of thousands of lives; Catholics accuse London as well as the local Protestant-dominated administration of discrimination, whereas Protestants accuse Catholics of seeking union with the Republic of Ireland. The cost of Northern Ireland to the UK, in terms of policing measures as well as economic assistance, has been enormous, but no investment has been enough to stabilize the situation. In 1996, still another effort at peacemaking involving all parties (including the government of the Republic of Ireland) was in progress. But geographic realities here form immense obstacles. The resource base is poor; a potential tourist industry has been stifled by the conflict. British infusions of capital and raw materials have stimulated the emergence of a diversified industrial base, but unemployment—and discontent—are still high. Any solution that would reduce British investment would worsen an already bleak economic picture and fuel new discontent.

The Republic of Ireland

All but the northeastern corner of the island of Ireland forms the territory of the Republic of Ireland, freed from British colonial rule just three generations ago. The island is shaped like a saucer with a wide rim broken toward the east, where lowlands open to the sea. Here lies Dublin, the capital; the general topography is revealed by the flow of streams (Fig. 1-15).

Excessive water, not high relief, is the inhibiting factor in farming in still strongly rural Ireland. Ireland's cool, moist environments are well suited to potato farming, and following its introduction from America in the 1600s the potato quickly took the place of other crops. But even the potato could not withstand the large amount of precipitation that fell year after year during the 1840s. Having become the nutritive basis for Ireland's population (which by 1830 had reached 8 million), the potato crop failed repeatedly, ravaged by blight and rotting in the ground. Over 1 million Irish died in the resulting famines, and nearly twice that number left the country. Today, the entire island of Ireland has a population of only 5.3 million, and the Republic of Ireland only 3.6 million.

The decline continues but now for other reasons. Economic causes are driving young Irish people away (unemployment here ranks among Europe's highest); emigrants also cite the conservatism of Irish cultural and religious life. Given the employment situation, one would expect foreign companies to establish themselves in Ireland to take advantage of low labor costs, tax breaks, and proximity to the European Union. During the 1980s many such companies did so, but the resulting upsurge of light-industry manufactures was cut short by labor problems (insufficient skills, union strictures) and infrastructure obstacles. In the 1990s, the effects of a long recession were evident almost everywhere. Nonetheless, a growing telecommunications service industry has taken root, and Ireland today is Europe's leading call center for the realm's rapidly expanding toll-free telephone market. Another bright spot is tourism: Ireland's legendary beauty attracts a large flow of tourists every year.

Ireland's regional development is also being influenced by events in England's Affluent South. Southeastern Ireland, notably the area centered on Waterford, enjoys a somewhat sunnier climate than the remainder of the island, and lower real estate prices (compared to England and Western European countries) have contributed to a Sunbelt phenomenon that is transforming towns and villages into vacation and retirement sites. Such population movements are made easier by EU membership: Ireland (unlike the United Kingdom) is a strong and referendum-approved supporter of the Union, and has benefited from the economic upturn of its southeastern segment.

❖ NORTHERN (NORDIC) EUROPE

North of Europe's Western European core area lies a disconnected group of six countries that exemplify core-periphery contrasts. Northern Europe is a region of difficult environments: generally cold climates, poorly developed soils, limited mineral resources, long distances. Together, the six countries—Sweden, Norway, Denmark, Finland, Estonia, and Iceland—contain just over 25 million inhabitants, which is a lower total than that of Benelux and only one-seventh that of Western Europe. The overall land area, on the other hand, is almost the size of the entire European core. Here in peripheral *Norden*, as the people call their northerly domain, national core areas lie in the south: note the location of the capitals of Helsinki, Stockholm, and Oslo at approximately the same latitude (Fig. 1-16).

Northern Europe's peripheral situation is more than environmental. As viewed from Europe's core area, Norden is on the way to nowhere. No major shipping lanes lead from Western Europe past Norden to other productive areas of the world, limiting interaction of the sort that ties the British Isles to the mainland. Moreover, except for relatively small Denmark and Estonia, all of Norden lies separated from the European mainland by water. At all levels of spatial generalization, isolation is a pervasive reality in Norden. Raw-material reserves are limited, although the

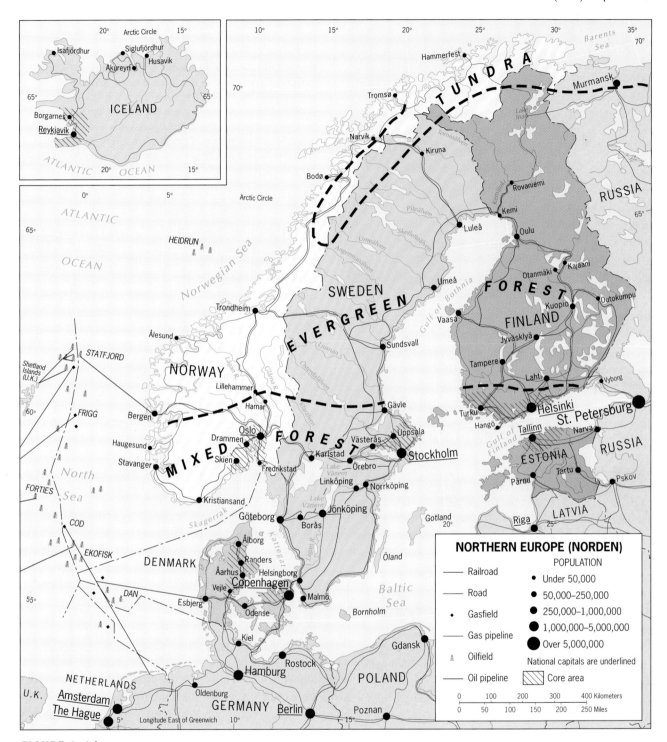

FIGURE 1-16

high relief does provide opportunities for hydroelectric power generation.

Northern Europe's remoteness, isolation, and environmental severity also have had positive effects for this region. The countries of the Scandinavian Peninsula lay removed from the wars of mainland Europe (although Norway was overrun by Nazi Germany during World War II). The three major languages—Danish, Swedish, and Norwegian—are mutually intelligible, which creates one of the criteria on which this region is delimited. Another regional criterion is the overwhelming adherence to the same Lutheran church in each of the Scandinavian countries (Norway, Sweden, and Denmark), Iceland, and Finland. Furthermore, democratic and representative governments emerged early, and individual rights and well-being have long been carefully protected.

From the Field Notes

"High relief makes Norway a string of comparatively poorly connected coastal cities, towns, and villages. Our drive to the capital, Oslo, left us with a sense of remoteness and distance from the large towns on the Atlantic coast, Trondheim and Bergen. Oslo looks southward along the waters of the Oslofjord to the Skagerrak from the head of the fjord, seen here below the twin towers of a government building. Like other Scandinavian cities, Oslo displays a very orderly urban life, cleanliness, a lack of urban blight and poverty, and a wealth of open-air public art."

Sweden

Territorially and demographically the largest Nordic country, Sweden also lies at the heart of the region (Fig. 1-16). The great majority of Swedes live south of the zone where evergreen forest yields to mixed deciduous and needleleaf forests. Here lie the capital, core area and, as Figure 1-5 shows, the main industrial districts; here, too, are the main agricultural areas that benefit from the lower relief, better soils, and milder climate.

Sweden long served as an exporter of raw or semi-finished materials to industrial countries, but today the Swedes are making finished products themselves, including automobiles, electronics, stainless steel, furniture, and glassware. Much of this production is based on local resources, including a major iron ore reserve at Kiruna in the far north (there is a steel mill at Luleå). Swedish manufacturing, in contrast to that of several Western European countries, is based in dozens of small and medium-sized towns specializing in particular products.

Norway

In 1994, when Sweden and Finland voted to join the European Union, Norway rejected that option. With its limited patches of cultivable soil, steep slopes, extensive forests, and lengthy, spectacularly fjorded coastline, Norway has found its economic opportunities on, in, and beneath the sea. Norway's merchant marine spans the world. Its fishing industry exploits extensive nearby waters. In addition, Norway's portion of the North Sea floor contains major reserves of oil and gas, boosting the country's econ-

omy. These assets persuaded a majority of Norwegians that remaining outside the EU would serve the country best.

As the map indicates, Norway has nothing to compare to Sweden's internal agricultural or industrial development. Its cities, from the capital of Oslo and the port of Bergen to the old national focus of Trondheim and Arctic Hammerfest, lie in the coastal zone and have difficult overland connections. The warmth of Atlantic waters makes life possible along Norway's high-latitude coastline, but the nation remains a string of clusters along its seaward margin.

Denmark

Territorially the smallest Nordic state, Denmark's population of 5.2 million ranks second in the region after Sweden. The country consists of the Jutland Peninsula and adjacent islands between Sweden and Western Europe. (Denmark also has ties to Greenland—see box titled ''Greenland and Europe.'') Denmark has a comparatively mild, moist climate; level land and good soils help sustain intensive agriculture over 75 percent of its area. Denmark exports dairy products, meats, poultry, and eggs, mainly to its chief trading partners, Germany, Sweden, and the United Kingdom.

Denmark's capital, Copenhagen, has long been a center where large quantities of goods are collected, stored, and transshipped. This is because it lies at the *break-of-bulk* point where many oceangoing vessels are prevented from entering the shallow Baltic Sea; conversely, ships with smaller tonnages ply the Baltic and bring their cargoes to this pivotal collecting station. Thus Copenhagen is an *entrepôt* whose transfer functions maintain the city's position as the lower Baltic's leading port.

Finland

Almost as large as Germany, Finland has just over 5 million residents, most of them concentrated in the triangle formed by the capital, Helsinki, the textile-producing center of Tampere, and the shipbuilding center of Turku (Fig. 1-16). A land of evergreen forests and glacial lakes, Finland has an economy that has long has been sustained by the export of wood and wood products. But the Finns, being a skillful and productive people, have developed a diversified economy in which the manufacture of machinery and the growing of staple crops play important roles.

Finland lost crucial lower-latitude territory (by Finnish standards) to the Soviet Union, now incorporated as a Russian "republic" under Moscow's control. Vyborg once was a Finnish town, but Finland avoided the fate of the Baltic states on the other side of the Gulf of Finland. The Finns also battled the Swedes, and Swedish stands alongside Finnish as an official language in the country. As Figure 1-6 reveals, the Finnish and Estonian languages are not closely related to those spoken in Scandinavia, and the origin of the Finns and their culture, and the question of how they came to this corner of Europe, continue to be topics of study and debate.

Following approval by Finland's voters, the country became a member of the European Union in 1995, leaving behind five decades of neutrality enforced by the Soviet Union. It was a crucial decision, because the collapse of the U.S.S.R. in 1991 severed trade links to the east that included the provision of much-needed oil by Moscow. EU membership will ease the reorientation of the Finnish economy.

Estonia

Ethnic and linguistic criteria support the inclusion of Estonia within the Northern European region. Not only does the Estonian language belong to the Finnic subfamily, but also the Estonians claim Nordic ties because their country was part of the Kingdom of Sweden when it was annexed by the Russians in 1710. During the Soviet period, Finland's fragile independence on the communist perimeter kept alive Estonia's dream of liberation and independence. Today, busy maritime traffic links the capital, Tallinn, to Helsinki directly across the Gulf of Finland.

Political and economic problems plague Estonia. About one-third of the population of 1.5 million is Russian, colonists who came here during the Soviet period. When Estonia became independent in 1991, its government tightened the rules of citizenship against this minority, and this action led to a deterioration of relations with Russia's new rulers. Meanwhile, long-term trade links with Russia degenerated, and the breakdown of the Russian economy severely affected Estonia as well. During the Soviet period, Estonia became industrialized through the exploitation of extensive oil shale deposits, which provided ample electricity. But the events of the 1990s disrupted the flow of raw materials and Estonia's once-guaranteed access to markets to the east. Recovery is not yet in sight.

Iceland

In the frigid waters of the North Atlantic, just south of the Arctic Circle, lies the volcanic, glacier-studded island of Iceland. Inhabited by people with Scandinavian ancestries (population: 269,000), Iceland and its small neighboring archipelago, the Westermann Islands, are of special scientific interest because they lie on the Mid-Atlantic Ridge, where the Eurasian and North American tectonic plates of the Earth's crust are diverging and new land can be seen to form.

Iceland's population is almost entirely urban, and the capital, Reykjavik, contains about half the country's population. The nation's economic geography is almost entirely oriented to the surrounding waters, whose seafood harvests give Iceland one of the world's highest standards of living—but at the risk of overfishing. Disputes over fish-

GREENLAND AND EUROPE

Greenland, the world's largest island, lies much closer to North America than to Europe. But its historical geography links it to Europe—specifically, Denmark. Peopled originally by Inuits whose communities clustered along the southwestern coast, Greenland was reached by Norwegian Vikings before A.D. 1000 and became a Norwegian dependency in 1261. After Norway united with Denmark in 1380, the frigid island came under Danish rule. When the union ended in 1814, Greenland continued under the Danish crown. In 1953, its status changed from colony to province, and in 1979, Greenland won home rule as a "Self-Governing Overseas Administrative Division" of Denmark with an Inuit name: *Kalaallit Nunaat*.

Greenland's 55,000 inhabitants, 85 percent of whom are Inuits or native-born Greenlanders of European ancestry, depend almost entirely on fishing, a declining industry. Mining operations (lead, zinc) failed, and tourism is yet a minor source of income. Grants from Copenhagen keep the territory going.

Economic problems and dependence on Denmark have not stifled Greenlanders' political liberties. When Denmark joined the (then) European Economic Community in 1973, Greenland entered the EEC as part of the Danish state. After achieving home rule in 1979, Greenlanders voted to secede, taking their territory out of the budding European Union. They were the first (but possibly not the last) to do so.

Still, the Scandinavian imprint on Greenland's Arctic cultural landscapes is indelible, as reflected by the capital, Nuuk (Godthab), with its Danish government buildings and brightly colored wooden houses, located on the comparatively moderate southwestern coast. The realm boundary shown in Figure I-1 confirms that while Greenland lies just a few miles east of Canada's Ellesmere Island, its functional ties are across the Atlantic, not across Baffin Bay.

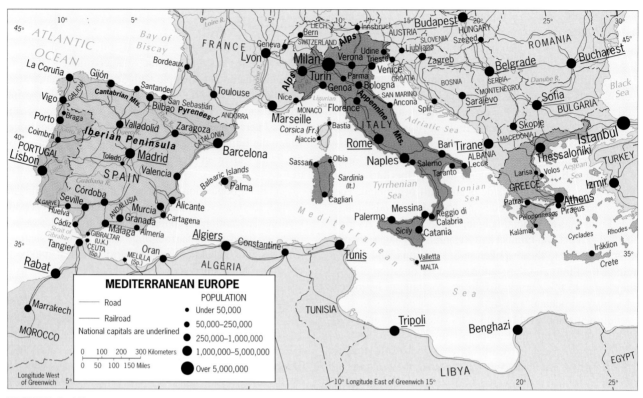

FIGURE 1-17

ing grounds and fish quotas have intensified in recent decades; the Icelanders argue that, unlike the Norwegians or the British, they have little or no alternative economic opportunities.

❖ MEDITERRANEAN EUROPE

On the southern side of Europe's core lie the five countries that constitute the Mediterranean region: Italy, Spain, Portugal, Greece, and the island microstate of Malta (Fig. 1-17). With this shift from near-polar to near-tropical Europe, we should expect strong contrasts, and there are many—but similarities also exist. Once again, this is a region of peninsulas. Like Northern Europe, Mediterranean Europe is a discontinuous region. Moreover, there (again) is separation from the European core, so that core-periphery contrasts are manifest here as well. A degree of cultural continuity, dating from Greco-Roman times, marks the region's religions and languages, lifeways and landscapes. As Figure I-7 shows, natural environments in this region are dominated by a climatic regime that bears its very name—Mediterranean. Dry, hot summers are the norm, so that moisture supply often is problematic during much of the growing season.

In terms of raw materials, Southern Europe (as the region is also called) is not nearly as well endowed as the

European core, a situation that is reflected by the map of industrial complexes (Fig. 1-5). Only northern Spain and northern Italy (the latter through massive imports of coal and iron ore) have become part of the core area's economic geography. Even then, Spain exports much of the minerals it mines directly to Western Europe. Forest resources also offer a discouraging picture: unlike Scandinavia and Finland, the Mediterranean region has been largely deforested. Despite its often rugged topography, Southern Europe's hydroelectric opportunities are constrained by limited and highly seasonal water supply.

A key contrast between Northern and Southern Europe lies in their populations. Although territorially smaller than Northern Europe, Mediterranean Europe has nearly five times as many people (118 million in 1997). Population distribution continues to reflect the agricultural bases of the preindustrial era, with large concentrations in coastal lowlands and fertile river basins, although the growth of major industrial centers, notably in northern Italy and northern Spain, has superimposed a new mosaic. Still, the level of urbanization in Southern Europe is far below that of Western or Northern Europe, or of the British Isles: only tiny Malta had reached 70 percent by 1997, and Portugal's 34 percent was one of the lowest in Europe. Other data indicate that living standards in Mediterranean Europe also lag well behind (Appendix A). This was, and remains, one of the challenges confronted by the European Union as its leaders strive for equality among the 15 members.

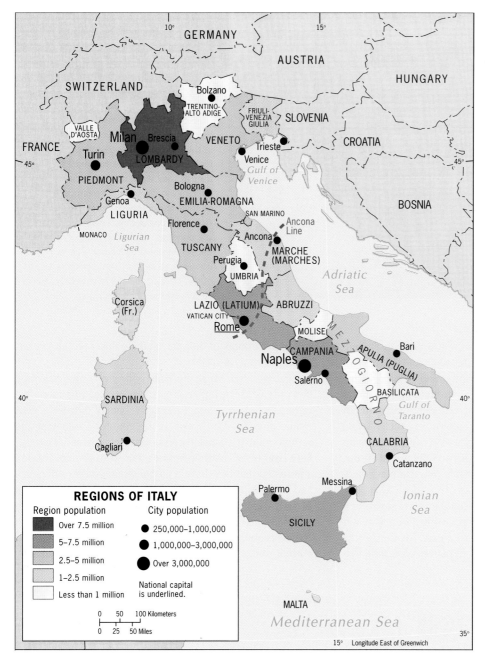

FIGURE 1-18

Italy

Centrally located in the Mediterranean region, most populous of the Mediterranean states, best connected to the European core, and economically most advanced is Italy (58 million), a charter member of Europe's Common Market.

Administratively, Italy is organized into 20 regions, many with historic roots dating back centuries (Fig. 1-18). Several of these regions have become powerful economic entities centered on major cities, such as Lombardy (Milan) and Piedmont (Turin); others are historic hearths of Italian culture, including Tuscany (Florence) and Veneto (Venice). These regions, in the northern half of Italy, stand in strong social, economic, and political contrast to such southern regions as Calabria (the "toe" of the Italian "boot"), Sicily, and Sardinia (Italy's two major Mediterranean islands). Not surprisingly, Italy is often described as two countries—a progressive north and a stagnant south, or *Mezzogiorno*.

North and south are bound by the ancient headquarters, Rome, which lies astride the narrow transition zone between Italy's contrasting halves. This zone is referred to as

From the Field Notes

"In Rome, ancient treasures coexist with modern urbanization as nowhere else in the world; commuters circle the Colosseum and other landmarks as though they were merely ordinary buildings in disrepair. The famous legacies of ancient Rome—the Forum, the great Arch of Constantine, Trajan's Column—are only part of this heritage. Walk along the streets of the city, and you will come upon places such as this, preserved parts of the classical city surrounded by, but as yet protected from, encroachment."

the *Ancona Line*, after the town on the Adriatic coast where the zone reaches the other side of the peninsula (Fig. 1-18). Whereas Rome remains Italy's capital and cultural focus, the functional core area of Italy has shifted northward into Lombardy in the basin of the Po River. Here lies Southern Europe's leading manufacturing complex, in which a large, skilled labor force and ample hydroelectric power from Alpine and Appennine slopes combine with a host of imported raw materials to produce a wide range of machinery and precision equipment. The Milan–Turin–Genoa triangle exports appliances, instruments, automobiles, ships, and many specialized products. Meanwhile, the Po Basin, lying on the margins of the region's dominant Mediterranean climatic regime, enjoys a more even pattern of rainfall dis-

tribution throughout the year, making this a productive agricultural zone as well.

Metropolitan Milan embodies the new, modern Italy. Not only is Milan Italy's largest city and leading manufacturing center—making Lombardy one of Europe's economic *motors*—but it also is the country's financial and service-industry headquarters. Privately owned factories here tend to be small operations, but so well managed and automated that they have a global reputation for efficiency and adaptability. Today, the Milan area, with just 9 percent of Italy's population, accounts for one-third of the country's national income. Its productivity has propelled the Italian economy past that of the United Kingdom and, if current trends continue, will push it past that of France by

High-rise, high-technology corporate campuses are familiar landmarks in the "edge cities" that cluster along freeways in the increasingly urbanized suburbs of large cities in the United States. The Europeans have been slower to adopt these automobile-generated activity complexes, but as we saw in the case of Paris's *La Défense* (p. 61), major business centers are now beginning to emerge outside the central-city core. This U.S.-style, high-tech complex—one of Europe's most advanced—is located on suburban Milan's *Tangenziale* (Beltway) at the interchange with the *autostrada* (expressway) that leads to the port city of Genoa.

■ AMONG THE REALM'S GREAT CITIES...

Rome

From a high vantage point, Rome seems to consist of an endless sea of tiled roofs, above which rise numerous white, ochre, and gray domes of various sizes; in the distance, the urban perimeter is marked by highrises fading in the urban haze. This historic city lives amid its past as perhaps no other as busy traffic encircles the Colosseum, the Forum, the Pantheon, and other legacies of Europe's greatest empire.

Founded about 3,000 years ago at an island crossing point on the Tiber River about 15 miles (24 km) from the sea, Rome had a high, defensible site. A millennium later, with a population some scholars estimate as high as 1 million, it was the capital of a Roman domain that extended from Britain to the head of the Persian Gulf, and from the shores of the Black Sea to North Africa. Rome's emperors endowed the city with magnificent, marble-faced, columned public buildings, baths, stadiums, obelisks, arches, and statuary; when Rome became a Christian city, the domes of churches and chapels added to its luster.

It is almost inconceivable that such a city could collapse, but that is what happened after the center of Roman power shifted eastward to Constantinople (now Istanbul). By the end of the sixth century, Rome probably had fewer than 50,000 inhabitants, and a mere 30,000 in the thirteenth. Papal rule and a Renaissance revival lay ahead, but in 1870, when Rome became the capital of newly united Italy, it still had a population of only 200,000.

Now began a growth cycle that eclipsed all previous records. As Italy's political, religious, and cultural focus (although not an industrial center to match), Rome grew to 1 million by 1930, to 2 million by 1960, and today approaches 3 million. The religious enclave of Vatican City, Roman Catholicism's headquarters, makes Rome a twin capital; the Vatican functions as an independent entity and has a global influence that Italy cannot equal.

Rome today remains a city whose economy is dominated by service industries: national and local government, finance and banking, insurance, retailing, and tourism employ three-quarters of the labor force. The new city sprawls far beyond the old, walled, traffic-choked center where the Roman past and Italian future come face to face.

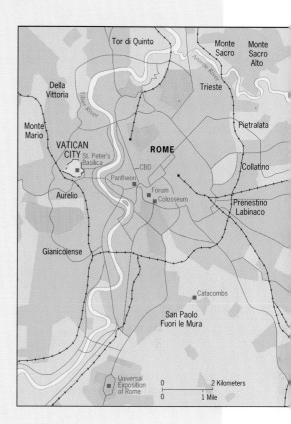

century's end. Italy's success can be read in Appendix A; its per capita GNP is by far the highest in Mediterranean Europe and rivals that of several countries in the European core.

This is all the more remarkable because Italy south of the Ancona Line is headed in the opposite direction. The highly advanced industrial north, dominated by private enterprises, stands in ever sharper contrast to the stagnant agricultural south, where large state-owned enterprises are common. The cost of supporting the poverty-stricken Mezzogiorno is enormous, not only in financial but also in social terms. Taxpayers in the north object to having their money poured into the "bottomless pit" of the Mezzogiorno; worse, an extremist reaction in the north (mainly in Lombardy) against the perceived North Africanization of the south is polarizing Italian politics. The south is seen as unyielding to modernizing efforts, unwilling to rid itself of organized crime clans that control village and countryside, and unable to stem the tide of immigration from North Africa (a migrant stream that is actually quite small in number).

What Italy needs is a stable and effective government; instead, it has had half a century of incompetence and corruption. More than 50 governments have formed and failed in just the past 50 years, leading to a huge national debt, a weak currency, and repeated failures to meet the criteria of the European Union in various economic and social spheres. As Italy's centrifugal forces grow, there is a real chance that weakness at the center may not only undo the economic gains made in the north, but also threaten the survival of the state itself.

Spain

At the western end of Mediterranean Europe lies the Iberian Peninsula, separated from France and Western Europe by the rugged Pyrenees and separated from North Africa

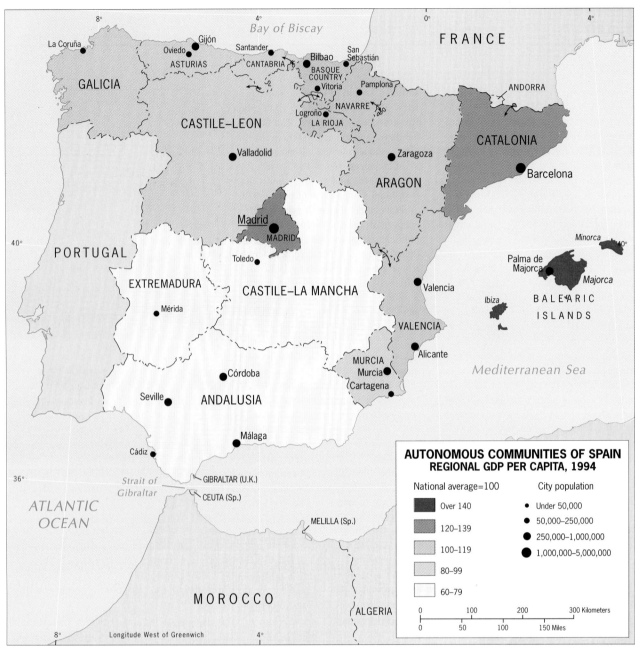

FIGURE 1-19

by the narrow Strait of Gibraltar (Fig. 1-17). Spain occupies most of this compact landmass, which is about the size of California plus Oregon.

Like Germany and France, Spain has reorganized and decentralized its internal politico-geographical structure, creating 17 subnational regions called *Autonomous Communities* (ACs) and given their historic names. (Sixteen are shown in Fig. 1-19; the Canary Islands AC lies in the nearby Atlantic.) All these ACs have their own parliaments and administrations that control planning, public works, cultural affairs, education, environmental policy-making, and, to some extent, international commerce. Each AC is

allowed to negotiate its own degree of autonomy, and those with the greatest economic strength and the most active separatist movements have gained the greatest concessions from the national government in Madrid. Among the latter, the Basque Country, a small AC facing the Bay of Biscay, has been most successful. The Basques, an ethnic minority whose passionate nationalism has been a centrifugal force in Spain for decades, have wrested considerable powers from the central government.

More consequential, however, is the case of Catalonia, the triangular AC in Spain's northeastern corner, adjacent to France (Fig. 1-19). Centered on prosperous, productive

Barcelona, Catalonia is Spain's leading industrial area, an AC with a population approaching 7 million (comparable to Sweden!). It is imbued with a fierce nationalism, endowed with its own language and culture, and economically is one of the driving *motors* of Europe. Note that while most of Spain's industrial raw materials lie in the northwest, its major industrial development has taken place in the northeast, where innovations and skills propel a high-technology-driven regional economy. In 1995, Catalonia, with 6 percent of Spain's territory and 17 percent of its population, produced 25 percent of all Spanish exports and nearly 40 percent of its industrial exports. Such economic strength translates into political power, and in Spain the issue of Catalonian separatism is never far from the surface.

As Figure 1-19 indicates, Spain's capital and largest city, Madrid, lies very near the geographic center of the state; it also lies along an economic-geographic divide. As ACs, Catalonia and Madrid are Spain's most prosperous; the contiguous group of five ACs between the Basque Country and Valencia rank next. Tourism (notably along the Mediterranean coast) and winegrowing (especially in La Rioja) contribute importantly here. Ranking below this cluster of ACs are the four regions of the northwest: Galicia, Castile-Leon, and industrialized Asturias and Cantabria. Incomes here are below the national average as industrial obsolescence, dwindling raw material sources, and emigration afflict the economy. Worst off, however, are the three large ACs to the south of Madrid: Extremadura, Castile-La Mancha, and Andalusia. Drought, inadequate land reform, scarce resources, and remoteness from Spain's fast-growing northeast are among the factors that inhibit development here. Low local wages have attracted some companies to set up factories in Seville, Córdoba, and Málaga, but until now the hopes raised by Spain's participation in the European Union have not materialized. Again, a spurt of growth has been followed by a slowdown that has had the effect of intensifying the regional disparities that weaken the fabric of the state. Even as EU members seek to reduce the differences among themselves, individual countries face the geographic consequences of focused growth that deepen internal divisions.

Portugal

The southwestern corner of the Iberian Peninsula is occupied by the state of Portugal (population: 10 million), a comparatively poor country that benefited enormously from its admission to the EU. One of the rules of EU membership is that the richer members provide assistance to the poorer ones, and Portugal shows the results in a massive renovation project in the capital, Lisbon, and in the modernization of surface transport routes.

Unlike Spain, which has major population clusters on

From the Field Notes

"We climbed up to the ramparts of the fort built by the Moors, who took the city in 711 A.D., to get a sense of the layout of Málaga. Tourism and seasonal residency have transformed the waterfront of this otherwise industrial city, now the major center of the crowded Costa del Sol. The bullring, we were told, serves more frequently as a site for rock concerts than as a bullfighting venue; 'anyway,' said a colleague, 'the tourists, not the locals, are who keep this disgusting spectacle going.' Highrises accommodate sun-seeking visitors and part-time residents, but this area suffers from growing water shortages, a concern for the future."

its interior plateau as well as its coastal lowlands, the Portuguese are concentrated along and near the Atlantic coast. Lisbon and the second city, Porto, are coastal cities; the best farmlands lie in the moister western and northern zone of the country. But farms here are small and inefficient, and despite the fact that Portugal remains dominantly rural, it must import as much as half of its foodstuffs. Exporting textiles, wines, corks, and fish, and running up an annual deficit, the Portuguese economy remains a far cry from those of other European countries of similar dimensions.

Greece

The peninsulas and islands that form the territory of Greece also constitute the easternmost outlier of both Mediterranean Europe and the European Union. Greece has land boundaries with Turkey, Bulgaria, Macedonia, and Albania; as Figure 1-17 reveals, it owns islands just offshore from mainland Turkey. Altogether, the Greek archipelago numbers some 2,000 islands ranging in size from Crete

■ AMONG THE REALM'S GREAT CITIES . . .

Athens

Take a map of Greece. Draw a line to encompass its land area as well as all the islands. Now find the geographic center of this terrestrial and maritime territory, and you will find a major city nearby. That

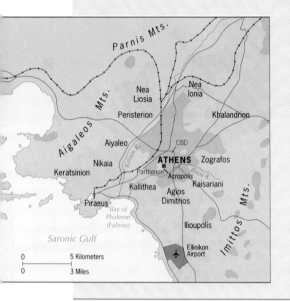

city is Athens, capital of modern Greece, culture hearth of ancient Greece, source of Western civilization.

Athens today is only part of a greater metropolis that sprawls across, and extends beyond, a mountain-encircled, arid basin that opens southward to the Bay of Phaleron, an inlet of the Aegean Sea. Here lies the port of Piraeus, Greece's largest, linked by rail and road to the capital. Other towns, from Keratsinion in the west to Agios Dimitrios in the east, form part of an urban area that not only lies at, but *is*, the heart of modern Greece. With a population of almost 4 million, most of the country's major industries, its densest transport network, and a multitude of service functions ranging from government to tourism, Greater Athens is a well-defined national core area.

Arrive at Ellinikon Airport south of Athens and drive into the city, and you can navigate on the famed Acropolis, the 500-foot (150-m) high hill that was ancient Athens' sanctuary and citadel, crowned by one of humanity's greatest historic treasures, the Parthenon, temple to

the goddess Athena. It is only one among many relics of the grandeur of Athens 25 centuries ago, now engulfed by mainly low-rise urbanization and all too often obscured by smog generated by factories and vehicles and trapped in the basin. Conservationists decry the damage done by air pollution to the city's remaining historic structures, but in truth the greatest destruction was wrought by the Ottoman Turks, who plundered the country during their occupation of it, and the British and Germans who carted away priceless treasures to museums and markets in Western Europe.

Travel westward from the Middle East or Turkey, and Athens will present itself as the first European city on your route. Travel eastward from Europe, however, and your impression of its cultural landscape may differ: behind the modern avenues and shopping streets are bazaar-like alleys and markets where the atmosphere is unlike any other in this geographic realm. And beyond the margins of Greater Athens lies a Greece that lags far behind this bustling, productive coreland.

(3,218 square miles) to small specks of land in the Cyclades.

Ancient Greece was a cradle of Western civilization; later it was absorbed by the expanding Roman Empire. For some 350 years from the mid-fifteenth century, Greece was under the sway of the Ottoman Turks. Independence was regained in 1827, but not until nearly a century later did Greece push its boundaries to their present position in a series of Balkan Wars. During World War II, Nazi Germany occupied and ravaged the country, and in the postwar period the Greeks have quarreled with the Turks, the Albanians, and the newly independent Macedonians. This remains a volatile part of the world.

Modern Greece is a nation of more than 10 million centered on historic Athens, one of the realm's great cities. With its port of Piraeus, metropolitan Athens contains about 40 percent of the Greek population, making this one of Europe's most congested and polluted urban areas. Athens is the quintessential primate city, its cultural landscape still dominated by the monumental architecture of ancient Greece. The Acropolis and other prominent landmarks at-

tract a steady stream of visitors; tourism is one of Greece's leading sources of foreign revenues, and Athens is only the beginning of what the country has to offer.

Greece's limestone-dominated topography has been sculpted under Mediterranean climatic conditions into stark, often barren scenery that forms a dramatic backdrop for the whitewashed towns and villages that dot the countryside. In the days of ancient Greece, extensive forests draped the landscape, but centuries of exploitation have denuded all the Mediterranean countries. None of this makes farming any easier, because soils have been washed away and arable land amounts to less than 30 percent of the total area. Still a strongly agrarian country, Greece is self-sufficient in staple foods and imports only livestock products; Mediterranean fruits (grapes, olives, figs, citrus) go to EU markets.

Greece occupies a pivotal position on the European periphery, adjacent to Islamic Turkey in one direction and to fractious Eastern Europe in another. In per capita GNP terms, it remains the poorest EU country, burdened by unstable and often ineffective government that has been unable to put Greece's financial affairs in order. Major loans from the EU have had strings attached that Greece has broken, and the economic reforms required under EU agreements have not materialized. As a result, Greece's points of modernization are mired in a morass of impediments, testing the promises of EU membership.

❖ EASTERN EUROPE

Of Europe's five regions, Eastern Europe is the most complex, troublesome, and transitory. The name Eastern Europe has for centuries (indeed, millennia) been synonymous with fragmentation, instability, and strife. Rugged topography, crucial rivers, isolated lowlands, and strategic valleys have witnessed tumultuous migrations, epic battles, and the rise and fall of states and empires. Slavic, Germanic, Turkish, Hungarian, Romanian, Albanian, and other peoples have converged on this region from near and far. Ethnicities, religions, alphabets, and lifeways have kept them in conflict.

Geographers call this region a **shatter belt**, a zone of chronic political splintering and fracturing. Other areas of the world merit this designation, but none can compare to Eastern Europe. Geographic terminology is replete with words that describe the breakup of established order and that have their roots in this part of the world. One of them is **balkanization** (see the box titled ''Balkanization''). A more recent term is *ethnic cleansing*, the forcible ouster of entire populations from their homelands by a stronger power bent on taking their territory. The term may be new, but the process is as old as Eastern Europe itself.

For part of the twentieth century, it appeared that greater stability might be in the offing. After World War I, in 1919, a group of combative peoples along the Adriatic Sea were combined in a single country called Yugoslavia and placed under the monarchy of Serbia. After World War II, Yugoslavia emerged as a communist dictatorship without falling apart. At the same time, the Soviet Union established its dominance over most of the region, imposing an enforced order through the communist parties of postwar Poland, East Germany, Czechoslovakia, Hungary, Romania, and Bulgaria. As Figure 1-20 shows, none of this came about easily. The whole territory of Poland was shifted westward. German Prussia along the Baltic was eliminated, and the Germans were driven off. Czechoslovakia lost its eastern extremity, Romania its entire northeast to the expansionist Soviet Union. But in truth, Figure 1-20 displays what for Eastern Europe passes as comparative stability. Not long before, an Austro-Hungarian Empire was challenging an empire built across the Balkans by the Muslim Turks, Poland had disappeared altogether, and nation-states such as those on the map had hardly begun to form.

BALKANIZATION

— ∎ ∎ ∎ —

The southern half of Eastern Europe is sometimes referred to as the Balkans or the Balkan Peninsula. This refers to the triangular landmass whose points are at the southern tip of the Greek mainland, the head of the Adriatic Sea, and the northwestern corner of the Black Sea. The name itself comes from a mountain range in Bulgaria, but it has also become a concept, born of the reputation for division and fragmentation in this portion of Eastern Europe. Any good dictionary carries a definition of the verb *balkanize*: for example, "to break up (as a region) into smaller and often hostile units," to quote Webster.

— ∎ ∎ ∎ —

The collapse of the Soviet Union, soon after the beginning of the 1990s, tested this newfound stability—and Eastern Europe failed. In 1990, civil war broke out in Yugoslavia, a conflict that was to cost hundreds of thousands of lives and dislocate millions in some of the most vicious ethnic-cleansing campaigns the world had ever seen. Out of Colorado-sized Yugoslavia came five new countries, one of which (Bosnia) was headed for partition. In 1993, the Czechoslovakian state broke up into two countries; this peaceful separation later generated tensions between the new neighbors. Meanwhile, Bulgarian government policies led to a mass exodus of Turks, and Hungarians living in Romania found themselves under pressure from the post-communist regime in Bucharest.

The collapse of Soviet communism had another, potentially even more significant impact on Eastern Europe. Former constituent republics of the U.S.S.R. turned their gaze from east to west, from Moscow toward Strasbourg and Brussels. They harbored hopes of joining the European Union, and some wanted to join NATO. Most viewed a closer association with Europe as a safeguard against the Russian giant next door. And so, even as the ''old'' Eastern Europe returned to its Balkanizing ways, a ''new'' Eastern Europe was in the making (Fig. 1-21). The realm boundary shifted eastward from Poland and Romania to Latvia and Ukraine, and five new states were added to the region: Latvia and Lithuania in the north, Ukraine and Moldova in the south, and, potentially, Belarus in the middle. In the late 1990s, Belarus remained the most closely tied of all the former Soviet republics to Moscow, but reformers in its capital, Mensk continued to promote European links.

As delimited in Figure 1-21, today's Eastern Europe displays several important geographic qualities. It is territorially and demographically Europe's largest region; it adjoins three of the four other European regions; it contains more countries than any other European region; it includes

FIGURE 1-20

Europe's largest state in area, Ukraine; it incorporates Europe's poorest country, Albania. In the late 1990s, none of its states could meet the criteria for membership in the European Union. And Eastern Europe reaches into a zone that Russia regards as part of its sphere of influence, not least because sizeable Russian minorities still live in all five of the former Soviet republics now realigning themselves toward Europe.

The Politico-Geographical Framework

Eastern Europe extends from the Baltic Sea in the north to the Black Sea and the Adriatic Sea in the south, and from the Russian border in the east to the German boundary in the west. To help us better understand this complex region, we group its 17 countries by geographic area:

FIGURE 1-21

1. *Countries Facing the Baltic Sea.* These countries are Poland, Lithuania, Latvia, and also Belarus; although Belarus does not have a Baltic Sea coast itself, it borders all three of the Baltic coast states and has a direct link to the Russian exclave of Kaliningrad.

2. *The Landlocked Center.* These are the countries at the heart of the old Eastern Europe: the Czech Republic (westernmost and most Western of the region's states), Slovakia, and Hungary.

3. *Countries Facing the Adriatic Sea.* This is a complex and changing group of countries devolved from the former Yugoslavia: Slovenia, Croatia, Bosnia, Serbia, and Macedonia; adjoining Albania is also included.

4. *Countries Facing the Black Sea.* This cluster of countries consists of Bulgaria and Romania of the old Eastern Europe as well as Moldova and Ukraine of the new. As Figure 1-21 shows, Moldova is technically landlocked but does border two Black Sea states.

Countries Facing the Baltic Sea

Poland

The northwestern corner of Eastern Europe is dominated by Poland. Look again at Figure 1-20 and you will note that Warsaw, the capital and primate city of Poland, once lay near the center of the national territory (outlined by the gray prewar boundary) but today lies in its eastern quadrant. (Berlin, not far from present-day Poland's western border, once was much more central to a greater Germany too.) Situated on the North European Lowland, Poland has traditionally been an agrarian country. Large-scale industrialization was introduced by the communist planners of the post-World War II period, who exploited the substantial coal and mineral supplies contained in the south. Along the boundary with the Czech Republic lie the Sudeten Mountains, and just to the east of them, in Silesia, Poland's industrial heartland emerged (Fig. 1-21). Katowice, Wroclaw, and Krakow came to symbolize the new era introduced by the communists. Unchecked environmental degradation, including the pollution of air, streams, and groundwater, accompanied this transformation of the south.

From the Field Notes

"Turbulent history *and* prosperous past are etched on the cultural landscapes of Poland's port, Gdynia, and the old Hanseatic city of Gdansk nearby. Despite major wartime destruction, much of the old architecture survives, and we found restoration underway throughout these twin cities. Gdansk was the stage for the rebellion of the Solidarity labor union against Poland's communist regime; today city and country struggle through the difficult transition to a new economic and social order. Attracting foreign visitors is one way to raise revenues, and the historic old city on the Baltic is becoming a tourist draw."

Poland's best farmlands also lie in the south, where farming is intensive and wheat is the leading crop. In central Poland, thin and often rocky soils that developed on glacial deposits support rye and potatoes; farther north, the Baltic rim has such poor soils that cultivation gives way to pastureland and moors. Farming in Poland suffered severely during the communist period. The communist planners attempted to collectivize all farms but without making investments comparable to those made on state-owned factories. This created a severe economic-geographic imbalance, so that Polish farming today copes with many problems. Farmers still work with tools and equipment that would be in an historical museum in any Western European country.

Warsaw lies near the head of navigation on the Vistula River in a productive agricultural area. As Figure 1-21 demonstrates, this focus of Polish history and culture lies at the hub of a radiating network of transport routes that reaches all parts of the country. One routeway links Warsaw to the port city of Gdansk on the Baltic Sea coast. There, in the docks and shipyards, began the political events of the 1980s that were to put Poland in the vanguard of Eastern Europe's political transformation, as the Solidarity labor union defied the communist regime. Today, Poland is on a difficult course of economic and political reform. The cities reflect the new market era, but the villages and countryside lag badly. In a stunning political development in 1995, a revived Communist Party (now called the Democratic Left Alliance) teamed up with the Peasant's party to regain power in the capital.

In their struggle to establish a democratic and economically viable Poland, the country's nearly 40 million citizens have several advantages. Poland is a nation-state with only minuscule minorities. One religion (Roman Catholicism) prevails overwhelmingly; one language unites the Poles. Urban-rural contrasts and present regional disparities notwithstanding, Poland has the potential to advance in an atmosphere of unity that is denied to so many Eastern European countries.

Lithuania

The Grand Duchy of Lithuania once dominated Eastern Europe from the Baltic coast to the Black Sea and from Warsaw to Tula, incorporating such major cities as Minsk (Mensk) and Kiev (Kyyiv). Eventually the Russians, Poles, and Germans pushed the Lithuanians back, and in 1940 the Soviet Union annexed their country, making it a Soviet republic. At the Potsdam Conference in 1945, very much at Lithuania's expense, the victorious Soviets were awarded the former German base at Königsberg as well as a Connecticut-sized hinterland; the Russians renamed this entity Kaliningrad (see box titled ''Kaliningrad: Russia on the Baltic''). This left Lithuania with a mere 50 miles (80 km) of coastline including the small port of Klaipeda (Fig.

KALININGRAD: RUSSIA ON THE BALTIC

■■■

The Russian *exclave* of Kaliningrad lies wedged between Lithuania and Poland, facing the Baltic Sea through its gigantic naval port (Fig. 1-21). Soon after the Soviets acquired this German base in 1945, virtually all ethnic Germans were expelled. Not much was left of their cultural landscape: relentless British bombing had devastated the place.

Russians replaced the departed Germans, and today Kaliningrad's population approaches 1 million people, 90 percent of whom are Russian. In the Soviet political scheme of things, Kaliningrad had the status of an *oblast* (see Chapter 2), and it was made part of the Russian Republic—as it remains today.

Uncertainty now prevails in Kaliningrad. Independence is not an issue, but that could change. Despite the destruction of much of seven centuries of German heritage, the German past still can be felt: the renowned philosopher Immanuel Kant lived and worked here and lies buried in the ruins of the German cathedral. The city's German name, Königsberg, appears now and then in advertising (e.g., for German tourists).

Kaliningrad's post-Soviet years have been difficult. Rumors flew—for example, that the exclave would be used to resettle ethnic Germans from Central Asia, that Russia's president had offered it for sale to Germany. Locals hoped to establish a free trade area, envisioning Kaliningrad as a future Hong Kong of the Baltic. Rising nationalism and militarism in Russia put an end to all this. In the late 1990s, Kaliningrad was reverting to its old role as a crucial military base for a country with few warm-water ports—a naval bulwark, potentially, in an Eastern Europe into which Western military alliances, chiefly NATO, are spreading.

■■■

1-21). But Klaipeda would be of little importance: it was not even connected by rail to the interior. As a Soviet republic, Lithuania was induced to look toward Moscow, not to the Baltic Sea and beyond. Hence the capital, Vilnius, is situated in the interior, near the Belarus border, rather than on the coast.

Lithuania's declaration of independence in 1990 produced a Soviet economic blocade, followed in 1991 by a tank and paratrooper attack on radio and television facilities. Moscow acknowledged Lithuania's sovereignty just months before the U.S.S.R. collapsed at the end of 1991. But independence left this small country with large economic problems, despite self-sufficiency in electric power from Soviet-era (and aging, unsafe) nuclear plants. The Soviet Union was Lithuania's chief source of raw materials for industry as well as its leading market, and the disruption of those linkages has produced a severe economic setback. Lithuania refused to comply with Russian demands that it provide unlimited military access to Kaliningrad; in response, Moscow imposed heavy duties on Lithuanian prod-

ucts entering Russia. Meanwhile, Lithuania sought security by becoming an associate partner in the Western European Union and applying to join NATO.

Latvia

Approximately the size of West Virginia, Latvia, Lithuania's northern neighbor, has just 2.5 million inhabitants, about one-third of whom are Russians. Latvians, in fact, constitute only a bare majority (52 percent) in their own country.

Latvia is Eastern Europe's northernmost country, and contrasts between the cultural landscapes of its capital (Riga) and that of neighboring, Nordic Estonia (Tallinn) confirm these limits. Latvia's historic and cultural links are to the south; like the Lithuanians, the Latvians lost their independence to the Soviets in 1940 and regained it in 1991. But unlike Lithuania, whose Russian minority is a mere 9 percent, Latvia copes with a major social as well as economic problem. Without resources other than its good soils and stands of forest, Latvia during Soviet times was industrialized to produce a wide range of specialized products such as television sets, refrigerators, and railroad cars. Dependence on the former Soviet Union for raw materials and markets, coupled with tensions in Riga as well as Moscow over the fate of Latvia's nearly 900,000 Russians, has cast doubt over Latvia's economic future.

Belarus

Between the Baltic states to the north and giant Ukraine to the south lies Kansas-sized Belarus, Moscow's closest ally during the Soviet period and still strongly linked to Russia today. Centered on the capital and largest city, Mensk (Minsk), Belarus has a population of just over 10 million of whom 80 percent are Belarussians (''White'' Russians). Like the Poles and the Slovaks, the Belarussians are a West Slavic people. Only 13 percent of the population is (East Slavic) Russian.

Belarus lies in a shatter belt that took the brunt of World War II. Its limited agricultural and industrial economies were smashed by eastward-moving German armies, which made Belarus their first Soviet victim. One-quarter of the entire population was killed, destruction was nearly complete, and recovery was slow and painful. Loyal cooperation with Soviet authorities made Belarus a favorite in Moscow, although its role in the overall Soviet economy was overshadowed by other republics, notably Ukraine.

The country's resource base is limited to peat (for fuel) and potash (for fertilizer production), and its soils are not as rich as Ukraine's. During the Soviet period, nonetheless, Belarus became a major machine-fabricating center, and Mensk a large industrial city producing tractors, trucks, and machine tools. Although some oil has been discovered in Belarus, the country's relative location between the heart

From the Field Notes
"This monument to the struggle for independence stands in the capital of Latvia, Riga. Our visit was a learning experience as a large demonstration nearby displayed not only written signs denouncing lingering Russian influence in Latvia, but also maps that equated Latvia with internal Russian republics such as Karelia and Chechnya."

of Russia and the rest of Eastern Europe assures it a plentiful supply: both the Druzba oil pipeline and the Northern Lights gas pipeline cross Belarus, and these supply the country's needs. Overall, economic reforms here have gone slowly during the post-Soviet period, and in the late 1990s there were signs that Belarus might become the first ex-Soviet republic to return to Moscow's fold. While some economic ties with Europe had strengthened, political and financial links with Russia (including parity and joint use of the Russian and Belarussian ruble) suggested an eastward orientation unique in the new Eastern Europe.

The Landlocked Center

It is no overstatement to say that the Czech Republic and Slovakia (the two components of former Czechoslovakia) as well as Hungary are pivotal countries in the changing Eastern Europe. In 1956, Hungarians rose in revolt against Soviet domination, and Russian tanks left no doubt regarding the true relationship between Moscow and this Eastern European colony. In Czechoslovakia in 1968, what began as a political movement toward greater freedom—the so-called Prague Spring—also was crushed by Soviet arms, again with much loss of life. These uprisings failed, but they revealed Soviet rule for what it was and contributed to the endurance of national identities and aspirations. In 1989, communist rule was swept away, and a new era opened for this heart of Eastern Europe. The ''velvet divorce'' between the Czechs and the Slovaks in 1993 complicated the map, creating a third entity in the Landlocked Center.

The Czech Republic

The Czech Republic not only is Eastern Europe's westernmost country; it also is the region's most Westernized state. Bohemia, the mountain-enclosed core area that contains the capital, Prague, always has been cosmopolitan in character and Western in its exposure, outlook, development, and linkages. A major manufacturing center, Prague lies in the lowland of the Elbe River, Bohemia's historic outlet to northern Germany and the North Sea. Prague is a classic primate city, its cultural landscapes totally faithful to Czech traditions; but the city also lies at the heart of the country's greatest concentration of wealth and productive capacity. The encircling mountains contain many valleys in which lie small industrial towns that specialize, Swiss-style, in fabricating numerous high-quality goods. In Eastern Europe the Czechs always have been the leaders in technology and engineering skills; even during the communist period their products found markets in foreign countries near and far.

As Figure 1-21 shows, the Czech Republic also incorporates an eastern zone named Moravia, after an old kingdom but signifying a zone of communist-era, heavy industrial development today. Here, Poland's Silesian manufacturing region overlaps into the Czech Republic, and local raw materials lie astride the gap between the Sudeten Mountains and the Carpathians (Fig. 1-2). This gap is called the Moravian Gate because of its vital importance as a passageway between the Danube valley to the south and the North European Lowland to the north. During the communist period, Moravia and the cities of Ostrava and Brno grew into important industrial centers producing steel, other metals, and chemicals. Today, many of those inefficient state-supported industries are obsolete and uncompetitive, posing a major challenge to the democratic Czech government.

To the often-asked question as to which Eastern European country might be the first to qualify for membership in the EU, the answer may well be the Czech Republic with its more than 10 million inhabitants, established trade links (the EU is already its chief partner), reforming economy, and diversified resource base. Still, as Appendix A shows, per capita GNP data put this country in the lower-middle income category, a long way below even Portugal and Greece. Despite its landlocked situation, the Czech Republic has the potential to lead the region into the EU.

Slovakia

For nearly three generations, there was the familiar form of Czechoslovakia on the map (Fig. 1-20)—through Nazi encroachment, war, communist rule, and the Soviet collapse. But liberation from Moscow's dominance undid more than 70 years of multicultural nation-building. The 5.4 million Slovaks in the east, their ethnic consciousness and nationalism fueled by the new freedom, opted to separate from their Czech neighbors to create their own republic. It was a negotiated secession based not only on ethnic and cultural issues but also on economic ones. The Slovaks did not share the Czechs' determination to pursue rapid economic reforms. Thus the Slovak city of Bratislava, on the Danube River just 50 miles (80 km) downstream from Vienna, became a national capital. In the division of assets, the Slovaks got about one-third of the wealth.

The breakup of Czechoslovakia left the Slovaks with serious problems. Trade links, access to markets outside Slovakia, finances, and boundary definition all caused difficulties. But other problems were worsened by the Slovaks themselves. Unlike the Czech Republic, Slovakia is a multicultural state: note, in Figure 1-22, that the Czech-Slovak boundary leaves a small Czech pocket in northwestern Slovakia. More important is the large (ca. 600,000) Hungarian minority in the south, constituting about 11 percent of the population. In late 1995, Slovakia's parliament passed a language law that severely restricts the use of the Hungarian language in the country over the bitter opposition of leaders of the Hungarian community. Meanwhile, international Slovak-Hungarian tensions have been heightened by the 1993 completion of the Gabcikovo Dam on the Danube, which diverts the river's water through a channel inside Slovakia, to which the Hungarians (on environmental and economic grounds) object.

Slovakia always was the least developed, most rural part of the former Czechoslovakia; it was more compliant during the communist era and more closely tied to the Soviet economic system than Bohemia. Deteriorating relationships with the Czechs, economic chaos in its old hinterland, ethnic frictions, and autocratic behavior on the part of its political leaders combine to cloud this new republic's future.

Hungary

On the threshold of the Balkan Peninsula lies Hungary, land of the middle Danube Basin, cornerstone of an historic empire, ally of Nazi Germany during World War II, and restive Soviet colony until 1989. Hungary is a nation-state of 10 million, but there also are Hungarians in Romania, Slovakia, Slovenia, Croatia, and Serbia, remnants of a time when Hungarian power was wider in this region (Fig. 1-22). In Hungary, these people across the borders are not

FIGURE 1-22

forgotten, and the Hungarian government from time to time reminds its neighbors of its interest in these nationals. These minority Hungarians, in return, look to the mother country for support.

Such behavior, when a state appeals to a regionally concentrated minority of ethnic cohorts in an adjacent state and seeks to protect it, is termed **irredentism**. The term derives from an old issue of this kind: attempts in Italy to achieve the incorporation of a neighboring Italian-speaking part of Austria. The Italians called their objective *Italia Irredenta* (Unredeemed Italy). Today, with numerous mi-

norities across Eastern European borders (Fig. 1-22), irredentism is a common pursuit in this region.

Eastern Europe's great river, the Danube, touches many countries in the region but serves mostly as a divider, not a unifier. But the Danube and its middle basin did function as a spatial bond for the Hungarians (Magyars), who arrived here more than a thousand years ago. A people of distant Asian origin, the Magyars have neither Slavic nor Germanic roots. They have converted their fertile lowland into a thriving state while retaining their cultural and linguistic identity even while vying for power in this turbulent region. The twin-cities capital astride the Danube, Buda and Pest (better known as Budapest), is a primate city nearly 10 times the size of the next largest town, reflecting the continuing rural character of the country (still less than 65 percent urbanized in the late 1990s).

Hungary's potentially productive rural economy has suffered a series of setbacks. Land fragmentation after 1919 (when its large estates were carved up), the destruction of World War II, and the communist collectivization campaign all impaired production, and today the privatization effort is causing difficulties as well. But Hungary not only has remained self-sufficient in food: it is the region's only net food exporter. A stabilizing economy will finally allow agriculture to come into its own as a major contributor to the national economy.

Industrialization in Hungary gained momentum during the Soviet period in the communist tradition of large state enterprises. Local coal, as well as iron ore sent by barge up the Danube, formed the basis for steel production. Metallurgical and engineering industries were developed in the Budapest area, but Hungary never matched (then) Czechoslovakia in manufacturing. Rather, Hungary continued to export millions of tons of raw materials, especially from its large bauxite (aluminum ore) reserves, to other countries.

Today, Hungary's best hope is considered to lie in smaller, privately owned, precision-manufacturing plants that require modest quantities of imported raw materials and will exploit the skills of Hungary's well-educated labor force. Such a transformation, however, takes time and determination. In the mid-1990s, public impatience and even a longing for the comparatively secure days of communist rule appeared to slow Hungary's transition.

Countries Facing the Adriatic Sea

As recently as 1990, only two Eastern European countries faced the Adriatic Sea: Yugoslavia and Albania (Fig. 1-20). Today Albania survives, but Yugoslavia has splintered into five countries, one of which may not survive the ongoing catharsis.

Former Yugoslavia

The breakup of Yugoslavia is the great human tragedy of Europe during the second half of the twentieth century. Long-dormant, long-contained centrifugal forces broke apart a multinational, multicultural state that had survived seven decades of turmoil and war. Yugoslavia ("Land of the South Slavs") lay between the Adriatic Sea to the west and Romania to the east, and between Austria and Hungary to the north and Bulgaria and Greece to the south. This was a country thrown together on maps after World War I, a

land of 7 major and 17 smaller ethnic and cultural groups (Fig. 1-22). The north, where Slovenes and Croats prevailed, was Roman Catholic; the south, Serbian Orthodox. Several million Muslims lived in Christian-surrounded enclaves. Two alphabets were in use. First the Royal House of Serbia and then a communist dictatorship personified by war hero Marshal Tito held Yugoslavia together. But when the communist system collapsed in Eastern Europe and the Soviet Union, so did the Yugoslav state.

Communist social planners unwittingly laid the groundwork for the disaster to come. They subdivided Yugoslavia into six internal ''republics'' based on the Soviet model, each dominated (except Bosnia) by one major group. Since all these ''republics'' inevitably incorporated minorities, the state guaranteed their rights, albeit through autocratic means.

When the communist system collapsed, individual ''republics'' proclaimed their independence—that is, the *majorities* in these entities did so. What remained of the machine of state, still dominated by the Serbs, tried to halt this disintegration. That effort soon failed, and new countries named Slovenia, Croatia, Bosnia, Macedonia, and, by subtraction, Serbia appeared on the map. Minorities in these new states, however, objected and, in Croatia and Bosnia, rose against those who were advancing statehood. The result was a catastrophic conflict at the very time that grandiose Euro-unification schemes were under way. Europe, which twice during the twentieth century had plunged the world into war and which vowed that genocide would ''never again'' cloud European horizons, failed this first test of its declarations. EU members, led by Germany, quickly accorded official recognition to the post-Yugoslavian states, even before the concerns of the frightened minorities could be considered. Then, European powers stood by as more than 250,000 people were killed, perhaps a million more were injured, ethnic cleansing and mass executions depopulated entire areas, refugees streamed into neighboring countries, and historic treasures were demolished. In the meantime, the conferees in Maastricht and Brussels poured champagne to herald the progress of the European Union.

In late 1996, the evolving map (Fig. 1-21) of former Yugoslavia consisted of five components: Slovenia, Croatia, Bosnia, Serbia, and Macedonia.

Slovenia The first to secede and proclaim independence, Slovenia is ethnically the most homogeneous of the republics but territorially small with only 8 percent (2 million) of former Yugoslavia's 23 million citizens. The most Western-appearing, this Alpine-mountain state produced as much as 20 percent of the former Yugoslavia's GNP from metallurgy, electronics, and motor vehicles. Slovenia is a well-defined nation-state with its own language, cultural landscape, primate city (Ljubljana), and social atmosphere.

Croatia This crescent-shaped territory with prongs along the Hungarian border and the Adriatic coast only partially reflects the distribution of Croats in the former Yugoslavia. In addition to more than 4 million Croats in the republic, another 800,000 are in neighboring Bosnia (Fig. 1-22). Substantial Serb minorities in Croatia made conflict inevitable when Croatia followed Slovenia and declared independence in 1991. Germany shortly thereafter recognized Croatia—an ally of Nazi Germany during World War II that fought against anti-Nazi Serbia. Based on communist-era industrialization and self-sufficiency in food, this was the most prosperous former Yugoslav republic after Slovenia, but the economy has been badly damaged by civil war and disruption of trade links.

Bosnia Centrally positioned and virtually landlocked, Bosnia is a cauldron of conflict, with 3.5 million inhabitants, of whom 49 percent are identified by the Muslim faith, 31 percent are Serbs, and 17 percent are Croats. The embattled, severely damaged capital of Sarajevo, holy place to Muslims and Orthodox Serbs and thus the Jerusalem of Bosnia, lies on the eastern margin of a wedge-shaped territorial remnant where Muslims remain in the majority. Croats and Muslims have been allied against the

The drama in former Yugoslavia continues. Much of Bosnia's infrastructure was destroyed in the early years of the conflict. When an agreement was reached to allow "peacekeeping" forces to enter, new bridges and roads had to be constructed to make this possible. United States Army engineers built this pontoon bridge spanning 700 yards (600 m) across the Sava River in just one week, allowing personnel and vehicles to enter Bosnia from Croatia. The bridge in the background was rendered unserviceable by bombing, and will require years for repairs. This photograph represents many similar situations in former Yugoslavia, where huge investments in reconstruction are needed.

Serbs who, at one time, had conquered 70 percent of Bosnia amid brutal ethnic cleansing. It was the Muslim-led government that declared independence in 1992, resulting in Serbian rebellion aided by the Serb-dominated state to the east. The civil war that followed claimed an estimated 250,000 lives and destroyed much of Bosnia's infrastructure and cultural heritage. Not until late 1995 did a U.S.-brokered agreement, reached in Dayton, Ohio, produce a shaky truce, policed by United Nations forces including more than 20,000 U.S. troops (see photo this page). But the old, newly-revived animosities may not be buried yet, and meanwhile Bosnia lies devastated, dependent on imported food and material, its industries (including the communist-era weapons industry) collapsed.

Serbia Largest and most populous of the republics (11 million), Serbia goes by several names, including Serbia-Montenegro (Montenegro is the coastal territory of the republic inhabited by about 700,000 Montenegrins, allies of the Serbs) and Yugoslavia, the Serbs having adopted the name of the defunct federation. Also part of Serbia are Vojvodina in the north, home to a large Hungarian minority, and Kosovo in the south, dominantly Albanian-Muslim and ruled oppressively by Serbian authorities (Fig. 1-20). The notion of a Greater Serbia has propelled a war of domination supported by this republic's 9 million Serbs that has pitted Serb against Slovene, Serb against Croat, Serb against Bosnian Muslim. International sanctions against the regime in the capital, Belgrade, have reduced this role; but, as noted, the conflicts go on. Serbia's economy, capable of producing a wide range of agricultural products and industrially developed on the communist heavy-industry model, has suffered severely, not only because of the nearby war and associated sanctions but also because the long-term system of interdependence with the neighboring republics has been destroyed.

Macedonia The independence of Macedonia, proclaimed in 1991, produced retaliation in Greece, which claimed that ''Macedonia'' is a Greek regional name and must not be used by any external area. The nearby Greek port of Thessaloniki was closed to Macedonian traffic, further damaging the already collapsing economy of this, the poorest of all the former Yugoslav republics. The matter was not resolved until mid-1995, but by then Macedonia had been reduced to a life of virtual subsistence. United Nations (including U.S.) troops were stationed in the capital, Skopje, to guard against any spillover effect of the conflicts raging to the north, but ultimately Macedonia's greatest challenge is likely to be internal. Its population of 2.2 million is 65 percent Macedonian and 21 percent Albanian (see Fig. 1-22), many of whom are Muslims. In this as in other parts of the world, a declining economy coupled with strong ethnic and cultural contrasts spells trouble.

Albania

Because of the severity of the strife that has engulfed former Yugoslavia, world attention has focused on that Adriatic country's disastrous devolution. But this subregion contains another potentially consequential country, Albania, an anachronism even in turbulent Eastern Europe. Alone among remnants of the Turkish Ottoman Empire, Albania became independent with a large (70 percent) Muslim majority. Alone among postwar European countries, it turned to China, not the Soviet Union, for its communist ideology. Alone among communist states, it remained tightly closed and retained its communist dogma long after its neighbors had opened their doors and altered their course.

Today, Albania still ranks lowest in all of Europe on many indices of economic well-being. A majority of its 3.6 million people (Albania has by far the fastest rate of population growth in the realm) subsist on livestock herding and farming on the one-seventh of this mountainous, earthquake-prone country that can be cultivated. But Albania's potential importance may not be measurable in economic terms. Figure 1-22 reveals why Albania may matter a great deal in the future: nearly half again as many Albanians as live in their country occupy areas of adjacent Serbia as well as Macedonia (Greece was moved to close its border with Albania when emigration mushroomed). Many of those external Albanians have ties to the motherland, and any conflict between Serbs or Macedonians with their Albanian minorities would quickly involve Albania itself. In 1996, the Albanian capital, Tirane, remained a virtually unaltered relic of another Eastern European era. As Yugoslavia's unraveling has proven however, change can come very rapidly in this volatile subregion.

Countries Facing the Black Sea

Four countries form Eastern Europe's Black Sea quadrant: Bulgaria, Romania, Moldova, and Ukraine (Fig. 1-21). All except Moldova have coastlines on the Black Sea, but none of their capital cities or core areas lies on the coast. This reflects an inward orientation that is characteristic of the subregion as a whole.

Bulgaria

In the south, bordering Turkey and Greece, lies Bulgaria, about the size of Tennessee, dominated by rugged topography except in the Danube lowland shared with Romania in the north. The Balkan Mountains, the east-west range that forms Bulgaria's backbone and separates the Danube and Maritsa drainage, has given the southern half of Eastern Europe its infamous name.

A Bulgarian state did not appear until 1878, when the Russian tsar's armies drove the Turkish rulers out of this area. The Slavic Bulgars, who constitute 86 percent of the population of nearly 8.4 million, never forgot who their liberators were, and during the Soviet period they proved to be compliant allies of Moscow. After the Soviet collapse, they reelected ex-communists to their government.

As elsewhere in the Balkans, Turkish minorities remain in Bulgaria, comprising about 9 percent of the population. The Bulgarians have not treated these people kindly: in 1974, the government issued a decree prohibiting use of the Turkish language in teaching; another closed the country's 1,300 mosques. In 1984, the Turks were forced to give up their Turkish names and take Slavic ones. This so-called assimilation policy led to hunger strikes and clashes in which Turks were killed. In 1989, when changing times opened the Turkish-Bulgarian border, more than 300,000 Turks fled across it as refugees. Another fertile field for irredentism had been plowed.

Until mid-century Bulgaria was essentially a peasant society, with more than three-quarters of the population engaged in farming. During the Soviet period, central planning, rapid industrialization on the communist model, and the collectivization of agriculture transformed the economy, which achieved self-sufficiency in food production, manufactured machinery and equipment for the Soviet Union, and exploited its bauxite, copper, and coal deposits. The post-Soviet period has been difficult for Bulgaria because its aging industries are failing and economic reform lags.

As the map shows, Bulgaria lies in a strategic location: near the entrance to the Black Sea, flanking the Danube River, and on the key land routes between Europe and Southwest Asia. Sofia, the old mountain-protected capital, lies on the railroad from Vienna and Belgrade to Istanbul. This relative location may in time become one of Bulgaria's leading assets.

Romania

More than twice as large and with more than twice the population of Bulgaria, Romania also is richer in raw materials and better connected to the outside world. None of this, however, has translated into a stronger economy or wealthier society. Romania is the ultimate tragedy of Soviet-dominated Eastern Europe; here communist totalitarianism reached its destructive extreme in the person of a dictator, Nicolae Ceausescu, who squandered Romania's wealth, destroyed much of the country's architectural heritage, and plunged the state into a social abyss from which it will take generations to recover.

Romania has a wide window on the Black Sea and a long border on the Danube River (although the capital, Bucharest, lies on neither). As Figure 1-2 shows, the country has a varied physiography. The resource-laden Carpathian Mountains sweep into Romania from the northwest in a broad arc. The Transylvanian Alps delimit the lower Danube Basin. Along their southern flank lie oil reserves (now nearly exhausted) that long made Romania an energy exporter.

As its name suggests, Romania was a Roman province in ancient times; Romanian is an outlier of the Romance subfamily of Indo-European languages, encircled by Slavic tongues (Fig. 1-6). The ethnic map (Fig. 1-22) reveals that Romania has substantial numbers of minorities, chiefly Hungarians in Transylvania (2 million, or about 8 percent) as well as smaller numbers of Ukrainians and Germans. The German minority was much larger at the end of World War II but was driven off during the postwar period.

Romania's economic geography is wretched. Despite the exhaustion of most of its oil reserves, natural gas still is available among a range of other raw materials including coal and iron ore; soil and climate could support substantial farm exports; timber remains in plentiful supply. But corruption and mismanagement were rife in Romania even before the postcommunist transition, and today Romania's reform program falters as the nation continues to suffer.

If Romania's economic geography is bleak, its social geography is tragic. During the communist period, much of the country's urban and rural cultural landscape was swept away to make room for massive public buildings and endless rows of featureless apartment blocks; in the villages, the drive to collectivize destroyed the fabric of society. In the cities, tens of thousands of unwanted orphans languish with minimal care in institutions, products of Ceausescu's repugnant population policies that required all women to bear five children. Once known as Eastern Europe's most civilized society, its capital the Paris of the region, Romania today is the basket case of the Balkans.

Moldova

A barely landlocked territory about twice the size of New Jersey, Moldova (formerly Moldavia) and its dominantly Romanian population were taken from Romania by the Soviets in 1940 and made into a Soviet Socialist Republic of the U.S.S.R. Soviet planners deprived Moldova of a Black Sea coast by leaving a narrow strip of coastline in Ukrainian hands (Fig. 1-21). Thus delimited, the country, independent since 1991, lies in the hinterland of the Ukrainian port of Odesa (Odessa).

Without industrial raw materials, Moldova remains a dominantly agrarian country known for its specialized crops and products such as fruits, tobacco, and wines. The capital, Chisinau (formerly Kishinev), is the center of what modest industrialization the Soviets introduced; apart from food processing, all else depends on raw-material imports.

But it is political geography that dominates life in Moldova today. After having struggled to get the Romanian language and its alphabet recognized by the Soviets, the

Moldovans overwhelmingly voted against reattachment to Romania in the post-Soviet period. Romanians now constitute only 65 percent of the population of 4.4 million, with about 14 percent Ukrainians and 13 percent Russians. Russians and Ukrainians are concentrated in the country's extreme east, between the Ukrainian border and the Dniester River (here called the Nistru). They have refused to participate fully in Moldova's political life, preferring talk of a Trans-Dniester Republic that would secede from Moldova. In 1996, this remained one of Eastern Europe's unresolved geopolitical dilemmas, risking stability in yet another part of the region.

Ukraine

As part of the new Eastern Europe, Ukraine is the region's most populous country, with 52 million inhabitants; it also is the largest territorially in the entire European realm. Its capital, Kiev (Kyyiv), is a major historic, political, and cultural focus. Briefly independent before the communist takeover in Russia, Ukraine regained its sovereignty as a much-changed republic in 1991. A land of famously fertile soils and Ukrainian farmers, it emerged from the Soviet period with a huge industrial complex and a diversified economy—as well as a large (22 percent) Russian minority.

Ukraine's border also changed during the Soviet era. In 1954, a Soviet dictator, as a "reward" for Ukraine's contribution to the Soviet Union's well-being, capriciously transferred the Crimea Peninsula from the Russian Soviet Republic to Ukraine. This important area (see Fig. 1-21) had been the historic home of descendants of Russia's Mongol invaders, the Crimean Tatars. They were accused of Nazi collaboration during World War II, and in 1945 they were ruthlessly expelled to Central Asia. Replacing them were more than a million Russians who, in 1954, found themselves under the jurisdiction of the Ukrainian S.S.R.

At the time, it seemed to matter little; both Russia and Ukraine, after all, were parts of the greater Soviet Union. But at the end of 1991 these Russians—as well as several million more living in Ukraine's industrialized east—suddenly found themselves residing in a foreign country. In truth, many had voted to approve of independence for Ukraine, but the implications of 1991 emerged in subsequent years. Before long, an autonomy movement arose in the Crimea Peninsula, with sympathizers in urban-industrial, Russified eastern Ukraine. Moscow hinted that a "revision" of the 1954 boundary might have to be sought.

In reality there are two Ukraines, and the challenge facing the government in Kiev is to keep them united and economically healthy. To get an idea of the regional contrasts, the much-dammed Dnieper River (Fig. 1-21) is a good geographic reference: to its west lies agrarian, rural,

The amount of environmental deterioration in the former Soviet Union became clear when former President Gorbachev's *glasnost* (openness) policy brought it into the open. Groundwater, stream water, seas, soils, and air had suffered severely from pollution ranging from chemical to nuclear. This view over an old residential district of the Ukrainian city of Donetsk symbolizes what has happened in the former U.S.S.R.: slag heaps now tower over the homes in this once-leafy suburb; smokestacks belching pollutants render the air acrid and gray.

Roman Catholic Ukraine, while east of it lies industrial, urban, Russian Orthodox Ukraine. Links with Russia in the west are few and indirect; links with Russia in the east continue to be numerous and active.

The economic dimension of this division already had evolved by the end of the nineteenth century, when coal mining in the east produced some 90 percent of the tsarist empire's high-quality coal. The Donets Basin (*Donbas* for short), where all this coal was mined, lay less than 200 miles (320 km) from vast deposits of high-quality iron ore at Kryvyy Rih (Krivoy Rog), and the industrial development of this zone was a high priority of communist planners. A cluster of major cities evolved, making eastern Ukraine a Soviet Ruhr.

As time went on, the highest quality reserves of Donbas-area coal and iron became depleted, and fuels and ores from Russia increasingly stoked the furnaces of eastern Ukraine. Indeed, the economic geographies of Russia and Ukraine were closely interwoven. Ukraine's shortage of oil and natural gas was met from Soviet supplies of these fossil fuels; Russia's chronic need for food stimulated large farm exports from Ukraine.

Ukraine's frontage on the Black Sea meant little to Soviet Ukraine other than as a break-of-bulk for imports and as a base (at the Crimean port of Sevastopol) of the joint military fleet. In the Soviet system, Ukraine looked and traded inward, not outward to the world. But today, Ukraine's Black Sea outlet is one of its leading geographic assets: the government in Kiev seeks to combine a loosening of the Russian connection with a strengthening of Ukraine's European links. Some Ukrainian leaders envisage Ukrainian membership in the European Union; significantly, the country is in the position of being able to reach the EU by land and sea without having to go through Russian territory.

Ukraine's determined reform efforts continue against some ominous backdrops, however. Dependence on Russian raw materials will long continue. The Russian minority remains a formidable force, always better educated and disproportionately influential in the republic's affairs. The energy problem is symbolized by the nuclear accident at Chornobyl (Chernobyl) north of Kiev in 1986, which wafted deadly radiation as far away as Poland and Norway, killed dozens, and poisoned thousands. At the time, Moscow's stonewalling and the Soviets' incompetent handling of the crisis galvanized anti-Russian feelings in Ukraine. Now the risk is rising that the measures taken to contain the reactor were inadequate and may lead to a repetition.

Massive, populous, and crucially located Ukraine is a potential juggernaut in the new Eastern Europe—and in the future of Europe as a whole.

With 582 million inhabitants in 38 countries, including some of the world's highest-income economies, a politically stable and economically integrated Europe would be a superpower in the twenty-first century. But Europe's political geography is anything but stable, as devolutionary forces and cultural conflict continue to trouble the realm. And economic integration so far involves less than half its countries, a process that will become more difficult as the European Union confronts applications from marginally qualifying states in the east. Europe always has been a realm of revolutionary change, and it remains so today.

■ PRONUNCIATION GUIDE ■

Abidjan (abb-ih-JAHN)
Acropolis (uh-CROP-uh-liss)
Adriatic (ay-dree-ATTIC)
Aegean (uh-JEE-un)
Agios Dimitrios (AH-jee-ohss deh-MEE-
    tree-ohss)
Allemanni (alla-MAH-nee)
Ancona (ahng-KOH-nuh)
Andalusia (unda-looh-SEE-uh)
Antwerp (ANN-twerp)
Appennines (APP-uh-neinz)
Arc de Triomphe (arc duh
    tree-AWMFF)
Ardennes (ar-DEN)
Asturias (uh-STOOR-ree-uss)
Athena (uh-THEENA)
Autostrada (AU-toh-strah-duh)
Avignon (ah-veen-YAW)
Azerbaijan (ah-zer-bye-ZHAHN)
Baden-Württemberg (BAHDEN
    VEERT-um-bairk)
Baku (bah-KOO)
Balkan (BAWL-kun)
Barcelona (bar-suh-LOH-nuh)
Basque (BASK)
Bavaria (buh-VAIRY-uh)
Belarus (bella-ROOSE)
Belarussia (bella-RUSH-uh)
Bern (BAIRN)
Biafra (bee-AH-fruh)
Bilbao (bil-BAU)
Bohemia (boh-HEE-mee-uh)
Bordeaux (boar-DOH)
Bosnia-Herzegovina (BOZ-nee-uh
    hert-suh-go-VEE-nuh)
Brandenburg (BRAHN-den-boorg)
Bratislava (BRUDDIS-lahva)
Brazzaville (BRAHZ-uh-veel)
Bremen (BRAY-mun)
Brno (BURR-noh)
Bucharest (BOO-kuh-rest)
Budapest (BOODA-pest)
Bulgaria (bahl-GHAIR-ree-uh)
Byzantine (BIZZ-un-teen)
Calabria (kuh-LAH-bree-uh)
Calais (kah-LAY)
Cantabria (kahn-TAH-bree-uh)
Carpathians (kar-PAY-theons)
Cartagena (karta-HAY-nuh)
Castile-La Mancha (kuh-STEEL
    luh-MAHN-chuh)
Castile-Leon (kuh-STEEL lay-OAN)
Catalan (katta-LAHN)

Catalonia (katta-LOH-nee-uh)
Caucasian (kaw-KAY-zhun)
Caucasus (KAW-kuh-zuss)
Ceausescu, Nicolae (chow-SHESS-koo,
    NICK-oh-lye)
Celt(ic) (KELT [ick])
Centrifugal (sen-TRIFFA-gull)
Centripetal (sen-TRIPPA-tull)
Champagne (sham-PANE)
Champs Elysées (SHAWZ elly-ZAY)
Charisma (kuh-RIZZ-muh)
Charleroi (SHARL-rwah)
Chisinau (kee-shih-NAU)
Chornobyl (CHAIR-noh-beel)
Conurbation (konner-BAY-shun)
Copenhagen (koh-pen-HAHGEN)
Córdoba (KORR-doh-buh)
Corsica (KORR-sih-kuh)
Crete (KREET)
Crimea (cry-MEE-uh)
Croat (KROH-aht/KROH-at)
Croatia (kroh-AY-shuh)
Cyclades (SIK-luh-deeze)
Cyrillic (suh-RILL-ick)
Czech (CHECK)
Czechoslovakia (check-uh-sloh-VAH-
    kee-uh)
Dakar (duh-KAHR)
Dalmatia (dall-MAY-shuh)
Danube (DAN-yoob)
De Gaulle, Charles (duh-GAWL,
    SHARL)
Département (day-pahrt-MAW)
Devolution (dee-voh-LOOH-shun)
Dinaric (dih-NAHR-rick)
Dnieper (duh-NYEPP-er)
Dniester (duh-NYESS-truh)
Dnipropetrovsk (duh-nep-roh-puh-
    TRAWFSSK)
Donbas (DAHN-bass)
Donets (duh-NETTS)
Donetsk (duh-NETTSK)
Druzba (DROOZE-bah)
Duchy (DUTCH-ee)
Edinburgh (EDDIN-burruh)
Eire (AIR)
Elbe (ELB)
Ellesmere (ELZ-mear)
Ellinikon (eh-LINNY-konn)
England (ING-glund)
Entrepôt (AHNTRA-poh)
Estonia (eh-STONY-uh)
Eurasia (yoo-RAY-zhuh)

Extremadura (ess-truh-muh-
    DOORA)
Fjord (FYORD)
Frankfurt (FRUNK-foort)
Gabcikovo (gahb-CHEE-kuh-voh)
Gaelic (GALE-ick)
Galicia (guh-LEE-see-uh)
Garonne (guh-RON)
Gdansk (guh-DAHNSK)
Gdynia (guh-DINNY-uh)
Geneva (jeh-NEE-vuh)
Genoa (JENNO-uh)
Glasgow (GLASS-skau)
Gneiss (NICE)
Godthab (GAWT-hawb)
Göteborg (GOAT-uh-borg)
Grande Arche (GRAWND-ARSH)
Greenwich (GREN-itch)
Guadalquivir (gwahddle-kee-
    VEER)
Gulag (GHOO-lahg)
Guiana (ghee-AH-nah)
Hague (HAIG)
Hamburg (HAHM-boorg)
Hanseatic (han-see-ATTIC)
Hegemony (heh-JEH-muh-nee)
Helsinki (hel-SINKEE)
Hesse (HESS)
Iberia (eye-BEERY-uh)
Igneous (IG-nee-uss)
Île de France (EEL duh-FRAWSS)
Île de la Cité (EEL duh-la-see-TAY)
Inuit (IN-yoo-it)
Inverness (IN-ver-ness)
Irrendentism (irruh-DEN-tism)
Islington (IZZ-ling-tunn)
Istanbul (iss-tum-BOOL)
Jutland (JUT-lund)
Kalaallit Nunaat (kuh-LAHT-lit
    noo-NAT)
Kaliningrad (kuh-LEEN-in-grahd)
Kattegat (KAT-ih-gat)
Kenyatta, Jomo (ken-YUTTA, JOH-
    moh)
Keratsinion (keh-rut-SINNY-onn)
Kiev (KEE-yeff)
Kiruna (kih-ROONA)
Kishinev (KISH-uh-neff)
Klaipeda (KLYE-puh-duh)
Königsberg (KAY-nix-bairk)
Kosovo (KAW-suh-voh)
Krivoy Rog (krih-voy-ROAG)
Kyyiv (KAY-eff)

La Défense (lah-day-FAWSS)
Land (LAHNT)
Länder (LEN-derr)
Latvia (LATT-vee-uh)
Le Havre (luh-HAHV)
Leipzig (LYPE-sik)
Leith (LEETH)
Liechtenstein (LIK-ten-shtine)
Liège (lee-EZH)
Lingua franca (LEEN-gwuh
    FRUNK-uh)
Lithuania (lith-oo-AINY-uh)
Ljubljana (lee-oo-blee-AHNA)
Loess (LERSS)
Loire (luh-WAHR)
Lombardy (LOM-bar-dee)
Louvre (LOOV)
Luleå (LOO-lee-oh)
Luxembourg (LUX-em-borg)
Lyon (lee-AW)
Maastricht (mah-STRICT)
Macedonia (massa-DOH-nee-uh)
Madagascar (madda-GAS-kuh)
Madrid (muh-DRID)
Magyar (MAG-yahr)
Main (MINE)
Málaga (MAHL-uh-guh)
Mao Zedong (MAU zee-DUNG)
Maritsa (muh-REET-suh)
Marseille (mar-SAY)
Meseta (meh-SAY-tuh)
Metamorphic (metta-MOR-fik)
Meuse/Maas (MERZZ/MAHSS)
Mezzogiorno (met-soh-JORR-noh)
Milan (muh-LAHN)
Milano fiori mee-LAHNO-fyor-ree
Moldavia (moal-DAY-vee-uh)
Moldova (moal-DOH-vuh)
Montenegro (mon-teh-NEE-groh)
Montpellier (maw-pell-YAY)
Moravia (more-RAY-vee-uh)
Moselle (moh-ZELL)
Munich (MYOO-nik)
Muslim (MUZZ-lim)
Nehru, Jawaharlal (NAY-roo,
    juh-WAH-hur-lahl)
Nigeria (nye-JEERY-uh)
Nistru (NEE-stroo)
Nord-Pas de Calais (NORD pah-
    duh-kah-LAY)
Notre Dame (NOH-truh DAHM)
Nouméa (noo-MAY-uh)
Nuuk (NOOK)

Oblast (OB-blast)
Ochre (OAKER)
Oder-Neisse (OH-der NICE)
Odesa (oh-DESSA)
Oise (WAHZ)
Oslo (OZ-loh)
Oslofjord (OZ-loh-fyord)
Ossies (OH-seez)
Ostrava (AW-struh-vuh)
Papeete (pahp-ee-ATE-tee)
Parliament (PAR-luh-ment)
Peloponnesus (pelloh-puh-
    NEEZE-uss)
Peron, Juan (puh-ROAN, WAHN)
Phaleron (fuh-LEER-un)
Phidias (FIDDY-uss)
Piraeus (puh-RAY-uss)
Provence (pro-VAHNSS)
Prague (PRAHG)
Prut (PRROOT)
Pyrenees (PEER-unease)
Randstad (RUND-stud)
Réunion (ray-yoon-YAW)
Reykjavik (RAKE-yah-veek)
Rhine (RYNE)
Rhône-Saône (ROAN say-OAN)
Riga (REEGA)
Rio de Janeiro (REE-oh day zhah-
    NAIR-oh)
Rioja (ree-OH-hah)
Romania (roh-MAIN-yuh)
Romansch (roh-MAHN-ssh)
Rouen (roo-AW)
Ruhr (ROOR)
Saarland (ZAHR-lunt)
Saigon (sye-GAHN)
Saint-Denis (san-deh-NEE)
Sarajevo (sahra-YAY-voh)
Sardinia (sahr-DINNY-uh)
Sava (SAH-vuh)
Saxony (SAX-uh-nee)
Seine (SENN)
Serbia (SER-bee-uh)
Sevastopol (seh-VASS-toh-paul)
Seville (suh-VILL)
Sicily (SISS-uh-lee)
Silesia (sye-LEE-zhuh)
Skagerrak (SKAG-uh-rak)
Skopje (SKAWP-yay)
Slav(ic) (SLAHV [ick])
Slovakia (slow-VAH-kee-uh)
Slovenia (slow-VEE-nee-uh)
Sofia (SOH-fee-uh)

Stockholm (STOCK-hoam)
Strasbourg (STRAHSS-boorg)
Stuttgart (SHTOOT-gart)
Sudeten (soo-DAYTEN)
Tahiti (tuh-HEE-tee)
Tallinn (TALLEN)
Tampere (TAHM-puh-ray)
Tangenziale (tahn-jen-tsee-AHL-lay)
Tatar (TAHT-uh)
Tatra (TAHT-truh)
Thames (TEMZ)
Thamesmead (TEMZ-meed)
Thessaloniki (thess-uh-luh-NEE-kee)
Tiber (TYE-ber)
Tibet (tuh-BETT)
Tilbury (TILL-buh-ree)
Tirane (tih-RAHN-uh)
Tito, Josip Broz (TEE-toh, yossip
    broz)
Train à grande vitesse (TRAN ah-
    grawnd-vee-TESS)
Trondheim (TRAHN-hame)
Tula (TOO-luh)
Tunisia (too-NEE-zhuh)
Turin (TOOR-rin)
Turku (TOOR-koo)
Tuscany (TUSS-kuh-nee)
Tyrol (tih-ROLL)
Ukraine (yoo-CRANE)
Ural (YOOR-ull)
Utrecht (YOO-trekt)
Valencia (vuh-LENN-see-uh)
Veneto (VENN-uh-toh)
Versailles (vair-SYE)
Vienna (vee-ENNA)
Vilnius (VILL-nee-uss)
Vistula (VIST-yulluh)
Vojvodina (VOY-vuh-deena)
Von Thünen, Johann Heinrich
    (fon-TOO-nun, YOH-hahn
    HINE-rish)
Vyborg (VEE-borg)
Wallonia (wah-LOANY-uh)
Walloon (wah-LOON)
Weber (VAY-buh)
Weser (VAY-zuh)
Westphalia (west-FAIL-yuh)
Westermann (VESS-tair-munn)
Yonne (YAHN)
Yugoslavia (yoo-goh-SLAH-vee-uh)
Zeeland (ZAY-lund)
Zuider Zee (ZYDER ZEE)
Zürich (ZOOR-ick)

CHAPTER 2

Russia's Fractious Federation

copyright © Rand McNally, 1993

The name *Russia* evokes cultural-geographic images of the past: terrifying tsars, conquering Cossacks, Byzantine bishops, rousing revolutionaries, clashing cultures. Russia repulsed the Tatar hordes, expanded into a vast contiguous empire, defeated Napoleon, succumbed to communism, and, when the communist system failed, lost most of its imperial domain. Today, Russia is in a tenuous—not to say dangerous—transition.

Russia is a land of vast distances, bitter cold, impenetrable forests, treacherous mountains, isolated outposts, and remote frontiers. Precommunist Russian culture was a culture of strong nationalism, resistance to change, political despotism, bejeweled aristocrats, desperate serfs. Great authors lamented the plight of the poor; major composers celebrated the indomitable Russian people and, as Tchaikovsky did in his *1812 Overture*, commemorated their victories over foreign foes.

Under the tsars, Russia grew from nation into empire. The insatiable demands of these rulers for wealth, territory, and power sent Russian armies across the plains of Siberia, through the deserts of interior Asia, and into the mountains along Russia's rim. Russian pioneers ventured even farther, entering Alaska, traveling down the Pacific coast of North America, and planting the Russian flag near San Francisco in that year of triumph, 1812. As Russia's empire expanded, its internal weaknesses gnawed at the power of the tsars. Serfs rebelled. Unpaid (and poorly fed) armies mutinied. When the tsars tried to initiate reforms, the aristocracy objected. The empire at the beginning of the twentieth century was ripe for revolution, which began in 1905 (see box titled ''The Soviet Union, 1924–1991'').

The tsar was overthrown in 1917, and a struggle for control followed. The victorious communists led by V. I.

THE SOVIET UNION, 1924–1991

■ ■ ■

For 67 years Russia was the cornerstone of the *Soviet Union*, the so-called Union of Soviet Socialist Republics (U.S.S.R.). The Soviet Union was the product of the Revolution of 1917, when more than a decade of rebellion against the rule of Nicholas II led to the tsar's abdication. Russian revolutionary groups were called *soviets* ("councils"), and they had been active since the first workers' uprising in 1905. In that crucial year, thousands of Russian workers marched on the tsar's palace in St. Petersburg in protest, and soldiers opened fire on them. Hundreds were killed and wounded. Russia descended into chaos.

The tsar's abdication in 1917 was forced by a coalition of military and professional men. After the tsar, Russia was ruled briefly by a Provisional Government. In November 1917, the country held its first democratic election ever—and, as it turned out, the last for more than 70 years to come.

The Provisional Government allowed the return to Russia of exiled activists in the Bolshevik camp (there were divisions among the revolutionaries): Lenin from Switzerland, Trotsky from New York, and Stalin from internal exile in Siberia. In the political struggle that ensued, Lenin's Bolsheviks gained control over the revolutionary soviets, and this ushered in the era of communism. In 1924, the new communist empire was formally renamed the Union of Soviet Socialist Republics, or Soviet Union in shorthand.

Lenin the organizer (who died in 1924) was succeeded by Stalin the tyrant, and many of the peoples under Moscow's control suffered unimaginably. In pursuit of communist reconstruction, Stalin and his elite starved millions of Ukrainian peasants to death, forcibly relocated entire ethnic groups, and exterminated "uncooperative" or "disloyal" peoples. The full extent of these horrors will never be known. Many of the country's most creative people were executed.

On December 25, 1991, the inevitable occurred: the Soviet Union ceased to exist, its economy a shambles, its political system shattered, the communist experiment a failure. The last Soviet President, Mikhail Gorbachev, resigned, and the Soviet hammer-and-sickle flying on the flagstaff atop the Kremlin was lowered for the last time, immediately replaced by the Russian tricolor. A new and turbulent era commenced, but Soviet structures and systems will long cast their shadows over transforming Russia.

■ ■ ■

Lenin soon swept away many vestiges of the Russia of the past. The Russian flag disappeared. The last tsar and his entire family were executed. The old capital of Russia, St. Petersburg, was renamed Leningrad in honor of the revolutionary leader. The interior city of Moscow was chosen as the new capital for a country with a new name, the *Soviet Union*. Eventually this Union came to consist of 15 political entities, each a Soviet Socialist Republic. Russia was just one of these republics, and so the name Russia disappeared from the international map.

Russia remained first among equals, however. Not for nothing had the communist revolution been known as the *Russian* Revolution. The Soviet Empire was the legacy of the tsars' expansionism, and the new communist rulers were Russians first and foremost. As the new Union was laid out, Russians moved by the millions to the fringes of the empire—the fringes where lay the "republics" formed from the tsars' colonies (or conquered later by the Red Army). The Russification of the Soviet Empire proceeded just as the British and French and Belgians and Portuguese were also moving in large numbers to their colonies. The Soviet Union was a Russian colonial empire.

The world's great colonial empires did not endure, and (as we predicted in earlier editions of this book) neither did the Soviet Union. Non-Russian peoples stirred in opposition to Moscow, their nationalism or ethnic consciousness mobilized by memories of a long-suppressed past. Lithuanians, Ukrainians, Georgians, and other peoples enmeshed in the Soviet Empire moved to throw off the communist yoke. And in Russia itself, nationalism stirred as well—in opposition not to other peoples of the empire but to the communist system that for nearly 70 years had bound them together.

Like the last tsar long ago, the last communist dictator, Mikhail Gorbachev, tried to control and channel the forces of change that swept through his country. And like his predecessor, he failed as events overtook him (unlike Tsar Nicholas II, however, Gorbachev lived to tell the tale in books and lectures). And so today, Russia's national flag flies once again over the Kremlin, the empire is no more, and the challenge facing Russia involves political stability and economic redirection, not Cold War politics and superpower competition.

Fragments of the Soviet Empire now tend to divert their gaze away from Moscow. The six westernmost Soviet Socialist Republics (although to a lesser degree Belarus) are looking westward, to Europe, as they seek a new place in the international community. The five Central Asian republics now constitute a distinct geographic region, *Turkestan*, where Islam is reviving and links to the Muslim world are strengthening. The three republics between the Black and the Caspian seas, wedged between Russia to the north and Turkey and Iran to the south, lie in the Transcaucasian Transition Zone, part of neither the Russian geographic realm nor the Islamic realm to the south (Fig. I-12).

MAJOR GEOGRAPHIC QUALITIES OF RUSSIA

■ ■ ■

1. Russia is by far the largest territorial state in the world. Its area is nearly twice as large as the next-ranking country (Canada).

2. Russia is the northernmost large and populous country in the world; much of it is very cold and/or very dry. Extensive rugged mountain zones separate Russia from warmer subtropical air, and the country lies open to Arctic air masses.

3. Russia was one of the world's major colonial powers. Under the tsars, the Russians forged the world's largest contiguous empire; this empire was taken over and expanded by the Soviet rulers who succeeded the tsars.

4. For so large an area, Russia's population of less than 150 million is comparatively small. The population remains heavily concentrated in the westernmost one-fifth of the country.

5. Development in Russia is concentrated in the western part of the country, west of the Ural Mountains; here lie the major cities, leading industrial regions, densest transport networks, and most productive farming areas. National integration and economic development east of the Urals extend mainly along a narrow corridor that stretches from the southern Urals region to the southern Far East around Vladivostok.

6. Russia is a multicultural state with a complex domestic political geography. Twenty-one internal republics, originally based on ethnic clusters, continue to function as politico-geographical entities.

7. Its large territorial size notwithstanding, Russia suffers from land encirclement within Eurasia; it has few good and suitably located ports.

8. Regions long part of the Russian and Soviet empires are realigning themselves in the postcommunist era. Eastern Europe and the heavily Muslim Southwest Asia realm are encroaching on Russia's imperial borders.

9. The failure of the Soviet communist system left Russia in economic disarray. Many of the long-term components described in this chapter (food-producing areas, railroad links, pipeline connections) broke down in the transition to the post-communist order.

10. Russia long has been a source of raw materials but not a manufacturer of export products, except weaponry. Very few Russian (or Soviet) automobiles, televisions, cameras, or other consumer goods reach world markets.

■ ■ ■

■ ■ ■ ■ ■ ■ ■ ■ ■ ■ ■ **FOCUS ON A SYSTEMATIC FIELD** ■ ■ ■ ■ ■ ■ ■ ■ ■ ■ ■

Climatology

Climate and weather are much in the news these days. Worldwide concern over long-term global warming has made the *greenhouse effect* a household word. The possibility that human pollution of the atmosphere is trapping more heat, preventing its escape, and enhancing this greenhouse function has produced international efforts to control the sources. As the number and range of weather extremes reported from around the world increase, scientists look for evidence that the global environmental system shows signs of instability. Unusual droughts in Africa and floods in the Americas are linked to the behavior of an ocean current off South America, and by extension the tropical Pacific Ocean, in what has become known as ENSO, the *El Niño/Southern Oscillation*.

All this is part of a branch of physical geography called **climatology**. Climatologists analyze the distribution of climatic conditions over the Earth's surface and investigate the processes that underlie this spatial arrangement. They trace environmental changes of the past and predict what may lie ahead for our planet. Glaciation, desertification, and hurricane tracking are some of the topics they pursue. As geographers, climatologists study the relationships between human societies and the climatic environments that prevail where they live. Understanding the regional climatology of a country can be a key to explaining its population distribution, cultural landscapes, and economic geography. For all our technological progress, climate and weather still are powerful influences over our lives.

The distinction between *weather* and *climate* is important. The term *weather* refers to atmospheric conditions occurring at a given place at a particular time: temperature, precipitation, humidity, wind factors. The *climate* of that place is a composite of these weather data, recorded over a long term as a set of averages. Such records allow climatologists to draw maps at various levels of generalization, from the global (such as Fig. I-7) to the regional and the local. A careful reading of the map and its legend reveals the strong contrasts between *humid equatorial*, *humid temperate*, and *cold polar* climates.

Figure I-7, however, is a kind of cartographic snapshot, today's version of a constantly changing distribution. Just 10,000 years ago, the map would have looked quite different. Icesheets were then receding from North America, and the Midwest remained a frigid tundra environment. If theories about enhanced greenhouse warming are proven correct, the map may change radically over the next century. Computer models are used to predict what future maps may look like.

To interpret the global climatic map, we must understand the processes that create it. Atmosphere and ocean are linked in a giant system that redistributes warmth (solar radiation) from the tropics to higher latitudes. The Earth's rotation sets up circulation cells in the oceans (*gyres*) and in the atmosphere (pressure-driven *highs* and *lows*). Globe-encircling wind currents move the atmospheric cells from east to west in tropical latitudes (the *trade winds*) and from west to east in the middle latitudes (the *westerlies*). The irregular outlines of the continents, and the variable relief of the landmasses, complicate the system. But high above the ground, high-speed *jet streams*, tube-like rivers of air that snake their way around the globe, function to compensate for congestion in the lower atmosphere.

Figure I-6 reflects a critical feature of the global environmental system: air contains moisture, and air masses driven by prevailing winds carry moisture onto the landmasses. Where humid air collides with high topography and rises to cross the mountains, its cooling is accompanied by condensation and precipitation. When the air mass moves on into the interior of the continent, it already has lost much of its moisture and can produce much less rain or snow. Interiors of continents, especially outside the tropics and walled off by mountains, thus tend to be dry.

Now turn to Figure I-7 and read on the map about Russia's climatic misfortune. By the time the humid air masses that bring plentiful moisture to Western Europe reach the longitude of St. Petersburg, they have lost much of their water. Western Russia is the realm's best-watered zone nevertheless, receiving between 20 and 40 inches (50 to 100 cm) of precipitation annually (Fig. I-6). Eastward, beyond the Ural Mountains, Russia is hemmed in between polar cold and inner Asian deserts. No protective mountain barriers shield Russia from Arctic air masses, and in the south, tall mountains prevent moist air from the Indian Ocean from reaching its interior. Only in the area between the Black and the Caspian seas are small parts of Russia warm enough to be mapped as humid temperate climate. Everywhere else, coldness, short summer growing seasons, and long winter nights dominate life in this most populous of high-latitude countries.

The climatic map goes far to explain why the majority of the Russian population today remains concentrated in the western one-fifth of Russia's gigantic territory, why agriculture remains a critical problem, and why, east of the Urals, the population extends ribbon-like along the southernmost margins of the country, always in the shadow of the region whose name is synonymous with climatic extremes—Siberia. Russian tsars fought for warm-water ports because the country's high-latitude harbors froze up for several months each year; later, Soviet communist planners tried to divert entire rivers to bring cultivation to the empire's desert lands. Russia is a storehouse of mineral wealth, but it faces a host of environmental problems that can be summarized under one rubric: its climatology.

■ ■ ■

DEFINING THE RUSSIAN REALM

RUSSIA'S DIMENSIONS

Even without its 14 former Soviet co-republics, territorially Russia is by far the largest country in the world. From the Bering Sea near Alaska to the Gulf of Finland, Russia stretches across 11 time zones. From well inside the Arctic Circle, Russia reaches southward to the latitude of Salt Lake City. It is a land of almost unimaginable dimensions, more than twice as large as the United States or China.

Russia also is a land of vast empty spaces. The country's population of 145 million is far outnumbered by China's 1.3 billion and India's 960-plus million. And, as Figure I-9 reveals, the Russian population still is strongly concentrated in what is sometimes called ''European'' Russia—Russia west of the Ural Mountains, which divide the west from Siberia (see map, p. 108). The name Siberia, which means ''sleeping land,'' is appropriate for a vast area in which the population still lives in isolated clusters and along discontinuous ribbons. If the Russia of the tsars and the Soviet Union of the communists had something in common, it was their desire to populate the east, to strengthen the Russian presence in that remote frontier. Over time an eastward vanguard of settlements did emerge, the largest ones along two railroads: one built by the tsars before the revolution and the other constructed by the communists afterward. The suitably named city of Novosibirsk is such a center, one of many trans-Ural places whose names start with *novo* (''new'').

Russia is the northernmost large and populous country on Earth, and its land lies unprotected by natural barriers against the onslaught of Arctic air masses. Moscow lies farther north than Edmonton, Canada, and St. Petersburg (the former Leningrad) lies at latitude 60° North—the latitude of the southern tip of Greenland. Winters are long, dark, and bitterly cold in most of Russia; summers are short and growing seasons limited. To make things worse, rainfall ranges from modest to minimal because the warm, moist air carried from the North Atlantic Ocean across Europe loses most of its warmth and much of its moisture by the time it reaches western Russia. (The term **continentality** is used to describe such an inland climatic environment remote from moderating maritime influences.) Drought, variable rainfall, and temperature extremes, therefore, have plagued Russian farmers as long as they have tilled the land.

The tsars embarked on their imperial conquests in part because of Russia's relative location: Russia always lacked warm-water ports. Their southward push might have reached the Persian Gulf or even the Mediterranean Sea, had the Revolution not intervened. Tsar Peter the Great

envisaged a Russia open to trading with the entire world; he developed St. Petersburg into Russia's leading port. But in truth, Russia's historical geography is one of remoteness from the mainstreams of change and progress, and of self-imposed isolation. Not even a string of warm-water ports would have been likely to transform Russia into an outward-looking, trading state. The tsars' objectives were primarily strategic, not economic.

From the Field Notes

"Walking the streets of St. Petersburg in 1994, more than 30 years after our first trip to the city, brought home the magnitude of the changes after 1991. Not only Nevsky Prospekt, the city's main street seen here, but other streets as well, are crowded with cars and pedestrians. Advertisements are everywhere. Small makeshift stalls selling goods ranging from books to family memorabilia line the sidewalks. Luxury goods and services have made their appearance, and the newly wealthy shop at recently opened, swank stores. Also in evidence are the newly destitute and desperate; just down the street from this scene a little girl stood playing the violin all day, her father leaning against the wall nearby, waiting to collect the coins passersby gave her. Crime has escalated as well, and extortion threatens many new small businesses. Like other Russian cities, St. Petersburg ('Peter' as locals call it) is in post-Soviet ferment."

AN IMPERIAL MULTINATIONAL STATE

Centuries of Russian expansionism did not confine itself to empty land or unclaimed frontiers. The Russian state became an imperial power that annexed and incorporated peoples of many nationalities and cultures. This was done by force of arms, by the overthrow of uncooperative rulers, by annexation, and by stoking the fires of ethnic conflict. By the time the ruthless Russian regime began to face revolution among its own people, the tsars held sway over the largest contiguous empire on Earth. Tsarist Russia was a hearth of **imperialism**, and its empire contained peoples representing more than 100 nationalities. The winners in the ensuing revolutionary struggle—the communists who forged the Soviet Union—did not liberate these subjugated peoples. Rather, they changed the empire's framework, binding the peoples colonized by the tsars into a new system that would in theory give them autonomy and identity. In practice, it doomed those peoples to bondage and, in some cases, extinction.

A Eurasian Heritage

The historical geography of this turbulent realm focuses on Russia west of the Urals and also involves Ukraine and other neighboring areas. Although our knowledge of Russia before the Middle Ages is only fragmentary, it is clear that peoples moved in great migratory waves across the plains on which the modern state eventually was to emerge. The dominant direction seems to have been from east to west; many groups came from interior Asia and left their imprints on the makeup of the population. Scythians, Sarmatians, Goths, and Huns came, settled, fought, and were absorbed or driven off. Eventually, the Slavs emerged as the dominant people in what is today Ukraine. From this base, on the productive soils north of the Black Sea, they expanded their domain.

Important centers in the evolution of Slavic states were Novgorod, situated on Lake Ilmen in the north and linked to the Hanseatic trade network, and Kiev (Kyyiv) on the Dnieper River in the heartland of the Slavs and in the zone of contact between the forests of middle Russia and the grassland steppes to the south. Kiev also had centrality, serving as a meeting place for Scandinavian and Mediter-

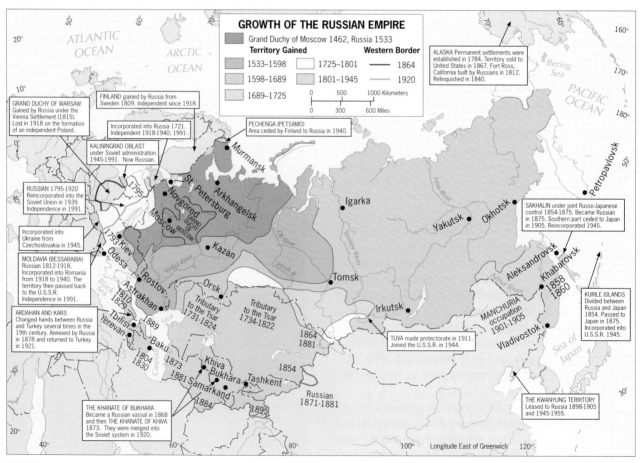

FIGURE 2-1

ranean Europe. Novgorod, like Kiev, was a princedom known locally as a *Rus*. During the eleventh and twelfth centuries, the Kievan Rus and the Novgorod Rus united to form a large and comparatively prosperous regional state.

Mongol Invasion

Prosperity, however, attracts competition, and the Kievan Rus suffered from internal division and external invasion. The external threat came from the Mongol Empire far to the east in interior Asia, which had been building under Gengis Khan. The Tatar hordes rode into the Kievan Rus, and the city as well as the state fell in 1240. Many Russians fled into the forests, where the horsemen of the open steppes were far less threatening. In forested middle Russia, a number of weak feudal states arose, many of them ruled by princes who paid tribute to the Tatars in order to be left in relative peace.

Moscow lay at the center of one of these Russes, a place with superior locational advantages on the river route to Novgorod and at the center of other Russian settlements. The Moscow River gave the site defensibility, and when the Mongols challenged Moscow late in the fourteenth cen-

tury and the Muscovites prevailed, Moscow's primacy was assured.

Soon Moscow's geographic advantages began to be felt. During the fifteenth century, Moscow's ruler was able to take control of Novgorod. But Muscovy became a real tsardom under the reign of Ivan (IV) the Terrible (1547–1584), who began Russia's advances into non-Russian territory; this campaign did not end until well after the tsarist regime was terminated by the Revolution of 1917. Ivan first established control over the entire basin of the important Volga River to the east, inflicted heavy defeats on the Muslim Tatars, and then pushed eastward across the Urals into Siberia.

Cossacks

This eastward expansion of Russia was spearheaded by a relatively small group of seminomadic peoples who came to be known as Cossacks and whose original home was in present-day Ukraine. Opportunists and pioneers, they sought the riches of the eastern frontier, chiefly fur-bearing animals, as early as the sixteenth century. By the middle of the seventeenth century they reached the Pacific Ocean, defeating Tatars in their path and consolidating their gains by constructing *ostrogs*, strategic fortified waystations along river courses. Before the eastward expansion halted in 1812 (Fig. 2-1), the Russians had moved across the Bering Strait to Alaska and down the western coast of North America into what is now northern California (see box titled "Russians in North America").

RUSSIANS IN NORTH AMERICA

■ ■ ■

The first white settlers in Alaska were Russians, not Western Europeans, and they came across Siberia and the Bering Strait, not across the Atlantic and North America. Russian hunters of the sea otter, valued for its high-priced pelt, established their first Alaskan settlement at Kodiak Island in 1784. Moving southward along the North American coast, the Russians founded additional villages and forts to protect their tenuous holdings until they reached as far as the area just north of San Francisco Bay, where they built Fort Ross in 1812.

But the Russian settlements were isolated and vulnerable. European fur traders began to put pressure on their Russian competitors, and Moscow found the distant settlements a burden and a risk. In any case, American, British, and Canadian hunters were decimating the sea otter population, and profits declined. When U.S. Secretary of State William Seward offered to purchase Russia's holdings in 1867, St. Petersburg quickly agreed—for $7.2 million. Thus Alaska, including its lengthy southward coastal extension, became United States territory and, eventually, the 49th State. Although Seward was ridiculed for his decision—Alaska was called "Seward's Folly" and "Seward's Icebox"—his reputation was redeemed when gold was discovered there in the 1890s. The twentieth century has proved Seward's action one of great wisdom, strategically as well as economically. At Prudhoe Bay off Alaska's northern Arctic slope, large oil reserves are still being exploited. And like Siberia, Alaska probably contains other yet-unknown riches.

■ ■ ■

Tsar Peter the Great

By the time Peter the Great took over the leadership of Russia (he ruled from 1682 to 1725), Moscow already lay at the center of a great empire—great, at least, in terms of the territories under its control. The Islamic threat had been ended with the defeat of the Tatars, who were now reduced to remnant strongholds. The Russian Orthodox Church was represented in the cultural landscape by distinctive religious architecture and in the ruling elite by powerful bishops.

Peter consolidated Russia's gains and hoped to make a modern, European-style state out of his loosely knit country. He built St. Petersburg as a **forward capital**, placing it on the doorstep of Swedish-held Finland, fortifying it with major military installations on the nearby island of Kronstadt, and making it Russia's leading port.

Tsar Peter the Great was an extraordinary leader, in many ways the founder of modern Russia. In his desire to remake Russia—to pull it from the forests of the interior to the waters of the west, to open it to outside influences, to relocate its population—he left no stone unturned. Prominent merchant families were forced by law to move from other cities to St. Petersburg. Ships and wagons entering the city had to bring building stones as an entry toll. The tsar himself, aware that the future of Russia as a major force lay in strength at sea as well as power on land, went to the Netherlands to work as a laborer in the famed Dutch shipyards to learn how ships were most efficiently built. Meanwhile, the tsar's forces continued to conquer people and territory: Estonia was incorporated in 1721, widening Russia's window to the west, and major expansion occurred south of the city of Tomsk (Fig. 2-1).

Tsarina Catherine the Great

Under Tsarina Catherine the Great, who ruled during most of the second half of the eighteenth century, Russia's empire in the Black Sea area grew at the expense of the Ottoman Turks. The Crimean Peninsula, the port city of Odesa (Odessa), and the whole northern coastal zone of the Black Sea fell under Russian control. Also during this period, the Russians made a fateful move: they penetrated the area between the Black and Caspian seas, the mountainous Caucasus with dozens of ethnic and cultural groups, many of them Islamized. The cities of Tbilisi (now in Georgia), Baki (Baku) in Azerbaijan, and Yerevan (Armenia) were captured. Eventually the Russian push toward an outlet on the Indian Ocean was halted by the British, who held sway in Persia (modern-day Iran), and the Turks. But Catherine the Great had made Russia a colonial power.

The Nineteenth Century

Russian expansionism was not yet satisfied. While extending the empire southward, the Russians also took on the Poles, old enemies to the west, and succeeded in taking most of what is today the Polish state, including the capital of Warsaw. To the northwest, Russia took over Finland from the Swedes in 1809. During most of the nineteenth century, however, the Russians were preoccupied with Central Asia—the region between the Caspian Sea and western China—where Toshkent (Tashkent) and Samarqand (Samarkand) came under St. Petersburg's control (Fig. 2-1). The Russians here were still bothered by raids of nomadic horsemen, and they sought to establish their authority over the Central Asian steppe country as far as the edges of the high mountains that lay to the south. Thus Russia gained a considerable number of Muslim subjects, for this was Islamic Asia they were penetrating. Under tsarist rule, these people acquired a sort of ill-defined protectorate status while retaining some autonomy. Much farther to the east, a combination of Japanese expansionism and a decline of Chinese influence led Russia to annex from China several provinces to the east of the Amur River. Soon thereafter, in 1860, the port of Vladivostok on the Pacific was founded.

Now began the course of events that was to lead, after five centuries of almost uninterrupted expansion and consolidation, to the first setback in the Russian drive for territory. In 1892, the Russians began building the Trans-Siberian Railroad in an effort to connect the distant frontier more effectively to the western core. As the map shows, (p. 114), the most direct route to Vladivostok was across northeastern China (Manchuria). The Russians wanted China to permit the construction of the last link of the railway across their territory, but the Chinese resisted. Taking advantage of the Boxer Rebellion in China in 1900 (see Chapter 9), Russia responded by annexing Manchuria and occupying it. This brought on the Russo-Japanese War of 1905, in which the Russians were disastrously defeated; Japan even took possession of southern Sakhalin Island (which they called Karafuto).

Colonial Legacy

Thus Russia—recipient of British and European innovations in common with Germany, France, and Italy—expanded by **colonialism** too. Yet where other European powers traveled by sea, Russian influence traveled overland into Central Asia, Siberia, China, and the Pacific coastlands of the Far East. What emerged was not the greatest empire but the largest *territorially contiguous* empire in the world. It is tempting to speculate what would have happened to this sprawling realm had European Russia (for such it still was) developed politically and economically in the manner of the other European power cores. At the time of the Japanese war, the Russian tsar still ruled over more than 8.5 million square miles (22 million sq km), just a tiny fraction less than the area of the Soviet Union after the 1917 Revolution. Thus the communist empire, to a very large extent, was the legacy of St. Petersburg and European Russia, not the product of Moscow and the socialist revolution.

THE PHYSICAL STAGE

Russia's climatic and biotic environments, we noted earlier, reflect the country's interior location, its high-latitude position, its exposure to Arctic air masses, and its encirclement by mountains on its landward side. Low temperatures, short growing seasons, and limited water supplies combine to challenge Russian farmers.

While the Soviet Union existed and control was exercised from Moscow, the communist planners initiated major irrigation projects to bring water to the warmer areas then under their control, especially the arid Muslim republics in Central Asia east of the Caspian Sea. Some of these schemes were successful, but others led to environmental disaster. Surface streams were diverted, groundwater supplies dwindled, pesticides caused widespread chemical pol-

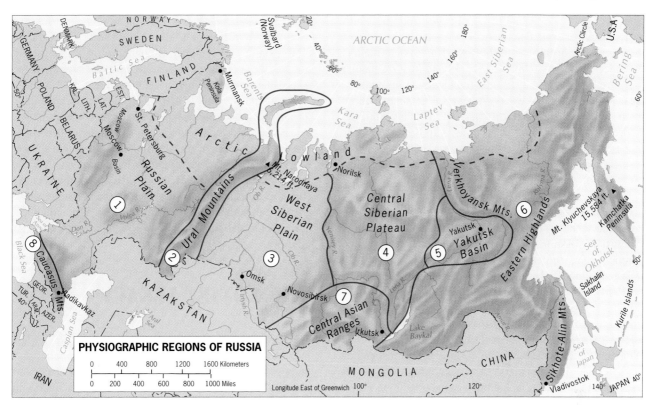

FIGURE 2-2

lution, and social costs (especially deterioration of public health) were high. The Aral Sea, on the border between the former Kazak and Uzbek republics, has lost more than half its water surface since 1960 because streams feeding it were diverted for irrigation.

Climate is a major reason why the independence of Ukraine has been of such great concern to the Russians. Ukraine has precious areas of *Cfa* and *Dfa* climate (the *f* signifying year-round precipitation and the *a* meaning comparatively warm summers), and it also contains the most southerly portion of the still-moderate *Dfb* climate (Fig. I-7). In an average year, Ukraine produced as much as 45 percent of all farm produce harvested in the entire Soviet Union. Much of that production was consumed in Russia—at prices set by the communist system, not by world markets. The prospect that this bounty would be lost to Russia following the dissolution of the Soviet Union was a major stimulus for the creation of the successor Commonwealth of Independent States in December 1991 (see box titled ''The Soviet Union, 1924–1991''). Russia's president Boris Yeltsin took the lead; to his Russian Republic, keeping Ukraine in the fold was a most urgent matter. Ultimately this campaign failed, and Ukraine not only is no longer in the Russian fold; relations between the two countries have become difficult.

Whatever the political manipulations, however, the climatic environment (and the biotic and soil conditions that go with it) will for many years continue to make Russia dependent on outside food sources. Expensive grain imports were a major problem for the old Soviet Union, Ukraine's productivity notwithstanding. The new Russia will have it no easier.

Physiographic Regions

Some of the forces shaping Russia's harsh environments are revealed in the map of its physiography (Fig. 2-2). South of the Russian state lie mountains—the Caucasus ⑧ in the west, the Central Asian Ranges ⑦ in the center, and the Eastern Highlands ⑥ facing the Pacific from the Bering Strait to the North Korean border. Warm subtropical air has little or no opportunity to penetrate here, while cold Arctic air masses can sweep southward without interruption. Russia's Arctic fringe is a lowland sloping gently toward the frigid Arctic Ocean. Overall, we can discern eight physiographic regions in Figure 2-2.

The *Russian Plain* ① is the eastward continuation of the North European Lowland, and here the Russian state formed its first core area. At its heart lies the Moscow Basin. Travel northward from Moscow, and the countryside soon is covered by needleleaf forests like those of Canada; to the south lie the grain fields of southern Russia and, beyond, those of Ukraine. Note the Kola Peninsula and the

Barents Sea in the far north: warm water from the North Atlantic comes around northern Norway and keeps the port of Murmansk open most of the year. The Russian Plain is bounded on the east by the *Ural Mountains* ②; though not a very high range, it is topographically prominent because it separates two extensive plains. The range is more than 2,000 miles (3,200 km) long and reaches from the shores of the Kara Sea to the border with Kazakstan. It is not a barrier to east-west transportation, and its southern end is quite densely populated. Here the Urals yield a variety of minerals as well as fossil fuels.

East of the Urals lies Siberia. The *West Siberian Plain* ③ has been described as the world's largest unbroken lowland; this is the vast basin of the Ob and Irtysh rivers. Over the last 1,000 miles (1,600 km) of its course toward the Arctic Ocean, the Ob falls less than 300 feet (90 m). In Figure 2-2, note the dashed line that extends from the Finnish border to the East Siberian Sea and offsets the Arctic Lowland. North of the line, water in the ground is permanently frozen; this *permafrost* creates yet another obstacle to permanent settlement. Looking again at the West Siberian Plain, we see that the north is permafrost-ridden and the central zone is marshy. The south, however, carries significant settlement, including such major cities as Omsk and Novosibirsk, within the corridor of the Trans-Siberian Railroad.

East of the West Siberian Plain the country begins to rise, first into the *Central Siberian Plateau* ④, another sparsely settled, remote, permafrost-afflicted region. Here winters are long and extremely cold, and summers are short; the area still remains barely touched by human activity. Beyond the *Yakutsk Basin* ⑤, the terrain becomes mountainous and the relief high. The *Eastern Highlands* ⑥ are a jumbled mass of ranges and ridges, precipitous valleys, and volcanic mountains. Lake Baykal lies in a trough that is over 5,000 feet (1,500 m) deep—the deepest rift lake in the world (see Chapter 7). On the Kamchatka Peninsula, volcanic Mount Klyuchevskaya reaches nearly 15,600 feet (4,750 m).

The northern part of region ⑥ is Russia's most inhospitable zone, but southward along the Pacific coast the climate is less severe. Nonetheless, this is a true frontier region. The forests provide opportunities for lumbering, a fur trade exists, and there are gold and diamond deposits. To help develop this promising area, the Soviets constructed the new Baykal-Amur Mainline (BAM) Railroad, a 2,200-mile-long (3,540 km) route that roughly parallels the aging Trans-Siberian line to the south (see Fig. 2-5).

The southern margins of Russia are also marked by mountains: the *Central Asian Ranges* ⑦ from the Kazak border in the west to Lake Baykal in the east, and the *Caucasus* ⑧ in the land corridor between the Black and Caspian seas. The Central Asian Ranges rise above the snow line and contain extensive mountain glaciers. The annual melting brings down alluvium to enrich the soils on the lower slopes and water to irrigate them. The Caucasus form an extension of Europe's Alpine mountain system and exhibit a similar topography (see photo, p. 150), but they do not provide convenient passes. Here Russia's southern border is sharply defined by topography.

As the physiographic map suggests, the more habitable terrain in Russia becomes latitudinally narrower from west to east; beyond the southern Urals, settlement becomes a discontinuous ribbon. As we will note later in this chapter, isolated towns did develop in interior Siberia (such as Yakutsk on the Lena River and Norilsk near the Yenisey River). Yet, as in Canada, the population is clustered markedly in the southern, most livable zone of the country, a narrow belt that widens only on Russia's Far Eastern, Pacific rim.

THE SOVIET LEGACY

The era of communism may have ended in the Soviet Empire, but its effects on Russia's political and economic geography will long remain. Seventy years of centralized planning and implementation cannot be erased overnight; regional reorganization toward a market economy cannot be accomplished in a day.

While the world of capitalism celebrates the failure of the communist system in the former Soviet realm, it will do well to remember why communism found such fertile ground in the Russia of the 1910s and 1920s. In those days Russia was infamous for the wretched serfdom of its peasants, the cruel exploitation of its workers, the excesses of its nobility, and the ostentatious palaces and riches of its rulers. Ripples from the Western European Industrial Revolution introduced a new age of misery for those laboring in factories. There were workers' strikes and ugly retributions, but when the tsars finally tried to better the lot of the poor, it was a case of too little too late. There was no democracy, and the people had no way to express or channel their grievances. Europe's democratic revolution passed Russia by, and its economic revolution touched the tsars' domain only slightly. The vast majority of Russians, and tens of millions of non-Russians under the tsars' control, faced exploitation, corruption, starvation, and harsh subjugation. When the people began to rebel in 1905, there was no hint of what lay in store; even after the full-scale Revolution of 1917, Russia's political future hung in the balance.

The Russian Revolution was no unified uprising. There were factions and cliques; the Bolsheviks (''Majority'') took their ideological lead from Lenin, while the Mensheviks (''Minority'') saw a different, more liberal future for their country. The so-called ''Red'' army factions fought against the ''Whites,'' while both battled the forces of the tsar. The country stopped functioning; terrible deprivations visited the people in the countryside as well as in the cities. Most Russians (and other nationalities within the empire, as well) were ready for radical change.

That change came when the Revolution succeeded and the Bolsheviks bested the Mensheviks, most of whom were exiled. In 1918, the capital was moved from Petrograd (as St. Petersburg had been renamed in 1914, to remove its German appellation) to Moscow. This was a symbolic move, the opposite of the forward-capital principle: Moscow lay deep in the Russian interior, not even on a major navigable waterway (let alone a coast), amid the same forests that much earlier had afforded the Russians protection from their enemies. The new Soviet Union would look inward, and the communist system would achieve with Soviet resources and labor the goals that had for so long eluded the country. The chief political and economic architect of this effort was the revolutionary leader who prevailed in the power struggle: V.I. Lenin (born Vladimir Ilyich Ulyanov).

The Political Framework

Russia's great expansion had brought a large number of nationalities under tsarist control; now it was the turn of the revolutionary government to seek the best method of organizing this heterogeneous ethnic mosaic into a smoothly functioning state. The tsars had conquered, but they had done little to bring Russian culture to the peoples they ruled. The Georgians, Armenians, Tatars, and residents of the Muslim khanates of Central Asia were among dozens of individual cultural, linguistic, and religious groups that had not been "Russified." The Russians themselves in 1917, however, constituted only about one-half of the population of the entire country. Thus it was impossible to establish a Russian state instantly over the whole of this vast political region, and these diverse national groups had to be accommodated.

The question of the nationalities became a major issue in the young Soviet state after 1917. Lenin, who brought the philosophy of Karl Marx to Russia, talked from the beginning about the "right of self-determination for the nationalities." The first response by many of Russia's subject peoples was to proclaim independent republics, as was done in Ukraine, Georgia, Armenia, Azerbaijan, and even in Central Asia. But Lenin had no intention of permitting any breakup of the state. In 1923, when his blueprint for the new Soviet Union went into effect, the last of these briefly independent units was fully absorbed into the sphere of the Moscow regime. Ukraine, for example, declared itself independent in 1917 and managed to sustain this initiative until 1919. But in that year the Bolsheviks set up a

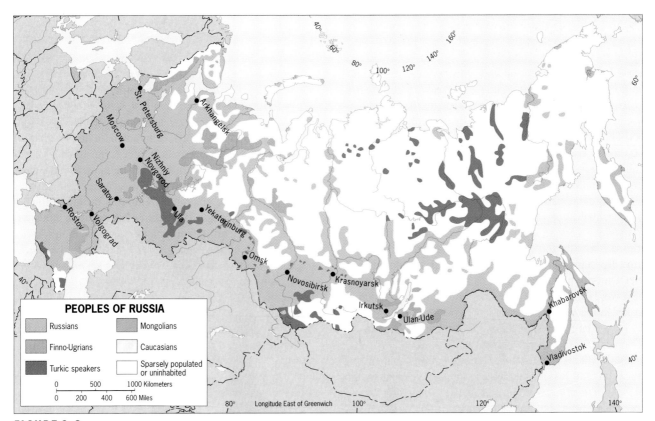

FIGURE 2-3

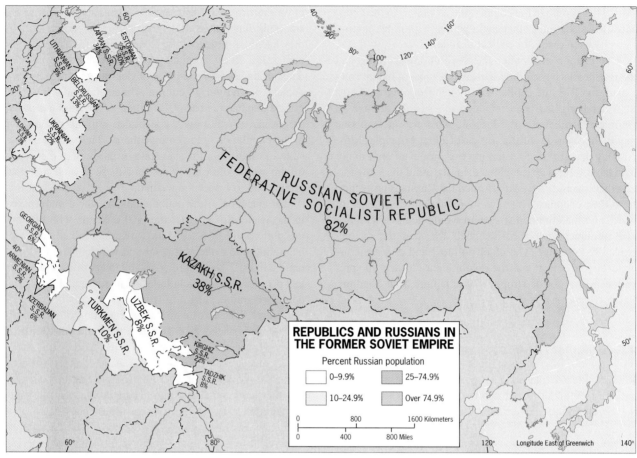

FIGURE 2-4

provisional government in Kiev, thereby ensuring the incorporation of the country into Lenin's Soviet framework.

The Bolsheviks' political framework for the Soviet Union was based on the ethnic identities of its numerous incorporated peoples. Given the size and cultural complexity of the empire, it was impossible to allocate territory of equal political standing to all the nationalities; the communists controlled the destinies of well over 100 peoples, including large nations as well as small isolated groups. It was decided to divide the vast realm into *Soviet Socialist Republics (S.S.R.s)*, each of which was delimited to correspond broadly to one of the major nationalities. At the time, Russians constituted about half of the developing Soviet Union's population, and, as Figure 2-3 shows, they also were (and still are) the most widely dispersed ethnic group in the realm. The Russian Republic, therefore, was by far the largest designated S.S.R., comprising just under 77 percent of total Soviet territory.

Within the S.S.R.s, smaller minorities were assigned political units of lesser rank. These were called *Autonomous Soviet Socialist Republics (A.S.S.R.s)*, which in effect were republics within republics; other areas were designated *Autonomous Regions* or other nationality-based units. It was

a complicated, cumbersome, often poorly designed framework, but in 1924 it was launched officially under the banner of the *Union of Soviet Socialist Republics (U.S.S.R.)*.

Eventually, the Soviet Union came to consist of 15 S.S.R.s (shown in Fig. 2-4), including not only the original republics of 1924 but also such later acquisitions as Moldova (formerly Moldavia), Estonia, Latvia, and Lithuania. The internal political layout often was changed, sometimes at the whim of the communist empire's dictators. But no communist *apartheid*-like system of segregation could accommodate the shifting multinational mosaic of the Soviet realm. The republics quarreled among themselves over boundaries and territory. Demographic changes, migrations, war, and economic factors soon made much of the layout of the 1920s obsolete. Moreover, the communist planners made it Soviet policy to relocate entire peoples from their homelands to better fit the grand design, and to reward or to punish—sometimes, it would appear, capriciously. The overall effect, however, was to move minority peoples eastward and to replace them with Russians. As time went on, the Russification of the Soviet Empire also produced a map of substantial ethnic Russian minorities in all the non-Russian republics (Fig. 2-4).

The Russian Republic, though only one of the 15 S.S.R.s, was the Soviet Union's dominant entity—the centerpiece of a tightly controlled **federation** (see box titled "The Soviet Federation and Its Legacy"). With half the population, the capital city, the realm's core area, and over three-quarters of the Soviet Union's territory, Russia was the empire's nucleus. In other republics, "Soviet" often was simply equated with "Russian"—it was the reality with which the lesser republics lived. Russians came to the other republics to teach (Russian was taught in the colonial schools), to organize (and often dominate) the local Communist Party, and to implement Moscow's economic decisions. This was colonialism, but somehow the communist guise and the contiguous spatial nature of the empire made it appear to the rest of the world as something else. Indeed, on the world stage, the Soviet Union became a champion of oppressed peoples, a force in the decolonization process. It was an astonishing contradiction that would, in time, be fully exposed.

The Soviet
Economic Framework

The geopolitical changes that resulted from the founding of the Soviet Union were accompanied by a gigantic economic experiment: the conversion of the empire from a capitalist system to communism. From the early 1920s onward, the country's economy would be *centrally planned*; that is, all decisions regarding **economic planning** and development were made by the communist leadership in Moscow. Soviet planners had two principal objectives: (1) to speed industrialization and (2) to collectivize agriculture. To accomplish these objectives, the entire country was mobilized, with a national planning commission (*Gosplan*) at the helm. For the first time ever on such a scale, and for the first time in accordance with Marxist-Leninist principles, an entire country was organized to work toward national goals prescribed by a central government.

The Soviet planners believed that agriculture could be made more productive by organizing it into huge state-run enterprises. The holdings of large landowners were expropriated, private farms were taken away from the farmers, and land was consolidated into collective farms. Initially, it was intended that all such land would be part of a *sovkhoz*, literally a grain-and-meat factory in which agricultural efficiency, through maximum mechanization and minimum labor requirements, would be at its peak. But the Soviets ran into opposition from many farmers who tried to sabotage the program in various ways, hoping to retain their land.

The fate of farmers and peasants who obstructed the communists' grand design was dreadful; it is one of the factors that galvanized anti-Russian nationalism in the now-independent republics. In the 1930s, for instance, Stalin confiscated Ukraine's agricultural output and then ordered part of the border between the Russian and Ukrainian republics sealed—thereby creating a famine that killed several million farmers and their families. In the Soviet Union

RUSSIA: MAJOR URBAN CENTERS AND SURFACE COMMUNICATIONS

POPULATION

- Under 50,000
- 50,000–250,000
- 250,000–1,000,000
- 1,000,000–5,000,000
- Over 5,000,000

Railroad built before 1917
Railroad built 1917–1991
Road

National capitals are underlined

0   200   400   600   800 Kilometers
0   100 200 300 400 500 Miles

FIGURE 2-5

under communist totalitarianism, the ends justified the means, and untold hardship came to millions who had earlier suffered under the tsars. (In his book *Lenin's Tomb*, author David Remnick estimates that between 30 and 60 million people lost their lives from imposed starvation, purges, Siberian exile, and other causes.)

It was a human tragedy of incalculable dimensions, but it was hidden from the world by the secretive and covert character of Soviet officialdom. Only after the last Soviet president, Mikhail Gorbachev, initiated his *glasnost* (openness) campaign did meaningful data on such events begin to emerge. One result has been the rehabilitation of the reputations of many prominent Russians and others who protested these actions and who were executed during Stalin's dictatorial rule (1929–1953).

While dissidents were ruthlessly eliminated, Soviet planners did realize that a smaller-scale collective farm, the *kolkhoz*, would be more acceptable and more efficient than the *sovkhoz*. After 15 years of reform, just before the outbreak of World War II in 1939, about 90 percent of all farms, ranging from large estates to peasant holdings, had been collectivized.

Productivity, however, did not increase as the Soviet planners had hoped. Farmers like to tend their own land, and they take better care of it than they do the land of the state. Resentment over the harsh imposition of the program undoubtedly played a role also, but ultimately the two persistent problems were poor management and weak incentives for farmers to do their best. Add to these the recurrent weather problems caused by the country's disadvantageous relative location, and it is not surprising that Soviet farm yields were below expectations—and below requirements. Throughout the three generations of communist rule, the Soviets had to turn to the outside world for food supplies.

Even as collectivization proceeded, the Soviets brought millions of acres of new land into cultivation through ambitious irrigation schemes. During the 1950s they launched the Virgin and Idle Lands Program in Kazakstan, turning pasturelands into wheatfields. The program transformed parts of the Kazakh S.S.R., but success was linked to environmental disaster as diverted streams dried up, groundwater was poisoned by pesticides, and large numbers of people became ill. Again, the grand design created new problems as it solved old ones.

The Soviet planners hoped that collectivized and mechanized agriculture would free hundreds of thousands of workers to labor in the industries they wanted to establish. Enormous amounts of money were allocated to the development of manufacturing; the Soviets knew that national power would be based on the country's factories. The

Novorossiysk lies in Russia's window on the Black Sea, between Georgia and Ukraine. This port, also the terminus of oil pipelines from the interior, has grown in importance (to Russia) since the collapse of the Soviet Union. The facilities shown here are part of the oil-transfer installations. The Russians would like to link this terminal to oilfields in Azerbaijan, thereby enhancing their influence over that oil-rich former Soviet republic; but an alternate route via Georgia to Turkey's Mediterranean coast may outflank this plan. Another oil-connection opportunity may be to western Kazakstan as that former S.S.R.'s rich Tengiz Basin oilfields begin major production.

transport networks were extended; a second rail route to the Far East (the BAM) was built in the 1970s and 1980s, and such remote places as Almaty (formerly Alma-Ata) and Qaraghandy (formerly Karaganda) were connected to the system (Fig. 2-5). Energy development was given priority; compared to agriculture, the U.S.S.R.'s industrialization program had good results. Productivity rose rapidly, and when World War II engulfed the empire, the Soviet manufacturing sector was able to generate the equipment and weapons needed to repel the German invaders.

Yet even in this context, the Soviet grand design held liabilities for the future. Because they could ignore market pressures and certain cost factors, Soviet planners assigned the production of particular manufactured goods to particular places, often disregarding the locational considerations of economic geography. For example, the manufacture of railroad cars might be assigned (as indeed it was) to a factory in Latvia. No other factory anywhere else would be permitted to produce this particular equipment—even if supplies of raw materials would make it much cheaper to build them near, say, Volgograd 1,200 miles away. Such practices made manufacturing in the U.S.S.R. extremely expensive, and the absence of competition made managers complacent and workers less productive than they could be.

The Soviet system tightly bound the economic geography of the republics to the center—and to each other. Each republic and other administrative entity in the Soviet Union depended for raw materials, energy, or other needs on another part. Assignment by Moscow, rather than market forces, ensured the development of territories and places. The city of Bratsk, for example, began as a penal colony in Siberia; once dams and roads had been built, it was made a factory town and grew into a major regional center of state enterprises. The lack of viability of many such enterprises in the post-Soviet period created a major impediment to economic reform.

The devolution of the Soviet Empire began even before the reform-minded Gorbachev took office in 1985. A costly military campaign in Afghanistan had failed, and this "Soviet Vietnam" took its toll on Russian society. Soviet economic, political, and social systems were showing signs of failure. Stirrings of democracy were growing in communist-dominated Eastern Europe. Gorbachev knew that the great Soviet communist experiment was ending in failure, but he worked to control the forces that speeded the collapse. That the dissolution of the old order occurred with minimal loss of life will be a permanent monument to the leader who opened Soviet society to the world—and to itself.

The shrinkage of the Soviet Empire and sphere of influence is reminiscent of the decolonization of Africa. The French flag once flew across much of West and Equatorial Africa, a political continuity across the Mediterranean.

When France lost its colonies, the state seemed much diminished. Russia lost two giant domains: first, most of Eastern Europe, where communist parties had ensured Moscow's dominance, and second, its crescent of 14 colonies from the Baltics to Turkestan. What was left of the Soviet Empire was Russia, still the world's largest state territorially, but now ranked sixth in terms of population (Pakistan will overtake it by the turn of the century).

Russia lost more than territory and population (about 25 million Russians remain stranded in the 14 colonies of the former empire); it also lost resources and raw materials and faced a spatially dismembered economy. Before the collapse, the Soviet Union was the world's leading oil producer. Many of the oil reserves lie outside Russia proper, and the Russians are trying to keep some control over the exploitation and export of this resource in what they call their **Near Abroad**, the 14 other countries of the former Union. Soviet investments built the industry's infrastructure, they argue, giving Moscow legitimate claim to share in the proceeds. Russia also has put pressure on the governments of former Soviet republics not to allow the laying of pipelines so as to avoid Russian territory. Like other colonial powers, the Russians do not want to let go of their former possessions without maintaining a presence.

RUSSIA'S CHANGING POLITICAL GEOGRAPHY

Political maps sometimes show a "European" Russia (bounded on the east by the Urals) and an "Asian" Russia beyond. No geographic justification exists for this division. True, as Figure 2-5 shows, the great majority of Russia's nearly 150 million population is concentrated in the western one-fifth of the national territory. But as the map of ethnic groups (Fig. 2-3) reveals, Russians—who make up about 83 percent of the population—are as dominant in the vast eastern regions of the country as they are in the western heartland. The Far Eastern cities of Komsomolsk and Khabarovsk are no less "European" Russian than Rostov or Nizhniy Novgorod (formerly Gorkiy). In short, the Russian geographic realm extends from St. Petersburg to Vladivostok and from Murmansk to the Georgian border.

Russia is nonetheless a patchwork of nations and ethnic groups, and even after nearly 75 years of communist domination and enforced Russification, many non-Russian cultures remain vibrant and strong. As discussed earlier, when the Soviet planners were confronted by the U.S.S.R.'s cultural diversity, they created a politico-geographical framework that would, they hoped, satisfy ethnic demands and pressures by establishing smaller political units within the

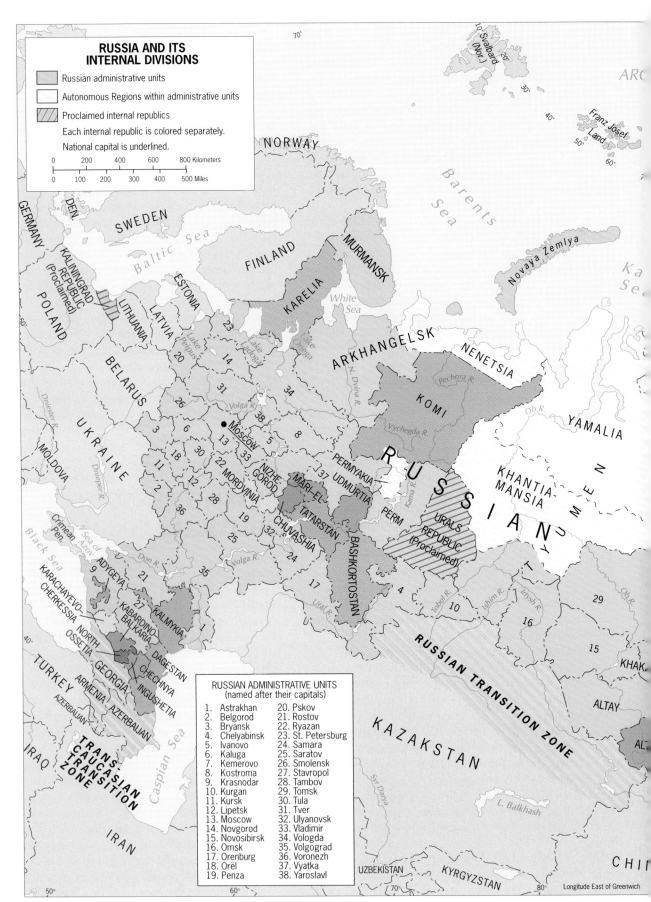

RUSSIA AND ITS INTERNAL DIVISIONS

Russian administrative units

Autonomous Regions within administrative units

Proclaimed internal republics

Each internal republic is colored separately.

National capital is underlined.

0   200   400   600   800 Kilometers

0   100   200   300   400   500 Miles

NORWAY

SWEDEN

DEN.

GERMANY

POLAND

KALININGRAD REPUBLIC (Proclaimed)

LITHUANIA

LATVIA

ESTONIA

BELARUS

UKRAINE

MOLDOVA

Baltic Sea

FINLAND

KARELIA

MURMANSK

White Sea

ARKHANGELSK

Barents Sea

Novaya Zemlya

Svalbard (Nor.)

Franz Josef Land

ARC

NENETSIA

KOMI

YAMALIA

KHANTIA-MANSIA

R U S S I A N

PERMYAKIA

UDMURTIA

PERM

URALS REPUBLIC (Proclaimed)

Moscow

NIZHE-GOROD

MORDVINIA

MARI-EL

TATARSTAN

CHUVASHIA

BASHKORTOSTAN

RUSSIAN TRANSITION ZONE

KAZAKSTAN

Crimean Pen.

Sea of Azov

Black Sea

ADYGEYA

KARACHAYEVO-CHERKESSIA

KABARDINO-BALKARIA

NORTH OSSETIA

GEORGIA

ARMENIA

AZERBAIJAN

KALMYKIA

INGUSHETIA

CHECHNYA

DAGESTAN

Caspian Sea

TURKEY

IRAQ

IRAN

TRANS-CAUCASIAN TRANSITION ZONE

UZBEKISTAN

KYRGYZSTAN

L. Balkhash

Syr Darya

TOMSK (29)

OMSK (16)

NOVOSIBIRSK (15)

ALTAY

KHAK

ALT

CHI

Longitude East of Greenwich

RUSSIAN ADMINISTRATIVE UNITS
(named after their capitals)

1. Astrakhan
2. Belgorod
3. Bryansk
4. Chelyabinsk
5. Ivanovo
6. Kaluga
7. Kemerovo
8. Kostroma
9. Krasnodar
10. Kurgan
11. Kursk
12. Lipetsk
13. Moscow
14. Novgorod
15. Novosibirsk
16. Omsk
17. Orenburg
18. Orël
19. Penza

20. Pskov
21. Rostov
22. Ryazan
23. St. Petersburg
24. Samara
25. Saratov
26. Smolensk
27. Stavropol
28. Tambov
29. Tomsk
30. Tula
31. Tver
32. Ulyanovsk
33. Vladimir
34. Vologda
35. Volgograd
36. Voronezh
37. Vyatka
38. Yaroslavl

FIGURE 2-6

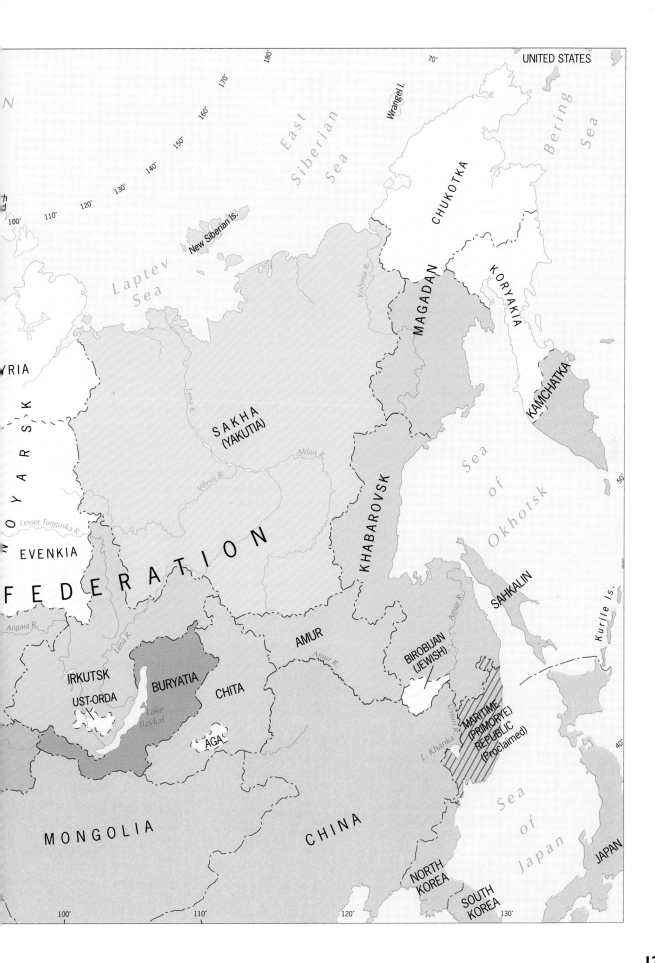

N

RIA

N O Y A R S K

EVENKIA

F E D E R A T I O N

East
Siberian
Sea

Laptev
Sea

New Siberian Is.

Wrangel I.

CHUKOTKA

MAGADAN

KORYAKIA

Bering
Sea

KAMCHATKA

Lena R.

SAKHA
(YAKUTIA)

Aldan R.

Vilvuy R.

Kolyma R.

KHABAROVSK

Sea
of
Okhotsk

Lesser Tunguska R.

Angara R.

Lena R.

IRKUTSK

UST-ORDA

BURYATIA

Lake
Baykal

CHITA

AGA

AMUR

Amur R.

SAHKALIN

Amur R.

BIROBIJAN
(JEWISH)

Kurile Is.

L. Khanka

Ussuri R.

MARITIME
(PRIMORYE)
REPUBLIC
(Proclaimed)

M O N G O L I A

C H I N A

NORTH
KOREA

SOUTH
KOREA

Sea
of
Japan

JAPAN

Soviet Socialist Republics. Within the Transcaucasus republic of Georgia, for instance, they laid out two so-called Autonomous Soviet Socialist Republics (A.S.S.R.s) and one Autonomous Region (see pp. 149–152).

The most complicated part of this framework lay in Russia itself. The Soviet Russian Republic was administratively and politically divided into 6 Territories, 49 Regions, 16 A.S.S.R.s, 5 Autonomous Regions, and 10 Autonomous Areas. The 6 Territories and 49 Regions were designed as administrative units, mainly for the purposes of national economic planning, census taking, and so forth. However, the original 16 A.S.S.R.s and 5 Autonomous Regions were designed to award recognition to some of the nearly 40 substantial minorities on Russian soil. One of these Regions, for example, was the Jewish Autonomous Region in the remote Soviet Far East, a rugged and near-isolated corner of the country not much larger than Vermont (and never populated by more than 30,000 Russians of Jewish descent).

Even before the Soviet Union fell apart, this unwieldy politico-geographical framework was failing. While the 5 Autonomous Regions held a higher status than the 10 Autonomous Areas, the peoples and leaders of several of the Autonomous Areas had begun to play a more prominent role in national affairs than those of some of the regions. (The Jewish Region, for instance, was never a factor.) But these issues were submerged when the devolution of the whole Soviet Union gathered momentum. When Russia emerged as a newly independent republic at the end of 1991, it took the official name *Russian Federation* in acknowledgment of its inherited, multi-tiered political geography and the diversity of its cultural geography.

The Russian Federation inherited the complicated administrative structure the Soviets had created, and naturally it was impossible to change that structure overnight. Yet something had to be done, because some of the republics *within* Russia, and even some other administrative units, were themselves agitating for independence or greater autonomy.

On March 31, 1992, most of Russia's nearly 90 internal republics, autonomous regions, *oblasts*, and *krays*, (all territorial components of the administrative hierarchy) signed the Russian Federation Treaty. Holdouts included Tatarstan, whose president insisted that his territory should be as independent as Latvia or Georgia, and a republic in the Caucasus zone, Chechnya-Ingushetia, which experienced a political crisis and split into two separate republics, Chechnya and Ingushetia (Fig. 2-6). After initially agreeing to sign the treaty, Chechnya reconsidered, and in mid-1996 neither Chechnya nor Ingushetia had formally acceded to the treaty. In the case of Chechnya, Russian military intervention to force the republic into Moscow's fold led to a prolonged and violent conflict with disastrous consequences for Chechnya's people and for Russia's political leadership.

The Federal Framework

The spatial framework of the evolving Russian Federation is no simpler than that of the Russian Soviet Federative Socialist Republic during communist times. In 1996, the federation consisted of 89 entities: 21 Republics, 11 Autonomous Regions (*okrugs*), 49 Provinces (*oblasts*), 6 Territories (*krays*), and 2 Autonomous Federal Cities.

Evidence of continuing change abounds. By January 1, 1996, a half dozen entities of lower rank had proclaimed themselves republics (three of these are shown on Fig. 2-6), including the cities of Moscow and St. Petersburg! Although the Russian Parliament had not gotten around to dealing with these assertions, none was withdrawn, and some of the aspirants, notably the Maritime ''republic'' and the Urals ''republic,'' loudly insisted on their prerogative. Some of the claims seem more reasonable than others. Kaliningrad, the Baltic-Russian exclave, has a case because of its isolation and separation from the main territory. On the other hand, it is difficult to take seriously the proclamation by two provinces, Amur (population about 1 million) or Vologda (a rural area on the forest margin 400 miles [700 km] east of St. Petersburg).

As we note in the discussion that follows, most of the recognized republics have substantial minority populations; the newly self-proclaimed republics do not. But the most important geographic point made by Figure 2-6 is this: subtract the internal republics, and what is left of Russia is an elongated, fragmented, perforated strip of territory that constitutes not much more than half the total area of the federation. Most of the country's western heartland is ethnically Russian, but east of Moscow lies a cluster of six republics that extends into Russia's very core area. Several of these republics contain energy installations, weapons factories, crucial railroad segments, and other facilities important to the country as a whole. Russia's federal map is a blueprint for trouble.

On the basis of relative location, we group Russia's republics as follows:

1. *The Republics of the Heartland*
   Mordvinia, Chuvashia, Mari-el, Tatarstan, Udmurtia, Bashkortostan (plus the newly proclaimed Urals or Sverdlovsk Republic).
2. *The Republics of the Caucasian Periphery*
   Dagestan, Chechnya, Ingushetia, North Ossetia, Kabardino-Balkaria, Karachayevo-Cherkessia,* Adygeya,* Kalmykia
3. *The Republics of the North*
   Karelia, Komi, Sakha (Yakutia)
4. *The Republics of the Southeast*
   Altaya,* Khakassia,* Buryatia, Tuva (plus the newly proclaimed Maritime or Primorye Republic).

*Formerly Autonomous Regions or Autonomous Areas.

REPUBLICS OF THE RUSSIAN REALM

REPUBLICS OF THE HEARTLAND

Mordvinian Republic (Mordvinia)

Travel southeastward from Moscow, and the first republic you will reach is Mordvinia (Fig. 2-7), formerly known as Mordovia. Once under the sway of the Tatars of Kazan, Mordvinia came under Russian control in the sixteenth century and was proclaimed a Soviet A.S.S.R. in 1934. Its population grew beyond 1 million only recently.

The Mordvinians, a people with Finnish and Ugric (Hungarian) ancestries, speak a language belonging to the Ural-Altaic family, which has not only Finnish and Hungarian but also Persian roots. Although Mordvinian culture has been fading, more than one-third of the population remains ethnic Mordvinian (Russians make up about 60 percent). In recent years, Mordvinian national costumes, elaborately embroidered tunics and dresses, have made a comeback in the cultural landscape.

Unlike many other republics, Mordvinia was never given a key role in the former Soviet economic system. The local economy remains chiefly agricultural; beekeeping is an old and continuing tradition here. Manufacturing, most of it of the light variety, concentrates in and near the capital, Saransk.

Chuvash Republic (Chuvashia)

Western Russia's great Volga River flows eastward north of Moscow until, in the vicinity of the Tatar city of Kazan, it bends southward on its journey to the Caspian Sea. Before the bend, on its right bank, lies the small, New Jersey-sized Chuvash Republic (Fig. 2-7).

The Chuvash are a Turkic people, sedentary farmers who inhabited this part of the middle Volga valley before the tsar's armies swept across their land in the sixteenth century. Farming and livestock raising continue to be the dominant way of life here; to this day, the Chuvash outnumber Russians in their republic by 70 to 25 percent.

Chuvashia became an A.S.S.R. in 1925. Under Soviet administration, some industrialization took place, and the capital, Cheboksary, situated on the Volga River, became a minor manufacturing center. The Chuvash, who are distantly related to the Bulgars, largely converted from Islam to the Russian Orthodox faith before the 1917 Revolution;

religion has not been an issue in Chuvash-Russian relations. During the collapse of the U.S.S.R. in 1991, Chuvash leaders reaffirmed their republic's "sovereignty."

Mari Republic (Mari-el)

Across the Volga River from the Chuvash Republic, about 400 miles (650 km) east of Moscow, lies the small Mari Republic (Fig. 2-7), a Massachusetts-sized enclave whose historical geography has been closely linked to that of its neighbors. The Mari, like the Komi and other minorities in east-central and northern Russia, have ancient ties to the Finns, the Hungarians, and even the Persians. At one time, they were subjugated under the Islamic Tatar khanate, but when Tsar Ivan the Terrible conquered the Tatar state in the sixteenth century, the Mari domain became a Russian colony. After the Revolution it became an *oblast* (province), and in 1936 it was proclaimed an A.S.S.R.

Today, about 45 percent of the population of nearly 800,000 has Mari ethnic ancestry. Nearly half of the people are Russians. Tatars (7 percent) still make up a noticeable minority.

Timber-based industries are important here, but under Soviet economic planning the Mari Republic was assigned machine building as its principal role. Still, logs from Mari forests continue to float down the Volga, to be made into furniture, prefabricated homes, and paper in the factories located in large Russian industrial cities.

Mari ethnic consciousness revived during the Soviet Union's disintegration, and the republic's leaders are demanding greater autonomy. The course of events in the neighboring Tatar Republic will be a key to Mari's future—as was the case in the past.

Tatar Republic (Tatarstan)

Russia's Tatars are scattered remnants of the Mongol invaders, the "Golden Horde" that swept across the realm nine centuries ago. Between 5 and 6 million ethnic Tatars survive, more than half of them clustered in the middle Volga valley (Fig. 2-7). Here lies the Tatar Republic, established in 1920.

The Tatar Republic lies where the Kazan khanate once did, a powerful Muslim state centered on the city of Kazan. In 1552, Tsar Ivan the Terrible conquered the khanate, destroying some 400 mosques (even as he ordered the construction of St. Basil's Cathedral in Moscow). Sporadic Muslim uprisings followed, always crushed with great loss

FIGURE 2-7

of life; but Islam did not die here. When the Revolution erupted, the Kazan Tatars joined it, hoping for a better life after the tsars' overthrow. They were rewarded with their republic even before the U.S.S.R. was formally constituted.

Under Soviet rule, atheism was state policy, and Islam continued to suffer. In addition, the republic (somewhat smaller than Maine) lay in the path of communist-planned industrial expansion. Agricultural and pastoral Tatarstan was ''endowed'' with truck factories, chemical industries, and engineering plants. For a time it was the former Soviet Union's largest oil producer; the first well was drilled in 1943, and growth was rapid. All this industrial activity

attracted Russian workers, and the population balance shifted. The Tatars began to object, their anger already aroused by cultural Russification.

Today, ethnic Tatars constitute around 48 percent and Russians 44 percent of the population of about 4 million. The Tatars in the republic are trying to revive their remaining mosques, rekindle their Arabic language, and stem the tide of Slavic culture. In 1991, the republic's president demanded the right to sign any new treaties as an equal partner with former Soviet Socialist Republics, and in 1992 he refused to sign the Russian Federation Treaty. Tatar nationalism remains a potent force here.

Udmurt Republic (Udmurtia)

Udmurtia on its south adjoins the Tatar Republic and is the fifth of the cluster of republics situated east of Moscow (Fig. 2-7). About the size of Estonia, the Udmurt Republic is the ancestral home of still another Finnic people, the Udmurts. Their autonomy was proclaimed in 1934. Today, ethnic Udmurts constitute barely over 30 percent of the population; they are far outnumbered by Russians (nearly 60 percent). A Tatar minority (7 percent) also remains here, as well as some Mari.

During the Soviet era, Udmurtia became heavily industrialized. Weapons, steel, machines, tools, engines, and vehicles were produced in factories in Izhevsk, the capital, and other cities. Lumbering, sawmilling, leatherworking, and food processing also take place; grain cultivation and livestock raising further contribute to a varied economic geography. All this explains the large Russian population in this republic. The overall population now approaches 2 million.

Undoubtedly reflecting the large Russian majority, the political scene in Udmurtia has been relatively calm, although local leaders have given notice that they want to strengthen their ability to negotiate with Moscow over "exports."

Bashkort Republic (Bashkortostan)

Located north of Kazakstan and extending westward from the slopes of the southern Ural Mountains (Fig. 2-7), Illinois-sized Bashkortostan has a population of 4.3 million and an economy based on rich mineral resources (including oil and natural gas) and productive agriculture. Of major significance is Bashkortostan's oil-refining capacity: huge facilities have been built, making this republic one of the most important and centralized nodes in Russia's oil pipeline network.

The Bashkorts, a Turkic people, settled in this area during the Mongol invasions. The Russians conquered this land during the rule of Ivan the Terrible (1552), and the capital and largest city, Ufa, was founded shortly thereafter.

In 1919, Muslim Bashkort leaders, enticed by Bolshevik promises of autonomy, joined the "Reds" in the Revolution. When the acreage of the large estates was redistributed, however, Russians—not Bashkorts or Tatars also living there—got most of it. A local rebellion was ruthlessly put down after the Bashkort A.S.S.R. was established (also in 1919).

Today, Bashkorts (24 percent) are outnumbered by Russians (40 percent) and Tatars (25 percent) in their own republic. Major industrialization has occurred in the Urals portion of the republic, based not only on oil production and refining but also on natural gas, timber, iron, manganese, and copper ores.

Non-Russians remain in the majority, and Islam and Orthodox Christianity are being revived. Bashkortostan's leaders, using the importance of the republic's economy to Russia as leverage, have demanded greater autonomy in the new Russian framework.

REPUBLICS OF THE CAUCASIAN PERIPHERY

Dagestan Republic (Dagestan)

At the eastern end of the Caucasus Mountains, between the crestline of the main range and the Caspian shore, lies the Dagestan Republic (Fig. 2-8), about half the size of Virginia. Dagestan (the name means "mountain country") contains some 30 distinct nationalities among its 2 million people, many of them confined to isolated valleys. The Russians penetrated this territory in the fifteenth century, forcing the Persians (whose frontier it was) to yield in 1723. Formal annexation occurred in 1813, but the locals continued to fight back until 1877. In 1921, Dagestan was made an Autonomous Republic, a congregation of minorities in which the largest ethnic group is the Avartsy (27 percent) and the second largest the Dargintsy (16 percent). Russians account for only 11 percent.

Dagestan lies in a sensitive area, between Shi'ite Muslim Azerbaijan to the south and the multicultural Kalmyk Republic to the north, and adjacent to war-torn Chechnya in the west. When Russian forces invaded Chechnya in 1994 to establish Moscow's control over that rebellious republic, there was much sympathy for the Chechens in Dagestan, and refuge was given to revolutionaries who crossed the border. But Dagestan itself is divided over relations with Moscow. There is potential for trouble here because this republic contains sizeable oil and gas reserves on its Caspian coastal plain near the capital, Makhachkala.

Chechen Republic (Chechnya)

Until mid-1992, Chechnya (Fig. 2-8) was part of the Chechen-Ingush Republic. Old Muslim enemies, the Chechen and the Ingush heroically stood together and repelled the Cossacks fighting for the tsars. Not until 1870 did the Russians prevail. Forcibly united by the Soviet-communist planners in 1936, the two ethnic groups were awarded status as an Autonomous Republic in 1957.

When the Soviet Union collapsed, the Chechens moved to install their leader as president of the republic. Before

FIGURE 2-8

long, the old adversaries were in conflict again, pitting the Chechens (who made up more than half of the population) against the minority Ingush (12 percent). The solution was to divide the republic into two. The larger eastern part, the heartland of the Chechens, became Chechnya; the smaller west was designated Ingushetia.

As early as 1991, Chechen leaders had vowed not to submit to Moscow's rule. The new Chechnya made good on that pledge by establishing a rebel regime in the capital, Groznyy, after the separation. When negotiations failed, Moscow in 1994 sent in its armed forces, and a devastating conflict ensued with echoes of Afghanistan and unprecedented scenes of destruction and desperation (busloads of

Russian mothers came to the war zone in search of their missing soldier sons). Tens of thousands of casualties, the almost total devastation of Groznyy, rising ethnic and cultural tensions in neighboring republics, and a protracted guerrilla war all were part of the price paid by Moscow for its subjugation of the tiny republic.

Chechnya consists of three well-defined geographic regions. North of the Terek River the land is flat, the people are farmers, and Russian domination has long been an accepted fact. South of the Terek and north of the mountains lies a middle zone that contains the capital, the oil installations and pipelines, and limited industrial development. Here, too, lie the larger towns beside Groznyy. And south of this middle zone lie the foothills, deep valleys, and towering peaks of the Caucasus, sheltering the villages where rebel Chechens are now based.

During Soviet times, Chechnya became an important entity in the empire's economic geography. Oil and natural gas went by pipeline from Groznyy as far north as Ukraine and central Russia. By 1990 about 30 percent of Chechnya's population was Russian, mostly concentrated in the middle zone. Today the question for Moscow is how to fit a defeated but still rebellious Chechnya into the political geography of the Russian Federation.

Ingush Republic (Ingushetia)

Caucasus peaks over 14,000 feet (4,200 m) high tower over this tiny republic on Russia's Transcaucasian rim (Fig. 2-8). The Ingush, one of Russia's many small ethnic groups (numbering under 300,000), have occupied mountain valleys here for centuries and have suffered dreadfully for it. In the 1940s, the Soviet dictator Josef Stalin sliced off part of the historic Muslim Ingush domain and put it under the control of neighboring, pro-Soviet North Ossetia. Already, the other flank of Ingush society lay in a Chechen-dominated Autonomous Republic.

In 1991, Chechens seized political control of (then) Chechnya-Ingushetia, but worse was to come. In 1992, Ingush Muslims living in North Ossetia faced a campaign of "ethnic cleansing" by North Ossetian soldiers helped by Russian forces. Hundreds of Ingush villages were destroyed, and thousands of Ingush families were driven into the icy mountains, where many perished. When the Russian Parliament declared a separate republic for the Ingush, it was carved out of western Chechnya-Ingushetia, but it did *not* include any Ingush land in North Ossetia. A narrow sliver of territory, the Ingush Republic is the smallest and weakest of all of Russia's 21 internal republics, with little infrastructure and less cohesion. Nazran, the capital, is a small, stagnant, isolated place.

Moscow's contradictory actions—creating an Ingush Republic on the one hand while abetting the killing of Ingush citizens in North Ossetia on the other—probably arise from a new regional concern here: the Muslim Confederation of Caucasian Mountain Peoples, an organization that seeks to unite Islamic groups on the Russian side of the Caucasus with the principal aim of breaking away from the Russian Federation.

North Ossetian Republic (North Ossetia)

As Figure 2-8 shows, there are two Ossetian Republics: one in Russia (North Ossetia) and the other in Georgia (South Ossetia). When the Soviet Union prevailed, this division was a technicality: both republics were part of the empire. But when the U.S.S.R. collapsed and Georgia became an independent state, the Ossetians found themselves separated by an international boundary.

North Ossetia's 3,000 square miles (8,000 sq km) of mountainous terrain lie on the northern flank of the Caucasus Mountains (Fig. 2-8). Critical road and rail links cross the Caucasus via this small but pivotal republic.

Ossetian historical geography is long and turbulent. The Ossetians lived in this area as long as 2,500 years ago. The Tatars incorporated Ossetia into their empire in the thirteenth century; the Turks contested the area; and then the Russians colonized it in the late 1700s. Today, Ossetians still constitute half the population of about 700,000; Muslim Ingush number about 50,000. Russians make up 35 percent. The Republic was proclaimed in 1936.

Ossetian ethnic consciousness was raised by the Soviet collapse and by Georgia's push for independence, which the South Ossetians perceived as a threat. Voices favoring reunification of the Ossetian domain were heard, and on the Georgian side, an insurrection arose with reunification as its goal. That uprising was put down by Georgia's army, but the north was preoccupied with another issue: the presence of the Ingush minority and the campaign against it. While North Ossetians historically have taken a pro-Russian position, this stance did not deter them from changing the name of their capital from the Soviet Ordzhonikidze to the precommunist Vladikavkaz.

Kabardino-Balkar Republic (Kabardino-Balkaria)

Located in the central segment of the Caucasus cluster of internal republics (Fig. 2-8), Kabardino-Balkaria has some of Russia's most spectacular scenery, including Mount Elbrus (18,510 feet/5,640 m) in its southwestern corner (see photo, p. 150). Glaciers, swift-running streams, alpine meadows, forests, and lower plains form a highly varied topography, despite the republic's small area (about the size of Connecticut).

About half of the 800,000 inhabitants are Kabardinians, people of the lower plains who allied themselves with the Russians as early as the sixteenth century. One-third of the population is Russian, many of Cossack descent. The Balkars of the mountains (10 percent) fought the Russian incursions for a long time. During World War II, the communists alleged, the Balkars sympathized with the German invaders of the U.S.S.R., and Stalin ordered their deportation as punishment. Later, many returned to the republic, which had been officially constituted in 1936.

Small but important deposits of valuable minerals (gold, chromium, nickel, tungsten, molybdenum) make mining an important industry. Soviet economic planners assigned the manufacture of oil-drilling equipment to this republic as part of their grand design—although Kabardino-Balkaria has no oil.

Karachayevo-Cherkess Republic (Karachayevo-Cherkessia)

The spectacular scenery of this small, Connecticut-sized republic is reminiscent of Switzerland: jagged peaks of rock and ice tower over forested slopes and alpine meadows. The southern boundary of Karachayevo-Cherkessia (Fig. 2-8) coincides with the crestline of the Caucasus Mountains, and here elevations exceed 13,000 feet (4,000 m).

This republic exemplifies the Soviet administration of minorities. A combined Autonomous Region for the Turkic Karachay and Christian Cherkess peoples was created in 1922, but a few years later this unit was divided to give each group an *oblast* (administrative area). This arrangement was dissolved in 1943 when the Karachay were accused of supporting the Germans during World War II; as a result, many Karachay were exiled. In 1957, the present territory was reestablished as an Autonomous Region to accommodate both ethnic groups. Today it is identified officially as a republic of the Russian Federation.

Livestock herding continues on the northern plains, but mining, lumbering, and some manufacturing have developed. The total population remains around 500,000. The capital, Cherkessk, lies in the non-Turkic part of the republic.

Adygey Republic (Adygeya)

The Adyghian people are related to the Cherkess, and their republic was first created as an Autonomous Region in 1922. Somewhat smaller than Puerto Rico, Adygeya lies in the hinterland of the city of Krasnodar in the Russian Federation's southwestern corner (Fig. 2-8), where the plains to the north merge into the foothills of the Caucasus to the south. Today, Adyghians make up about one-quarter of the total population of 480,000. Agriculture dominates, and a wide variety of farm produce is exported; the area is famous for its flowers, especially its "Crimean" roses.

Kalmyk Republic (Kalmykia)

Kalmykia does not lie on the slopes of the Caucasus, but it adjoins Dagestan to the south. The Kalmyks, a nomadic people of Mongol ancestry, migrated to their present home in the lowlands northwest of the Caspian Sea during the seventeenth century (Fig. 2-8). They brought with them their Tibetan Buddhist faith (some Kalmyks are Muslims, however); their yurts (tents of felt fabric on a lattice frame, easily moved); and their livestock (horses, cattle, goats, sheep, and camels as beasts of burden). Overrun by the Russians in the eighteenth century, they found themselves in a Soviet *oblast* in 1920; in 1935, they were awarded status as an Autonomous Republic. Stalin abolished Kalmykia in 1943, alleging that Kalmyks were collaborating with the invading Germans; after the war he exiled the entire population (with much loss of life) to Soviet Central Asia. After Stalin's death in 1953, the Kalmyks were allowed to return to Kalmykia, and in 1958 their Autonomous Republic was restored.

About as large as South Carolina, Kalmykia flanks the lower Volga River as well as the Caspian Sea, leaving a narrow Russian corridor to the Caspian shore (Fig. 2-8). It was Soviet policy to permanently settle the traditionally nomadic Kalmyks, who continue to depend on their livestock. Sheep breeding is a major source of income. Russian farmers moved into the arable lands of Kalmykia, and today the population of 400,000 is about 43 percent Kalmyk, 42 percent Russian, and 15 percent other ethnic minorities.

In 1993, Kalmykia's high-profile president, touting his republic as "Europe's only Buddhist state," demanded greater autonomy for his country and "international recognition" from abroad. The Russians show little enthusiasm for such aspirations.

REPUBLICS OF THE NORTH

Karelian Republic (Karelia)

In the far northwestern corner of Russia lies a republic larger in area than California but smaller in total population than San Francisco: Karelia (Fig. 2-7). To get there, you go north from St. Petersburg or south from Murmansk, but no matter what your route, expect cold weather, clear lakes, dense forests, and very little surface relief. The glaciers scoured the north of Karelia, and they deposited their till (unconsolidated materials) over the south as they melted back.

An area as remote and forbidding as Karelia would not be expected to have a checkered politico-geographical past, but this republic has been a changeable frontier. The east has been part of Russia since the early fourteenth century, but the west has seen many boundary changes involving Russia's neighbor, Finland. The Karelian A.S.S.R. was proclaimed in 1923, and after the war between the Soviets and Finland in 1940, territory gained by the U.S.S.R., including the port of Vyborg, was attached to Karelia. In the same year, Karelia was made a full-scale Soviet republic. In 1946, however, Karelia's coastline was taken away and attached to the Russian Republic, and the status of the republic reverted to A.S.S.R. within the boundaries on the map today.

The Karelians are a Finnic people, not Slavs, but they constitute only 11 percent of the current population. Russians make up 70 percent, and Belarussians about 8 percent. Karelia's economic geography is based on its vast timber resources. During the communist era, it was a feared place of exile for criminals and political dissidents.

Komi Republic (Komi)

The vast northeastern corner of Russia west of the Urals is the territory of Komi, land of the Arctic Circle (Fig. 2-7). From the crestline of the mountains in the east (sometimes rising about 6,000 feet), the surface drops into the basin of the Pechora River on its way to the Arctic Ocean. This is a country of tundra and taiga (snowforest), of frigid winters and short cool summers.

Komi became a Russian domain as early as the fourteenth century, and fur hunting drew settlers who made Syktyvkar, now the capital, a major base of operations. An A.S.S.R. was established in 1936 expressly to recognize the Komi ethnic group, a Finnic people who today constitute a mere 25 percent of the population of nearly 1.5 million.

In the 1940s, the Pechora Railway was built diagonally across the entire republic from near Syktyvkar in the southwest to the extreme northeastern corner. The economic geography of Komi centers on its extensive forest resources but importantly also on its coal, natural gas, and oil. In terms of value, mining is the leading activity. A nuclear power plant near the town of Pechora provides electricity—and, being of Chornobyl vintage and construction, worries local residents.

Sakha Republic (Yakutia)

Territorially, the Yakut Republic (renamed *Sakha* in 1993) in Russia's far northeast (Fig. 2-6) is larger than all the other internal Russian republics combined, but its population barely exceeds 1 million. Sakha is nearly one-third the size of Canada, and its environments are even harsher: in fact, this republic has the harshest climate in all the inhabited world.

The Yakuts, traditionally nomadic pastoralists, moved northward from central Asia into the basin of the Lena River; their ancestral connections are Mongol and Turkic, but their history is uncertain. They arrived here more than a thousand years ago, and despite the difficult climate they continued to herd cattle as well as reindeer. In the first half of the seventeenth century, Russian pioneers arrived in their domain, and the present capital, Yakutsk, was founded in 1632. The Yakut A.S.S.R. was established in 1922.

Sakha is a permafrost-ridden, isolated, topographically varied land; surface communications remain primitive. But the republic's mineral potential is enormous. Oil and gas have been found as well as gold and diamonds. Timber resources are huge. Russians (52 percent) outnumber ethnic Yakuts (38 percent) in the population, but in the 1990s the Yakuts presented a united front, demanding that any

"foreigners" exploiting or exploring for Yakutia's raw materials pay Yakutsk for the right to do so—Russian outsiders included. Japan has expressed interest in Yakutia's resources, but who will strike the deal: Moscow or Yakutsk?

REPUBLICS OF THE SOUTHEAST

Altay Republic (Altaya)

Altaya is situated at the sensitive crossroads where Russia, Kazakstan, China, and Mongolia meet (Fig. 2-6). About the size of Maine, it is home to a small population (roughly 250,000), most of whom still depend on cattle breeding for their livelihood; the mining of gold and coal also contributes to the economy.

The original designation of the Autonomous Region in 1922 was based on the local people's distinct language and traditional culture. This republic has had various names on older maps, including Oirot and Gorno-Altay.

Khakass Republic (Khakassia)

Khakassia began its existence in 1930 as an *oblast* in Krasnoyarsk Territory, a West Virginia-sized area in the upper basin of the Yenisey River (Fig. 2-6). The heart of this republic, including the capital, Abakan, lies in the valley of the Abakan River, a mountain-encircled basin with an arid climate and steppe vegetation. The grasslands were converted into grain and potato fields during the Virgin Lands campaign of the 1950s, but livestock herding continues to be an important industry. Pine forests clothe the mountain slopes, and timber is an important export.

The Khakass people who first occupied this area were Turkic nomads who drove their herds across the plains, but today they are greatly outnumbered by ethnic Russians. Copper ores first attracted the Russian immigrants; later, iron, gold, coal, and ferroalloys were discovered. The republic is linked by rail to the Trans-Siberian mainline.

The population remains about 600,000, but Khakassia lies in a sensitive zone near Mongolia's northern border. Some current Chinese texts describe Khakassia as part of an area where territorial issues remain "unsettled."

Buryat Republic (Buryatia)

Montana-sized, mountainous Buryatia flanks Lake Baykal and extends westward to border the Tuva Republic (Fig. 2-6). As the map shows, a comparatively narrow Russian corridor separates Sakha (Yakutia) to the north from Buryatia on the Mongolian border.

Mineral-rich and scenically spectacular, Buryatia was settled by nomadic Buddhist Mongols who kept cattle as well as camels. The territory was annexed from China by the Russians during the seventeenth century, and colonization began. Gold and furs were the early objectives, but the Buryats offered strong opposition. Then the Trans-Siberian Railroad was built across Buryat Territory during the late nineteenth century, and industrial and urban expansion accelerated. Valuable ferroalloys as well as coal and iron are found here. Mining and metallurgy dominate the economy, and pastoralism is still important.

Reflecting the economic significance of Buryatia, many Russians settled in the area. The population of about 1.1 million is more than 70 percent Russian today. Less than one-quarter are ethnic Buryats in this important prize of tsarist expansion.

Tuva Republic (formerly Tannu Tuva)

The Tuva Republic, about the size of North Dakota, lies wedged between Mongolia to the south, Buryatia to the east, Altaya to the west, and Siberian Russia to the north (Fig. 2-6). Its varied environments range from high snow-capped mountains and dry steppe-like basins to dense pine forests. Population is sparse (about 330,000).

This area was part of China's empire from the mid-eighteenth century until 1911, when the Russians incited an anti-Chinese uprising among the Turkic Tuvans. In 1914, the tsar established a ''protectorate'' for the area, but after the Revolution the Tuvans declared themselves independent from both China and Russia and formed the Tannu Tuva People's Republic. The Soviets annexed their small neighbor in 1944 and proclaimed an A.S.S.R. there in 1961. In 1971, the name was changed to the Tuva Republic.

Tuvans (or Tuvinians) still constitute about 60 percent of the population (Russians 35 percent), and pastoralism remains the chief economic activity. Some gold has been found, along with cobalt and asbestos.

There are signs that China has not forgotten Russia's takeover of Tuva. Some Chinese maps show Tuva as disputed territory.

Primorye (Maritime) Republic [Proclaimed]

As Figure 2-6 shows, a territory in the Russian Far East, bounded by China and North Korea and facing the Sea of Japan, has proclaimed itself a republic, the Primorye or Maritime Republic. This is a significant development because it reflects more than the mere desire on the part of this territory's representatives to enhance its status in Moscow. Rather, it mirrors an impatience with Russia's political and economic rules and regulations (including taxation), here in an outpost about as far from the country's core area as one can get.

The capital is Vladivostok, which until 1991 was a naval base closed to all foreigners; Nakhodka, with its oil and container terminals at Vostochni, is the main commercial port. The ruling group in the capital sometimes is described as a freewheeling syndicate that tolerates no opposition; it elects to ignore Moscow when it wishes. Some of its members even speak of outright independence, but this territory's greatest opportunities lie in its situation at the end of the Trans-Siberian Railway and across the sea from a Japan covetous of Siberia's raw materials. More sensible talk is of a Hong Kong of the north, a threshold to Russia and a gateway to the Pacific Rim. Economic success is more likely than political posturing to lead to republic status.

REPUBLICS AND REGIONS

As we noted earlier, the Russian Federation, in addition to its 21 republics, contains nearly 70 regions and other designated territories (including 11 Autonomous Regions). The end of the Soviet period presented an opportunity for a simplification of the politico-geographical map; instead, things became even more complicated. Under the federal system, the 21 republics enjoy more rights than any of the other jurisdictions. Half the seats in the Federation Council are reserved for the republics, even though these make up under 30 percent of Russia's territory and account for just 16 percent of its population. Regions, therefore, aspire to republic status because this will give them more influence in federal affairs. Republics, on the other hand, are reluctant to approve such elevation because this will dilute their influence in the affairs of state. In the still-formative arena of Russian democracy powerful individual leaders have given certain regions greater visibility and influence in the capital than their status merits, but Russia's first president, Boris Yeltsin, frequently asserted that he favored giving all regions rights equal to those of the republics.

In 1996, Russia's complex federal structure still was in transition. The republics, however, continue to hold center stage in the process, standing as legacies of a time when communist planners rewarded minorities with territory and punished them with exile.

RUSSIA'S PROSPECTS

When Peter the Great began his campaign to reorient his country toward Europe and the outside world, he envisaged a Russia with warm-water ports, a nation no longer encircled by Swedes, Lithuanians, Poles, and Turks, and a force in continental affairs. Core-periphery relationships always have been crucial to Russia, and they are just as important today. With 25 million Russians living in the former Soviet republics on Russia's rim, without the effective protection they enjoyed during the communist period, and with conflicts raging among several of these neighbors from Georgia to Tajikistan, Russia worries anew about its periphery, and the geographic term *Near Abroad* has become a household word in Moscow. The Russians argue that, the collapse of their Soviet Union notwithstanding, they have the right to be involved, and if necessary to intervene, in their former colonial empire.

Heartlands and Rimlands

Russia's relative location has long been the subject of study and conjecture by geographers. Nearly a century ago, the British geographer Sir Halford Mackinder (1861–1947) argued in a still-discussed article entitled ''The Geographical Pivot of History'' that western Russia and Eastern Europe enjoyed a combination of natural protection and resource wealth that would someday propel its occupants to world power. The protected core area, he reasoned, far overshadowed the exposed periphery. Eventually, this pivotal interior region of Eurasia, which he later called the *heartland*, would become a stage for world domination.

Mackinder's **heartland theory** was published at a time when Russia was a weak, lagging, revolution-ripe society, but Mackinder stuck to his guns. When Russia, now in control of Eastern Europe and with a vast colonial empire, emerged from the Second World War as a superpower, Mackinder's conjectures seemed accurate.

But not all geographers agreed. Probably the first scholar to use the term *rimland* (today Pacific Rim is part of everyday language!) was Nicholas Spykman, who in 1944 countered Mackinder by calculating that Eurasia's periphery, not its core, held the key to global power. Spykman foresaw the rise of rimland states as superpowers and viewed Japan's emergence as just the beginning of that process.

Internal Problems

In the late 1990s, however, Russia is reorganizing rather than pursuing global strategic objectives. Its political spatial structure, as we have noted, continues to be modified.

Its economic geography also is in transition, a vast reorganization that is inflicting much pain on a fractured and sometimes bewildered people. The Soviet Union always was a country of huge regional disparities, inadequate infrastructures, and deep social contradictions. Enormous problems confront the Russian Federation in its aftermath.

Comparisons between the pre-1991 Soviet Russia and the present state are difficult to make, because the information published during Soviet times often was skewed for political or propaganda purposes. For decades the Soviet Union presented itself as a Second World alternative to the capitalist First World, equal in most respects, ahead in some (such as the space race during the 1950s), and comparable based on official statistics. The Soviet ruble was valued as high as U.S. $1.30; census data ranging from the ethnic composition of the republics to the productivity of agriculture, and from energy consumption to health conditions, presented a positive picture of the workers' paradise.

But Gorbachev's *glasnost* campaign during the late 1980s exposed flaws that the secretive system had hidden, and the post-Soviet decade of the 1990s revealed the full extent of the misrepresentation in which the Soviets at home, and the world at large, had believed. The Russian ruble fell to more than 1,000 to the dollar. Health conditions turned out to be far worse than indicated. There were reports of a dramatic drop in male life expectancy, from

Economic dislocation in Russia during "restructuring" has had an especially severe impact on agriculture. Farm output has plummeted in a country heavily dependent on its own production. Privatization of collective farms has gone haltingly, has been resisted in many areas of the country, and has led to labor problems as farmers abandoned the land and moved to the cities. The government, alarmed at the effect of all this on production, has resorted to the use of the military to harvest crops. Here, soldiers are harvesting turnips on a farm in the Vologda Region.

over 70 to under 60 years (the 1995 figure is 59). In fact, while male life expectancy did indeed decline, it probably never rose as high as census data had suggested. Based on World Bank criteria (see map, p. 29), the Soviet Union should have been classified as a lower-middle-income

economy. Significantly, Russia's current status prompted the Bank in 1996 to make that reclassification, demoting the country from the upper-middle-income designation it had held since these rankings were first published.

So many problems beset the Russian economy that recovery will take decades. In agriculture, the reprivatization of farming is a slow and dangerous process, risking dislocation and lowered production; and as always, the harsh natural environment is playing its role. In 1994, an unusually cold summer stunted Russia's crops, and then heavy rains washed some of them away. The production of meat, milk, vegetables, and grain has declined every year since 1989.

In industrial production, the Russian economy is suspended between the ponderous, inefficient state enterprises still operating and the brutal requirements of the free market. The country's extraction and sale of raw materials (including oil) is hampered by the continuing inefficiencies and breakdowns of the infrastructure inherited from Soviet times.

All these problems have political implications. Should the federation's tenuous stability be damaged by disorder arising from unemployment, pension-fund shortages, separatist movements, food shortages, or other causes, Russia's transition to democracy could be halted. Former communists and strident nationalists would enlarge their already substantial role in national affairs. And despite its disastrous campaigns in Afghanistan and Chechnya, the Russian army still is a potential political force to be reckoned with. Russia inherited from the Soviet Union an enormous arsenal of nuclear weapons, and its armed forces still have the capacity to intervene at home and to imperil the world.

BORDERLANDS OF THE SENSITIVE EAST

FIGURE 2-9

Russia's boundaries with its neighbors in many areas are sensitive and disputed. In response, these borders have been demarcated and reinforced. This electrified fence, running through the center of an off-limits border zone, marks the contact between Russia and China near the western corner of Mongolia.

External Challenges

Take the train eastward from Moscow, and for nearly a week a seemingly changeless landscape unfolds: vast stretches of forest; villages of wooden houses with muddy, unpaved streets; bleak cities with rows of gray tenements and clusters of smog-belching smokestacks. Beyond Novosibirsk the scenery becomes more varied, and after Irkutsk and the shores of Lake Baykal it is in places spectacular. Here, too, it gets colder, and the ride seems longer than ever. For good reason: the Trans-Siberian Railroad now makes a giant bend past Khabarovsk before turning south for Vladivostok. A direct route across China would cut more than a day off the trip. But no such link exists.

The routes of Russia's eastern railroads reflect the troubled relationships the tsars as well as the communists had with their Chinese neighbors. Tsarist Russia's plan to annex China's Northeast (Manchuria) and to build a line between Irkutsk and Vladivostok was foiled; nor did such a route materialize during the days when Soviet advisers counseled their Chinese communist comrades. Rather, China reminded Moscow that Russia's eastern flank was captured from the Chinese during the Manchu period (see the map on p. 426). In the eighteenth century, China's sphere of influence extended far beyond the Amur River.

Today, much of the Amur River forms the boundary between China and Russia (Fig. 2-9).

In 1995, when the ruling clique in Vladivostok summarily ousted the hundreds of Chinese street vendors who had come to ply their wares in the city, Chinese newspapers printed maps reminding their readers of the area's historical geography. The Chinese traders, it was argued, were on their "own" territory. Indeed, China and Russia have intermittently disputed (and skirmished about) the boundaries along the Amur and Ussuri rivers. After the collapse of the Soviet Union, Russia took the initiative to secure treaty agreements on sections of these borders, and in 1992 settlements were reached on more than a dozen disputed segments (others remain unresolved). But for the future, the larger question will involve China's "lost" territories. As Figure 2-9 shows, Russia's eastern flank just touches North Korea, virtually landlocking China's historic industrial heartland. National security will match economic development as a challenge for Moscow in this remote arena.

Russia's other external challenges are more immediate but probably more surmountable as well. They include:

1. *Reasserting influence in the Near Abroad.* Three issues confront Russia here: the fate of Russians now under the jurisdiction of new governments (see box titled "The Russians are Leaving!"); the suppression of conflict in the newly independent states; and the sharing of oil and gas revenues from reserves in neighboring countries.

2. *Sustaining the Commonwealth of Independent States.* Upon the collapse of the Soviet Union, Russia took the lead in establishing a multinational union of former Soviet republics that would form a successor to the U.S.S.R. in international and global economic affairs. By 1996, the organization seemed to have lost most of its relevance; its main outcome appeared to be a closer relationship between co-founder Belarus and Russia.

3. *Reaffirming a role in Eastern Europe.* Russia is a dominantly Slavic state, and Slavic peoples in Eastern Europe look to Moscow for support. During the war of the 1990s in former Yugoslavia, Russia proclaimed its support for Serb causes and threatened intervention when NATO attacked Serbian positions. Moscow may use future disputes in Eastern Europe to reestablish its influence there.

Bounded by 14 countries and flanked by several more from Sweden in the west to the United States and Japan in the east, occupying the core of the world's greatest landmass and sharing it with the world's two most populous states, Russia remains a strong, if diminished, force in international affairs. The nature of the temporary "New World Order," now in the making, will depend substantially on the course of events in the land forged by the tsars.

THE RUSSIANS ARE LEAVING!

■ ■ ■

While Moscow frets about the fate of ethnic Russians in the dominantly non-Russian Near Abroad, many of those Russians are making their own decision: they are leaving. And they are not the only ones to abandon the newly independent republics on Russia's periphery. Others without long-term ancestral roots in the area are also departing—if they can. Kazakhstan alone, according to United Nations observers, has seen an exodus of nearly 1.2 million ethnic Russians. At least another 500,000 ethnic Germans have also left. Estimates vary, but available data suggest that as many as 3.5 million Russians may have left the Near Abroad since 1991.

It is a classic case of what geographers call *push factors* causing a migration stream (see box, p. 165). As the new governments install local languages as official, relegating Russian to secondary status, change the education system, deprive Russians of the privileges and priorities they once enjoyed, revive local traditions and reform the economic system in their constituents' favor, Russians feel excluded—if not threatened. Those who can (the well-educated and employable, the better-off) are leaving in large numbers. Left are the elderly and the poor, reducing the likelihood of cultural clashes as their strength as a community wanes. Still, the Russian population of the Near Abroad exceeds 20 million in the late 1990s, and Moscow's concern is not unwarranted.

■ ■ ■

REGIONS OF THE RUSSIAN REALM

So vast is Russia, so varied its physiography, and so diverse its cultural landscape, that regionalization requires a small-scale perspective and a high level of generalization. Figure 2-10 creates a four-region framework within Russia proper, with a fifth region, the Transcaucasian Transition Zone, in the contentious area between the Black and Caspian seas.

❖ RUSSIAN CORE

The heartland of a state is its **core area**. Here a large part of the population is concentrated, and leading cities, major industries, dense transport networks, intensively cultivated lands, and other essentials of the country cluster within a relatively small region. Core areas of long standing carry the imprints of culture and history especially strongly. The Russian core area, broadly defined, is the region that extends from the western border of the Russian realm to the Ural Mountains in the east (Fig. 2-10). This is the Russia of Moscow and St. Petersburg, of the Volga River and its industrial cities, of farms and forests. Here, the Muscovy Russians asserted their power and began the formation of

the Russian Empire, forerunner to the now-defunct Soviet Union.

Central Industrial Region

At the heart of the Russian Core lies the Central Industrial Region (Fig. 2-11). The precise definition of this subregion varies, for all regional definitions are subject to debate. Some geographers prefer to call this the Moscow Region, thereby emphasizing that for over 250 miles (400 km) in all directions from the capital, regional orientations are toward this historic focus of the state. As the map shows, Moscow has maintained its decisive **centrality**: roads and railroads radiate in all directions to Ukraine in the south; to Mensk (Belarus) and the rest of Eastern Europe in the west; to St. Petersburg and the Baltic coast in the northwest; to Nizhniy Novgorod (formerly Gorkiy) and the Urals in the east; to the cities and waterways of the Volga Basin in the southeast (a canal links Moscow to the Volga, Russia's most important navigable river); and even to the subarctic northland that faces the Barents Sea (see box titled ''Facing the Barents Sea'').

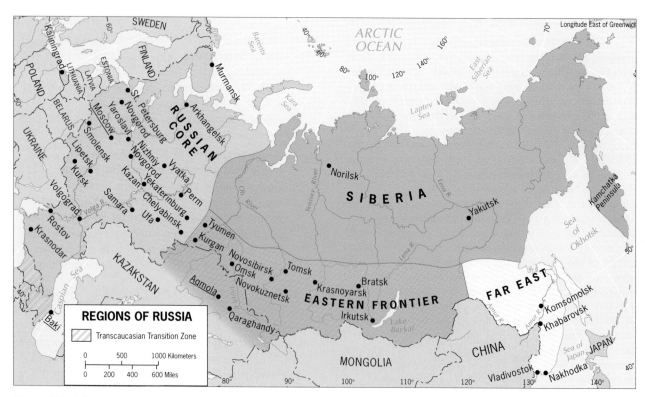

FIGURE 2-10

Major Cities of the Realm

City	Population* (in millions)
Baki, Azerbaijan	1.9
Irkutsk, Russia	0.7
Kazan, Russia	1.1
Moscow, Russia	9.3
Nizhniy Novgorod, Russia	1.5
Novosibirsk, Russia	1.5
St. Petersburg, Russia	5.1
Tbilisi, Georgia	1.4
Vladivostok, Russia	0.7
Volgograd, Russia	1.0
Yekaterinburg, Russia	1.4
Yerevan, Armenia	1.4

*Based on 1997 Population Estimates.

Moscow itself is a transforming metropolis (1997 population: 9.3 million), with high-rise apartment complexes increasingly dominating the residential landscape. Although this construction has helped alleviate the capital's severe housing shortage, Moscow—like nearly all Russian cities—remains badly overcrowded, with most people forced to accept unreasonably cramped personal living spaces.

Moscow is also the focus of an area that includes some 50 million inhabitants (one-third of the country's total population), many of them concentrated in such major cities as Nizhniy Novgorod, the automobile-producing "Soviet Detroit"; Yaroslavl, the tire-producing center; Ivanovo, the heart of the textile industry; and Tula, the mining and metallurgical center where lignite (brown coal) deposits are worked.

St. Petersburg (the former Leningrad) remains Russia's second city, with a population of 5.1 million. There was a time when St. Petersburg was the focus of Russian political and cultural life and Moscow was a distant second city. Today, however, St. Petersburg has none of Moscow's locational advantages, at least not with respect to the domestic market. It lies well outside the Central Industrial Region near the northwestern corner of the country, 400 miles (650 km) from Moscow. Neither is it better off than Moscow in terms of resources: fuels, metals, and foodstuffs must all be brought in, mostly from far away. The former Soviet emphasis on self-sufficiency even reduced St. Petersburg's asset of coastal location, because some raw materials could have been imported much more cheaply across the Baltic Sea from foreign sources than from do-

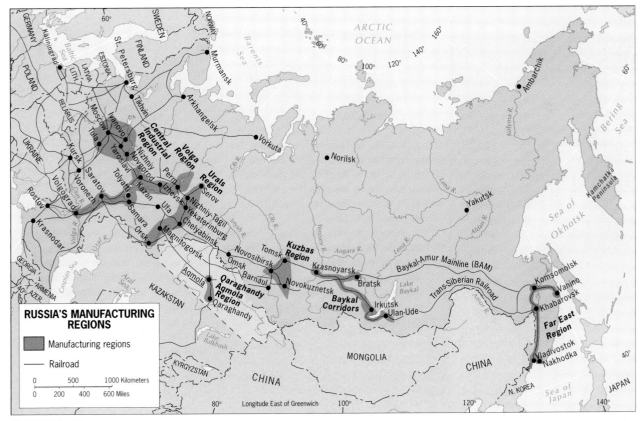

FIGURE 2-11

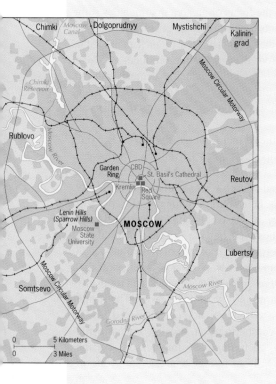

Moscow

In the vastness of Russia's territory, Moscow, capital of the federation, seems to lie far from the center, close to its western margin. But in terms of Russia's population distribution, Moscow's centrality is second to none among the country's cities. Moscow lies at the heart of Russia's primary core area and at the focus of its circulation systems.

On the banks of the meandering Moscow River, Moscow's skyline of onion-shaped church domes and modern buildings rises from the forested flat Russian Plain like a giant oasis in a verdant setting. Archeological evidence points to ancient settlement of the site, but Moscow enters recorded history in the middle of the twelfth century. Forest and river provided defenses against Tatar raids, and when a Muscovy force defeated a Tatar army in the late fourteenth century, Moscow's primacy was assured. A huge brick Kremlin (citadel; fortress) with walls more than a mile in length and with 18 towers was built to ensure the city's security. From this base, Ivan the Terrible expanded Muscovy's domain and laid the foundations for Russia's vast empire.

The triangular Kremlin and the enormous square in front of it (Red Square of revolutionary times), flanked by St. Basil's Cathedral and overlooking the Moscow River, is the center of old Moscow and is still the heart of the city. From here, avenues and streets radiate in all directions to the Garden Ring and beyond. Until the 1970s, the Moscow Circular Motorway encircled most of the built-up area, but today the city sprawls far outside this beltway. For all its size, Moscow never developed a world-class downtown skyline. Communist policy was to create a "socialist city" in which neighborhoods—consisting of apartment buildings, workplaces, schools, hospitals, shops, and other amenities—would be clustered so as to obviate long commutes into a high-rise city center.

Moscow may not be known for its architectural appeal, but the city contains noteworthy historic as well as modern structures, including the Cathedral of the Archangel and the towers of Moscow State University.

St. Petersburg

Tsar Peter the Great and his architects transformed the islands of the Neva Delta, at the head of the Gulf of Finland, into the Venice of the North, its palaces, churches, waterfront facades, bridges, and monuments giving St. Petersburg a European look unlike any other city in Russia. Having driven out the Swedes, Peter laid the foundations of the Peter and Paul fortress on the bank of the wide Neva River in 1703, and the city was declared the capital of Russia in 1712.

Peter's "window on Europe" (named after the saint, not the tsar) was to become a capital to match Paris, Rome, and London. During the eighteenth century, St. Petersburg acquired a magnificent skyline dominated by tall, thin spires and ornate cupolas, and graced by baroque and classical architecture. The Imperial Winter Palace and the adjoining Hermitage Museum at the heart of the city are among a host of surviving architectural treasures.

Revolution and war lay in St. Petersburg's future. In 1917, the Russian Revolution began in the city (then named Petrograd), and following the communist victory it lost its capital functions to Moscow and its name to Lenin. As Leningrad, it suffered through an 872-day Nazi siege during World War II, holding out heroically through endless bombardment and starvation that took nearly 1 million lives and severely damaged many of its buildings.

The communist period witnessed the neglect and destruction of some of Leningrad's most beautiful churches, the intrusion and addition of crude monuments and bleak apartment complexes, and the rapid industrialization and growth of the city. Immediately after the collapse of the Soviet Union, it regained its original name, and a new era opened. The Russian Orthodox church revived, building restoration went forward, and tourism boomed. But the transformation from communist to capitalist ways has been accompanied by social problems that include crime and poverty.

FACING THE BARENTS SEA

■ ■ ■

North of the latitude of St. Petersburg, the region we have named the Russian Core takes on Siberian properties. However, the Russian presence is much stronger in this remote northland than it is in Siberia proper. Two substantial cities, Murmansk and Arkhangelsk, are road- and rail-connected outposts in the shadow of the Arctic Circle.

Murmansk lies on the Kola Peninsula not far from the border with Finland (Fig. 2-7). In its hinterland lie a variety of mineral deposits, but Murmansk is particularly important as a naval base. During World War II, Allied ships brought supplies to Murmansk; the city's remoteness shielded it from German occupation. After the war, it became a base for nuclear submarines. This city is also an important fishing port and a container facility for cargo ships.

Arkhangelsk is located near the mouth of the Northern Dvina River where it reaches an arm of the White Sea (Fig. 2-7). Its site was chosen by Ivan the Terrible during Muscovy's early expansion; the tsar wanted to make this the key port on a route to maritime Europe. Yet Arkhangelsk, mainly a port for lumber shipments, suffers from a more restricted ice-free season than Murmansk, whose port can be kept open with the help of the North Atlantic Drift ocean current (and by icebreakers when the need arises).

Nothing in Siberia east of the Urals rivals either of these cities—yet. But their very existence and growth prove that Siberian barriers to settlement can be overcome.

■ ■ ■

From the Field Notes

"Although the former Soviet Union was a prodigious producer of equipment, almost all of this production was for domestic use; little was exported to the non-communist world. One exception was weapons, and Russia today continues to market arms worldwide. Here in the port of St. Petersburg, cannons and armored cars await transportation, reportedly to India. Also note the red farm tractors ready for transfer, the large container facility, and St. Petersburg's high-rise apartment buildings from the Soviet era in the background."

mestic sites in distant Central Asia (only bauxite deposits lie nearby, at Tikhvin).

Yet St. Petersburg was at the vanguard of the Industrial Revolution in Russia, and its specialization and skills have remained. Today, the city and its immediate environs contribute about 10 percent of the country's manufacturing, much of it through the building of high-quality machinery. In addition to the usual association of industries (metals, chemicals, textiles, and food processing), St. Petersburg has major shipbuilding plants and, of course, its port and naval station. This productive complex, though not enough to maintain its temporary pre-Revolution advantage over Moscow, kept the city in the forefront of modern Soviet development and will continue to do so in the new Russia.

Povolzhye: The Volga Region

A second region lying within the Russian Core is the *Povolzhye*, the Russian name for an area that extends along the middle and lower valley of the Volga River. It would be appropriate to call this the Volga Region, for that greatest of Russia's rivers is its lifeline, and most of the cities that lie in the *Povolzhye* are situated on its banks (Fig. 2-11). In the 1950s, a canal was completed to link the lower

Volga with the lower Don River (and thereby the Black Sea), extending this region's waterway system still further.

The Volga River was an important historic route in old Russia, but for a long time neighboring regions overshadowed it. The Moscow area and Ukraine were far ahead in industry and agriculture. The Industrial Revolution that came late in the nineteenth century to the Moscow Region left the *Povolzhye* little affected. Its major function remained the transit of foodstuffs and raw materials to and from other regions.

This transport function is still important, but things have changed in the *Povolzhye*. First, World War II brought furious development because the Volga River, located east of Ukraine, was protected by distance from the German armies that invaded from the west. Second, in the postwar era the Volga-Urals region proved to be the greatest source

It is unlikely that you have ever seen a Russian car in the street, but Russia does have automobile factories. This view of part of the LADA auto assembly plant in Tolyatti, on the Volga River south of Kazan, shows vehicles ready for shipment.

of petroleum and natural gas in the entire Soviet Union. From near Volgograd (formerly Stalingrad) in the southwest to Perm on the Urals' flank in the northeast lies a belt of major oilfields (Fig. 2-12). These deposits were once believed to constitute the former Soviet Union's largest reserve, but as the map shows, later discoveries in western Siberia indicate that even more extensive oil- and gasfields lie beyond the Urals. Still, because of their location and size, the Volga Region's fossil-fuel reserves retain their importance.

Third, the transport system has been greatly expanded. The Volga-Don Canal directly connects the Volga waterway to the Black Sea; the Moscow Canal extends the northern navigability of this river system into the very heart of the Central Industrial Region; and the Mariinsk canals provide a link to the Baltic Sea. Today, the Volga Region's population exceeds 25 million, and the cities of Samara (formerly Kuybyshev), Volgograd, Kazan, and Saratov all have populations in the range of 1.0 to 1.3 million. Manufacturing has also expanded into the middle Volga Basin, emphasizing more specialized engineering industries. The huge Fiat-built auto assembly plant in Tolyatti, for example, is one of the world's largest of its kind.

The Urals Region

The Ural Mountains form the eastern limit of the Russian Core. These mountains are not particularly high; in the north they consist of a single range, but southward they broaden into a zone of hilly topography. Nowhere do they form an obstacle to east-west transportation. An enormous storehouse of metallic mineral resources located in and near the Urals has made this area a natural place for industrial development. Today, the Urals Region, well connected to the Volga and Central Industrial region, extends from Serov in the north to Orsk in the south (Fig. 2-11).

The Urals Region rose to prominence during World War II and its aftermath when it benefited from its remoteness from the German invaders, allowing its factories to support the war effort without threat of destruction. Coal had to be imported, but the problem of energy supply here has been considerably relieved by the discovery of oilfields in the zone lying between the Urals and the Volga Region (Fig. 2-12). The more recent West Siberian oil and gas exploitation (from fields lying northeast of the Urals) increasingly supports development in the Urals as well.

The Central Industrial, Volga, and Urals regions form the anchors of the Russian core area. For decades they have been spatially expanding toward one another, their interactions ever more intensive. These regions of the Russian Core stand in sharp contrast to the comparatively less developed, forested, Arctic north and the remote upland south between the Black and Caspian seas. Thus even within this Russian coreland, frontiers still await growth and development.

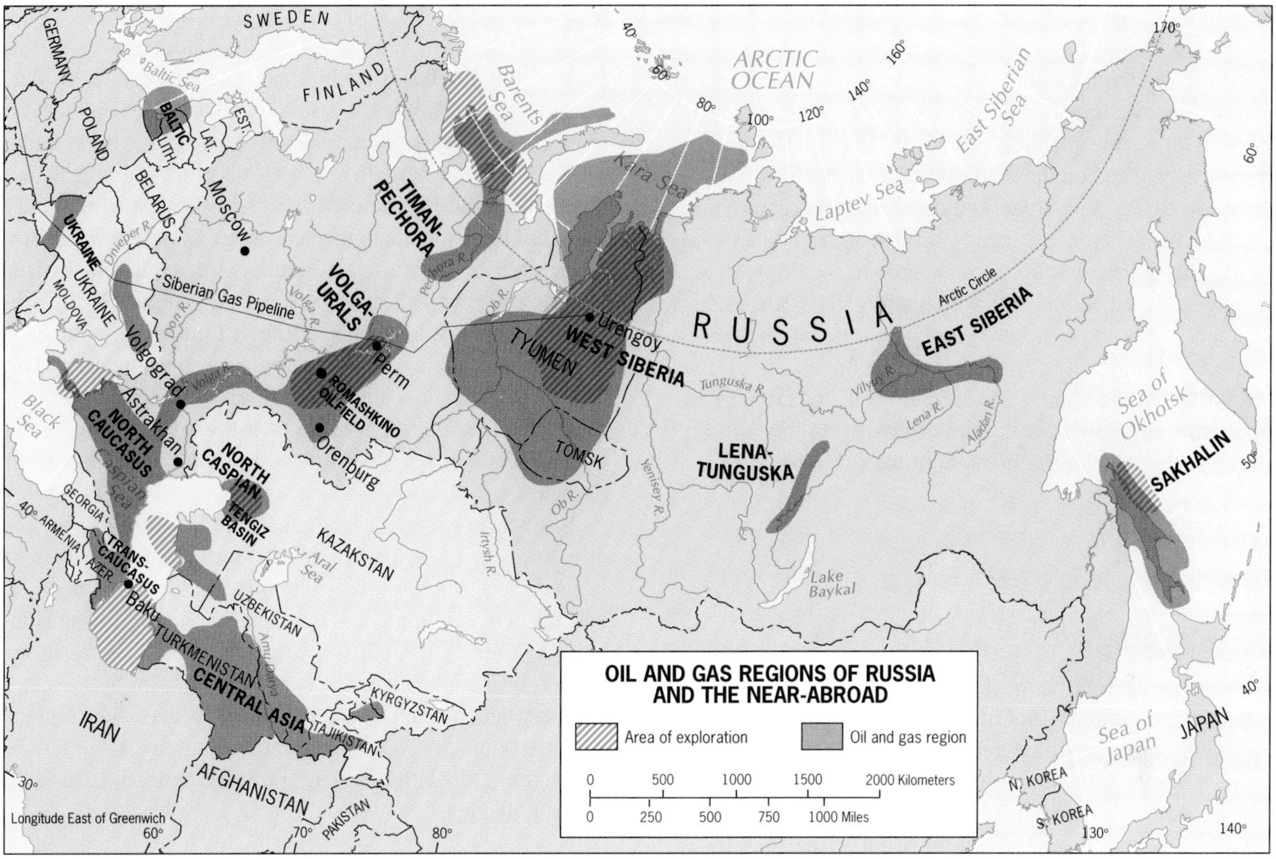

FIGURE 2-12

❖ THE EASTERN FRONTIER

From the eastern flanks of the Ural Mountains to the headwaters of the Amur River, and from the latitude of Tyumen to the northern zone of neighboring Kazakstan, lies Russia's vast Eastern Frontier Region, product of a gigantic experiment in the eastward extension of the Russian Core (Fig. 2-10). As the maps of cities and surface communications suggest, this eastern frontier is more densely peopled and more fully developed in the west than in the east; at the longitude of Lake Baykal, settlement has become linear, marked by ribbons and clusters along the east-west railroads. Three subregions dominate the geography: the Kuznetsk Basin in the west, the Lake Baykal area in the east, and the Qaraghandy-Aqmola zone in northern Kazakstan in the south.

The Kuznetsk Basin (Kuzbas)

Some 900 miles (1,450 km) east of the Urals lies another of Russia's primary regions of heavy manufacturing resulting from the communist period's national planning: the Kuznetsk Basin, or *Kuzbas* (Fig. 2-11). In the 1930s, it was opened up as a supplier of raw materials (especially coal) to the Urals, but that function steadily diminished in importance as local industrialization accelerated. The original plan was to move coal from the Kuzbas west to the Urals and allow the returning trains to carry iron ore east to the coalfields. However, good-quality iron ore deposits were subsequently discovered in the vicinity of the Kuznetsk Basin itself. As the new resource-based Kuzbas industries grew, so did its urban centers. The leading city, located just outside the region, is Novosibirsk, which stands at the intersection of the Trans-Siberian Railroad and the Ob River as the very symbol of Russian enterprise in the vast eastern interior. To the northeast lies Tomsk, one of the oldest Russian towns in all of Siberia, founded three centuries before the Bolshevik uprising and now caught up in the modern development of the Kuzbas region. Southeast of Novosibirsk lies Novokuznetsk, a city that specializes in the production of steel for the region's machine and metalworking plants; aluminum products using Urals bauxite are also manufactured here.

Impressive as the concentration of coal, iron, and other resources may be in the Kuznetsk Basin, the industrial and urban development that has taken place there must in large measure be attributed once again to the ability of the communist state and its planners to promote this kind of expansion, notwithstanding what capitalists would see as excessive investment. In the absence of such financial constraints, the state planners were able to push the country vigorously ahead on the road toward industrialization, with the hope that certain areas would successfully reach ''takeoff'' levels. Then these areas would require progressively fewer direct investments because growth would to an ever greater extent become self-perpetuating. The Kuzbas, for instance, was expected to grow into one of the Soviet Union's leading industrial agglomerations, with its own important market and with a location, if not favorable to the Urals and points west, then at least fortuitous with respect to the developing markets of the Far East.

Qaraghandy-Aqmola (formerly Karaganda-Tselinograd)

The eastward march of Soviet development knew no boundaries, certainly not internal political ones. It spilled over into northern Kazakstan, which, during the Soviet period, became almost totally Russified. As Figure 2-11 shows, northern Kazakstan lay directly in the path of Soviet eastward economic expansion toward the Kuznetsk Basin, and transport routes constructed across it connected this area more closely to Russia than it was linked to the rest of Kazakstan. This helps explain why Kazakstan, a Soviet republic established for the Kazaks and other Muslim minorities in this part of Central Asia, had the largest Russian minority of all the Soviet republics: 38 percent of the total (Kazaks made up 40 percent). The ethnic spatial pattern that evolved saw Russians concentrated in the north, while Kazaks and others dominated in the rest of the republic.

The Qaraghandy-Aqmola area (Fig. 2-11) reflects this sequence of events. Qaraghandy (Karaganda), once a small Kazak town, became an iron- and steel-producing center, a Russified city that exchanged raw materials and finished products with other Russian production zones, especially the Urals. Similarly, Aqmola (Tselinograd) emerged in the 1950s as an administrative focus and entrepôt for the disastrous Virgin and Idle Lands Program.

Today, Russians are leaving Kazakstan in large numbers. Independence, Kazak nationalism, and Islamization have combined to make Russians second-class citizens in this part of the Near Abroad, especially in the remote capital of Almaty which is being shifted to Aqmola in 1998. But many workers in the north are unable to join the more than 1 million Russians (and over 500,000 ethnic Germans) who had left by mid-1996. The political geography has changed, but northern Kazakstan's economic geography—and associated cultural landscapes—have not. Here, certainly, the regional name **frontier** is appropriate.

The Lake Baykal Area

East of the Kuzbas, development becomes more insular, and distance becomes a stronger adversary (Fig. 2-11). North of the Tuva Republic and eastward around Lake Baykal, larger and smaller settlements cluster along the two railroads to the Pacific coast (Fig. 2-5). West of the lake, these rail corridors lie in the headwater zone of the Yenisey River and its tributaries. A number of dams and hydroelectric projects serve the valley of the Angara River, particularly the city of Bratsk. Mining, lumbering, and some farming sustain life here, but isolation dominates it. The city of Irkutsk, near the southern end of Lake Baykal, is the principal service center for a vast Siberian region to the north and for a lengthy east-west stretch of southeastern Russia.

Eastward beyond Lake Baykal, the Eastern Frontier really lives up to its name: this is southern Russia's most rugged, remote, forbidding country. Settlements are few and far between; many of them are mere camps. The Buryat Republic (Fig. 2-6) is part of this zone; the territory bordering it to the east was taken from China by the tsars and may become an issue in the future. Where the Russian-Chinese boundary turns southward, along the Amur River, the region called the Eastern Frontier ends and Russia's Far East begins.

❖ SIBERIA

Before we assess the potentials of Russia's Pacific Rim, we should remind ourselves that the ribbons of settlement just discussed hug the southern perimeter of this giant country, avoiding the vast Siberian region to the north (Fig. 2-10). Siberia extends from the Ural Mountains to the Kamchatka Peninsula—a vast, bleak, frigid, forbidding land. Larger than the conterminous United States but inhabited by only an estimated 15 million people, Siberia quintessentially symbolizes the Russian environmental plight: vast distances, cold temperatures worsened by strong Arctic winds, difficult terrain, poor soils, and limited options for survival.

But Siberia is also a region of resources. From the days of the first Russian explorers and Cossack adventurers, word of Siberia's riches filtered back to the west. Gold, diamonds, and other precious minerals were found. Later, metallic ores including iron and bauxite were discovered. Still more recently, the Siberian interior proved to contain sizeable quantities of oil and natural gas (Fig. 2-12) and began to contribute significantly to the country's energy supply.

As the physiographic map (Fig. 2-2) shows, several major rivers—the Ob, Yenisey, and Lena—flow gently northward across Siberia and the Arctic Lowland into the Arctic

Ocean. Hydroelectric power development in the basins of these rivers has generated electricity used in the extraction and refinement of local ores, and in the lumber mills that have been set up to exploit the vast Siberian forests. Soviet planners even gave serious consideration to grandiose schemes to reverse the flow of Siberian streams (to divert Siberian water onto parched farmlands far to the southwest)—ideas their Russian successors are likely to discard.

The human geography of Siberia is fragmented, and much of the region is for all intents and purposes uninhabited (Fig. 2-3). Ribbons of Russian settlement have developed; the Yenisey River, for instance, can be traced on this map of Soviet peoples (a series of small settlements north of Krasnoyarsk), and the upper Lena valley is similarly fringed by ethnic Russian settlement. Yet these ribbons and other islands of habitation are separated by hundreds of miles of empty territory. During the worst excesses of the Stalinist period (and long afterward as well), dissidents and criminals were exiled to Siberian mining and lumbering camps to serve their terms at hard labor under the harshest of conditions; many never returned.

The political geography of eastern Siberia is marked by the growing identity of Sakha (the Yakut Republic). As additional resources are discovered here (including oil and natural gas), the importance of this republic, centered on the capital, Yakutsk, will increase.

Siberia, Russia's freezer, is stocked with goods that may become mainstays of future national development. Already, precious metals and mineral fuels are contributing importantly to the Russian economy. As time passes, we may expect Siberian resources to sustain the development of the Eastern Frontier. As that happens, this southern zone of development will expand northward as well as eastward, bringing more of Siberia into the national sphere. One step in that direction has already been accomplished with the completion of the BAM (Baykal-Amur Mainline) Railroad during the 1980s. This route, lying north of and parallel to the Trans-Siberian Railroad, extends 2,200 miles (3,540 km) eastward from Tayshet (near Krasnoyarsk) directly to Komsomolsk (Fig. 2-5). To date, however, the BAM has faced problems, and only portions of it are in regular use. Nonetheless, it is still expected to become the axis of a major new industrial corridor in the twenty-first century.

❖ THE FAR EAST

Russia has about 5,000 miles (8,000 km) of Pacific coastline—more than the United States (including Alaska). However, most of its coastal margin lies north of the latitude of the State of Washington, and from the point of view of coldness, the Russian coastline lies on the ''wrong'' side of the Pacific. The port of Vladivostok, the original eastern terminus of the Trans-Siberian Railroad, lies at a latitude

Oil and natural gas are among the former Soviet Union's—and Russia's—most valuable resources, and during the Soviet period the U.S.S.R. was the world's leading producer, mainly for domestic consumption. Equipment and facilities, however, were subject to breakdowns and inadequate maintenance. Oil spills and other accidents went unreported in secretive Soviet society. Today, the full extent of these problems is becoming known, and pollution events are reported and investigated. Here is an example: the disastrous 1994 oil spill in the Usinsk Region in northern Russia, which created huge pools of oil in the taiga. Russian authorities insisted that "only" 13,000 tons of oil had escaped. Independent investigators estimated the spill at 280,000 tons.

about midway between San Francisco and Seattle, but it must be kept open throughout the winter by icebreakers—something unheard of in these American ports. The climate, to say the least, is harsh. Winters are long and bitterly cold; summers are cool.

Nonetheless, the Russians have been determined to develop their Far East Region (Fig. 2-10) as intensively as possible. Their resolve has recently been spurred by ideological differences between democratizing Russia and hardline-communist China. In the Asian interior, high mountains and empty deserts mark most of the Russo-Chinese borderland zone; south of the Kuznetsk Basin and Lake Baykal, the Republic of Mongolia functions as a sort of buffer between Chinese and Russian interests. But in the Pacific area, the Russian Far East and Northeast China confront each other directly across an uneasy river boundary (along the Amur and its tributary, the Ussuri). Not only the Russian tsars but also the Soviet communists always wanted to consolidate their distant Pacific outposts. The Soviets even offered financial and other incentives to western residents willing to move east.

As the map of climates (Fig. I-7) reminds us, the Russian Far East is no Garden of Eden. The proximity of the

Vladivostok today is a city that personifies the Russian "Wild East." During Soviet times this was a closed port, a huge naval military complex accessible only to authorized Communist Party members of high standing. Today the harbor brims with freighters bringing in contraband ranging from used cars to videotapes; the Russian clique that rules the city envisages a "Maritime Republic" and pays little heed to Moscow's dictums; and the once-proud Russian fleet lies rusting at the docks. Now Vladivostok's leaders want to wrest commercial and fishing primacy away from nearby Nakhodka, the port built to accommodate trading activities while Vladivostok was closed to outsiders. It is a wild time on the Russian Pacific Rim.

Pacific Ocean is not very effective in moderating the cold; summers are only slightly longer and milder than they are in Siberia. Most of the Far East, with its rugged terrain and vast wilderness, remains a sparsely peopled region. Lumbering centers and fishing villages only occasionally break the emptiness of the countryside; seafood is still the leading product here. From Vladivostok and points north, Russian fishing fleets sail the Sea of Okhotsk and the northern Pacific in search of salmon, herring, cod, and mackerel to be frozen, canned, and shipped via rail to far-off western markets.

The mineral potential of the Far East, like that of eastern Siberia, is still only partially known. A deposit of high-quality coal lies in the Bureya River valley (a tributary of the Amur), and there is iron ore near Komsomolsk, now the region's first steel-producing center. Tin deposits near Komsomolsk are especially important because they are Russia's main source.

The major axis of urban and industrial development in the Far East has emerged along the Amur-Ussuri river system, from Vladivostok in the south (with its huge military installations, shipbuilding, and fish-processing plants) to Komsomolsk in the north (Fig. 2-11). Near the confluence of the two rivers lies Khabarovsk, the city with the greatest advantages of centrality. At Khabarovsk, machine and metal-working industries process iron and steel from Komsomolsk; chemical industries depend on nearby Sakhalin's oil, and furniture factories use the area's ubiquitous timber.

Until now, very little development has taken place in Vanino, the coastal terminus of the new BAM Railroad. There is a ferry to Kholmsk on the island of Sakhalin, and Vanino and its neighbor, Gavan, are home to small fishing fleets. Until the opening of Vladivostok, the more significant growth had been taking place at the new coastal endpoint of the Trans-Siberian Railroad, which now goes 55 miles (90 km) beyond Vladivostok to Nakhodka. As Figure 2-5 shows, this rapidly growing port is the southernmost city of the Russian Far East and faces the Sea of Japan; across the water lies one of the world's most powerful—and most raw-material-deficient—economies.

Why did Soviet-Japanese economic interactions in this area fail to develop, given its regional geography? The circumstances are ironic. Japan and the former Soviet Union never signed a treaty ending the war between them (World War II) because the Soviets occupied and annexed four small Japanese islands in the Kurile chain, northeast of Hokkaido (see Fig. 9-15). The Japanese demanded their return, the Soviets refused, and there the matter stood, holding up not only the treaty but also economic relations between the two powers. In 1990, then-Soviet President Gorbachev met with Japanese leaders, who offered a (U.S.) $26 billion deal for the islands, a proposal that included exploration in the Russian Far East and eastern Siberia, the purchase of raw materials, and assistance in the development of the region's infrastructure. This negotiation was abrogated by the collapse of the Soviet Union, and when the new Russian president Yeltsin faced the Japanese, he vowed not to yield ''one square meter'' of Russian territory. Despite some diplomatic efforts since then, the matter still stands there, holding up the growth of an entire region and delaying Russia's entry into the economic sphere of the East Asian Pacific Rim.

The Russian Far East today sometimes is compared to the American Wild West of another era. Thievery, violent crime, corruption, and high-handed actions by authorities make this a difficult place in which to do business. Add to this the unreliability of train transport to and from the interior and the slow workings of the Vostochni container

FIGURE 2-13

port at Nakhodka, and it is evident that the advantages of relative location can be negated by inaction and mismanagement. The Russian Far East is a region of great potential, but internal as well as external obstacles stand in the way. Not only is the Japanese opportunity being lost, but looming on the landward side is a China whose best port of exit would be . . . Vladivostok. Nothing the Russians of the Far East have done would suggest any interest in fostering Russian-Chinese cooperation, however. And so the Maritime Republic, if it ever achieves approval in Moscow, is likely to have a turbulent youth.

❖ TRANSCAUCASIAN TRANSITION ZONE

Between the Russian geographic realm and the North Africa/Southwest Asian realm lies a geographic region of enormous physiographic, historical, and cultural complexity. Physiographically, it is dominated by the Caucasus

Mountains, in whose valleys countless struggles have been waged. Historically, this has been—and is—a battleground for Christians and Muslims, Russians and Turks, Armenians and Persians. Culturally, the region has been a jigsaw of languages, religions, and traditions. It is, by many measures, the Balkans of Asia.

The geographic name for this turbulent region signifies its changeable character: Transcaucasian Transition Zone (TTZ). As Figure 2-13 shows, the TTZ presently consists of three political entities: Georgia, Armenia, and Azerbaijan. All three are former republics of the Soviet Union, and all three are part of what Russia conceives of as its Near Abroad

Before we examine the regional geography of the TTZ, we should consider its relative location. To the north lies the Russian Federation—but not Russia itself. Almost all the way across the TTZ's northern flank, from the Caspian to the Black Sea, lies the tier of Russia's internal republics discussed earlier: some are stable and quiet, others volatile and active (Fig. 2-8). To the south lie two Muslim countries, each with links to the region: Iran, with historic ties to Azerbaijan, and Turkey, with a long history of adversarial relationships with Armenians. These outside forces continue to influence the course of events in Transcaucasia—just as external powers have long done in Eastern Europe's Balkan **shatter belt**.

Where does the TTZ, as a region, belong in the global spatial framework? That question remains unanswered in the late 1990s. Georgia, the country that faces the Black Sea, has proclaimed its ''Europeanness,'' and its embattled government has stated its intention to seek closer ties to Europe. Armenia, landlocked and fragmented, is a Christian country nearly surrounded by Muslim domains. And Azerbaijan, as we see in Figure 2-13, is actually the northern half of a larger province, the southern part of which lies in Iran today. So pressing and numerous are the unresolved territorial issues in the TTZ that we discuss the region separately here, awaiting indications of its future disposition. Transcaucasia is in transition; its three republics are small and comparatively weak, and major contenders for influence in the region (and perhaps for control over parts of it) lie nearby.

Georgia

Of the three republics in the TTZ, only Georgia has a Black Sea coast. Somewhat smaller than South Carolina, Georgia is a country of high mountains and fertile valleys, with a complicated political geography. Its population of 5.4 million people is more than 70 percent Georgian but also includes Armenians (8 percent), Russians (6 percent), Azerbaijanis or Azeris (6 percent), and smaller numbers of Ossetians (3 percent) and Abkhazians (2 percent). Within the republic lie three minority-based autonomous entities, the Abkhazian and Ajarian Autonomous Republics, and the South Ossetian Autonomous Region (Fig. 2-13).

Sakartvelo, as the Georgians call their country, has a long and turbulent history. Tbilisi, the capital for 15 centuries, lay at the core of a major empire around the turn of the thirteenth century, but the Mongol invasion ended this era. Next, the Christian Georgians found themselves in the path of the Islamic wars between Turks and Persians. Turning northward for protection, the Georgians soon were annexed by the Russians, who were looking for warm-water ports. Like other peoples overpowered by the tsars, the Georgians seized on the Russian Revolution to reassert their independence; but the Soviets reincorporated Georgia in 1921 and proclaimed a Georgian Soviet Socialist Republic in 1936. Josef Stalin, the communist dictator who succeeded Lenin, was a Georgian.

Georgia is renowned for its scenic beauty, warm and favorable climates, varied agricultural production (especially tea), timber, manganese, and other products. Georgian wines, tobacco, and citrus fruits are much in demand. The diversified economy has the potential to support Georgia's consolidation as a viable state.

From the Field Notes

"One of the most arduous trips in memory, this ride in an old bus along the Georgian Military Highway from Yerevan in Armenia via Tbilisi in Georgia and on to Pyatigorsk in Russia, all across the Caucasus Mountains. Near the Georgia-Russia border we stopped in sight of Mount Elbrus (18,510 feet; 5,642 m), tallest peak in the Caucasus and sometimes mistakenly called the highest mountain in Europe. An Eastern Orthodox Church stood on a high mountain nearby. Along the route were numerous small villages, isolated and hidden away in valleys, some with Islamic crescents painted on doors."

Unfortunately, Georgia's political geography is loaded with **centrifugal forces**. After Georgia declared its independence in 1991, factional fighting destroyed its first elected government. But worse was to come. In the Autonomous Region of South Ossetia, which borders the Russian internal republic of North Ossetia, conflict broke out over local demands for self-determination; many of Georgia's Ossetians preferred unification with Russia's Ossetians over minority status in the new Georgian state. Even as this costly civil strife continued, one of Georgia's two internal republics, Abkhazia (in the country's northwest corner) proclaimed its independence, ousting Georgians and setting up a separatist regime.

In time, the Abkhazian uprising would have the direst consequences for Georgia. Initially, Russian army leaders supported the separatist movement, for reasons having less to do with strategy than with comfort and spite: comfort because the Abkhazian Black Sea coast had long been a favorite resort for Russian (Soviet) army brass, and spite because of a lingering animosity toward the Georgian leader, Eduard Shevardnadze, former Soviet foreign minister who was blamed for the Soviet debacle. With Russian help, the Abkhazians, a minority in their own republic, drove 250,000 Georgians from their homes.

However, then Russian attitudes changed—but at a price. In return for permanent Russian military bases on Georgian soil, Moscow ended its support for the Abkhazians and even blockaded the common border between Abkhazia and Russia. When Abkhazia's rebel leaders stood fast, Georgia lost on two counts: unable to regain control over Abkhazia, it lost sovereignty as Russian troops were stationed within its borders.

In the meantime, Georgia's promising economy suffered severely; the tourist industry was moribund, and the country's famed subtropical health resorts (those not destroyed in the conflict) stood empty. If there was a positive side to the political geography of Georgia, it involved the southwestern republic within its borders, Ajaria. No secession attempt was made here; more than 80 percent of the population is Georgian, and of the remainder, 10 percent are Russians and 5 percent Armenians. That bright spot, however, was not enough to give much hope to a country whose post-Soviet hopes were soon dashed by civil war.

Armenia

Armenia, Georgia's landlocked southern neighbor, fared even worse than Georgia politically and far worse environmentally. Armenia, as the atlas map (p. 108) shows, occupies some of the most rugged and mountainous terrain in the earthquake-prone Transcaucasus. Territorially the smallest of all former Soviet republics, Maryland-sized Armenia extends southeastward in a narrow corridor that separates a part of neighboring Azerbaijan from its main body (Fig. 2-13). Moreover, an outlier of Christian Armenian settlements lies surrounded by Muslim Azerbaijan to the east, an **exclave** that was engulfed by ethnic conflict soon after independence came to these countries.

This spatial recipe for trouble was engineered by Soviet sociopolitical planners in the 1920s. Armenians, who adopted Christianity 17 centuries ago, for more than a millennium have tried to establish a secure homeland here on the margin of the Muslim world. During World War I, the Ottoman Turks turned on their Armenian minority in a ruthless campaign of destruction and expulsion, driving the survivors into the Transcaucasus. At the end of the war, an independent Armenia briefly arose in the area of the present country, but it lasted only two years. In 1920, it was taken over by the Soviets; in 1936, it was proclaimed one of the 15 constituent republics of the Soviet Union. The collapse of the Soviet Empire gave Armenia what it had lost three generations earlier: independence.

But independence was no guarantee of stability or prosperity. Indeed, Armenia, on the outer periphery of the former U.S.S.R., benefited more than most Soviet republics from Moscow's control. Protection against Islamic adversaries and considerable economic investment, along with a relaxed communism that came with distance from the Soviet core, gave Armenia security and relative well-being. The Soviet map, however, contained the seeds of trouble. After independence, with Moscow's enforced security gone, Armenia soon was embroiled in a bitter conflict with Azerbaijan over its exclave of Nagorno-Karabakh (Fig. 2-13). Supply lines of oil and gas from Russia across neighboring Georgia stopped functioning, and Azerbaijan cut the pipeline from the Caspian Sea. Armenia's economy, already fragile, collapsed.

And yet Armenia is not without assets. Its population of nearly 4 million is unusually homogeneous for a former Soviet (especially Transcaucasian) entity; well over 90 percent of the people are ethnic Armenians and adhere to the Christian faith. Mineral-rich and producing a variety of subtropical fruits in well-watered valleys, Armenia has economic potential, enhanced by plentiful hydroelectric power derived from Soviet-era installations. The capital, Yerevan, is a major metropolis containing more than one-third of the country's population; it lies within sight of the Turkish border. Today, however, Armenians, as so often in their history, suffer from the aftermath of environmental disaster (a 1988 earthquake killed tens of thousands) and from the vagaries of conflict arising from the actions of neighbors near and far.

Azerbaijan

Whereas Georgia and Armenia, on the western flank of the Transcaucasian Transition Zone, are Christian societies, Azerbaijan, to the east and facing the Caspian Sea, is an old Islamic society with close historic ties to neighboring Iran on its southern border. As we will see in Chapter 6, there also is an Iranian province called Azerbaijan; the split resulted from the Russian southward advance during the nineteenth century. Official Soviet communist atheism notwithstanding, the (now) nearly 8 million *Azeris* never abandoned their Shi'ite faith while Azerbaijan was a Soviet republic, and after independence Islam revived vigorously.

Azerbaijan's territorial morphology as well as its relative location spell trouble. As Figure 2-13 shows, the country is divided into two parts by Armenia's corridor, which separated Naxcivan from the main body; moreover, Muslim Azerbaijan encircles the Christian-Armenian *enclave* of Nagorno-Karabakh. The costly war of the early 1990s over this enclave was a severe setback for both countries, and led not only to thousands of casualties but also to substantial refugee movements. Armenia wanted to link its exclave to its main territory via a corridor; Azerbaijan wanted to drive the Armenians out and establish control over its problem enclave. Mediation by Turkey and other countries reduced the violence, but the issues were never satisfactorily resolved.

One reason for the slowdown of strife in the interior was the emergence of a bigger problem for Azerbaijan: its relations with Moscow in light of its large oil production. On paper, Azerbaijan would seem to have a bright economic future based on the sale of its oil (produced from reserves in the vicinity of the capital, Baki [Baku]) and on the prospective exploitation of major new reserves found under Caspian waters. But Azerbaijan's ability to sign agreements with foreign countries for the sale of this oil is being thwarted by Russian interference in this part of the Near Abroad. First, there is the matter of ownership. Russia's government has repeatedly asserted that any oil found under the Caspian Sea is the common property of all post-Soviet states and can only be exploited under joint agreement. Second, Azerbaijan presently depends for its oil exports on a pipeline that goes through Dagestan and Groznyy (Chechnya) to the Black Sea port of Novorossiysk. Russia has said that it will not tolerate the building of competing pipelines to coastal Turkey or Iran. To emphasize that they are serious, the Russians have taken a hand in Azeri politics, allegedly helping conspirators launch an (unsuccessful) coup against the country's uncooperative president in 1995.

With military bases in Georgia and economic involvement in Azerbaijan, Russia clearly regards the TTZ as a crucial part of the Near Abroad. As Figure 2-4 reveals, Russian minorities here never were large. Thus it is not the protection of Russian expatriates that motivates Russian involvement in this transitory region: national political and economic motives drive Russia's actions. Here, Russia faces two of the world's leading Islamic countries; here, Russia contends with restive Muslim minorities on its own side of the border; and here, Russians built an oil and gas production complex on which much of southern Russia still depends.

And so, among the regions of Russia proper and of the Near Abroad, the Transcaucasian Transition Zone is transitional in several ways. This region still displays many of the frontier qualities that marked it in the past. Russian influences still radiate southward just as Islamic pressures diffuse to the north. Old animosities still divide Turks and Armenians; old cultural legacies link Azeris in Azerbaijan and Iran. At some future time, the regional geography of this area may justify the delimitation of an inter-realm boundary here. But as the turn of the century looms, that time appears a long way off.

■ PRONUNCIATION GUIDE ■

Abakan (ahb-uh-KAHN)
Abkhazia (ahb-KAHZ-zee-uh)
Adygeya (ah-duh-GAY-uh)
Adyghian (ah-duh-GAY-un)
Ajaria (uh-JAR-ree-uh)
Alma-Ata (ahl-muh-uh-TAH)
Almaty (AHL-mah-tee)
Altaic (al-TAY-ik)
Altay (AL-tye)
Altaya (al-TYE-uh)
Amur (uh-MOOR)
Angara (ahng-guh-RAH)
Apartheid (APART-hate)
Apparatchik (appa-RAH-chick)
Aqmola (AHK-moh-luh)
Aral (ARREL)
Archangel (ARK-ain-jull)
Arkhangelsk (ahr-KAN-jelsk)
Armenia (ar-MEENY-uh)
Astrakhan (ASTRA-kahn)
Avartsy (ah-VAHR-tsee)
Azerbaijan (ah-zer-bye-JAHN)
Azeri (ah-ZAIRY)

Baki (Baku) (bah-KOO)
Barents (BARRENS)
Bashkort (BAHSH-kort)
Bashkortostan (bahsh-kort-oh-STAHN)
Basil (BAY-zull)
Bauxite (BAWKS-site)
Baykal (bye-KAHL)
Baykalsk (bye-KAHLSK)
Belarus (bella-ROOSE)
Bolshevik (BOAL-shuh-vick)
Bratsk (BRAHTSK)
Bureya (buh-RAY-yuh)
Buryat(ia) (boor-YAHT [ee-uh])
Byzantine (BIZZ-un-teen)
Caspian (KASS-spee-un)
Caucasus (KAW-kuh-zuss)
Cheboksary (cheb-ahk-SAH-ree)
Chechen (CHEH-chen)
Checheno-Ingushetia (cheh-CHENNO in-goo-SHETTY-uh)
Chechnya (CHETCH-nee-uh)
Cherkessk (cher-KESK)
Chornobyl (CHAIR-nuh-beel)

Chukotka (chuh-KAHT-kuh)
Chuvash(ia) (choo-VAHSH [ee-uh])
Coup d'état (koo-day-TAH)
Crimean (cry-MEE-un)
Dagestan (dag-uh-STAHN)
Dargintsy (dar-GHINT-see)
Dnieper (duh-NYEPPER)
Donets (duh-NETTS)
Dvina (duh-vee-NAH)
El Niño (ell-NEEN-yoh)
Finno-Ugric (finno-YOO-grick)
Gavan (guh-VAHN)
Georgia (GEORGE-uh)
Glasnost (GLUZZ-nost)
Gorbachev, Mikhail (GOR-buh-choff, meek-HYLE)
Gorkiy (GORE-kee)
Gorno-Altay (gore-noh-AL-tye)
Gorod (guh-RAHD)
Groznyy (GRAWZ-nee)
Gyre (JYER)
Hegemony (heh-JEH-muh-nee)
Hokkaido (hah-KYE-doh)

Ingush (in-GOOSH)
Ingushetia (in-goo-SHETTY-uh)
Iran (ih-RAN/ih-RAHN)
Irkutsk (ear-KOOTSK)
Irtysh (ear-TISH)
Islam (iss-LAHM)
Ivanovo (ee-VAH-nuh-voh)
Izhevsk (EE-zheffsk)
Kabardinian (kabber-DIN-ee-un)
Kabardino-Balkar(ia) (kabber-DEE-noh
    bawl-KAR [ree-uh])
Kalmyk (KAL-MIK)
Kalmykia (kal-MIK-ee-uh)
Kamchatka (kahm-CHUT-kuh)
Kara (KAHR-ruh)
Karachay (kah-ruh-CHYE)
Karachayevo-Cherkessia (kahra-CHAH-
    yeh-vuh cheer-KESS-ee-uh)
Karaganda (karra-gun-DAH)
Karelia (kuh-REE-lee-uh)
Kazak (kuzz-uck)
Kazakstan (KUZZ-uck-STAHN/
    KUZZ-uck-stahn)
Kazan (kuh-ZAHN)
Khabarovsk (kuh-BAHR-uffsk)
Khakassia (kuh-KAHSS-ee-uh)
Khan, Genghis (KAHN, JING-guss)
Khanate (KAHN-ate)
Kholmsk (KAWLMSK)
Kiev(Kyyiv) (KEE-yeff)
Kievan (kee-EVAN)
Kirghiz (keer-GEEZE)
Klyuchevskaya
    (klee-ooh-CHEFF-skuh-yuh)
Kodiak (KOH-dee-ak)
Kolkhoz (KOLL-koze)
Kolyma (koh-LEE-mah)
Komi (KOH-mee)
Komsomolsk (komm-suh-MAWLSK)
Koryakia (kor-YAH-kee-uh)
Krasnodar (KRASS-nuh-dahr)
Krasnovodsk (krass-nuh-VAUGHTSK)
Krasnoyarsk (krass-nuh-YARSK)
Kronstadt (KROAN-shtaht)
Kura (KOOR-uh)
Kurile (CURE-reel)
Kuybyshev (KWEE-buh-sheff)
Kuzbas (kooz-BASS)
Kuznetsk (kooz-NETSK)
Kyrgyzstan (KEER-geeze-stahn)
Kyyiv [See Kiev]
Lena (LAY-nuh)
Lenin (LENNIN)
Mackinder, Halford (muh-KIN-der,
    HAL-ferd)

Makhachkala (muh-kahtch-kuh-LAH)
Mari(-el) (MAH-ree [el])
Mariinsk (muh-ree-EENTSK)
Menshevik (MEN-shuh-vick)
Meshketian (mesh-KETTY-un)
Moldova (moal-DOH-vuh)
Mordvinia (mord-VINNY-uh)
Moscow (MOSS-kau)
Murmansk (moor-MAHNTSK)
Muscovy (muh-SKOH-vee)
Muslim (MUZZ-lim)
Nagorno-Karabakh (nuh-GORE-noh
    KAH-ruh-bahk)
Nakhodka (nuh-KAUGHT-kuh)
Naxcivan (nah-kee-chuh-VAHN)
Nazran (nahz-RAHN)
Neva (NAY-vuh)
Nevsky Prospekt (NEFF-skee
    PROSS-spekt)
Nizhniy Novgorod (NIZH-nee NAHV-
    guh-rahd)
Norilsk (nuh-REELSK)
Novgorod (NAHV-guh-rahd)
Novokuznetsk (noh-voh-kooz-NETSK)
Novorossiysk (noh-voh-ruh-SEESK)
Novosibirsk (noh-voh-suh-BEERSK)
Oblast (OB-blast)
Odesa (Odessa) (oh-DESSA)
Oirot (AW-ih-rut)
Okhotsk (oh-KAHTSK)
Okrug (uh-KROOG)
Ordzhonikidze (or-johnny-KIDD-zuh)
Ossetia (oh-SEE-shuh)
Ossetians (oh-SEE-shunz)
Pechora (peh-CHORE-ruh)
Perestroika (perra-STROY-kuh)
Perm (PAIRM)
Permyakia (pairm-YAH-kee-uh)
Petrograd (PETTRO-grahd)
Povolzhye (puh-VOLL-zhuh)
Primoriye (pree-MOHR-ree-eh)
Pyatigorsk (pea-ah-tee-GORSK)
Qaraghandy (kah-rah-GONDY)
Rus (ROOSE)
Sakartvelo (sah-KART-vuh-loh)
Sakha (SAH-kuh)
Sakhalin (SOCK-uh-leen)
Samara (suh-MAH-ruh)
Samarqand (Samarkand) (sah-mahr-
    KAHND)
Saransk (sun-RAHN-tsk)
Saratov (suh-RAHT-uff)
Sarmatian (sahr-MAY-shee-un)
Scythian (SITH-ee-un)
Serov (SAIR-roff)

Shevardnadze, Eduard (sheh-vart-
    NAHD-zeh, ED-wahrd)
Shi'ite (SHEE-ite)
Siberia (sye-BEERY-uh)
Slavic (SLAH-vick)
Sovkhoz (SOV-koze)
Spykman (SPIKE-mun)
Stalin (STAH-lin)
Sverdlovsk (sferd-LOFFSK)
Syktyvkar (sik-tiff-KAR)
Taiga (TYE-guh)
Tajikistan (tah-JEEK-ih-stahn)
Tannu Tuva (tan-ooh-TOO-vuh)
Tashkent (Toshkent) (tahsh-KENT)
Tatar (TAHT-uh)
Tatarstan (TAHT-uh-STAHN)
Tayshet (tye-SHET)
Tbilisi (tuh-BILL-uh-see)
Tchaikovsky (chye-KOFF-skee)
Terek (TEH-rek)
Tianshan (tyahn-SHAHN)
Tikhvin (TIK-vun)
Tolyatti (tawl-YAH-tee)
Transcaucasia (tranz-kaw-KAY-zhuh)
Tsar (SAHR)
Tsarina (sah-REE-nuh)
Tselinograd (seh-LEENO-grahd)
Tula (TOO-luh)
Turkmenistan (terk-MEN-uh-stahn)
Tuva (TOO-vuh)
Tuvinian (too-VINNY-un)
Tyumen (tyoo-MEN)
Udmurt(ia) (ood-MOORT [ee-uh])
Ufa (oo-FAH)
Ukraine (yoo-CRANE)
Ulyanov (ool-YAH-noff)
Ural (YOOR-ull)
Usinsk (oo-SINSK)
Ussuri (ooh-SOOR-ree)
Uzbekistan (ooze-BECK-ih-stahn)
Vanino (VAH-nih-noh)
Varangian (vuh-RANGE-ee-un)
Vladikavkaz (vlad-uh-kuff-KAHZ)
Vladivostok (vlad-uh-vuh-STAHK)
Volgograd (VOLL-guh-grahd)
Vologda (VAW-lug-duh)
Vostochni (vaw-STOTCH-nee)
Yakut(sk) (yuh-KOOT [sk])
Yakutia (yuh-KOOTY-uh)
Yaroslavl (yar-uh-SLAHV-ull)
Yekaterinburg (yeh-KAHTA-rin-berg)
Yeltsin, Boris (YELT-sinn, BAW-reese)
Yenisey (yen-uh-SAY)
Yerevan (yair-uh-VAHN)
Yurt (YOORT)

CHAPTER 3

North America: The Postindustrial Transformation

The North American realm consists of two countries that are alike in many ways. In the United States and in Canada, European cultural imprints prevail. Indeed, the realm is often called *Anglo-America*: English is the dominant language of the United States and officially shares equal status with French in Canada. The overwhelming majority of churchgoers adhere to Christian faiths. Most (but not all) of the people trace their ancestries to various European countries. In the arts, architecture, and other modes of cultural expression, European norms prevail.

North American society is the most highly urbanized of the world's realms, and nothing symbolizes it quite as strongly as the skyscraper panoramas of New York, Toronto, or Chicago. North Americans also are hypermobile, with networks of superhighways, commercial air lanes, and railroads efficiently interconnecting the realm's far-flung cities and regions. Commuters stream into and out of suburban business centers and central-city downtowns by the millions each working day. And every year at least one out of every six individuals changes his or her residence.

In the 1990s, North America has entered a new age, the third since the arrival of Columbus in the New World more than 500 years ago. The first four centuries were dominated by agriculture and rural life; the second age—industrial urbanization—has endured over the past century but is now ending. In its place, the United States and Canada are today experiencing the maturation of a **postindustrial society and economy**, which is dominated by the production and manipulation of information, skilled services, and high-technology manufactures, and operates within a global-scale framework of business interactions. As blue-collar rapidly yields to white-collar employment and as the automated office becomes the dominant workplace, fundamental dislocations are felt throughout North America. The aging and declining Manufacturing Belt of the northeastern quadrant of the United States and adjacent southeastern

IDEAS & CONCEPTS

Postindustrial society and economy	Megalopolitan growth
Urban geography	Eras of intraurban structural evolution
Cultural pluralism	Suburban downtown
Time-space convergence	Urban realms model
Physiographic province	Von Thünen model
Rain shadow effect	Economies of scale
Migration	Historical inertia
Culture hearth	
Epochs of metropolitan evolution	

REGIONS

Continental Core	The South
New England/Atlantic Provinces	The Southwest
French Canada	Marginal Interior
Agricultural Heartland	West Coast

Canada is now often referred to as the *Rustbelt*. But the U.S. southern-tier states (the so-called *Sunbelt*) contain growing numbers of glamorous, high-prestige locales, led by suburban San Francisco's Silicon Valley and its high-technology counterparts in Texas, North Carolina, Florida, Arizona, and Southern California.

Not surprisingly, the human geography of the United States and Canada is undergoing a parallel transformation

THE MAJOR GEOGRAPHIC QUALITIES OF NORTH AMERICA

■ ■ ■

1. North America encompasses two of the world's biggest states territorially (Canada is the second largest in size; the United States is fourth).

2. Both Canada and the United States are federal states, but their systems differ. Canada's is adapted from the British parliamentary system and is divided into 10 provinces and 2 territories. The United States separates its executive and legislative branches of government, and it consists of 50 States, the Commonwealth of Puerto Rico, and a number of island territories under U.S. jurisdiction in the Caribbean Sea and the Pacific Ocean.

3. Both Canada and the United States are plural societies. Although ethnicity is increasingly important, Canada's pluralism is most strongly expressed in regional bilingualism. In the United States, major divisions occur along racial/ethnic lines.

4. A majority of Quebec's French-speaking citizens supports a strong movement that seeks independence for the province. The die may have been cast in the 1995 referendum in which (minority) non-French speakers were the difference in the narrow defeat of separation;

after the next referendum, Canada as we now know it may well disappear.

5. By world standards, this is a rich realm where high incomes and high rates of consumption prevail. North America possesses a highly diversified resource base, but nonrenewable raw materials are consumed prodigiously and domestic energy prospects remain uncertain.

6. North America's population, not large by international standards, is the most highly urbanized and mobile among the world's geographic realms.

7. North America is home to one of the world's great manufacturing complexes. The realm's industrialization generated its unparalleled urban growth, but a new postindustrial society and economy are rapidly maturing in both countries.

8. The North American Free Trade Agreement is linking the economies of the United States, Canada, and Mexico ever more tightly; international barriers to trade and investment flow are being dismantled among the three countries.

■ ■ ■

as dynamic new locational forces surface. As new regions emerge, the older ones struggle to reinvent themselves. Simultaneously, at the intrametropolitan scale, the industrial cities turn inside out and provide major new opportunities for their surrounding suburbs (which now contain more than half the U.S. population) as well as the urbanizing countryside beyond. Whatever the outcome, the winners and losers in the current scramble to adapt to the changing spatial infrastructure will shape the geography of these two leading countries well into the next century.

DEFINING THE REALM

TWO HIGHLY ADVANCED COUNTRIES

Although Canada and the United States share a number of historical, cultural, and economic qualities, the two countries differ in important ways. The differences, in fact, have firm spatial dimensions. The United States, somewhat smaller territorially than Canada, occupies the heart of the North American continent and, as a result, encompasses a greater environmental range. The U.S. population is dispersed across most of the country, forming major concentrations along both (the north-south-trending) Atlantic and Pacific coasts; the overwhelming majority of Canadians, however, live in an east-west corridor that lies across southern Canada, mostly within 200 miles (320 km) of the U.S. border. The United States also encompasses North America's northwestern extension, Alaska. (Offshore Hawaii, however, belongs to the Pacific realm.) Thus, unlike Canada, the United States is a *fragmented state*, a discon-

■■■■■■■■■■■■ FOCUS ON A SYSTEMATIC FIELD ■■■■■■■■■■■

Urban Geography

Urban geography is concerned with the spatial interpretation of city-centered population concentrations that exhibit a high-density, continuously built-up settlement landscape. No major world realm is more heavily urbanized than North America—a dimension of vital importance for understanding the human geography of the United States and Canada.

Contemporary urban geography is organized into four leading components. The first is an appreciation for the *historical evolution* of urban society. The second and third involve the study of contemporary spatial patterns at two distinct levels of generalization: macroscale or *interurban geography*, which treats cities as a system of interacting points that serve large surrounding areas, and microscale or *intraurban geography*, which focuses on the internal structure and functioning of individual metropolitan (central city-suburban ring) complexes. The fourth component deals with *planning and policy-making*, applications of urban-geographical knowledge to help define and solve spatial problems.

Urban evolution is concerned with the historical geography of cities, emphasizing the forces that produced urban growth during each of the major periods of human development since the city was innovated in Southwest Asia's Mesopotamia (modern-day Iraq) about 7,000 years ago. As urbanization spread outward from this source area, it was incorporated into the cultures of various realms—as we observe in each of this book's chapters.

Interurban geography deals with the broad scope of national-scale urban systems, viewing individual cities as points in a network that interact with one another and serve hinterlands whose territories are commensurate with the size and functional diversity of each city. The study of city types, functions, and spheres of influence within such urban systems has led to the development of useful classifications and models. One is *central place theory*, which describes some of the basic rules that govern the locations of and relationships among settlements within an urban system.

Intraurban geography is concerned with the internal spatial organization of the metropolis, emphasizing structural form as well as the distributions of diverse population groups and activities. Many models of urban structure have been devised, the most successful appearing between 1920 and 1945. Those based on concentric zones of rising affluence, sectors associated with radial transport routes, and (pre-World War II) activity clusters linked by automobile still provide a useful generalization of the land-use organization in today's large, industrial-era, central cities. But the contemporary metropolis has spilled out of its central-city confines in the final third of this century, and such models are unable to accommodate the new urban reality wherein the suburbs have become the essence of the American city. (A more recent model of this metropolitan transformation is discussed on pp. 172–173.)

The spatial arrangement of people within the metropolis is the domain of *urban social geography*. The residential territorial mosaic is the focus of attention and is studied from a number of perspectives, including ethnicity, race, socioeconomic status, community formation, housing-market operations, and the dynamics of intraurban migration. The intrametropolitan distribution of nonresidential activities is the sphere of *urban economic geography*. Today this involves major new suburban business centers as well as the central city's downtown central business district (CBD) and its tributary commercial zones.

The discipline's concern with the current human condition is expressed in the deepening involvement of urban geographers with the fourth leading component, *planning and policy-making*, which provides the opportunity for employing spatial concepts and methods to help resolve social, economic, and environmental problems in metropolitan areas. Since the planning process is fundamentally a spatial one, geographers have much to contribute through the practical application of such advanced skills as computer mapping, interpretation of satellite imagery, and the management of geographic information systems. Urban geographers also participate in policy-making by demonstrating the likely outcomes of various policy options. Many practitioners argue that the achievement of social justice means the elimination of spatial inequalities that still deeply divide the metropolitan population mosaic. In the United States, these geographic disparities are intensifying as new forces continue to reshape the metropolis.

■■■

tinuous country whose national territory consists of two or more individual parts separated by foreign territory and/or international waters.

Population Contrasts

Differences also become apparent when population totals and composition are examined. The 1997 population of the United States was approximately 267 million; Canada's was 30 million, just over one-tenth as large. Although comparatively small, Canada's population is divided by culture and tradition, and this division has a pronounced regional expression. English is the home language of 62 percent of Canada's citizens, French is spoken by 24 percent, and other languages are used by 14 percent of the population; Amerindians and Inuit peoples (formerly called Eskimos) make up only about 2 percent of the total.

This view of the downtown Chicago skyline reminds us that even the wealthiest countries contain areas of grinding poverty. This street corner is located in the middle of the sprawling ghetto on the city's southwest side, amid a sea of crumbling and abandoned dwellings studded with now-inefficient factories and warehouses that have outlived their usefulness. Except for Sears Tower (crowned by the pair of white masts), North America's tallest skyscraper, there is nothing unique about this scene—which is depressingly typical of the inner central city of the large, industrial, U.S. metropolis.

Canada's multilingual situation is accentuated by the strong spatial clustering of French speakers in the second most populous of the country's 10 provinces—Quebec (which also contains the second biggest Canadian metropolis, Montreal). More than 85 percent of Quebec's population is French Canadian, and Quebec is the historic, traditional, and emotional focus of French culture in Canada. Quebec straddles the central and lower St. Lawrence River valley, the leading route of access into North America for the early French settlers. In recent decades a strong nationalist movement has emerged in Quebec, and today demands nothing less than outright separation from the rest of Canada (independence was very narrowly defeated in the latest referendum held in 1995).

Internal regionalism also affects Canada to the west of *Francophone* (French-speaking) Quebec. The country's core area lies mainly in Ontario, Canada's most populous province, centered on metropolitan Toronto; Ontario's French-speaking minority, constituting only about 5 percent of the population, is clustered mostly in the east near the Quebec border. Farther west lie the interior Prairie Provinces of Manitoba, Saskatchewan, and Alberta, where the French join the Germans and Ukrainians to form small minorities ranging from 4 to 7 percent of each provincial population. The French cause is weakest in Canada's westernmost province, British Columbia, which is third in population and focuses on Greater Vancouver; here, over 80 percent of the population claims English ancestry, whereas only about 1.5 percent are ethnic French—well behind the province's rapidly growing Asian community.

No multilingual divisions affect the unity of the North American realm's other federation, but **cultural pluralism** of another kind prevails south of the border in the United States. More persistent in the U.S. cultural mosaic than language or ethnicity is the division between peoples of European descent (just over 80 percent of the population)

and those of African origin (12 percent in 1995). Despite the significant progress of the modern civil rights movement, which decidedly weakened *de jure* racial segregation in public life, whites overwhelmingly refuse to share their immediate living space with blacks; thus *de facto* residential segregation is all but universal. Although separatist regional thinking on the provincial/state scale of Quebec does not exist, a compelling case can be made that persistent local racial segregation has, in effect, produced two societies—one white and one black—that are both separate and unequal.

Economic Affluence and Influence

The United States and Canada rank among the most highly advanced countries of the world by every measure of national development, possessing two of the highest living standards on Earth. Yet the good life is not shared equally by all of North America's residents. Deprivation is surprisingly widespread, with notable spatial concentrations in the United States inside the inner-ring slums of big cities and on rural reservations containing Native Americans or Inuit. In the worst of these poverty pockets, malnutrition is commonplace, although its severity pales in comparison to the daily misery experienced by the poor of most developing countries.

North America's highly developed societies have clearly achieved a global leadership role, which arose from a combination of history and geography. Presented with a rich abundance of natural and human resources over the past 200 years, Americans and Canadians have brilliantly converted these productive opportunities into continent-wide affluence and worldwide influence as their booming

Industrial Revolution surpassed even Europe's by the early decades of this century.

Perhaps the greatest triumph of the North American realm was its hard-won victory over a vast and difficult natural environment. By persistently improving transportation and communications, the United States and Canada were finally able to spatially organize their entire countries across great distances (exceeding 2,500 miles/4,000 km) in the needed east-west direction across a terrain in which the grain of the land—particularly mountain barriers—is consistently oriented north-south. As breakthroughs in rail, highway, and air-transport technology succeeded each other over the past 150 years, geographic space-shortening—or **time-space convergence**—allowed distant places to become ever nearer to one another and constantly enabled people and activities to disperse more widely.

Although they continue to have their share of differences, the United States and Canada maintain close and cordial relations, and the border between them is by far the longest open international boundary on Earth (about 80 million people cross it each year). Clearly, the two countries are firmly locked together in a mission of world leadership. What is now happening in the North American realm—and why—matters enormously to every other country. The United States and Canada constitute the most advanced realm in the world; wherever it goes in the new postindustrial age, many of the rest will eventually follow.

NORTH AMERICA'S PHYSICAL GEOGRAPHY

Before we examine the human geography of the United States and Canada more closely, it is important to consider the physical setting in which they are rooted. The North American continent extends from the Arctic Ocean to Panama, but we will confine ourselves here to the territory north of Mexico—a geographic realm that still stretches from the near-tropical latitudes of southern Florida and Texas to the subpolar lands of Alaska and Canada's far-flung northern periphery. The remainder of the North American continent, which constitutes a separate realm (together with the island chains of the Caribbean Sea), comprises *Middle America* and will be treated in Chapter 4.

Physiography

North America's physiography is characterized by its clear, well-defined division into physically homogeneous regions called **physiographic provinces**. Each region is marked by a certain degree of uniformity in relief, climate, vegetation, soils, and other environmental conditions, resulting

in a scenic sameness that comes readily to mind. For example, we identify such regions when we refer to the Rocky Mountains, the Great Plains, and the Appalachian Highlands. However, not all the physiographic provinces of North America are so easily delineated.

The complete layout of the continent's physiography is seen in Figure 3-1. The most obvious aspect of this map of North America's physiographic provinces is the north-south alignment of the continent's great mountain backbone, the Rocky Mountains, whose rugged topography dominates the western segment of the continent from Alaska to New Mexico. The major feature of eastern North America is another, much lower chain of mountain ranges called the Appalachian Highlands; these uplands also trend approximately north-south and extend from Canada's Atlantic Provinces to Alabama. The orientation of the Rockies and Appalachians is important because, unlike Europe's Alps, they do not form a topographic barrier to polar or tropical air masses flowing southward or northward, respectively, across the continent's interior.

Between the Rocky Mountains and the Appalachians lie North America's vast interior plains, which extend from the Mackenzie delta on the Arctic Ocean to the coast of the Gulf of Mexico. These can be subdivided into several provinces: (1) the great Canadian Shield, which is the geologic core area containing North America's oldest rocks; (2) the Interior Lowlands, covered largely by glacial debris laid down by meltwater and wind during the Late Cenozoic glaciation; and (3) the Great Plains, the extensive sedimentary surface that slowly rises westward toward the Rocky Mountains. Along the southern margin, these interior plainlands merge into the Gulf-Atlantic Coastal Plain, which extends from southern Texas along the seaward margin of the Appalachian Highlands and the neighboring Piedmont until it ends at New York's Long Island.

On the western side of the Rocky Mountains lies the zone of Intermontane Basins and Plateaus. This physiographic province includes: (1) the Colorado Plateau in the south, with its thick sediments and spectacular Grand Canyon; (2) the lava-covered Columbia Plateau in the north, which forms the watershed of the Columbia River; and (3) the central Basin-and-Range country (Great Basin) of Nevada and Utah, which contains several extinct lakes from the glacial period as well as the surviving Great Salt Lake.

This province is called *intermontane* because of its position between the Rocky Mountains to the east and the Pacific coast mountain system to the west. From the Alaskan Peninsula to Southern California, the west coast of North America is dominated by an almost unbroken corridor of high mountain ranges whose origins stem from the contact between the North American and Pacific Plates (Fig. I-3). The major components of this coastal mountain belt include California's Sierra Nevada, the Cascades of Oregon and Washington, and the long chain of highland

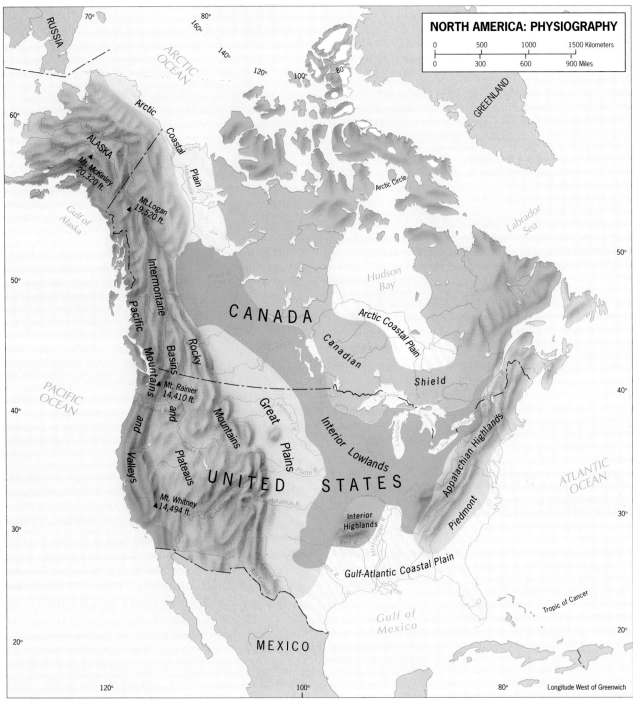

FIGURE 3-1

massifs that line the British Columbian and southern Alaska coasts. Three broad valleys—which contain dense populations—are the only noteworthy interruptions: California's Central (San Joaquin-Sacramento) Valley, the Puget Lowland of Washington State that extends southward into western Oregon's Willamette valley, and the lower Fraser valley which slices through southern British Columbia's coast range.

Climate

The various climatic regimes and regions of North America are clearly depicted on the world climate map (Fig. I-7). In general, temperature varies latitudinally—the farther north one goes, the cooler it gets. Local land-and-water-heating differentials, however, distort this broad pattern. Because land surfaces heat and cool far more rapidly than

THE RAIN SHADOW EFFECT
■ ■ ■

The dryness of most of the western interior of the United States and Canada is a result of the relationship between climate and physiography. Despite the prevailing wind direction from west to east, moisture-laden air moving onshore from the Pacific is unable to penetrate the center of the continent because high mountains stand in the way. When eastward-moving Pacific air reaches the foot of these north-south-aligned ranges, it is forced to rise up the western (windward) slope in order to surmount the topographic barrier. As the air rises toward the summit level, it steadily cools. Since cooler air is less able to hold moisture, falling temperatures produce substantial rainfall (and snowfall), in effect "squeezing out" the Pacific moisture as if the air were a sponge filled with water. This altitude-induced precipitation is called *orographic* (mountain) rainfall.

Although precipitation totals are quite high on the upper windward slopes of these West Coast mountain ranges, the major impact of the orographic phenomenon is felt to the east. Robbed of most of its moisture content by the time it is pushed across the summit ridge, the eastward-moving air now rushes down the leeward (downwind) slopes of the mountain barrier. As the air warms, its capacity to hold moisture greatly increases; the result is a warm, dry wind known as a *chinook* that can blow strongly for hundreds of miles inland. This widespread existence of semiarid (and in places even truly arid) environmental conditions is known as the **rain shadow effect**—with mountains quite literally creating a leeward "shadow" of dryness.

■ ■ ■

From the Field Notes
"We were able to land at the St. Louis Airport, not affected by the disastrous flood, but almost immediately upon leaving the terminal we got caught up in snarled traffic and disrupted surface routes. The Great Midwestern Flood of 1993 not only threatened (and inundated part of) the city itself; in its hinterland, farmhouses, fields, forests, and all else lay swamped under many feet of water. From an embankment we looked in the direction of the confluence of the Mississippi and Missouri rivers, where thousands of farm animals drowned, crops were ruined, and homes damaged if not destroyed. Human engineering had failed to hold nature in check, but after the waters receded, the people came back to live again, and to face the risk again, on these fertile soils."

water bodies, yearly temperature ranges are much larger where *continentality* is greatest.

Precipitation generally tends to decline toward the west—except for the Pacific coastal strip itself—as a result of the **rain shadow effect**, whereby most Pacific Ocean moisture is effectively screened from the continental interior (see box titled "The Rain Shadow Effect"). This broad division into Arid (western) and Humid (eastern) America, however, is marked by a fuzzy boundary that is best viewed as a wide transitional zone. Although the separating criterion of 20 inches (50 cm) of annual precipitation is easily mapped (see Fig. I-6), that generally north-south *isohyet* (the line connecting all places receiving exactly 20 inches per year) can and does swing widely across the drought-prone Great Plains from year to year because highly variable warm-season rains from the Gulf of Mexico come and go in unpredictable fashion.

On the other hand, precipitation in Humid America is far more regular. The prevailing westerly winds (blowing from west to east—winds are always named for the direction *from which they come*), which normally come up dry for the large zone west of the 100th meridian, pick up con-

siderable moisture over the Interior Lowlands and distribute it throughout eastern North America. A large number of storms develop here on the highly active weather front between tropical Gulf air to the south and polar air to the north. Even if major storms do not materialize, local weather disturbances created by sharply contrasting temperature differences are always a danger. There are more tornadoes (nature's most violent weather) in the central United States each year than anywhere else on Earth. And in winter, the northern half of this region receives large amounts of snow, especially around the Great Lakes.

Figure I-7 shows the absence of humid temperate (*C*) climates from Canada (except along the narrow Pacific coastal zone) and the prevalence of cold in Canadian environments. East of the Rocky Mountains, Canada's most *moderate* climates correspond to the *coldest* of the United States. Nonetheless, Southern Canada does share the environmental conditions that mark the Upper Midwest and Great Lakes areas of the United States, so that agricultural productivity in the Prairie Provinces and in Ontario is substantial. Canada is a leading food exporter (chiefly wheat), as is the United States, in spite of its comparatively short growing season.

The broad environmental partitioning into Humid and Arid America is also reflected in the distribution of the realm's soils and vegetation. For farming purposes there is usually sufficient soil moisture to support crops where annual precipitation exceeds the critical 20 inches; where the yearly total is less, soils may still be fertile (especially in the Great Plains), but irrigation is often necessary to achieve their full agricultural potential. Vegetation patterns are displayed on the world map (Fig. I-8), but the enormous human modification of the environment in North America's settled zones has reduced much of this regionalization scheme to the level of the hypothetical. Nonetheless, the Humid/Arid America dichotomy is again a valid generalization: the natural vegetation of areas receiving more than 20 inches of water annually is *forest*, whereas the drier climates give rise to a *grassland* cover.

Los Angeles is infamous for its atmospheric pollution, underscored here in contrasting views of the air-quality extremes over the central business district. Smog shrouds this sprawling city on many days of the year, despite efforts to curb the production of pollutants from automobile exhaust, factories, and other sources.

Hydrography (Water)

Surface water patterns in North America are dominated by the two major drainage systems that lie between the Rockies and the Appalachians: (1) the five Great Lakes (Superior, Michigan, Huron, Erie, and Ontario) that drain into the St. Lawrence River, and (2) the mighty Mississippi-Missouri river network, fed by such major tributaries as the Ohio, Tennessee, and Arkansas rivers. Both are products of the last episode of Late Cenozoic glaciation, and together they amount to nothing less than the best natural inland waterway system in the world. Human intervention has further enhanced this network of navigability, mainly through the building of canals that link the two systems as well as the St. Lawrence Seaway.

Elsewhere, the northern east coast of the continent is well served by a number of short rivers leading inland from the Atlantic. In fact, many of the major northeastern seaboard cities of the United States—such as Washington, D.C., Baltimore, and Philadelphia—are located at the waterfalls that marked the limit to tidewater navigation (hence their designation as *fall line cities*). Rivers in the Southeast and west of the Rockies at first offered little practical value owing to orientation and navigability problems. In the Far West, however, the Colorado and Columbia rivers have become supremely important as suppliers of drinking and irrigation water as well as hydroelectric power.

THE UNITED STATES

The broad outline of North America we have just sketched will be useful in developing the regionalization scheme that appears in the final section of this chapter. To fully appreciate the realm's internal regional organization, however, it is first necessary to examine in some detail the changing human geography of each country. We begin with the United States, and it may be helpful to take a few moments to review and further familiarize yourself with the map of its basic contents (Fig. 3-2). Our focus is on contemporary urban, cultural, and economic geography, and we start by tracing the evolution of that most essential of all human-geographical expressions—the map of population distribution.

Population in Time and Space

The current population distribution of the United States is shown in Figure 3-3. It is important to note that this map is the latest "still" in a motion picture, one that has been unreeling for nearly four centuries since the founding of the first permanent European settlements on the northeastern coast. Slowly at first, then with accelerating speed after

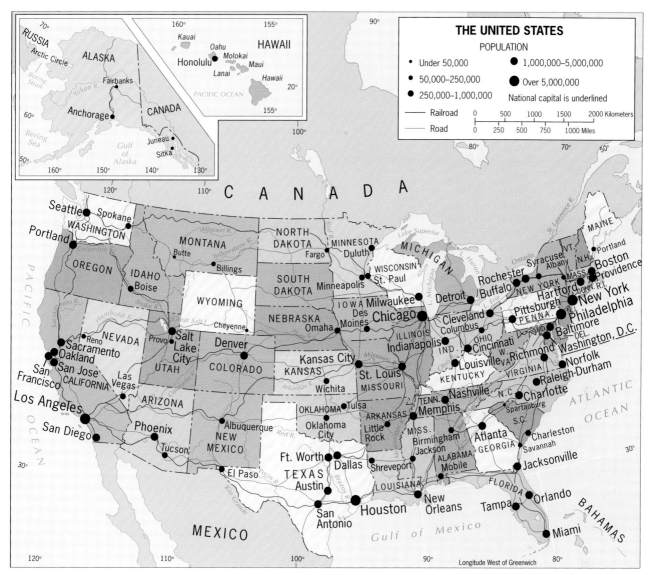

FIGURE 3-2

1800, as one major technological breakthrough followed another, Americans (and Canadians) took charge of their remarkable continent and pushed the settlement frontier westward to the Pacific. The swiftness of this expansion was dramatic, but Americans have long been the world's most mobile people. In fact, migrations continue to reshape the United States in the 1990s, perhaps the most significant being the persistent drift of people and livelihoods toward the south and west (the so-called Sunbelt), away from the north and east. The 1980 census revealed that, for the first time, the geographic center of the U.S. population had crossed west of the Mississippi River, capping a 200-year movement in which every decennial census reported a westward shift. The 1990 census found that this center had shifted an additional 40 miles (65 km) southwestward dur-

ing the 1980s. Undoubtedly, the 2000 census will report a further movement in this direction.

To understand the contemporary population map, we need to review the major forces that have shaped, and continue to shape, the distribution of Americans and their activities. Since its earliest days, the United States has been perceived as the world's premier "land of opportunity," and it has attracted a steady influx of immigrants who were rapidly assimilated into the societal mainstream (see box titled "The Migration Process"). Within the country, people have sorted themselves out to maximize their proximity to existing economic opportunities, and they have shown little resistance to relocating as the nation's changing economic geography has successively favored different sets of places over time.

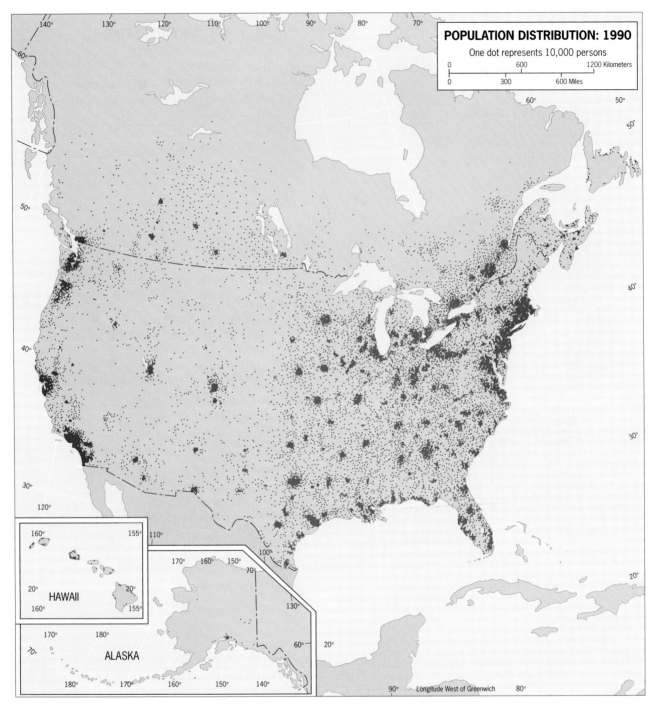

FIGURE 3-3

During the past century these transformations have spawned a number of major migrations. Although the American frontier closed 100 years ago, the westward shift of population continues—but now with that pronounced deflection to the south. The explosive growth of cities, triggered by the Industrial Revolution in the final decades of the nineteenth century, launched a rural-to-urban migratory stream that lasted into the 1960s. The middle decades of

this century also witnessed significant migration from south to north—particularly by blacks—but that, too, has ended and even reversed as substantial numbers of blacks have returned to the southeastern United States. This is part of the much larger north-to-south movement toward the Sunbelt mentioned above—a migratory stream that remains strong for the following reasons: (1) the U.S. economy and its higher-paying jobs continue to shift in that

THE MIGRATION PROCESS

▪ ▪ ▪

The United States as well as Canada is the product of **migration**. After tens of thousands of years of native settlement, European explorers reached the shores of North America beginning with Columbus in 1492 (and possibly centuries earlier than that). The first permanent colonies were established in the early seventeenth century, and from them evolved the modern United States of America. The Europeanization of North America doomed the continent's aboriginal societies, but this was only one of many areas around the world where local cultures and foreign invaders came face to face. Between 1835 and 1935, perhaps as many as 75 million Europeans departed for distant shores—most of them bound for the Americas (Fig. 3-4). Some sought religious freedom; others escaped poverty and famine; still others simply hoped for a better life.

Studies of the *migration decision* indicate that migration flows—then as now—vary in size with: (1) the perceived degree of difference between one's home, or source, and the destination; (2) the effectiveness of the information flow, that is, the news about the destination sent back by those who migrated to those who stayed behind waiting to decide; and (3) the distance between the source and the destination (shorter moves attracting many more migrants than longer ones). More than a century ago, the British social scientist Ernst Georg Ravenstein studied the migration process; many of his conclusions remain valid today. For example, every migration stream from source to destination produces a counter-stream of returning migrants who are unable to adjust, are unsuccessful, or are otherwise persuaded to go back home.

Studies of migration also conclude that several factors are at work in the migration process. *Push factors* motivate people to move away; *pull factors* attract them to new destinations. To those early Europeans, the United States was a new frontier, a place where one might acquire a piece of land, some livestock. The opportunities were reported to be unlimited.

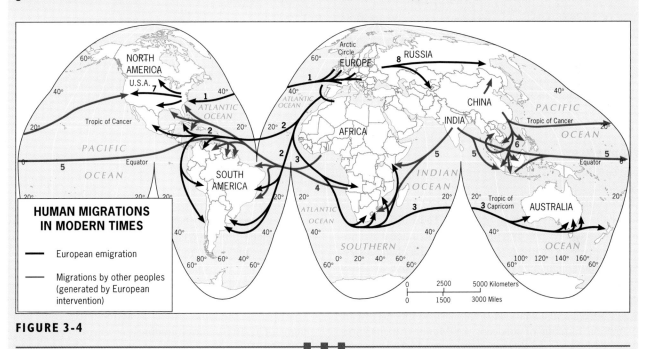

FIGURE 3-4

direction; (2) the retirement migration of the affluent elderly to such states as Florida and Arizona, though slowing, shows little sign of ending; and (3) much of the new wave of immigration from Middle America and eastern Asia is directed, respectively, toward the zone adjacent to the southern border and California's coastal cities.

Let us now look more closely at the historical geography of these changing population patterns, considering first the initial rural influence and then the decisive impacts of industrial urbanization.

Pre-Twentieth-Century Population Patterns

The spatial distribution of the U.S. population is rooted in the colonial era of the seventeenth and eighteenth centuries that was dominated by England and France. The French were concerned with penetrating the continental interior in order to establish a lucrative fur-trading network. The other major (and larger) colonizing force, the English, concentrated their settlement efforts along the coast of what is

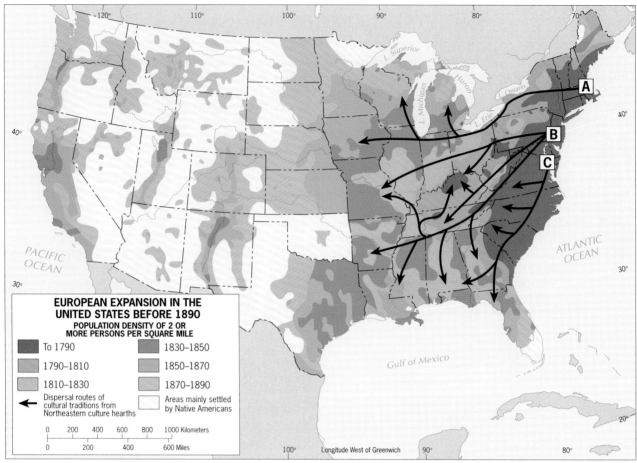

FIGURE 3-5

today the northeastern U.S. seaboard. These British colonies quickly became differentiated in their local economies, a diversity that was to endure and later shape American cultural geography. The northern colony of New England (Massachusetts Bay and environs) specialized in commerce; the southern Chesapeake Bay colony (Tidewater Virginia and Maryland) emphasized the large-scale plantation farming of tobacco; the Middle Atlantic area lying in between (southeastern New York, New Jersey, eastern Pennsylvania) was home to a number of smaller, independent-farmer colonies.

As these three adjacent colonial regions thrived and yearned to expand after 1750, the British government kept the inland frontier closed. It also exerted tightening economic controls, thereby unifying the three groups of colonies in their growing dislike for the increasingly heavy-handed mother country. By 1775, escalating tensions produced open rebellion and the onset of the eight-year-long Revolutionary War. This resulted in a resounding English defeat and independence for the newly formed United States of America.

The western frontier of the fledgling nation now swung

open, and the old British Northwest Territory (Ohio-Michigan-Indiana-Illinois-Wisconsin) was promptly settled. Propelling this rise of the trans-Appalachian West was the discovery that the soils (and climate) of the Interior Lowlands were especially favorable for farming—a remarkable improvement over the relatively infertile seaboard soils. That sparked the rapid growth of agriculture and the widening of coastal-interior trading ties across ever greater distances. New interregional complementarities now emerged, a cornerstone of an economy whose spatial organization was assuming national-scale proportions.

By the time the westward-moving frontier swept across the Mississippi valley in the 1820s, it was clear that the three former seaboard colonies (**A**, **B**, and **C** in Fig. 3-5) had become separate **culture hearths**—primary source areas and innovation centers from which migrants carried cultural traditions into the central United States (as shown by the arrows in Fig. 3-5). The northern half of this vast interior space soon became well unified as its infrastructure steadily improved. By 1860 the railroad had replaced earlier plank roads and canals, providing a more efficient network that offered fast and cheap long-distance transportation, thereby producing significant time-space convergence among once-distant regions within the northeastern quadrant of the United States. The American South, however, did not wish to integrate itself economically with the North to any great degree, preferring instead to export tobacco and cotton from its plantations to overseas markets. This divergent regionalism, together with its insistence on preserving slavery, soon led the South into secession and the ruinous Civil War (1861–1865); in the dismal aftermath, it took the South decades to rebuild.

The second half of the nineteenth century also saw the frontier cross the western United States, although at first it was a *hollow frontier*: parts of the Pacific coast were settled before the dry steppelands of the Great Plains region far to the east (Fig. 3-5). As early as 1869, agriculturally booming California was linked to the rest of the nation by transcontinental railroad, and these same steel tracks also began to open up the bypassed Great Plains. Those semiarid grasslands, it turned out, were quite fertile after all, particularly for raising wheat on a large scale if limited water supplies were carefully husbanded.

By the time the U.S. frontier closed in the 1890s, today's rural settlement pattern was firmly in place, anchored to a set of enduring national agricultural regions (discussed later in this chapter). The closure of the frontier also coincided with an accelerating shift in the distribution of the American population: the census of 1900 revealed that the newly developing metropolitan areas were growing more than twice as fast as the countryside. Clearly, an urban revolution was accompanying the Industrial Revolution that had taken hold after 1870. To be sure, cities had been important since colonial times, but the degree and extent of urbanization by 1900 was approaching massive proportions. In retrospect, of course, the demographic tilt from rural to urban America was just beginning: by 1920, 51 percent of the U.S. population resided in cities and towns; by 1950, 64 percent; and by 1997, 75 percent.

Twentieth-Century Industrial Urbanization

In the United States the Industrial Revolution occurred almost a century later than in Europe, but when it finally did cross the Atlantic in the 1870s, it took hold so successfully and advanced so robustly that only 50 years later America was surpassing Europe as the world's mightiest industrial power. Thus the far-reaching social and economic changes that Europe's industrializing countries experienced were accelerated in the United States, heightened all the more by the arrival of nearly 25 million European immigrants—who overwhelmingly headed for jobs in the major manufacturing centers—between 1870 and 1914.

The impact of industrial urbanization occurred simultaneously at two levels of spatial generalization. At the national or *macroscale*, a system of new cities rapidly emerged, specializing in the collection, processing, and distribution of raw materials and manufactured goods, linked together by an efficient web of railroad lines. Within that urban network, at the local or *microscale*, individual cities prospered in their new roles as manufacturing centers, generating an internal structure that still forms the geographic framework of most of the central cities of America's large metropolitan areas. We now examine the urban trend at both of these scales.

Evolution of the U.S. Urban System The rise of the national urban system in the late nineteenth century was based on the traditional external role of cities: providing goods and services for their hinterlands in exchange for raw materials. Because handicrafts and commercial activities were already agglomerated in existing (preindustrial) cities, the emerging industrialization movement tended to favor such locations. Moreover, these urban centers contained concentrations of labor and investment capital, provided a market for finished goods, and possessed superior transport and communications connections. Importantly, these cities could also absorb the hordes of in-migrants who would cluster by the thousands around new factories built within and just outside the municipal boundaries. Their growing incomes, in turn, permitted industrially intensifying cities to invest in a bigger local infrastructure of private and public services as well as housing—and thereby convert each round of industrial expansion into a new stage of urban development. Once generated, this whole process unfolded so swiftly that planning was impossible, and by the turn of the twentieth century a number of large cities had unexpectedly materialized on the U.S. landscape.

The rise of the national urban system, unintended though it may have been, was a necessary by-product of industrialization; without it the rapid economic development of the United States could not have taken place. Even though it emerged during the Industrial Revolution of the 1870–1920 period, the national urban system had been in the process of formation for several decades preceding the Civil War. The evolutionary framework of the system over the past two centuries can be summarized within the multistage model developed by John Borchert, which consists of five **epochs of metropolitan evolution** based on transportation technology and industrial energy.

- The *Sail-Wagon Epoch* (1790–1830), marked by primitive overland and waterway circulation. The leading cities were the northeastern ports of Boston, New York, and Philadelphia, none of which had yet emerged as the primate city. They were more heavily oriented to the European overseas trade than to their still rather inaccessible western hinterlands—though the Erie Canal, linking the coast and the Great Lakes, was completed as the epoch came to a close.

- The *Iron Horse Epoch* (1830–1870), dominated by the arrival and spread of the steam-powered railroad, which steadily expanded its network from east to west until the transcontinental line was opened in 1869. A nationwide transport system had now been forged, coal-mining centers boomed (to keep locomotives running), and—aided by the easier movement of raw materials—small-scale urban manufacturing began to disperse outward from its New England hearth. The national urban system started to take shape: New York advanced to become the primate city by 1850, and the next level in the hierarchy was occupied by booming new industrial cities such as Pittsburgh, Detroit, and Chicago.

- The *Steel-Rail Epoch* (1870–1920), which spanned the U.S. Industrial Revolution. Several forces now shaping the growth and full establishment of the national metropolitan system included: the rise of the pivotal steel industry along the Chicago-Detroit-Pittsburgh axis; the increasing scale of manufacturing that necessitated greater agglomeration in the most favored raw-material and market locations for industry; and the use

of steel in railway construction, which permitted significantly higher speeds and longer hauls of heavy commodities.

- The *Auto-Air-Amenity Epoch* (1920–1970), which encompassed the later stage of U.S. industrial urbanization and the maturation of the national urban hierarchy. The key innovation was the gasoline-powered internal combustion engine, which permitted ever greater automobile- and truck-based regional and metropolitan dispersal. As technological advances in manufacturing led to the increasing automation of blue-collar jobs, the American labor force steadily shifted toward domination by white-collar, personal and professional services to manage the industrial economy. This kind of productive activity began to respond to the growing locational pull of *amenities* (pleasant environments) available in suburbs and the outlying Sunbelt states in a nation now fully interconnected by jet travel and long-distance communications networks.

- The *Satellite-Electronic-Jet Propulsion Epoch* (1970–   ), shaped by the newest advancements in global-scale communications, computer technologies,

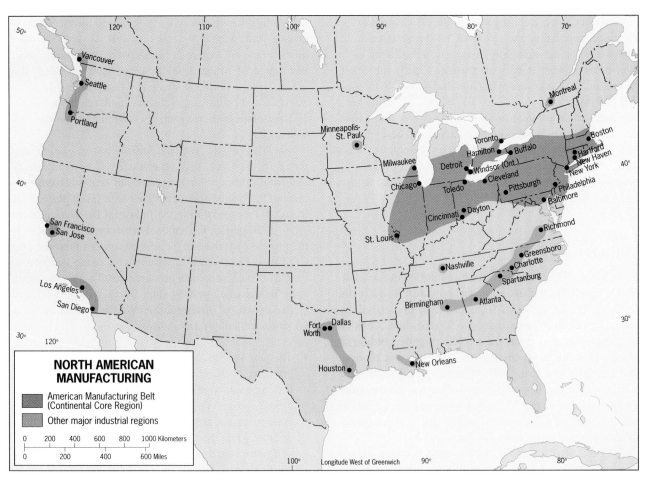

FIGURE 3-6

and transoceanic travel. The key trends working to re-structure the U.S. urban system today are: increasing international interaction; research and development related to information management; the heightened lure of recreational amenities; the constraints of more costly energy supplies; and the cutback in military ex-penditures. The most successful metropolises are those with direct intercontinental air connections, expand-ing overseas trade and business linkages, and thriving high-technology research communities; they include a number of urban areas along the Pacific and Atlantic coasts as well as in Texas. Among the now-disadvan-taged metropolises are the older manufacturing cen-ters, particularly those that have long depended on defense-industry contracts.

The growth of the U.S. urban system and its industrial-based economy produced dramatic spatial changes as pop-ulations relocated to keep pace with shifting employment opportunities. The most notable regional transformation was the early-twentieth-century emergence of the conti-nental *core area*, or American Manufacturing Belt, which contained the lion's share of industrial activity in both the United States and Canada. As Figure 3-6 shows, the geo-

graphic form of the core region—which includes southern Ontario—was a great rectangle whose four corners were Boston, Milwaukee, St. Louis, and Baltimore. However, because manufacturing is such a spatially concentrated ac-tivity, the core area should *not* be thought of as a contin-uous factory-dominated landscape. In truth, well under 1 percent of the territory of the Manufacturing Belt is actu-ally devoted to industrial land use. Most of its mills and foundries are tightly clustered into a dozen districts that center on metropolitan Boston; Hartford-New Haven; New York-northern New Jersey; Philadelphia; Baltimore; Buffalo; Pittsburgh-Cleveland; Detroit-Toledo; Chicago-Milwaukee; Cincinnati-Dayton; St. Louis; and Ontario's Toronto-Hamilton-Windsor.

At the subregional scale, as transportation break-throughs permitted progressive urban decentralization and **megalopolitan growth**, the expanding peripheries of ma-jor cities soon coalesced to form a number of conurbations. The most important of these by far is the *Atlantic Seaboard Megalopolis* (Fig. 3-7), the 600-mile (1,000-km) urbanized northeastern coastal strip extending from southern Maine to Virginia that contains metropolitan Boston, New York, Philadelphia, Baltimore, and Washington. This was the economic heartland of the U.S. core region, the seat of

F

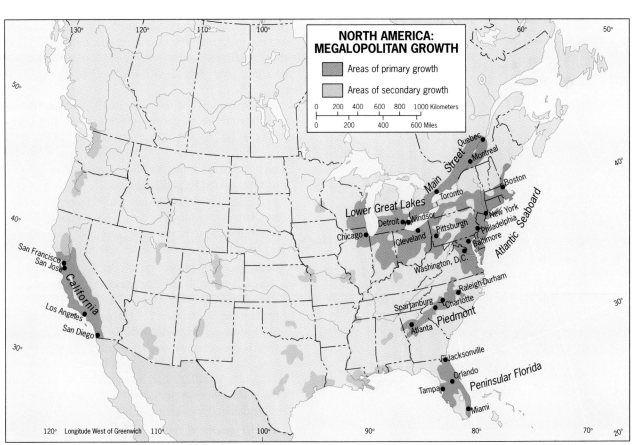

FIGURE 3-7

American government, business, and culture, as well as the trading "hinge" between the United States and the Old World across the Atlantic. Four other primary conurbations also emerged—*Lower Great Lakes* (Chicago-Detroit-Pittsburgh), *California* (San Diego-Los Angeles-San Francisco), *Peninsular Florida* (Jacksonville-Orlando-[Tampa]-Miami), and *Piedmont* (Atlanta-Charlotte-Raleigh/Durham)—and are expected to continue growing, along with a number of secondary megalopolitan concentrations (Fig. 3-7). (And note, for later reference, that in Canada the same forces unleashed a parallel episode of rapid industrial urbanization and created a nationally predominant conurbation—known as *Main Street*—linking Quebec City, Montreal, Toronto, and Windsor.)

The Changing Structure of the U.S. Metropolis The internal structure of the metropolis reflected the same mixture of forces that shaped the national urban system. As an industrial city, however, it represented a departure from its European parentage. Whereas Europe's major cities were historically centers of political and military power—onto

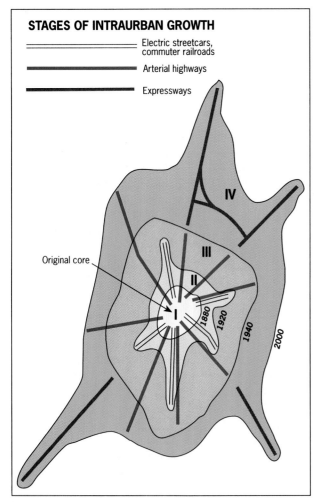

FIGURE 3-8

which industrialization was grafted almost as an afterthought—a large number of U.S. cities came into being as economic machines to produce the goods and services required to sustain an ongoing Industrial Revolution. Thus, right from the start, the performances of America's cities were judged mainly in terms of their profit-making abilities. The chief social function of the city was to receive and process immigrants (domestic-rural as well as foreign) for assimilation into the mainstream American nativist culture, which increasingly concentrated in the rapidly growing suburbs after 1920.

At the intrametropolitan scale, too, transportation technology was a decisive force in shaping geographic patterns. Rails—in this instance, lighter street rail lines—once again shaped spatial structure, as horse-drawn trolleys were succeeded by electric streetcars in the late nineteenth century. The arrival of the automobile after World War I changed all that, and America began to turn from building compact cities to the widely dispersed metropolises of the post–World War II highway era. By 1970, the new intraurban expressway network had equalized location costs throughout the metropolis, setting the stage for suburbia to swiftly transform itself from a residential preserve into a complete *outer city* with amenities and new prestige that proved highly attractive to the business world. As the newly urbanized suburbs started to capture major economic activities and thereby gain a surprising degree of functional independence, many large cities saw their status diminish to that of coequal, their once thriving central business districts (CBDs) all but reduced to serving the less affluent populations that now dominated the central city's neighborhoods.

Intrametropolitan growth can also be organized into **eras of intraurban structural evolution** based on transportation, as described by John Adams (Fig. 3-8).

- The initial Stage I, the *Walking-Horsecar Era* (prior to 1888), entailed a pedestrian city, slightly augmented after 1850 when horse-drawn trolleys began to operate. Urban structure was dominated by compactness (everything had to be within a 30-minute walk), and very little land-use specialization could occur.

- The invention in 1888 of the electric traction motor, a device that could easily be attached to horsecars, launched the *Electric Streetcar Era* (1888–1920). Speeds of up to 20 miles (32 km) per hour enabled the 30-minute travel radius and the urbanized area to expand considerably along new outlying trolley corridors (Stage II in Fig. 3-8), spawning streetcar suburbs and helping to differentiate space within the older core city. In the older core city, the CBD, industrial, and residential land uses emerged in their modern form.

- Stage III was the *Recreational Automobile Era* (1920–1945), when the initial impact of cars and highways steadily improved the accessibility of the outer metro-

From the Field Notes

"Monitoring the urbanization of U.S. suburbs for the past quarter-century has brought us to Tyson's Corner on many a field trip and data-gathering foray. It is now hard to recall from this mid-1990s view that less than 30 years ago this place was merely a near-rural crossroads. But as nearby Washington, D.C. steadily decentralized, *Tyson's* capitalized on its unparalleled regional accessibility (its Capital Beltway location at the intersection with the radial Dulles Airport Toll Road) to attract a seemingly endless parade of high-level retail facilities, office complexes, and a plethora of supporting commercial services. We would rank it today among the three largest and most influential suburban downtowns (or 'edge cities') in North America."

politan ring, thereby launching a wave of mass sub-urbanization that further extended the urban frontier (and the half-hour time-distance radial). Meanwhile, the still-dominant central city experienced its economic peak and the partitioning of its residential space into neighborhoods sharply defined by income level, ethnicity, and race.

- The final Stage IV, the *Freeway Era* (1945 to the present), saw the full impact of automobiles as high-speed expressways pushed suburban development and the time-distance limit in certain sectors to more than 30 miles (50 km) from downtown.

The social geography of the evolving industrial metropolis over the past 150 years has been marked by the development of a residential mosaic that exhibited the congregating of ever-more-specialized groups. Before the arrival of the electric streetcar, which introduced ''mass'' transit affordable by every social class, the heterogeneous city population had been unable to sort itself into ethnically uniform neighborhoods. Immigrants pouring into the industrializing cities were forced to reside within walking distance of their workplaces in crowded tenements and row houses that were filled in the order that their tenants arrived in town. When the United States sharply curtailed foreign immigration in the 1920s, industrial managers soon discovered the large African-American population of the rural South—increasingly unemployed there as cotton-related agriculture declined—and began to recruit these workers by the thousands for the factories of Manufacturing Belt cities. This new migration had an immediate impact on the social geography of the industrial city because whites refused to share their living space with the racially different newcomers. The result was the involuntary segregation of these newest migrants, who were channeled into geographically separate, all-black areas. By 1945, these areas became large expanding *ghettos*, speeding the departure of many white central-city communities in the postwar era and helping to create a racially divided society.

The suburban component of the intraurban residential mosaic, which now contains just over 50 percent of the U.S. population, is home to the more affluent residents of the metropolis. As noted above, suburbia has also acquired a massive infusion of nonresidential activities since 1970, transforming it into a low-density, ring-like city that surrounds the old metropolitan core. A quarter of a century

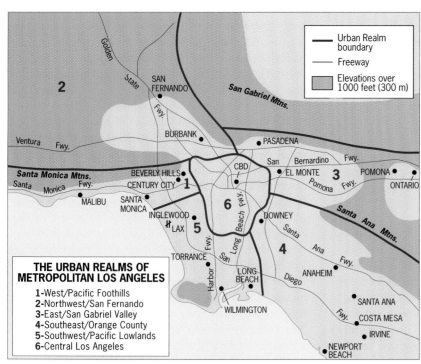

IDEAL FORM OF MULTICENTERED URBAN REALMS MODEL

FIGURE 3-9

F

multipurpose activity nodes often developed around big regional shopping centers, whose prestigious images attracted scores of industrial parks, office campuses and high-rises, hotels, restaurants, entertainment facilities, and even major league sports stadiums and arenas, which together formed burgeoning new **suburban downtowns** that are an automobile-age version of the CBD (see photo p. 171). As suburban downtowns flourished, they attracted tens of thousands of local residents to organize their lives around them—offering workplaces, shopping, leisure activities, and all the other elements of a complete urban environment. In this way, they further loosened ties not only to the central city but also to other sectors of the suburban ring.

These newest spatial elements of the contemporary metropolis are assembled in the model displayed in Figure 3-9. The rise of the outer city has today produced a *multicentered* metropolis consisting of the traditional CBD, as well as a set of increasingly coequal suburban downtowns, with each activity center serving a discrete and self-sufficient surrounding area. James Vance defined these new tributary areas as **urban realms**, recognizing in his studies that each such realm maintains a separate, distinct economic, social, and political significance and strength. The urban realms model is applied to Los Angeles in Figure 3-10; a similar regionalization scheme could easily be drawn for other large American metropolises.

The position of the central city within the new multinodal metropolis of realms is eroding. No longer the dom-

ago, the suburbs surpassed the central cities in total employment; by the late 1990s, even Sunbelt metropolises were experiencing the suburbanization of a critical mass of jobs (greater than 50 percent of the urban-area total).

As this outer city grew rapidly, the volume and level of interaction with the central city constantly lessened and suburban self-sufficiency increased. The shift toward functional independence was heightened as new suburban nuclei sprang up to serve the new local economies, the major ones locating near key freeway interchanges. These

THE URBAN REALMS OF METROPOLITAN LOS ANGELES

1-West/Pacific Foothills
2-Northwest/San Fernando
3-East/San Gabriel Valley
4-Southeast/Orange County
5-Southwest/Pacific Lowlands
6-Central Los Angeles

FIGURE 3-10

inant metropolitanwide center for goods and services, the CBD increasingly serves the less affluent residents of the innermost realm and those working downtown. As manufacturing employment declined precipitously, many large cities adapted successfully by shifting toward service industries. Accompanying this switch is downtown commercial revitalization, which has been widespread since 1970; but in many cities, for each shining new skyscraper that goes up, several old commercial buildings are abandoned. Residential reinvestment has also occurred in many downtown-area neighborhoods, usually taking the form of *gentrification*—the upgrading of communities by new higher-income settlers. To succeed, however, this development usually requires the displacement of established lower-income residents, an emotional issue that has sparked many a conflict. Beyond the CBD zone, the vast inner city remains the problem-ridden domain of low- and moderate-income people, with most forced to reside in ghettos. Financially ailing big-city governments are unable to fund adequate schools, crime-prevention programs, public housing, and sufficient social services, and so the downward spiral, including abandonment in the aged industrial cities, will continue unabated into the next century.

Cultural Geography

In the United States, over the past two centuries, the contributions of a wide spectrum of immigrant groups have shaped—and continue to shape—a rich and varied cultural mosaic. Great numbers of these newcomers were willing to set aside their original cultural baggage in favor of assimilation into the emerging culture of their adopted homeland, which itself was a hybrid nurtured by constant infusions of new influences. For most upwardly mobile immigrants, this plunge into the much-touted "melting pot" promised a ticket for acceptance into mainstream American society. Yet for millions of other would-be mainstreamers, the melting pot has proven to be a lumpy stew. Whether or not by choice, they continue to stick together in ethnic communities (especially within the inner cities of the Manufacturing Belt), with a large proportion of those neighborhoods inhabited by racial minorities who do not participate fully in the national culture.

American Cultural Bases

As the nativist culture of the United States matured, it came to develop a set of powerful values and beliefs: (1) love of newness; (2) a desire to be near nature; (3) freedom to move; (4) individualism; (5) societal acceptance; (6) aggressive pursuit of goals; and (7) a firm sense of destiny. Brian Berry has discerned these cultural traits in the behavior of people throughout the evolution of urban America. A "rural ideal" has prevailed throughout U.S. history

and is still expressed in a strong bias against residing in cities. When industrialization made urban living unavoidable, those able to afford it soon moved to the emerging suburbs (*newness*) where a form of country life (*close to nature*) was possible in a semiurban setting. The fragmented metropolitan residential mosaic, composed of a myriad of slightly different neighborhoods, encouraged frequent *unencumbered mobility* as middle-class life revolved around the *individual* nuclear family's *aggressive pursuit* of its aspirations for *acceptance into the next higher stratum of society*. These accomplishments confirmed to most Americans that their goals could be attained through hard work and perseverance, and that they possessed the ability to realize their *destiny* by achieving the "American Dream" of homeownership, affluence, and total satisfaction.

Language

Although linguistic variations play a far more important role in Canada, no less than one-eighth of the U.S. population spoke a primary language other than English in 1990. Differences in English usage are also evident at the subnational level in the United States, where regional variations (*dialects*) can still be noted despite the recent trend toward a truly national society. The South and New England immediately come to mind as areas that still possess distinctive accents. An even closer connection between language and landscape is established through *toponymy*, the naming of places. U.S. place-name geography provides important clues to the past movements of cultural influences and national groups. For instance, the preponderance of such Welsh place names as Cynwyd, Bryn Mawr, and Uwchlan just to the west of Philadelphia is the final vestige of an erstwhile colony of Celtic-speaking settlers from Wales.

Religion

North America's Christian-dominated kaleidoscope of religious faiths contains important spatial variations. Many major Protestant denominations are clustered in particular regions, with Baptists localized in the southeastern quadrant of the United States, Lutherans in the Upper Midwest and northern Great Plains, and Mormons in Utah and southern Idaho. Roman Catholics are most visibly concentrated in two locations: (1) the Manufacturing Belt metropolises and nearby New England, which received a huge influx of Catholic Europeans over the past century; and (2) the entire Mexican borderland zone that is home to a burgeoning Hispanic-American population. Judaism is the nation's most highly agglomerated major religious group; its largest congregations are clustered in the cities and suburbs of Megalopolis, Southern California, South Florida, and the Midwest.

Ethnicity

Ethnicity (or nationality) was a decisive influence in shaping American culture, which in turn also reshaped the cultural traditions of newcomers who were assimilated into mainstream society. The current U.S. ethnic tapestry is, as always, a complex mosaic (see box titled "The Ethnic Tapestry of the United States"). The spatial distribution of ethnic minorities is mapped in Figure 3-11, and also includes the Native American population of the conterminous United States. Because Native Americans largely occupy tribal lands on reservations ceded by the federal government, they are undergoing very little distributional change. The Hispanic population, on the other hand, is growing rapidly through in-place natural increase as well as in-migration from Middle America (much of it illegal). Spatially, Hispanics are both increasing in density along the southwestern border and fanning out toward large metropolitan areas to the north and east. Blacks are not as mobile as Hispanics, undoubtedly constrained by racially segregated housing markets. Nonetheless, upwardly mobile blacks are increasing their presence both in suburbia and, via return migration, in the more affluent areas of the South—although settling in communities that remain highly segregated. The newest elements in the ethnic tapestry (as indicated in the box) are Asian Americans, who heavily cluster in major West Coast metropolitan areas.

The Mosaic Culture

American cultural geography continues to evolve. What is now taking place is a new fragmentation into a *mosaic culture*, an increasingly heterogeneous complex of separate, uniform "tiles" that cater to more specialized groups than ever before. No longer based solely on such broad divisions as income, race, and ethnicity, today's residential communities of interest are also forming along the dimensions of age, occupational status, and especially lifestyle. Their success and steady proliferation reflect an obvious satisfaction on the part of a majority of Americans. Yet such balkanization—fueled by people choosing to interact only with others exactly like themselves—threatens the survival of important democratic values that have prevailed throughout the evolution of U.S. society.

The Changing Geography of Economic Activity

The economic geography of the United States today is the product of all of the foregoing, as bountiful environmental, human, and technological resources have cumulatively blended together to create one of the world's most advanced economies. Perhaps the greatest triumph was over-

THE ETHNIC TAPESTRY OF THE UNITED STATES

■ ■ ■

The diversity and complexity of America's ancestral makeup was reported by the 1990 census, which showed that nearly 85 percent of the U.S. population identified with one of more than 140 different national backgrounds. There were more Americans of full or partial English descent than the total population of England, and more than one-half as many Americans of German stock as currently reside in Germany; the Irish Americans outnumber the population of the Republic of Ireland by a ratio of almost 13 to 1. In addition, so many Americans listed themselves as African-Americans that only four Subsaharan African states recorded a larger national population total. Other significant ethnic ancestries include French, Italian, Scottish, Polish, and Mexican. Those of Native American background accounted for 0.8 percent of the U.S. total.

Geographically, the dominant European ancestries—French, Irish, and those originating in the United Kingdom—were dispersed throughout the United States. Most of the others showed an affinity for a particular region: Italians, Portuguese, and Russians clustered in the Northeast, whereas Scandinavians and Czechs were localized in the north-central states. California, the most populous state, best exhibited the nation's diverse ethnicity (see Fig. 3-11), with more Americans of English, German, Irish, French, Scottish, Dutch, Swedish, and Danish origin concentrated there than in any other state.

Although the African-American population remains fairly stable in relative size (12 percent of the U.S. total in 1995), two other minority groups—Asians and Hispanics—are substantially increasing their presence in the national ethnic tapestry through in-place growth and sizeable immigration. During the 1980s, the number of Asians and Pacific Islanders more than doubled, for the most part remaining clustered in the urban areas of the West Coast. The biggest single Asian group today is the Chinese, followed by the Filipinos, Japanese (South Asian) Indians, Koreans, and Vietnamese. The Hispanic population, which is expected to surpass the black population to become the nation's largest minority by 2005, grew by nearly 55 percent between 1980 and 1990. Disaggregated by national background, the Mexican component accounts for 60 percent of the Hispanic population, Central and South Americans 15 percent, Puerto Ricans 12 percent, and Cubans 5 percent. As Figure 3-11 indicates, Hispanics are most heavily concentrated in the southwestern quadrant of the United States.

■ ■ ■

coming the "tyranny" of distance, as people and activities were organized into a continentwide spatial economy that took maximum advantage of agricultural, industrial, and urban development opportunities. Yet, despite these past achievements, American economic geography at the end of the twentieth century is once again in upheaval as the transition is being completed from industrial to postindustrial society.

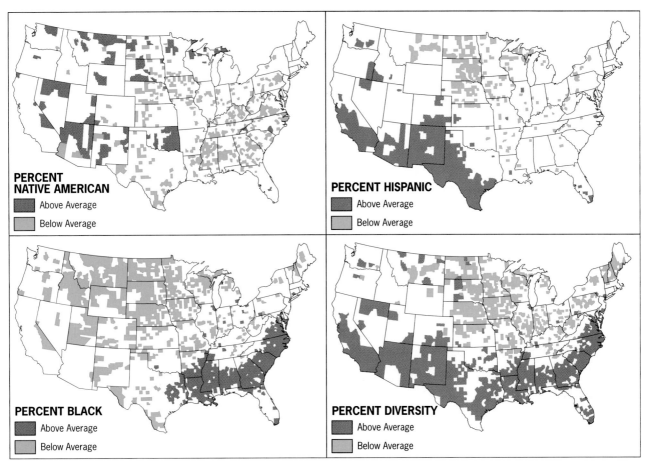

FIGURE 3-11

Major Components of the Spatial Economy

Economic geography is mainly (though not exclusively) concerned with the locational analysis of productive activities. Four major sets may be identified:

- **Primary activity**: the extractive sector of the economy in which workers and the environment come into direct contact, especially in *mining* and *agriculture.*
- **Secondary activity**: the *manufacturing* sector, in which raw materials are transformed into finished industrial products.
- **Tertiary activity**: the *services* sector, including a wide range of activities from retailing to finance to education to routine office-based jobs.
- **Quaternary activity**: today's dominant sector, involving the collection, processing, and manipulation of information; a subset, sometimes referred to as **quinary activity**, is the managerial activity associated with decision-making in large organizations.

Historically, each of these activities has successively dominated the American labor force for a time over the past 200 years, with the quaternary sector now dominating the economy for the foreseeable future. Agriculture was dominant until late in the nineteenth century, giving way to manufacturing by 1900. The steady growth of services after 1920 finally surpassed manufacturing in the 1950s but now shares a dwindling portion of the limelight with the still-rising quaternary sector. The approximate breakdown by major sector of employment in the U.S. labor force today is agriculture, 2 percent; manufacturing, 15 percent; services, 18 percent; and quaternary, 65 percent (with about 10 percent in the quinary sector). We now sequentially treat these major productive components of the spatial economy in the following coverage of resource use, agriculture, manufacturing, and the postindustrial revolution.

Resource Use

The United States (and Canada) was blessed with abundant deposits of mineral and energy resources. Fortunately, these resources were usually concentrated in sufficient quantities to make long-term extraction an economically feasible proposition, and most of the richest raw material sites are still the scene of major drilling or mining opera-

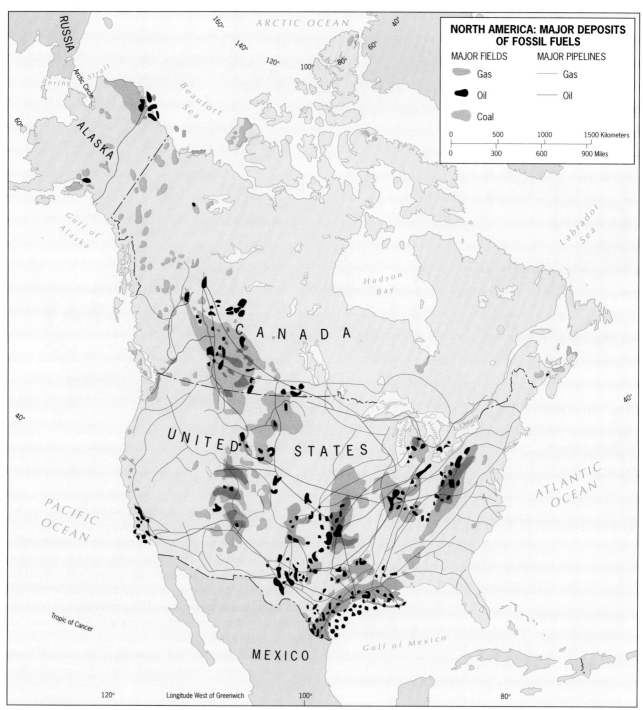

FIGURE 3-12

tions. Moreover, the continental and offshore mineral/fuel storehouse may yet contain significant undiscovered resources for future exploitation.

Mineral Resources North America's rich mineral deposits are localized in three zones: the Canadian Shield north of the Great Lakes, the Appalachian Highlands, and scattered areas throughout the mountain ranges of the West. The Shield's most noteworthy minerals are iron ore (Minnesota's Mesabi Range just west of Lake Superior and the Labrador Trough lying along the Quebec-Labrador border); nickel (localized in Sudbury, Ontario, northern Manitoba, and the central coastland of Labrador); and gold, uranium, and copper (northwestern Canada). Besides vast

Agriculture has not been excluded from the technological revolution that is sweeping across every corner of North America's spatial economy. Global Positioning Satellite (GPS) technology is now becoming so accurate a tool that farmers soon will be applying all-important fertilizer with precision equipment guided by a computerized map of each field's fertility variations. The process is being perfected at this field station of the University of Illinois near the main campus at Urbana—and will shortly make the farms of the surrounding Corn Belt even more productive.

deposits of soft (bituminous) coal, the Appalachian region also contains hard (anthracite) coal in northeastern Pennsylvania and iron ore in central Alabama. The western mountain zone contains significant deposits of coal, copper, lead, zinc, molybdenum, uranium, silver, and gold.

Fossil Fuel Energy Resources   The realm's most strategically important resources are its coal, petroleum (oil), and natural gas supplies—the *fossil fuels*, so named because they were formed by the geologic compression and transformation of plant and tiny animal organisms that lived hundreds of millions of years ago. These energy supplies, mapped in Figure 3-12, reveal abundant deposits and distribution networks, particularly in the conterminous United States and Alaska.

The realm's *coal* reserves are among the greatest anywhere on Earth; the U.S. portion alone contains at least a 400-year supply. Three coal regions are evident: (1) Appalachia, still the largest region but declining steadily because its high-sulfur coal must be expensively treated to meet federal clean-air standards; (2) the Western coal region, centered on the northern U.S. Great Plains and southern Alberta, which continues to expand because its many low-sulfur, near-surface coal beds can easily be strip-mined; and (3) the Midcontinent coalfields, an arc of high-sulfur, strippable deposits centered on southern Illinois and western Kentucky, which is also declining as the West surges.

The leading *oil*-production areas of the United States are located along and offshore from the Texas-Louisiana Gulf Coast (where huge additional reserves were discov-ered below the gulf's floor in the mid-1990s); in the Midcontinent district, extending through western Texas-Oklahoma-eastern Kansas; and along Alaska's central North Slope facing the Arctic Ocean. Lesser oilfields are also found in Southern California (major untapped reserves lie offshore), west-central Appalachia, the northern Great Plains, and the southern Rockies. Canada's major oilfields lie in a wide crescent curving southeastward from the Athabasca tar-sand deposits in northern Alberta (which alone produce more than 25 percent of the country's crude oil) to the prairies of southern Manitoba.

The distribution of *natural gas* deposits resembles the geography of oilfields because petroleum and energy gas are usually found in similar geologic formations (the floors of ancient shallow seas). Accordingly, major gasfields are located in the Gulf, Midcontinent, and Appalachian districts. However, when subsequent geologic pressures are exerted on underground oil, the liquid is converted into natural gas. This has happened frequently in mountainous zones so that western gas deposits in and around the Rockies tend to stand apart from oilfields.

Agriculture

Despite the twentieth-century emphasis on urbanization and the development of the nonprimary sectors of the spatial economy, agriculture remains an important element in America's human geography. Because it is the most extensive (space-consuming) economic activity, vast expanses of the U.S. (and Canadian) landscape are clothed with fields of grain. In addition, great herds of livestock are

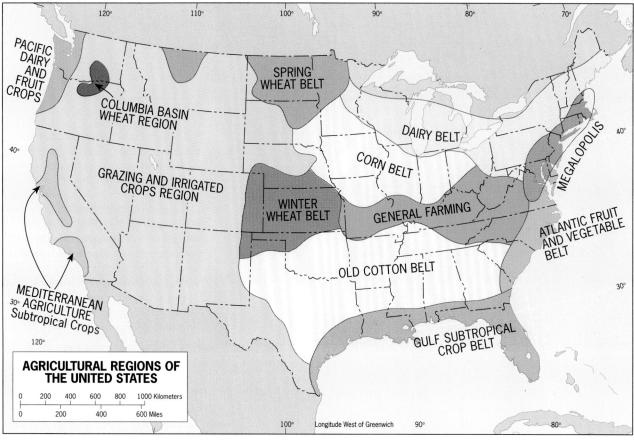

FIGURE 3-13

sustained by pastures and fodder crops because this wealthy realm can afford the luxury of feeding animals from its farmlands—and demands huge quantities of red meat in its diet. The growing application of high-technology mechanization to farming has steadily increased both the volume and value of total agricultural production. It also has been accompanied by a sharp reduction in the number of those actively engaged in agriculture: in 1995, only a dwindling 1.7 percent of the U.S. population still lived on farms. At the same time, the increasing size and efficiency of the individual producing unit are continuing to diminish the number of farms—to less than 1.9 million today, the lowest total since the 1850s.

The regionalization of U.S. agricultural production is shown in Figure 3-13, its spatial organization developing largely within the framework of the **von Thünen model** (see pp. 52–54). As in Europe (Fig. 1-4), the early nineteenth-century, original-scale model of town and hinterland—in a classical demonstration of time-space convergence—expanded outward (with constantly improving transportation technology) from a locally "isolated state" to encompass the entire continent by 1900. As this macrogeographic structure formed, the greatly enlarged original Thünian production zones (Fig. 1-3) were modified: (1) the

first ring was now differentiated into an inner fruit/vegetable zone and a surrounding belt of dairying; (2) the forestry ring was displaced to the outermost limits of the regional system because railroads could now transport wood quite cheaply; (3) the field crops ring subdivided into an inner, mixed crop-and-livestock ring to produce meat (the Corn Belt, as it came to be called) and an outer zone that specialized in the mass production of wheat grains; and (4) the ranching area remained in its outermost position, a grazing zone that supplied young animals to be fattened in the meat-producing Corn Belt as well as supporting an indigenous sheep-raising industry. The "supercity" anchoring this macro-Thünian regional system was the northeastern Megalopolis, well on its way toward coalescence and already the dominant food market and transport focus of the entire country.

Although the circular rings of the model are not apparent in Figure 3-13, many spatial regularities can be observed (remember that von Thünen, too, applied his model to reality and thereby introduced several distortions of the theoretically ideal pattern). Most significant is the sequence of farming regions as distance from the national market increased, especially westward from Megalopolis toward central California, which was the main directional thrust of the historic inland penetration of the United States. The Atlantic Fruit and Vegetable Belt, Dairy Belt, Corn Belt, Wheat Belts, and Grazing Region are indeed consistent with the model's logical structure, each zone successively farther inland astride the main transcontinental routeway. Deviations from the scheme may be attributed to irregularities in the environment or to unique conditions. For example, the nearly year-round growing seasons of California and the Gulf Coast-Florida region permit those distant areas (with the help of efficient refrigerated transport) to produce fruits and vegetables in competition with the regional system's innermost zone.

Manufacturing

The geography of North America's industrial production has long been dominated by the Manufacturing Belt (Fig. 3-6). As noted earlier, the internal structure of the Belt is organized around a dozen urban-industrial districts interlinked by a dense transportation network. This highly agglomerated spatial pattern developed because of the economic advantages of clustering industries in cities, which contained the largest concentrations of labor and capital, constituted major markets for finished goods, and were the most accessible nodes on the national transport network for assembling raw materials and distributing manufactures. As these industrial centers expanded, they also achieved **economies of scale**, savings accruing from large-scale production in which the cost of manufacturing a single item was further reduced as individual companies mechanized assembly lines, specialized their workforces, and purchased raw materials in huge quantities.

This efficient production pattern served the nation well throughout the remainder of the industrial age. Because of the effects of **historical inertia**—the need to continue using hugely expensive manufacturing facilities for their full, multiple-decade lifetimes to cover initial investments—the Manufacturing Belt is not about to disappear for a long time to come. However, as aging factories are retired, a process well under way, the distribution of American industry will change. As transportation costs equalize among U.S. regions, as energy costs now favor the south-central oil- and gas-producing states, as high-technology manufacturing advances reduce the need for lesser skilled labor, and as locational decision-making intensifies its attachment to noneconomic factors, industrial management has increasingly demonstrated its willingness to relocate to regions it perceives as more desirable in the South and West.

The Decline of Traditional Industry   The decline of traditional heavy industry has been a major trend in the Manufacturing Belt since the 1970s. Problems of obsolescence, economic decline, and foreign competition are widespread, and the region's aging physical plant is seldom worth reinvesting in; on the blighted landscape of the inner industrial city, in particular, the abandoned manufacturing complex is a depressing, all-too-familiar sight. Competing, high-quality industrial imports from abroad have grabbed a major share of the U.S. market during the past two decades. Once based exclusively on cheap labor, America's chief foreign competitors in the western Pacific Rim and Europe now possess superior technologies and productive efficiencies in a growing list of industries.

The United States' reduced global position is well illustrated in the case of steel, one of the most basic heavy

The trend toward modernization of the steel industry is not limited to the Manufacturing Belt: this is the North Star Steel Company's new mini-mill at Beaumont on the Gulf Coast of Texas.

industries and the very foundation on which the Manufacturing Belt was built. American steelmakers led the world in production until about 30 years ago, but then fared so poorly that by 1986 sales had declined by over 40 percent and employment by nearly 60 percent. That decline was due to lowered worldwide demand and the steady rise of steel imports (whose lower prices found many U.S. customers). Yet the American steel industry also fell behind its competitors because it was too slow to modernize its old-fashioned manufacturing methods.

The Revitalization Effort Belatedly, U.S. manufacturers are learning from their pacesetting foreign rivals. A major effort has been under way for the past decade to create high-technology "factories of the future," with much of this activity concentrated in the Midwest portion of the Manufacturing Belt. High-skilled jobs have been created by the tens of thousands in research and development and new equipment manufacturing, emphasizing robotics and other state-of-the-art automation technologies. Significantly, the steel industry has been in the forefront, with many companies more than doubling their productivity since 1990. Among the most successful ventures have been nonunionized *mini-mills* built by smaller firms (mostly in the Midwest and South), which manufacture steel by using electric power to melt down scrap metal instead of beginning with ore and coal.

There is a downside to this progress, however: the more successful high-tech manufacturing becomes, the fewer the factory jobs to be filled. Thus it is *not* the American manufacturing sector that is declining; indeed, industrial production in the mid-1990s was more than 60 percent higher than in 1970. Rather, it is the rapid shrinkage of blue-collar employment opportunities that has caused most of the pain, a still ongoing human problem because the dynamics driving the postindustrial revolution continue to mandate change in the structure and composition of the U.S. labor force.

The Postindustrial Revolution

The signs of postindustrialism are visible throughout the United States and Canada today, and they are popularly grouped under such names as "the information society" and "the electronic era." The term *postindustrial* by itself, of course, tells us mainly what the theme of the American economy is no longer. Yet many social scientists also use the term to refer to a specific set of societal traits that signal an historic break with the recent past (see box titled "The Lineaments of Postindustrial Society"). Many of the urban spatial expressions of the currently transforming society have already been highlighted; we focus here on some broader economic-geographical patterns.

High-technology, white-collar, office-based activities are the leading growth industries of the postindustrial economy. Most are spatially footloose and are therefore quite

THE LINEAMENTS OF POSTINDUSTRIAL SOCIETY

■ ■ ■

In his classic work, *The Coming of Postindustrial Society*, sociologist Daniel Bell sketched the distinguishing features of the emerging economy and its impact on American life. The new society and economy are marked by a fundamental change in the character of technology use—from fabricating to processing—in which telecommunications and computers are vital for the exchange of knowledge. Information becomes the basic product, the key to wealth and power, and is generated and manipulated by an intellectual rather than a machine technology. Yet postindustrial society does not displace industrial society—just as manufacturing did not obliterate agriculture. Instead, in Bell's words, "the new developments overlie the previous layers, erasing some features and thickening the texture of society as a whole."

Several hallmarks of postindustrialism can be identified. Knowledge is central to the functioning of the economy, which is led by science-based and high-tech industry. The technical/professional or "knowledge" class increasingly dominates the workforce: quaternary and quinary activity already employed 40 percent of the U.S. labor force in 1975, a proportion that has surpassed 65 percent in the 1990s. The nature of work in the information society focuses on person-to-person interaction rather than person-product or person-environment contacts. A meritocracy now prevails in which individual advancement is based on education and acquired job skills.

Among the many spatial implications of this socioeconomic transformation, Bell observed that postindustrial occupations would gravitate toward five major types of locations: economic enterprises, government, universities, social service complexes, and the military. Central North Carolina's well-known *Research Triangle Park*, located near the center of the "triangle" circumscribed by the nearby cities of Raleigh, Durham, and Chapel Hill, is an outstanding example of a prestigious Sunbelt high-technology manufacturing and research complex. Even a small sampling of the list of tenants fully supports Bell's hypothesis: (1) among its business enterprises are such corporate giants as IBM, TRW, and the headquarters of Burroughs Wellcome Pharmaceuticals; (2) government is represented by the U.S. Environmental Protection Agency, the Southern Growth Policies Board, and North Carolina's Board of Science and Technology; (3) three local universities are prime movers in the Park's research operations—Duke, North Carolina, and North Carolina State—which have also attracted the National Humanities Center and the Sigma Xi Scientific Research Society; (4) social services are the business of the National Center for Health Statistics Laboratory and the International Fertility Research Program; and (5) the military presence is embodied by the U.S. Army Research Office.

■ ■ ■

responsive to such noneconomic locational forces as geographic prestige, local amenities, and proximity to recreational opportunities. Northern California's *Silicon Val-*

ley—the headquarters of the U.S. microprocessor industry, which produces components for computers and electronic appliances—epitomizes the blend of locational qualities that attract a critical mass of high-tech companies to a given locality. According to a recent study, they include: (1) a nearby major university (Stanford) that offers a top-notch graduate engineering program; (2) close proximity to a cosmopolitan urban center (San Francisco); (3) a large local pool of skilled and semiskilled labor; (4) year-round pleasant weather and recreational opportunities; (5) high-quality nearby housing; and (6) a local economic climate especially conducive to corporate prosperity. Not surprisingly, Silicon Valley's success has spawned a number of similar complexes, most notably in the outer suburban cities of California's San Diego and Los Angeles, Texas's Austin and Dallas, and North Carolina's Raleigh-Durham.

Beyond these thriving complexes, the geographic impact of the postindustrial revolution has been uneven over the past two decades. States offering high-amenity environments have generally fared best, especially California, Florida, Texas, North Carolina, Oregon, and Washington. As we might expect, the Manufacturing Belt states as a group performed weakly, although the Midwest has enjoyed a resurgence since 1985. Most interesting has been the overall performance of the highly touted Sunbelt.

The notion that the southern tier of states is a uniformly booming area is a myth, and stark contrasts in development abound. Even at the state level, significant internal variations are obvious, and all too often the burgeoning economy of one county leaves a dormant neighbor completely untouched. It is worth pointing out, too, that even prosperous corners of the Sunbelt are not immune to changing economic currents—as Texas, Oklahoma, and Louisiana discovered when the oil industry entered a long slump in the 1980s. Yet, however spotty the overall economic development pattern, it is still very likely that many Sunbelt metropolises will continue to capture an expanding share of major U.S. business activity.

CANADA

Like the United States, Canada is a federal state, but it is organized differently. Canada is divided administratively into 10 provinces and 2 federal territories (Fig. 3-14). The two territories, the Yukon and the Northwest Territories,* together occupy a massive area half the size of Australia—

*Figure 3-14 also shows a third northern territory, Nunavut, which is planned to come into existence in 1999. This proposed new territory is the outcome of a major native land claim agreement between the Inuit and the Canadian government, and encompasses all of Canada's eastern Arctic. When it becomes part of the official map, Nunavut will cover one-fifth of Canada's land, an area larger than that of any other province or territory. Since 80 percent of its approximately 22,000 residents are Inuit, Nunavut will be the first territory to have a substantial degree of native self-government.

but are inhabited by fewer than 100,000 people. The 10 provinces range in territorial size from tiny, Delaware-sized Prince Edward Island to sprawling Quebec, more than twice the area of Texas. Beginning in the east, the four Atlantic Provinces are Nova Scotia, New Brunswick, Prince Edward Island, and Newfoundland. To their west lies Quebec, and to *its* west Ontario—Canada's two biggest provinces. Most of western Canada (which is what the Canadians call everything west of Lake Superior) is covered by the three Prairie Provinces—Manitoba, Saskatchewan, and Alberta. In the far west, beyond the Canadian Rockies and facing the Pacific, lies the tenth province, British Columbia.

In terms of population (in descending order), Ontario (10.9 million) and Quebec (7.4 million) are again the leaders; British Columbia ranks third with 3.8 million; next come the three Prairie Provinces with a combined total of 5 million; the Atlantic Provinces are the four smallest, together containing 2.8 million. The sparsely populated territories in the far north contain only about 95,000 residents. Canada's total population of 30 million is only slightly larger than one-tenth the size of the U.S. population. Spatially, as we noted at the outset of the chapter, most Canadians reside within 200 miles (325 km) of the United States border.

Population in Time and Space

The map showing the distribution of Canada's population (Fig. 3-3, p. 164) immediately tells us that only a fraction of this enormous country is effectively settled—about one-eighth, to be more precise. Even within that most heavily populated strip along the southern border, people tend to cluster in a pattern that resembles a strung-out island chain in a vast ocean. Five such clusters can be identified on the map, the largest by far being Main Street (home to more than 6 out of every 10 Canadians). As we noted earlier (Fig. 3-7), this is the conurbation that stretches across southeasternmost Quebec and Ontario, from Quebec City on the lower St. Lawrence River through Montreal and Toronto to Windsor on the Detroit River. The four lesser clusters are: (1) the Saint John-Halifax crescent in central New Brunswick and Nova Scotia; (2) the area surrounding Winnipeg in southeastern Manitoba; (3) the Calgary-Edmonton corridor in south-central Alberta; and (4) the southwestern corner of British Columbia, focused on Canada's third-largest metropolis, Vancouver.

Pre-Twentieth-Century Canada

Canada's population map evolved more slowly than that of the United States. Compared to European expansion in the United States (Fig. 3-5), penetration of the Canadian interior (Fig. 3-15) lagged several decades behind. In 1850, for example, when the eastern United States and much of

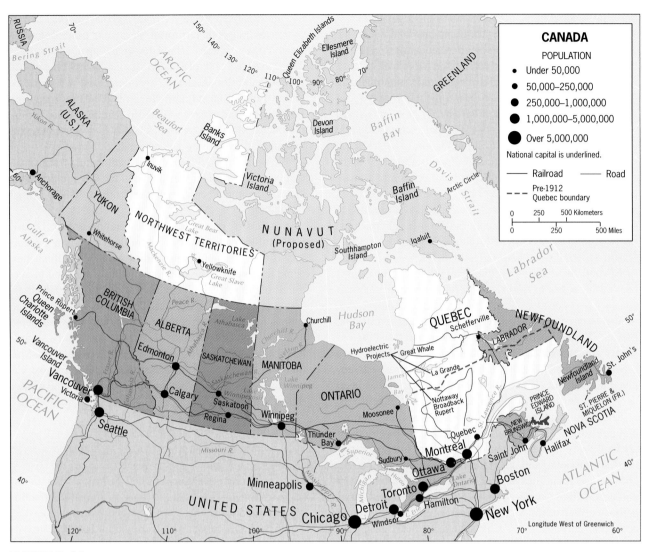

FIGURE 3-14

the Pacific coast had already been settled, Canada's frontier had only reached the shores of Lake Huron. The numbers of westward-moving Canadian pioneers, of course, were always far fewer than the totals of their counterparts to the south. In terms of political geography, Canada did not unify before the last third of the nineteenth century—and then only reluctantly because of fears the United States was about to expand in a northerly direction. Moreover, the Canadian push to the west was undoubtedly delayed by major physical obstacles, including the barren Canadian Shield north and west of Lakes Huron and Superior, and the rugged highlands dividing the Great Plains from the Pacific—which lacked convenient mountain passes.

The evolution of modern Canada, as well as its contemporary cultural geography (as we will see in the next section), is also deeply rooted in the bicultural division discussed at the outset of this chapter. Its origin lies in the fact that it was the French, not the British, who were the

first European colonizers of present-day Canada, beginning in the 1530s. *New France*, during the seventeenth century, grew to encompass the St. Lawrence Basin, the Great Lakes region, and the Mississippi valley. In the 1680s a series of wars between the English and French began, ending with France's defeat and the cession of New France to Britain in 1763. By the time London took control of this new possession, the French had made considerable progress in their North American domain. French laws, the French land-tenure system, and the Roman Catholic Church prevailed, and substantial settlements (including Montreal on the St. Lawrence) had been established. The British, anxious to avoid a war of suppression and preoccupied with problems in their other American colonies, gave former French Quebec—the region extending from the Great Lakes to the mouth of the St. Lawrence—the right to retain its legal and land-tenure systems as well as freedom of religion.

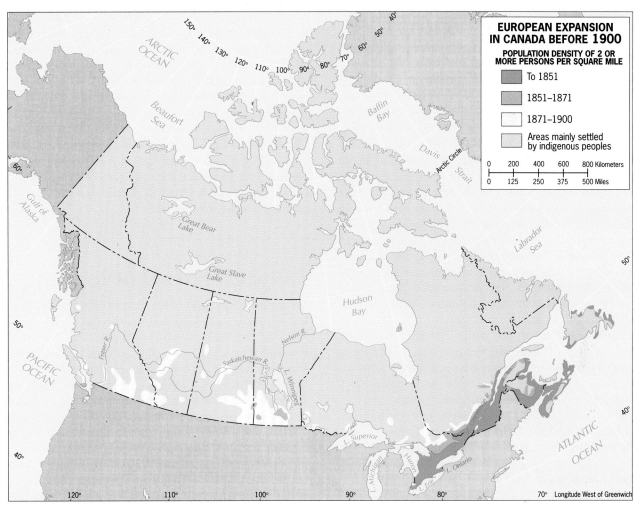

FIGURE 3-15

After the American War for Independence, London was left with a region it called British North America (the name Canada—derived from an aboriginal word meaning settlement—was not yet in use), but whose cultural imprint still was decidedly French. The war drove many thousands of English refugees northward, and soon difficulties arose between them and the French in British North America. In 1791, heeding appeals by these new settlers, the British Parliament divided Quebec into two provinces: Upper Canada, the region upstream from Montreal centered on the north shore of Lake Ontario, and Lower Canada, the valley of the St. Lawrence. Upper and Lower Canada became, respectively, the provinces of Ontario and Quebec (Fig. 3-14). By parliamentary plan, Ontario would become English-speaking and Quebec would remain French-speaking.

This earliest cultural division did not work well, and in 1840 the British Parliament tried again. This time it reunited the two provinces through the Act of Union under which Upper and Lower Canada would have equal representation in the provincial legislature. That, too, was a failure, and efforts to find a better system finally led to the 1867 British North America Act that established the Canadian federation (consisting initially of Upper and Lower Canada, New Brunswick, and Nova Scotia, later, between 1870 and 1949, to be joined by the other provinces and territories). Under the 1867 Act, Ontario and Quebec were once again separated, but this time Quebec was given important guarantees: the French civil code was left unchanged, and the French language was protected in Parliament and in the courts.

Twentieth-Century Canada

By the close of the nineteenth century, the fledgling Canadian federation was making major strides toward regional development and the spatial integration of a continentwide economy. In 1886, the transcontinental Canadian Pacific Railway had finally been completed to Vancouver, a condition of British Columbia's entry into the federation

15 years earlier. This new lifeline soon spawned the settlement of not only the far west but also the fertile Prairie Provinces, whose wheat-raising economy expanded steadily as immigrants from the east and abroad poured in during the early years of this century (Fig. 3-15). Industrialization also began to increasingly affect Canada at this time, and by 1920 manufacturing had surpassed agriculture as the leading source of national income. As noted earlier (Fig. 3-6), the dominant zone of industrial activity is the Toronto-Hamilton-Windsor corridor of southern Ontario (which also functions as one of the 12 districts of the American Manufacturing Belt). That has been the case since the Industrial Revolution took hold in Canada during World War I (1914–1918), when the country was heavily involved in Britain's war effort on the European mainland.

As industrial intensification proceeded, it was accompanied by the expected parallel maturation of the national urban system. Along the lines of Borchert's five epochs of American metropolitan evolution, Maurice Yeates has constructed a similar multistage model of urban-system development that divides the briefer Canadian experience into three eras. The initial *Frontier-Staples Era* (prior to 1935) encompasses the century-long transition from a frontier-mercantile economy to one oriented to staples (production of raw materials and agricultural goods for export), with increasing manufacturing activity in the budding industrial heartland. By 1930, Montreal and Toronto, reflecting their different cultural constituencies, had emerged as the two leading cities atop the national urban hierarchy (thus Canada has no single primate city).

Next came the *Era of Industrial Capitalism* (1935–1975), during which Canada achieved U.S.-style prosperity. However, most of this development—involving the tremendous growth of manufacturing, tertiary activities, and urbanization—took place after 1950 because the Great Depression lasted until World War II, which was followed by a long recession. A major stimulus was the investment of U.S. corporations in Canadian branch-plant construction, especially in the automobile industry in Ontario near the automakers' Detroit-area headquarters. In western Canada, the rapid growth of oil and natural gas production fueled Alberta's urban development, but new agricultural technologies reduced farm labor needs and sparked a rural out-migration in neighboring Saskatchewan and Manitoba. The postwar period also saw the ascent of Main Street, which on less than 2 percent of Canada's land quickly came to contain more than 60 percent of its people, two-thirds of its national income, and nearly 75 percent of its manufacturing jobs.

The third stage, ongoing since 1975, is the *Era of Global Capitalism*, signifying the rise of additional foreign investment from the western Pacific Rim and Europe. This, of course, is also the era of transformation into a postindustrial economy and society, and in the process Canada is experiencing many of the same upheavals as the United States. Interestingly, while U.S. urban growth has recently leveled off at about 75 percent of the total population, Canada still continues to urbanize (reaching 77 percent in 1997). Most of this development is occurring in the form of new suburbanization. Yet even though suburban business complexes are springing up outside Toronto and West Edmonton's megamall continues to be the world's biggest shopping center, it would be a mistake to presume that Canada's intraurban geography mirrors that of the United States. On the contrary, the Canadian metropolis has experienced far less automobile-generated deconcentration, and the central city has retained much of its middle-income population and economic base.

Cultural/Political Geography

The historic cleavage between Canada's French- and English-speakers, supposedly resolved by the British North America Act, has resurfaced in the past three decades to dominate the country's cultural geography. By the time the Canadian federation observed its centennial in 1967, it had become evident that Quebecers regarded themselves as second-class citizens; that to them bilingualism meant that French-speakers had to learn English but not vice versa; and that they perceived that Quebec was not getting its fair share of the country's wealth.

Since the 1960s, the intensity of ethnic feelings in Quebec has risen in surges despite the federal government's efforts to satisfy the province's demands. During the

From the Field Notes

"Automobile license plates in Quebec send a cultural-political message: 'I remember.' What will not be forgotten is Quebec's Frenchness, its aspirations to be 'itself' in a greater Canada—or outside of it. But Anglophone Canadians living in Quebec may not *souvenir* anything of the sort, and in 1995 they, and a small minority of Francophone *Quebecois*, combined to outvote (but barely) those who opted for independence. This kind of one-sided political message on a public decal is representative of the cultural irritations that have driven Quebecers apart, and may ultimately splinter the Canadian federation."

1970s, while a separatist political party came to power in Quebec, a new federal constitution was drawn up in Ottawa. In 1980, Quebec's voters solidly rejected sovereignty when given that choice in a referendum. But the new constitution did *not* satisfy the Quebecers, and throughout the 1980s and early 1990s the federal government struggled unsuccessfully to devise a plan, acceptable to all the provinces, that would keep Quebec in the Canadian federation.

By late 1995, with Canada's interest in constitutional reform utterly exhausted, a second referendum on Quebec's sovereignty could no longer be delayed. With the reenergized separatist party again leading the way, the Francophone-dominated electorate almost approved independence (which attracted an astonishing 49.4 percent of the vote). In the aftermath, many Canadians, both inside and outside Quebec, have come to believe that their country will soon cease to exist, perhaps as early as the end of the century. Indeed, as of early 1996, efforts were moving forward in Quebec toward organizing a third referendum that presumably would turn narrow defeat into separatist victory.

If sovereignty should be approved in another referendum, several issues would be raised. To begin with, the federal government may refuse to negotiate Quebec's separation if the vote does not go far beyond 50 percent (a 67 percent vote to break away has been mentioned as a minimum). And if negotiations should proceed, certain provinces insist that Quebec be treated toughly. This is particularly true for oil-rich Alberta and export-booming British Columbia; both have threatened to secede and forge their *own* independent state if more concessions are made to Quebec.

From a geographic standpoint, perhaps the most important issue of all is that the French linguistic region does *not* coincide with Quebec's territorial boundaries. As can be seen in Figure 3-16, ''French Canada'' is not only limited to the area centered on the middle and lower St. Lawrence valley, but also spills across provincial borders to the east and west as well as the U.S. border to the southeast. Moreover, a number of English-speaking communities remain inside the Francophone region, including many large towns (such as Hull across the Quebec border from Ottawa) and even sizeable portions of Montreal. Although younger inhabitants of these communities are leaving, many others have not only declared their intention to stay but asserted that they have the same right to secede from Quebec that Quebec has to secede from Canada!

B

Finally, the culture-based Quebec crisis has stirred the ethnic feelings of Canada's 1.3 million native peoples. Their assertions have also received a sympathetic hearing in Ottawa—as the planned creation of Nunavut demonstrates (see footnote p. 181). Most of all, the aboriginal peoples—who claim they constitute Canada's ''first nations''—want their rights protected by the federal government against the provinces. Among these will be the Native

Canadians of Quebec's northern frontier, the Cree, whose historic domain covers more than half of the province of Quebec as it appears on current maps. Administration of the Cree was assigned to Quebec's government in 1912, a responsibility that may well be invalidated by a move toward independence. In any case, the Cree would likely be empowered to seek independence as well. This would leave the French-speaking remnant of Quebec with only about 45 percent of the province's present area. As the map (Fig. 3-14) shows, the territory of the Cree is no unproductive wilderness; it contains vital facilities of the James Bay Hydroelectric Project, a massive scheme of dikes, dams, and artificial lakes that has transformed much of northwestern Quebec and generates electrical power for a huge market within and outside the province. Clearly, the stakes of independence are high and the risks are great.

Economic Geography

As in the United States, the growth of Canada's spatial economy has been supported by a diversified, high-quality *resource base*. We noted earlier that the Canadian Shield is endowed with rich mineral deposits and that oil is extracted in sizeable quantities in Alberta to the west of the Shield. Canada has long been a leading *agricultural* producer and exporter, especially of wheat and other grains from its breadbasket in the fertile Prairie Provinces. New technologies keep productivity high, but labor requirements are diminishing to the point where farm workers now account for less than 3 percent of the national workforce. Postindustrialization has caused substantial employment decline in the *manufacturing sector*, including the loss of at least 300,000 factory jobs during the recession of the early 1990s. These sharp cutbacks have most severely affected the Southern Ontario industrial heartland, which will continue to be plagued by unemployment and retraining problems as manufacturing plunges from 14 percent of the Canadian labor force (in 1990) to an expected mere 8 percent by 2000. On the other hand, Canada's already robust *tertiary and quaternary sectors* (which employed an expanding 67 percent of the total workforce in 1990) are creating a host of new economic opportunities. Fortunately, Southern Ontario is benefiting in this postindustrial transformation because the country's leading high-technology, research-and-development complex has grown up around Waterloo and Guelph, a pair of university towns about 60 miles (100 km) west of Toronto.

Canada's economic future is also going to be strongly affected by the continuing development of its trading relationships. The landmark U.S.–Canada Free Trade Agreement, signed in 1989, initiated a 10-year-long phasing out of all tariffs and investment restrictions between the

■ AMONG THE REALM'S GREAT CITIES...

Toronto

Toronto, the capital of Ontario and the country's largest metropolis (4.7 million), is the historic heart of English-speaking Canada. The landscape of much of its center is dominated by exquisite Victorian-era architecture, and surrounds a thriving downtown that comprises Canada's leading economic center. Landmarks abound in this CBD, and include the Skydome stadium, the famous City Hall with its pair of facing curved high-rises, and mast-like CN Tower, which at 1,815 feet (553 m) is the world's tallest freestanding structure. Toronto also is a leading port and manufacturing complex, with miles of factories, docks, and railroad yards lining the bustling shoreline of Lake Ontario, whose outlet accesses the Atlantic via the St. Lawrence Seaway.

Livability is one of the first labels that Torontonians would apply to their city, which has experienced less suburbanization than central cities of its size in the United States. *Diversity* is another leading characteristic because this is North America's richest urban ethnic mosaic. Among the largest and most vigorous of more than a dozen major ethnic communities are those dominated by Italians, Portuguese, Chinese, Greeks, French, and Ukrainians. Perhaps Toronto works so well because it pioneered the realm's first metropolitan government federation in 1954. After 40-plus years, the central city and the suburbs continue to confront tough problems and work harmoniously toward their resolution.

NORTH AMERICAN FREE TRADE AGREEMENT (NAFTA)

■ ■ ■

The economies of Canada, the United States, and Mexico formalized their interactions under the North American Free Trade Agreement (NAFTA) that took effect in 1994, and plans were launched in 1995 to add to include South America's Chile. This grand economic alliance has forged the world's largest trading bloc, a (U.S.) $7-trillion-plus market of 410 million consumers (surpassing the European Union's 373 million even without Chile). Between now and its completion date of 2009, NAFTA will integrate the four economies through a number of steps. Thousands of individual tariffs, quotas, and import licenses are scheduled to be removed or rolled back to eliminate most trade barriers for agricultural and manufactured goods as well as tertiary and quaternary services. Moreover, restrictions on the flow of investment capital across international borders will be lifted; regulations concerning foreign ownership of productive facilities are also being liberalized.

The architects of NAFTA cite the opportunities the maturing pact will bring. Business investment in all four countries will surge as unimpeded exports flow throughout the vast new marketplace. At the same time, intensified competition will give consumers access to a wider variety of quality products at reduced prices. Canada—which joined NAFTA mainly to safeguard the concessions it received in its 1989 Free Trade Agreement with the United States—faced the most skeptical citizenry at the outset, with opinion polls showing a majority of Canadians believed the 1989 pact caused them

more economic harm than good. Those doubts soon evaporated, however, as Canadian exports—and new jobs—under NAFTA rose substantially during the mid-1990s. Trade with the United States, already heightened under the bilateral 1989 agreement, propelled much of this growth, especially the southward export of goods whose tariffs had been lowered.

If NAFTA realizes its goals, it will require each member country to concentrate on the production of those goods and services for which it has a comparative advantage (in research and development, technological skills, managerial know-how, and the like). As workers move from less productive to more productive industries, jobs will be redistributed—a process that would be occurring anyway as these countries continue to adapt to continental and global economic change.

NAFTA's success has attracted the intense interest of all the other countries in the Western Hemisphere. Although most of them are involved in their own regional efforts aimed at greater economic integration (as we shall see in Chapters 4 and 5), the great majority desire to become part of an expanded NAFTA. To that end, ambitious plans were launched at summit meetings in the mid-1990s to create the *Free Trade Area of the Americas (FTAA)*, a hemisphere-encompassing free trade zone that would establish a single market of more than 800 million consumers. If FTAA is implemented by its target date of 2005, it may accelerate the completion of NAFTA—as well as NAFTA's evolution into a much larger organization.

■ ■ ■

two countries. This alone will amount to a monumental achievement, given that the annual flow of goods and services across the U.S.-Canada border is already the largest in the international trade arena. Moreover, in 1994 these tightening economic linkages were further reinforced by the implementation of the North American Free Trade Agreement (NAFTA), which consolidates the gains of the 1989 pact and opens major new opportunities for both countries by adding Mexico (and soon Chile) to the trading partnership (see box titled "North American Free Trade Agreement [NAFTA]").

At a recent conference of Canadian scholars and public officials, participants were asked to focus on the present and near-future consequences of the free trade pacts for their nation's economy. Most predicted a spatial reorientation of Canada, whereby traditional east-west linkages would weaken as north-south ties strengthened within a framework of transnational regions straddling the U.S.-Canadian border. Thus several parallel north-south relationships—many already well developed because of geographical and historical commonalities—can be expected to intensify: the Atlantic Provinces with neighboring New England, Quebec (even if independent) with New York State, Ontario with Michigan and surrounding Midwestern states, the Prairie Provinces with the Upper Midwest, and British Columbia with the (U.S.) Pacific Northwest. These international interactions are also well established in the overall regional configuration of North America, to which we now turn in the concluding section of this chapter.

REGIONS OF THE NORTH AMERICAN REALM

The ongoing transformation of North America's human geography is fully reflected in its internal regional organization. As new forces uproot and redistribute people and activities, a number of old locational rules no longer apply. However, the varied character of the realm's physical, cultural, and economic landscapes assures that meaningful regional differences will persist. The current areal arrangement of the United States and Canada will now be examined within a framework of eight regions (Fig. 3-16).

❖ THE CONTINENTAL CORE

The Continental Core region (Fig. 3-6)—synonymous with the American Manufacturing Belt—has been discussed earlier. Serving as the historic workshop for the linked spatial economies of the United States and Canada, this region was the unquestioned leader and centerpiece during the century between the Civil War and the close of the industrial age (1865–ca. 1970). The onset of postindustrialization has reduced that linchpin regional role, and today the Continental Core increasingly shares a number of major functions with still-emerging areas to the west and south.

Unfortunately for the Manufacturing Belt, its sharp recent decline in blue-collar employment was not accompanied by the creation of meaningful retraining programs or sufficient replacement jobs in other sectors. Thus parts of the region have become home to the chronically unemployed—a problem most apparent in the inner cities and innermost suburbs of Detroit, Chicago, Philadelphia, Cleveland, Baltimore, and Pittsburgh.

Manufacturing remains a highly important activity within the transformed American economy, but the pro-

Major Cities of the North American Realm

City	Population* (in millions)
Atlanta	4.1
Boston	3.2
Chicago	8.3
Dallas-Ft. Worth	5.2
Detroit	4.5
Houston	5.1
Los Angeles	10.0
Montreal	3.4
New York	17.9
Philadelphia	5.1
San Francisco-Oakland	4.2
Seattle-Tacoma	3.9
Toronto	4.7
Vancouver	1.9
Washington, D.C.	4.4

*Based on 1997 estimates.

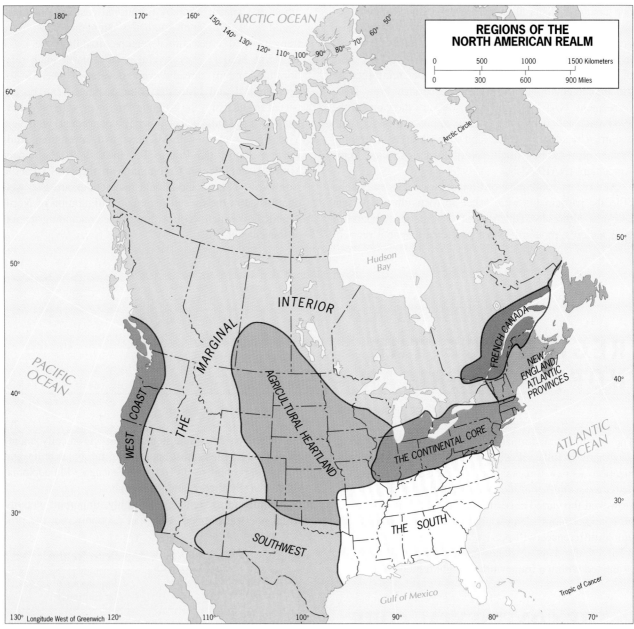

FIGURE 3-16

ductivity and obsolescence problems that have increased production costs in the Core region present obstacles in an era when its traditional industries are better able to respond to locational forces that operate in favor of newer regions. As we noted earlier, the Midwest portion of the Manufacturing Belt has responded successfully to this challenge by pursuing the high-tech upgrading of its aged industrial base.

In other parts of the Core region, where manufacturing has historically played a lesser role, postindustrial development has produced a number of new growth centers. The major metropolitan complexes of the northeastern Megalopolis—already the scene of much quaternary and quinary economic activity—are adjusting fairly well. Though still mired in the aftermath of recession, the Boston area, well endowed with research facilities, has attracted innovative high-tech businesses that are focused on the Route 128 freeway corridor that girdles the central city. New York City remains the national leader in finance and advertising, and it houses the broadcast media. But even in these spheres the city increasingly shares its once-exclusive decision-making leadership with its huge outer suburban city,

Chicago

Chicago lies near the southern end of Lake Michigan not far from the geographic center of the conterminous United States. Its centrality and crossroads location were evident to its earliest indigenous settlers, who developed the site as a portage where canoes could be hauled from the lake to the headwaters of a nearby stream that led to the Mississippi River. Centuries later, when the modern U.S. spatial economy emerged, Chicago became the leading hub on the continental transport network. As a freight-rail node, it still reigns supreme; and even though most long-distance passengers have switched to jet planes, Chicago remained the quintessential hub city as O'Hare International became the world's busiest airport more than a generation ago.

Poet Carl Sandburg described Chicago as ''the city of big shoulders,'' underscoring its personality and prowess as the leading U.S. manufacturing center during the century following the Industrial Revolution of the 1870s and 1880s. Among the myriad products that emanated from Chicago's industrial crucible were the first steel-frame skyscraper, elevated railway, refrigerated boxcar, cooking range, electric iron, atomic reactor, and window

envelope. Chicago also became a major commercial center, spawning a skyscraper-dominated Central Business District second only to Manhattan's in size and influence.

Chicago today (metro population: 8.3 million) is trying hard to make the transition to the postindustrial age. Its diversified economy has proven to be a precious asset in this quest, and the city is determined to reinvent itself as a first-order service center. The new forces it must contend with, however, are symbolized in the fate of emptying Sears Tower, the city's (and the nation's) tallest building: Sears & Roebuck recently abandoned the structure for a new headquarters campus along Interstate-90 in the far northwestern suburb of Hoffman Estates. This shift acknowledges the rise of Chicagoland's thriving outer city, one of the country's largest and most successful. Around O'Hare, a huge and still-expanding suburban office/industrial-park complex has truly become Chicago's ''Second City.'' It suggests that major airports will themselves play a leading role as urban growth magnets in the ever more globalized spatial economy of the dawning twenty-first century.

New York

New York is much more than the largest city of the North American realm. It is one of the most famous places on Earth; it is the hemisphere's gateway to Europe and the rest of the Old World; it is the seat of the United Nations; it is one of the globe's most important financial centers— a true ''world city'' in every sense of that overused term.

New York City consists of five boroughs, centered by the island of Manhattan which contains the CBD south of 59th Street (the southern border of Central Park). Here is a skyscrapered landscape unequaled anywhere, studded with more fabled landmarks, streets, squares, and commercial facilities than any other city except Paris. This is also the cultural and media capital of the United States, which is another way of saying that the city's influence constantly radiates across the planet thanks to New York-based television networks, newspapers and magazines, book publishers, fashion and design leaders, and artistic trendsetters.

At the metropolitan scale, New York centers a vast urban region, 150 miles (250 km) square (population: 17.9 million), which sprawls across parts of three States in the heart of Megalopolis. That outer city has become a giant in its own right with its population of more than 10 million, massive business complexes, and burgeoning suburban downtowns. Thus, despite its global connections, New York is very much caught in the currents that affect U.S. central cities today. Its ghettos of disadvantaged populations steadily grow; its aging port and industrial base continue to shrink; its corporate community is increasingly stressed by the high cost of doing business in ultra-expensive Manhattan. New York has not yet found the solutions but is struggling to overcome these problems by reinventing itself in order to become a leading economic player in the twenty-first century. Given its reputation, dynamism, and always confident spirit, a successful future cannot be ruled out.

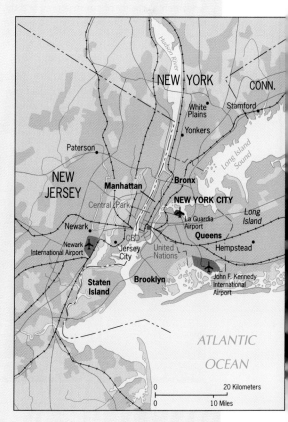

which more than a quarter-century ago surpassed Manhattan in total number of corporate headquarters facilities.

Perhaps the Core region metropolis that has gained the most from postindustrialization is Washington, D.C. As the information-producing and managerial sectors have blossomed, and as the U.S. federal government extended its ties to the private sector, the District of Columbia and its surrounding outer city of affluent Maryland and Virginia suburbs has amassed an enormous complex of office, research, trade-organization, lobbying, and consulting firms. The suburbanization of facilities has been heightened by the lack of space in the District's center, avoidance of poor inner-city neighborhoods near downtown, and particularly the lure of the Capital Beltway—a 66-mile (105-km) freeway that encircles Washington and connects its most prestigious suburbs. Indeed, this thriving curvilinear outer city (see photo p. 171) underscores the fact that nonresidential suburbia constitutes the healthiest category of economic subareas throughout the Continental Core.

From the Field Notes

"The older residential areas of Boston, between the Charles River and the high-rise CBD, is a legacy of a time when cities were more compact, and travel downtown from far-flung suburbs was still in the future. High-density, relatively low-rise buildings mixed residences and businesses together, and neighborhoods were defined by churches and their parishes. Many cities lost large parts of this heritage when suburbanization and modernization arrived, but Boston retains a substantial legacy. A local urban geographer told us that the fundamental quality of construction, and the relative location of this area close to the burgeoning downtown, combined to revive it."

❖ NEW ENGLAND/ATLANTIC PROVINCES

New England, one of the realm's historic culture hearths, has retained a strong regional identity for almost 400 years. The urbanized southern half of New England has been the northeastern anchor of the Continental Core since the mid-nineteenth century—where it must be regionally classified. However, the six New England states (Maine, New Hampshire, Vermont, Rhode Island, Massachusetts, and Connecticut) still share many common characteristics. Besides this overlap with the Manufacturing Belt (and Megalopolis) in its south, the New England region also extends northeastward across the Canadian border to encompass most of the four Atlantic Provinces of New Brunswick, Nova Scotia, Prince Edward Island, and Newfoundland (Fig. 3-16).

A long association based on economic and cultural similarities has tied northern New England to Atlantic Canada. Both are rural in character, possess rather difficult environments in which land resources are limited, and were historically bypassed in favor of more fertile inland areas. Thus economic growth in New England has always lagged behind the rest of the realm. Development has centered on primary activities, mainly fishing the (once) rich offshore banks of the nearby North Atlantic, forestry in the uplands, and farming in the few fertile valleys available. Recreation and tourism have boosted the regional economy in recent times, with scenic coasts and mountains attracting millions from the neighboring Core region. The growth of skiing has also helped, extending the tourist season through the harsh winter months.

A rich sense of history and tradition permeates the close cultural affinity of New England and the Atlantic Provinces. Both possess homogeneous British-based cultures that have long withstood persistent Francophone incursions from adjacent Quebec, and both have given rise to a staunchly self-sufficient, pragmatic, and conservative population. Village settlement is overwhelmingly preferred, in sharp contrast to the dispersed rural population of the North American interior that clings to its individual farmsteads.

In the past decade, New England's economy has experienced a swing from one extreme to the other. The middle and late 1980s were years of remarkable growth, fueled by the computer industry and surging federal expenditures on military research. But without warning, the region's fortunes nosedived during the early 1990s as smaller computers (mainly manufactured outside New England) quickly captured the market and the end of the Cold War signaled substantial cutbacks in defense spending. Moreover, these developments coincided with a nationwide recession that was felt so strongly in New England that this region was declared to be the U.S. region with the most severe economic downturn since the Great Depression of the 1930s. By the mid-1990s recovery had barely begun; this painfully slow process could last into the next century.

The Atlantic Provinces have also experienced hard economic times in the 1990s. Most adversely affected has been the groundfish industry as offshore stocks of flounder, haddock, and cod became severely depleted through overfishing. (This situation also affects the industry in Massachusetts, resulting in the 1994 closure of fishing on parts of the Georges Bank off Cape Cod.) And disagreements over Canada's maritime fishing boundaries have resulted in the seizure of vessels and heightened tensions with other nations, including the United States. In 1996, this situation remained a potential threat to the good relationships that have for so long marked international connections in the realm's northeastern corner.

❖ FRENCH CANADA

Francophone Canada constitutes the effectively settled, southern portion of Quebec, which straddles the central and lower St. Lawrence valley from where that river crosses the Ontario-Quebec boundary just upstream from Montreal to its mouth in the Gulf of St. Lawrence. Also included are sizeable concentrations of French speakers who reside just across the provincial borders with Ontario and New Brunswick (where they are called Acadians) as well as the U.S. border in northernmost Maine (Fig. 3-16). The Old World charm of Quebec's cities is matched by an equally unique rural settlement landscape introduced by the French: narrow rectangular farms, known as *long lots*, are laid out in sequence perpendicular to the St. Lawrence and other rivers, allowing each farm access to the waterway.

The economy of French Canada is no longer rural (although dairying remains a leading agricultural pursuit), exhibiting urbanization rates similar to those of the rest of the country. Industrialization is widespread, supported by cheap hydroelectric energy generated at huge dams in northern Quebec. Tertiary and postindustrial commercial activities are centered in Montreal; tourism and recreation are also important.

As we have seen, Quebec's heightened nationalism has had a major impact on the Canadian federation since 1970. Many important changes have also occurred within the province. During the 1980s, the provincial government enacted new laws that strengthened the French language and culture throughout Quebec. With English domination ended, the French Canadians have been channeling their energies into developing Quebec's economy. A by-product of that effort was the rapid urbanization of young families and a loosening of their ties to Roman Catholicism. The outcome was a sharp decline in the provincial birth rate, which the Quebec government now tries to offset by encouraging (with cash payments) large families and Francophone immigration (mostly from former colonies of France around the world).

❖ THE AGRICULTURAL HEARTLAND

In the heart of North America, agriculture becomes the predominant feature of the landscape. Whereas the innermost fruit-and-vegetable and dairying belts in the U.S. macro-Thünian regional system (Fig. 3-13) are in competition with other economic activities in the Continental Core, by the time one reaches the Mississippi valley, meat and grain production prevail for much of the 1,000 miles (1,600 km) west to the base of the Rocky Mountains. Because the eastern half of the Agricultural Heartland lies in Humid America and closer to the national food market on the northeastern seaboard, mixed crop-and-livestock farming wins out over less competitive wheat raising, which is relegated to the fertile but semiarid environment of the central and western Great Plains on the dry side of the 100th meridian. The latter area also contains Canada's agricultural heartland north of the 49th-parallel border. Although these fertile Prairie Provinces are less subject to serious drought than the U.S. high plains to the south, they are situated at a more northerly latitude and thus have a shorter growing season.

■ AMONG THE REALM'S GREAT CITIES . . .

Montreal

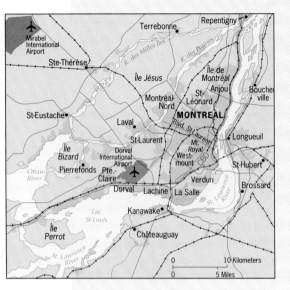

Montreal lies on a large triangular island in the St. Lawrence River, its historic core laid out along the base of what Montrealers call "the mountain"—flat-topped Mount Royal, from which the city takes its name. Culturally, Montreal is the largest French-speaking city in the world after Paris, and therein lie its opportunities and challenges. Without a doubt, this is one of the realm's most cosmopolitan cities, a rich European flavor pervading its stately CBD, its vibrant neighborhoods, and its throbbing street life. To escape the long harsh winters here, much of downtown Montreal has become an underground city with miles of handsome passageways lined with shops, restaurants, and movie theaters, all of it served by one of the most modern and attractive subway systems.

Yet, despite the sparkle and continental atmosphere, Montreal has not been a happy place in recent times. With the es-calation of tensions between Francophone and Anglophone Canada over the past three decades, it is understandable that the confrontation should overshadow life in Quebec's largest metropolis (3.4 million), and so the city has become deeply divided along ethnolinguistic lines. Spatially, this polarization is especially evident along the Boulevard St. Laurent, which marks the boundary between the French-speakers in Montreal's East End (who constitute about 70 percent of the city's population) and the English speakers who cluster in the West End. Montreal's economy hangs in the balance, too, and has already been adversely affected as several Anglophone companies withdrew from the province altogether. How all of present-day Canada ultimately resolves the Quebec issue is of critical importance to this jewel of a city—whose future prospects remain clouded by uncertainty.

The distribution of North American corn and wheat production is shown in Figure 3-17, which neatly defines the boundaries of the Agricultural Heartland region. The Corn Belt to the east is focused on Illinois and Iowa, with extensions into neighboring states (see Fig. 3-13). The area of north-central Illinois, which displays a major cluster of corn farming, is a classic example of transition along a regional boundary: the Agricultural Heartland/Continental Core dividing line connecting Milwaukee and St. Louis passes right through this zone (Fig. 3-16), which contains a number of sizeable urban-industrial concentrations interspersed by some of the most productive corn-producing land on Earth. On its western margins, the Corn Belt quickly yields to the dryland Wheat Belts. The growth of foreign wheat-sale opportunities in recent years has led to the intensification of production in the Winter Wheat Belt through the use of center-pivot irrigation, a means of overcoming the moisture deficit where groundwater supplies are sufficient (aquifers are most abundant in central Nebraska, southwestern Kansas, and the panhandles of Texas and Oklahoma).

Throughout the Agricultural Heartland, nearly all activity is oriented toward farming. Its leading metropolises— Kansas City, Minneapolis-St. Paul, Winnipeg, Omaha, and even Denver—are major processing and marketing points for pork and beef packing, flour milling, and soybean and canola oil production. People in this region are generally of Northern European ancestry and conservative. Yet rapid technological advances require that farmers keep abreast of increasingly scientific agricultural techniques and business methods if they are to survive in the hypercompetitive atmosphere of present-day agriculture.

❖ THE SOUTH

Of the realm's eight regions, none has undergone more change, during the past quarter-century, than the American South. Choosing to pursue its own sectional interests practically from the advent of nationhood, this region experienced an even deeper economic and cultural isolation from the rest of the United States during the long and bitter aftermath of the ruinous Civil War. For over a century, the South languished in economic stagnation; but the 1970s witnessed a reassessment in the nation's perception of the

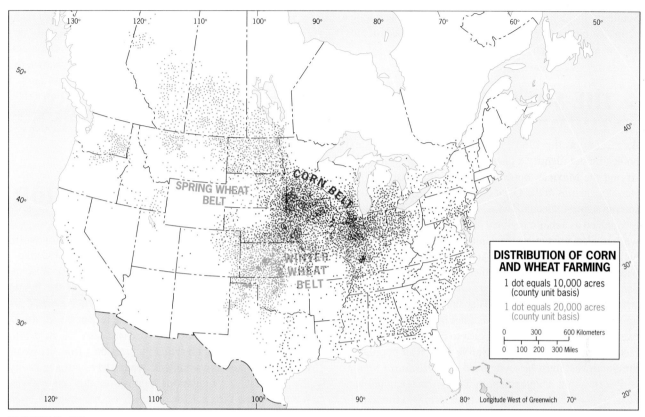

FIGURE 3-17

region that launched a still-ongoing wave of growth and change unparalleled in Southern history.

Propelled by the forces that created the Sunbelt phenomenon, people and activities streamed into the urban South. Cities such as Atlanta, Houston, Miami, Tampa, and Charlotte turned into booming metropolises practically overnight, and conurbations swiftly formed in such places as southern and central Florida, the Carolina Piedmont, and along the Gulf Coast from Houston through New Orleans to Mobile, Alabama. This ''bulldozer revolution'' was matched in the most favored rural areas by an agricultural renaissance that stressed such higher-value commodities as beef, soybeans, poultry, lumber, and, most recently, resurging cotton. On the social front, institutionalized racial segregation was dismantled three decades ago. Although certain problems in minority relations persist, inequalities are no worse in the South today than in other parts of the country.

Yet for all the growth that has taken place since 1970, the South remains a region beset by many economic problems because the geography of its development has been decidedly uneven. Although certain metropolitan and farming areas have benefited, many others containing large populations have not. Even places adjacent to boom areas

have often been left unaffected, and the juxtaposition of progress and backwardness is frequently encountered in the Southern landscape. The geographic selectivity of nonagricultural growth is noteworthy in another regard: with the exception of centrally located Atlanta and the I-85 corridor extending northeastward into North Carolina, much of the leading development has occurred near the region's periphery in the Washington suburbs of northern Virginia, the Houston area, and Florida's central and southern-coast corridors. As the checkerboard development pattern intensifies, even places within boom areas are now falling behind. This is especially true of the large Southern central cities, whose shrinking prospects are converging with those of their northern counterparts as burgeoning outer suburban cities around Atlanta, Houston, Miami, Tampa, Orlando, Charlotte, Raleigh-Durham, and New Orleans capture ever larger shares of total metropolitan employment.

In retrospect, recent Southern development has produced much beneficial change and has helped to remove the region's longstanding inferiority complex. But the South still has a long way to go, and even today much of the region lags behind the times. One of its foremost students, the historian C. Vann Woodward, offers this revealing insight about the so-called New South: ''The old

Southern distinction of being a people of poverty among a people of plenty lingers on. There is little prospect of closing the gap overnight.''

❖ THE SOUTHWEST

As recently as the 1960s, North America geography textbooks did not identify a distinct southwestern region, classifying the Mexican borderland zone into separate southward extensions of the Great Plains, Rocky Mountains, and Intermontane Plateaus. Today, however, this area must be recognized as a major regional entity. The emerging Southwest is also unique in the United States because it is a bicultural regional complex. Atop the crest of the Sunbelt wave, in-migrating Anglo-Americans join a rapidly expanding Mexican-American population anchored to the sizeable long-time resident Hispanic population, which traces its origins to the Spanish colonial territory that once extended from Texas's Gulf Coast to central California. In fact, when we add the large Native American population, the Southwest must be recognized as a *tricultural* region.

Recent rapid development in Texas, Arizona, and New Mexico is essentially built on a three-pronged foundation: (1) availability of huge amounts of *electricity* to power air conditioners through the long, brutally hot summers; (2) sufficient *water* to supply everything from swimming pools to irrigated crops so that large numbers may reside in this dry environment; and (3) the *automobile*, so that affluent newcomers may spread themselves out at much-desired low densities. The first and third of these foundation prongs have been rather easily attained because the eastern flank of the Southwest is abundantly endowed with oil and natural gas (Fig. 3-12). The future of water supplies, however, is far more problematical. For instance, Phoenix and Tucson could not keep growing at their current swift pace without the massive new Central Arizona Project canal, the state's last major source (the Colorado River) of drinking water.

D

So far, the generation of new wealth in this growing region makes it an undeniable success. Since much of this achievement is linked to the fortunes of the oil business, however, the Southwest may well be entering a future beset by unpredictable economic turns if the experiences of the energy industry since 1970 are an indicator. In the 1980s, much of Texas was staggered by the falling fortunes of oil, and it rebounded only slowly during the early 1990s. But the postindustrial revolution is also prominently represented in this region—with a specialized activity complex of electronic and space-technology facilities located in the eastern Texas triangle formed by connecting Houston, San Antonio, and Dallas-Fort Worth. (The state capital of Austin, near the triangle's center, has particularly benefited and

is now one of the nation's leading centers of high-technology research.) The Southwest's newest economic opportunities revolve around NAFTA-spawned development. The Texas sector of the U.S.-Mexico border has shown the most promise so far, especially within a new transnational growth corridor that extends for more than 500 miles (800 km) southward from Dallas through Austin and San Antonio to Monterrey 150 miles (250 km) inside Mexico.

❖ THE MARGINAL INTERIOR

The Marginal Interior is an appropriate name for North America's largest region by far. It covers most of Canada, all of Alaska, the northern salients of Minnesota-Wisconsin-Michigan, New York's Adirondacks, and the inland U.S. West between (and including) the Sierra Nevada-Cascades and the Rocky Mountains. There can be no doubt about the interior position of this vast slice of the North American realm; its marginal nature stems from its isolation and rugged environment, which have attracted only the sparsest of populations relative to the other seven regions. In the U.S. portion of the region, for example, even after counting substantial recent growth, population density is still only about 12 per square mile (5 per sq km) in contrast to 72 per sq mi (30 per sq km) for the country as a whole. Yet, its disadvantages aside, the Marginal Interior contains great riches because it is one of the Earth's major storehouses of mineral and energy resources.

Not surprisingly, this region's history has been a frustrating one of boom-and-bust cycles, determined by technological developments, economic fluctuations, and corporate and government decisions made in the Continental Core. The events of the past several years underscore this bittersweet development process. Following the oil shortage of the early 1970s, there was a virtual invasion in the search for additional petroleum and natural gas supplies, with efforts concentrating on the Arctic North Slope of Alaska and adjacent Canada as well as southwestern Wyoming, northwestern Colorado, and parts of Alberta. With the opening of the Trans-Alaska Pipeline in 1977, optimism reached an all-time high, and western Wyoming and Colorado were inundated by tens of thousands of migrants who taxed local community facilities to the breaking point. But, as has happened so many times before, the boom soon fizzled out (when the oil shortage ended and fuel prices plummeted).

Elsewhere in the region, recent economic recessions negatively affected the mining industry: many iron mines were closed in Minnesota, northern Ontario, and Labrador, and once-productive copper pits in Montana, Utah, and Arizona were faced with unprecedented slowdowns. Nevertheless, the future still beckons brightly. Rocky Mountain coal- and uranium-extraction operations were relatively untouched by these economic misfortunes, and the mining of silver, lead, zinc, and nickel endures at scattered sites throughout western North America and on the Canadian Shield to the east. And surely, as the most easily extractable fossil-fuel supplies decline worldwide, oil and gas drillers will one day return in force to the central Rockies and its vast deposits of oil shale.

The recent sharp swings of the extractive economy also mask a steadier influx of population and nonprimary activities to other parts of the Marginal Interior. In fact, from 1970 to 1995, the U.S. component of the region grew by more than 70 percent. The guiding force was the search for high-amenity locations, with the clean, stress-free, wide-open spaces of certain parts of the Rockies and Intermontane Plateaus a particular attraction. Accordingly, Nevada, Arizona, Idaho, Colorado, Utah, New Mexico, and Montana all ranked among the 10 fastest-growing states in the mid-1990s. In addition, business and industrial firms are streaming into the Marginal Interior, lured to its booming metropolitan areas by amenities, comparatively low wages and taxes, and skilled labor. Utah has benefited the most, especially *Software Valley*, the 40-mile (65-km) corridor

between Salt Lake City and Provo along the base of the Wasatch mountain front. Here lies the world's second largest concentration of information technology (after Silicon Valley), built up around two universities, a highly educated and dedicated workforce, and a postindustrial complex specializing in the production of word-processing and computer-simulation software.

❖ THE WEST COAST

The Pacific coast of the conterminous United States and southwesternmost Canada has been a powerful lure to migrants since the Oregon Trail was pioneered more than 150 years ago. Unlike the remainder of the North American west, the narrow strip of land between the Sierra Nevada-Cascade mountain wall and the sea receives adequate moisture. It also possesses a far more hospitable environment, with generally delightful weather south of San Francisco, highly productive farmlands in California's Central Valley, and such scenic glories as the Big Sur coast, Washington State's Olympic Mountains, and the spectacular waters surrounding San Francisco-Oakland, San Diego, Seattle-

■ AMONG THE REALM'S GREAT CITIES . . .

Los Angeles

It is hard to think of another city that in the past few years has been in more headlines than Los Angeles. Much of this publicity has been negative. But despite its earthquakes, mudslides, brushfires, pollution alerts, showcase trials, and riots, the glamorous image of "LA" has hardly been dented. This is compelling testimony to the city's resilience and vibrancy and to its unchallenged position as the Western world's entertainment capital.

The plane-window view during the descent into Los Angeles International Airport, almost always across the heart of the metropolis, gives a good feel for the immensity of this urban landscape. It not only fills the huge natural amphitheater known as the Los Angeles Basin, but it also oozes into adjoining coastal strips, mountain-fringed valleys and foothills, and even the margins of the Mojave Desert more than 50 miles (80 km) inland. This quintessential spread city, of course,

could only have materialized in the automobile age. In fact, most of it was built rapidly over the past 40 years, propelled by the swift expansion of a high-speed freeway network unsurpassed anywhere in metropolitan America. In the process, Greater Los Angeles became so widely dispersed that today it has reorganized within a geographic framework of six sprawling urban realms (Fig. 3-10).

The metropolis as a whole is home to 10 million Southern Californians, constitutes North America's second largest urban agglomeration, and forms the southern anchor of the huge conurbation that lines the Pacific coast from San Francisco southward to the Mexican border (Fig. 3-7). It also is the West Coast's leading trade, manufacturing, and financial center. In the global arena, Los Angeles is the eastern Pacific Rim's biggest city and the origin of the greatest number of transoceanic flights and sailings to Asia.

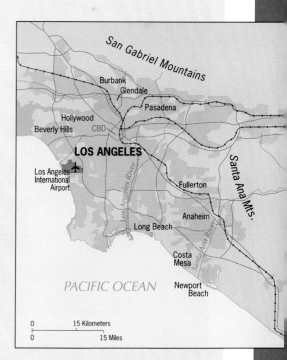

Tacoma, and Vancouver. Most major development here took place during the post-World War II era, accommodating enormous population and economic growth, and the West Coast is just now beginning to face the less pleasant consequences of regional maturity.

Fifty years of unrelenting growth have especially taken their toll in California. The massive development of America's most populated state (33 million in 1997) has been overwhelmingly concentrated in the teeming conurbation extending south from San Francisco through San Jose, the San Joaquin valley, the Los Angeles Basin, and the southwestern coast into San Diego at the Mexican border. Environmental hazards bedevil this entire corridor, including inland droughts, coastal-zone flooding, mudslides, brushfires, and earthquakes—with the ominous San Andreas Fault practically the axis of megalopolitan coalescence. To all this, humans have added their own abuses of California's fragile habitat, from overuse of water supplies (requiring vast aqueduct systems to import water from hundreds of miles away) to the appalling air pollution of Los Angeles caused by the emissions of more than 8 million motor vehicles, which are vital to movement within this particularly dispersed metropolis of 10 million.

D

Lost somewhere in the shuffle is yesterday's glamorous image of Southern California—communicated so stunningly in the motion pictures of the postwar period—which is based on relaxed outdoor living in luxurious horticultural suburbs in one of the most agreeable climates to be found anywhere on Earth. Undeniably, economic development has brought California prosperity, and the state has decidedly established its leadership as a national innovator. But this affluence has also been subject to sudden shifts because much of the California economy is tied to the aerospace, weapons, and other volatile industries. Today that dependency is lessening, particularly in Southern California, which was battered by the recession of the early 1990s. More than half of the 350,000 aerospace jobs the state recorded in 1990 have disappeared, but new Pacific Rim-oriented service industries (most notably involving entertainment) are rapidly emerging to take up some of the slack. Clearly, however, the years of unbridled optimism are over, and the future is likely to bring new struggles to

Vancouver, leading city of Canada's British Columbia, has been favored by its location on the Pacific Rim. Large numbers of Asian immigrants, particularly from Hong Kong since 1990, have come to settle in the Vancouver area, intensifying the city's already strong ties to its Transpacific trading partners.

maintain the state's still-considerable advantages in the face of growing competition from would-be new Californias elsewhere on the continent—not to mention the Pacific Rim.

The northern portion of the West Coast region includes the Pacific Northwest—focused on Oregon's Willamette valley and the nearby Cowlitz-Puget Sound lowland of western Washington—and the British Columbia coastal zone of southwesternmost Canada. Originally built on timber and fishing (primary activities that still thrive here), this area found its impetus for industrialization in the massive Columbia River dam projects of the 1930s and 1950s, which generated cheap hydroelectricity. This in turn attracted aluminum and aircraft manufacturers (and gave rise to the huge Boeing aerospace complex around Seattle). Unique environmental amenities—zealously safeguarded here—also have lured hundreds of growth companies, and the Pacific Northwest is adjusting smoothly to the maturing postindustrial economy.

Perhaps the greatest potential advantage of the Pacific Northwest, shared with urban California to the south, is its location as a gateway to the western Pacific Rim—which continues to emerge as one of the world's most dynamic economic arenas. As North America becomes increasingly enmeshed in the global economy, the eastern edge of Asia will be a critically important marketplace. Both Seattle and Portland, Oregon are bound to benefit because their airports are the closest to East Asia of any in the conterminous United States. North of the Canadian border, Vancouver enjoys the same geographic advantages, as well as a head start in forging trade and investment linkages to the western Pacific Rim thanks to its large and growing Asian population (almost 20 percent of the city's residents are of Chinese descent). The uncertainties surrounding the reunification of Hong Kong with China in 1997 have swelled these ranks as thousands of affluent citizens of the former crown colony have resettled in Vancouver, creating their own residential and (Asian-tied) business communities. As a result, more than 35 percent of British Columbia's exports now flow to the countries of the western Pacific Rim, a proportion far higher than in any other Canadian province or U.S. State. ◀|

■ **PRONUNCIATION GUIDE** ■

Adirondack (add-ih-RONN-dak)	Mesabi (meh-SAH-bee)	Ravenstein (RAVVEN-steen)
Boeing (BOH-ing)	Mojave (moh-HAH-vee)	San Joaquin (san-wah-KEEN)
Boulevard St. Laurent (boo-leh-VAHR san-law-RAH)	Montreal (mun-tree-AWL)	Saskatchewan (suss-KATCH-uh-wunn)
	Monterrey (mon-teh-RAY)	*Souvenir* (soo-veh-NEER)
Bryn Mawr (BRINN-mar)	Newfoundland (NYOO-fun-lund)	Tertiary (TER-shuh-ree)
Calgary (KAL-guh-ree)	Nova Scotia (nova-SKOH-shuh)	Toponymy (toh-PONN-uh-mee)
Caribbean (kuh-RIB-ee-un/karra-BEE-un)	Nunavut (NOON-uh-voot)	Tucson (TOO-sonn)
	Ottawa (OTTA-wuh)	Ungava (ung-GAH-vuh)
Cenozoic (senno-ZOH-ik)	Puerto Rico (pwair-toh-REE-koh)	Uwchlan (YOO-klan)
Cynwyd (KIN-widd)	Puget (PYOO-jet)	Valdez (val-DEEZE)
Francophone (FRANK-uh-foan)	Quaternary (kwuh-TER-nuh-ree)	Von Thünen (fon-TOO-nun)
Guelph (GWELF)	Quebec (kwih-BECK)	Wasatch (WAW-satch)
Inuit (IN-yoo-it)	Quebecer (kwih-BECK-er)	Willamette (wuh-LAMM-ut)
Isohyet (EYE-so-hyatt)	Quebecois (kay-beh-KWAH)	Winnipeg (WIN-uh-peg)
Megalopolis (mega-LOPP-uh-liss)	Quinary (KWYE-nuh-ree)	Yeates (YATES)

Scale 1:16 000 000; one inch to 250 miles. Polyconic Projection
Elevations and depressions are given in feet

copyright © Rand McNally, 1993

A-530000-76 -9 23
COPYRIGHT BY
RAND McNALLY & COMPANY
MADE IN U.S.A.

Scale 1:1 000 000
0   10 Miles
0 4 8 12 16 Kilometers

Bahía de Panamá

Caribbean
Sea

PANAMA

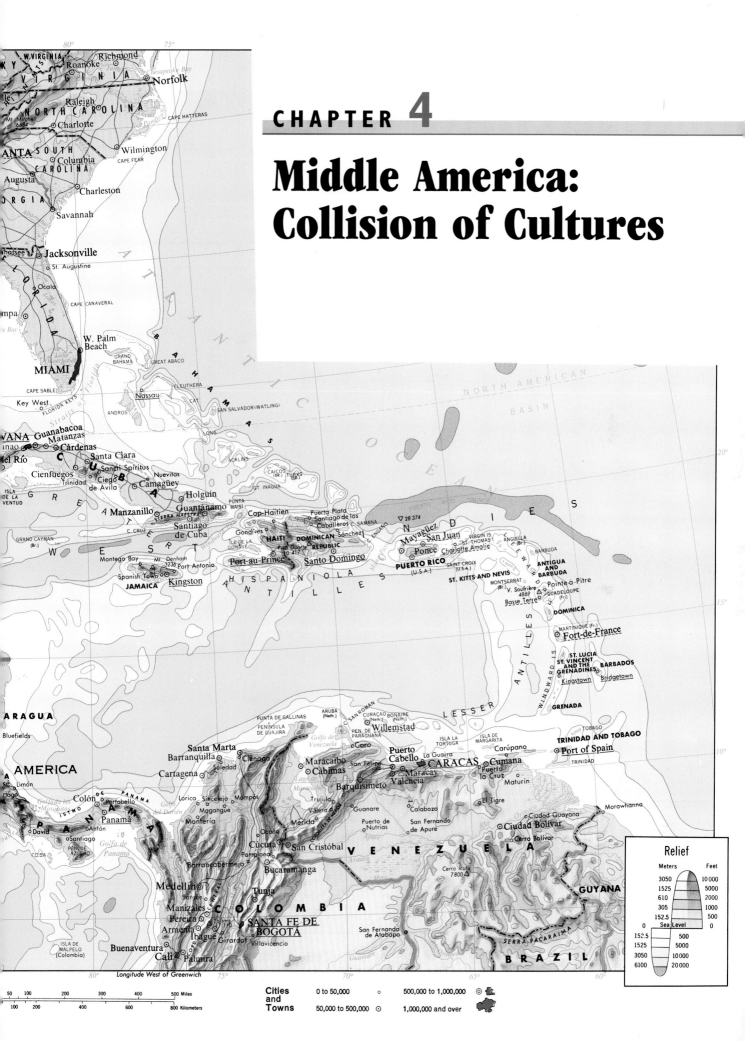

CHAPTER 4

Middle America: Collision of Cultures

IDEAS & CONCEPTS		◈ **REGIONS**
Historical geography	Plantation	Caribbean Basin
Cultural landscape	Plural societies	Greater Antilles
Land bridge	Transculturation	Lesser Antilles
Culture hearth	Maquiladora	Mexico
Mainland-Rimland framework	Altitudinal zones	Central America
Hacienda	Tropical deforestation	

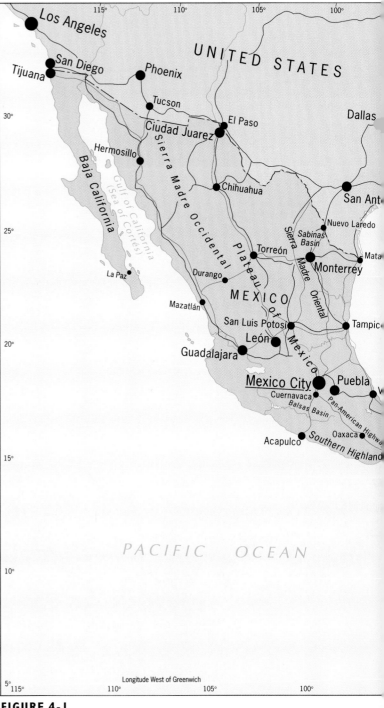

FIGURE 4-1

Middle America is a realm of vivid contrasts, turbulent history, political turmoil, and an uncertain future. The realm's diversity, visible in Figure 4-1, comprises all the lands and islands between the United States to the north and South America to the south. This includes the substantial landmass of Mexico, the narrowing strip of land to its southeast that extends from Guatemala to Panama, and the many large and small islands of the Caribbean Sea to the east. Middle America is a realm of soaring volcanoes and forested plains, of mountainous islands and flat coral cays. Moist tropical winds sweep in from the sea, watering windward (wind-facing) coasts while leaving leeward (wind-protected) areas dry. Soils vary from fertile volcanic to desert barren. Spectacular scenery abounds, and tourism is one of the realm's leading industries.

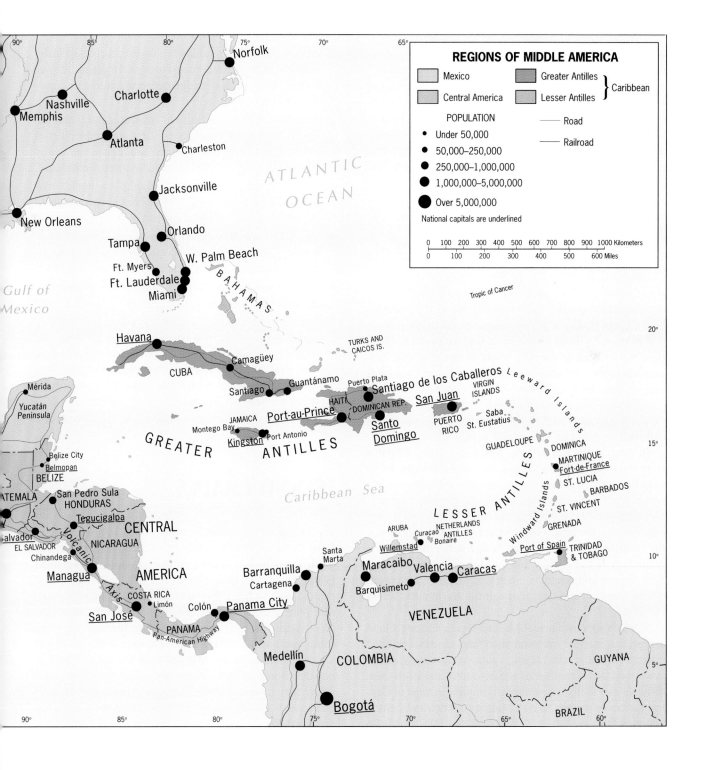

DEFINING THE REALM

Is Middle America a discrete geographic realm? Some geographers combine Middle and South America into a realm they call "Latin" America, citing the dominant Iberian (Spanish and Portuguese) heritage and the prevalence of the Roman Catholic religion. But these criteria apply far more strongly to South America. In Middle America, large populations exhibit African and Asian as well as European ancestries. Nowhere in South America has native, Amerindian culture contributed to modern civilization as strongly as it has in Mexico. The Caribbean Basin is a patchwork of independent states, territories in political transition, and residual colonial dependencies. The Dominican Republic speaks Spanish, adjacent Haiti uses French, Dutch is spoken in Curaçao and on neighboring islands, while English is spoken in Jamaica. Middle America thus gives vivid definition to concepts of cultural-geographical pluralism.

A LAND BRIDGE

Compared to the landmass of South America, the Middle American realm is divided and fragmented. Even its funnel-shaped mainland, a 3,800-mile (6,000-km) connection between the North and South American realms, narrows to a slim 40-mile (65-km) ribbon of land in Panama. Here, this strip of land, or *isthmus*, bends eastward so that Panama's orientation is east-west (with the Panama Canal cutting it northwest-southeast). On the map, continental (mainland) Middle America looks like a bridge between the Americas, and this is exactly what physical geographers call such an isthmian link: a **land bridge**.

In their geological history, North and South America were not always connected this way. Land bridges form, exist for a time, and then disappear, either through geologic processes or because sea levels rise and submerge them. If you examine a globe, you can see other present and former land bridges: the Sinai Peninsula between Asia and Africa, the (now-broken) Bering land bridge between northeasternmost Asia and Alaska, and the shallow waters between New Guinea and Australia. Such land bridges have played crucial roles in the dispersal of animals and humans across the globe.

Even though continental Middle America forms a land bridge, its internal fragmentation has always inhibited movement. Mountain ranges, volcanoes, swampy coastlands, and dense rainforests make contact and interaction difficult, especially where they all combine in most of Central America (see box titled "Middle America and Central America").

THE MAJOR GEOGRAPHIC QUALITIES OF MIDDLE AMERICA

■ ■ ■

1. Middle America is a fragmented realm that consists of all the mainland countries from Mexico to Panama and all the islands of the Caribbean Sea to the east.

2. Middle America's mainland constitutes a crucial barrier between Atlantic and Pacific waters. In physiographic terms, this is a land bridge between continental landmasses.

3. Middle America is a realm of intense cultural and political fragmentation. The political geography defies unification efforts, but countries and regions are finally beginning to work together to solve mutual problems.

4. Middle America's cultural geography is complex. African influences dominate the Caribbean, whereas Spanish and Amerindian traditions survive on the mainland.

5. The realm contains the Americas' least-developed territories. New economic opportunities may help to alleviate Middle America's endemic poverty.

6. In terms of area, population, and economic potential, Mexico dominates the realm.

7. Mexico is reforming its economy and has experienced major industrial growth. Its hopes for continuing this development are heavily tied to overcoming its present economic crisis and to expanding trade with the United States, Canada, and soon Chile under the North American Free Trade Agreement.

■ ■ ■

ISLAND CHAINS

The islands of the Caribbean Sea stretch in a lengthy arc from Cuba to Trinidad, with numerous outliers outside (such as Barbados) and inside (for example, the Cayman Islands) the main chain. The four large islands (Cuba, Hispaniola, Puerto Rico, Jamaica) are called the *Greater Antilles*; the smaller islands are referred to as the *Lesser Antilles*. Again, the map gives a hint of the physiography: the entire *archipelago* (island chain) consists of the crests and tops of mountain chains that rise from the floor of the Caribbean. Some of these crests are relatively stable; elsewhere they contain active volcanoes. Almost everywhere, earthquakes are a hazard. In the islands as well as on the mainland, the crust is unstable where the Caribbean, Cocos, and North American Plates come together (Fig. I-3). Add to this the seasonal risk of hurricanes spawned and nurtured by the realm's abundant tropical waters, and the Middle American environment ranks among the world's most hazardous.

LEGACY OF MESOAMERICA

Mainland Middle America was the scene of the emergence of a major ancient civilization. Here lay one of the world's true **culture hearths** (see Fig. 6-2), a source area from which new ideas radiated and whose population could expand and make significant material and intellectual progress. Agricultural specialization, urbanization, and transport networks developed, and writing, science, art, architecture, religion, and other spheres of achievement saw major advances. Anthropologists refer to the Middle American culture hearth as *Mesoamerica*, which extended southeast from the vicinity of present-day Mexico City to central Nicaragua. What is especially remarkable about its development is that it occurred in very different geographic environments, each presenting obstacles that were overcome in order to unify and integrate large areas. First, in the low-lying tropical plains of what is now Honduras, Guatemala, Belize, and Mexico's Yucatán Peninsula, and perhaps simultaneously in Guatemala's highlands, the Maya civilization arose more than 3000 years ago, building cities such as Mirador between 1000 and 500 B.C. Later, on the high plateau of central Mexico, the Aztecs founded a major civilization centered on the largest city ever to arise in pre-Columbian times. Other states with complex cultures arose

From the Field Notes
"We had seen the twin Piton volcanic peaks during the last part of the long drive south from Castries on the Caribbean island of St. Lucia. But only when we crossed the last ridge did we get a view of the site of the town of Soufrière, dangerously located at the base of Petit Piton, nearest of the two volcanic peaks. The Pitons have threatened Soufrière in the past, causing the town's evacuation; the population lives with constant risk. Volcanic eruptions, earthquakes related to plate movement, and hurricanes make this one of the world's most hazardous regions for its inhabitants."

elsewhere, making Mesoamerica one of the crucibles of statecraft and knowledge.

MIDDLE AMERICA AND CENTRAL AMERICA

■ ■ ■

Middle America, as we define it, includes all the mainland and island countries and territories that lie between the United States and the continent of South America. Sometimes the term *Central America* is used to identify the same realm, but Central America is actually a region within Middle America. Central America comprises the republics that occupy the strip of mainland between Mexico and Panama: Guatemala, Belize, Honduras, El Salvador, Nicaragua, and Costa Rica. Panama itself is regarded here as belonging to Central America as well. However, it should be noted that many Central Americans still do not consider Panama to be part of their region because that country was, for most of its history, a part of South America's Colombia.

■ ■ ■

The Lowland Maya

As we note later, the Maya civilization is the only one on the world culture map that arose substantially in lowland tropics. Its great cities, with stone pyramids and other elaborate buildings, still yield archeological information today. Maya culture over three millennia experienced successive periods of glory and decline, reaching its zenith—the Classic Period—from the third to the tenth centuries A.D.

The Maya civilization unified an area larger than any of the modern Middle American states except Mexico. Its population was probably somewhere between 2 and 3 million; the Maya language—the local *lingua franca*—and some of its related languages remain in use in the area to this day. The Maya state was a theocracy (ruled by religious leaders) with a complex religious hierarchy, and the great cities with their imposing structures that now lie in ruins served primarily as ceremonial centers. We also know that Maya culture produced skilled artists, writers, mathe-

The ancient Maya founded one of the world's great culture hearths in the tropical zone of Middle America extending from the Yucatán Peninsula into present–day Mexico as far as Chiapas and into Guatemala. The Maya domain was endowed with major stone-built centers, massive rock-carved figures, elaborate tombs, detailed inscriptions (hieroglyphs), and numerous structures still being yielded by the dense forests of the interior. This pyramid at Palenque, in the Mexican State of Chiapas, is known as the Temple of the Inscriptions.

maticians, and astronomers, but it had a practical side too. In agriculture and trade, these people achieved a great deal. They grew cotton, created a rudimentary textile industry, and even exported finished cotton cloth by seagoing canoes to other parts of Middle America in return for, among other things, the cacao (raw material of chocolate) they prized so highly that these beans were used as money, obsidian for tools and weapons, and quetzal feathers for fashion.

The Highland Aztecs

In what is today the intermontane highland zone of Mexico, significant cultural developments were taking place as well. Here lay Teotihuacán, the first true urban center in the Western Hemisphere, founded around the beginning of the Christian era. For nearly seven centuries it thrived, its population reaching over 125,000, its urban landscape dominated by great pyramids, its influence radiating into Maya culture. Later the Toltecs migrated into this area from the north, conquered and absorbed the local Amerindian peoples, and formed a powerful state. The Toltecs' period of hegemony was relatively brief, lasting less than three centuries after their rise to power around A.D. 900, but they conquered parts of the Maya domain, absorbed many Mayan innovations and customs, and introduced them on the plateau. When the Toltec state was in turn penetrated by new elements from the north, it was already in decay, but its technology was readily adopted and developed by the conquering Aztecs.

The Aztec state, the pinnacle of organization and power in pre-Columbian Middle America, is thought to have orig-

inated in the early fourteenth century with the founding of a settlement on an island in one of the many lakes that lay in the *Valley of Mexico* (the area surrounding present-day Mexico City). This urban complex, a functioning city as well as a ceremonial center, named Tenochtitlán, was soon to become the greatest city in the Americas and the capital of a large and powerful state. Through a series of alliances with neighboring peoples, the Aztecs gained control over the entire Valley of Mexico, the pivotal geographic feature of Middle America that is still the heart of the modern state of Mexico. This 30-by-40-mile (50-by-65-km) region is, in fact, a mountain-encircled basin positioned about 8,000 feet (nearly 2,500 m) above sea level. Both elevation and interior location affect its climate; for a tropical area, it is quite dry and very cool. The area's lakes formed a valuable means of internal communication, and the Aztecs connected several of them by canals. This fostered a busy canoe traffic, bringing agricultural produce to the cities and tribute paid by their many subjects to the headquarters of the ruling nobility.

Throughout the fourteenth century, the Aztec state strengthened its military position, and by the early fifteenth century the conquest of neighboring peoples had begun. The Aztec drive to expand the empire was directed primarily eastward and southward, and little resistance was encountered. The Aztecs' objective was neither the acquisition of territory nor the spreading of their culture, but rather the subjugation of peoples and towns in order to extract taxes and tribute. As Aztec influence spread throughout Middle America, the goods streaming back to the Valley of Mexico included gold, cacao beans, and cotton cloth. The state grew ever richer, its population mush-

■■■■■■■■■■■■ FOCUS ON A SYSTEMATIC FIELD: ■■■■■■■■■■■

Historical Geography

Although our survey of world realms is less than half complete, it has already become evident that the geographic present is a product of the past. The study of how a region evolves and continues to change is the domain of **historical geography**. Because it treats the all-encompassing dimension of time, the concepts and methods of historical geography—like those of regional geography—are applicable to every branch of the discipline.

Historical geographers approach their subdiscipline from a number of different perspectives. The first of these is the study of the *geographic past*, wherein earlier spatial patterns are researched, reconstructed, and compared for time periods that were crucial in the formation of the present regional structure. The second perspective focuses on *landscape evolution*, the historical analysis and interpretation of existing **cultural landscapes**. This search for ''the past in the present'' can involve the study of individual relics and comparative regional analysis of landscape artifacts such as house types. A third approach to historical geography can be called *perception of the past*, the application of modern behavioral concepts to recreate and study the attitudes of previous generations toward their environments.

One of the most active areas of historico-geographical research is the study of *settlement geography*, the facilities people build while occupying a region. Because these facilities usually survive beyond the era of their original functions, they often provide some of the clearest expressions of the past in the contemporary landscape. The colonial landscapes of the New World have received considerable attention, especially the layout of Spanish colonial towns in Middle America. Shown in Figure 4-2, this layout's historico-geographical significance is described by Charles Sargent:

The morphology, *or form, of a city reflects its past, and the clearest reflections are found in the street pattern, the size of city blocks, the dimensions of urban lots, and the surviving colonial architecture. In the typical Span-*

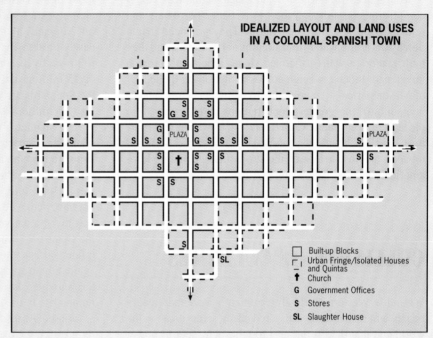

IDEALIZED LAYOUT AND LAND USES IN A COLONIAL SPANISH TOWN

☐ Built-up Blocks
⌐╵ Urban Fringe/Isolated Houses and Quintas
✝ Church
G Government Offices
S Stores
SL Slaughter House

FIGURE 4-2

ish American city, the colonial past is on view in what is now the city core, which is commonly laid out in a gridiron of square or rectangular blocks subdivided into long, narrow lots along equally narrow streets.

This particular spatial pattern hardly came about by accident—it was decreed by royal regulations down to the smallest details—and was repeated throughout Hispanic America. Because most Spaniards were urban dwellers, it was hardly surprising that they chose to build towns in their colonial territories. But why *this* particular town form? The answer rests with that other major characteristic of an urban settlement—function.

The Spanish colonial town possessed several functions that were best suited to the compact, gridiron-street-pattern layout. It is often said that the threefold aims of the rulers of New Spain were ''God, glory, and gold.'' Because the main function of the settlement was administrative,

town sites were chosen to maximize accessibility to trade routes and sources of tribute from local Amerindians.

This control function also extended to the internal structure of the town: everything was tightly focused on the central *plaza*, or market square, under the watchful eye of government authorities in adjacent buildings (**G** symbols in Fig. 4-2). Another leading town function was expressed in the central role of the church, which always faced the plaza as well. The Roman Catholic church sought to convert as many Amerindians to its faith as possible, and the easiest way to do this was to forcibly resettle the dispersed indigenous population in Spanish towns, where the collection of tribute, the recruitment of mine workers, and the farming of land surrounding the town were also facilitated. Because of the gridiron street plan, any insurrections by the resettled Amerindians could be contained by having a small military force seal off the affected blocks and root out the troublemakers.

roomed, and its cities expanded (Tenochtitlán peaked at a population in excess of 100,000). These cities were not just ceremonial centers but true cities with a variety of economic and political functions and large populations, including specialized, skilled labor forces.

The Aztecs produced a wide range of impressive accomplishments, although they were better borrowers and refiners than they were innovators. They practiced irrigation by diverting water from streams to farmlands, and they built elaborate walls to terrace slopes where soil erosion threatened. Indeed, when it comes to measuring the legacy of Mesoamerica's Amerindians to their successors (and to humankind), the greatest contributions surely came from the agricultural sphere. Corn (maize), the sweet potato, various kinds of beans, the tomato, squash, cacao tree, and tobacco are just a few of the crops that grew in Mesoamerica when the Europeans first made contact.

COLLISION OF CULTURES

We in the Western world all too often believe that history began when the Europeans arrived in some area of the world and that the Europeans brought such superior power to the other continents that whatever existed there previously had little significance. Middle America bears out this perception: the great, feared Aztec state fell before a relatively small band of Spanish invaders in an incredibly short time (1519–1521). Let us not lose sight, however, of a few facts. At first the Spaniards were considered to be ''White Gods,'' whose arrival had been predicted by Aztec prophecy. Having entered the Aztecs' territory, the earliest Spanish visitors could see that great wealth had been amassed in Aztec cities. And Hernán Cortés, for all his 508 soldiers, did not singlehandedly overthrow this powerful empire: he sparked a rebellion by peoples who had fallen under Aztec domination and who had seen their relatives carried off for human sacrifice to Aztec gods. Led by Cortés with his horses and guns, these indigenous peoples rose against their Aztec oppressors and joined the band of Spaniards headed toward Tenochtitlán. There, thousands of these Amerindians died in combat against the Aztec warriors; Tenochtitlán would not have fallen so easily to the Spaniards without their valor.

The defeat of Middle America's dominant indigenous state opened the door to Spanish penetration and supremacy. None of Middle America's other local states could withstand the European invasion either. Only some communities protected by geography—deep in the forests or high in the mountains—survived the onslaught for a time. But eventually the soldiers, the priests, and the Spaniards' diseases reached them too.

Effects of the Conquest

In Middle America the confrontation between Hispanic and native cultures spelled disaster for the Amerindians in every conceivable way: a drastic decline in population, rapid deforestation, pressure on vegetation from grazing animals, substitution of Spanish wheat for maize (corn) on cropland, and the concentration of Amerindians into newly-built towns. The quick defeat of the Aztec state was followed by a catastrophic decline in population. Of the 15 or 25 million native inhabitants of Middle America when the Spaniards arrived (estimates vary), only a century later just 2.5 million survived.

The Spaniards were ruthless colonizers, but not more so than other European powers that subjugated other cultures. True, the Spaniards first enslaved the Amerindians and were determined to destroy the strength of indigenous society. But biology accomplished what ruthlessness could not have achieved in so short a time. Nowhere in the Americas did the native peoples have immunity to the diseases the Spaniards brought: smallpox, typhoid fever, measles, influenza, and mumps. Nor did they have any protection against the tropical diseases that the Europeans introduced through their African slaves, which took enormous tolls on human life in the hot, humid lowlands of Middle America.

Middle America's cultural landscape—its great cities, its terraced fields, its dispersed aboriginal villages—was thus drastically modified. The Amerindian cities withered. Having destroyed Tenochtitlán, the Spaniards did recognize the attributes of its site and situation and chose to rebuild it as their mainland headquarters. Whereas the Amerindians had used stone almost exclusively as their building material, the Spaniards employed great quantities of wood and used charcoal for heating, cooking, and smelting metal. The onslaught on the forests was immediate, and expanding rings of deforestation quickly formed around the Spanish towns. The Spaniards also introduced livestock, whose numbers multiplied rapidly and made increasing demands not only on the existing grasslands but on the cultivated crops as well. Cattle and sheep quickly became sources of wealth, and the owners of the herds benefited. But the livestock now competed with the people for available food. This required the opening up of vast areas of marginal land in higher and drier locations, thereby contributing to a major disruption of the region's food-production balance. Moreover, the Spaniards brought in their own crops (notably wheat) and farming equipment, and soon large fields of wheat began to encroach upon the small plots of corn that the natives cultivated. Neither were their irrigation systems spared: the Spaniards needed water for their fields and hydropower for their mills, and they had the technology to take over and modify regional drainage and irrigation systems. This they did, leaving the Amerindians' fields insufficiently watered, thereby diminishing even further these peoples' food supply.

The most far-reaching changes in the cultural landscape introduced by the Spaniards had to do with their traditions as town dwellers. To facilitate domination, they relocated the Amerindians from their land into nucleated villages and towns the Spaniards established and laid out (see Fig. 4-2). In these settlements, the Spaniards could exercise the kind of government and administration to which they were accustomed; the focus of each new urban center was the plaza, site of the Catholic Church. Each town was located near what was thought to be good agricultural land (which more than occasionally turned out not to be the case), so that the Amerindians could go out each day and work in the fields. Packed tightly into these towns and villages, the Amerindians came face to face with Spanish culture. Here they learned the Europeans' Roman Catholic religion and paid their taxes and tribute to a new master. Nonetheless, the nucleated indigenous village survived. Its administration was also taken over by the Spaniards (and later by their postcolonial successors). Today it is still a key landscape feature of remote Amerindian areas in southeastern Mexico and inner Guatemala, where native languages still prevail over Spanish (Fig. 4-3).

Once the indigenous population was conquered and resettled, the Spaniards were able to pursue another primary goal in their New World territory: the exploitation of its wealth for their own benefit. Lucrative trade, commercial agriculture, livestock ranching, and especially mining were the avenues to affluence. The mining of gold held the greatest initial promise, and the Spaniards simply took over the Amerindian miner workforce already in operation. But when accessible gold deposits soon diminished, Spanish prospectors began searching for other valuable minerals. They were quickly successful, finding enormously profitable silver and copper deposits, particularly in a wide frontier zone north of the Valley of Mexico (see Fig. 4-6). The development of these resources set into motion a host of changes in this part of Middle America because mining towns required labor forces, extractive equipment, food supplies, and the transportation connections to handle the intensifying flows of people, goods, and services. Thus was born a new urban system, one that not only integrated and organized the Spanish domain in Middle America but also extended effective economic control over some far-flung parts of New Spain. Mining truly became the mainstay of colonial Middle America.

Whatever the arena of Spanish domination—towns, farms, mines, indigenous villages—the Roman Catholic church was the supreme cultural force transforming Amerindian society. Jesuits and soldiers pushed the Spanish frontier northward and westward, the church acquiescing in the often brutal occupation of Amerindian domains by the state. In the wake of conquest, it was the church that controlled, pacified, organized, and acculturated the Amerindian peoples.

MAINLAND AND RIMLAND

In Middle America outside Mexico, only Panama, with its twin attractions of interoceanic transit and gold deposits, became an early focus of Spanish activity (the Spaniards founded Panama City in 1519). Apart from their use of the approximate route of the modern Panama Canal as an Atlantic-Pacific link, their main interest lay on the Pacific side of the isthmus, a base from which Spanish influence steadily began to extend northwestward into Central America. The leading center of Spanish activity, however, remained in what is today central and southern Mexico.

The major arena of international competition in Middle America lay not on the Pacific side but on the islands and coasts of the Caribbean Sea. Here the British gained a foothold on the mainland, controlling a narrow, low-lying coastal strip that extended from Yucatán to what is now Costa Rica. As the colonial-era map (Fig. 4-4) shows, in the Caribbean the Spaniards faced the British, French, and Dutch, all interested in the lucrative sugar trade, all searching for instant wealth, and all seeking to expand their empires. Later, after centuries of European colonial rivalry in the Caribbean Basin, the United States entered the picture and made its influence felt in the coastal areas of the mainland—not through colonial conquest but through the introduction of widespread, large-scale, banana plantation agriculture.

The effects of these plantations were as far-reaching as the impact of colonialism on the Caribbean islands. The economic geography of the Caribbean coastal zone was transformed as hitherto unused alluvial soils in the many

AMERINDIAN LANGUAGES OF MIDDLE AMERICA

0   250   500   750 Kilometers
0   250   500 Miles

Gulf of Mexico

MEXICO

PACIFIC OCEAN

BELIZE

GUATEMALA   HONDURAS

EL SALVADOR   NICARAGUA

COSTA RICA

Areas where native languages are spoken by more than half of the people

FIGURE 4-3

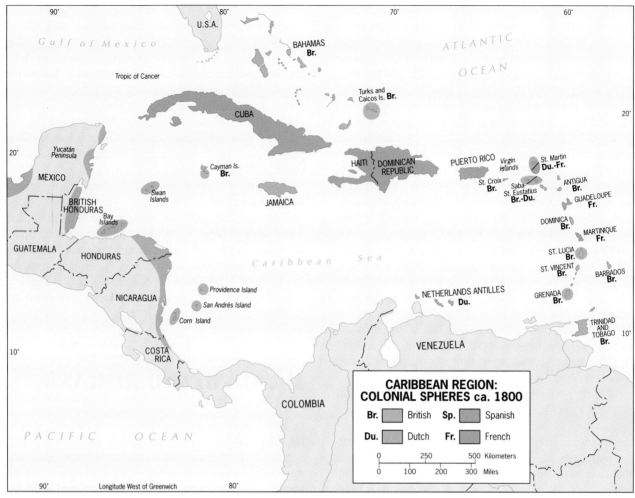

FIGURE 4-4

river lowlands were planted with thousands of acres of banana trees. Because the diseases the Europeans had brought to the New World had been most rampant in these hot and humid areas, the Amerindian population that survived was too small to provide a sufficient labor force.

The comparatively small populations of indigenous peoples on the Caribbean islands succumbed even more rapidly than the mainland Amerindians to the harsh treatment by the European colonists and to their imported diseases. In little more than half a century they were wiped out, creating a labor shortage and generating the trans-Atlantic slave trade from Africa that transformed the Caribbean's demography (see Fig. 7-5). When labor was needed on continental Middle America's Caribbean coast, tens of thousands of black laborers were brought to the mainland from Jamaica and other islands (many more came when the Panama Canal was dug between 1904 and 1914), completely altering the demographic complexion here as well.

These contrasts between the Middle American highlands on the one hand, and the coastal areas and Caribbean islands on the other, were conceptualized by John Augelli

into the **Mainland-Rimland framework** (Fig. 4-5). Augelli recognized (1) a Euro-Amerindian **Mainland**, consisting of mainland Middle America from Mexico to Panama, with the exception of the Caribbean coast from mid-Yucatán southeastward; and (2) a Euro-African **Rimland**, which included this coastal zone and the islands of the Caribbean. The terms *Euro-Amerindian* and *Euro-African* underscore the cultural heritage of each region: on the Mainland, European (Spanish) and Amerindian influences are paramount; in the Rimland, the heritage is European and African.

As Figure 4-5 shows, the Mainland is subdivided into several areas on the basis of the strength of the Amerindian legacy. In southern Mexico and Guatemala, Amerindian influences are prominent; in northern Mexico and parts of Costa Rica, those influences are limited; between these areas lie sectors with moderate Amerindian influence. The Rimland, too, is subdivided. The most obvious division is between the mainland-coastal plantation zone and the islands. But the islands themselves can be classified according to their cultural heritage. Thus there is a group of islands with Spanish influence (Cuba, Puerto Rico, and the

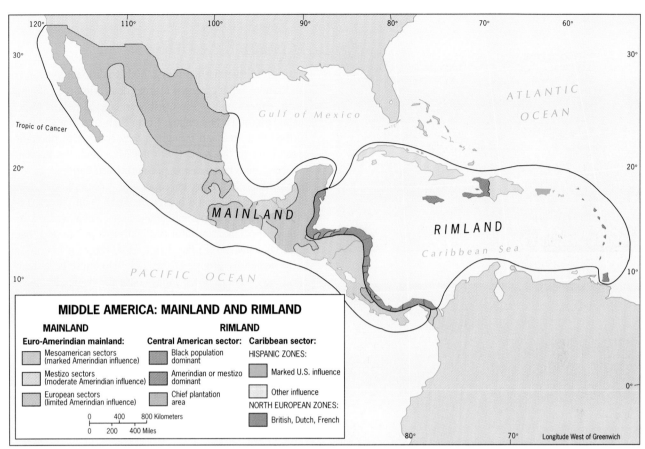

FIGURE 4-5

Dominican Republic on old Hispaniola) and another group with other European influences, including the former British West Indies, the various French islands, and the Netherlands Antilles.

These contrasts of human habitat are supplemented by regional differences in outlook and orientation. The Rimland was an area of sugar and banana plantations, of high accessibility, of seaward exposure, and of maximum cultural contact and mixture. The Mainland, being farther removed from these contacts, was an area of greater isolation. The Rimland was the region of the great plantation, and its commercial economy was therefore susceptible to fluctuating world markets and tied to overseas investment capital. The Mainland was the region of the hacienda, more self-sufficient and considerably less dependent on external markets.

The Hacienda

This contrast between plantation and hacienda land tenure in itself constitutes strong evidence for a Rimland-Mainland division. The hacienda was a Spanish institution, but the modern plantation, Augelli argued, was the concept of Europeans of more northerly origin. In the **hacienda**, Spanish landowners possessed a domain whose productiv-

ity they might never push to its limits: the very possession of such a vast estate brought with it the social prestige and comfortable lifestyle they sought. The native workers lived on the land—which may once have been *their* land—and had plots where they could grow their own subsistence crops. Traditions survived in the Amerindian villages incorporated into the early haciendas, in the methods of farming, and in the means of transporting produce to markets. All this is written as though it is mostly in the past, but the legacy of the hacienda, with its inefficient use of land and labor, is still visible throughout mainland Middle America as well as parts of South America.

The Plantation

The **plantation**, in contrast, was conceived as something entirely different. In *Middle America: Its Lands and Peoples* (1989), Robert West and John Augelli list five characteristics of Middle American plantations that clearly illustrate the differences between hacienda and plantation: (1) plantations are located in the humid tropical coastal lowlands of the realm; (2) plantations produce for export almost exclusively—usually a single crop; (3) capital and skills are often imported so that foreign ownership and an outflow of profits occur; (4) labor is seasonal—needed in

Contrasting land uses in the Middle American Rimland give rise to very different rural cultural landscapes. Huge stretches of the realm's best land continue to be controlled by (often absentee) landowners whose haciendas yield export or luxury crops, or foreign corporations that raise fruits for transport and sale on their home markets. The banana plantation shown here lies in northern Honduras and is owned by United Brands Company. The vast fields of banana plants stand in strong contrast to the lone peasant who ekes out a bare subsistence from small cultivable plots of land, often in high-relief countryside where grazing some goats or other livestock is the only way to use most of the land.

large numbers during the harvest period but often idle at other times—and such labor has been imported because of the scarcity of Amerindian workers; and (5) with its "factory-in-the-field" operation, the plantation is more efficient in its use of land and labor than the hacienda. The objective was not self-sufficiency but profit, and wealth rather than social prestige is a dominant motive for the plantation's establishment and operation.

Legacy of Land Tenure Traditions

During the twentieth century, both traditional systems of land tenure have changed a great deal. Massive U.S. corporate investment in the Caribbean coastal belt of Guatemala, Honduras, Nicaragua, Costa Rica, and Panama transformed that area and brought a new concept of plantation agriculture to the region. On the Mainland, the hacienda has been under increasing pressure from national governments that view it as an economic, political, and social liability. Indeed, some haciendas have been parceled out to small landholders, and others have been pressed into greater specialization and productivity. In Mexico, hacienda land has been placed in *ejidos*, where it is communally owned by groups of families. Although its remnants continue to break down today, both the hacienda and the plantation have for centuries contributed to the different social and economic directions that gave the Mainland and Rimland their respective regional personalities.

POLITICAL DIFFERENTIATION

Continental Middle America today is fragmented into eight different countries, all but one of which have Hispanic origins. (That exception—Belize, the former British Honduras—is now being transformed as thousands of post–1980, Spanish-speaking immigrants from neighboring countries will soon form the majority of the tiny Belizean population.) Largest of them all is Mexico, the giant of Middle America, whose 762,000 square miles (1,972,500 sq km) constitute more than 70 percent of the realm's entire land area (the Caribbean Basin included) and whose nearly 100 million people outnumber those of all the other countries and islands of Middle America combined.

The cultural variety in Caribbean Middle America is much greater. Here Cuba dominates: its area is almost as large as that of all the other islands put together, and its population of 11.3 million is well ahead of the next-ranking country (the Dominican Republic, with 8.2 million). As we have seen, however, the Caribbean is hardly an area of exclusive Spanish influence. For example, whereas Cuba has an Iberian heritage, its southern neighbor, Jamaica (population 2.5 million, mostly black), has a legacy of British involvement, and in nearby Haiti (7.5 million, overwhelmingly black) the strongest imprints have been African and French. The crowded island of Hispaniola (population 15.7 million) is shared between Haiti and the Dominican Republic, where Hispanic culture prevails. Spanish influence also predominates in the nearby, remaining island of the Greater Antilles, Puerto Rico (3.7 million), a commonwealth of the United States.

The Lesser Antilles, too, exhibits great cultural diversity. There are the (once Danish) U.S. Virgin Islands; French Guadeloupe and Martinique; a group of British-influenced islands, including Barbados, St. Lucia, St. Vincent, and Grenada; and Dutch St. Maarten (shared with the French), Saba, St. Eustatius, and the A-B-C islands—Aruba, Bonaire, Curaçao—off the northwestern Venezuelan coast. Standing apart from the Antillean arc of islands (off northeastern Venezuela) is Trinidad, another former British dependency that, along with its smaller neighbor Tobago, became a sovereign state in 1962.

Independence movements stirred Middle America at an early stage. In the Greater Antilles region, where Spain held Cuba and Puerto Rico and the British controlled Jamaica, Hispaniola's Afro-Caribbean population mounted a rebellion against their French slave masters. They succeeded, in 1804, in creating the Republic of Haiti. On the mainland, revolts against Spanish authority (beginning in 1810) achieved independence for Mexico by 1821, and for the Central American republics later in the decade of the 1820s.

The colonial map of the Caribbean was a patchwork of holdings by British, Dutch, French, Spanish, and Danish interests (Fig. 4-4). All were competitors in the sugar boom made possible by African labor. The United States, concerned over European designs in the region, in 1823 proclaimed the Monroe Doctrine, which was intended to deter any European power from reasserting its authority in newly independent republics or from further expanding its existing domains. The U.S. as yet had no holdings in Middle America, but the emerging republics on its doorstep presented obvious economic and strategic opportunities.

By the end of the nineteenth century, the United States had become a major factor in the realm. The Spanish–American War of 1898 made Cuba independent and put

Puerto Rico under the U.S. flag; soon afterwards, the Americans were in Panama, constructing the Panama Canal. Meanwhile, U.S. corporations were riding a boom based not on sugar, but on huge banana plantations. Central American republics were U.S. colonies in all but name.

Independence came to the Greater and Lesser Antilles in fits and starts. Afro-Caribbean Jamaica as well as Trinidad and Tobago, where the British had brought a large South Asian population, attained full sovereignty from Britain in 1962; other British islands (Barbados, St. Vincent, Dominica) became independent later. But France retains Martinique and Guadeloupe as overseas *départements* of the French Republic, and the Dutch islands are at various stages of autonomy. The U.S. Virgin Islands, purchased from Denmark in 1917, have no prospect of independence.

Major Cities of the Middle American Realm

City	Population* (in millions)
Guadalajara, Mexico	3.3
Guatemala City, Guatemala	1.0
Havana, Cuba	2.3
Managua, Nicaragua	1.3
Mexico City, Mexico	24.0
Monterrey, Mexico	2.9
Panama City, Panama	1.0
Port-au-Prince, Haiti	1.4
San José, Costa Rica	1.0
San Juan, Puerto Rico	1.1
San Salvador, El Salvador	0.6
Santo Domingo, Dominican Rep.	2.8
Tegucigalpa, Honduras	0.8

*1997 Population Estimates

REGIONS OF THE MIDDLE AMERICAN REALM

❖ THE CARIBBEAN

Caribbean America today is a land crowded with so many people that, as a region (encompassing the Greater and Lesser Antilles), it is the most densely populated part of the Americas. It is also a place of grinding poverty and, in all too many localities, of unrelenting misery, for most,

with little chance for escape. In some respects, U.S.-affiliated Puerto Rico—which in 1993 elected to maintain its commonwealth status—and rebounding communist Cuba (see box titled "Castro's Cuba in the Post-Soviet Era") constitute exceptions to any such generalization made about Caribbean America. On most of the other islands, however, life for the average person is difficult, often hopeless, and tragically short.

All this is in jarring contrast to the early period of riches based on the sugar trade. Yet that initial wealth was gained while an entire ethnic group (the Amerindians) was being wiped off the Caribbean map and while another (the Africans) was being imported in bondage. The sugar reve-

nues, of course, always went to the planters, not the laborers. Subsequently, the regional economy faced rising competition from other tropical sugar-producing areas; it soon lost its monopoly of the European market, and difficult times prevailed. Meanwhile, just as they did in other parts of the world, the Europeans helped stimulate the rapid growth of the island population by improving sanitation and medical standards. Death rates were lowered, but birth rates remained high and explosive population increases resulted.

With the decline of the sugar trade, millions of people were pushed into a life of subsistence, malnutrition, and hunger. Many sought work elsewhere. Tens of thousands of Jamaican laborers went to the plantations of the Rimland coast; large numbers of British West Indians went to Eng-

CASTRO'S CUBA IN THE POST-SOVIET ERA

■ ■ ■

Located only 90 miles (145 km) from southernmost Florida, Cuba, the Caribbean's largest island, has been under the dictatorial rule of Fidel Castro since 1959. A convert to Soviet-style communism soon after taking power, Castro spent the next three decades as the Western Hemisphere's pariah, constantly seeking to export his socialist revolution to other parts of the Americas. But all that began to change after 1991 in the wake of the collapse of the Soviet Union. Cuba's economy was devastated by the loss of most of its former benefactor's market and imported raw materials. To alleviate this hardship and to shore up his popular support, Castro was forced to introduce free market reforms, while stubbornly clinging to the "socialist principles" of his revolution of so long ago. Among these changes were the rapid growth of private businesses, massive layoffs of government workers, and the encouragement of foreign investment.

Foreign joint ventures have mushroomed in the 1990s, and today involve European-backed tourist resorts and international banks, Japanese automobile dealerships, Israeli textile plants, a state-of-the-art telephone system developed by Mexican entrepreneurs, and much more. With the drive toward economic integration intensifying across the Americas, countries throughout the hemisphere are now eager to build trading ties with Cuba because they recognize that the post-Soviet

transformation has eliminated Castro as an ideological threat. These new relationships notwithstanding, the United States refuses to deal with Cuba (one of only four such countries left in the hemisphere) and continues its longstanding trade embargo—which is widely ignored in Middle America.

Given the island's favorable geographic position and resource base, these recent developments suggest that Cuba has the potential to become a Caribbean economic tiger. Conditions for commercial agriculture rank among the best in the Antillean archipelago, and the crop base is diversifying. Mineral resources include high-quality nickel, copper, and iron ores, and oil deposits may also exist. Skilled manufacturing has been nurtured since 1960; among others, a respectable biotechnology industry has emerged (which successfully produced a vaccine for meningitis). Moreover, higher-education facilities are the most advanced in the Caribbean Basin, and the communications infrastructure is approaching world-class standards.

Importantly, all this progress is taking place with the end of the Castro regime in sight. Even if what is left of communist rule should persist into the twenty-first century, Cuba is poised to become a major player in the economic geography of Middle America.

■ ■ ■

The Caribbean is a region of sharp, often searing contrasts. Haiti is the Western Hemisphere's poorest country, its city slums (here in the capital, Port-au-Prince) the most desperate of human environments. Tourism and wealthy waterfront living create another face of the region, generating landscapes of conspicuous consumption—but producing some of the limited economic opportunities in the Caribbean Basin.

land in search of a better life; Puerto Ricans and (more recently) Dominicans streamed to New York. But this outflow has failed to stem the tide of regional population growth: today there are over 37 million people on the Caribbean islands, a total expected to increase by up to 40 percent over the next three decades.

The Caribbean islanders simply have not had many alternatives in their search for betterment. Their dispersed habitat is fragmented by both water and mountains; the total amount of flat cultivable land is only a small fraction of the far-flung Antillean archipelago. Although some diversification of production has occurred in the region, agriculture remains a significant economic activity. Sugar is still a leading product and continues to head the export lists of the Dominican Republic and Cuba; in Haiti, coffee has become the leader among exported primary products. In the Lesser Antilles, sugar has retained a somewhat less prominent position, having been supplanted by such crops as bananas, limes, spices, and sea-island cotton. Even here, however, sugar still predominates in some places, particularly in Barbados and St. Kitts-Nevis. And tourism, the region's single growth industry, is a mixed blessing.

Problems of Widespread Poverty

All the crops grown in the Caribbean—Haiti's coffee, Jamaica's bananas, the Lesser Antilles' fruits as well as the pivotal sugar industry—constantly face severe competition from other parts of the world and have still not become established at a scale that could begin to improve standards of living. Those minerals that do exist in this region—Jamaica's bauxite (aluminum ore), Trinidad's oil, Cuba's nickel—do not support any significant industrialization within the Caribbean Basin itself. As in other lower-income parts of the world, these resources are exported for use elsewhere. Thus the Caribbean countries, dependent on uncertain foreign markets for much of their revenues, find themselves trapped in an international economic order they cannot change.

Given these conditions, the vast majority of the people in this region continue to eke out a precarious living from small plots of ground, are mired in poverty and are threatened by disease. Food supplies are frequently inadequate on Caribbean islands, with the best land always used to raise cash crops for the export trade rather than staples for local consumption (many countries—most notably Haiti—are chronically food-deficient). Cultivation methods have undergone little change over the generations. Land inheritance customs have subdivided peasant families' plots until they have become so small that the owner must sharecrop some other land or seek work on a plantation in order to supplement the meager harvest. Furthermore, soil erosion constantly threatens: much of the Jamaican countryside is scarred by gulleys and ravines, and Haiti's land has become so ravaged by agricultural overuse and deforestation that the entire country is turning into an ecological wasteland (see photo p. 212).

With such problems, it would be unlikely for Caribbean America to have many large cities. Nonetheless, the urbanization of the region's population surpasses 60 percent, far exceeding the global average of 43 percent. And the four leading urban agglomerations—the Dominican Republic's Santo Domingo, Cuba's Havana, Haiti's Port-au-

Prince, and Puerto Rico's San Juan—contain more than 20 percent of the Caribbean's total population. Often, however, towns and cities exhibit even more miserable living conditions than the poorest rural areas. The slums of Port-au-Prince are among the world's worst (see photo p. 213); we should not be surprised that such abysmal conditions drive away the most desperate Haitians as "boat people" in search of a better life elsewhere.

African Heritage

The human geography of the Caribbean islands is also a legacy of Africa, and there are places where the region's cultural landscapes strongly resemble those of West and Equatorial Africa. In the construction of village dwellings, the operation of rural markets, the role of women in rural life, the preparation of certain kinds of food, the methods of cultivation, the nature of the family, artistic expression, and in an abundance of other traditions, the African heritage can be read throughout the Caribbean-American scene.

Nevertheless, in general terms it may still be argued that the European or white person is in the best position in this island chain, politically and economically; the *mulatto* (mixed white-black) ranks next; and the black person ranks lowest. In Haiti, for instance, where 95 percent of the population is "pure" black and only 5 percent mulatto, this mulatto minority has held a disproportionate share of

From the Field Notes

"Tourism has become an important source of jobs and revenues in Caribbean-island countries where economic opportunities are few. Miami has become the world's leading cruise port, and from there (and from other Florida ports) dozens of cruise ships ply Caribbean waters and visit islands in the Greater and Lesser Antilles. Here the Royal Viking Sun lies alongside in a bay on the west coast of Dominica. The ship's more than 700 passengers are ashore taking tours and visiting shops and restaurants."

power. On neighboring Jamaica, the 15 percent "mixed" sector of the population has played a role of prominence in island politics far out of proportion to its numbers. In the Dominican Republic, the pyramid of power puts the white, Spanish-European sector (15 percent) at the top, the mixed group (70 percent) in the middle, and the black population (15 percent) at the bottom. In Cuba, too, the 11 percent of the population that is black has found itself less favored than the white sector (37 percent), the mulatto sector (30 percent), and the *mestizo* (mixed white-Amerindian) population (21 percent).

The composition of the population of the islands is further complicated by the presence of Asians from both China and India. During the nineteenth century, the emancipation of slaves and ensuing local labor shortages brought some far-reaching solutions. Some 100,000 Chinese emigrated to Cuba as indentured laborers (today they still constitute 1 percent of the island's population); and Jamaica, Guadeloupe, Martinique, and especially Trinidad saw nearly 250,000 East Indians arrive for similar purposes. To the African-modified forms of English and French heard in the Caribbean, therefore, can be added several Asian languages. Hindi is particularly strong in Trinidad, whose overall population is now 41 percent South Asian. The ethnic and cultural variety of the **plural societies** of Caribbean America is indeed endless.

Tourism: The Irritant Industry

The resort areas, scenic treasures, and historic locales of Caribbean America attract up to 25 million visitors annually, making this region one of the world's most popular tourist destinations. About 10 million of these tourists travel on cruise ships (which are based in Miami and Fort Lauderdale), and many of their ports of call and nearby coastal strips have enjoyed spectacular growth since 1980. Jamaica is a good example. The recent development of such places as Ocho Rios, Port Antonio, and Montego Bay has boosted the tourist industry to where it now accounts for one-sixth of the country's gross domestic product and employs more than one-third of the Jamaican labor force.

Certainly, Caribbean tourism is a prospective money-maker for most of the islands, but it has serious drawbacks. The invasion of overtly poor communities by affluent, sometimes raucous tourists contributes to a rising sense of local anger and resentment. At the same time, tourism can have the effect of debasing local culture, which often is adapted to suit the visitors' tastes (as at hotel-staged "culture" shows). And while tourism does generate income in the Caribbean where alternatives are few, the intervention of island governments and multinational corporations removes opportunities from local entrepreneurs in favor of large operators and major resorts.

Tourism, then, is a mixed blessing for the developing

Caribbean Basin. Given the region's limited options, it provides revenues and jobs where otherwise there would be none. Yet there is a negative cumulative effect that intensifies contrasts and disparities: gleaming hotels tower over substandard housing, luxury liners glide past poverty-stricken villages, opulent meals are served in places where, down the street, children suffer from malnutrition. Clearly, the tourist industry contributes positively to island economies but strains the fabric of the local communities involved.

Regional Cooperation

Many of the world's realms and regions, as we noted in the chapters on Europe and North America, are today exploring new forms of international cooperation. The Caribbean Basin is no exception, the diverse cultural and political backgrounds of its island-nations notwithstanding. Prompted by shared economic interests and a growing sense of regional identity, all 16 independent states in 1994 concluded an agreement to create the Association of Caribbean States (ACS). This initiative was spearheaded by CARICOM (Caribbean Community and Common Market), a 20-year-old customs union of former British colonies, which wanted to expand by building ties to nearby Spanish-speaking countries. These new linkages came to include not only the Dominican Republic and Cuba (as well as Creole-speaking Haiti) in the Caribbean, but all of the realm's mainland countries except Panama, plus Venezuela and Colombia in northern South America. In all, 25 member-states constitute the ACS, with the only Caribbean nonparticipants being non-self-governing states affiliated with France, the Netherlands, the United Kingdom, or the United States.

Whereas the ACS marks solid progress toward the broad goal of achieving closer ties among Caribbean-area states, more immediate regional issues also played a role in the organization's formation. Most important was the launching of the North American Free Trade Agreement (NAFTA), which threatened the existing duty-free access of Caribbean exports (especially clothing) to the U.S. market. These privileges had been granted as part of the Caribbean Basin Initiative following the U.S. ouster of Grenada's communist regime in 1983, and resulted in substantial job increases in several Caribbean countries as their exports to the United States surged. With its entry into NAFTA in 1994, Mexico became a much stronger competitor on the U.S. market for these Caribbean goods because the tariffs on Mexican trade were set on a downward course toward elimination. But by banding together under the new ACS banner, the Caribbean states successfully exerted collective pressure on the United States and extracted an agreement from the U. S. that guaranteed tariff parity with Mexico into the late 1990s.

THE CONTINENTAL MOSAIC

Continental Middle America consists of two regions, Mexico and Central America, the former constituted by a single country and the latter by seven. Mexico is Middle America's giant, a region by virtue of its physical size, population, cultural qualities, resource base, and relative location. Mexico's geographic position adjacent to the United States has been a blessing as well as a curse—a blessing because it has facilitated economic interaction, a curse because it cost Mexico huge tracts of territory and, more recently, has led to neighborly friction over massive illegal emigration. Much of that cross-border population outflow was triggered by economic conditions in Mexico. These conditions may now improve, for the North American Free Trade Agreement (NAFTA), despite Mexico's initial problems, should stimulate industrial and other forms of regional development.

The seven republics of Central America cannot match Mexico in terms of total population or territory. But they constitute a geographic region nonetheless—a region plagued by armed conflict (with recent big-power involvement), leadership struggles, economic stagnation, rapid population growth, and environmental crisis. We focus first on Mexico and then turn to Central America.

❖ MEXICO AND ITS UNFINISHED REVOLUTION

Mexico is the colossus of Middle America, with a 1997 population of 97.8 million—exceeding the combined total of all the other countries and islands of the realm by 27 million—and a territory more than twice as large (Fig. 4-6). Indeed, in all of Spanish-influenced Middle and South America, no country has even half as large a population as Mexico. Moreover, in 1998 Mexico will become the eleventh of the world's countries to surpass the 100-million mark. It has grown so swiftly that its population has *doubled* since 1970, and it is on track to double again by 2030.

The physiography of Mexico resembles the western United States, although environments are more tropical. Figure 4-1 shows several prominent features: the elongated Baja (Lower) California Peninsula in the northwest, separated from the mainland by the Gulf of California (which Mexicans call the Sea of Cortés); the far eastern Yucatán Peninsula, jutting out into the Gulf of Mexico; and the Isthmus of Tehuantepec in the southeast, where Mexico's landmass tapers. Here in the southeast, Mexico most resembles Central America physiographically; a mountain backbone forms the isthmus, curves southeast into Guatemala, and extends northwest toward Mexico City (Fig.

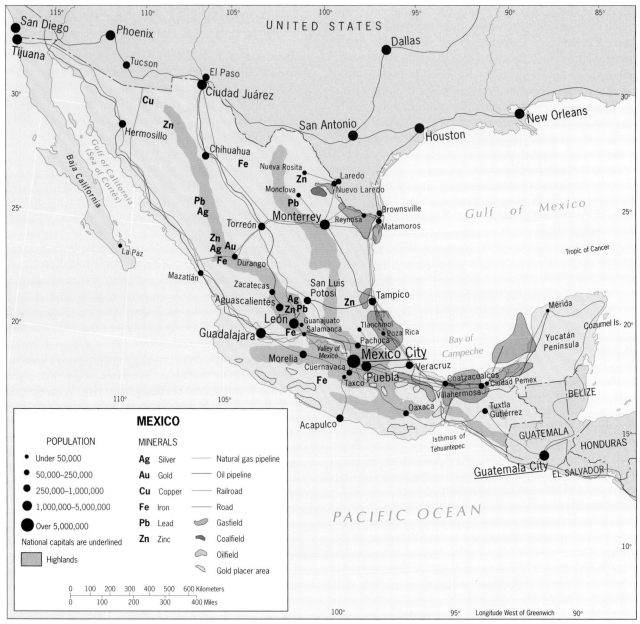

FIGURE 4-6

4-6). Shortly before reaching the capital, this mountain range divides into two chains, the Sierra Madre Occidental (in the west) and Sierra Madre Oriental (in the east). These diverging ranges frame the funnel-shaped Mexican heartland, the center of which consists of the extensive Plateau of Mexico (the Valley of Mexico lies near its southeastern end). This rugged tableland is some 1,500 miles (2,400 km) in length and up to 500 miles (800 km) wide. The plateau is highest in the south, near Mexico City, where it is about 8,000 feet (2,450 m) in elevation; from there it gently declines to the northwest toward the Rio Grande. As Figure I-7 reveals, Mexico's climates are marked by dryness, particularly in the broad, mountain-flanked north. Most of

the better-watered areas lie in the southern half of the country where the major population concentrations have developed.

The distribution of population across Mexico's 31 internal States is shown in Figure 4-7. The largest concentrations containing more than half the Mexican people, extends across the densely populated "waist" of the country from Veracruz State on the eastern Gulf Coast to Jalisco State on the Pacific. The center of this corridor is dominated by the most populous State, Mexico, at whose heart lies the Federal District of Mexico City—which is about to become the largest urban agglomeration on Earth. In the dry and rugged terrain to the north of this central corridor

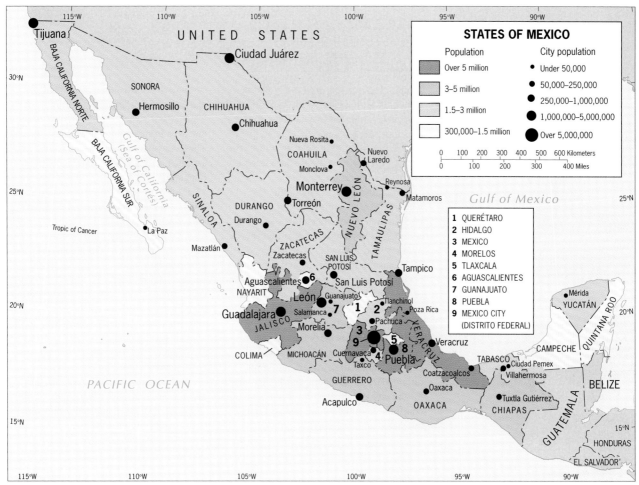

FIGURE 4-7

lie Mexico's least-populated subregions, among them the six States bordering the United States—Tamaulipas, Nuevo León, Coahuila, Chihuahua, Sonora, and Baja California Norte. Southern Mexico also exhibits a sparsely peopled periphery in the hot and humid lowlands of the Yucatán Peninsula, but most of the highlands of the continental spine southeast of Mexico City contain sizeable populations, particularly in the States of Guerrero, Oaxaca, and Chiapas.

Another major feature of Mexico's population map is urbanization, driven by the *pull* of the cities (with their perceived opportunities for upward mobility) in tandem with the *push* of the economically-stagnant countryside. Today, no less than 71 percent of the Mexicans reside in towns and cities—up from 59 percent as recently as 1970. This is a surprisingly high proportion for a developing country (the current U.S. figure is 75 percent). Undoubtedly, these numbers are affected by the explosive recent growth of metropolitan Mexico City conurbation, which now totals 24 million and is home to an astonishing 25 percent of the national population. Among the other lead-

ing cities in the central population corridor are Guadalajara, Puebla, and León (Fig. 4-7). Outside this heartland, some of the country's fastest-growing cities are found in the U.S. border zone: both Tijuana and Ciudad Juárez have just surpassed 1 million, and Monterrey (at nearly 3 million) is booming to the extent that Nuevo León State now ranks higher than its neighbors in Figure 4-7. Urbanization rates at the other end of Mexico, however, are at their lowest in those remote uplands where Amerindian society has been least touched by modernization.

Nationally, the Amerindian imprint on Mexican culture remains quite strong. Today, 60 percent of all Mexicans are mestizos and 30 percent are "pure" Amerindians; only about 8 percent are Europeans. There is a Mexican saying that Mexicans who do not have Amerindian blood in their veins nevertheless have the Amerindian spirit in their minds. Certainly the Mexican Amerindian has been Europeanized, but the Amerindianization of modern Mexican society is so powerful that it would be inappropriate here to speak of one-way, European-dominated *acculturation*. Clearly, what took place in Mexico is **transculturation**—

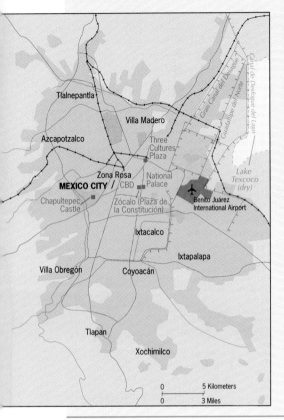

Mexico City

Middle America has only one great metropolis: Mexico City. With 24 million inhabitants, Mexico City is home to one-fourth of Mexico's population and grows by a half million each year. In a few years Mexico City will be the world's largest urban agglomeration.

Lakes and canals marked this site when the Aztecs built their city of Tenochtitlán here seven centuries ago. The conquering Spaniards made it their headquarters, and following independence the Mexicans made it their capital. Centrally positioned and well connected to the rest of the country, Mexico City, hub of the national core area, became the quintessential primate city.

Vivid contrasts mark the cityscape. Historic plazas, magnificent palaces, churches, and villas, superb museums, ultramodern skyscrapers, and luxury shops fill the city center. Beyond lies a zone of comfortable middle-class and struggling but stable working-class neighborhoods. Outside this belt, however, lies a ring

of more than 500 slums plus countless, and even poorer, *ciudades perdidas*, the ''lost cities'' where newly arrived peasants live in miserable poverty and squalor. These squatter settlements may contain as much as *one-third* of Mexico City's population.

Environmental problems match the social troubles. The surface waters have long since dried up, and water shortages now prevail; the city must be served by pipelines from across the mountains. Air pollution here ranks among the world's worst as 3.5 million cars and 40,000 factories create smog in Mexico City's thin, high-altitude air that sometimes reaches 100 times the acceptable level. Add to this a set of geologic hazards: land subsidence, earthquakes, and the risk of volcanic activity nearby.

And yet the great city beckons, and every year hundreds of thousands of the desperate and the dislocated arrive with hope—and little else.

---

the two-way exchange of culture traits between societies in close contact. In the southeastern periphery (Fig. 4-3), several hundred thousand Mexicans still speak only an Amerindian language, and perhaps another 5 million still use these languages in everyday conversation even though they also speak Mexican Spanish. The latter has been strongly influenced by Amerindian languages, but this is only one aspect of Mexican culture that has received an Amerindian impress. Uniquely Mexican modes of dress, foods and cuisine, sculpture and painting, architectural styles, and folkways also vividly reflect the Amerindian contribution. The fusion of these Spanish and Amerindian heritages gives modern Mexico a distinctiveness that it alone possesses in Middle and South America, and is the product of an upheaval that began to reshape the country nearly a century ago.

Revolution and Its Aftermath

Modern Mexico was forged in a revolution that began in 1910 and set into motion a sequence of events that is still unfolding today. At its heart, this revolution was about the redistribution of land, an issue that had not been resolved

after Mexico freed itself from Spanish colonial control in the early nineteenth century. As late as 1900, more than 8,000 haciendas blanketed virtually all of Mexico's good farmland, and about 96 percent of all rural families owned no land whatsoever and toiled as *peones* (landless, constantly indebted serfs) on the haciendas. Led by the legendary Emiliano Zapata, the triumphant revolution produced a new constitution in 1917 that launched a program of expropriation and parceling out of the haciendas to rural communities.

Since 1917, about half the cultivated land of Mexico has been redistributed, mostly to peasant communities consisting of 20 families or more. Such lands are called *ejidos*; the government holds title to the land, and use rights are parceled out to villages and then individuals for cultivation. Most of the *ejido* lands carved out of haciendas lie in central and southern Mexico, where Amerindian traditions of landownership and cultivation survived and where the adjustments were most successfully made. Despite an understandable temporary decline in agricultural productivity during the transition, the miracle is that land reform was carried off without major dislocation and that the power of the wealthy land-owning aristocracy could be broken without ruin to the state. Although considerable malnutrition and poverty persisted in the countryside, it was widely recognized that Mexico, alone among the realm's countries with large Amerindian populations, had made major strides toward solving the land question. Just how much farther Mexico has to go, however, became evident at the outset of 1994 when, unexpectedly, the issue resurfaced at the heart of a rebellion that broke out in southeasternmost Chiapas State and made global headlines on the very day that NAFTA went into operation.

Chiapas, the poorest of the 31 States, lies wedged against the Guatemalan border and is more reminiscent of Central America than Mexico. The complexion of its rapidly growing population of 3.9 million is heavily Amerindian (largely of Mayan background) and is dominated by families of peasant farmers who eke out a precarious existence on hilly, poor-quality farmlands. The rest of the productive soils have for centuries been incorporated into the estates of the large landholders, a system that endures virtually unaffected by the land redistribution that reshaped many other parts of rural Mexico. In response to this lack of development, the Mexican government has in recent years pledged to introduce new services and job-creation programs in Chiapas, but its main effort was a coffee-raising scheme in the 1980s that soon came to naught when world coffee prices plummeted. This experience intensified the longstanding bitterness of the Chiapans, and a radical group of Mayan peasant farmers quietly began to organize and build support to resume the historic struggle of Amerindian *peones* to gain land, fair treatment, and escape from extreme poverty.

On January 1, 1994, this organization, now calling itself the Zapatista National Liberation Army (ZNLA), ignited a guerrilla war with coordinated attacks on several Chiapan towns. By timing this uprising to coincide with the launching of NAFTA—and by taking the legendary name of Zapata—the ZNLA ensured maximum publicity for its agenda that included land reform, access to greater economic opportunity, heightened cultural identity, and local autonomy. The Mexican military reacted to these events by quickly driving the insurgents out of the towns, but eased their heavy-handed pursuit when the press exposed a number of human rights violations. The guerrillas regrouped in the mountains and have remained a force to be reckoned with as negotiations replaced armed conflict later in 1994. As of mid-1996, the fighting had not resumed but tensions persisted as the dispute remained unresolved.

The Chiapas rebellion raises a number of broader issues that have important implications for Mexico's immediate future. First, it emphasizes that certain areas of the country do not participate in the ongoing NAFTA-related development thrust. In spatial terms, the Chiapas situation is a classic case of core-periphery confrontation, with the outlying population demanding its share of the growing national pie. A second issue, which now affects so many of the world's countries, is devolution. The ZNLA demand for "autonomy" is modeled after Spain's autonomous

From the Field Notes
"Mexico's high-relief west coast marks the collision zone between the North American Plate and the Cocos Plate, a belt of spectacular scenery and high-risk living. Beaches, scenery, and climate have engendered one of the world's most famous resort areas here, but this photograph is a reminder that Acapulco (in Guerrero State) also serves other functions—including a naval base—and has a storied commercial past. This was colonizing Spain's point of departure and return to and from distant Pacific holdings in the Philippines; at this latitude, as the map shows, Mexico is comparatively narrow, facilitating an overland link to the Atlantic (Gulf) port of Veracruz. But times (and functions) change, and tourism, not trade, dominates the economy of Acapulco today."

communities (see p. 90), and involves not secession but the decentralization of powers from the federal to the State government that allows the latter more local control, particularly over cultural affairs. A third issue raised is vital to Mexico's social geography. The ZNLA crusade elicited wide sympathy among indigenous populations in four other southern States (Guerrero, Oaxaca, Puebla, and Michoacán), and could well spark a nationwide civil rights movement for Amerindians. As in Caribbean societies, darkness of skin color is directly related to a person's social status, a linkage that in Mexico further extends to an individual's degree of "Indianness."

Economic-Geographic Transformation

In recent decades, Mexico has made important progress in several productive sectors. During the early 1990s, its economy was further transformed by the boom preceding the implementation of NAFTA, which by 2010 will bind the economies of Mexico, the United States, Canada, and Chile into a single free trade zone and market of nearly

470 million people. The launching of NAFTA in 1994, however, was followed by a number of unexpected shocks to the Mexican political establishment and economic system. The most serious of these shocks was a substantial devaluation of the peso in late 1994. As a result, Mexico plunged into deep economic recession in 1995, thereby testing the faith of its workers, business leaders, and foreign investors, and challenging the government to correct the problems and restart the boom. We now review the changing geography of Mexico's economic activity against the background of these events.

Agriculture

Whereas traditional subsistence agriculture has not changed a great deal in the poorer areas of rural Mexico, commercial agriculture has diversified during the past quarter-century and made major strides with respect to both the home market and export. The greatest productivity still emanates from private cultivators, although much of the land involved has been subdivided into *ejidos*. The central plateau is geared mainly to the domestic production of food crops, but in the arid north, major irrigation projects have been built on streams flowing down from the interior highlands. Along the booming northwest coast, mechanized large-scale cotton production supplies the domestic market as well as the thriving export trade. For the home market, wheat and winter vegetables are grown; but fruit and vegetable cultivation are attracting the investments of foreigners, especially to such crops as bananas and sugarcane. Cattle raising is another leading pursuit and continues to expand onto the Gulf Coast lowlands from its long-time base in the northern interior.

Energy Resources

While Mexico's metal mining industries are less important today than they once were, the country has since 1970 enjoyed the advantages—and suffered the problems—of being a major petroleum producer. Centered on the southern Gulf Coast's Bay of Campeche around Villahermosa in Tabasco State, huge oilfields brought Mexico abundant revenues when the world oil price was high in the late 1970s and serious economic difficulties when the price fell after 1980. Discoveries of massive oil and natural gas reserves in this area have made Mexico self-sufficient in these energy resources, adding to already substantial reserves located in oilfields lying along the Gulf Coast to the northwest and inner Yucatán to the northeast (Fig. 4-6). The high petroleum prices of the 1970s stimulated the beginnings of Mexico's economic-geographic transformation, but the oil crash of the 1980s thwarted the momentum for a decade. This crash occurred because the government found itself without the expected oil revenues it needed to pay the interest on the huge foreign loans it had taken to finance domestic development programs. Such a disaster could recur if a future oil boom again tempts Mexican leaders to take such risks.

Industrialization

Manufacturing is the centerpiece of Mexico's latest development episode, but this economic sector actually got its start almost a century ago. Blessed with a wide range of raw materials, many of which are located in the north (Fig. 4-6), the country began to industrialize in 1903 with the completion of an iron and steel plant (now abandoned) in

The border between Mexico and the United States occasionally provides some stunning contrasts. Here, looking westward, we observe the opposing economic geographies that mark the border landscape of the Imperial valley east of the urban area formed by Mexicali, Mexico and Calexico, California. The larger town of Mexicali (on the left) sprawls eastward along the border, with a cluster of maquiladora assembly plants, visible in the left foreground, surrounded by high-density housing. On the U.S. side, lush irrigated croplands blanket the otherwise dry countryside, fed by the All-American Canal that taps the waters of the Colorado River before they cross into Mexico. At this point, the canal swings northward to bypass border-adjoining Calexico (right rear) before rejoining the international boundary beyond the town's western edge.

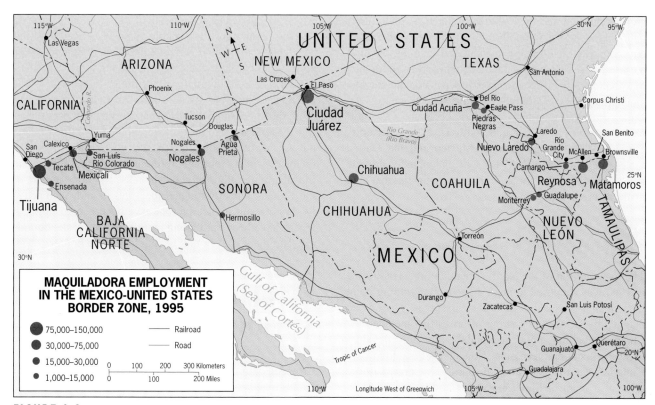

**MAQUILADORA EMPLOYMENT
IN THE MEXICO-UNITED STATES
BORDER ZONE, 1995**

● 75,000–150,000   —— Railroad

● 30,000–75,000   —— Road

● 15,000–30,000   0   100   200   300 Kilometers

• 1,000–15,000   0   100   200 Miles

FIGURE 4-8

the northeastern city of Monterrey. A second steel complex was built at nearby Monclova in the 1950s, a period that saw the spreading of factories across many parts of central Mexico, particularly in and around the capital. The growth of the industrial sector since that time has been steady and for the past two decades has employed at least one-sixth of the Mexican labor force.

The most significant new development in Mexico's manufacturing geography is the growth of maquiladora plants in the northern border zone. The **maquiladoras** are foreign-owned factories (mainly by large U.S. companies) that assemble imported, duty-free components and raw materials into finished industrial products. At least 80 percent of these goods are then reexported to the United States, whose import tariffs are limited to the value added to the products during their Mexican fabrication stage (under NAFTA these tariffs will soon be phased out). All parties benefit from this industrial system: the Mexicans gain a significant number of jobs, and the foreign owners benefit from Mexico's wage rates, which average less than 20 percent of those north of the border. Although this development program was initiated in the 1960s, the number of maquiladoras grew only to a modest 588 (with 122,000 employees) by 1982. Suddenly, however, this innovation took flight, and by the early 1990s about 1,800 assembly plants were employing some 500,000 workers. Among the goods being assembled were electronic equipment, electri-

cal appliances, auto parts, clothing, plastics, and furniture. By 1995, more than 2,300 maquiladoras with over 600,000 employees were operating all along the northern border (Fig. 4-8), and accounted for a robust one-fifth of Mexico's industrial labor force and more than 5 percent of its gross domestic product.

The Mexican government is trying to capitalize on the success of the maquiladora program by encouraging the industrial development of other parts of the country. It would most like to see industrial firms create complete manufacturing complexes deeper within Mexico rather than limit their investments to assembly plants that hug the U.S. border. In recent years, several multinational corporations did undertake such ventures, most notably the three leading U.S. automakers and four of their Japanese and German competitors (the automobile industry today is Mexico's biggest employer and export leader). To speed the opening of the rest of the country, the government is giving priority to programs aimed at upgrading Mexico's infrastructure to world-class standards, especially its telecommunications, superhighway, and electrical-power networks.

Burgeoning Monterrey—150 miles (250 km) inside Mexico yet close enough to the Texas border to have benefited from all the recent development trends—is frequently singled out as a model of what a future Mexican growth center should be. Among this city's assets are a highly educated and well-paid labor force, a stable international business community, and a thriving high-technology complex of ultramodern industrial facilities that has attracted blue-chip multinational companies. In addition, a new expressway link to the Rio Grande is helping to forge an international corridor of growth between Monterrey and Dallas–Fort Worth, which is already becoming the dominant axis of cross-border trade with Texas. Furthermore, as trading restrictions between Mexico and the United States are eliminated, NAFTA will undoubtedly provide an additional impetus for loosening the spatial ties that bind manufacturers to the border zone.

Continuing Challenges

The inauguration of the North American Free Trade zone in 1994 was supposed to mark an economic turning point for Mexico. Instead, NAFTA's first two years were plagued by an unending string of crises that included the Chiapas rebellion, political scandals, monetary devaluation, and economic recession. Nonetheless, exports to the United States rose substantially during this period (more than 80 percent of Mexican exports now go to the U.S., which is also the source for about 70 percent of all imported goods). When the peso devaluation crisis struck in late 1994, the United States unhesitatingly played a pivotal role in the effort to rescue the Mexican economy. As Mexico pursues economic recovery and the resumption of its growth boom, we should keep in mind that NAFTA is the first reciprocal trade agreement among high–income and upper-middle-income countries. As events since 1993 have demonstrated, there are bound to be internal as well as external problems during the transition to the new supranational economic order.

As Mexico strives to join the ranks of the world's ad-vantaged states, the movements of its inhabitants indicate that the country is still a wobbly giant. Well over a million Mexicans annually migrate to the United States border zone, lured by the promise of jobs in agriculture and manufacturing. Most migrants would prefer to keep on going north, and hundreds of thousands do indeed manage to cross the U.S. boundary each year. Large numbers of frustrated Mexicans also illegally attempt to leave the country, but are intercepted and deported by U.S. authorities; undoubtedly, many more get through. In the mid-1990s, this illegal immigration triggered increased public support in the United States for tougher measures to curb such activity. As a result, police patrols have been beefed up, new fences have been built along many stretches of the boundary, and there is even talk about closing the border.

❖ THE CENTRAL AMERICAN REPUBLICS

Crowded onto the narrow segment of the Middle American land bridge between Mexico and the South American continent are seven countries collectively known as the Central American republics (Fig. 4-9). Territorially, they are all quite small: only one, Nicaragua, is larger than the Caribbean island of Cuba. Populations range from Guatemala's 11.3 million down to Panama's 2.7 million in the six Hispanic republics, whereas the sole former British territory, Belize, has only about 240,000 inhabitants.

The narrowing land bridge on which these republics are located consists of a highland belt flanked by coastal lowlands on both the Caribbean and Pacific sides. From the earliest times, the people have been concentrated in the upland *tierra templada* (temperate) zone. Here tropical

In the path of NAFTA: if the agreement takes full effect, Monterrey in the Mexican State of Nuevo León is most advantageously situated.

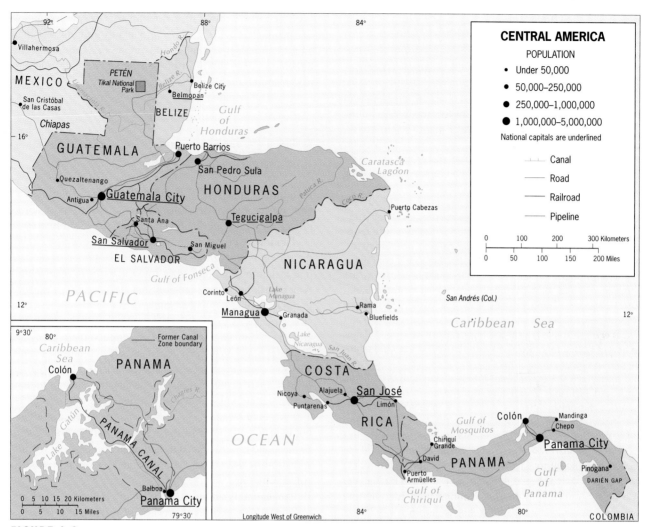

FIGURE 4-9

temperatures are moderated by elevation (see box titled ''Altitudinal Zonation''), and rainfall is adequate for the cultivation of a variety of crops. As noted earlier, the Middle American highlands are studded with volcanoes, and local areas of fertile volcanic soils are scattered throughout the region. The old Amerindian agglomerations were located in these more fertile parts of the highlands, and this human distribution persisted during the Spanish period.

The distribution of population within Central America, apart from its concentration in the region's uplands, also exhibits greater densities toward the Pacific than toward the Caribbean coastlands (Fig. I-9). El Salvador, Belize, and (to some degree) Panama are exceptions to the rule that people in continental Middle America are concentrated in the *templada* zone. Most of El Salvador is *tierra caliente*, and the majority of its 6.2 million people are crowded (710 to the square mile—a population density approaching India's) onto the intermontane plains lying less than 2,500 feet (750 m) above sea level. In Nicaragua, too, the Pacific-

side areas are the most densely populated; the early Amerindian centers lay near Lake Managua, Lake Nicaragua, and in the adjacent highlands. The frequent activity of volcanoes in this Pacific zone is accompanied by the emission of volcanic ash, which settles over the countryside and quickly weathers into fertile soils.

By contrast, the Caribbean coastal lowlands—hot, wet, and awash in leached soils—support comparatively few people. In the most populous republic, Guatemala, the heartland also has long been in the southern highlands. Although the large majority of Costa Rica's population is concentrated in the central valley around San José, the Pacific lowlands have been the scene of major in-migration since banana plantations were first established there. Even in Panama, there is a strong Pacific orientation. More than half of all Panamanians (and this means about two-thirds of the rural population) live in the southwestern lowlands and on adjoining mountain slopes; another 25 percent live and work in the Panama Canal corridor; and of the re-

ALTITUDINAL ZONATION

■ ■ ■

Continental Middle America and the western margin of South America are areas of high relief and strong local contrasts. People live in clusters in hot tropical lowlands, in temperate intermontane valleys, and even on high plateaus just below the snow line in the Andes. In each of these various zones, distinct local climates, soils, crops, domestic animals, and modes of life prevail. Such **altitudinal zones** are known by specific names as if they were regions with distinguishing properties—as, in reality, they are.

The lowest vertical zone, from sea level to 2,500 feet (about 750 m), is known as the *tierra caliente*, the "hot land" of the coastal plains and low-lying interior basins where tropical agriculture (including banana plantations) predominates. Above this lowest zone lie the tropical highlands containing Middle and South America's largest population clusters, the *tierra templada* of temperate land reaching up to about 6,000 feet (1,850 m). Temperatures here are cooler; prominent among the commercial crops is coffee, while corn (maize) and wheat are the staple grains. Still higher, from about 6,000 feet to nearly 12,000 feet (3,600 m) is the *tierra fría*, the cold country of the higher Andes where hardy crops such as potatoes and barley are the mainstays. Only small parts of the Middle American highlands reach into the *fría* zone, but in South America this environment is much more extensive in the Andes. Above the tree line, which marks the upper limit of the *tierra fría*, lies the *puna* (also known as the *páramos*); this fourth altitudinal zone, extending from about 12,000 to 15,000 feet (3,600 to 4,500 m), is so cold and barren that it can support only the grazing of sheep and other hardy livestock. The highest zone of all is the *tierra helada*, or "frozen land," a zone of permanent snow and ice that reaches to the peaks of the loftiest Andean mountains.

These elevation ranges are for highlands lying in the equatorial latitudes. Of course, as one moves poleward of the tropics beyond 15 degrees of latitude, the sequence of five vertical zones is ratcheted downward, with the breaks occurring at progressively lower altitudes.

■ ■ ■

mainder, a majority (many of them descendants of black immigrants from the Caribbean) live on the Caribbean or Rimland side of the isthmus.

Middle America's smaller republics face the same problems as the less developed parts of Mexico, only more so; they also share many of the difficulties confronting the poorer Caribbean islands. No present or future challenge, however, is greater than Central America's overpopulation. The region's population explosion began around mid-century, expanding from a base of 9.3 million people in 1950 to a total of 34.4 million in 1997. Unlike Mexico, which has substantially reduced its rate of natural increase since 1980, Central America (except for Costa Rica and Panama)

is on a course that will see a doubling of today's population to 70 million by 2020. This amounts to nothing less than an onrushing demographic catastrophe in a region already unable to cope with most of its social, economic, and natural resource problems.

Emergence From a Turbulent Era

Inequities, repressive governments, external interference, and the frequent unleashing of armed forces have destabilized Central America for much of its modern history.

The roots of these upheavals are old and deep and today the region is struggling to emerge from a period of turmoil that lasted through the 1980s into the early 1990s. Central America is not a large region, but because of its physiography it contains many isolated, comparatively inaccessible locales. Conflicts between Amerindian population clusters and mestizo groups are endemic to the region, and contrasts between the privileged and the poor are especially harsh. Dictatorial rule by local elites followed authoritarian rule by Spanish colonizers, and the latest episode of violent confrontation was but one in a series. One unprecedented side effect of these recent conflicts has been an intermittent but enormous flow of refugees. These people, by the tens of thousands, were forced to flee from combat and "death-squad" terrorism, often leaving behind broken families and shattered lives.

Despite the challenges that remain, battle-scarred Central America's prospects are finally brightening. The expanding opportunities of individual republics are highlighted in the country profiles that follow. An important breakthrough is also occurring at the supranational level, where a new spirit of cooperation is forging a sense of regional identity that has barely existed in the past. What began in 1993 as an exploratory effort to resuscitate local trade within the framework of the 30-year-old Central American Common Market rapidly escalated into a series of pacts to create a more meaningful economic union. At the same time, free trade agreements were negotiated with several nearby countries outside Central America, and all seven republics became founding members of the Association of Caribbean States (which includes most of the realm's countries). By 1995, their new interrelationships were functioning so smoothly that all the republics except tiny Belize announced their intention to apply as a group for full membership in NAFTA.

Guatemala: Still in Turmoil

Westernmost of Central America's seven republics, Guatemala has more land neighbors than any other. Straight-line boundaries across the tropical forest mark much of the border with Mexico, creating the box-like region of Petén between Chiapas State and Belize to the east (Fig. 4-9). Also to the east lie Honduras and El Salvador.

Guatemala, heart of the ancient Maya Empire and still strongly infused by Amerindian culture and tradition, has just a small window on the Caribbean but a longer Pacific coastline. It was still part of Mexico when the Mexicans threw off the Spanish yoke, and although independent from Spain after 1821, Guatemala did not become a separate republic until 1838. Mestizos, not the Amerindian majority, secured the country's independence.

Most populous of all seven Central American republics with more than 11 million inhabitants (mestizos are now in the majority with 55 percent, Amerindians 45 percent), Guatemala has seen much conflict. Repressive regimes made deals with U.S. and other foreign economic interests that stimulated development, but at a high social cost. Over the past half-century, military regimes have dominated political life. The deepening split between the wretchedly poor Amerindians and the better-off mestizos, who call themselves *ladinos*, generated a civil war that started in the 1950s and has since claimed more than 100,000 lives. Most of the victims have been of Mayan descent; the mestizos control the government, army, and land-tenure system.

In the rugged highlands of the south, where the capital of Guatemala City is located, and in the remote plains on the Mexican border, physiography and remoteness provide refuge for insurgents, and for the ladinos the war was unwinnable. International outrage over human-rights violations perpetrated by the armed forces also contributed to negotiations that, in the mid-1990s, seemed to be moving Guatemala in the direction of compromise and, possibly, stability.

The tragedy of Guatemala is that its economic geography has considerable potential, long destroyed by its internal conflicts. Elevations in the Sierra Madre, the range that forms the south's mountain backbone, reach over 13,000 feet (4000 m), soils are fertile, and moisture is ample over areas large enough to produce a wide range of crops including excellent coffee. The country's mineral wealth probably is only partially known, but includes nickel in the highlands and oil on the north side of the mountains. Peace and security would engender a major tourist industry based on Guatemala's incomparable Maya heritage (both past and present), spectacular scenery, and magnificent beaches.

Stability also would enable Guatemala to build an infrastructure it badly needs. The incipient core area around Guatemala City still lies poorly connected to the outside world; the road (no railroad!) to Puerto Barrios on the Caribbean coast is a tortuous, slow trip, and little port development has occurred on the Pacific coast. If the challenge of ethnic division can be overcome, the rewards for Guatemala would be great.

Belize: Changing Identity

Strictly speaking, Belize is not a Central American republic in the same tradition as the other six. This country, a wedge of land between northern Guatemala, Mexico's Yucatán Peninsula, and the Caribbean, was a British dependency (called British Honduras) until 1981. For much of the post-independence period, Guatemala refused to recognize Belize's existence (claiming an 1859 treaty with Britain had "stolen" the territory from Guatemala), but the dispute has now been settled. Slightly larger than Massachusetts and with a minuscule population of only 240,000 (many of African descent), Belize has been more reminiscent of a Caribbean island than of a continental Middle American state.

In the late 1990s, all that is changing as migrations reshape the demographic complexion of Belize. Thousands of residents of African descent (the Creoles, predominantly English speakers) have recently emigrated. Most went to the United States, and their departure precipitated a drop in the proportion of Belize's Creole population from 40 percent in 1980 to less than 30 percent in 1995. They have been replaced by more than 50,000 Spanish-speaking immigrants—mostly escapees from the strife in nearby Guatemala, El Salvador, and Honduras—who saw *their* proportion of the Belizean population rise from 33 to 43 percent between 1980 and 1991. By 2000, the newcomers will be in the majority, Spanish will have become the *lingua franca*, and Belize's cultural geography will exhibit an expansion of the Mainland at the expense of the Rimland.

The Belizean transformation extends to the economic sphere as well. No longer just an exporter of sugar, Belize is producing new commercial crops and has a thriving seafood processing industry that have become leading revenue earners. Also important is eco-tourism, based on the natural attractions of the country's still pristine environment. Belize is also developing a reputation as one of the region's major "offshore" banking centers—a financial haven for foreign companies and individuals who want to avoid paying taxes in their home countries. (This is one of the stimuli spurring the growth of coastal Belize City; the capital, interior Belmopan, has a mere 4,000 inhabitants.) A not unrelated new pursuit is Belize's designation as the world's leading transactional center for the computerized gambling that takes place on the Internet.

Honduras: Mired in Poverty

As the map (Fig. 4-9) shows, Honduras, in direct contrast to Guatemala, has a lengthy Caribbean coastline and a small window on the Pacific. Honduras occupies a critical place in the political geography of Central America, flanked as it is by Nicaragua, El Salvador, and Guatemala—all grappling with the aftermath of years of internal conflict. During the war-torn 1980s, refugees streamed into the country, and both American and Soviet arms crossed Honduran territory and waters bound for neighboring Nicaragua and El Salvador.

Honduras has a democratically elected government, but the military wields considerable power. With 5.8 million inhabitants, about 90 percent mestizo, Honduras is the region's poorest, least developed country. Agriculture, livestock, forestry, and some mining (lead and zinc) form the mainstays of the economy. The familiar Central American products—bananas, coffee, apparel—earn most of the external income. During the 1970s, there was some promising development of light industry around San Pedro Sula near the northwestern coast. However, Honduras fell victim to regional strife, foreign investors were afraid to risk their funds, and tourism abated. The economic outlook has not significantly improved. This is a tragedy because social and ethnic divisions, which so strongly mark Honduras's neighbors, are not serious here, and the gap between rich and poor, though evident, is not as wide.

El Salvador: Postwar Reconstruction

El Salvador is Central America's smallest country territorially, smaller even than Belize; yet, with a population 25 times as large (6.2 million), it is the most densely peopled. Again, like Belize, it is one of only two continental republics that do not have coastlines on the Caribbean as well as Pacific sides. El Salvador adjoins a narrow Pacific coastal plain in a chain of volcanic mountains. The country's heartland lies behind those mountains in the interior, where the capital, San Salvador, is located. North of this core area, at the Honduran border, lies another zone of mountains, which has always contained areas beyond effective governmental control.

Unlike neighboring Guatemala, El Salvador's population is quite homogeneous (89 percent mestizo and just 10 percent Amerindian). Yet ethnic homogeneity did not translate into social or economic equality or even opportunity. Other Central American countries were called "banana republics"; El Salvador was a coffee republic, and the coffee was produced on the huge landholdings of a few landowners and on the backs of a subjugated peasant labor force. The military supported this system and repeatedly suppressed violent and desperate peasant uprisings.

From 1980 to 1992, El Salvador was ripped by a devastating civil war that was worsened by outside arms supplies from the United States (supporting the government) and Nicaragua (aiding the rebel forces). Following the negotiated end to that war, in which some 75,000 people, mostly peasants, were killed, an effort has been under way to prevent a recurrence. But El Salvador, its historical geo-

graphy one of division and disintegration, will have difficulty overcoming its legacy of searing inequality.

Perversely, the civil war did have one positive result: wealthy citizens who left the country and did well in the U.S. and elsewhere are sending substantial funds back home, a boost to the economy. This money has helped stimulate such urban industries as textile and footwear manufacturing, as well as food processing. Certainly the economy is recovering, but El Salvador's biggest problems lie in its marginal, interior zones, where peasant families on overpopulated lands say that they have been forgotten in the national reconstruction effort. El Salvador's future still hangs in the balance.

Nicaragua: Deluged by Disasters

When studying Nicaragua, one does well to look again at the map (Fig. 4-9), which underscores the country's pivotal position in the heart of Central America. The Pacific coast follows a southeasterly direction, but the Caribbean coast is oriented north-south so that Nicaragua forms a triangle of land with its lakeside capital, Managua, located in a great longitudinal valley on the mountainous, Pacific side. Indeed, the core of Nicaragua always has been in this western zone. The Caribbean side, where the mountains and valleys give way to a coastal plain of (disappearing) rainforest, pine savanna, and swampland, has for centuries been home to Amerindian peoples such as the Miskito who have been relatively remote from the focus of national life.

Nicaragua was the typical Central American republic, ruled by a dictatorial government and exploited by a wealthy land-owning minority, its export agriculture dominated by huge plantations owned by foreign corporations. It was a situation ripe for insurgency, and in 1979 the leftist Sandinista rebels overthrew the government in Managua. However, Sandinista rule quickly produced its own excesses, resulting in civil war for most of the 1980s. The conflict ended in 1990, and in the free elections that followed a more democratic, anti-Sandinista regime was voted into office. Nonetheless, Sandinistas continue to wield great influence in the running of the country, especially the armed forces.

A leading casualty of this turmoil is the Nicaraguan economy, and during the past few years the country has become continental Middle America's poorest. Besides massive problems of poverty, unemployment, and public services provision, the government is bedeviled by a land redistribution crisis of nightmarish proportions as it tries to sort out thousands of pre- and postwar property claims (most of them complicated by the ongoing breakup of huge Sandinista-era collective farms). Somehow, the government has also kept the country on a steady transitional course toward a free market economy, assisted by significant foreign aid from the United States. Nicaragua possesses few comparative advantages in the coming economic competition against the region's more prosperous republics, but some development possibilities do exist. One is a much-discussed proposal to build a new ''dry'' canal across the flatlands of central Nicaragua, with high-speed trains ferrying freight-filled containers between Caribbean and Pacific ports that can be directly loaded onto ultramodern ships too large to fit through the isthmus's aging Panama Canal. Economic opportunity, however, must also be weighed against looming demographic disaster: unless Nicaragua can rein in the high natural-increase rate in its population (which now even exceeds Haiti's by nearly 20 percent), the country's living standards will not have a chance to improve in the foreseeable future.

Costa Rica: Durable Democracy

As if to confirm what was said about Middle America's endless variety and diversity, Costa Rica differs in significant ways from its neighbors—and from the norms of Central America. Bordered by two volatile countries (Nicaragua to the north and Panama to the east), Costa Rica is a nation with an old democratic tradition and, in this cauldron, *no* standing army! The country is, in fact, the oldest democracy in Middle and South America, enjoying a freely elected government (except for two brief periods) since 1889. Although the initial Hispanic imprint here was similar to that found elsewhere on the Mainland, early independence, the good fortune to lie remote from regional strife (which fostered an enduring posture of neutrality), and a leisurely pace of settlement allowed Costa Rica the luxury of concentrating on its economic development and on public education, which is universal and free. Perhaps most important, internal political stability has prevailed over much of the past 175-odd years. The last brush with conflict 50 years ago left the nation resolved to avoid further violence, and the armed forces were abolished in 1948 (along with a military establishment, so often the source of trouble throughout Central America).

Costa Rica, like its neighbors, is also divided into environmental zones that parallel the coasts. The most densely settled is the central highland zone, lying in the cooler *tierra templada*. Volcanic mountains prevail throughout much of this zone, but the heartland is the Valle Central (Central Valley), a fertile 40-by-50-mile (65-by-80-km) basin that contains the leading population cluster focused on the capital city and the country's main coffee-growing area. The capital, San José, is atypical of Middle America—a clean and virtually slumless city that is the most cosmopolitan urban center between Mexico City and South America.

To the east of the highlands are the hot and rainy Caribbean lowlands, a sparsely populated segment of Rimland where many plantations have now been abandoned and replaced by subsistence farmers. Between 1930 and 1960, the (U.S.-based) United Fruit Company shifted most of the country's banana plantations from the crop-disease-ridden Caribbean littoral to Costa Rica's third zone—the plains and gentle slopes of the Pacific coastlands. This gave the Pacific zone a major boost in economic growth (although banana production now is moving to Panama, where wages are lower). Today it is the scene of diversifying and expanding commercial agriculture (often requiring irrigation) as well as successful colonization schemes in previously undeveloped valleys and basins. Additional Pacific-coast development is spurred by new container-port construction at Puerto Caldera.

The long-term development of Costa Rica's economy has given it the region's highest standard of living, literacy rate, and life expectancy. Agriculture continues to dominate, with coffee, bananas, ornamental plants, pineapples, and beef the leading exports. As the country pays off the enormous foreign debt it accumulated during the 1980s, considerable investment capital from abroad is again flowing in. Much of it has financed new tourist facilities. These advances notwithstanding, the country's veneer of development cannot mask serious problems. In terms of social structure, almost one-quarter of the population is trapped in an unending cycle of poverty, and the huge gap between the poor and the affluent is constantly widening. And in the environmental sphere, recent progress has come at the expense of the tropical forest, whose rapid destruction portends ecological disaster (see box titled ''Tropical Deforestation'').

Politically, Costa Rica remains quite stable. Despite its proximity to the region's trouble spots, it has resisted involvement because the overwhelming majority of its peace-loving people prefer the country to maintain its neutrality as ''the Switzerland of Central America.''

Panama: Strategic Canal, Reorganizing Corridor

The Republic of Panama owes its birth to an idea: the construction of an artificial waterway to connect the Atlantic and Pacific oceans and thereby avoid the lengthy circumnavigation of South America. In the 1880s, when Panama was still an extension of neighboring Colombia, a French company tried and failed to build such a canal here; thousands of workers died of yellow fever, malaria, and other tropical diseases, and the company went bankrupt. By the turn of the century, U.S. interest in a Panama canal

TROPICAL DEFORESTATION

■ ■ ■

As the world vegetation map (Fig. I-8) indicates, two-thirds of Central America was covered by tropical rainforests. We deliberately use the word *was* because destruction of this precious woodland resource has proceeded so swiftly that a typical landscape scene today shows not a rainforest but a scraggly, tree-cleared wasteland ripe for environmental disaster (see photo). **Tropical deforestation** in continental Middle America began with the Spanish colonial era in the sixteenth century. The pace has accelerated incredibly in recent decades; since 1950, over 80 percent of Central America's forests have been decimated. Today, more than 3 million acres of woodland in Central America and Mexico disappear *each year*, an area equivalent to one-third the size of the Netherlands. El Salvador has already lost 98 percent of its forests, and most of the six other republics will reach that stage within a decade.

The causes of tropical deforestation are related to the persistent economic and demographic problems of disadvantaged countries. In Central America, the leading cause has been the need to clear rural lands for cattle pasture as many countries (notably Costa Rica) attempt to become meat producers and exporters. Although some gains have been recorded, the price of environmental degradation has been enormous. Because tropical soils are so nutrient-poor, newly deforested areas can function as pastures only for a few years at most. These fields are then abandoned for other freshly cut lands and quickly become the devastated landscape seen in the photograph. Without the protection of trees, local soil erosion and flooding immediately become problems, affecting still-productive nearby areas. A second cause of deforestation is the rapid logging of tropical woodlands as the timber industry increasingly turns from the exhausted forests of the mid-latitudes to harvest the rich tree resources of the equatorial zones, responding to accelerating global demands for new housing, paper, and furniture. The third major contributing factor is related to the region's population explosion: as more and more peasants are required to extract a subsistence from inferior lands, they have no choice but to begin cutting the forest for both firewood and additional crop-raising space, and their intrusion prevents the trees from regenerating (Haiti is the extreme example—see photo p. 212).

Although deforestation is a depressing event, tropical pastoralists, farmers, and timber producers do not apparently consider it to be life-threatening, and perhaps it even seems to offer some short-term economic advantages. Why, then, should there be such an outcry from the scientific community? And why should the World Resources Institute call this "the world's most pressing land-use problem"? The answer is that unless immediate large-scale action is taken, soon after the turn of the century the tropical rainforest will be reduced to two large patches—the western Amazon Basin of South America and the middle Zaïre (Congo) Basin of Equatorial Africa.

The tropical forest, therefore, must be a very important

This scene in Costa Rica shows how badly the land can be scarred in the wake of deforestation. Without roots to bind the soil, tropical rains swiftly erode the unprotected topsoil.

part of our natural world—and indeed it is. Biologically, the rainforest is by far the richest, most diversified arena of life on our planet: even though it covers only a shrinking 5 percent of the Earth's land area, it contains well over half of all plant and animal species. Its loss would mean not only the extinction of millions of species, but also the *end of birth* because the evolutionary process that produces new species would be terminated. Because the rainforests already yield countless valuable medicinal, food, and industrial products, how many potential disease-combatting drugs or new crop varieties to feed undernourished millions would be irretrievably lost?

Another consequence of rainforest removal would be its impact on the planet's climates. Many environmental scientists hypothesize that a global warming trend would be enhanced as the burning of the remaining forests adds large quantities of carbon dioxide to the air, forming a thickening layer of the gas that would prevent excess heat from escaping the atmosphere. This trapping of warm air—the so-called *greenhouse effect*, because of the obvious analogy—would be further intensified by the loss of the carbon dioxide-absorbing trees and could lead to higher temperatures at every latitude, polar-icecap melting, and rising sea levels that would imperil the world's crowded coastal zones.

■ ■ ■

From the Field Notes

"The Panama Canal remains an engineering marvel more than 80 years after it opened in August, 1914. The parallel lock chambers each are 1000 feet long and 110 feet wide, permitting vessels as large as the Queen Elizabeth II to cross the isthmus. Ships are raised by a series of locks to Gatun Lake, 85 feet above sea level. We watched as tugs helped guide the QE-II into the Pedro Miguel Lock, the upper (of three) locks on the Pacific side. Gaillard Cut, the most spectacular part of the Canal, is in the background; behind us is Miraflores Lake, from which the double Miraflores Locks lead down to sea level. Note that the water level on the far right, where the QE-II will enter, already has been raised to the level of the Gaillard Cut. The lock doors are 65 feet wide and seven feet thick, and range in height from 47 to 82 feet. The motors that move them are recessed in the walls of the lock chambers. Once inside the locks, ships are pulled by powerful locomotives called *mules* that ride on rails that ascend and descend the system. Traversing the Panama Canal took us about 12 hours from pilot station to pilot station; it was one of the most fascinating days imaginable."

(which would shorten the sailing distance between the East and West Coasts by 8,000 nautical miles) rose sharply, and the United States in 1903 proposed a treaty that would permit a renewed effort at construction across Colombia's Panamanian isthmus. When the Colombian Senate refused to go along, Panamanians rebelled, and the United States supported this uprising by preventing Colombian forces from intervening. The Panamanians, at the behest of the United States, declared their independence from Colombia, and the new republic immediately granted the United States rights to the Canal Zone, averaging about 10 miles (16 km) in width and just over 50 miles (80 km) in length.

Soon canal construction commenced, and this time the project succeeded as American engineering and technology—and medical advances—triumphed over a formidable set of obstacles. The Panama Canal (see inset map, Fig. 4-9) was opened in 1914, a symbol of U.S. power and influence in Middle America. The Canal Zone was held by the United States under a treaty that granted it "all the rights, powers, and authority" in the area "as if it were the sovereign of the territory."

4.1

Such language might suggest that the United States held rights over the Canal Zone in perpetuity, but the treaty nowhere stated specifically that Panama permanently yielded its own sovereignty in that transit corridor. In the 1970s, as the canal was transferring more than 14,000 ships per year (that number is now down to less than 13,000, but the tonnage is up) and generating hundreds of millions of dollars in tolls, Panama sought to terminate U.S. control in the Canal Zone. Delicate negotiations began. In 1977, an agreement was reached on a staged withdrawal by the United States from the territory, first from the Canal Zone and then—on January 1, 2000—from the Panama Canal itself. This agreement took the form of two treaties, and following the signing by Presidents Jimmy Carter and Omar Torrijos, they were ratified in spite of stubborn opposition in the U.S. Senate.

In the late 1990s, with Panama already in control of most of the Canal Zone, international concerns were being raised over Panama's long-term capacity to operate and maintain the canal. These concerns were spurred by the deteriorating condition of buildings and facilities in the Canal Zone already vacated by the United States. Panama is not a rich country, and the canal requires major investments for its upkeep.

Panama reflects some but not all of the usual geographic features of the Central American republics. Its population of 2.7 million is about 70 percent mestizo, but it also contains substantial black, white, and Amerindian minorities. Spanish is the official (and majority) language, but English is in widespread use. Ribbon-like and oriented east-west, Panama's topography is mountainous and hilly, with some mountains reaching higher than 10,000 feet (3,000 m). Eastern Panama, especially Darien Province adjoining Colombia, is densely forested, and here is the only remaining gap in the otherwise complete, intercontinental Pan American Highway. Most of the rural population lives in the uplands west of the canal (much of the urban population is concentrated in the vicinity of the waterway, anchored by the cities at each end of the canal, Panama City and Colón). There, Panama produces bananas, rice, sugarcane, and coffee—and from the sea, shrimp and fishmeal.

The Panama Canal, despite its age and inability to accommodate today's largest ships, remains Panama's focus, its lifeline, its future. It has now been augmented by an oil pipeline that crosses the isthmus in the Far west, linking the Pacific port of Armuelles with the Caribbean terminal at Chiriquí Grande. But breakdowns afflict the railroad paralleling the canal; plans for a four–lane expressway between Colón and Balboa are on hold; and concerns over freshwater supply for interior Lake Gatún, which feeds the locks, is rising as deforestation scars its catchment basin.

While Colón, at the Caribbean end of the canal, is a rather typical Middle American city now benefiting from a free trade zone, it is Panama City, on the Pacific coast (and the only coastal capital in mainland Middle America), that raises eyebrows. A high-rise skyline unlike any other in the realm suggests Singapore or Miami rather than San Salvador or San José. This world-class skyline towers over an urban area that anchors a country of a mere 3 million people. What generates this glass-tower cityscape? The official answer: international banking. The geographic answer suggests the power of relative location, including a new linkage to drug-plagued Colombia, the country on which Panamanians nearly a century ago turned their backs.

■ PRONUNCIATION GUIDE ■

Acapulco (ah-kah-POOL-koh)
Antigua (an-TEE-gwuh)
Antilles (an-TILL-eeze)
Archipelago (ark-uh-PELL-uh-go)
Armuelles (ahr-MWAY-yace)
Aruba (uh-ROO-buh)
Augelli (aw-JELLY)
Baja (BAH-hah)
Balsas (BAHL-suss)
Barbados (bar-BAY-dohss)
Bauxite (BAWKS-site)
Belize (beh-LEEZE)
Belmopan (bell-moh-PAN)
Bering (BERRING)
Bonaire (bun-AIR)
Cacao (kuh-KAY-oh/kuh-KOW)
Campeche (kahm-PAY-chee)
Caribbean (kuh-RIB-ee-un/
    karra-BEE-un)
CARICOM (CARRY-komm)
Castries (CASS-treez)
Cay (KEE)
Chiapas (chee-AHP-uss)
Chihuahua (chuh-WAH-wah)
Chiriquí Grande (chih-rih-KEE
    GRAHN-day)
Ciudad Juárez (see-you-DAHD
    WAH-rez)
Ciudades perdidas (see-you-DAH-dayss
    pair-DEE-duss)
Coahuila (koh-uh-WEE-luh)
Cocos (KOH-kuss)
Colón (kuh-LOAN)
Cortés, Hernán (kor-TAYSS, air-NAHN)
Costa Rica (koss-tuh-REE-kuh)
Curaçao (koor-uh-SAU)
Darien (dar-YEN)
Dominica (duh-MIN-ih-kuh)
Ejido[s] (eh-HEE-doh[ss])
El Paso (ell-PASSO)
Fonseca (fahn-SAY-kuh)
Fort-de-France (for-duh-FRAWSS)
Gaillard (gil-YARD)
Gatún (guh-TOON)
Grenada (gruh-NAY-duh)
Guadalajara (gwah-duh-luh-HAHR-uh)
Guadeloupe (GWAH-duh-loop)

Guatemala (gwut-uh-MAH-lah)
Guerrero (geh-RARE-roh)
Hacienda (ah-see-EN-duh)
Haiti (HATE-ee)
Hegemony (heh-JEH-muh-nee)
Hispaniola (iss-pahn-YOH-luh)
Honduras (hon-DURE-russ)
Isthmus/isthmian (ISS-muss/
    ISS-mee-un)
Jalisco (huh-LISS-koh)
Jamaica (juh-MAKE-uh)
Jesuit (JEH-zoo-it)
Junta (HOON-tah)
Ladino (luh-DEE-noh)
León (lay-OAN)
Lingua franca (LEEN-gwuh FRUNK-uh)
Littoral (LIT-oh-rull)
Maize (MAYZ)
Managua (mah-NAH-gwuh)
Maquiladora (mah-kee-luh-DORR-uh)
Martinique (mahr-tih-NEEK)
Maya[n] (MY-uh[un])
Mesoamerica (MEZZOH-america)
Mestizo (meh-STEE-zoh)
Michoacán (mee-chuh-wah-KAHN)
Mirador (meera-DOAR)
Miraflores (meera-FLAW-rayss)
Miskito (mih-SKEE-toh)
Monterrey (mahnt-uh-RAY)
Mulatto (moo-LAH-toh)
Nassau (NASS-saw)
Nevis (NEE-vuss)
New Guinea (noo-GHINNY)
Nicaragua (nick-uh-RAH-gwuh)
Norte (NOR-tay)
Nuevo León (noo-AY-voh lay-OAN)
Oaxaca (wuh-HAH-kuh)
Ocho Rios (oh-choe REE-ohss)
Páramos (PAH-ruh-mohss)
Palenque (puh-LENG-kay)
Peón (pay-OAN)
Peones (pay-OH-nayss)
Petén (peh-TEN)
Petit Piton (peh-TEE pea-TAW)
Philippines (FILL-uh-peenz)
Piton (pea-TAW)
Placer[ing] (PLASS-uh[ring])

Port-au-Prince (por-toh-PRANSS)
Puebla (poo-EBB-luh)
Puerto Barrios (pwair-toh bah-REE-uss)
Puerto Caldera (pwair-toh kahl-DERRA)
Puerto Rico (pwair-toh REE-koh)
Puna (POONA)
Quetzal (kay-DZAHL)
Saba (SAY-buh/SAH-buh)
Sabinas (sah-BEE-nuss)
Salinas (sah-LEE-nuss)
Sandinista (sahn dee-NEE-stuh)
San José (sahn hoe-ZAY)
San Juan (sahn HWAHN)
San Pedro Sula (sahn pay-droh-SOO-
    luh)
Sierra Madre (see-ERRA MAH-dray)
    Occidental (oak-see-den-TAHL)
    Oriental (aw-ree-en-TAHL)
St. Eustatius (saint yoo-STAY-shuss)
St. Lucia (saint LOO-shuh)
St. Maarten (sint MAHRT-un)
Sonora (suh-NORA)
Soufrière (soo-free-AIR)
Tabasco (tuh-BAH-skoh)
Tamaulipas (tah-mau-LEEPUS)
Tampico (tam-PEEK-oh)
Taxco (TAHSS-koh)
Tegucigalpa (tuh-goose-ih-GAHL-puh)
Tehuantepec (tuh-WHAHN-tuh-pek)
Tenochtitlán (tay-noh-chit-LAHN)
Teotihuacán (tay-uh-tee-wah-KAHN)
Tierra caliente (tee-ERRA
    kahl-YEN-tay)
Tierra fría (tee-ERRA FREE-uh)
Tierra helada (tee-ERRA ay-LAH-dah)
Tierra templada (tee-ERRA
    tem-PLAH-dah)
Tijuana (tee-WHAHN-uh)
Tobago (tuh-BAY-goh)
Torrijos (tor-REE-hohss)
Valle Central (VAH-yay sen-TRAHL)
Veracruz (verra-CROOZE)
Villahermosa (vee-yuh-air-MOH-suh)
Yucatán (yoo-kuh-TAHN)
Zaïre (zah-EAR)
Zapata, Emiliano (zuh-PAH-tuh,
    eh-mee-lee-AH-noh)

HAVANA

CUBA

Tropic of Cancer

Gulfo de Campeche

PEN DE YUCATÁN

HISPANIOLA

San Juan

PUERTO RICO (U.S.A.)

GUADELOUPE (Fr.)

MARTINIQUE (Fr.)

NORTH AMERICAN BASIN

ATLANTIC

JAMAICA

Golfo de Honduras

WEST INDIES

CARIBBEAN SEA

BARBADOS

OCEAN

CENTRAL

Lago de Nicaragua

PUNTA DE GALLINAS

Venezuela

TRINIDAD AND TOBAGO

AMERICA

Barranquilla
Cartagena

Maracaibo

Port of Spain

Panamá
IST DE PAN.

LLANOS

Valencia

CARACAS

Mérida

Ciudad
Bolívar

GUYANA

Georgetown

Paramaribo

Golfo de Panamá

ISLA DEL COCO
(Costa Rica)

ISLA DE MALPELO
(Colombia)

Medellín

Cerro Icutú
7800

SANTA FE
DE BOGOTA

VENEZUELA

Boa Vista do
Río Branco

GUIANA
HIGHLANDS

SURINAME

Cayenne

FR.
GUIANA

Nevado del Tolima
17,110

COLOMBIA

ILHA DE MARAJÓ

ROCEDOS SÃO PEDRO
E SÃO PAULO
(Brazil)

ARCHIPIÉLAGO DE COLÓN
(GALÁPAGOS ISLANDS)
(Ec.)

Quito
Cotopaxi

ECUADOR
Chimborazo
20,561

Japurá

Río Negro

Río Amazonas

Manaus
(Manáos)

Belém
(Pará)

São Luís
(Maranhão)

Equator

Guayaquil

Iquitos

Leticia

Río Solimões (Amazonas)

Fortaleza
(Ceará)

Golfo de Guayaquil

Chiclayo

PERU

Purús

Juruá

Madeira

Tapajós

Xingu

Tocantins

Teresina

ARQUIPÉLAGO
FERNANDO DE NORONHA
(Brazil)

CABO DE SÃO ROQUE

Natal

João Pessoa (Paraíba)

Trujillo

ANDES

Río Branco

Pôrto
Velho

BRAZIL

SERRA DO PIAUÍ

RECIFE (Pernambuco)

Nevs Huascaran
22,205

LIMA

Callao

Cuzco

CHAPADA DE
MATO GROSSO

Brasília

Maceió

Volcán Misti
19,098

Arequipa

Mollendo

La Paz
Nev Illimani
21,151

BOLIVIA

Cuiabá

BRAZILIAN

Salvador
(Bahia)

Sucre
Potosí

Diamantina

HIGHLANDS

SERRA DO ESPINHAÇO

Iquique

GRAN
CHACO

Belo Horizonte

Pico da Bandeira
9482

Antofagasta

DESIERTO DE ATACAMA

Salta

PARAGUAY

Vitória

ISLA DE SAN FÉLIX
(Chile)

ISLA DE SAN AMBROSIO
(Chile)

Cerro Azufre (Copiapó)
19,947 Vol.

Tucumán

Asunción

SÃO PAULO

Santos

CABO FRIO

RIO DE JANEIRO

Copiapó

Corrientes

Iguaçú Falls

Tropic of Capricorn

ARGENTINA

Coquimbo

Santa Fe

Salto

Paraná

Florianópolis

Córdoba

Rosario

Cerro Aconcagua
22,834

CHILE

Valparaíso

Mendoza

ANDES

URUGUAY

Pôrto Alegre

SANTIAGO

BUENOS AIRES

La Plata

MONTEVIDEO

Río Grande

ISLAS DE JUAN FERNÁNDEZ
(Chile)

Concepción

PAMPAS

Río de la Plata

Valdivia

Bahía Blanca

Puerto Montt

Viedma

Golfo San Matías

ISLA DE CHILOE

Comodoro Rivadavia

ARCHIPIÉLAGO
DE LOS CHONOS

Golfo San Jorge

WELLINGTON

Monte
Valentin
1314

FALKLAND IS.
(ISLAS MALVINAS)
(Br.)

HANOVER

Río Gallegos

Stanley

DESOLACIÓN

Punta Arenas

Estrecho de Magallanes

TIERRA DEL FUEGO

ISLA DE LOS ESTADOS

Mt. Sarmiento
8100

CABO DE HORNOS
(CAPE HORN)

SOUTH GEORGIA
(Falkland Is.)

PACIFIC

OCEAN

Drake Passage

SOUTH
SANDWICH
ISLANDS
(Falkland Is.)

SOUTH SHETLAND
ISLANDS
(U.K.)

SOUTH ORKNEY IS.
(U.K.)

JOINVILLE

ATLANTIC

OCEAN

A-540000-76 4 13
COPYRIGHT BY
RAND McNALLY & COMPANY
MADE IN U.S.A.

ANTARCTIC PENINSULA

JAMES ROSS

Longitude West of Greenwich

Antarctic Circle

| 0 | 200 | 400 | 600 | 800 | 1000 Miles |

| 0 | 400 | 800 | 1200 | 1600 Kilometers |

copyright © Rand McNally, 1993

Scale 1:40 000 000; one inch to 630 miles. Lambert's Azimuthal, Equal Area Projection
Elevations and depressions are given in feet

Relief

Meters		Feet
3050		10 000
1525		5000
610		2000
305		1000
0	Sea Level	
152.5		500
1525		5000
3050		10 000
6100		20 000

South America: Continent of Contrasts

Of all the continents South America has the most familiar shape—that giant triangle connected by mainland Middle America's tenuous land bridge to its sister continent in the north. What we realize less often about South America is that it lies not only south but also mostly east of its northern counterpart as well. Lima, the capital of Peru—one of the continent's westernmost cities—lies farther east than Miami, Florida. Thus South America juts out much more prominently into the Atlantic Ocean than does North America, and South American coasts are much closer to Africa and even to southern Europe than are the coasts of Middle and North America. Lying so far eastward, South America's western flank faces a much wider Pacific Ocean than does North America; the distance from its west coast to Australia is nearly twice that from California to Japan.

As if to reaffirm South America's northward and eastward orientation, the western margins of the continent are rimmed by one of the world's longest and highest mountain ranges, the Andes, a gigantic wall that extends from Tierra del Fuego near the southern tip of the triangle to Venezuela in the far north (Fig. 5-1). Every map of world physical geography clearly reflects the existence of this mountain chain—in the alignment of isohyets (lines connecting

IDEAS & CONCEPTS	
Economic geography	Urbanization
Agricultural systems	Rural-to-urban migration
Land alienation	The "Latin" American city
Isolation	Forward capital
South American culture	Growth-pole concept
spheres	Elongated state
Complementarity	

❖ REGIONS	
Brazil	The Andean West
The Caribbean North	The Mid-Latitude South

places of equal precipitation totals; see Fig. I-6), in the elongated zone of highland climate (Fig. I-7), and in the regional distribution of vegetation (Fig. I-8). Moreover, as Figure I-9 reveals, South America's largest population clusters are located along the eastern and northern coasts, overshadowing those of the Andean west.

DEFINING THE REALM

Long characterized by regional disparities, political turmoil, and developmental inertia, South America is today entering a new era of opportunity. Its major countries, heretofore accustomed to going their separate ways, are discovering the benefits of forging closer individual and multinational ties, especially through free trade pacts. Juntas and their remnants are being swept off the political landscape as military forces retreat to their barracks in favor of more democratic forms of government. New transport routes are opening new settlement frontiers in many once-remote parts of the continent. South America now leads the world in new mining ventures: coal and oil in Colombia; iron ore and gold in the wide fringes of the vast Amazon Basin; copper in southern Peru and northern

FIGURE 5-1

Chile; silver, zinc, and lead in the Andes of Peru. The perception in the late 1990s is that, finally, things are improving and that this realm is at the threshold of a period of unprecedented economic growth.

This optimism, however, must be tempered by the recognition that serious challenges remain to be confronted and overcome. As certain areas of South America make real and lasting progress, too many others continue to be plagued by infrastructure shortcomings, inefficiency and corruption, and endless rounds of no-gain, boom/bust cycles. Most importantly, throughout the realm the gulf between the rich and the poor steadily widens: since 1980, the proportion of the continent's population living below the poverty level has jumped from 27 to 33 percent. Put another way, according to the World Bank, the richest 20 percent of the population controls *67 percent* of South

America's wealth while the poorest 20 percent controls only (an astonishingly small) 2 percent. These numbers reveal the greatest gap between affluence and poverty to be found in any geographic realm. They also remind us— recent advances notwithstanding—that South America, above all else, is still a continent of awesome contrasts.

THE HUMAN SEQUENCE

Although modern South America's largest populations are situated in the east and north, there was a time during the height of the Inca Empire when the Andes Mountains contained the most densely peopled and best organized state on the continent. Although the origins of Inca civilization are still shrouded in mystery, it has become generally accepted that the Incas were descendants of ancient peoples who came to South America via the Middle American land bridge (possibly following earlier migrations from Asia to North America via the Bering land bridge). Thus for thousands of years before the Europeans arrived in the sixteenth century, indigenous Amerindian communities and societies had been developing in South America.

About a thousand years ago, a number of regional cultures thrived in Andean valleys and basins and at places along the Pacific coast. The llama had been domesticated; religions flourished; sculpture, painting, and other art forms were practiced. Over these cultures the Incas extended their authority from their headquarters in the Cuzco Basin of the Peruvian Andes, beginning late in the twelfth century, to forge the greatest empire in the Americas prior to the coming of the Europeans. (Nothing to compare with these cultural achievements of the central Andean zone existed anywhere else in South America.)

The Inca Empire

When the Inca civilization is compared to that of ancient Mesopotamia, Egypt, the old Asian civilizations, and the Mexican Aztec Empire, it quickly becomes clear that this civilization was an unusual achievement. Everywhere else, rivers and waterways provided avenues for interaction and the circulation of goods and ideas. Here, however, an empire was forged from a series of elongated basins (called *altiplanos*) in the high Andes, created when mountain valleys between parallel and converging ranges filled with erosional materials from surrounding uplands. These *altiplanos* are often separated from one another by some of the world's most rugged terrain, with high snowcapped mountains alternating with precipitous canyons. Individual *altiplanos* accommodated regional cultures; the Incas themselves were first established in the intermontane basin of Cuzco (Fig. 5-3). From that hearth, they conquered and extended their authority over the peoples of coastal Peru and other *altiplanos*.

More impressive than the Incas' military victories was their subsequent capacity to integrate the peoples and regions of the Andean domain into a stable and efficiently functioning state. The odds would seem to have been against them, because as they progressed their domain became ever more elongated, making effective control much more difficult. The Incas, however, were expert road and bridge builders, colonizers, and administrators, and in an incredibly short time they consolidated these new territories—which extended from southern Colombia southward to central Chile (brown zone, Fig. 5-3).

At its zenith, the Inca Empire may have counted more than 20 million subjects. Of course, the Incas themselves were always in a minority in this huge state, and their position became one of a ruling elite in a rigidly class-structured society. The Incas, representative of the emperor in Cuzco, formed a caste of administrative officials who implemented the decisions of their monarch by organizing all aspects of life in the conquered territories. The life of the empire's subjects was strictly controlled by this bureaucracy of Inca administrators, and there was very little personal freedom. So highly centralized was the state and so complete the subservience of its tightly controlled population that a takeover at the top was enough to gain power over the entire empire—as the Spaniards quickly proved in the 1530s.

Economic Geography

Economic geography is concerned with the diverse ways in which people earn a living, and how the goods and services they produce are expressed and organized spatially. In Chapter 3, we identified four sets of productive activities as the major components of the *spatial economy*: (1) primary activities (agriculture, mining, and other extractive industries); (2) secondary activities (manufacturing); (3) tertiary activities (services); and (4) quaternary activities (information and de-cision-making). That discussion empha-sized the three latter activities, particularly their geographic dimensions in highly de-veloped countries. Here the focus is on *agriculture*, the dominant livelihood in South America (and in the world).

The map of South America's **agricul-tural systems** is displayed in Figure 5-2 and reveals an unusual pattern: here com-mercial (for-profit) and subsistence (mini-mum-life-sustaining) farming exist side by side to a greater degree than in any other realm (where one of the two always geo-graphically dominates the other). This, of course, does not represent a planned "bal-ance" between the two. Rather, it reflects the continent's deep internal cultural and economic divisions, on which this chapter elaborates.

The commercial agricultural side of South America is expressed in (1) a huge cattle-ranching zone near the coasts (Cate-gory 9) that stretches southwest from northeastern Brazil to Patagonia; (2) Ar-gentina's wheat-raising Pampa (Category 3), which is comparable to the U.S. Great Plains; (3) a Corn Belt-type crop and live-stock zone (Category 2) in northeastern Argentina, Uruguay, southern Brazil, and south-central Chile; (4) a number of sea-board tropical plantation strips (Category 6) located in eastern Brazil, the Guianas, Venezuela, Colombia, and Peru; and (5) a Mediterranean-type agricultural zone (Cat-egory 5) in central Chile.

In stark contrast to these commercial systems, subsistence farming covers the rest of South America's arable land. Prim-itive shifting cultivation (Category 8) oc-

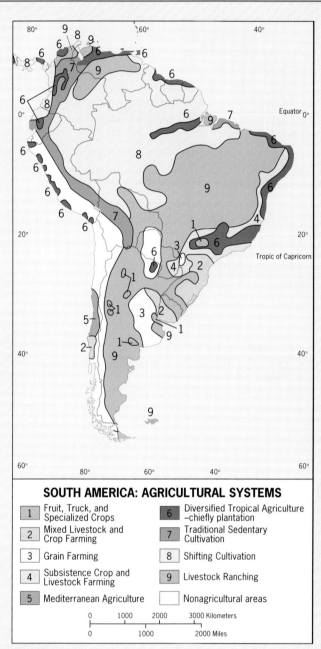

SOUTH AMERICA: AGRICULTURAL SYSTEMS

1	Fruit, Truck, and Specialized Crops	**6**	Diversified Tropical Agriculture –chiefly plantation
2	Mixed Livestock and Crop Farming	**7**	Traditional Sedentary Cultivation
3	Grain Farming	**8**	Shifting Cultivation
4	Subsistence Crop and Livestock Farming	**9**	Livestock Ranching
5	Mediterranean Agriculture		Nonagricultural areas

0   1000   2000   3000 Kilometers

0   1000   2000 Miles

FIGURE 5-2

curs in the rainforested Amazon Basin and its hilly perimeter; rudimentary sedentary cultivation (Category 7) dominates the Andean plateau country from Colombia in the north to the Bolivian *Altiplano* in the south; and a ribbon of mixed subsistence farming (Category 4) courses through most of east-central Brazil between the coastal plantation and interior grazing zones.

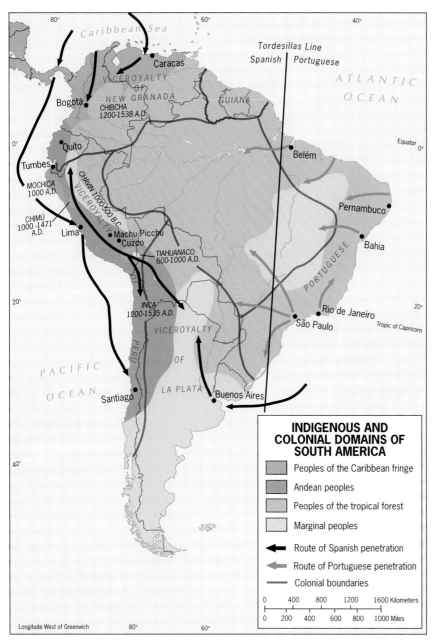

FIGURE 5-3

The Inca Empire, which had risen to greatness so rapidly, disintegrated abruptly under the impact of the Spanish invaders. Perhaps the swiftness of its development contributed to its fatal weakness, but the empire was also ripe for internal revolt in the early sixteenth century as the vaunted administrative framework was increasingly stressed by the Incas' push into the northern Andes. At any rate, apart from spectacular ruins such as those at Peru's Machu Picchu, the empire left behind social values that have remained a part of Amerindian life in the Andes to this day and still contribute to fundamental divisions between the Hispanic and Amerindian populations in this part of South America. For

example, the Inca state language, Quechua, was so firmly rooted that it is still spoken by millions of Amerindians living in the highlands of Peru, Ecuador, and Bolivia.

The Iberian Invaders

In South America as in Middle America, the location of indigenous peoples determined to a considerable extent the direction of the thrusts of European invasion. The Incas, like Mexico's Maya and Aztec peoples, had accumulated gold and silver in their headquarters, possessed productive

FIGURE 5-4

farmlands, and constituted a ready labor force. Not long after the 1521 defeat of the Aztecs, the Spanish conquerors crossed the Panamanian isthmus and sailed southward along the continent's northwestern coast. On his first journey in 1527, Francisco Pizarro heard of the existence of the Inca Empire and soon withdrew to Spain to organize its overthrow. Four years later he returned to the Peruvian coast with 183 men and two dozen horses at a time when the Incas were preoccupied with problems of royal succession and strife in the northern provinces. The events that followed are well known, and in 1533 the party rode victorious into Cuzco.

At first, the Spaniards kept intact the Incan imperial structure by permitting the crowning of an emperor who was in fact under their control. But soon the land- and gold-hungry invaders were fighting among themselves, and the

breakdown of the old order began. The new order that eventually emerged in western South America placed the indigenous peoples in serfdom to the Spaniards. Great haciendas were formed by **land alienation** (the takeover of former Amerindian lands), taxes were instituted, and a forced-labor system was introduced to maximize the profits of exploitation. As in Middle America, most of the Spanish invaders had little status in Spain's feudal society, but they brought with them the values that prevailed in Iberia: land meant power and prestige, gold and silver meant wealth.

Lima, the west coast headquarters of the Spanish conquerors, was founded by Pizarro in 1535, about 375 miles (600 km) northwest of the Andean center of Cuzco. Before long Lima was one of the richest cities in the world, its wealth based on the exploitation of vast Andean silver deposits. The city quickly became the capital of the viceroyalty of Peru, as the authorities in Spain integrated the new possession into their colonial empire (Fig. 5-3). Subsequently, when Colombia and Venezuela came under Spanish control and, later on, when Spanish settlement began to expand in the coastlands of the Rio de la Plata estuary in what is now Argentina and Uruguay, two additional viceroyalties were added to the map: New Granada in the north and La Plata in the south.

Meanwhile, another vanguard of the Iberian invasion was penetrating the east-central part of the continent, the coastlands of present-day Brazil. This area had become a Portuguese sphere of influence, because Spain and Portugal had agreed in the Treaty of Tordesillas (1494) to recognize a north-south line (drawn by Pope Alexander VI) 370 leagues west of the Cape Verde Islands as the boundary between their New World spheres of influence. This border ran approximately along the meridian of 50°W longitude, thereby cutting off a sizeable triangle of eastern South America for Portugal's exploitation (Fig. 5-3).

A brief look at the political map of South America (Fig. 5-4), however, shows that the 1494 treaty did not succeed in limiting Portuguese colonial territory to the east of the agreed-upon 50th meridian. True, the boundaries between Brazil and its northern and southern coastal neighbors (French Guiana and Uruguay) both reach the ocean near 50°W; but then the Brazilian boundaries bend far inland to include almost the entire Amazon Basin as well as a good part of the Paraná-Paraguay Basin in the south. Thus Brazil came to be only slightly smaller in territorial size than all the other South American countries combined—and today also accounts for just under half the total population of the continent. The successful, enormous westward thrust was the work of many Brazilian elements, including missionaries in search of converts and explorers in search of quick wealth. No group, however, did more to achieve this penetration than the so-called *Paulistas*, the settlers of São Paulo who needed Amerindian slave labor to run their highly profitable plantations.

The Africans

As Figure 5-3 shows, the Spaniards initially got very much the better of the territorial partitioning of South America—not just quantitatively but qualitatively as well. There were no rich Amerindian states to be conquered and looted east of the Andes, and no productive agricultural land was under cultivation. The comparatively few eastern Amerindians constituted no usable labor force. It has been estimated that the entire area of present-day Brazil was inhabited by no more than 1 million aboriginal people.

When the Portuguese finally began to develop their New World territory, they turned to the same lucrative activity that their Spanish rivals had pursued in the Caribbean—the plantation cultivation of sugar for the European market.

From the Field Notes

"The city of Salvador, capital of Bahia State, is one of the most historic places in the Americas. Great wealth was accumulated here during the sixteenth and seventeenth centuries, and behind the bluff above the port on All Saints Bay there arose one of the world's most magnificent cities. The sixteenth-century cathedral, still standing, is one of the world's finest of the time. Hundreds of churches and public buildings and numerous elaborate villas mark the era of whaling, estate agriculture based on slave labor from Benin in West Africa, and political power. After Brazil's capital was transferred to Rio de Janeiro and the whaling industry collapsed, Salvador declined. When we first studied it in 1982, the old town lay in ruins, and a new, modern city was growing nearer the mouth of the bay. But United Nations World Heritage designation and international assistance enabled Bahia to begin restoration, and in 1995 old Salvador was one vast construction site."

And they, too, found their labor force in the same source region, as millions of Africans were brought in slavery to the tropical Brazilian coast north of Rio de Janeiro. Not surprisingly, Brazil now has South America's largest black population, which is still heavily concentrated in the country's poverty-stricken northeastern states. Today, with the overall population of Brazil at 163 million, 6 percent (almost 10 million) of the people are "pure" black and another 39 percent (63 million) are of mixed African, white, and Amerindian ancestry. Africans, then, definitely constitute the third major immigration of foreign peoples into South America (see Figs. 3-4 and 7-5).

Longstanding Isolation

Despite their common cultural heritage (at least insofar as their European-mestizo population is concerned), their adjacent location on the same continent, their common language, and their shared national problems, the countries that arose out of South America's Spanish viceroyalties have existed in a considerable degree of **isolation** from one another. Distance and physiographic barriers reinforced this separation, and the realm's major population agglomerations still adhere to the coast, mainly the eastern and northern coasts (Fig. I-9). Compared with other world realms, South America may be described as underpopulated not just in terms of its modest total for a landmass of its size but also because of the resources available for or awaiting development. This continent never drew as large an immigrant European population as did North America. The Iberian Peninsula could not provide the numbers of people that western and northwestern Europe did, and Spanish colonial policy had a restrictive effect on the European inflow.

The New World viceroyalties existed primarily for the purpose of extracting riches and filling Spanish coffers. In Iberia there was little interest in developing the American lands for their own sake. Only after those who had made Spanish and Portuguese America their permanent home and who had a stake there rebelled against Iberian authority did things begin to change, and then very slowly. South America was saddled with the values, economic outlook, and social attitudes of eighteenth-century Iberia—not the best tradition from which to begin the task of forging modern nation-states.

Independence

Certain isolating factors had their effect even during the wars for independence. Spanish military strength was always concentrated at Lima, and those territories that lay farthest from this center of power—Argentina and Chile—were the first to establish their independence from Spain (in 1816 and 1818, respectively). In the north, Simón Bolívar led the burgeoning independence movement, and in 1824 two decisive military defeats there spelled the end of Spanish power in South America. Thus, in little more than a decade, the Hispanic countries had fought themselves free. But this joint struggle did not produce unity: no fewer than nine countries emerged from the three former viceroyalties.

It is not difficult to understand why this fragmentation took place. With the Andes intervening between Argentina and Chile and the Atacama Desert between Chile and Peru, overland distances seem even greater than they really were, and these obstacles to contact proved quite effective. Hence the countries of South America began to grow apart—a separation process sometimes heightened by uneasy frontiers. Friction and even wars have been frequent, and dozens of boundary disputes remain unresolved to this day.

Brazil attained independence from Portugal at about the same time that the Spanish possessions in South America were struggling to end overseas domination, but the sequence of events was quite different. In Brazil, too, there

had been revolts against Portuguese control. But the early 1800s, instead of witnessing a decline in Portuguese authority, actually brought the Portuguese government (headed by Prince Regent Dom João) from Lisbon to Rio de Janeiro! Incredibly, Brazil in 1808 was suddenly elevated from colonial status to the seat of empire. It owed its new position to Napoleon's threat to overrun Portugal, which was then allied with the British. Although many expected that a new era of Brazilian progress and development would follow, things soon fell apart. By 1821, with the Napoleonic threat removed, Portugal decided to demote Brazil to its former colonial status. But Dom Pedro, Dom João's son and the new regent, defied his father and led a successful struggle for Brazilian independence.

The postindependence relationships between Brazil and its Spanish-influenced neighbors have been similar to the limited interactions among the individual Hispanic republics themselves. Distance, physical barriers, and cultural contrasts served to inhibit positive contact and interaction. Thus Brazil's orientation toward Europe, like that of the other republics, remained stronger than its involvement with the countries on its own continent. (As noted earlier, only in the 1990s are the countries of the South American realm finally beginning to recognize the mutual advantages of increasing cooperation.)

CULTURAL FRAGMENTATION

When we speak of the "orientation" or "interaction" of South American countries, it is important to keep in mind just who does the orienting and interacting, for there is a tendency to generalize the complexities of these countries away. The fragmentation of colonial South America into 10 individual republics and the nature of their subsequent relationships was the work of a small minority of the people in each country. People of African descent in Brazil, at the time of independence, had little or no voice in the course of events; the Amerindians in Peru, numerically a vast majority, could only watch as their European conquerors struggled with each other for supremacy. It would not even be true to say that the European minorities *in toto* governed and made policy: the wealthy, landholding, upper-class elite determined the posture of the state.

So complex and heterogeneous are the societies and cultures of Middle and South America that practically every generalization has to be qualified. Take the one so frequently used—the term *Latin* America. Apart from the obvious exceptions that can be read from the map, such as Jamaica, Guyana, and Suriname, which are clearly not "Latin" countries, it may be improper even to identify some of the Spanish-influenced republics as "Latin" in their cultural milieu. Certainly the white, wealthy upper

FIGURE 5-5

classes are of Latin European stock, and they have the most influence at home and are most visible abroad; they are the politicians and the businesspeople, the writers and the artists. Their cultural environment is made up of the Spanish (and Portuguese) language, the Roman Catholic church, and the picturesque Mediterranean-style architecture of Middle and South America's cities and towns. These things provide them with a common bond and strong ties to Iberian Europe. However, in the mountains and villages of Ecuador, Peru, and Bolivia are millions of people to whom the Spanish language is still alien, to whom the white people's religion is another unpopular element of acculturation, and to whom decorous Spanish styles of architecture are meaningless when a decent roof and a solid floor are still unattainable luxuries.

South America, then, is a continent of *plural societies*,

where Amerindians of different cultures, Europeans from Iberia and elsewhere, blacks from western tropical Africa, and Asians from India, Japan, and Indonesia have produced a cultural and economic kaleidoscope of almost endless variety. Certainly, calling this human spatial mosaic "Latin" America is not very useful. Is there a more meaningful approach to a regional generalization that would better represent and differentiate the continent's cultural and economic spheres? John Augelli, who also developed the Rimland-Mainland concept for Middle America, made such an attempt. His map (Fig. 5-5) shows that five **South American culture spheres**—internal cultural regions—blanket the realm. This scheme is quite useful if one keeps in mind that South America is undergoing economic change in certain areas today and that these culture spheres are generalized and subject to further modification as well.

Tropical-Plantation Region

The first culture sphere, the *tropical-plantation* region, in many ways resembles the Middle American Rimland. It consists of several separated areas, of which the largest lies along the northeastern Brazilian coast, with four others along the Atlantic and Caribbean coastlands of northern South America. Location, soils, and tropical climates favored plantation crops, especially sugar. The fact that the aboriginal population was small led to the introduction of millions of African slave laborers, whose descendants today continue to dominate the racial makeup and strongly influence the cultural expression of these areas. The plantation economy later failed, soils became exhausted, slavery was abolished, and the people were largely reduced to poverty and subsistence—socioeconomic conditions that now dominate most of the region mapped as tropical-plantation.

European-Commercial Region

The second region on Augelli's map, identified as *European-commercial*, is perhaps the most truly "Latin" part of South America. Argentina and Uruguay, each with a population that is at least 85 percent "pure" European and with a strong Hispanic cultural imprint, constitute the bulk of the European-commercial region. Two other areas also lie within it: most of Brazil's core area and the heartland of central Chile. Southern Brazil shares the temperate grasslands of the Argentine Pampa and Uruguay (see Fig. I-8), and this area is important as a zone of livestock raising as well as corn production. Middle Chile is an old Spanish settlement zone, home to the approximately one-sixth of the Chilean population who claim pure Spanish ancestry. Here, in an area of Mediterranean climate (Fig. I-7), cattle and sheep pastoralism as well as mixed farming are practiced. In general, then, the European-commercial region is economically more advanced than the rest of the continent. A commercial economy rather than a subsistence way of life prevails, living standards are better, literacy rates are higher, transportation networks are superior, and (as Augelli has pointed out) the overall development of this region surpasses that of parts of Europe itself.

Amerind-Subsistence Region

The third region is identified as *Amerind-subsistence*, and it forms an elongated zone along the length of the central Andes from southern Colombia to northern Chile/northwestern Argentina, closely approximating the area occupied by the old Inca Empire. The feudal socioeconomic structure that was established here by the Spanish conquerors still survives. The Amerindian population forms a large, landless peonage, living by subsistence or by working on haciendas (or smaller farms called minifundias) far removed from the Spanish culture that forms the primary force in the national life of their country. This region includes some of South America's poorest areas, and what commercial activity there is tends to be in the hands of whites or mestizos. The Amerindian heirs to the Inca Empire live, often precariously, at high elevations (as much as 12,500 feet/3800 m) in the Andes. Poor soils, uncertain water supplies, high winds, and bitter cold make farming a constantly difficult proposition.

Mestizo-Transitional Region

The fourth region, *mestizo-transitional*, surrounds the Amerind-subsistence region, covering coastal and interior Peru and Ecuador, most of Colombia and Venezuela, much of Paraguay, and large parts of Argentina, Chile, and especially Brazil (including the developing portions of the Amazon Basin). This is the zone of mixture between European and Amerindian—or African in coastal sections of Venezuela, Colombia, and northeastern Brazil. The map thus reminds us that such countries as Bolivia, Peru, and Ecuador are dominantly Amerindian and mestizo. In Ecuador, for instance, these two groups make up about 80 percent of the total population, and less than 10 percent can be classified as European. The term *transitional* has an economic connotation, too, because (as Augelli puts it) this region "tends to be less commercial than the European sphere but less subsistent in orientation than dominantly [Amerindian] areas."

Undifferentiated Region

The fifth region on the map is marked as *undifferentiated* because its characteristics are hard to classify. Some of the Amerindian peoples in the interior of the Amazon Basin had remained almost completely isolated from the momentous changes in South America since the days of Columbus. Although remoteness and lack of change are still two notable aspects of this subregion, the ongoing development of Amazonia is reversing that situation. The most remote Amazon backlands as well as the Chilean and Argentinean southwest also are sparsely populated and exhibit only very limited economic development; inaccessibility continues to contribute to the unchanging nature of these areas.

The framework of the five culture spheres just described is necessarily a broad generalization of a rather complex geographic reality. But it does underscore the diversity of South America's peoples, cultures, and economies—a quality that lately has been overshadowed by a still-emerging, continentwide spirit of international cooperation.

ECONOMIC INTEGRATION

All across South America today, the separatism that has for so long marked international relations is giving way as countries discover the benefits of forging new partnerships with one another. The distribution of the realm's resources has always presented opportunities for building interregional **complementarities**, and the resurgence of civilian, democratic rule is finally making that development possible on a continental scale. With mutually advantageous trade the catalyst, international cooperation is blossoming in every sphere. Periodic flareups of boundary disputes now rarely escalate into prolonged conflicts—witness the defusing of the 1995 armed confrontation between Ecuador and Peru or the peaceful settlement of more than 20 territorial disputes between Chile and Argentina since 1990. Cross-border rail, road, and pipeline projects, stalled for years, are suddenly proceeding apace. In southern South America, five formerly contentious nations are working to complete the *Hidrovia*, a system of river locks that will soon open most of the huge Paraná-Paraguay Basin to barge transport. And investments now flow freely from one country to another, particularly in the agricultural sector: Brazilian farmers already raise most of Paraguay's soybean crop and are developing huge tracts of fertile land in Brazil-adjoining portions of Bolivia and Uruguay.

Speaking at Miami's 1994 Summit of the Americas, Argentine President Carlos Menem captured the essence of the attitude that now prevails among the continent's leaders: "Free trade can solve an infinity of problems that have dragged us down for more than a century." To ensure the institutionalization of that guiding principle, South American governments are actively pursuing several avenues of regional economic integration. In 1996, South America's republics were affiliated with no less than four major trading blocs:

- **Mercosur** Launched in 1995, this Southern Cone Common Market established a free trade zone and customs union linking Brazil, Argentina, Uruguay, and Paraguay.
- **Andean Group** Formed as the Andean Pact in 1969 but restarted in 1995 as a far more effective customs union with common tariffs for imports, its members are Venezuela, Colombia, Peru, Ecuador, and Bolivia.
- **Group of Three (G-3)** Launched in 1995, this free trade agreement involves Mexico, Venezuela, and Colombia, and aims to phase out all internal tariffs by 2005.
- **North American Free Trade Agreement (NAFTA)** Launched by the United States, Canada, and Mexico in 1994, NAFTA is expanding into South America to include Chile; it aims to phase out all internal tariffs

and complete the formation of a free trade zone no later than 2010.

These organizations represent only an intermediate step toward a much grander goal: the creation of the *Free Trade Area of the Americas (FTAA)*, a single market of more than 800 million consumers that would extend from the Arctic shores of Alaska and Canada to Chile's southernmost Cape Horn. This hemisphere-spanning free trade zone—by far the world's largest—is to be formed by combining 24 existing multinational agreements (including the four listed above) into one. Negotiators will undoubtedly face some formidable obstacles in assembling this gigantic trading bloc; as of 1996, its architects were optimistic that FTAA would meet its target completion date of 2005.

URBANIZATION

As in most other developing countries, residents of South America's republics are leaving the land and moving to the cities. This **urbanization** process intensified sharply after mid-century, and it persists so strongly that South America's urban population percentage now ranks with those of Europe and the United States. In 1925, about one-third of South America's peoples lived in cities and towns, and as recently as 1950, the percentage was just over 40. But by 1975 the continentwide figure had surpassed 60 percent, and today 73 percent of the South American population resides in urban areas. Of course, these percentages mask the actual numbers, which are even more dramatic. Between 1925 and 1950, the realm's towns and cities grew by about 40 million residents as the urbanized percentage rose from 33 to 42. Then between 1950 and 1975, more than 125 million people crowded into the teeming metropolitan areas—more than *three times* the total for the previous quarter-century—and an additional 110-plus million since 1975 have swelled the continental total to its current 241 million.

Nowhere is South America's population of 330 million (25 percent greater than that of the United States) increasing faster than in the towns and cities. We usually assume that the populations of rural areas grow more rapidly than urban areas because farm families traditionally have more children than city dwellers. Yet, overall, the urban population of South America has grown annually by about 5 percent since 1950, while the rural areas increased yearly by less than 2 percent. These figures underscore the dimensions of the **rural-to-urban migration** from the countryside to the cities—still another migration process that affects modernizing societies throughout the world.

The generalized spatial pattern of South America's urban transformation is seen in Figure 5-6, which shows a *cartogram* of the continent's population.* Here we see not only the realm's 13 countries in population-space relative to each other, but also the proportionate sizes of individual large cities within their total national populations.

Regionally, the realm's highest urban percentages are found in southern South America. Today in Argentina, Chile, and Uruguay, at least 85 percent of the people reside in cities and towns (statistically on a par with Europe's most urbanized countries). Ranking next among the most heavily urbanized populations is Brazil at 77 percent (2 percent higher than the United States), now growing so rapidly that it records an increase of about 1 percent each year. That may not sound very significant, but once again percentages mask the real numbers; in a population of 163 million, 1 percent indicates that an additional 1,630,000 annually are somehow squeezed into Brazil's badly over-crowded urban centers.

The next highest group of countries, averaging 73 per-cent urban, border the Caribbean in the north. Here, Venezuela leads with 84 percent, and Colombia follows with 68 percent. Not surprisingly, the Amerind-subsistence-dominated Andean countries constitute the realm's least urbanized zone. Peru, because of its strong Spanish im-print, is the region's leader by far, with 70 percent of its population agglomerated in towns and cities. Ecuador, Bo-livia, and transitional Paraguay, on the other hand, lag well behind the rest of the continent, with all three exhibiting urban proportions in the 51 to 60 percent range. Figure 5-6 also tells us a great deal about the relative positions of major metropolises in their countries. Six of them—Bra-zil's São Paulo and Rio de Janeiro, Argentina's Buenos Aires, Peru's Lima, Colombia's Bogatá, and Chile's San-tiago—rank among the world's megacities (whose popu-lations exceed 5 million).

In South America, as in Middle America, Africa, and Asia, people are attracted to the cities and driven from the poverty of the rural areas. Both *pull* and *push factors* are at work. Rural land reform has been slow in coming, and every year tens of thousands of farmers simply give up and leave, seeing little or no possibility for economic advance-ment. The cities lure them because they are perceived to provide opportunity—the chance to earn a regular wage. Visions of education for their children, better medical care, upward social mobility, and the excitement of life in a big city draw hordes to places such as São Paulo and Caracas. Road and rail connections continue to improve, so that ac-cess is easier and exploratory visits can be made. City-based radio and television stations beckon the listener to the locale where the action is.

But the actual move can be traumatic. Cities in devel-oping countries are surrounded and often invaded by

*A *cartogram* is a specially transformed map in which countries and cities are represented in proportion to their populations. Those containing large numbers are "blown up" in population-space, while those containing les-ser numbers are "shrunk" in size accordingly.

From the Field Notes

"Look one way between the towering skyscrapers of water-front Rio de Janeiro, and you see one of the world's most famous beaches. Look landward, and luxury gives way to deprivation. Rio's *favelas* are vast and poor. Still, this photo-graph gives rise to hope. Ten years ago this was a miser-able shantytown whose residents lived in shacks. Most of those dwellings have been replaced by more permanent housing; sanitary conditions may not be adequate, but there is electricity, walls are bricked, windows and doors the norm. Some *favela* residents have done well enough to build sub-stantial homes. That is the hope for the future: that stability and rising incomes among *favela* residents will transform shantytown into suburb."

squalid slums, and this is where the uncertain urban im-migrant most often finds a first (and frequently permanent) abode in a makeshift shack without even the most basic amenities and sanitary facilities. Many move in with rela-tives who have already made the transition but whose dwelling can hardly absorb yet another family. And un-employment is persistently high, often exceeding 25 per-cent of the available labor force. Jobs for unskilled workers are hard to find and pay minimal wages. But the people still come, the overcrowding in the shantytowns worsens, and the threat of epidemic-scale disease (and other disas-ters) rises. It has been estimated that in-migration accounts for over 50 percent of urban growth in some developing countries, and in South America the poorest urban popu-lations exhibit high natural growth rates as well.

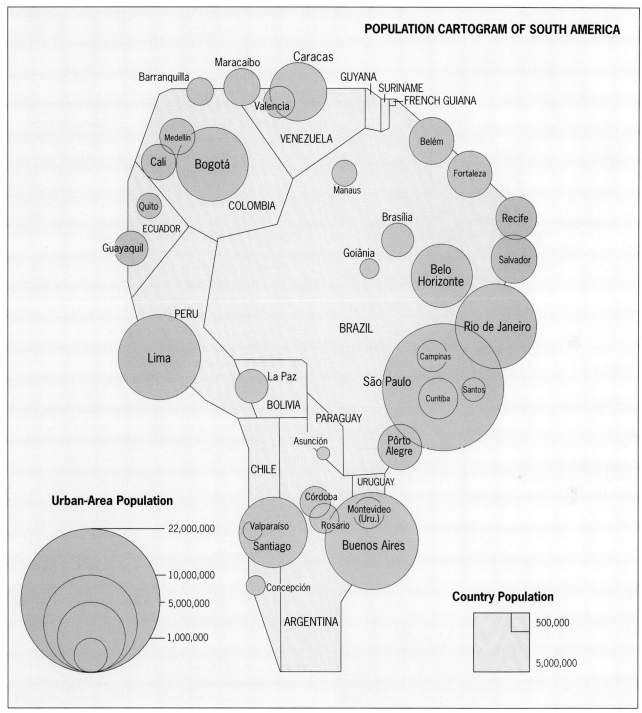

POPULATION CARTOGRAM OF SOUTH AMERICA

FIGURE 5-6

The "Latin" American City

Although the urban experience has been varied in the South and Middle American realms because of diverse historical, cultural, and economic influences, there are many common threads that have prompted geographers to search for useful generalizations. One is the model of the intraurban spatial structure of the **"Latin" American city** proposed by Ernst Griffin and Larry Ford (Fig. 5-7), which may have wider application to cities in developing countries in other geographic realms.

The basic spatial framework of city structure, which blends traditional elements of South and Middle American culture with modernization forces now reshaping the urban scene, is a composite of radial sectors and concentric zones. Anchoring the model is the thriving CBD, which, like its

A GENERALIZED MODEL OF LATIN AMERICAN CITY STRUCTURE

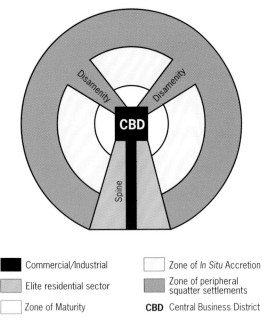

Commercial/Industrial

Elite residential sector

Zone of Maturity

Zone of *In Situ* Accretion

Zone of peripheral squatter settlements

CBD Central Business District

FIGURE 5-7

European counterpart, remains the primary business, employment, and entertainment focus of the surrounding metropolitan agglomeration. The landscape of the CBD contains many modern high-rise buildings but also mirrors its colonial beginnings. As shown in Figure 4-2, when the Spanish colonizers laid out their New World cities, they created a central square, or *plaza*, dominated by a church and flanked by imposing government buildings (see photo below). Lima's *Plaza de Armas*, Bogotá's *Plaza Bolívar*,

and Buenos Aires' *Plaza de Mayo* are classic examples. Early in the South American city's development, the plaza formed the hub and focus of the city, surrounded by shopping streets and arcades. Eventually the city outgrew its old center and new commercial districts formed elsewhere within the CBD, leaving the plaza to serve as a largely ceremonial link with the past.

Emanating outward from the urban core along the city's most prestigious axis is the commercial *spine*, which is surrounded by the *elite residential sector* (shown in green in Fig. 5-7). This widening corridor is essentially an extension of the CBD, featuring offices, shopping, high-quality housing for the upper and upper-middle classes, restaurants, theaters, and such amenities as parks, zoos, and golf courses that give way to wealthy adjoining suburbs, which carry the elite sector beyond the city limits.

The three remaining concentric zones are home to the less fortunate residents of the city (who comprise the great majority of the urban population), with income levels and housing quality decreasing markedly as distance from the city center increases. The *zone of maturity* in the inner city contains the best housing outside the spine sector, attracting middle-class urbanites who invest sufficiently to keep their solidly built but aging dwellings from deteriorating. The adjacent *zone of in situ accretion* is one of much more modest housing interspersed with unkempt areas, representing a transition from inner-ring affluence to outer-ring poverty. The residential density of this zone is usually quite high, reflecting the uneven assimilation of its occupants into the social and economic fabric of the city.

The outermost *zone of peripheral squatter settlements* is home to the impoverished and unskilled hordes that have recently migrated to the city from rural areas. Although housing in this ring consists mainly of teeming, high-density shantytowns, residents here are surprisingly optimistic about finding work and eventually bettering their living conditions. Many achieve their first economic success by becoming part of the "informal sector." This primitive form of capitalism is now common in many devel-

From the Field Notes

"We walked from the *Plaza de Armas*, Lima's most impressive plaza, to the *Plaza de San Martin*, the city's other major square. The *Plaza de Armas* is best seen from street level, with its eighteenth-century cathedral and wooden balconies and porticos. The *Plaza de San Martin*, on the other hand, seemed to demand a more panoramic view, but it was not easy to get permission to step out on a rooftop; the political situation was tense and armed military were much in evidence. Eventually we were allowed to view the *Plaza de San Martin* (he is the subject of the equestrian statue left of center) from atop the Columbus Hotel. One of Lima's major avenues, the *Avenida Nicolas de Pierola*, can be seen to reach the plaza in the left background; in the "Latin" American city, major square and major thoroughfare tend to intersect this way."

oping countries, taking place beyond the control—and especially the taxation—of the government. Participants are unlicensed sellers of homemade goods (such as arts and crafts, clothing, and food specialties) and services (auto repair, odd jobs, and the like). Their willingness to engage in this hard work has transformed many a shantytown into a beehive of activity that can propel resourceful residents toward a middle-class existence elsewhere in the metropolis.

A final structural element of many South American cities is the *disamenity sector* that contains relatively unchanging slums, known as *barrios* or *favelas*. The worst of these poverty-stricken areas often include sizeable numbers of people who are so poor that they are forced literally to live in the streets. Thus the realm's cities present sharp contrasts between poverty and affluence, squalor and comfort—a harsh juxtaposition all too frequently observed in the cityscape (see photo on p. 244).

As with all geographic models, this particular construct can be criticized for certain shortcomings. Among the criticisms directed at this generalization are the need for a sharper sectoring pattern and a stronger time element involving a multiple-stage approach. On the other hand, many scholars view this work as a successful abstraction not only of the "Latin" American city but also of cities in any developing country.

REGIONS OF THE SOUTH AMERICAN REALM

We turn now to the countries of South America and the principal characteristics of their regional geography. In general terms it is possible to group the realm's countries into regional units (Fig. 5-4) because several have qualities in common. To begin with, Brazil by itself constitutes a geographic region because it contains half the continent's land and people. The northernmost Caribbean countries, Venezuela and Colombia, form a second regional unit that includes neighboring Guyana, Suriname, and French Guiana. On the basis of their Amerindian cultural heritage, Andean physiography, and modern populations, the western republics of Ecuador, Peru, and Bolivia constitute a third regional entity. Finally, in the south, Argentina, Uruguay, and Chile have a common regional identity as the realm's mid-latitude, most strongly Europeanized states; Paraguay, a transitional country between the Andean west and these southern republics, is also included with the latter.

❖ BRAZIL: GIANT OF SOUTH AMERICA

By any measure, Brazil is South America's giant. It is so large that it has common boundaries with all the realm's other countries except Ecuador and Chile (Fig. 5-4). Its tropical and subtropical environments range from the equatorial rainforest of the Amazon Basin to the humid temperate climate of the far south. Territorially, Brazil occu-

Major Cities of the Realm

Cities	Population* (in millions)
Asunción, Paraguay	0.6
Belo Horizonte, Brazil	4.2
Bogotá, Colombia	6.0
Buenos Aires, Argentina	11.2
Caracas, Venezuela	3.1
Guayaquil, Ecuador	1.9
La Paz, Bolivia	1.3
Lima, Peru	7.9
Manaus, Brazil	1.3
Montevideo, Uruguay	1.4
Quito, Ecuador	1.3
Rio de Janeiro, Brazil	10.1
Santiago, Chile	5.3
São Paulo, Brazil	21.7

*Based on 1997 Estimates

pies just under 50 percent of South America and is exceeded on the global stage only by Russia, Canada, China, and the United States. In population size, it also accounts for just below half the continental total and is again the world's fifth-largest country (surpassed only by China, India, the United States, and Indonesia). The Brazilian economy is now the ninth largest on Earth, and its modern industrial base the tenth largest; the country is likely to continue rising in the international ranks and is poised to become a world force in the twenty-first century.

Population Patterns

Brazil's population of 163 million is as diverse as that of the United States. In a pattern quite familiar in the Americas, the indigenous inhabitants of the country were decimated following the European invasion (no more than 275,000 Amerindians now survive deep in the Amazonian interior). Africans came in great numbers, too, and today there are about 10 million blacks in Brazil. Significantly, however, there was also much racial mixing, and nearly 65 million Brazilians have combined European, African, and minor Amerindian ancestries. The remaining 90 million—now barely in the majority at 55 percent—are mainly of European origin, the descendants of immigrants from Portugal, Italy, Germany, and Eastern Europe. The complexion of the population was further diversified by the arrival of Lebanese and Syrians—and, since 1908, the growth of

From the Field Notes

"Near the waterfront in Belém lie the now-deteriorating, once-elegant streets of the colonial city built at the mouth of the Amazon. Narrow, cobblestoned, flanked by tiled frontages and arched entrances, this area evinces the time of Dutch and Portuguese hegemony here. Mapping the functions and services here, we recorded the enormous diversity of activities ranging from carpentry shops to storefront restaurants and from bakeries to clothing stores. Dilapidated sidewalks were crowded with shoppers, workers, and people looking for jobs (some newly arrived, attracted by perceived employment opportunities in the growing city). The diversity of population in this and other tropical South American cities reflects the varied background of the region's peoples and the wide hinterland from which these urban magnets have drawn their inhabitants."

the largest community of ethnic Japanese outside Japan. The latter, now numbering over a million, are concentrated in São Paulo State where they have risen to the topmost ranks of Brazil's easygoing immigrant society as farmers, urban professionals, business leaders—and traders with Japan.

Brazilian society, to a greater degree than is true elsewhere in the Americas, has made progress in dealing with its racial divisions. To be sure, blacks are still the least advantaged among the country's major population groups, and community leaders continue to complain about discrimination. But ethnic mixing in Brazil is so pervasive that hardly any group is unaffected, and official census statistics about "blacks" and "Europeans" are meaningless. What the Brazilians do have is a true national culture, expressed in an adherence to the Catholic faith (this is the world's largest Roman Catholic country), in the universal use of a modified form of Portuguese as the common language, and in a set of lifestyles in which vivid colors, distinctive music, and a growing national consciousness and pride are fundamental ingredients.

One of the most noteworthy trends in Brazil's population geography has been the recent lowering of the country's high rate of natural increase. Over the past quarter-century, the birth rate has been halved (to a manageable 1.7 percent) and the average number of children born to a Brazilian woman plummeted from 4.4 in 1980 to 2.1 in 1995. Surprisingly, this impressive slowdown in population increase took place in the absence of an active birth control policy by the Brazilian government—and in direct contradiction to the teachings of the decreasingly influential Catholic church. Demographers believe that the decline is the result of three factors: a rapid spread in contraceptive usage, the negative influence on family formation of a period of economic stagnation in the 1980s and early 1990s, and newly widespread access to television (which is reshaping the attitudes and aspirations of the Brazilian people).

Development Prospects

Brazil is richly endowed with mineral resources, including enormous iron and aluminum ore reserves, extensive tin and manganese deposits, and major oil- and gasfields (Fig. 5-8). Other significant energy developments involve massive new hydroelectric facilities and the successful substitution of sugarcane-based alcohol (gasohol) for gasoline—allowing more than half of all cars to use this fuel instead of costly imported petroleum. Besides these natural endowments, Brazil's soils sustain a bountiful agricultural output that makes the country a global leader in the production and export of soybeans, orange juice concentrate, and coffee. Commercial agriculture, in fact, is now the fastest growing economic sector, propelled by mechanization and the opening of a major new farming frontier in the fertile grasslands of southwestern Brazil.

As noted earlier, Brazil ranks tenth among the world's industrial powers. Much of the momentum for this continuing development was unleashed in the early 1990s after the government opened the country's long-protected industries to international competition and foreign investment. These new policies are proving to be effective because productivity has risen by a third since 1990 as Brazilian manufactures attain world-class quality. Industrial goods also surpassed (in value) the agricultural sector during the mid-1990s to become the nation's export leader. Trade with Argentina (now Brazil's primary trading partner) is at the forefront, a relationship expected to intensify as Mercosur reaches full operation.

Brazil's current economic performance notwithstanding, the country's overall development during the past three decades has been marked by a roller coaster ride of boom and bust. The prospects for keeping economic growth on a more stable footing are heightened by the continent's new free trade environment, but serious internal problems of social and regional inequality must be mitigated as well. In a realm exhibiting the world's sharpest division between affluence and poverty, Brazil stands out with the widest income gap in South America. Today, the richest 10 percent of the population own two-thirds of the land and control more than half of the country's wealth. At the other end of the economic spectrum, poverty has increased by an astonishing 45 percent since 1980. With the poorest 20 percent of the population now living in the most squalid conditions to be found anywhere on the planet, too little is being done to improve their housing, education, and health standards (over half of all Brazilians suffer from chronic malnutrition). Even though considerable regional development is taking place, the benefits are overwhelmingly channeled toward already affluent individuals and powerful corporations, thereby further widening that enormous gulf between modern and traditional Brazil. Unless the country's resources and access to its opportunities are redistributed more equitably, Brazil's forward progress will increasingly be clouded by the negative consequences of uneven development.

Subregions

Brazil is a federal republic consisting of 26 States and the federal district of the capital, Brasília (Fig. 5-8). As in the United States, and for similar reasons, the smallest States lie in the northeast and the larger ones farther west. The State of Amazonas is the biggest, twice the size of Texas with over 600,000 square miles (1.6 million sq km), but its huge area contains less than 4 million people. At the other extreme, Rio de Janeiro State (17,000 square miles/ 44,000

sq km) on the southeastern coast has a population of more than 15 million. But the State with the largest population by far is São Paulo, now exceeding 35 million and growing phenomenally.

Although Brazil is about as large as the 48 contiguous United States, it does not possess the clear physiographic regionalism familiar to us. It has no Andes Mountains, and the countryside consists mainly of plateau surfaces and low hills (Fig. 5-1). Even the lower-lying Amazon Basin, which covers almost 60 percent of the country, is not entirely a plain: between the tributaries of the great river lie low but extensive tablelands. In the southeast, the plateau surface of the Brazilian Highlands rises slowly eastward toward the Atlantic coast. Along the shoreline itself, there is a steep escarpment leading from plateau surface to sea level that leaves almost no living space along its base. Thus, with relatively little coastal plainland, Brazil is indeed fortunate to possess several very good natural harbors. Given the country's physiographic ambiguity, the six subregions discussed next have no absolute or even generally accepted boundaries. In Figure 5-8, those boundaries have been drawn to coincide with the borders of States, making identifications easier.

The Northeast

The Northeast was Brazil's source area, its culture hearth. The plantation economy took root here at an early date, attracting Portuguese planters who soon imported the country's largest group of African slaves to work in the sugar fields. But the ample and dependable rainfall that occurs along the coast soon gives way to lower and more variable patterns in the interior, and today most of the Northeast is poverty-stricken, hunger-afflicted, overpopulated, and subject to devastating droughts. In fact, extended dry periods occur so regularly that this region (home to about 50 million people) has become known as the *Polygon of Drought*. Ironically, there is ample groundwater, but the region's huge population of peasant farmers cannot afford to drill the necessary wells. Without such irrigation, human and animal pressures on the land are combining to deplete the natural vegetation and thereby hasten the encroachment of aridity.

Sugar still remains the chief crop along the moister coast, and livestock herding prevails in the drier inland backcountry—the *sertão*—with beef cattle in the better grazing zones and goats elsewhere. The comparatively small areas of successful commercial agriculture—cotton in Rio Grande do Norte, sisal in Paraíba, and sugarcane in Pernambuco—stand in sharp contrast to the countless patches of shifting subsistence agriculture located nearby. The key is water, and when agribusiness or government can afford the investment, the land can become highly productive. For example, the lower São Francisco valley in

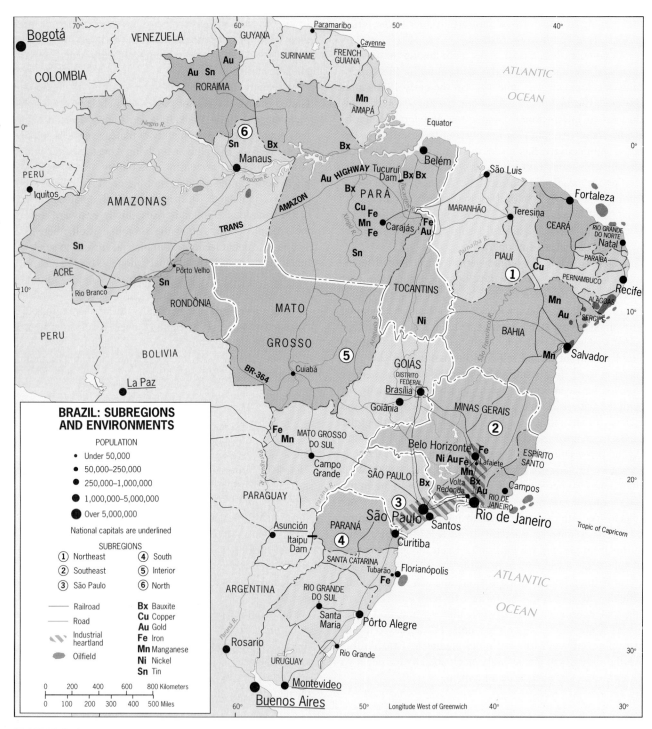

FIGURE 5-8

Bahia State has in recent years been transformed into South America's fastest growing area of irrigated farming, specializing in tropical fruits.

The Northeast today is Brazil's great contradiction. In the cities of Recife and Salvador, the architecture still bears the imprint of an earlier age of wealth, but thousands of peasants without hope—driven from the land by deterio-

rating conditions—constantly arrive to expand the usual surrounding shantytowns. As yet, few of the generalizations about emerging Brazil apply here, but there are some bright spots beyond irrigated-agriculture projects. A huge petrochemical complex was built near Salvador in the late 1980s, creating thousands of jobs, luring further (including foreign) investment, and boosting the region's industrial

■ AMONG THE REALM'S GREAT CITIES . . .

Rio de Janeiro

Say "South America" and the first image most people conjure up is Sugar Loaf Mountain, Rio de Janeiro's landmark sentinel that guards the entrance to beautiful Guanabara Bay. Nicknamed the "magnificent city" because of its breathtaking natural setting, Rio became Brazil's capital in 1763, a function that endured for two centuries until the federal government shifted its headquarters to interior Brasília in 1960. Rio de Janeiro's primacy also suffered another blow in the late 1950s: São Paulo, its hated urban rival 250 miles (400 km) to the southwest, surpassed Rio to become Brazil's largest city—a gap that has been widening ever since. Although these events triggered economic decline, Rio (10.1 million) remains a major entrepôt, air travel and tourist hub, and center of international business. As a focus of cultural life, however, Rio de Janeiro is unquestionably Brazil's leader, and the *cariocas* (as Rio's residents call themselves) continue to set the national pace with their entertainment industries, universities, museums, and libraries.

On the darker side, this city's reputation is increasingly tarnished by the widening abyss between Rio's affluent and poor populations—symbolizing inequities that rank among the world's most extreme. All great cities experience problems, and Rio de Janeiro is currently bedeviled by the drug use and crime waves emanating from its most desperate hillside *favelas* (slums) that continue to grow explosively.

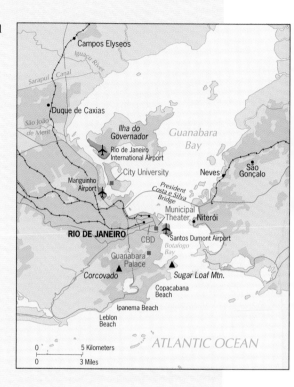

base in general. Tourism has begun to boom along the entire Northeast coast, whose new beachside resorts are attracting hordes of vacationing Europeans (flying time to Western Europe is only about seven hours). Most promising of all is the ongoing development of Fortaleza and surrounding Ceará State: a thriving new clothing industry has already put the city on the global map, and the social programs of local government have improved education, health care, and employment so substantially that they are cited as a model for all of urban Brazil.

The Southeast

In the State of Bahia, a transition occurs toward the south. The coastal escarpment becomes more prominent, the plateau higher, and the terrain more varied; annual rainfall increases and is seasonally more dependable. The Southeast has been modern Brazil's core area, with its major cities and leading population clusters. Gold first drew many thousands of settlers, and other mineral finds also contributed to the influx (Rio de Janeiro itself served as the terminus of the "Gold Trail"), but ultimately the region's agricultural possibilities ensured its stabilization and growth. The mining towns needed food, and prices for foodstuffs were high; farming was stimulated, and many farmers came (with their slave workers) to Minas Gerais State to till the soil. Eventually a pastoral industry came to predominate, with large herds of beef cattle grazing on planted pastures.

The post–World War II era brought another mineral age to the region, based not on gold or diamonds but on the iron ores around Lafaiete and the manganese and limestone carried to the steelmaking complex at Volta Redonda (Fig. 5-8). Iron mining has now become one of Brazil's leading economic activities, and since 1990 the country has been in the topmost ranks of the world's producers and exporters. Minas Gerais produces 77 percent of Brazil's iron ore and more than 40 percent of its steel. From this base, industrial diversification in the Southeast has proceeded apace, with the metallurgical center of Belo Horizonte paving the way. That booming city has also become the endpoint of a rapidly developing manufacturing corridor that stretches 300 miles (500 km) southwest to metropolitan São Paulo—which in the mid-1990s experienced growth rates on a par with the highest in the Pacific Rim. At the

■ AMONG THE REALM'S GREAT CITIES...

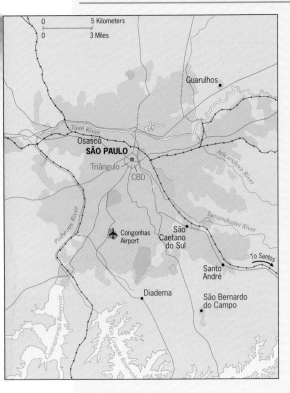

São Paulo

São Paulo, which lies on a plateau 30 miles (50 km) inland from its Atlantic outport (Santos), possesses no obvious locational advantages. Yet here on this site we find the third largest metropolis on Earth, whose population has multiplied so uncontrollably that São Paulo has doubled in size (from 11 to 22 million) since 1977.

Founded in 1554 as a Jesuit mission, modern São Paulo was built on the nineteenth-century coffee boom. It has since grown steadily as both an agricultural processing center (soybeans, sugar, and orange juice besides coffee) and a manufacturing complex (that now accounts for more than half of all of Brazil's industrial jobs), and has become Brazil's primary focus of commercial and financial activity. Today São Paulo's distinctive, high-rise CBD is the very symbol of urban South America and attracts the continent's largest flow of foreign investment.

The huge recent growth of this megacity has been accompanied by massive problems of overcrowding, pollution, and congestion. Staggering poverty—on a par with Mexico City's—is the most pressing crisis as the ever-expanding outer ring of shantytowns tightens its grip on the burgeoning metropolis. With a doubling time of only two decades, where will Greater São Paulo be 20 years from today?

same time, with an economy equivalent in size to Chile's in 1994, Minas Gerais has surpassed Rio de Janeiro to become Brazil's second largest State in industrial output.

São Paulo

The leading industrial producer is the State of São Paulo, long a part of the Southeast subregion, which today claims its own regional identity because it is the focus of ongoing Brazilian development. This economic-geographic powerhouse accounts for nearly 50 percent of the country's gross domestic product and more than 60 percent of its manufacturing activity. With an economy that now exceeds Argentina's in overall size, burgeoning São Paulo State has been a magnet for migrants (especially from the Northeast). Not surprisingly, it is growing prodigiously and already contains almost one-quarter of the Brazilian population.

The wealth of São Paulo State was built on its coffee plantations (known as *fazendas*), and up to a generation ago coffee accounted for at least half of Brazil's exports. Although it remains an important crop (Brazil is still the world's leading producer), coffee today claims only about 3 percent of the exports and has been eclipsed by other farm commodities. One of them is orange juice concentrate; São Paulo State now produces more than double the annual output of Florida, thanks to a climate free of winter freezes, ultramodern processing plants, and a fleet of specially equipped tankers to ship the concentrate to markets in the United States—and more than 50 other countries ranging from Japan to Russia. Another leading pursuit is soybeans; since 1990 Brazil has ranked first or second among the world's producers. The central and western portions of São Paulo State have benefited the most from this recent agricultural expansion, which also includes the raising of sugarcane for gasohol production, cotton, and rubber trees (Brazil is expected to become a global leader again within a decade).

Matching this agricultural prowess is the State's industrial strength. The revenues derived from the coffee plantations provided the necessary investment capital, ores from Minas Gerais supplied the vital raw materials, hydroelectric power from the slopes of the coastal escarpment produced the needed energy, the nearby outport of Santos facilitated access to the ocean, and immigration from Europe, Japan, and rural Brazil contributed the increasingly skilled labor force. As the capacity of the domestic market grew, the advantages of central location and agglomeration secured São Paulo's primacy—a position reinforced by the subsequent growth of international trade. As a result, metropolitan São Paulo is now the country's (and South America's) leading manufacturing complex, with Brazil's foremost industry—automobiles—based in the city of São Paulo's southern suburbs. As Brazilian and foreign automakers continue to pour in investments, that industry has become one of the world's largest, recently overtaking the output of Italy, the former republics of the Soviet Union, and the United Kingdom.

The South

Three States, whose combined population approaches 25 million, make up the southernmost Brazilian subregion: Paraná, Santa Catarina, and Rio Grande do Sul. The contribution of recent European immigrants to the agricultural development of southern Brazil, as in neighboring Uruguay and Argentina, has been considerable. Many came not to the coffee country of São Paulo State but to the available lands farther south. Here the newcomers introduced their advanced farming methods to several areas. Portuguese rice farmers clustered in the valleys of Rio Grande do Sul; the Germans, specialists in raising grain and cattle, occupied the somewhat higher areas to the north and in Santa Catarina; and the Italians selected the highest slopes, where they established thriving vineyards. These fertile lands proved highly productive, and with expanding markets in the large urban areas to the north, this tristate region became Brazil's most affluent corner.

European-style standards of living match the diverse European heritage that is reflected in the towns and countryside, where the German and Italian languages are still spoken alongside Portuguese. In places, zealous preservation of Old World cultural life has led to hostility against non-European Brazilians. Many communities now actively discourage poor, job-seeking migrants from the north by offering to pay return bus fares or even blocking their household-goods-laden vehicles. Moreover, several extremist groups have recently arisen to openly espouse the secession of the South from Brazil. The largest, calling itself ''The South Is My Country,'' promotes the creation of a *Republic of the Pampas* and already displays its intended flag. Accusing the separatist movement of subversion and racism, State governments have moved to thwart its progress. They claim the central issue is an economic one, complaining to Brasília that the South receives only about a 60 percent return on the federal taxes it pays. But with the polls in the mid-1990s showing that at least one-third of the population is sympathetic to the idea of secession, officialdom may well have a bigger struggle on its

No country seems to be immune to devolutionary forces. Canada has its Quebec, Britain its Scotland, France its Corsica, Spain its Basques and Catalans. Now Brazil faces a separatist movement in its three southernmost States, which constitute the region called The South in Figure 5-8, and correspond to the zone where what we have mapped as "European-commercial" extends into Brazil (Fig. 5-5). With more than 22 million people, an area the size of France, and a strong economy, Brazil's southerners, who argue that they are overtaxed and underrepresented in Brasília, have generated a drive to secede from mismanaged Brazil. Proponents of secession call their proposed country the *Republic of the Pampas*. In this photograph, taken in 1995, a leader of the secessionists, Irton Marx, holds the flag he hopes will some day wave over his nation.

hands in preventing the Brazilian South from succumbing to the devolutionary forces that are weakening nation-states around the globe.

Economic development in the South is not limited to the agricultural sector. Coal from Santa Catarina and Rio Grande do Sul, shipped north to the steel plants of Minas Gerais, was a crucial element in Brazil's industrial emergence. Local manufacturing is growing as well, especially in Pôrto Alegre and Tubarão, where South America's single largest steelmaking facility opened in 1983. More recently, an internationally significant center of the computer software industry was established in Florianópolis, the island city and State capital just off Santa Catarina's coast. Known as *Tecnópolis*, this complex continues to grow by capitalizing on its amenities, skilled labor force, superior air-travel and global communications linkages, and government and private-sector incentives to support new companies. Only the sparsely populated inland portion of the South has lagged behind the rest of the region, despite the new presence of massive Itaipu Dam in western Paraná State (the dam supplies São Paulo and surroundings with huge amounts of hydroelectric power, but did not stimulate the growth of its local area). Further to the north, however, interior Brazil has begun to awaken with a vengeance.

The Interior

The subregion of Interior Brazil—constituted by the three States of Goiás, Mato Grosso, and Mato Grosso do Sul—is also known as the Central-West (or *Centro-Oeste*). This is the subregion that Brazil's developers have long sought to make a part of the country's productive heartland, and in the 1950s the new capital of Brasília was deliberately situated on its margins (Fig. 5-8). By locating the new capital city in the untapped wilderness 400 miles (650 km) inland from its predecessor, Rio de Janeiro, the nation's leaders dramatically signaled the opening of Brazil's age of development and a new thrust toward the west. Brasília is also noteworthy in another regard because it represents what political geographers call a **forward capital**. There are times when a state will relocate its capital to a sensitive area, perhaps near a peripheral zone under dispute with an unfriendly neighbor, in part to confirm its determination to sustain its position in that contested zone. (An example was Pakistan's decision in the 1960s to move its capital from coastal Karachi to northern Islamabad, a "forward" position near the disputed territory of Kashmir lying in the border zone with India.) Brasília does not lie near a contested area, but Brazil's interior was an internal frontier to be conquered by a growing nation. In that drive, the new capital occupies a decidedly forward position.

Despite the subsequent growth of Brasília since 1960 (from zero to 1.9 million inhabitants today), it was not until the 1990s that the tristate Interior became economically integrated with the rest of Brazil. The catalyst was the exploitation of the vast *cerrado*—the fertile savannas that blanket the Central-West and make it one of the world's most promising agricultural frontiers (more than 80 percent of its arable land still awaits development). As with the U.S. Great Plains, the flat terrain of the *cerrado* is one of its main advantages because it facilitates the large-scale mechanization of farming with a minimal labor force. Another advantage is rainfall, which is more prevalent than in the Great Plains or Argentine Pampa (Fig. I-6). The leading crop is soybeans, whose output per acre here exceeds even that of the U.S. Corn Belt. Other grains and cotton are also expanding across the farmscape of the Centro-Oeste, but the current pace of regional development is inhibited by a major accessibility problem that is now in the process of being resolved. Unlike the Great Plains and Pampa, the growth of an efficient transportation network did not accompany the opening of this farming frontier. As a result, the Interior's products must travel along poor roads and intermittent railroads to reach the markets and ports of the Atlantic seaboard. Today several projects are under way to alleviate these bottlenecks, particularly the construction of a new, privately financed railway (the Ferronorte, popularly called the Soy Railroad) that will link São Paulo-Santos with the Cuiabá area—and eventually, perhaps, the central Amazon Basin (Fig. 5-8).

The Amazonian North

The largest and most rapidly developing subregion—whose population has tripled since 1980 to about 15 million—is also the most remote from the core of Brazilian settlement: the seven States of the Amazon Basin (Fig. 5-8). This was the scene of the great rubber boom at the turn of the century, when the wild rubber trees in the *selvas* (tropical rainforests) produced huge profits and the central Amazon city of Manaus enjoyed a brief period of wealth and splendor. The rubber boom ended in 1910, however, when plantations elsewhere (notably in Southeast Asia) began to produce rubber more cheaply, efficiently, and accessibly. For most of the seven decades that followed, Amazonia was a stagnant hinterland, but all that changed dramatically during the 1980s. New development has been spawned throughout this awakening region, which has become the scene of the world's largest migration into virgin territory as over 200,000 new settlers arrive every year. Most of this activity is occurring south of the Amazon River, in the tablelands between the major waterways and along the Basin's wide rim.

Two ongoing development schemes are especially worth noting because they are quintessential expressions of what is occurring here. The first is the *Grande Carajás Project* in eastern Pará State, a huge, multifaceted scheme centered on one of the world's largest known deposits of iron ore in the Serra dos Carajás hills (Fig. 5-8). In addition to a vast mining complex, other new construction here includes the Tucuruí Dam on the nearby Tocantins River and a 535-mile (850-km) railroad to the Atlantic port of São Luis. This ambitious development project also emphasizes further mineral exploitation (including bauxite, manganese, and copper), cattle raising, crop farming, and forestry. What is taking place is an application of the **growth-pole concept**. The term is almost self-explanatory: a growth pole is a location where a set of industries, given a start, will expand and generate widening ripples of development in the surrounding area. Growth plans for Grande Carajás are aimed at just such a stimulation of its immediate hinterland, and if they come to fruition the scheme could one day cover one-sixth of all Amazonia.

Understandably, tens of thousands of settlers have descended on this part of the Amazon Basin in recent years.

The end of the equatorial rainforest: deforestation in Amazonas State. In the mid-1990s, between 20,000 and 30,000 square miles (52,000 and 78,000 sq km) of forest fell annually in Brazil to chainsaw and fire. As roads and settlements penetrate the once-remote rainforest areas, and miners wash the soil away, streams are poisoned and clogged with sediment and ecosystems suffer irreversible damage. This photograph, however, could have been taken in areas other than Brazil's Amazonas. In Equatorial Africa and Southeast Asia (notably Borneo) the dreadful scenario is the same.

Those seeking business opportunities have been in the vanguard, but they have been followed by masses of lower-income laborers and peasant farmers in search of jobs and land ownership. The initial stage of this colossal enterprise has boosted the fortunes of many towns, most notably the spectacular revival of Manaus northwest of Carajás. An entire industrial complex has now emerged in the free trade zone that adjoins this city, specializing in the production of TVs, VCRs, and myriad other electronic goods; the advantages for manufacturers here are cheap assembly costs, outstanding air-freight operations, and central location with respect to all of northern South America (further enhanced by the 1995 completion of a new highway north to Caracas, Venezuela). Many problems have also arisen as the tide of pioneers rolled across central Amazonia, none more tragic than the plight of the Yanomami, one of the hemisphere's last remaining groups of indigenous people practicing a Stone Age culture. Following the 1987 discovery of gold deposits beneath their lands in Roraima State north of Manaus, the Yanomami homeland was overrun by thousands of claim-stakers who triggered many violent confrontations. The government was slow to intervene, and by the time it acted to halt the bloodshed the Yanomami's fragile way of life had fallen victim to the disruption and pollution of its habitat, the ravages of newly introduced diseases, and the inevitable modernization forced upon it by contact with other Brazilians.

The second leading development scheme, known as the *Polonoroeste Plan*, is located about 500 miles (800 km) to the southwest of Grande Carajás in the 1,500-mile-long Highway BR-364 corridor that parallels the Bolivian border and connects the western Brazilian towns of Cuiabá, Pôrto Velho, and Rio Branco (Fig. 5-8). Although the government had planned for the penetration of western Amazonia to proceed via the east-west Trans-Amazon Highway, the migrants of the 1980s and 1990s have preferred

to follow BR-364 and settle within the Basin's southwestern rim zone, mostly in Rondônia State. Agriculture has been the dominant activity here, attracting affluent growers and ranchers from the South, plantation workers from São Paulo and Paraná States displaced by agrarian mechanization, and subsistence farmers from all over the country. The common denominator has been the quest for land; bitter conflicts continue to break out between peasants and landholders as the Brazilian government cautiously pursues the persistent and volatile issue of land reform.

The usual pattern of settlement in this part of the Brazilian North is something like this. As main and branch highways are cut through the wilderness, settlers (enticed by cheap land) follow and move out laterally to clear spaces for farming. Crops, usually corn or upland rice, are planted, but within three years the heavy equatorial-zone rains leach out soil nutrients and accelerate surface erosion. As soil fertility declines, pasture grasses are then planted, and the plot of land is soon sold to cattle ranchers (most are associated with big agribusiness corporations based in Rio or São Paulo). The peasant farmers then move on to newly opened areas, clear more land for planting, and the cycle repeats itself. As long as open spaces remain, this is a profitable pursuit for all parties, but it assures the widespread establishment of a low-grade land use that will ultimately concentrate most of the earnings in the hands of large landowners. It also requires the clearing of enormous stands of tropical woodland: during the mid-1990s, more than 20,000 square miles (52,000 sq km) of rainforest were disappearing *annually* in Amazonia—an area almost the size of Pennsylvania. Moreover, Amazon woodland destruction has much wider implications: it accounts for over half of all tropical deforestation, an environmental crisis of global proportions whose future consequences could negatively affect most forms of terrestrial life (see box p. 228).

❖ THE NORTH: CARIBBEAN SOUTH AMERICA

As another look at Figure 5-5 confirms, the countries of South America's northern tier have something in common besides their coastal location: each has a tropical plantation zone signifying early European plantation development, the in-migration of black laborers, and the absorption of this African element into the population matrix. Not only did myriad black workers arrive; many thousands of South Asians also came to the realm's northern shores as contract laborers. The pattern is familiar: in the absence of large local labor sources, the colonists turned to slavery and indentured workers to serve their lucrative plantations. Between Spanish Venezuela and Colombia on the one hand and the non-Hispanic "three Guianas" on the other, there is this difference: in Venezuela and Colombia, the population center of gravity soon moved into the interior and the plantation phase was followed by a totally different economy. In the Guianas, coastal settlement and the plantation economy still predominate (see box titled "The Guianas").

Venezuela and Colombia have what the Guianas lack (Fig. 5-9): much larger territories and populations, more varied natural environments, and greater economic opportunities. Each also has a share of the Andes Mountains (Colombia's being larger), and each produces oil from reserves that rank among the world's major deposits.

Venezuela

Much of what is important in Venezuela is concentrated in the northern and western parts of the country, where the Venezuelan Highlands form the eastern spur of the north end of the Andes system. Most of Venezuela's 23 million people are concentrated in these uplands, which include the capital of Caracas, its early rival Valencia, the thriving commercial and industrial center of Barquisimeto, and San Cristóbal near the Colombian border.

THE GUIANAS

On the north coast of South America, three small countries lie east of Venezuela and next to northernmost Brazil (Fig. 5-9). In "Latin" South America, these territories are anomalies of a sort: their colonial heritage is Western European, and formerly they were known as British, Dutch, and French Guiana—and called the three *Guianas*. But two of them are independent now: Guyana, Venezuela's neighbor, and Suriname, the Dutch-influenced country in the middle. The easternmost territory, French Guiana, continues under colonial rule.

None of the three countries has a population over 1 million. Guyana is the largest with 920,000, Suriname has 445,000 inhabitants, and French Guiana a mere 170,000. Culturally and spatially, patterns here are Caribbean: peoples of South Asian and African descent are in the majority, with whites forming a small minority. In Guyana, Asians make up just over half the population, blacks and mixed Africans 43 percent, and the others, including Europeans, are tiny minorities. In Suriname, the ethnic picture is even more complicated, for the colonists brought not only Indians from South Asia (now 37 percent of the population) and blacks (31 percent) but also Indonesians (15 percent) to the territory in servitude. The remaining 10 percent of the Surinamese population consists of a separate category of black communities, peopled by descendants of African slaves (known as Bush blacks) who escaped from the coastal plantations and fled into the forests of the interior. French Guiana is the most European of the three countries. About three-quarters of its small population speak French, and the ethnic majority (66 percent) is of mixed African, Asian, and European ancestry—the *creoles*.

French Guiana, an overseas *département* of France, is the least developed of the three Guianas. There is a small shrimping industry, and some lumber is exported; however, food must be brought in from abroad. Suriname progressed more rapidly after independence in 1975, but unremitting political instability soon ensued. More than 100,000 residents emigrated to the Netherlands, and by the mid-1990s the stagnating economy neared bankruptcy. Nonetheless, Suriname not only remains self-sufficient in rice production but also exports surpluses of that staple crop along with bananas and shrimp. In most years, the leading Surinamese income earner is bauxite, mined in a zone across the middle of the country. British-influenced Guyana became independent in 1966 amid internal conflict that basically pitted people of African origins against people of Asian descent. Such problems have continued to plague this country, in which the great majority of the people live in small villages near the coast. Bauxite, gold from mineral-rich western Guyana, and seafood earn most of the country's annual revenues.

Today a major environmental crisis faces all three countries. With the economies of the two largest Guianas starved for funds, companies from Asia's Pacific Rim are offering to pay handsomely to exploit the most precious natural resource—the vast tracts of tropical rainforest that blanket the landscape beyond the populated coastal strips. Concessions have already been granted, but the battle has been joined by conservationists near and far, who are well aware that these same loggers have ravaged similar woodlands and caused the destructive erosion of hills and valleys throughout Southeast Asia (see top photo p. 24).

The Venezuelan Highlands are flanked by the Maracaibo Lowlands and Lake Maracaibo to the northwest and by a large region of savanna country called the *llanos* in the Orinoco Basin to the south and east. The Maracaibo Lowland, once a disease-infested, sparsely peopled coastland, is today one of the world's leading oil-producing areas. Much of the oil is drawn from reserves that lie beneath the shallow waters of the lake itself. Actually, Lake Maracaibo is a misnomer, for the ''lake'' is open to the ocean and is in fact a gulf with a very narrow entry. Venezuela's second city, Maracaibo, is the focus of the oil industry that transformed the Venezuelan economy in the 1970s. After 1985, however, the country suffered the consequences of the global oil depression. Because Venezuela had borrowed heavily against its future oil revenues, it now faces the same economic difficulties as Mexico in paying off its foreign debt—a burden increased and extended by a 1994 monetary devaluation and political crisis that produced a severe recession and widespread social unrest.

The *llanos* on the southern side of the Venezuelan Highlands and the Guiana Highlands in the country's southeast are two of those areas that contribute to South America's lingering image as ''underpopulated'' and ''awaiting development.'' Whereas the *llanos* are beginning to share in Venezuela's oil production (large reserves have been discovered here), the superior commercial agricultural potential of these savannas and of the *templada* areas (see box titled ''Altitudinal Zonation'' on p. 224) of the Guiana Highlands has yet to be realized. Economic integration of this interior zone with the rest of Venezuela has been encouraged by the discovery of rich iron ores on the northern flanks of the Guiana Highlands southwest of Ciudad Guayana. Local railroads connect with the Orinoco River, and from there ores are shipped directly to foreign markets.

Colombia

Colombia also has a vast area of *llanos* covering about 60 percent of the country. As in Venezuela, this is comparatively empty land far from the national core and much less productive than it could be. Eastern Colombia consists of the upper basins of major tributaries of the Orinoco and Amazon rivers; it lies partly under savanna and partly under rainforest vegetation. Since most of the country east of the Andes is controlled by guerrillas (who have driven out

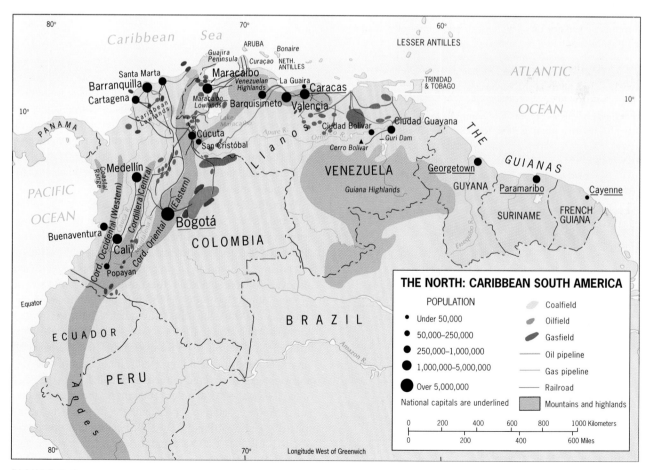

FIGURE 5-9

at least 500,000 settlers since 1990), it will be a long time before any part of eastern Colombia matches the Andean zone of the country or even the Caribbean Lowlands of the north. These two regions contain the vast majority of Colombians, and here lie the major cities and productive areas.

Western Colombia is dominated by mountains, but there is a regularity to this terrain. For the most part there are four parallel ranges, generally aligned north-south and separated by prominent valleys. The westernmost of these ranges is a coastal belt, less continuous and lower than the other three, which constitute the true Andean mountain chain of Colombia. In this country the Andes separate into three ranges: the Eastern (Oriental), Central, and Western (Occidental) Cordilleras (Fig. 5-9). The valleys between these Cordilleras open northward onto the Caribbean Lowland, where the three important Colombian ports of Barranquilla, Cartagena, and Santa Marta are located.

Colombia's population of 39 million consists of more than a dozen separate clusters, some of them in the Caribbean Lowland, others in the inter-Cordilleran valleys, and still others in the intermontane basins within the Cordilleras themselves. The capital city, Bogotá, was founded in one of the major basins in this Andean range at an elevation of 8,500 feet (2,700 m). For centuries the Magdalena valley between the Cordillera Oriental and the Cordillera Central was a crucial link in the key cross-continental route that began in Argentina and ended at the port of Cartagena, and Bogotá benefited from its position along this artery. Today, the Magdalena valley is still Colombia's leading transport corridor, but Bogotá's connections with much of the rest of Colombia still remain tenuous. Nonetheless, the capital centers a major metropolis that is home to 6 million.

Colombia's physiographic variety is matched by its demographic diversity. In the Andean south, it has a major cluster of Amerindian inhabitants, and here the country begins to resemble its southern neighbors. In the northern coastlands, traces remain of the plantation period and the African population it brought. Bogotá is a great Hispanic cultural headquarters, whose influence extends beyond the country's borders. In the Cauca valley between the Cordillera Central and Cordillera Occidental, the city of Cali is the commercial focus for a farming district that raises sugarcane, tobacco, and cattle. Farther north, the Cauca River flows through a region comprising the provinces of Antioquia and Caldas, whose urban focus is the textile manufacturing city of Medellín but whose greater importance to the Colombian economy has been the production of coffee.

With its extensive *templada* areas along the Andean slopes, Colombia in the mid-1990s was the world's second largest exporter of coffee (Brazil ranks first). But that is about to change because high production costs, low international prices, and worm infestations of coffee beans are combining to drive thousands of coffee growers out of business. The repercussions in this farming region are certain to be enormous as the coffee-production infrastructure is dismantled, and a social upheaval may well accompany the difficult transition to new livelihoods.

With the decline of coffee, oil and coal are becoming Colombia's two leading exports. Petroleum deposits have been worked for decades in the north, adjacent to the (still-disputed) Venezuelan border where the Maracaibo oilfields extend into Colombia. During the mid-1980s, significant discoveries were made in the northeastern province of Arauca (about 200 miles [320 km] south of Lake Maracaibo), a 500-mile pipeline was quickly built across the northern Andes to the Caribbean coast, and Colombia was turned into a net oil exporter overnight. In 1991, a much larger oil deposit (including a substantial quantity of natural gas) was discovered in the Cusiana district at the edge of the *llanos* about 100 miles (160 km) northeast of Bogotá, and proved to contain reserves that rank it among the hemisphere's leading oilfields. By 1995, petroleum production reached full operation here and the expectation is that Cusiana will become a major supplier to the gigantic U.S. market. Colombia also contains South America's richest

From the Field Notes

"From Venezuela all the way around western South America, the mountains of the Andes system create barriers to inland movement—and not only for transportation. Interior isolation and fragmentation lead to political problems as well. Remote from authority, rebels use the mountainous interior to stage revolts; interior peoples of Amerindian descent are almost literally walled off from the Europeans on the coast. Efforts to link the coast to the interior have produced some remarkable solutions. On this excursion, we found this cablecar system near El Cojo, that lifted us 3000 feet over the Cordillera del Litoral to Caracas. Along the way, we stopped at villages where we were joined by people, goats, and chickens; on the way down, we could barely see Caracas through the acrid smog that filled its valley."

coal deposits in its northernmost extremity around the base of the Guajira Peninsula. The mining complex at Cerrejón is one of the largest anywhere, and when the area's second major bituminous coal mine is completed, the country should rank as the world's third leading exporter (mainly to Europe).

Despite their great potential, the country's expanding energy industries are threatened by guerrilla organizations that force the military to tightly guard production facilities and pipelines. Insurgencies elsewhere on the continent have faded or disappeared, but Colombia's has stubbornly persisted for more than 30 years and still controls most of the territory to the east of the Andes. That part of Colombia is also well known as a major source of illicit drugs (see box titled "The Geography of Cocaine"), and guerrilla groups have in recent years abandoned political ideology

THE GEOGRAPHY OF COCAINE

Any geographical discussion of northwestern South America today must take note of one of its most widespread activities: the production of illegal narcotics. Of the enormous flow of illicit drugs that enter the United States each year, the most widely used substance undoubtedly is cocaine—all of which comes from South America, mainly Bolivia, Peru, and Colombia (the sources of more than 75 percent of the total world supply). Within these three countries, cocaine annually brings in billions of (U.S.) dollars and "employs" thousands of workers, constituting an industry that functions as a powerful economic force. The wealthy drug barons who operate the industry have accumulated considerable power through bribery of politicians, threats of terrorism, and alliances with guerrilla movements in outlying zones beyond governmental control. The cocaine industry itself is structured within a tightly organized network of territories that encompass the various stages of this drug's production.

The first stage of cocaine production is the extraction of coca paste from the coca plant, a raw-material-oriented activity that is located near the areas where the plant is grown. The coca plant was domesticated in the Andes by the Incas centuries ago; millions of their descendants still chew coca leaves for stimulation and brew them into coca tea, the area's leading beverage. The leading zone of coca-plant cultivation is along the eastern slopes of the Andes and adjacent tropical lowlands in Bolivia, Peru, and the western margins of Brazil's Amazonia. Today, three main areas dominate in the growing of coca leaves for narcotic production (Fig. 5-10): Bolivia's Chaparé district (in the marginal Amazon lowlands northeast of the city of Cochabamba), the Yungas Highlands (north of the Bolivian capital, La Paz), and north-central Peru's Huallaga valley (which may produce as much as half of the world's total supply). These areas thrive because they combine many favorable conditions: local environments are conducive to high leaf yields that produce more than two crops per year, and plants here have developed immunity to most diseases and insect ravages. The success of these areas has enabled their Amerindian populations to turn coca plants into a cash crop. Operations often reach the scale of plantations, thereby inducing subsistence farmers from other areas to migrate into these now-specialized coca-raising regions. In fact, profitable coca cultivation has led thousands of peasants to abandon the far less lucrative production of food crops, further reducing the capability of nutrition-poor Bolivia to feed itself. The

coca leaves harvested in these eastern Andean regions make their way to nearby centers—located at the convergence of rivers and trails—where coca paste is extracted and prepared.

The second stage of production involves refining the coca paste (about 40 percent pure cocaine) into cocaine hydrochloride (more than 90 percent pure), a lethal concentrate that is diluted with substances such as sugar or flour before being sold on the streets to consumers. Cocaine refining requires sophisticated chemicals, carefully controlled processes, and a labor force skilled in their supervision, and here Colombia has predominated. Most of this activity takes place in ultramodern processing centers in the southeastern lowlands where alliances between drug barons and guerrillas protect against the intrusion of the Bogotá government. Interior Colombia also possesses the geographic advantages of intermediate location, lying between the source areas to the south and the U.S. market to the north (a short hop for smugglers by air or sea across the Caribbean). Nonetheless, Colombia's involvement in cocaine refining is on the decline because, in the mid-1990s, the government successfully attacked a number of processing centers and captured or killed several leaders of the Medellín and Cali drug cartels. The slack has been taken up by new crime syndicates in Bolivia and particularly Peru, who are bypassing Colombia, refining their own cocaine, and dealing directly with counterparts in Mexico and elsewhere to market their product.

The final stage of production entails the distribution of cocaine to the U.S. marketplace, which depends on an efficient (but clandestine) transportation network. Most common is the use of private planes that operate directly out of remote airstrips near the refineries in places beyond effective government control. The transport of cocaine to the United States is often a two-step process, making use of such intermediate Middle American transshipment points as the Bahamas, Jamaica, Cuba, Panama, and particularly Mexico. Although South Florida is believed to be a leading port of illegal entry, other coastal points in the U.S. Southeast as well as cities along the Mexican border in the Southwest have recorded increases in cocaine trafficking since 1990. And judging by the size of recent drug seizures in Southern California, the overland route through mainland Middle America may now have become the smugglers' route of choice.

as they concentrate on highly lucrative operations as protectors and partners of the drug traffickers. This crime culture also affects the rest of the country in the form of widespread homicides, kidnappings, drug-gang shootouts, terrorist bombings, and other forms of violence. All this violence is widely reported by the international media and taints Colombia's external image.

Both Venezuela and Colombia exhibit a pronounced clustering of often-isolated populations, share a relatively empty interior, and depend on a small number of products for the bulk of their export revenues. The majority of the people in these countries practice subsistence agriculture and labor under the social and economic inequalities common to most of Iberian America.

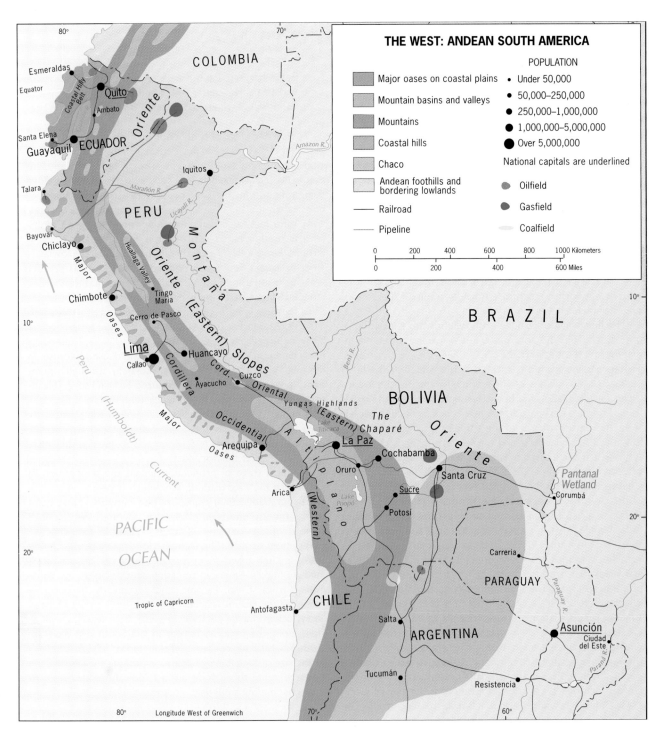

FIGURE 5-10

❖ THE WEST: ANDEAN SOUTH AMERICA

The third regional grouping of South American states—the Andean West (Fig. 5-10)—encompasses Peru, Ecuador, and Bolivia.* The map of culture spheres (Fig. 5-5) shows the Amerind-subsistence region extending along the Andes Mountains, indicating that these countries have large Amerindian components in their populations. Just over half the people of Peru are of Amerindian stock; in Bolivia the figure is 47 percent, and in Ecuador 40 percent. However, all these percentages are only approximate because it is often impossible to distinguish between Amerindian and ''mixed'' people of strong Amerindian character. There are, however, other similarities among these countries: their incomes are low; they are comparatively unproductive; and, unhappily, they exemplify the grinding poverty of the landless peonage.

Peru

In terms of territory as well as population (25 million), Peru is the largest of the three republics. Its half-million square

*Paraguay also exhibits many cultural characteristics of the Andean West, but is today turning southward in its economic-geographic orientation. We treat it as a transitional state and discuss it in the next regional section, the Mid-Latitude South.

miles (1.3 million sq km) divide both physiographically and culturally into three subregions: (1) the desert coast, the European-mestizo region; (2) the Andean highlands or Sierra, the Amerindian region; and (3) the eastern slopes and adjoining *montaña*, the sparsely populated Amerindian-mestizo interior (Fig. 5-10). It is symptomatic of the cultural division still prevailing in Peru that Lima, the capital whose population constitutes almost one-third of the national total, is not located centrally in a basin of the Andes but in the peripheral coastal zone.

At Lima in the heart of the desert coastal strip, the Spaniards avoided the greatest of the Amerindian empires, choosing a site a few miles inland from a suitable anchorage that became the modern outpost of Callao. From an economic point of view, the Spaniards' choice of a headquarters on the Pacific coast proved to be sound, for the coastal region has become commercially the most productive part of the country. A thriving fishing industry based on the cool productive waters of the Peru (Humboldt) Current offshore contributes significantly to the export trade. Irrigated agriculture in some 40 oases distributed all along the arid coast produces cotton, sugar, rice, vegetables, fruits, and wheat. The cotton and sugar are important export products, and raising the output of the other crops for sale in foreign markets is a major priority in Peru's current development plans.

The Andean (Sierra) region occupies about one-third of the country and contains the majority of Peru's Amerindian peoples, most of them Quechua-speaking. However, despite the size of its territory and population (nearly half of

■ AMONG THE REALM'S GREAT CITIES . . .

Lima

Even in a realm replete with primate cities, Lima (7.9 million) stands out. Here reside 32 percent of the Peruvian population—who produce over 70 percent of the country's gross domestic product, 90 percent of its collected taxes, and 98 percent of its private investments. Economically, at least, Lima *is* Peru.

Lima began as a modest oasis in a narrow coastal desert squeezed between the cold waters of the Pacific and the soaring heights of the nearby Andes. Near here the Spanish conquistadors discovered the best natural harbor on the western shoreline of South America, where they founded the port of Callao; but they built their city on a site 7 miles (11 km) inland where soils and water supplies proved

more beneficial. This city was named Lima and quickly became the Spaniards' headquarters for all their South American territories.

Since independence, Lima has continued to dominate Peruvian national life. Nonetheless, the city's population growth was manageable through the late 1970s. The past 20 years, however, have seen a disastrous doubling of the metropolitan population (though on a smaller numerical base than São Paulo's parallel growth explosion). The worst problems are localized in the peripheral squatter shantytowns that now house at least one-third of the urban population—which by South American standards are appalling (see photo p. 263).

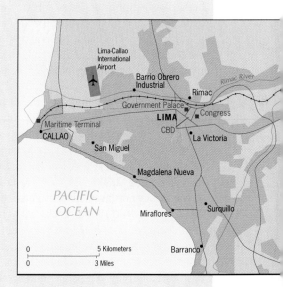

all Peruvians reside here), the political influence of this region is slight, as is its economic contribution (except for the mines). In the high valleys and intermontane basins, the Amerindian population is clustered either in isolated villages, around which people practice a precarious subsistence agriculture, or in the more favorably located and fertile areas where they are tenants, peons on white- or mestizo-owned haciendas. Most of these people never receive an adequate daily caloric intake or balanced diet of any sort; the wheat produced around Huancayo, for instance, is sent to Lima's European market and would be too expensive for the Amerindians themselves to buy. Potatoes, barley, and corn are among the subsistence crops grown here in the *tierra fría* zone, and in the *puna* of the higher basins the Amerindians graze their llamas, alpacas, cattle, and sheep (these vertical zones are discussed in the box on p. 224). The major mineral products from the Sierra are copper, silver, lead, and several other metallic minerals, with the largest mining complex centered on Cerro de Pasco.

Of Peru's three subregions, the *Oriente*, or East—the inland slopes of the Andes and the Amazon-drained, rainforest-covered *montaña*—is the most isolated. The focus of the eastern region, in fact, is Iquitos, a city that looks east rather than west and can be reached by oceangoing vessels sailing 2,300 miles (3,700 km) up the Amazon River across northern Brazil. Iquitos grew rapidly during the Amazon wild-rubber boom early in this century and then declined; now it is growing again and reflects Peruvian plans to begin development of the east. Petroleum was discovered west of Iquitos in the 1970s, and since 1977 oil has flowed through a pipeline built across the Andes to the Pacific port of Bayovar. Meanwhile, traders plying the navigable Peruvian rivers above Iquitos continue to collect such products as chicle, nuts, rubber, herbs, and special cabinet woods.

Peru today is emerging from nearly two decades of instability associated with the radical-communist *Sendero Luminoso* (Shining Path) insurgency. This guerrilla movement, which built itself by exploiting the desperate poverty of the Amerindians, established a territorial base in the central Andes. During the early 1990s, it spilled out of its mountain stronghold and for a while even threatened Lima. But the government finally responded strongly, and since 1994 the guerrillas have been in retreat—their leadership in disarray, their territorial control reduced to a few hardcore localities, and their offensive diminished to sporadic terrorist acts. With the revolutionary threat all but ended, investors and tourists have returned in force. The Peruvian economy, led by its reviving energy, mineral, and fishing industries, ranked among the world's fastest growing in 1995. The challenge to the country is to maintain such growth beyond the short-term period of recovery.

Ecuador

On the map, Ecuador, smallest of the three Andean West republics, appears to be just a corner of Peru. But that would be a misrepresentation because Ecuador possesses a full range of regional contrasts. It has a coastal belt; an Andean zone that may be narrow (under 150 miles or 250 km) but by no means of lower elevation than elsewhere; and an *Oriente*—an eastern region that is as sparsely populated and as underdeveloped as that of Peru. As in Peru, nearly half of the people of Ecuador are concentrated in the Andean intermontane basins and valleys, and the most productive region is the coastal strip. Here, however, the similarities end.

Ecuador's coastal zone consists of a belt of hills interrupted by lowland areas, of which the most important lies in the south between the hills and the Andes, drained by the Guayas River and its tributaries. The largest city and commercial center of the country (but not the capital), Guayaquil, forms the focus for this subregion. Ecuador's west coast lowland, moreover, is not a desert: it consists of fertile tropical plains not bedeviled by excessive rainfall. It is also far less Europeanized than Peru's Pacific-facing lowland because the white component of the total population of 12 million is only 10 percent. A sizeable proportion of whites are engaged in administration and hacienda ownership in the Andean interior, where most of the 25 percent of the Ecuadorians who are Amerindian also reside—and, not surprisingly, where land tenure reform is an explosive issue. Of the remaining national population, 10 percent is black and mulatto and 55 percent is mestizo (many with a stronger Amerindian ancestry).

The products of this coastal region differ as well from those of Peru. Ecuador is one of the realm's leading banana exporters; small farms owned by black and mulatto Ecuadorians and located in the north near Esmeraldas contribute to this total, as do farms on the eastern and northern margins of the Guayas Lowland. Cacao is another important lowland crop, and lucrative coffee is grown on the coastal hillsides as well as in the Andean *templada* zones. Cotton, rice, and oil palms are also cultivated, and cattle can be raised. Over the past quarter-century, petroleum production in the jungles of the *Oriente* region has reached substantial levels, and oil tops the country's export list. A trans-Andean pipeline was opened in 1972 to connect the rich interior oilfields to the port of Esmeraldas, but poor management contributed to a pollution nightmare of repeated oil spills and toxic-waste dumping along much of the 300-mile (480-km) route. A newly organized cleanup effort notwithstanding, this is likely to remain one of South America's worst environmental disasters for years to come.

Ecuador is not a poor country, and the coastal lowland has undergone vigorous development in recent years. But the Andean interior, where the white and mestizo administrators and hacienda owners are outnumbered by the Amerindians by about four or five to one, is a different story—or rather, a story similar to that of Andean regions in the other countries of western South America. Quito, the capital city, lies in one of the several highland basins in which the Andean population is clustered. Its functions remain primarily administrative, because productivity in the Andean region is too low to stimulate commercial and industrial development. The Ecuadorian Andes do differ from those of Peru in that they apparently lack major mineral deposits. So despite the completion of a railroad linking Quito to Guayaquil on the Pacific coast, the interior of Ecuador remains isolated—and, economically, comparatively inert.

Bolivia

From Ecuador southward through Peru, the Andes broaden until in Bolivia they reach a width of some 450 miles (720 km). In both the Cordillera Oriental and the Cordillera Oc-

From the Field Notes

"South of Lima we drove along the Pan American Highway to get a sense of the transition from urban center to countryside. The city had not reminded us nearly as strongly of the scarcity of water here: trees and gardens flank streets and homes in Lima's wealthier areas, but here the trees were few and the grass on the hillsides very sparse. A growing percentage of Lima's burgeoning population lives in shantytowns on the perimeter, and here people are at risk—as they are in other mushrooming cities. In 1992 a cholera epidemic, attributed to contaminated water, originated in this area and spread into tropical South and Middle America, claiming 10,000 lives and infecting about 1 million people."

cidental, peak elevations in excess of 20,000 feet (6,000 m) are recorded. Between these two great ranges lies the *Altiplano* proper (Fig. 5-10; as noted earlier, an *altiplano* is an elongated, high-altitude basin). On the boundary between Peru and Bolivia, freshwater Lake Titicaca—the highest large lake on Earth—lies at 12,507 feet (3,700 m) above sea level. Here, in its west, lies the heart of modern Bolivia; here, too, lay one of the centers of Inca civilization, and indeed of pre-Inca cultures. Bolivia's capital, La Paz, is also located on the *Altiplano* at 11,700 feet (3,570 m), one of the highest cities in the world.

Lake Titicaca is what helps make the *Altiplano* livable, for this large body of water ameliorates the coldness in its vicinity, where the snow line lies just above the plateau surface. On the surrounding cultivable land, grains have been raised for centuries in the Titicaca Basin to the extraordinary elevation of 12,800 feet (3,850 m). To this day, this area of Peru and Bolivia supports a major cluster of Amerindian subsistence farmers.

Modern Bolivia is the product of the European impact, an influence that bypassed many of the Amerindian population clusters. Of course, these indigenous Bolivians no more escaped the loss of their land than did their Peruvian or Ecuadorian counterparts, especially east of the *Altiplano*. What made the richest Europeans in Bolivia wealthy, however, was not land but minerals. The town of Potosí in the Cordillera Oriental became a legend for the immense deposits of silver in its vicinity; zinc, copper, and several alloys were also discovered there. Over most of the past century, Bolivia's tin deposits, which ranked among the world's richest, yielded much of the country's annual export income; but since 1980, declining tin reserves and falling world prices have forced much of the industry to shut down. Today, zinc has replaced tin as the leading metallic mineral export, and natural gas and oil are accounting for a growing share of foreign revenues. Bolivia now exports large quantities of gas to Argentina and Brazil. In return, Brazil is assisting in the development of the southeastern lowlands, where commercial agriculture—notably soybean production—is steadily expanding in the fertile savanna.

Bolivia has had a turbulent history. Apart from endless internal struggles for power, the country first lost its corridor to the Pacific coast in a disastrous conflict with Chile, then lost its northern territory of Acre to Brazil in a dispute involving the rubber boom in the Amazon Basin, and finally lost 55,000 square miles (140,000 sq km) of southeastern Gran Chaco territory to Paraguay. The most critical loss by far was that of its outlet to the sea over a century ago—which may yet be regained from Chile if longstanding negotiations can be concluded successfully.

Although Bolivia has rail connections to the Chilean ports of Arica and Antofagasta, it remains severely disadvantaged by its landlocked situation. Because the Cordillera Occidental and the *Altiplano* form the country's in-

hospitable western margins, one might suppose that the country would look eastward and that its *Oriente* might be somewhat better developed than that of Peru or Ecuador. As we noted, the opening of this interior subregion is under way, but the limited infrastructure still confines economic activity and population growth to a few pockets. The densest settlement clusters within Bolivia's dispersed population of 8 million are in the valleys and basins of the Cordillera Oriental, where agricultural opportunities are more favorable than on the barren *Altiplano* to the west. Regional development in this second poorest country of South America is also inhibited because an astonishing 70 percent of all Bolivians still live in poverty. The majority of the impoverished are the long-neglected indigenous peoples, who finally began to be recognized in Bolivian society in the mid-1990s when a newly-elected government took several steps to integrate Amerindian cultures into mainstream national life.

❖ THE SOUTH: MID-LATITUDE SOUTH AMERICA

South America's four southern countries—Argentina, Chile, Uruguay, and Paraguay—constitute the realm's final regional grouping (Fig. 5-11). Because the continent resembles an overstuffed ice cream cone on the map, this region also goes by its popular name—the *Southern Cone*. As we noted earlier, the Southern Cone Common Market (Mercosur) began operating here in 1995 as the hemisphere's second largest trading bloc after NAFTA (of which Chile is to be a member). Mercosur encompasses Argentina, Uruguay, Paraguay, and Brazil, and has expressed interest in adding Chile and Bolivia. This effort may prove to be unnecessary if the planned Free Trade Area of the Americas materializes after the turn of the century.

Regional ties in the Mid-Latitude South are built on its modern history as the heart of the European-commercial culture sphere, which also spills across the Brazilian border to include much of that country's core area (Fig. 5-5). Today Mercosur not only perpetuates those linkages, but has also drawn peripheral Paraguay into its orbit. Paraguay may be a less-developed state and transitional with respect to the adjoining Andean West region, but its neighbors to the south and east unhesitatingly welcomed Paraguay into their new free trade zone and customs union. Two of the largest dams on Earth (Yacyretá and Itaipu) already operate on the Paraná River along Paraguay's eastern border, and one of Mercosur's biggest ventures calls for the completion of the 2,150-mile (3,450-km)-long *Hidrovia* waterway that

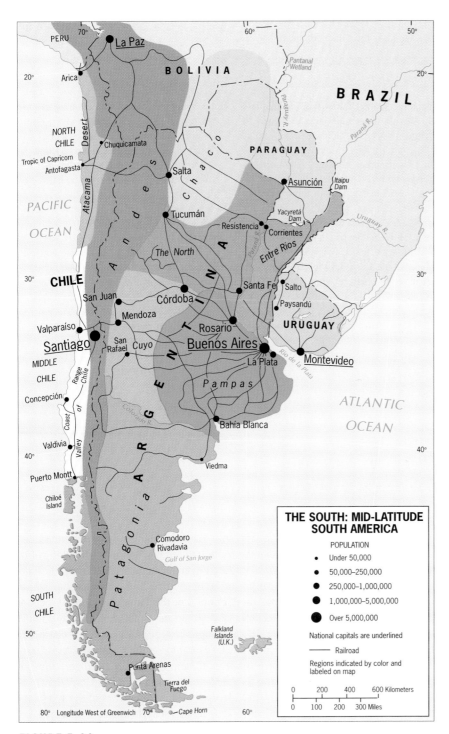

FIGURE 5-11

would open most of the Paraná-Paraguay river basin to barge transport. Intraregional ties are also being strengthened by many additional infrastructure improvement projects, including the construction of new routes over and under the Andes, the world's longest bridge across the Rio de la Plata between Buenos Aires and Uruguay, and hundreds of miles of new superhighways and railroads.

Argentina

The largest Southern Cone country by far, in terms of both population and territory, is Argentina, whose 1.1 million square miles (2.8 million sq km) and 36 million people rank second only to Brazil in this geographic realm. Argentina exhibits a great deal of physical-environmental va-

■ AMONG THE REALM'S GREAT CITIES...

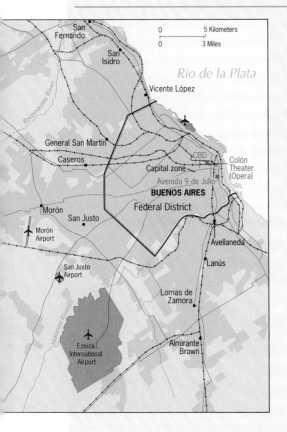

Buenos Aires

Its name means "fair winds," which first attracted European mariners to the site of Buenos Aires alongside the broad estuary of the muddy Rio de la Plata. The shipping function has remained paramount (see photo p. 267), and to this day the city's residents are known throughout Argentina as the *porteños* (the "port dwellers"). Modern Buenos Aires was built on the back of the nearby Pampa's grain and beef industry, and is often likened to Chicago and the Corn Belt because both cities are interfaces between immensely productive agricultural hinterlands and the rest of the world.

Buenos Aires (11.2 million) is another classic primate metropolis, housing nearly one-third of all Argentines, serving as the capital since 1880, and functioning as the country's economic core. Moreover, Buenos Aires is a cultural center of global standing, a monument-studded city that contains the world's widest boulevard (*Avenida 9 de Julio*). During the half-century between 1890 and 1940, the city was known as the "Paris of the South" for its architecture, fashion leadership, book publishing, and performing arts activities (it still has the world's biggest opera house). With the recent restoration of democracy, Buenos Aires is now trying to recapture its golden years. Besides reviving these cultural functions, the city has added a new one: the leading base of the hemisphere's motion picture and television industry for Spanish-speaking audiences.

riety within its boundaries, and the vast majority of the population is concentrated in the physiographic subregion known as the Pampa (a word meaning "plain"). Figure I-9 indicates the degree of clustering of Argentina's inhabitants on the land and in the cities of the Pampa. It also shows the relative emptiness of the other six subregions (mapped in Fig. 5-11)—the scrub-forest Chaco in the northwest, the mountainous Andes in the west (along whose crestline lies the boundary with Chile), the arid plateaus of Patagonia south of the Rio Colorado, and the undulating transitional terrain of intermediate Cuyo, Entre Rios (also known as "Mesopotamia" because it lies between the Paraná and Uruguay rivers), and the North.

The Argentine Pampa is the product of the past 150 years. During the second half of the nineteenth century, when the great grasslands of the world were being opened up (including those of the interior United States, Russia, and Australia), the economy of the long-dormant Pampa began to emerge. The food needs of industrializing Europe grew by leaps and bounds, and the advances of the Industrial Revolution—railroads, more efficient ocean transport, refrigerated ships, and agricultural machinery—helped make large-scale commercial meat and grain production in the Pampa not only feasible but also highly profitable. Large haciendas were laid out and farmed by tenant workers; railroads radiated ever farther outward from the burgeoning capital of Buenos Aires, and brought the entire Pampa into production. Yet the Pampa has hardly begun to fulfill its productive potential, which could easily double with more intensive agricultural practices. (For example, the land inheritance system too often fragments farms into inefficiently small units, and the use of fertilizers runs at perhaps one-twentieth the level in the U.S. Corn Belt.)

Over the decades, within the framework of the von Thünen model, several specialized agricultural areas appeared on the Pampa. As we would expect, a zone of fruit and vegetable production became established near the huge Buenos Aires conurbation located beside the estuary of the Rio de la Plata. To the southeast is the predominantly pastoral district, where beef cattle and sheep are raised. In the drier west, northwest, and southwest, wheat becomes the important commercial grain crop, but half the land remains devoted to grazing. Among the exports, cereals usually lead by value, followed by animal feed, vegetable oils, and meat products.

Argentina's affluence is reflected in its bustling cities,

which epitomize the European-commercial cultural character of the realm's southernmost countries. In fact, *87 percent* of the Argentine population may be classified as urbanized, a proportion typical of countries in Western Europe. More than 11 million (32 percent) reside in metropolitan Buenos Aires, which also contains most of the industries, many of them managed by Italians, Spaniards, and other immigrants. Córdoba, Argentina's second city with a much smaller population of 1.4 million, is another focus of industrial growth. Much of the manufacturing in the larger cities is associated with the processing of Pampa products and the production of consumer goods for the domestic market. Nearly one out of every three wage earners in the country is engaged in manufacturing—another indication of Argentina's somewhat advanced economic standing.

Argentina's population shows a high degree of clustering and a decidedly peripheral distribution. The Pampa subregion covers only a little more than 20 percent of Argentina's territory, but with more than three-fourths of the people concentrated here the rest of the country cannot be densely populated. Outside the Pampa, pastoralism is an almost universal pursuit, but the quality of the cattle is much lower than in the Pampa; in near-arid Patagonia sheep are raised, but ecological problems caused by overgrazing have resulted in a sharp decline since 1990. Some of the more distant areas contain potentially significant resources: in Patagonia an oilfield is in production near coastal Comodoro Rivadavia, and extensive reserves may also exist in the northern Chaco border zone with Paraguay. Yerba maté, a local tea, is produced in the North and Entre Rios subregions; and the quebracho-tree extract (for tanning leather) that both Paraguay and Argentina export comes from the central Paraguay-Paraná Basin. Northeastern Argentina is also an area targeted for industrial development, sparked by the massive Yacyretá Dam on the Paraná River. In both the North and Cuyo subregions, streams flowing off the Andes provide opportunities for irrigated agriculture; Tucumán is a major sugar-producing center, and Mendoza the heart of a productive district of vineyards and fruit orchards. But despite these sizeable near-Andean outposts and their good rail connections, effective Argentina remains the area within a radius of 350 miles (560 km) of Buenos Aires.

The 1980s was a particularly stressful decade for Argentina, involving the ''disappearances'' of thousands of nonrightist political activists, the unraveling of a repressive military junta, a severe economic crisis that accompanied the restoration of more democratic government, and a humiliating military defeat by the British in a brief war for the offshore Falkland Islands (whose unresolved sovereignty claims could trigger another conflict). The 1990s have been kinder to Argentina, beginning with a four-year growth spurt that revived the economy. Political stability has also returned, and the Argentines seem firm in their

From the Field Notes

"Whether you arrive in Buenos Aires by air or by sea, this city does not present a memorable site. It lies on the western side of the Plata estuary, not far south from the delta of the Paraná River. Sail upstream, and the low, flat banks of the Rio de la Plata narrow, presenting a string of coastal villages and some industry; then the high-rises of Buenos Aires come into view, seemingly standing in the muddy water (silting is a problem in the channel). As the photograph shows, Buenos Aires also lacks a dramatic CBD skyline; for so large a city, its vertical development is limited. Its areal sprawl, however, is considerable; there are few obstacles to expansion. Unremarkable architecturally, Buenos Aires has an Ibero-Italian atmosphere with some colonial remnants—in places. Historic preservation has not been a preoccupation here."

resolve to avoid the excesses and errors of the recent past. With its rich and plentiful resources and its promising new role in Mercosur, Argentina may finally be positioned to begin achieving its full potential.

Uruguay

Uruguay, unlike Argentina or Chile, is compact, small, and rather densely populated. This buffer state of old became a fairly prosperous agricultural country, in effect a smaller-scale Pampa (though possessing less favorable soils and topography). Figures I-4 and I-7 show the similarity of physical conditions on the two sides of the Plata estuary. Montevideo, the coastal capital, contains almost 45 percent of the country's population of 3.2 million; from here, railroads and roads radiate outward into the productive agricultural interior. In the immediate vicinity of Montevideo lies Uruguay's major farming area, which produces vegetables and fruits for the city as well as wheat and fodder crops. Most of the rest of the country is used for grazing cattle and sheep, with meat products, textiles, and hides dominating the export trade.

Uruguay, of course, is a small country, and its area of 68,000 square miles (176,000 sq km)—less territory even than Guyana—does not leave much room for population clustering. Nevertheless, a special quality of the land area of Uruguay is that it is rather evenly peopled right up to the boundaries with Brazil and Argentina. And of all the countries in South America, Uruguay is the most truly European, notably lacking the racial minorities that mark even Chile and Argentina, but with a sizeable non-Spanish European component in its population. As for future development prospects as Mercosur matures, Uruguay is in an excellent position to capitalize on its location between the Pampa to the south and Brazil's most dynamic areas to the north.

Paraguay

As Figure 5-4 indicates, Paraguay is one of those countries (Kazakstan is another) that lies across a regional transition zone. Although ranking slightly higher than Peru, Ecuador, and Bolivia, Paraguay's GNP per capita is more typical of the Andean West than the Mid-Latitude South. Ethnically, too, the pattern is more characteristic of the West: about 95 percent of Paraguay's 5 million people are mestizo, but with so pervasive an Amerindian influence that any white ancestry is almost totally submerged. In terms of language, Amerindian Guaraní is widely spoken alongside Spanish, and the country is probably the world's most completely bilingual. The physiography, however, is decidedly non-Andean because all of Paraguay lies to the east of the mountains in the center of the Paraná-Paraguay Basin; from here, the outlet to the sea has always been to the south. As the basin's waterways are improved—and as Mercosur drives the economic integration of the Southern Cone countries—that regional orientation is certain to strengthen.

Paraguay's landlocked position has had much to do with its modest economic development to date. Opportunities for pastoral and agricultural industries have long existed, but went largely unrealized because exports must be shipped through Buenos Aires—a long haul via the (as-yet-unimproved) Paraguay-Paraná waterway from the capital of Asunción. Nonetheless, soybean products, cotton, timber, hides, and meat reach foreign markets. Grazing in the dry Chaco to the west of the Paraguay River is an important commercial activity, but here cattle generally do not compare favorably to those of Argentina. A prosperous Mennonite agricultural colony in the remote heart of this zone is more successful, producing a large peanut crop and about half of the country's dairy products. In the northeast border zone, many Brazilians have crossed the Paraná to purchase fertile farmlands and raise lucrative soybeans. Further downstream, two ultramodern dams present possibilities for the growth of manufacturing (Fig. 5-11).

▮▶ Chile

For 2,500 miles (4,000 km) between the crestline of the Andes and the coastline of the Pacific lies the narrow strip of land that is the Republic of Chile. On the average just 90 miles (150 km) wide (and only rarely over 150 miles or 250 km in width), Chile is the textbook case of what political geographers call an **elongated state**, one whose shape tends to contribute to external political, internal administrative, and general economic problems. In the case of Chile, the Andes Mountains do form a barrier to encroachment from the east, and the sea constitutes an avenue of north-south communication. History has shown the country to be quite capable of coping with its northern rivals, Bolivia and Peru.

As Figures I-7 and 5-11 indicate, Chile is a three-subregion country. About 90 percent of its 15 million people are concentrated in what is called Middle Chile, where Santiago, the capital and largest city, and Valparaíso, the chief port, are located. North of Middle Chile lies the Atacama Desert, which is wider, drier, and colder than the coastal desert of Peru. South of Middle Chile, the coast is broken by a plethora of fjords and islands, the topography is mountainous, and the climate—wet and cool near the Pacific—soon turns drier and colder against the Andean interior. South of the latitude of Chiloé Island there are no permanent overland routes, and there is hardly any settlement. These three subregions are also clearly apparent on the map of culture spheres (Fig. 5-5), which displays a mestizo north, a European-commercial zone in Middle Chile, and an undifferentiated south. In addition, a small Amerind-subsistence zone in northern Chile's Andes is shared with Argentina and Bolivia.

Some intraregional differences exist between northern and southern Middle Chile, the country's core area. Northern Middle Chile, the land of the hacienda and of Mediterranean climate with its dry summer season, is an area of (usually irrigated) crops that include wheat, corn, vegetables, grapes, and other Mediterranean products. Livestock raising and fodder crops also take up much of the productive land but are giving way to the more efficient and profitable cultivation of fruits for export. Southern Middle Chile, into which immigrants from both the north and from Europe (especially Germany) have pushed, is a better-watered area where raising cattle has predominated. But here, too, more lucrative fruit, vegetable, grain, and other expanding food crops are changing the area's agricultural specializations.

Prior to the 1990s, the arid Atacama region in the north

From the Field Notes
"We stood off Chile's southernmost town, Punta Arenas across from Tierra del Fuego, waiting for the winds to subside and the waters to calm. Midsummer temperatures were in the low 50s and the Brunswick Peninsula beyond appeared bleak and barren. Once ashore, there was much in the cultural landscape to confirm this town's regional importance: a port with naval vessels to patrol Antarctic waters, a monument proclaiming Chile's hegemony over a wide Antarctic zone (including parts claimed by Argentina and the United Kingdom), a 'Patagonian Institute.' The southern end of the Southern Cone remains a contentious frontier."

accounted for most of Chile's foreign revenues. The Atacama contains the world's largest exploitable deposits of nitrates, and at first these provided the country's economic mainstay. But this mining industry declined after the discovery of methods of synthetic nitrate production early in the twentieth century. Subsequently, copper became the chief export (Chile possesses the world's largest reserves). It is found in several places, but the main concentration lies on the eastern margin of the Atacama Desert near the town of Chuquicamata, not far from the port of Antofagasta.

Chile today is in the midst of a development boom that is transforming its economic geography. For the past decade the country has been South America's greatest success story, and since 1990 no economy in the hemisphere has grown faster. Following the ouster of the repressive military government in 1989, Chile embarked on a program of reform that brought stable growth, dramatically lowered inflation and unemployment, reduced the poverty rate by more than 25 percent, and attracted massive foreign investment. The last is of particular significance because these new international connections enable the export-led Chilean economy to diversify and develop in important new ways. Copper remains the single leading export, but many other mining ventures (including gold) have been launched. In the agricultural sphere, fruit and vegetable production for export has soared because Chile's harvests coincide with the winter farming lull in the affluent countries of the Northern Hemisphere. Other primary economic activities that are increasing their foreign earnings are seafood and wood products, especially paper. Industrial expansion is occurring as well, with new factories manufacturing an array of goods that range from basic chemicals to computer software.

Chile's rapidly internationalizing economy has led the country to forge a major role for itself on the global trading scene. The Japanese expressed a particular interest in building ties, and during the early 1990s Japan was its leading trading partner. This relationship prompted the designation of Chile as the "economic tiger of the Andes," the first such burgeoning, beehive state on the South American shoreline of the sprawling Pacific Rim. By 1995, however, it became clear that Chile's Pacific Rim developments—at least for the time being—were going to be most heavily oriented to its own hemisphere. The Chileans quickly realized that regional economic integration offered the best opportunities for continued progress and eagerly accepted the invitation of the United States, Canada, and Mexico to join the North American Free Trade Agreement; as of mid-1996, Chile is first in line to join NAFTA. Among the foremost advantages of membership for Chile are unparalleled access to the huge NAFTA market, the accelerated inflow of long-term investment capital from Chile's new partners, and the technology transfers that will propel the sophisticated development of the industrial and service sectors.

Chile's ongoing transformation is widely touted today as the "economic model" for all of Middle and South America. That is probably an overstatement because few countries in these geographic realms can match the Chilean combination of natural and human resources and developmental circumstances. Yet all of those countries strongly aspire to follow in Chile's footsteps and for the first time in their modern histories are cooperating toward making that a reality. Their goal is the economic integration of all the Americas within the next decade—an achievement that could lead to a better life for every inhabitant of the Western Hemisphere. ◀◀

■ PRONUNCIATION GUIDE ■

Acre (AH-kray)
Altiplano (ahl-tee-PLAH-noh)
Amazonas (ahma-ZOAN-ahss)
Amerind (AMMER-rind)
Antioquia (ahn-tee-OH-kee-ah)
Antofagasta (untoh-fah-GAHSS-tah)
Arauca (ah-RAU-kah)
Arica (ah-REE-kah)
Asunción (ah-soohn-see-OAN)
Atacama (ah-tah-KAH-mah)
Augelli (aw-JELLY)
Avenida de 9 Julio (ah-veh-NEEDA day
    noo-AY-vay HOO-lee-oh)
Avenida Nicolas de Pierola (ah-veh-
    NEEDA NEEKO-lahss day pyair-
    ROH-lah)
Bahia (bah-EE-yah)
Barquisimeto (bar-key-suh-MAY-toh)
Barranquilla (bah-rahn-KEE-yah)
Barrio[s] (BAHR-ree-oh[ss])
Bauxite (BAWKS-site)
Bayovar (bye-YOH-vahr)
Belém (bay-LEM)
Belo Horizonte (BAY-loh
    haw-ruh-ZONN-tee)
Beni (BAY-nee)
Benin (beh-NEEN)
Bogotá (boh-goh-TAH)
Bolívar, Simón (boh-LEE-vahr,
    see-MOAN)
Brasília (bruh-ZEAL-yuh)
Brunswick (BROON-svik)
Buenos Aires (BWAY-nohss EYE-race)
Cacao (kuh-KAY-oh/kuh-KAU)
Caldas (KAHL-dahss)
Cali (KAH-lee)

Callao (kah-YAH-oh)
Caracas (kah-RAH-kuss)
Carioca (kah-ree-OH-kah)
Cartagena (kar-tah-HAY-nah)
Caste (CAST)
Cauca (KOW-kah)
Ceará (say-uh-RAH)
Central (sen-TRAHL)
Centro-Oeste (SENTRO oh-ESS-tee)
Cerrado (seh-RAH-doh)
Cerrejón (serray-HONE)
Cerro de Pasco (serro day PAH-skoh)
Chaco (CHAH-koh)
Chaparé (chah-pah-RAY)
Chibcha (CHIB-chuh)
Chile (CHILLI/CHEE-lay)
Chiloé (chee-luh-WAY)
Chuquicamata (choo-kee-kah-MAH-tah)
Ciudad Guayana (see-you-DAHD
    gwuh-YAHNA)
Cochabamba (koh-chah-BUM-bah)
Comodoro Rivadavia (comma-DORE-oh
    ree-vah-DAH-vee-ah)
Cordillera (kor-dee-YERRA)
Cordillera del Litoral (kor-dee-YERRA
    dale lee-toh-RAHL)
Córdoba (KORD-oh-bah)
Creole[s] (KREE-oal[z])
Cuiabá (koo-yuh-BAH)
Cusiana (koo-see-AHNA)
Cuyo (KOO-yoh)
Cuzco (KOO-skoh)
Département (day-part-MAW)
Ecuador (ECK-wah-dor)
El Cojo (el-KOH-hoh)
Entre Rios (en-truh-REE-ohss)

Esmeraldas (ezz-may-RAHL-dahss)
Essequibo (essa-KWEE-boh)
Falkland[s] (FAWK-lund[z])
Favela[s] (fah-VAY-lah[ss])
Fazenda (fah-ZENN-duh)
Ferronorte (feh-roh-NOR-tay)
Florianópolis (flaw-ree-ah-NOP-poh-
    leese)
Fortaleza (for-tuh-LAY-zuh)
Fría (FREE-uh)
Goiás (goy-AHSS)
Grande Carajás (GRUNN-dee
    kuh-ruh-ZHUSS)
Guajira (gwah-HEAR-ah)
Guanabara (gwah-nah-BAR-uh)
Guaraní (gwah-rah-NEE)
Guayaquil (gwye-ah-KEEL)
Guayas (GWYE-ahss)
Guiana[s] (ghee-AH-nah[z])
Guri (GOOR-ree)
Guyana (guy-AHNA)
Hacienda (ah-see-EN-duh)
Hidrovia (ee-DROH-vee-uh)
Huallaga (wah-YAH-gah)
Huancayo (wahn-KYE-oh)
Humboldt (HUMM-bolt)
Iberia (eye-BEERY-uh)
Iquitos (ih-KEE-tohss)
Isohyet (EYE-so-hyatt)
Islamabad (iss-LAHM-uh-bahd)
Itaipu (ee-TYE-pooh)
João, Dom (ZHWOW, dom)
Junta (HOON-tah)
Karachi (kuh-RAH-chee)
Kashmir (KASH-meer)
Kazakstan (kuzz-uck-STAHN)

Lafaiete (lah-fuh-YAY-tuh)
La Paz (lah-PAHZ)
Lima (LEE-muh)
Llama (LAH-muh)
Llano[*s*] (YAH-noh[ss])
Machu Picchu (MAH-choo PEEK-choo)
Magdalena (mahg-dah-LAY-nah)
Magellan (muh-JELLEN)
Manaus (muh-NAUSS)
Maracaibo (mah-rah-KYE-boh)
Mato Grosso (mutt-uh-GROH-soh)
Mato Grosso do Sul
    (mutt-uh-GROH-soh duh-SOOL)
Medellín (meh-deh-YEEN)
Mercosur (mair-koh-SOOR)
Mestizo (meh-STEE-zoh)
Minas Gerais (MEE-nuss zhuh-RICE)
Minifundia (minny-FOON-dee-uh)
Montaña (mon-TAHN-yah)
Montevideo (moan-tay-vee-DAY-oh)
Nigeria (nye-JEERY-uh)
Occidental (oak-see-den-TAHL)
Oriental (orry-en-TAHL)
Oriente (orry-EN-tay)
Orinoco (orry-NOH-koh)
Pampa (PAHM-pah)
Pará (puh-RAH)
Paraguay (PAHRA-gwye)
Paraíba (pah-rah-EE-buh)
Paraná (pah-rah-NAH)
Patagonia (patta-GOH-nee-uh)
Paulista[*s*] (pow-LEASH-tah[ss])
Pernambuco (pair-nahm-BOO-koh)
Pisac (pea-SAHK)
Pizarro, Francisco (pea-SAHRO,
    frahn-SEECE-koh)

Plata, Rio de la (PLAH-tah, REE-oh
    day lah)
Plaza de Armas (PLAH-sah day
    AR-mahss)
Plaza de Mayo (PLAH-sah day
    MYE-oh)
Plaza de San Martin (PLAH-sah
    day sahn mahr-TEEN)
Polonoroeste (POLLOH-nuh-roh-
    ESS-tee)
Porteño (por-TAIN-yoh)
Pôrto Alegre (POR-too uh-LEG-ruh)
Pôrto Velho (POR-too VELL-yoo)
Potosí (poh-toh-SEE)
Puna (POONA)
Punta Arenas (POON-tah
    ah-RAY-nahss)
Quebracho (kay-BROTCH-oh)
Quechua (KAYTCH-wah)
Quito (KEE-toh)
Recife (ruh-SEE-fuh)
Rio Branco (REE-oh BRUNG-koh)
Rio de Janeiro (REE-oh day
    zhah-NAIR-roh)
Rio Grande do Norte (REE-oh
    GRUN-dee duh NORTAH)
Rio Grande do Sul (REE-oh GRUN-dee
    duh SOOL)
Rondônia (roh-DOAN-yuh)
Roraima (raw-RYE-muh)
Salvador (SULL-vuh-dor)
San Cristóbal (sahn kree-STOH-bahl)
San Juan (sahn HWAHN)
San Rafael (sahn rah-fye-ELL)
Santa Catarina (SUN-tuh
    kuh-tuh-REE-nuh)

Santiago (sahn-tee-AH-goh)
Santos (SUNT-uss)
São Francisco (sau frahn-SEECE-koh)
São Luis (sau loo-EECE)
São Paulo (sau PAU-loh)
Sendero Luminoso (sen-DARE-oh
    loo-mee-NOH-soh)
Serra dos Carajás (SERRA doo
    kuh-ruh-ZHUSS)
Sertão (sair-TOWNG)
Sisal (SYE-sull)
Suriname (soor-uh-NAHM-uh)
Tanzania (tan-zuh-NEE-uh)
Tecnópolis (tek-NOH-poh-leece)
Templada (tem-PLAH-dah)
Tierra del Fuego (tee-ERRA dale
    FWAY-goh)
Titicaca (tiddy-KAH-kuh)
Tocantins (toke-un-TEENS)
Tordesillas (tor-day-SEE-yahss)
Tsunami (tsoo-NAH-mee)
Tubarão (too-buh-RAUNG)
Tucumán (too-koo-MAHN)
Tucuruí (too-koo-roo-EE)
Uruguay (OO-rah-gwye)
Valencia (vuh-LENN-see-uh)
Valparaíso (vahl-pah-rah-EE-so)
Venezuela (veh-neh-SWAY-lah)
Verde (VER-dee)
Volta Redonda (vahl-tuh rih-DONN-
    duh)
Von Thünen (fon-TOO-nun)
Yacyretá (yah-see-ray-TAH)
Yanomami (yah-noh-MAH-mee)
Yerba maté (YAIR-bah mah-TAY)
Yungas (YOONG-gahss)

CHAPTER 6

North Africa/Southwest Asia: The Energy of Islam

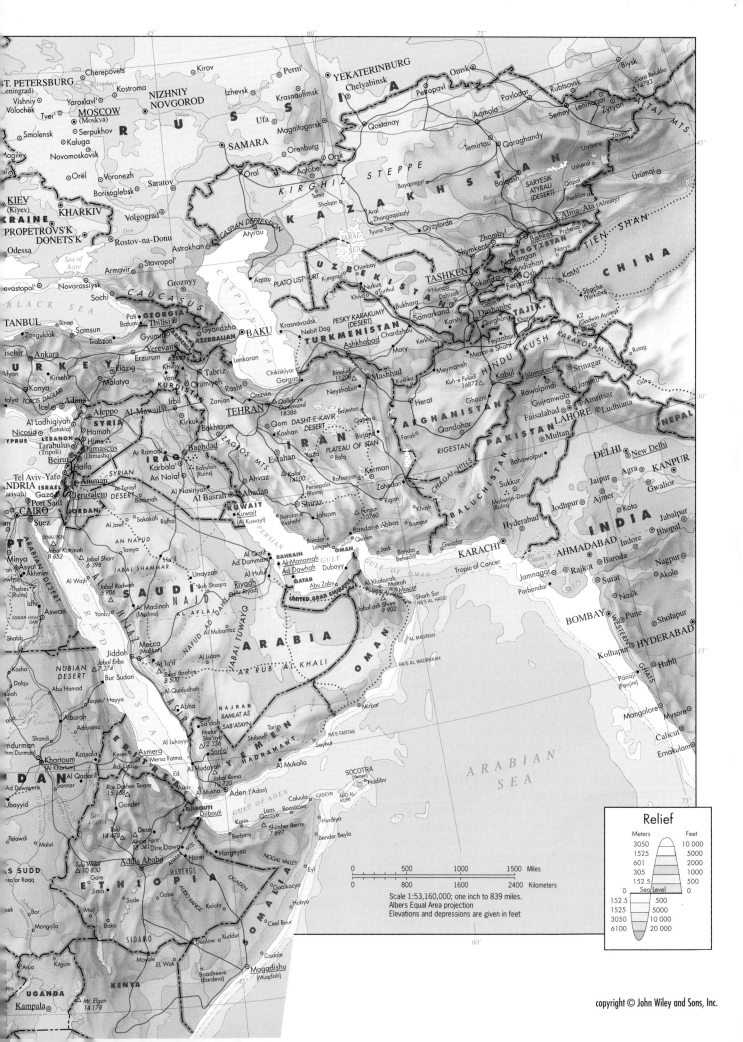

From Morocco on the shores of the Atlantic to the mountains of Afghanistan, and from the Horn of Africa to the steppes of inner Asia, lies a vast geographic realm of enormous cultural complexity. It lies at the crossroads where Europe, Asia, and Africa meet, and it is part of all three. Throughout history, its influences have radiated to these continents and to practically every other part of the world as well. This is one of humankind's primary source areas. On the Mesopotamian Plain between the Tigris and Euphrates rivers (in modern-day Iraq) and on the banks of the Egyptian Nile arose civilizations that must have been among the very earliest. In the soils, plants were domesticated that are now grown from the Americas to Australia. Along its paths walked prophets whose religious teachings are still followed by hundreds of millions of people. And in the final decade of the twentieth century, the heart of this realm is beset by some of the most bitter and dangerous conflicts on Earth.

IDEAS & CONCEPTS

Cultural geography	Relocation diffusion
Cultural landscape	Expansion diffusion
Popular culture	Contagious diffusion
Folk culture	Hierarchical diffusion
Culture hearth	Religious fundamentalism
Cultural diffusion	Geometric political
Cultural perception	boundaries
Culture conflict	Energy resources
Culture region	Cultural revival
Fertile Crescent	Basin irrigation
Hydraulic civilization theory	Perennial irrigation
Climate change theory	Nomadism
Spatial diffusion	Buffer state

REGIONS

Egypt and the	Middle East
Lower Nile Basin	Arabian Peninsula
Maghreb and Its Neighbors	Empire States
African Transition Zone	Turkestan

DEFINING THE REALM

It is tempting to characterize this geographic realm in a few words, and stress one or more of its dominant features. It is, for instance, often called the "Dry World," containing as it does the vast Sahara and Arabian deserts. But most of the realm's people live where there is water—in the Nile delta, along the Mediterranean coastal strip (or *tell*) of northwesternmost Africa, along the Asian eastern and northeastern shores of the Mediterranean Sea, in the Tigris-Euphrates Basin, in far-flung desert oases, and along the lower mountain slopes of Iran south of the Caspian Sea as well as those of Turkestan to the northeast. True, we know this world region as one where water is almost always at a premium, where peasants often struggle to make soil and moisture yield a small harvest, where nomadic peoples and their animals circulate across dust-blown flatlands, where oases are islands of sedentary farming and trade in a sea of aridity. But it also is the land of the Nile, the lifeline of Egypt, and the crop-covered *tell* of the northern coast of Morocco, Algeria, and Tunisia.

Before we investigate this realm further, it is useful to look again at Figure I-7, the map of the world climates. Note the considerable degree of coincidence between the extent of the *B* (desert and steppe) climates and the limits of both the North African and Southwest Asian segments. In the case of North Africa, the border between this realm and Subsaharan Africa comes close to matching the southern limits of the *BSh* zone. In Southwest Asia, deserts, steppes, and mountain climates dominate. Except for the Mediterranean coasts, this is a realm of very low and highly variable annual precipitation, of searing daytime heat and chilling nighttime cold, of strong winds and dust-laden air. Soils are thin; mountain slopes carry little vegetation. To all this, water brings exception—not only along coasts and rivers but also in oases and *qanats*, tunnels dug into water-bearing rock strata at an angle, so that the water drains to the surface. The population map (Fig. I-9) reminds us how scattered and isolated the population clusters are—as a matter of necessity.

An "Arab World"?

North Africa/Southwest Asia is also often referred to as the Arab World. Once again, this implies a uniformity that does not actually exist. In the first place, the name *Arab* is applied loosely to the peoples of this area who speak Ar-

THE MAJOR GEOGRAPHIC QUALITIES OF NORTH AFRICA/ SOUTHWEST ASIA

■■■

1. This realm contains several of the world's great ancient culture hearths and some of its most durable civilizations.

2. The North Africa/Southwest Asia realm is the source of several world religions, including Islam, Christianity, and Judaism.

3. This realm is predominantly but not exclusively Islamic (Muslim). That faith pervades cultures from Morocco in the west to Afghanistan in the east.

4. North Africa/Southwest Asia is the "Arab World," but significant population groups in this realm are not of Arab ancestry.

5. The population of North Africa/Southwest Asia is widely dispersed in discontinuous clusters.

6. Natural environments in this realm are dominated by drought and unreliable precipitation. Population concentrations occur where the water supply is adequate to marginal.

7. The realm contains a pivotal area in the "Middle East," where Arabian, North African, and Asian regions intersect.

8. North Africa/Southwest Asia is a realm of intense discord and bitter conflict, reflected by frequent territorial disputes and boundary frictions.

9. The end of Soviet rule and the revival of Islam in Turkestan have recently expanded this realm into central Asia.

10. Enormous reserves of petroleum lie beneath certain portions of the realm, bringing wealth to those favored places. But overall, oil revenues have raised the living standards of only a small minority of the total population.

■■■

abic and related languages, but ethnologists normally restrict it to certain occupants of the Arabian Peninsula—the Arab "source." In any case, the Turks are not Arabs, and neither (for the most part) are the Iranians or the Israelis. Moreover, although Arabic is spoken over a wide region that extends from Mauritania in the west across all of North Africa to the Arabian Peninsula, Syria, and Iraq in the east, in many areas within the realm it is not used by most of the people.

In Turkey, for example, Turkish is the major language, and it has Ural-Altaic rather than Semitic or Hamitic roots. In Iran, the Iranian language belongs to the Indo-European linguistic family. Other "Arab World" languages that have separate ethnological identities are spoken by the Jews of Israel, the Tuareg people of the Sahara, the Berbers of northwestern Africa, and the peoples of the transition zone between North Africa and Subsaharan Africa to the south.

An "Islamic World"?

Another name given to this realm is the World of Islam. The prophet Muhammad (Mohammed) was born in Arabia in A.D. 571, and in the centuries that followed his death in 632, Islam spread into Africa, Asia, and Europe. This was the age of Arab conquest and expansion. Their armies penetrated southern Europe, their caravans crossed the deserts, and their ships plied the coasts of Asia and Africa. Along these routes they carried the Muslim (Islamic) faith, converting the ruling classes of the states of the West African savanna, threatening the Christian stronghold in the highlands of Ethiopia, penetrating the deserts of inner Asia, and pushing into India and even the island extremities of Southeast Asia. Islam was the religion of the marketplace, the bazaar, the caravan. Where necessary, it was imposed by the sword, and its protagonists aimed directly at the political leadership of the communities they entered. Today, the Islamic faith with its more than 1 billion followers extends well beyond the limits of the realm discussed here (Fig. 6-1). It is the major religion in northern Nigeria, in Pakistan, and in Indonesia; it is influential in East Africa; and it still prevails in its old strongholds of Albania, Kosovo, and Bosnia in Eastern Europe. Moreover, in South Asia more than 100 million Muslims live in Hindu-dominated India, the world's largest religious minority.

On the other hand, the World of Islam is not entirely Muslim either. In Israel, Judaism is the prevailing faith; Christianity remains strong in Lebanon; and ancient Coptic Christian churches still exist in Egypt. Thus the connotation of World of Islam when applied to North Africa/ Southwest Asia is far from satisfactory: the religion prevails far beyond these areas, and within the realm there are a number of countries in which Islam is not the dominant faith.

"Middle East"?

Finally, this realm is frequently called the Middle East. That must sound quite odd to someone in, say, India, who might think of a Middle West rather than a Middle East! The name, of course, reflects the biases of its source: the "Western" world, which saw a "Near East" in Turkey, a "Middle" East in Egypt, Arabia, and Iran, and a "Far" East in China and Japan. Still, the term has taken hold, and

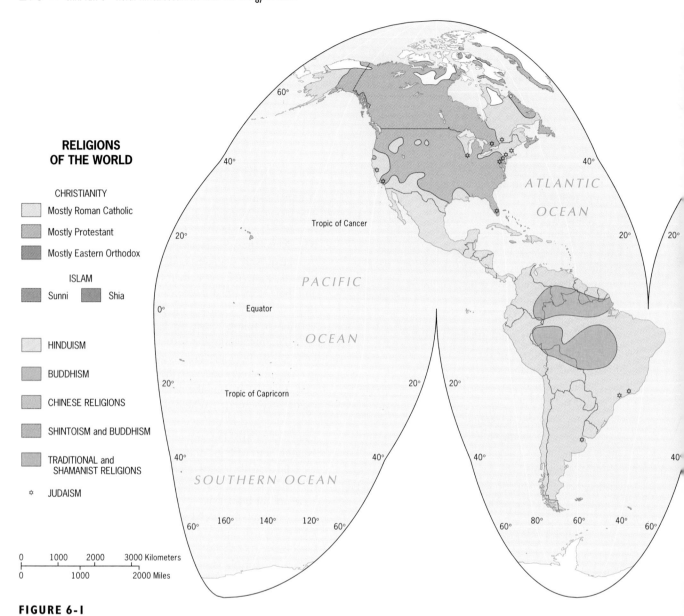

FIGURE 6-1

HEARTHS OF CULTURE

it can be seen and heard in everyday usage by journalists as well as members of the United Nations. In view of the complexity of this realm, its transitional margins, and its far-flung areal components, the name "Middle East" need be faulted only for being imprecise—it does not make a single-factor region of North Africa/Southwest Asia, as do the terms *Dry World*, *Arab World*, and *World of Islam*.

This geographic realm occupies a pivotal part of the world: here Eurasia, crucible of human cultures, meets Africa, source of humanity itself. A million years ago, the ances-

tors of our species walked from East Africa into North Africa and Arabia and spread from one end of Asia to the other. One hundred thousand years ago, *Homo sapiens* crossed these lands on the way to Europe, Australia, and, eventually, the Americas. Ten thousand years ago, human communities in what we now call the Middle East began to domesticate plants and animals, learned to irrigate their fields, enlarged their settlements into towns, and formed the earliest of states. One thousand years ago, the heart of the realm was stirred and mobilized by the teachings of Muhammad and the Koran, and Islam was on the march from North Africa to India. Today this realm is a cauldron of religious and political activity, weakened by conflict but empowered by oil, plagued by poverty but fired by a fundamentalist wave of religious revival.

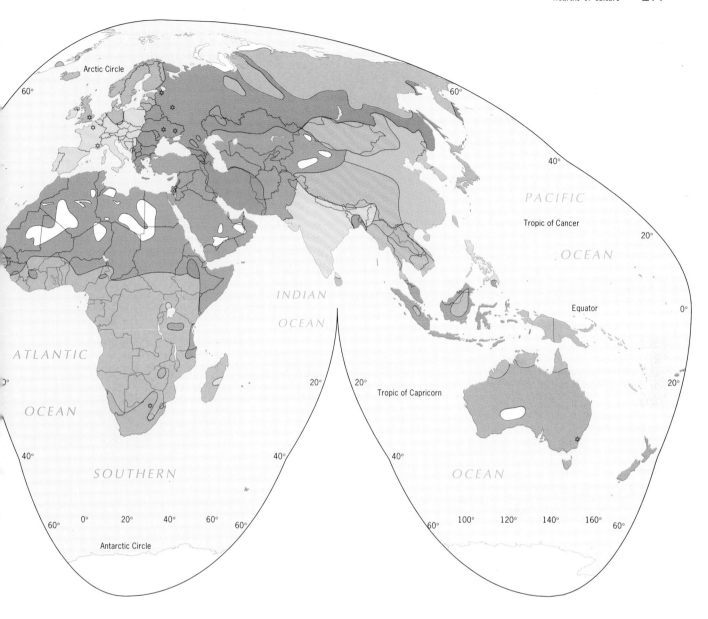

In the basins and valleys of the great rivers of this realm (the Tigris and Euphrates of modern-day Iraq and the Nile of Egypt) lay two of the world's earliest **culture hearths**—sources of innovations and ideas, traits and technologies, lifestyles and landscapes that spread far beyond their nuclei (Fig. 6-2). *Mesopotamia*, the "land between the rivers," was a zone of fertile alluvial soils, abundant sunshine, and ample water, and here, in the Tigris-Euphrates lowland between the head of the Persian Gulf and the uplands of present-day Turkey, arose one of humanity's first civilizations. Mesopotamia's agricultural know-how, which involved planned sowing and harvesting and the distribution of surplus grain for storage and later consumption, radiated to villages far away. Eventually the wider region was anchored by a **Fertile Crescent**, a zone of advanced farming that extended in a great arch from the Mediterranean coast to very near the Persian Gulf.

Irrigation was the key to prosperity and power in Mesopotamia, and urbanization was its reward. Among many settlements in the Fertile Crescent, some thrived, grew, enlarged their hinterlands, diversified socially and occupationally. Others failed. What determined success? One theory, the **hydraulic civilization theory**, holds that cities able to control irrigated farming over large hinterlands held power over others, used food as a weapon, and thrived. One such city, Babylon on the Euphrates River, endured for nearly 4,000 years (from 4100 B.C.). A busy port, its walled and fortified center endowed with temples, towers, and palaces, Babylon was the world's largest city.

Mesopotamia's culture hearth lay between that of the

■ ■ ■ ■ ■ ■ ■ ■ ■ ■ ■ ■ **FOCUS ON A SYSTEMATIC FIELD** ■ ■ ■ ■ ■ ■ ■ ■ ■ ■ ■ ■

Cultural Geography

We are constantly surrounded by the artifacts of our culture. From simple utensils to elaborate buildings, from food and clothing to superhighways and suburbs, we live in a world transformed by our products and preferences.

Scholars representing several disciplines study the phenomenon of culture from various vantage points. Geographers are especially interested in the *spatial expressions* of cultures, in the ways they are laid out and organized on the land, at every level of generalization. Another focus of interest has to do with the complex relationships between human cultures and the *natural environments* under which they have evolved. These are topics of the field of **cultural geography**, and in this chapter we have good reason to concentrate on it. Cultural imperatives and cultural conflicts are powerful forces in the realm under study.

Cultural geography is a wide-ranging and comprehensive field. Its major components focus on (1) cultural landscapes, (2) culture hearths, (3) cultural diffusion, (4) cultural environments, (5) cultural perception, and (6) culture regions.

As was noted in the Introduction, culture gives character to an area. The people of any particular culture transform their living space by building structures on it, creating routes of contact and communication, tilling the land, and channeling the water. This composite of artificial features is conceptualized as the **cultural landscape**, and the photographs in this chapter exhibit many of the major elements in this realm's expressions of it. The cultural landscape represents material (tangible) culture, but cultural geographers also study what they call **popular culture**, the behavior and preferences of large numbers of people influenced by mass communication in urbanized societies (rock music is an example), and **folk culture**, the durable ways of life still surviving in relatively

remote rural areas, where culture still is transmitted mainly by word of mouth in family and village settings.

North Africa and Southwest Asia, as a geographic realm, contained several of the world's great ancient **culture hearths**, crucibles of cultural maturation and achievement from which radiated ideas, innovations, and ideologies that changed the world beyond. Here lay the sources of several major civilizations. Here, too, emerged three major religions, Judaism, Christianity, and Islam. The first two arose near the eastern shore of the Mediterranean Sea, and the third in western Arabia. Reconstructing these and other culture hearths and determining their environmental settings are two of cultural geography's primary pursuits.

From ancient as well as modern culture hearths, knowledge and ideas spread far afield in a set of processes collectively called **cultural diffusion**. Geographic research into various diffusion processes now makes it possible to use modern techniques of reconstruction to determine old routes of diffusion, so that we are learning much about the long-distance contact of ancient civilizations. The diffusion of Islam, which began in the seventh century A.D. and continues today, is a prominent topic in this chapter.

The multiple relationships between human cultures and their physical settings are subsumed under the rubric of *cultural environments*, a branch of cultural geography also known as *cultural ecology*. Human cultures exist in long-term accommodation with (and adaptation to) their natural environments, seizing the opportunities these environments present and suffering from the extremes they sometimes impose. Even the most technologically advanced culture cannot completely evade the destructive forces of nature. But there is a more tendentious issue here. Just how does environment (mainly climate, but

also vegetation, relief, et al.) influence, over the long term, a society's collective attitudes, capacities, and energies? Does it, in fact, have any such influence at all? Questions of this kind have fueled some of geography's most intense and productive debates.

The recent and current tragedy in former Yugoslavia has reminded us of the significance of **cultural perception** as a factor in misunderstanding and conflict. The comfort we feel in a familiar cultural landscape (as many Americans report when they visit Australia or Britain), and the discomfort many of us experience when we are confronted with an unfamiliar place (say Khartoum or Damascus) is part of this. When the first Peace Corps volunteers went to Africa and other parts of the world, some suffered from "culture shock" and had to leave their posts. They could do so, but in former Yugoslavia, millions confronted the prospect that they would come under the control of cultures other than their own, as a result of political developments. This perception led to **culture conflict**, a condition, as we will see, of several states in North Africa and Southwest Asia as well.

The spatial extent of individual cultures can be mapped at all scales. A single city can consist of several cultural landscapes, each of which can be delimited. One country may contain several **culture regions**, including, in this realm, Sudan, Turkey, and several of the countries in Central Asia. At a smaller scale still, we are able to map the dozen world geographic realms of which North Africa/ Southwest Asia forms one. And, like the cultural landscapes of cities and the culture regions in countries, the great geographic realms are always changing, still affected by the flow of ideas, innovations, and interactions that continues to modify our world.

■ ■ ■

Nile valley to the west and the Indus valley (in present-day Pakistan) to the east. Egypt's cultural evolution may have started even earlier than Mesopotamia's, and the focus of

this civilization lay upstream from (south of) the Nile delta and downstream from (north of) the first of the Nile's series of rapids, or *cataracts* (Fig. 6-2). This part of the Nile

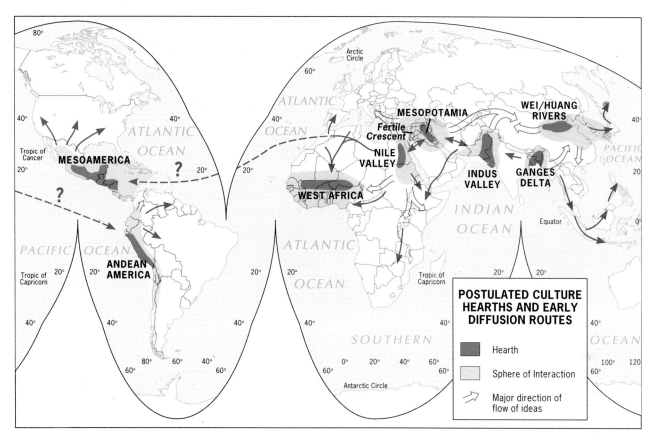

FIGURE 6-2

valley lies surrounded by inhospitable desert, and unlike Mesopotamia (which lay open to all comers), the Nile provided a natural fortress here. The ancient Egyptians converted their security into progress. The Nile was their highway of trade and interaction; it also supported agriculture by irrigation. The Nile's cyclical regime of ebb and flow was much more predictable than that of the Tigris-Euphrates river system. By the time Egypt finally fell victim to outside invaders (about 1700 B.C.), a full-scale urban civilization had emerged. Ancient Egypt's artist-engineers left a magnificent legacy in the form of massive stone monuments, some of them containing treasure-filled crypts of prominent rulers. These tombs have enabled archeologists to reconstruct the ancient history of this culture hearth.

To the east, separated from Mesopotamia by more than 1,200 miles (1,900 km) of mountain and desert, lies the *Indus valley* (see Fig. 6-2). By modern criteria, this eastern hearth lies outside the realm under discussion here; but in ancient times it had effective ties with the Tigris-Euphrates region. Mesopotamian innovations reached the Indus region early, and eventually the cities of the Indus became power centers of a civilization that reached far into present-day northern India.

Today, the world continues to benefit from the accomplishments of the ancient Mesopotamians and Egyptians. They domesticated cereals (wheat, rye, barley), vegetables (peas, beans), and fruits (grapes, apples, peaches); they also domesticated many animals (horses, pigs, sheep). They advanced not only irrigation and agriculture but also calendrics, mathematics, astronomy, government, engineering, metallurgy, and a host of other skills and technologies. As time went on, many of their innovations were adopted and then modified by other cultures in the Old World and eventually in the New World as well. Europe was the greatest beneficiary of these legacies of Mesopotamia and ancient Egypt, whose achievements constituted the very foundations of "Western" civilization.

Decline and Decay

Today, many of the early cities of this realm's culture hearths are archeological curiosities. In some instances, new cities have been built on the sites of the old, but the great cultural traditions of this realm went into deep decline after many centuries of continuity.

It cannot escape our attention that a large number of these ancient urban centers are located in what is now desert. Presuming that they were not built in the middle of these drylands, it is tempting to postulate a hypothesis that also could form an alternative to the hydraulic civilization theory: that of **climate change**.

The ancient Romans transformed the North African countryside by building cities, laying out irrigation systems, and farming fields that had been pastures for the herds of nomads. Their structures, from colonnaded avenues to tiered amphitheaters and from cobbled roadways to covered aqueducts, were built so well that parts of them endure to this day. These are the surviving remnants of a covered aqueduct that carried water for many miles across the undulating landscape of northeastern Tunisia. A cross-section of the "tube" atop the columned structure can be seen at the upper left; the extent of the system can be gauged at the right. Roman engineers sloped the aqueduct in such a way that water could be drawn off to various fields along its course. Not only was the system functional; the structure itself was designed to convey beauty as well as strength and effectiveness. Arched openings between the columns allowed strong Mediterranean winds to pass through, reducing the danger of weakening and collapse. It worked for centuries.

It may have been climate change that gave certain cities in the ancient Fertile Crescent an advantage over others, not a monopoly over irrigation techniques; and it may have been climate change, associated with shifting environmental zones in the wake of the last Pleistocene glacial retreat, that destroyed the last of the old civilizations. Perhaps overpopulation and human destruction of the natural vegetation contributed to the process. Indeed, some cultural geographers suggest that the momentous innovations in agricultural planning and irrigation technology were not "taught" by the seasonal flooding of the rivers, but impelled as a response to changing environmental conditions as the riverine communities tried to survive.

The scenario is not difficult to imagine. As outlying areas began to fall dry and farmlands were destroyed, people congregated in the already crowded river valleys—and every effort was made to increase the productivity of lands that could still be watered. Eventually overpopulation, destruction of the watershed, and perhaps reduced rainfall in the rivers' headwater areas combined to deal the final blow. Towns were abandoned to the encroaching desert; irrigation canals filled with drifting sand; remaining croplands dried up. Those who could migrated to areas reputed still to be productive. Others stayed, their numbers dwindling, increasingly reduced to subsistence.

As old societies disintegrated, new power emerged elsewhere. First the Persians, then the Greeks, and later the Romans imposed their imperial designs on the tenuous lands and disconnected peoples of North Africa/Southwest Asia. Roman technicians converted North Africa's farmlands into irrigated plantations whose products went by the boatload to Roman Mediterranean shores. Thousands of people were carried in bondage to the cities of the new conquerors. Egypt was quickly colonized, as was the area we now call the Middle East. One region that lay distant, and therefore remote from these invasions, was the Arabian Peninsula, where no major culture hearths or large cities had emerged, and where Arab settlements and nomadic routes remained unaffected by the turmoil.

STAGE FOR ISLAM

Islam today dominates the cultural geography of the realm under discussion, but long before Islam arose, other religions had emerged: Zoroastrianism in what is today Iran, and Judaism and later Christianity in the area of modern Israel. But the teachings of Zoroaster remained confined to Persia, Judaism was devastated by the Babylonians and later by the Romans, and Christianity became an alien faith as its center moved to Rome. Until the early seventh century A.D., none of the religions that had their origins in this realm became dominant within it.

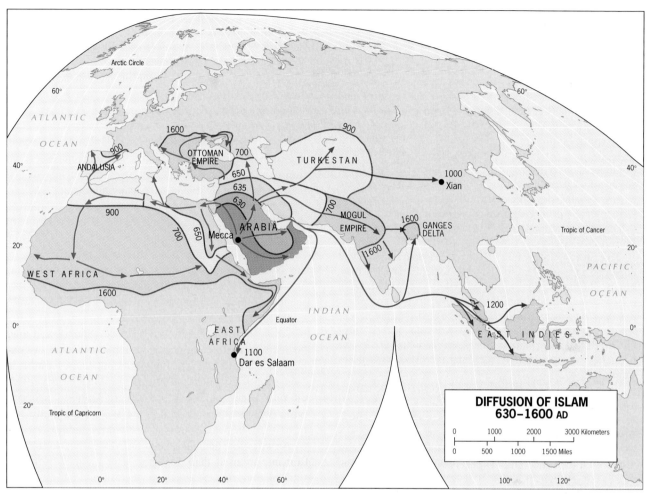

FIGURE 6-3

Muhammad the Prophet

In a remote place on the Arabian Peninsula, where Arab communities had been little affected by the foreign invasions of the Middle East, an event occurred early in the seventh century that was to change the course of history and affect the destinies of people in many parts of the world. In a town called Mecca, about 45 miles (70 km) from the Red Sea coast in the Jabal Mountains, a man named Muhammad in the year A.D. 613 began to receive a series of revelations from Allah (God). Muhammad (571–632) already was in his early forties, and he had barely 20 years to live. Convinced after some initial self-doubt that he was indeed chosen to be a prophet, Muhammad committed his life to the fulfillment of the divine commands he believed he had received. Arab society was in social and cultural disarray, but Muhammad forcefully taught Allah's lessons and began the transformation of his culture. His personal power soon attracted enemies, and he was forced to flee Mecca for the safer haven of Medina, where he continued his work. Mecca, of course, later became Islam's holiest place.

The precepts of Islam in many ways constituted a revision and embellishment of Judaic and Christian beliefs and traditions. There is but one god, who occasionally communicates through prophets; Islam acknowledges that Moses and Jesus were such prophets. What is earthly and worldly is profane; only Allah is pure. Allah's will is absolute; Allah is omnipotent and omniscient. All humans live in a world created for their use, but only to await a final judgment day.

Islam brought to the Arab World not only the unifying religious faith it had lacked but also a new set of values, a new way of life, a new individual and collective dignity. Islam dictated observance of the *Five Pillars*: (1) repeated expressions of the basic creed, (2) the daily prayer, (3) a month of daytime fasting (*Ramadan*), (4) the giving of alms, and (5) at least one pilgrimage to Mecca. And the faith prescribed and proscribed in other spheres of life as well. Alcohol, smoking, and gambling were forbidden. Polygamy was tolerated, although the virtues of monogamy were acknowledged. Mosques appeared in Arab settlements, not only for the (Friday) sabbath prayer, but also as social gathering places to bring communities closer to-

DIFFUSION PROCESSES

■ ■ ■

One of the most interesting areas of study in geography has to do with the way ideas, inventions, practices, and other phenomena spread through a population across space and time. A new musical style, a word or phrase, a new way of doing things, new fashions in clothing disseminate by a process called **spatial diffusion**. So do technological innovations, diseases, religious practices, and political ideas.

Understanding diffusion processes allows us to reconstruct the dispersal of cultural and technological ideas in the past. It also helps us predict such things as the impact of epidemics or the probability that an invention will find a market to justify the cost of production. It has many practical applications.

The Swedish geographer Torsten Hägerstrand in 1952 published a fundamental study on this topic entitled *The Propagation of Diffusion Waves*. Diffusion takes place in two types of processes: **expansion diffusion**, when propagation waves originate in a strong and durable source area and spread outward, affecting an ever larger area and population, and **relocation diffusion**, in which an idea, innovation (or, in the case of AIDS, a virus) is carried to a distant location and diffuses from there.

Islam, as Figure 6-3 shows, initially spread by rapid expansion diffusion from its western Arabian source. Virtually every village, town, and nomadic community was affected in a form of expansion diffusion shaped by local proximity; geographers call this process **contagious diffusion**. But Islam was also disseminated through the conversion of kings, chiefs, and other high officials, who in turn propagated it through their bureaucratic organizations. This is another form of expansion diffusion, called **hierarchical diffusion**, of which there are many variations; its main spatial expression is the downward filtering of the diffusing phenomenon from larger to smaller places within a national- or continental-scale urban hierarchy.

The map also shows that Islam spread far and wide by relocation diffusion, notably to the Ganges Delta and to the East Indies, where seafarers established remote new source areas. In East Africa it arrived by both expansion and relocation diffusion. Today the diffusion of Islam continues in many areas, including North America, where it arrived by relocation diffusion and now spreads by expansion diffusion.

■ ■ ■

gether. Mecca became the spiritual center for a divided, widely dispersed people for whom a collective focus was something new.

The Arab-Islamic Empire

Muhammad provided such a powerful stimulus that Arab society was mobilized almost overnight. The prophet died in 632, but his faith and fame continued to spread like wildfire. Arab armies carrying the banner of Islam formed, invaded, conquered, and converted where they went. As Figure 6-3 shows, by A.D. 700 Islam had reached far into North Africa, into Transcaucasia, and into most of Southwest Asia. In the centuries that followed, Islam penetrated Southern and Eastern Europe, central Asia's Turkestan, West Africa, East Africa, and South and Southeast Asia, even reaching China by A.D. 1000.

At the heart of this religious realm lay an Arab Empire. Its original capital was at Medina in Arabia, but as expansion continued its headquarters were moved first to Damascus and later to Baghdad. Meanwhile, Islam was carried ever farther outward by camel caravan and by pilgrim, by sailor, scholar, and sultan. Here was a manifestation of the process of **spatial diffusion** (see box titled ''Diffusion Processes''). And while Islam expanded, the Arab Empire that lay at the source matured and prospered. In such fields as architecture, mathematics, and science, the Arabs overshadowed their European contemporaries. The Arabs established institutions of higher learning in many cities including Baghdad, Cairo, and Toledo (Spain), and their distinctive cultural landscapes gave unity to their vast domain (see box titled ''The Flowering of Islamic Culture''). Non-Arab societies in the path of the Muslim drive were not only Islamized, but they also were Arabized, adopting other Arab traditions as well. The faith had spawned a culture; it still lies at the heart of that culture today.

As we noted, Islam's expansion eventually was checked in Europe, Russia, and elsewhere. But a map showing the total area under Muslim sway in Eurasia and Africa reveals the enormous dimensions of the domain affected by Islamization at one time or another (Fig. 6-4). The expansion of Islam, now mainly by relocation diffusion, continues. There are Islamic communities in places as widely scattered as Vienna, Singapore, and Cape Town, South Africa; Islam also continues to grow in the United States. With more than 1 billion adherents today, Islam remains a vigorous cultural force around the world.

ISLAM DIVIDED

For all its vigor and success, Islam did not escape fragmentation into sects. The earliest and most consequential division arose upon Muhammad's death. Who should be his legitimate successor? Some believed that only a blood relative should follow the prophet as leader of all Islam. Others, a majority, felt that any devout follower of Muhammad's was qualified. The first chosen successor was the father of Muhammad's wife (and thus not a blood relative). But this did not satisfy those who wanted to see a man named Ali, a cousin, made *caliph* (successor). When Ali's turn came, his followers, the Shi'ites, proclaimed that the first legitimate heir to the prophet had been installed. This offended the Sunnis, those who did not see a blood

THE FLOWERING OF ISLAMIC CULTURE

The conversion of a vast realm to Islam was not only a religious conquest: the rise of Islam was accompanied by a glorious explosion of Arab culture. In science, in the arts, in architecture, and in other fields, Arab society far outshone European society. While remnants of the Roman Empire languished, Arab energies soared.

When the wave of Islamic diffusion reached the Maghreb, the Arabs saw, on the other side of the narrow Strait of Gibraltar, an Iberia ripe for conquest and ready for renewal. An Arab-Berber alliance, the *Moors*, invaded Spain in 711, and before the end of the eighth century all but northern Castile and Catalonia were under Arab control.

It took seven centuries for Spain's Catholic armies to recapture all of the Arabs' Iberian holdings, but by then the Muslims had made an indelible imprint on the Spanish cultural landscape. The Arabs brought unity and imposed the rule of Baghdad, and their works soon overshadowed what the Romans had wrought. *Al-Andalus*, as this westernmost outpost of Baghdad was called, was endowed with thousands of magnificent mosques, castles, schools, gardens, and public buildings. The ultimately victorious Christians destroyed most of the less durable art (pottery, textiles, furniture, sculpture) and burned the contents of libraries, but the great Islamic structures survived, including the Alhambra in Granada, the Giralda in Seville, and the Great Mosque of Córdoba, three of the world's greatest architectural achievements. While Spanish culture became Hispanic-Islamic culture, the Muslims were transforming their cities from Turkestan to the Maghreb in the image of Baghdad. The lost greatness of a past era still graces those townscapes today.

Islam's penetration of Iberia permanently transformed the cultural landscapes of many Spanish cities. The Mesquita Mosque in Córdoba is only one of numerous Islamic legacies in Spain's southern province of Andalucía.

---

relationship as relevant in the succession. From the beginning of this disagreement, the numbers of Muslims who took the Sunni side far exceeded those who regarded themselves as Shia (followers) of Ali. The great expansion of Islam was largely propelled by Sunnis; the Shi'ites survived as small minorities scattered throughout the realm. Today, about 85 percent of all Muslims are Sunnis.

But the Shi'ites were masters of political manipulation who vigorously promoted their version of the faith. In the early sixteenth century their work paid off: the royal house of Persia (modern-day Iran) made Shi'ism the only legal religion throughout its vast empire. That domain extended from Persia into lower Mesopotamia (modern Iraq), into Azerbaijan, and into western Afghanistan and Pakistan. As the map of religions (Fig. 6-1) shows, this created for Shi'ism a large culture region and gave the faith unprecedented strength. Iran remains the bastion of Shi'ism in the realm today, and the appeal of Shi'ism continues to radiate into neighboring countries and even farther afield.

The differences between Sunni and Shia versions of Islam have had profound consequences for the realm. Sunni

Muslims believe in the effectiveness of family and community in the solution of life's problems, whereas Shi'ites follow their infallible *imams* (mosque officials who lead worshipers in prayer), the sole source of true knowledge. Sunnis tend to be comparatively reserved; Shi'ites often are passionate and emotional. For example, the death of Ali's son, Husayn, who would have been caliph, is commemorated annually with intense processions during which the marchers beat themselves with chains and cut themselves with sharp metal instruments. To many Sunnis, this is an unseemly demonstration of excess.

During the last decades of the twentieth century, Shi'ism gained unprecedented influence in the realm. In its heartland, Iran, a *shah* (king) tried to secularize the country and to limit the power of the imams; he provoked a revolution that cost him the throne and made Iran an Islamic republic—in fact, a Shi'ite republic. Before long, Iran was at war with Sunni-dominated Iraq, and Shi'ite parties and communities elsewhere were invigorated by the newfound power of Shi'ism. From Arabia to Africa's northwestern corner, Sunni-ruled countries warily watched their Shi'ite

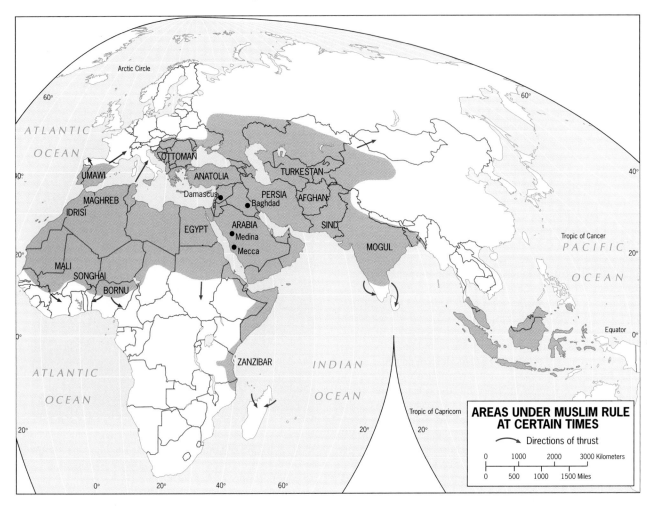

FIGURE 6-4

minorities, newly imbued with religious fervor. Mecca, the holy place for both Sunnis and Shi'ites, became a battle-ground during the week of the annual pilgrimage, and for a time the (Sunni) Saudi Arabian government denied entry to Iranian Shi'ite pilgrims. The split between Sunni and Shia is not the only fracture in Islam, but it is geographi-cally the most consequential.

Fundamentalism

Another cause of intra-Islamic conflict lies in the resur-gence of **religious fundamentalism**. In Shi'ite Iran, the imams wanted to reverse the shah's moves toward liber-alization and secularization: they wanted to (and did) recast society in traditional, fundamental Islamic molds. An *aya-tollah* (leader under Allah) replaced the shah in 1979; Is-lamic rules and punishments were instituted. Women, con-siderably liberated and educated during the shah's regime, resumed roles more in keeping with traditional Islam. Vestiges of Westernization, encouraged by the shah, dis-

appeared. The war against Iraq (1980–1990) began as a conflict over territory but became a holy war that cost more than a million lives.

Islamic fundamentalism did not rise in Iran alone, nor was it confined to Shi'ite communities. Many Muslims— Sunnis as well as Shi'ites—in all parts of the realm dis-approved of the erosion of traditional Islamic values, the corruption of society by European colonialists and later by Western modernizers, and the declining power of the faith in the secular state. While economic times were good, such dissatisfaction stayed beneath the surface. But when jobs were lost and incomes declined, the appeal of a return to fundamental Islamic ways increased.

This set Muslim against Muslim in all the regions of the realm. Fundamentalists fired the faith with a new militancy, challenging the status quo in countries ranging from Pa-kistan to Algeria. The militants forced their governments to ban ''blasphemous'' books, to resegregate the sexes in schools, to enforce traditional dress codes, to legitimize religious-political parties, and to heed the wishes of the *mullahs* (teachers of Islamic ways). Militant Muslims pro-

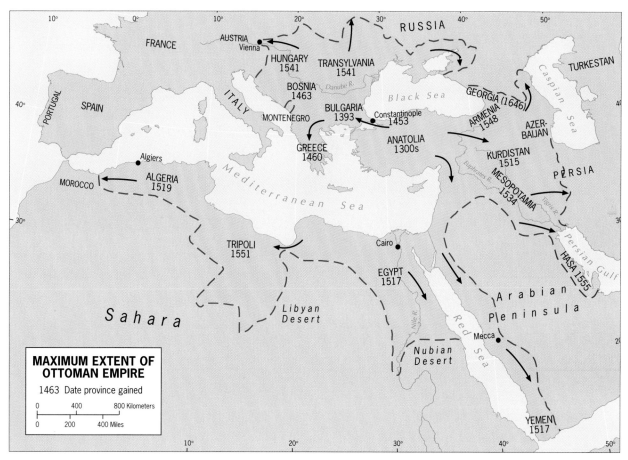

FIGURE 6-5

claimed that democracy inherited from colonialists and adopted by Arab nationalists was incompatible with the rules of the Koran.

In Algeria, the growing power of the fundamentalists led to a crisis. In late 1991, during democratic elections, the so-called Islamic Salvation Front (ISF) showed great strength. It was clear that the ISF would defeat the government in the second round of voting scheduled some weeks later; leaders of the ISF stated that they would transform Algeria into an Islamic republic. The government preempted the militants by resigning, leaving the door open for the armed forces to take control. Algeria was plunged into a long and deep crisis of confrontation and violence, ruining its social fabric, devastating its economy, killing tens of thousands, and spilling over into France, former colonial power and home to a large Algerian minority.

Already, Sudan, Pakistan, and of course Iran are officially Islamic republics. To many orthodox Sunnis who prefer to keep mosque and state separate, what has happened in Sudan is frightening. In 1989, a democratically elected government there was overthrown in a military coup. The army allowed the leaders of the National Islamic Front to institute Islamic laws; "nonbelievers" were

purged, and the *sharia* criminal code was introduced. (The sharia laws prescribe corporal punishment, amputations, stoning, and lashing for major and minor offenses.) Algeria's fundamentalists often referred to Sudan as their model as the elections approached, because Sudan proved that an Islamic state could indeed be established in a Sunni-dominated country. The specter of a Sudan-style regime in Algiers helped precipitate the crisis that began in 1991.

The rift between more liberal Muslims and fundamentalists poses a major challenge for the future. Militant Muslims confront the governments of Egypt, Jordan, Tunisia, and even Turkey. These governments have reacted in various ways (in Egypt, a combination of appeasement and containment; in Tunisia and Algeria, repression; in Jordan, co-option). Other countries, such as Morocco and Saudi Arabia, clearly are vulnerable. Hanging in the balance is the future of the realm.

We should not be led to believe that fundamentalism and religious militancy are exclusively Islamic phenomena. Such notions also infect Christian, Judaic, Hindu, and Buddhist societies. Nowhere, however, does the fundamentalist drive exhibit the intensity and vigor it displays in the Islamic realm.

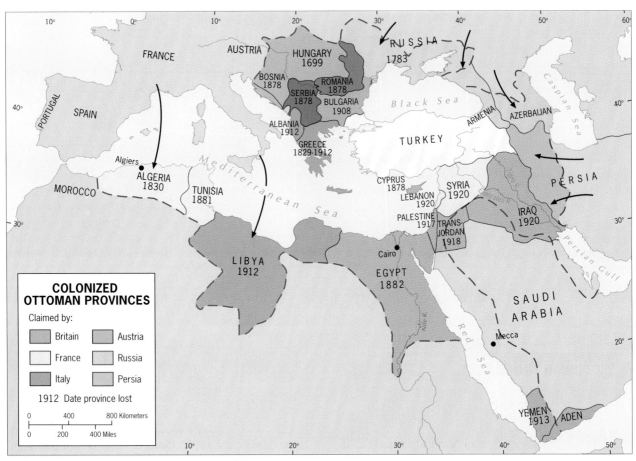

FIGURE 6-6

Islam and Other Religions

Two other major faiths had their sources in the *Levant* (the area extending from eastern Greece around the Mediterranean coast to northern Egypt), and both were older than Islam. Islam's rise overpowered many smaller Jewish communities, but it was the Christians, not the Jews, who waged centuries of holy war against the Muslims, seeking, through the Crusades, not only to drive Islam back but to reestablish Christian communities where they had been prior to the Islamic expansion. The aftermath of that campaign still marks the realm's cultural landscape today. A substantial Christian minority (about one-fourth of the population) remains in Lebanon, and Christian minorities also survive in Israel, Syria, Egypt, and Jordan. Strained relations between the long-dominant Christian minority in Lebanon and the Muslim majority (itself divided into five sects) contributed to the disastrous armed struggle that engulfed this country during the 1970s and 1980s.

But the most intense conflict in modern times has pitted the Jewish state, Israel, against its Islamic neighbors near and far. Israel's United Nations-sponsored creation in 1948 precipitated a half-century of intermittent strife and attempts at mediation; it also caused friction among Islamic states in the region. Jerusalem—holy city for Judaism, Christianity, *and* Islam—lies in the crucible of this confrontation.

The Ottoman Aftermath

It is a geographic twist of fate that Islam's last great advance into Europe resulted in the European occupation of Islam's very heartland. The Ottomans (named after their leader, Osman I), based in what is today Turkey, conquered Constantinople (now Istanbul) and pushed into Eastern Europe. Soon his forces were on the doorstep of Vienna; they also invaded Persia, Mesopotamia, and North Africa (Fig. 6-5). The Ottoman Empire under Suleyman the Magnificent, who ruled from 1522 to 1560, was the most powerful state of its time in all of western Eurasia.

The Ottoman Empire survived for more than four centuries, but it lost territory as time went on, first to the Hungarians, then to the Russians, and later to the Serbs until, after World War I, the European powers took over its provinces and made them colonies—colonies we now know by the names of Syria, Iraq, Lebanon, and Yemen (Fig. 6-6). As the map shows, the French and the British took large possessions; even the Italians claimed part of the Ottoman domain.

The boundary framework laid out by the colonial powers to delimit their holdings was anything but satisfactory. As Figure I-9 reminds us, this realm's population of nearly 500 million is clustered, fragmented, and strung out in river valleys, coastal zones, and crowded oases. Long stretches of boundary were laid out as ruler-straight lines across uninhabited territory; there seemed to be no need to adjust these **geometric political boundaries** to cultural or physical features in the landscape. Other boundaries, even some in desert zones, were poorly defined and never marked on the ground. Later, when the colonial powers had withdrawn and the colonies became independent states, such boundaries led to quarrels, even armed conflicts, among neighboring Muslim states.

THE POWER AND PERIL OF OIL

Travel through cities and towns of North Africa and Southwest Asia, talk to students and shopkeepers, taxi drivers and travelers, and you will hear the same refrain over and over again: ''Leave us in peace, let us do things our traditional way. Our problems—with each other, with the world—seem always to result from outside interference. The stronger countries of the world exploit our weaknesses and magnify our quarrels. We want to be left alone.''

That wish might have come closer to fulfillment were it not for two relatively recent events: the creation of the state of Israel and the discovery of some of the world's largest oil reserves. We defer our discussion of the founding of Israel until the regional section of this chapter. Here the focus is on the realm's most valuable export product: oil.

It is sometimes said that oil is the realm's most valuable *resource*. That, of course, is not accurate. The great majority of the people here continue to be farmers. The realm's most valuable resources are water and tillable soil. Oil is indeed a resource, an **energy resource**, and it is in demand throughout the world. But there was a Muslim world long before the first barrel of oil was taken from the ground, and there are tens of millions of people whose lives, to this day, are only indirectly affected by oil revenues. Egypt has very little oil (a few wells produce oil in the Sinai Peninsula). Morocco has almost none. Turkey depends heavily on oil imports.

Location of Known Reserves

In very general terms, oil (and associated natural gas) exists in this realm in three discontinuous zones (Fig. 6-7). The most productive of these zones extends from the southern and southeastern part of the Arabian Peninsula northwestward around the rim of the Persian Gulf, reaching into Iran and continuing northward into Iraq, Syria, and southeastern Turkey, where it peters out. The second zone lies across North Africa and extends from north-central Algeria eastward across northern Libya to Egypt's Sinai Peninsula, where it ends. The third zone begins on the margins of the realm in eastern Azerbaijan, continues under the Caspian Sea into Turkmenistan and Kazakstan, and also reaches into Uzbekistan, Tajikistan, and Krygyzstan.

The search for oil goes on, not only in this realm but also in many other areas of the world. As a result, estimates of the realm's reserves are subject to change as new discoveries are announced. Current assessments suggest that more than 65 percent of the world's known oil reserves lie in the North Africa/Southwest Asia realm.

In terms of production, Saudi Arabia long has been among the world leaders, and it is the undisputed top exporter. Two countries that rival Saudi Arabia as producers, the United States and Russia, also consume most of their output and do not export much oil. The United States, in fact, is the world's leading importer. Thus the oil production in Southwest Asia and North Africa is crucial to the rest of the world. As Figure I-11 indicates, oil wealth has elevated several of this realm's countries into the upper-middle and high-income categories. Oil wealth also has enmeshed these Islamic societies in world affairs, so that a threat to stability in an oil-rich country risks foreign intervention.

When the colonial powers laid down the boundaries that partitioned this realm among them, no one knew about the riches that lay beneath the ground. A few wells had been drilled, and production in Iran had begun as early as 1908 and in Egypt's Sinai Peninsula in 1913. But the major discoveries came later, in some cases after the colonial powers had already withdrawn. Some of the newly independent countries, such as Libya, Iraq, and Kuwait, found themselves with wealth undreamed of when the Turkish Ottoman Empire collapsed. As Figure 6-7 shows, however, others were less fortunate. A few countries had (and still have) potential. Among the smaller, weaker emirates and sheikdoms on the Arabian Peninsula there always was the fear that powerful neighbors would try to annex them. (Kuwait faced this prospect in 1990 when it was invaded by Iraq.) The unevenly distributed oil wealth, therefore, created yet another source of division and distrust among Islamic neighbors.

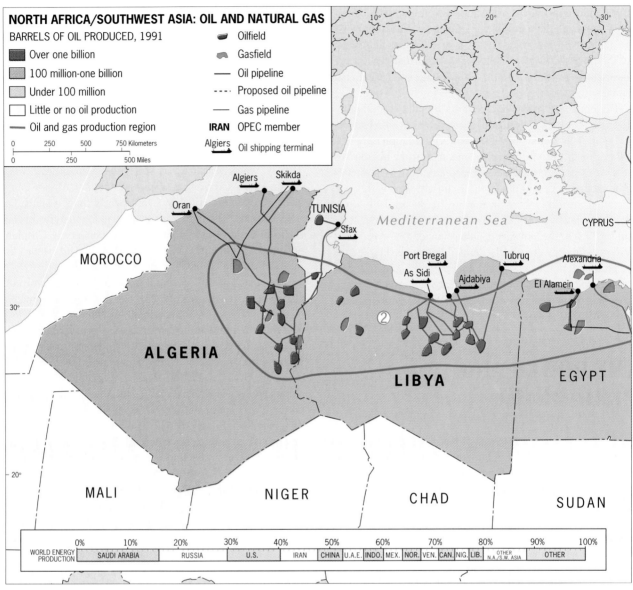

NORTH AFRICA/SOUTHWEST ASIA: OIL AND NATURAL GAS

BARRELS OF OIL PRODUCED, 1991

- Over one billion
- 100 million-one billion
- Under 100 million
- Little or no oil production
- Oil and gas production region

Oilfield
Gasfield
— Oil pipeline
---- Proposed oil pipeline
— Gas pipeline
IRAN OPEC member
Algiers Oil shipping terminal

0   250   500   750 Kilometers
0   250   500 Miles

FIGURE 6-7

A Foreign Invasion

The oil-rich countries of the realm had a coveted energy source, but they did not possess the equipment or skills to exploit it. These had to be introduced from the outside world and entailed what many tradition-bound Muslims feared most: the penetration of the vulgarities of Western ways. In his book *The Middle East*, geographer William Fisher reports that in Saudi Arabia there were two major cultural forces: Islam and Aramco, the joint Arab-American oil company. Yet while some diffusion of Western ideas and practices occurred, Islamic societies proved quite resistant to foreign acculturation. In the early 1990s, for example, when a group of women demanded the right to drive automobiles in Saudi Arabia, their campaign was firmly put down. Such Western ''excesses'' certainly have not taken root in this Islamic country.

In geographic terms, the impact of oil, its production and sale, in the exporting countries of this realm can be summarized as follows:

1. *High Incomes.* When oil prices on international markets were high, several countries in this realm

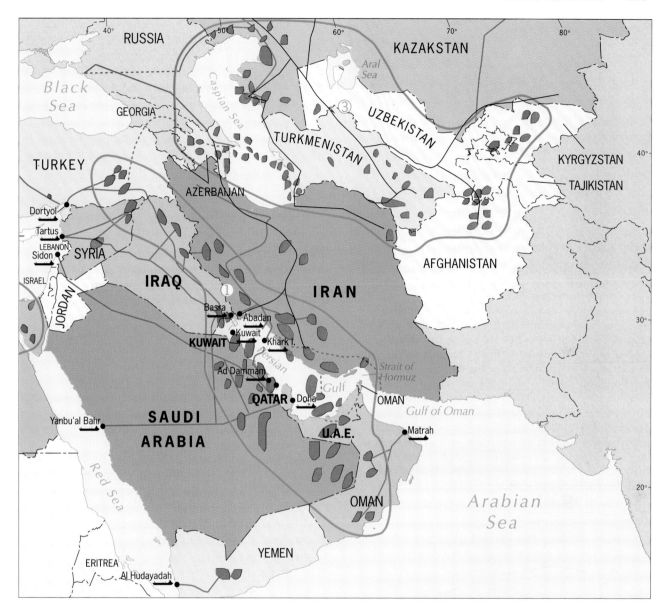

ranked as the highest income societies in the world. Even when oil prices declined, virtually all the oil-exporting states remained in the upper-middle-income category (Fig. I-11).

2. *Modernization.* Large oil revenues transformed cultural landscapes throughout the realm, producing a facade of modernization in the cultural landscape of ports and capitals. Gleaming glass-encased office buildings towered over mosques; superhighways crossed ancient camel paths; state-of-the-art port facilities handled oil and oil-funded trade.

3. *Industrialization.* Farsighted governments among those with oil wealth realize that oil reserves will not last forever, and have invested some of their income in industrial plants that will outlast the era of oil. Petrochemical industries, plastics fabrication, and desalinization plants are among these facilities.

4. *Intra-Realm Migration.* The oil wealth has allowed governments to attract millions of workers from less favored parts of the realm to work in the oilfields, in the ports, and in many other, mainly me-

From the Field Notes

"The impact of oil wealth is nowhere better illustrated than it is along the waterfront of the famous Creek of Dubai. We stopped on the bridge that links the two sectors of Dubai to observe the old *dhows* (most of these wooden boats now with motors as well as triangular sail) still at their centuries-old moorings, overlooked not by traditional Arab buildings but by modern skyscrapers. Oil may have transformed the local economy, but don't count the role of the dhows out just yet. They still carry trade and contraband. We went to the dock and asked what was being transported. 'Jeans and cassettes to Iran,' we were told, 'and caviar and carpets from Iran. It's a good business.' So the traditional role of the dhows goes on, even in the age of oil."

nial capacities. This has brought large numbers of Shi'ites to the countries of eastern Arabia; hundreds of thousands of Palestinians also are employed as laborers. Saudi Arabia, with a population of 20 million, has 5 million foreign workers.

5. *Inter-Realm Migration.* The willingness of workers from such countries as Pakistan, India, and Sri Lanka to work for wages even lower than those paid by the oil industry has attracted a substantial flow of temporary immigrants from outside the realm. These workers serve mostly as domestics, gardeners, refuse collectors, and the like.

6. *Regional Disparities.* Oil wealth and its manifestations in the cultural landscape create strong contrasts with areas not directly affected. The ultramodern east coast of Saudi Arabia is a world apart from its Red Sea zone in the west, where it becomes a land of desert, oasis, and camel, of vast distances, slow change, and isolated settlements. All oil-rich countries are to some degree affected by this phenomenon.

7. *Foreign Investment.* Oil-generated wealth has been invested, by governments and by Arab businesspeople, in foreign countries. These investments have created a network of international involvement that links many of this realm's countries not only to the economies of foreign states, but also to growing Arab (and thus Islamic) communities in those states.

The map (Fig. 6-7) contains a warning that came home to the oil-rich states (not only in this realm but elsewhere as well) in the 1980s when, after a period of high oil prices and huge revenues, the roof fell in and incomes plummeted. Even the power of a 12-member cartel, OPEC (Organization of Petroleum Exporting Countries), could not recover the lost advantage. Figure 6-7 shows a system of oil and gas pipelines that strongly resembles the exploitative interior-to-coast railroad lines in a mineral-rich colony of the past. Such a pattern spells disadvantage for the exporter, whether colony or independent country. Markets, not raw-material exporters, dominate international trade. Oil brought this realm into contact with the outside world in ways unforeseen just a century ago. Oil has strengthened and empowered some of its peoples; it has dislocated and imperiled others. It has truly been a double-edged sword.

REGIONS OF THE REALM

Identifying and delimiting regions in this vast geographic realm is a considerable challenge. Not only are population clusters widely scattered, but also cultural transitions—internal as well as external—make it difficult to discern a regional framework.

As we have noted on several earlier occasions, the world's regional geographic framework is subject to change. When Columbus sailed for the New World, the entire Balkan region of Eastern Europe was under the sway of the Muslim Ottoman Empire, and Islam cast its shadow over Vienna and Venice. In those times, this was not just a North African/Southwest Asian realm but an Eastern Eu-

Major Cities of the Realm

Cities	Population* (in millions)
Algiers, Algeria	4.1
Almaty, Kazakstan	1.3
Baghdad, Iraq	4.8
Beirut, Lebanon	2.1
Cairo, Egypt	10.2
Casablanca, Morocco	3.5
Damascus, Syria	2.2
Istanbul, Turkey	8.6
Jerusalem, Israel	0.6
Khartoum, Sudan	2.7
Riyadh, Saudi Arabia	2.9
Tehran, Iran	7.1
Tel Aviv, Israel	2.0
Toshkent, Uzbekistan	2.4
Tunis, Tunisia	2.2

*Based on 1997 estimates.

ropean one as well. Little more than a century ago, after the Austrian and Austro-Hungarian empires had wrested the upper Danube Basin from the Turks, the Muslims still ruled over much of Romania, Bulgaria, Serbia, Bosnia, Albania, and Greece. Not until the second decade of the twentieth century did the Ottomans lose the last of their European holdings, which finally erased the European Muslim region from the map. Ever since, Christian Greece and Islamic Turkey have been antagonistic neighbors.

In Asia, too, expanding Islam suffered setbacks. The Muslims spread northward from Persia into the Transcaucasian corridor between the Black and Caspian seas, reaching the northern slopes of the Caucasus Mountains and converting such peoples as the Ingush and the Chechens. But Russian armies stopped the advance and conquered the entire Muslim frontier. East of the Caspian Sea lay another arena of Islamic expansion, and much of central Asia became a mosaic of Muslim societies, a vast but sparsely peopled region where Islam traveled by caravan and took root in the oases. Once again the Russian tsars had other plans for this region, and the scattered sheikdoms and caliphates were no match for Moscow's armies. Later, when the Soviet communists inherited the Russian Empire, they divided Central Asia into five colonies and proceeded to try to extinguish Islam in favor of communism's official atheism.

Islam's defeats had different consequences in different regions. In Eastern Europe, old Christian traditions soon expunged Islam's imprints except in Albania, Kosovo, and Bosnia, where remnants of it survived, and in Bulgaria, where a small Turkish minority remained. But in Transcaucasia and in the five Soviet republics of Central Asia

(the region called *Turkestan*), Islam proved more durable. After the Soviet Empire collapsed in 1991, Islam quickly reasserted itself, and a wave of Muslim fervor swept over these old Islamic frontiers. Here was a classic case of **cultural revival**, the regeneration of a long-dormant culture through internal renewal and external infusion.

The realm's regional framework, therefore, has changed once again. We must now recognize the central Asian region as part of it. Change also is affecting Transcaucasia, where the freed former Soviet republic of Azerbaijan displays intense Muslim militancy. The borders of this realm have always been volatile; the 1990s are no exception.

The Regional Matrix

The vast North Africa/Southwest Asia realm is divided into seven regions (Fig. 6-8). As both Figures I-12 and 6-8 suggest, it is appropriate to recognize two sets of regions: those to the south of a line connecting the northeastern corner of the Mediterranean Sea and the head of the Persian Gulf, and those to the north. This is a line of contrast, conflict, and transition that not only separates Arab from non-Arab states but also divides different Islamic sects. Astride this line lies a nation without a state: the Kurds, victims of intraregional rivalries.

The following are the regional components of this geographic realm:

1. **Egypt and the Lower Nile Basin.** This region in many ways constitutes the heart of the realm as a whole. Egypt (together with Iran and Turkey) is one of the realm's three most populous countries. It is the historic focus of this part of the world and a major political and cultural force. It shares with its southern neighbor, Sudan, the waters of the lower Nile River.

2. **The Maghreb and Its Neighbors.** Western North Africa (the *Maghreb*) and the areas that border it also form a region, consisting of Algeria, Tunisia, and Morocco at the center and Libya, Chad, Niger, Mali, and Mauritania along the broad periphery. The last four of these countries also lie astride or adjacent to the broad transition zone where the Arab-Islamic realm of northern Africa merges into Subsaharan Africa.

3. **The African Transition Zone.** From southern Mauritania in the west to Somalia in the east, across the entire African landmass at its widest extent, the realm dominated by Islamic culture interdigitates with that of Subsaharan Africa. No sharp dividing line can be drawn here: people of African ethnic stock have adopted the Muslim faith and Arabic language and traditions. As a result, this is less a region than a zone of transition.

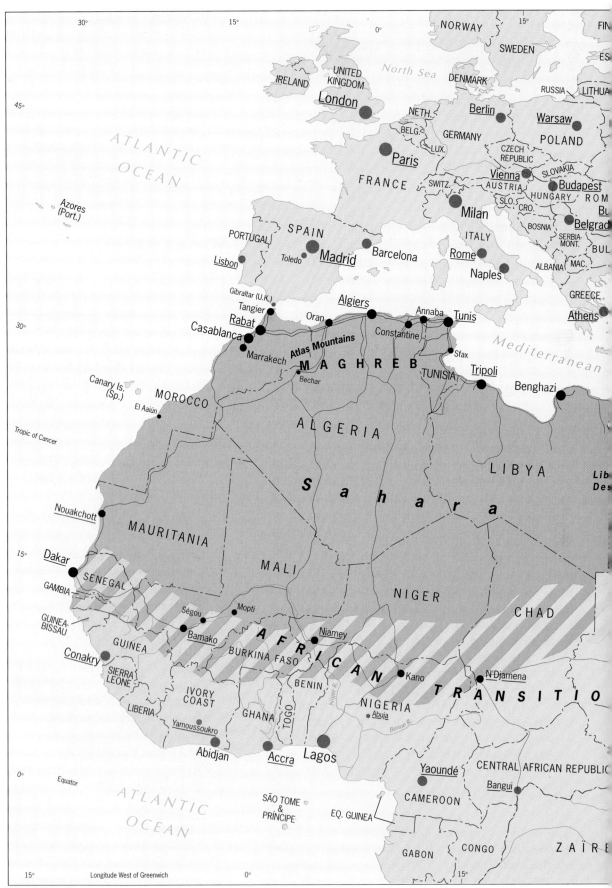

FIGURE 6-8

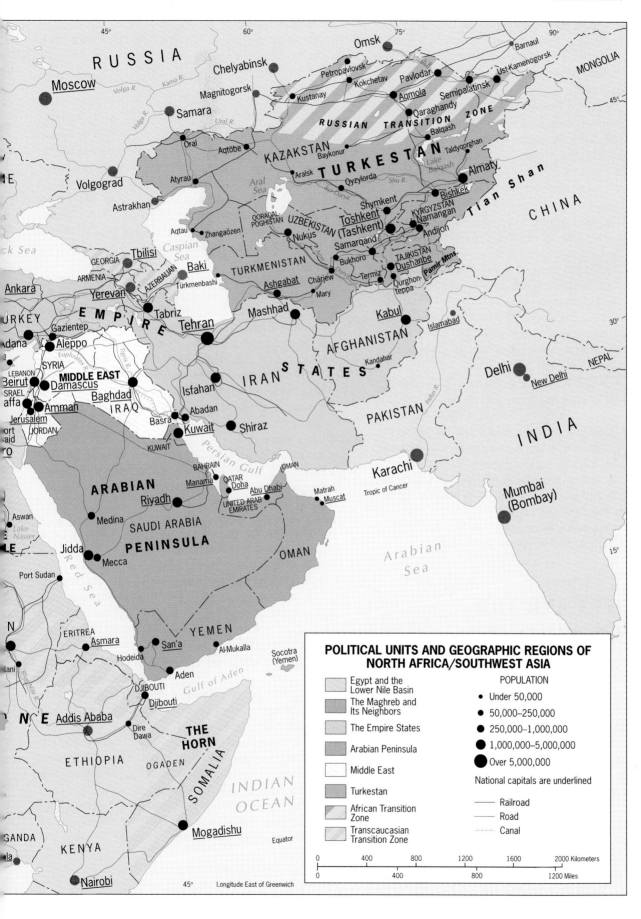

RUSSIA

Moscow

Chelyabinsk

Omsk

Barnaul

MONGOLIA

Magnitogorsk

Petropavlovsk

Kokchetav

Pavlodar

Ust-Kamenogorsk

Samara

Kustanay

Agmola

Semipalatinsk

RUSSIAN TRANSITION ZONE

Oral

Aqtöbe

KAZAKSTAN

Baykonur

Qaraghandy

Balqash

Taldyqorghan

TURKESTAN

Volga R.

Volga R.

Kama R.

Ural R.

45°

60°

75°

90°

45°

30°

15°

Volgograd

Atyrau

Aralsk

Qyzylorda

Shu R.

Lake
Balqash

Almaty

Tian Shan

CHINA

Astrakhan

Aral
Sea

Aqtau

Zhangaözen

QORAQAL-
POGHISTON

Shymkent

Toshkent
(Tashkent)

Bishkek

KYRGYZSTAN
Namangan

Andijon

ck Sea

Caspian
Sea

GEORGIA

Tbilisi

Baki

AZERBAIJAN

Türkmenbashi

UZBEKISTAN

Nukus

Samarqand

Bukhoro

Amu Darya

TAJIKISTAN
Dushanbe

Pamir Mtns.

ARMENIA

Ankara

Yerevan

Tabriz

TURKMENISTAN

Ashgabat

Chärjew

Mary

Termiz

Qurghon-
teppa

Syr-Darya

URKEY

EMPIRE

Tehran

Mashhad

Kabul

Islamabad

Delhi

New Delhi

NEPAL

Gazientep

adana

Aleppo

SYRIA

LEBANON

Beirut

SRAEL

affa

Amman

Jerusalem

ort

aid

JORDAN

MIDDLE EAST

Damascus

Baghdad

IRAQ

Euphrates R.

Tigris R.

Isfahan

Abadan

Basra

Kuwait

KUWAIT

Shiraz

IRAN

STATES

Kandahar

AFGHANISTAN

PAKISTAN

Indus R.

INDIA

Persian Gulf

BAHRAIN

Manama

QATAR
Doha

Abu Dhabi

UNITED ARAB
EMIRATES

OMAN

Matrah

Muscat

Karachi

Mumbai
(Bombay)

ARABIAN

Riyadh

Tropic of Cancer

Aswan

Lake
Nasser

Medina

SAUDI ARABIA

PENINSULA

Jidda

Mecca

OMAN

Arabian
Sea

Port Sudan

Red Sea

N

ERITREA

Asmara

San'a

YEMEN

Al-Mukalla

Socotra
(Yemen)

Hodeida

Aden

Gulf of Aden

DJIBOUTI

Djibouti

LE

Blue Nile R.

E

Addis Ababa

Dire
Dawa

THE
HORN

ETHIOPIA

OGADEN

SOMALIA

INDIAN
OCEAN

GANDA

Mogadishu

KENYA

Equator

45°

Longitude East of Greenwich

a

Nairobi

**POLITICAL UNITS AND GEOGRAPHIC REGIONS OF
NORTH AFRICA/SOUTHWEST ASIA**

Egypt and the
Lower Nile Basin

The Maghreb and
Its Neighbors

The Empire States

Arabian Peninsula

Middle East

Turkestan

African Transition
Zone

Transcaucasian
Transition Zone

POPULATION

● Under 50,000

● 50,000–250,000

● 250,000–1,000,000

● 1,000,000–5,000,000

● Over 5,000,000

National capitals are underlined

——— Railroad

——— Road

+++++ Canal

0   400   800   1200   1600   2000 Kilometers

0   400   800   1200 Miles

4. **The Middle East.** This region includes Israel, Jordan, Lebanon, Syria, and Iraq. In effect, it is the crescent-like zone of countries that extends from the eastern Mediterranean coast to the head of the Persian Gulf.

5. **The Arabian Peninsula.** Dominated by the large territory of Saudi Arabia, the Arabian Peninsula also includes the United Arab Emirates, Kuwait, Bahrain, Qatar, Oman, and Yemen. Here lies the source and focus of Islam, the holy city of Mecca; here, too, lie many of the world's greatest oil deposits.

6. **The Empire States.** Across the zone of mountains, highlands, and plateaus that extends from Turkey in the west across Iran in the center to Afghanistan in the east lies a region of five states dominated by two: the secular Turkish republic and the Islamic republic of Iran. Both have an imperial history, and the influence of each still extends into neighboring countries. In one of these countries, the island of Cyprus, the involvement is Turkish; in the other, Azerbaijan, the link is with Iran.

7. **Turkestan.** We revive an old name for a region where Islam is resurgent, where a critical period of redirection lies ahead. This is the old Soviet Central Asia, and while the majority of the population is nominally Muslim, Russian minorities continue to play a role here. Kazakstan, by far the largest state, contains a transition zone between its dominantly Muslim south and majority-Russian north (Fig. 6-8). The other components of this region are the most populous country, Uzbekistan, as well as Turkmenistan, Kyrgyzstan, and Tajikistan.

❖ EGYPT AND THE LOWER NILE BASIN

Egypt, anchor of the lower Nile Basin, occupies a pivotal location in the heart of a realm that extends for over 6,000 miles (9,600 km) longitudinally and some 4,000 miles (6,400 km) latitudinally. At the northern end of the Nile and of the Red Sea, at the eastern end of the Mediterranean Sea, in the northeastern corner of Africa across from Turkey to the north and Saudi Arabia to the east, adjacent to Israel, to Islamic Sudan, and to militant Libya, Egypt lies in the crucible of this realm. Because it owns the Sinai Peninsula (recently lost to and regained from Israel), Egypt, alone among states on the African continent, has a foothold in Asia, a foothold that gives it a coast overlooking the strategic Gulf of Aqaba. Egypt also controls the Suez Canal, vital link between the Indian and Atlantic

oceans and lifeline of Europe. It is hardly necessary to further justify Egypt's designation (together with northern Sudan) as a discrete region.

The Nile River

The Greek scholar Herodotus described Egypt as the gift of the Nile, but Egypt also was a product of natural protection. In the protective isolation of their culture hearth, the ancient Egyptians converted security into stability and progress. Modern Egypt's cultural landscape still carries the marks of antiquity's accomplishments in the form of great stone sculptures and pyramids that bear witness to the flowering of a continuous civilization that is now more than 5,000 years old.

Egypt's Nile is the aggregate of two great branches upstream: the White Nile, which originates in the streams that feed Lake Victoria in East Africa, and the Blue Nile, whose source lies in Lake Tana in Ethiopia's highlands. The two Niles converge at Khartoum in modern-day Sudan. Fly northward from Khartoum along the Nile toward Egypt, and you will realize how small and vulnerable this ribbon of water looks in the vast wasteland of the Sahara. And on this ribbon depend the lives of tens of millions of people.

About 95 percent of Egypt's 65 million people live within a dozen miles (20 km) of the great river's banks or in its delta (Fig. 6-9). It always has been this way: the Nile rises and falls seasonally, watering and replenishing soils and crops on its banks. In April and May, the river usually is at its lowest level, a trickle through the desert. Then during the summer months it rises, to reach flood stage at Cairo, the capital, in October. Then the Nile may be as much as 20 feet (6 m) above its lowest stage.

The ancient Egyptians used **basin irrigation**, building fields with earthen ridges and trapping the floodwaters with their fertile silt, to grow their crops. That practice continued for thousands of years until, during the nineteenth century, the construction of permanent dams made possible the **perennial irrigation** of Egypt's farmlands. These dams, with locks for navigation, controlled floods, expanded the country's cultivable area, and allowed the raising of more than one crop per year on the same field. In a single century, all of Egypt's farmland was brought under perennial irrigation.

The greatest of all Nile projects, the Aswan High Dam (which began operating in 1968), creates Lake Nasser, one of the world's largest artificial lakes (Fig. 6-9). As the map shows, the lake extends into Sudan, where 50,000 people had to be resettled to make way for it. The Aswan High Dam increased Egypt's irrigable land by nearly 50 percent and today provides the country with about 40 percent of its electricity.

Egypt's elongated oasis, just 3 to 15 miles (5 to 25 km) wide, widens north of Cairo across a delta anchored in the

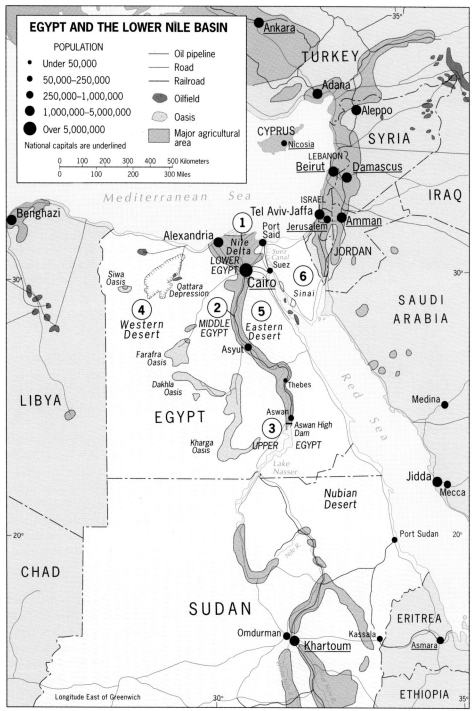

FIGURE 6-9

west by the great city of Alexandria and in the east by Port Said, gateway to the Suez Canal. The delta contains extensive farmlands, but it is a troubled area today. The ever more intensive use of the Nile's water and silt upstream is depriving the delta of much needed replenishment. And the low-lying delta is geologically subsiding, raising fears of salt-water invasion from the Mediterranean Sea that would damage soils here.

Egypt's millions of subsistence farmers, the *fellaheen*, still struggle to make their living off the land, as did the peasants of the Egypt of five millennia past. Rural landscapes seem barely to have changed; ancient tools are still in use, and dwellings remain rudimentary. Poverty, disease, high infant mortality rates, and low incomes prevail. And despite the Aswan High Dam and the expansion of irrigation, Egypt's available farmland per capita has steadily

declined during the past two centuries as population growth (still 2.3 percent per year) keeps nullifying gains in agricultural productivity.

Egypt's Regional Prominence

Despite all this, Egypt constitutes, in many ways, this realm's most important and influential country. Egypt lies spatially, culturally, and ideologically at the heart of the Arab World. Its population numbers are rivaled by those of Turkey and Iran, but Egypt is the dominant *Arab* country; neither Turks nor Iranians are Arabs. A secular state with a government that in theory is democratically elected, Egypt borders Israel to the east and Libya to the west,

controls a vital lifeline in the Suez Canal, and contains the realm's greatest city, Cairo, its capital. With leadership (in the peace process toward Israel, in its participation in the Gulf War of 1991 against Iraq) comes criticism, and there are concerns that Egypt's democratic institutions may not be strong enough to accommodate it. The government, seeking to limit the influence of the Muslim Brotherhood, sometimes uses intimidation to achieve that goal. The response is a growing militancy among the minority that would change Egypt into an Islamic Republic. An Algeria-type breakdown of order would be a calamity beyond measure.

Egypt has six subregions, mapped in Figure 6-9. The great majority of Egyptians live and work in Lower (i.e., northern) and Middle Egypt (regions ① and ②), the country's core area anchored by Cairo and flanked by the leading port and second industrial center, Alexandria. The economy has benefited from further oil discoveries in the Sinai (region ⑥ in Fig. 6-9) and in the Western Desert ④, so that Egypt now is self-sufficient and even exports some petroleum. Cotton and textiles form the other major source of external income, but the important tourist industry has been dealt a blow by Islamic extremists who have attacked visitors throughout the country. As the population mushrooms, the gap between food supply and demand widens, and Egypt must import grain. For many years, Egypt has been a major recipient of U.S. foreign aid.

Egypt in the late 1990s was at a crossroads in more than one way. Its planners know that reducing the high birth rate would improve the demographic situation, but fundamentalist Muslims object to any programs that promote family planning. Its accommodation with Israel helps ensure foreign aid but divides the people. Its government faces a fundamentalist challenge. Egypt's future, in this crucial corner of the realm, is uncertain.

❖ THE MAGHREB AND LIBYA

The countries of northwestern Africa are collectively called the *Maghreb*, but the Arab name for them is more elaborate than that: *Djezira-al-Maghreb*, or "Isle of the West," in recognition of the great Atlas Mountain range rising like a vast island from the waters of the Mediterranean Sea to the

From the Field Notes

"On the eastern edge of central Cairo we saw what looked like a combination of miniature mosques and elaborate memorials. Here lie buried the rich and the prominent of times past in what locals call the 'City of the Dead.' But we found it to be anything but a dead part of the city. Many of the tombs here are so large and spacious that squatters have occupied them. Thus the City of the Dead is now an inhabited graveyard, home to at least one million people. The exact numbers are impossible to determine; indeed, whereas metropolitan Cairo in 1997 has an official population of just over 10 million, many knowledgeable observers believe that 16 million is closer to the mark."

north and the sandy flatlands of the Sahara to the south.

The countries of the Maghreb (sometimes spelled *Maghrib*) are Morocco, last of the North African kingdoms; Algeria, a secular republic beset by the religious-political problems we noted earlier; and Tunisia, smallest and most Westernized of the three (Fig. 6-10). Libya, facing the Mediterranean between the Maghreb and Egypt, is unlike any other North African country: an oil-rich desert state whose population is almost entirely clustered in settlements along the coast.

Whereas Egypt is the gift of the Nile, the Atlas Mountains form the nucleus of the settled Maghreb. These high ranges wrest from the rising air enough orographic rainfall to sustain life in the intervening valleys, where good soils support productive farming. From the vicinity of Algiers

▪ AMONG THE REALM'S GREAT CITIES...

Cairo

Stand on the roof of one of the high-rise hotels in the heart of Cairo, and in the distance you can see the great pyramids, monumental proof of the longevity of human settlement in this area. But the present city was not founded until Muslim Arabs chose the site as the center of their new empire in A.D. 969. Cairo became and remained Egypt's primate city, situated where the Nile River opens into its delta, home today to almost one-sixth of the entire country's population.

Cairo ranks among the world's 20 largest urban agglomerations, and it shares with other cities of the poorer world the staggering problems of crowding, inadequate sanitation, crumbling infrastructure, and substandard housing. But even among such cities, Cairo is noteworthy for its stunning social contrasts. Along the Nile waterfront, elegant skyscrapers rise above carefully manicured, Parisian-looking surroundings. But look eastward, and the urban landscape extends gray, dusty, almost featureless as far as the eye can see. Not

far away, more than a million squatters live in the sprawling cemetery known as the City of the Dead (see photo above). On the outskirts, millions survive in overcrowded shantytowns of mud huts and hovels.

And yet Cairo is the dominant city not only of Egypt but for a wider sphere, the cultural capital of the Arab World, with centers of higher learning, splendid museums, world-class theater and music, and magnificent mosques and Islamic learning centers. Although Cairo always has been primarily a center of government, administration, and religion, it also is a river port and an industrial complex, a commercial center and, it sometimes seems, one giant *bazaar*. Cairo is the heart of the Arab World, a creation of its geography, and a repository of its history.

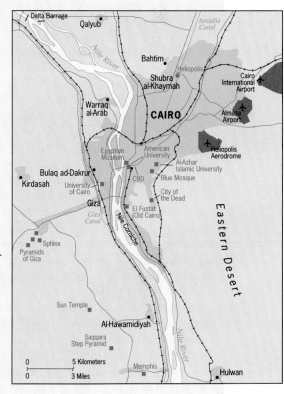

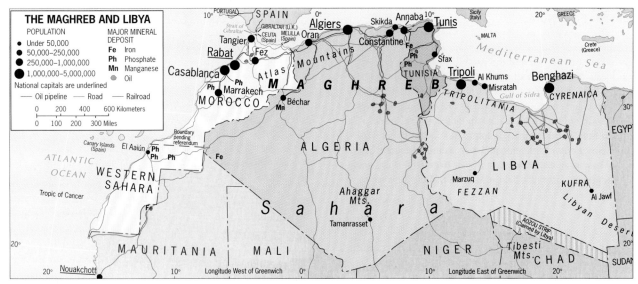

FIGURE 6-10

eastward along the coast into Tunisia, annual rainfall averages more than 30 inches (75 cm), a total more than three times as high as that recorded for Alexandria in Egypt's delta. Even 150 miles (240 km) inland, the slopes of the Atlas still receive over 10 inches (25 cm) of rainfall. The effect of the topography can be read on the world map of precipitation (Fig. I-6): where the highlands of the Atlas terminate, desert conditions immediately begin.

The Atlas Mountains are structurally an extension of the Alpine system that forms the orogenic backbone of Europe, of which Switzerland's Alps and Italy's Appennines are also parts. In northwestern Africa, these mountains trend southwest-northeast and commence in Morocco as the High Atlas, with elevations close to 13,000 feet (4,000 m). Eastward, two major ranges appear that dominate the landscapes of Algeria proper: the Tell Atlas to the north, facing the Mediterranean, and the Saharan Atlas to the south, overlooking the great desert. Between these two mountain chains, each consisting of several parallel ranges and foothills, lies a series of intermontane basins (analogous to South America's Andean *altiplanos* but at lower elevations), markedly drier than the northward-facing slopes of the Tell Atlas. In these valleys, the rain shadow effect of the Tell Atlas is reflected not only in the steppe-like natural vegetation but also in land-use patterns: pastoralism replaces cultivation, and stands of short grass and bushes blanket the countryside.

During the colonial era, well over a million Europeans came to settle in North Africa—most of them French, and a large majority to Algeria—and these immigrants soon dominated commercial life. They stimulated the renewed growth of the region's towns. Casablanca, Algiers, and Tunis rose to become the urban foci of the colonized territories. Although the Europeans dominated trade and commerce and integrated the North African countries with France and the European Mediterranean world, they did not confine themselves to the cities and towns. They recognized the agricultural possibilities of the favored parts of the *tell* (the lower Tell Atlas slopes and narrow coastal plains that face the Mediterranean) and established thriving farms. Not surprisingly, agriculture here is of the Mediterranean variety. Algeria soon became known for its vineyards and wines, citrus groves, and dates; Tunisia has long been one of the world's leading exporters of olive oil; and Moroccan oranges went to many European markets.

Oil and Emigrants

The fortunes of the Maghreb states since independence (four decades ago) have varied. Morocco became embroiled in a territorial conflict involving Western Sahara to its south, a former Spanish possession. A small but resilient group of Western Saharan residents resisted Morocco's absorption of this desert area, and Algeria and Mauritania gave them some support. This strained relations between Morocco and its neighbors, but in Morocco itself the campaign had a unifying effect at a time when the king faced growing antimonarchist sentiment. In Algeria, oil revenues replaced farm produce as the major export and revenue earner; reserves not only of oil but also of natural gas are linked by pipelines to Algerian as well as Tunisian ports for shipment to Europe (Fig. 6-10). For more than a decade, Algeria's income rose with world oil prices, but then a combination of dropping prices and diminishing reserves reversed the upward trend. Tunisia, too, went through an oil boom, but its supplies dwindled even faster, and other sources of income were needed. Fortunately these were available: Tunisia had not abandoned its agricultural economy (citrus fruits, olives and olive oil, textiles, leather goods) when the oil boom prevailed.

The other export of the Maghreb states has been people. North Africans emigrated to Europe, mainly to France, by the hundreds of thousands in search of work unavailable at home. Smaller emigration streams led to Spain and to Italy; in 1997, at least 2.5 million Algerians, Moroccans, and Tunisians resided in Europe.

The political landscape of the Maghreb is changing. Both Algeria and Tunisia face a rising tide of Islamic fundamentalism that strengthened as national economies weakened and poverty and frustration rose. Morocco, the leading conservative force in Arab affairs, has dealt severely with the fundamentalist challenge, but its nonoil economy (phosphates, fertilizers, and Mediterranean fruits are its chief income earners, along with tourism) is weak, and the market for militant Islam will thus widen.

The Desert Coast

Between the eastern terminus of the Atlas range (in Tunisia) and Egypt's Nile delta, the Sahara reaches the very shores of the Mediterranean Sea. Here lies Libya, small in population (5.6 million), large in area (nearly three times bigger than Texas), and rich in oil.

Almost rectangular in shape, Libya is a country whose four corners matter most (Fig. 6-10). What limited agricultural possibilities exist lie in the northwest in Tripolitania, centered on the capital, Tripoli, and in the northeast in Cyrenaica, where Benghazi is the urban focus. Between these two coastal clusters, which are home to 90 percent of the Libyan people, lies the Gulf of Sidra, a deep Mediterranean bay. Libya has claimed this gulf as its territorial sea, but this claim has not been accepted by other countries. Libya's two southern corners are the desert Fezzan, a mountainous area on the southwestern border with Algeria and Niger, and the sparsely populated Kufra Oasis in the southeast. Despite its huge size and tiny population, Libya has claimed a sector (the Aozou Strip) of its southern neighbor, Chad (Fig. 6-10).

Libya's long-time ruler, Muammar Gadhafi, used the revenues from major oil reserves in the country's northeastern quadrant for two purposes. The first was to improve the country's infrastructure. His most successful project has been ''the great man-made river,'' a gigantic, 600-mile (950-km)-long pipeline that carries water from aquifers beneath the Sahara in the south to the parched populations of Cyrenaica and Tripolitania. The second purpose was to promote revolutionary Arab-Islamic causes abroad. Despite armed retaliation and boycotts, that policy sustained for Libya a power position far greater than its population—less than one-tenth of Egypt's—would seem to merit.

❖ THE AFRICAN TRANSITION ZONE

Islam diffused by land and by sea. It crossed the Mediterranean Sea and the Indian Ocean; it gained footholds on many coasts from East Africa to Indonesia. In North Africa, Islam also crossed the desert to the south. It came up the Nile by boat and across the Sahara by caravan, and it reached the peoples of the interior steppes and savannas on the other side. There, along a wide stretch of Africa known today as the *Sahel* (an Arabic word for ''border'' or ''margin''), Islam's proselytizers converted millions to the Muslim faith. As we will discover in the next chapter, the zone between the desert and the forest was one of Africa's culture hearths, and here lay large and durable states. Their rulers and subjects adopted Islam, and soon huge throngs of West Africans marched along the savanna corridor eastward on their annual pilgrimages to Mecca.

Then came Europe's colonial powers, and the modern political map of Africa took shape. From Senegal to Sudan, Muslims and non-Muslims were thrown together in countries not of their making. Thus superimposed on the cultural landscape and on the political mosaic, a zone of transition marks the map (Figs. I-12 and 6-8). It is a zone in which peoples are ethnically African but culturally Arabized.

A Corridor of Instability

The African Transition Zone affects primarily (but not exclusively) Senegal, Mauritania, Mali, Burkina Faso, Niger, Nigeria, Chad, Sudan, and the countries of the *Horn of Africa* (Ethiopia, Eritrea, Djibouti, and Somalia). These are countries with one foot in each geographic realm, and none has escaped the consequences. From Senegal's efforts to merge with The Gambia to Somalia's pending fragmentation into two countries (Fig. 6-11), this is a corridor of culture conflict.

In the 1990s, tension and strife have afflicted Nigeria, Chad, Sudan, and the Horn. Nigeria has a Muslim north and a non-Muslim south, but Christian southerners have settled in northern cities to take jobs northerners would not. Intermittent clashes have cost thousands of lives there. Nigeria's northeastern neighbor, Chad, has a deep cultural chasm as well, its 45-percent Muslim population holding the north and center, the 55-percent non-Muslims the south. To the east, Sudan presents an even more complicated picture: a large population (30 million) in a crucial geographic location, divided into a Muslim north containing about 70 percent of all Sudanese and a mostly animist, traditional-African south that has for decades fought off the north's efforts to control it. This long-term conflict may

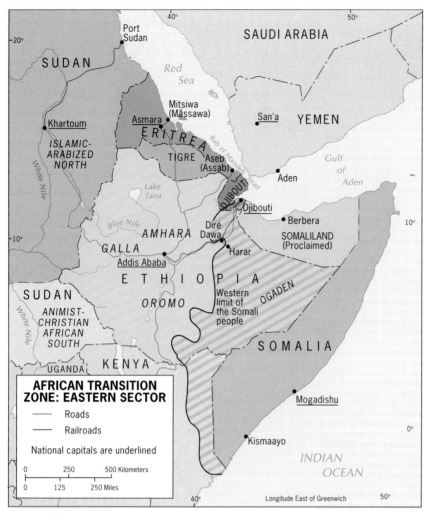

FIGURE 6-11

be the costliest, in terms of human lives, of all such wars to have broken out in Africa in the postcolonial period.

Culture conflict in the Horn of Africa has at times spilled over into Sudan, creating there some of the worst refugee crises on the continent. Ethiopia's highlands, many centuries ago, became the natural fortress for a Christian people, the Amhara, who not only resisted the Muslim advance successfully but also conquered a large Muslim domain in the lowlands surrounding their core area. This Ethiopian empire, centered on Addis Ababa, thus contained the seeds of war that sprouted in the 1980s. On the Red Sea coast, Muslim Eritrea broke away from Ethiopia in 1993, with the result that Ethiopia now is a landlocked state. And while the Amhara were forced to yield their monopoly on power in the new Ethiopian government, sharing it with a group of peoples known as the Oromo and others, the key issue still faces Addis Ababa: millions of Muslim Somali live within Ethiopia (Fig. 6-11). Even after the secession of all-Muslim Eritrea, Ethiopia's population of 60 million

still is about 45 percent Islamic. Here, the problems of location astride the African Transition Zone are far from over.

If there is one overriding geographic quality marking the African Transition Zone in addition to its cultural problems, it is its environmental plight. As Figure I-7 reveals, the African Transition Zone is almost coincident with the steppe (*BSh*) climatic region that extends from the Atlantic coast to the Horn. This is a physiographic belt of low and variable rainfall, cyclical desertification, and fragile vegetation; but it also is a zone of livestock herders. Cattle and goats contribute to the ecological damage done by human overpopulation; they worsen the effects of droughts by destroying what is left of the natural vegetation, and they delay its recovery later. Pastoralists are mobile people, and when drought limits the pastures in the steppe they tend to move southward, driving their herds into the fields of sedentary farmers. This creates still another source of friction in this troubled African Transition Zone.

❖ THE MIDDLE EAST

The regional term *Middle East*, as we noted earlier, is far from satisfactory, but it is in such common and general use that avoiding it creates more problems than are solved. It originated when Europe was the world's dominant realm and when things were ''near,'' ''middle,'' and ''far'' from Europe: hence a *Near* East (Turkey, Egypt, Libya), a *Far* East (China, Japan, Korea, and other countries of East Asia), and a *Middle* East (many of the countries in between). If you check definitions of the past, you will see that the terms were applied rather inconsistently: Syria,

Lebanon, Palestine, even Jordan sometimes were included in the ''Near'' East, and Persia and Afghanistan in the ''Middle'' East.

Today, the geographic designation *Middle East* has a more specific meaning. And at least half of it has merit: this region, more than any other, lies at the middle of the vast Islamic realm (Fig. 6-8). To the north and east of it, respectively, lie Turkey and Iran, with Afghanistan and Muslim Turkestan beyond the latter. To the south lies the Arabian Peninsula. And to the west lie the Mediterranean Sea and Egypt, and the rest of North Africa. This, then, is the pivotal region of the realm, the very heart of it.

Five countries form the Middle East (Fig. 6-12): Iraq,

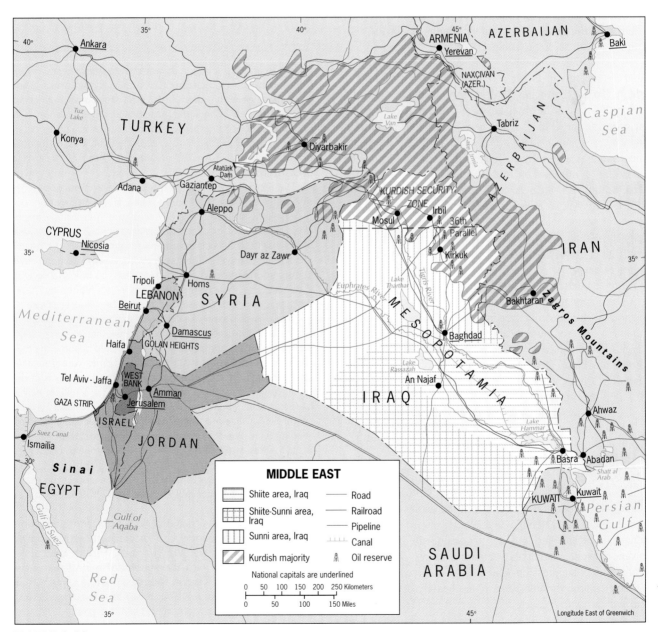

FIGURE 6-12

largest in population and territorial size, facing the Persian Gulf; Syria, next in both categories and fronting the Mediterranean; Jordan, linked by the narrow Gulf of Aqaba to the Red Sea; Lebanon, whose survival as a unified state has come into question; and Israel, Jewish nation in the crucible of the Muslim world.

Iraq

California-sized Iraq (22 million) comprises nearly 60 percent of the total area of the Middle East and has 40 percent of the region's population. With major oil reserves and large areas of irrigated farmland, Iraq also is best endowed with natural resources. Iraq is heir to the early Mesopotamian states and empires that emerged in the basin of the Tigris and Euphrates rivers, and the country is studded with significant archeological sites.

Today, Iraq is bounded by as many as six neighbors, and has recently had adversarial relationships with most of them. To the north lies Turkey, source of both its vital rivers. To the east lies Iran, target of a decade of destructive war during the 1980s. At the head of the Persian Gulf is Kuwait, invaded by Iraq's armies in 1990. To the south lies Saudi Arabia, ally of Iraq's adversaries. And to the west, Iraq is adjoined by Jordan and Syria. Jordan afforded Iraq an outlet to the sea via the port of Aqaba during the 1991 Gulf War in defiance of U.N. sanctions, but in the postwar period relations have soured. Syria lies functionally most remote from Iraq despite its long desert border with its embattled neighbor. Note, however, that the Euphrates River traverses Syria before entering Iraq. Water is a potent source of tension in this arid zone.

The regional geography of Iraq reveals the potent divisions within the country (Fig. 6-12). The heart of Iraq is the area centered on the capital, Baghdad, situated on the Tigris River amid the productive farmlands of the Tigris-Euphrates Plain. Here the great majority of the people are Sunni Muslims, who dominate the core area and the country's political machine.

As many as 10 million of Iraq's 22 million-plus citizens, however, are not Sunnis but Shi'ites, most of them concentrated in the populous south. Here lie some of Shia's holiest places, and here the Ayatollah Khomeini, escaping persecution by the shah of Iran, was given refuge before his eventual return to Tehran. But Iraq's Shi'ites do not have proportional representation in the country's government, and the 1991 Gulf War spelled disaster for them and their provinces. Fearing a Shi'ite uprising after their ouster from Kuwait, Baghdad's remaining forces unleashed a campaign of suppression against the south in which towns, villages, and religious shrines were heavily damaged and untold casualties occurred.

A close look at Figure 6-12 reveals the importance of this southern zone of Iraq. The Tigris and Euphrates join

to become the Shatt al Arab, Iraq's water outlet to the Persian Gulf. Over its last 50 miles or so, the Shatt al Arab waterway also becomes the boundary between Iraq and Iran. Iraq's territorial claim to land on the Iranian side of the waterway precipitated the decade of war between the two countries beginning in 1980.

Also note the situation of Iraq's most important southern city, Basra, and the position of the boundary with Kuwait. In this area lie several major oilfields; one, the Rumailah oilfield, extends from beneath Iraqi soil into Kuwait. In 1990, Iraq claimed that Kuwait was draining oil from this field by drilling slanted wells under Iraqi territory; that the boundary between the two countries was

A FUTURE KURDISTAN?

Political maps of the region do not show it, but where Turkey, Iraq, and Iran meet, the cultural landscape is not Turkish, Iraqi, or Iranian. Here live the Kurds, a fractious and fragmented nation of about 25 million (no certainty exists about their numbers). More Kurds live in Turkey than in any other country (perhaps as many as 12 million); possibly as many as 7 million in Iran; and smaller numbers in Syria, Armenia, and even Azerbaijan (see Fig. 6-12).

The Kurds have occupied this isolated, mountainous frontier zone for more than 3,000 years. They are a nation, but they have no state; nor do they enjoy the international attention that peoples of other stateless nations (such as the Palestinians) receive. Turkish and Iraqi repression of the Kurds, and Iranian betrayal of their aspirations, briefly make the news but are soon forgotten. Relative location has much to do with this: access to their landlocked domain is inhibited by its remoteness and by the obstacles created by ruling regimes.

Many Kurds dream of a day when their fractured homeland will be a nation-state. Most would agree that the city of Diyarbakir, now in southeastern Turkey, would become the capital. It is the closest any Kurdish town comes to a dominant city, although the largest urban concentration today is in the shantytowns of Istanbul, where as many as 3 million Kurds have migrated. After the 1991 Gulf War, the United Nations established a "security zone" for Kurds in northern Iraq, extending from the 36th parallel northward to the borders of Syria, Turkey, and Iran (Fig. 6-12). This was done to encourage refugee Kurds to return and to protect them from further Iraqi mistreatment. But this also may be the closest the Kurds will come, in the foreseeable future, to their dream of a free Kurdistan.

Given their numbers and regional concentration, some form of federal autonomy in a reconstituted Turkey might seem a logical objective. Unfortunately, Turkish cultural repression, combined with the actions of a small extremist Kurdish group, the Kurdish Workers Party, makes this an unlikely prospect. Nor will boundaries be moved to create an independent Kurdistan. The stakes are too great.

not agreed upon; and that Kuwait was failing to adhere to OPEC rules on oil production and pricing. The Shatt al Arab waterway still was filled with the wreckage of the Iran-Iraq War, so that the annexation of Kuwait would also give Iraq a new outlet to the Persian Gulf. This combination of justifications and potential returns persuaded Iraq to embark on its disastrous invasion of Kuwait in 1990.

The core area, centered on Baghdad, and the Shi'ite south are two of Iraq's major subregions; a third subregion lies in the north. Here most of the people are Sunni Muslims, but they are not Arabs: this is the land of the Kurds. Fewer than 4 million Kurds live in the mainly mountainous areas of northern Iraq, and they constitute perhaps 15 percent of the total Kurdish population in the realm. The Kurds are minorities in all the countries they inhabit, and governments like to undercount their minorities. Geographers estimate that there may be as many as 25 million Kurds, with the largest number in Turkey, the next largest minority in Iran, then in Iraq, Syria, and small clusters even in Armenia and Azerbaijan (Fig. 6-12).

As the map shows, Iraq's Kurds occupy a sensitive area of the country: the huge oil reserves on which the city of Kirkuk is situated lie on the margins of the Kurdish domain. During the Iran-Iraq War and again during the Gulf War, the Kurds rose against their Iraqi rulers and briefly succeeded in taking control over parts of their domain, but Iraq's response always ended such efforts, often with great loss of life. In the late 1980s, Baghdad even resorted to the use of cyanide and mustard gas against Kurdish villagers. At the end of the Gulf War, the United Nations established a security zone between the 36th parallel and Iraq's northern border to encourage refugees who had crossed into Turkey and Iran to return. However, as Figure 6-12 shows, a large part of Iraq's Kurdish domain lies outside this safety zone. Once again, the Kurds are what they have always been: victims of more powerful neighbors (see box titled ''A Future Kurdistan?'').

Iraq's infrastructure and economy were shattered during the Gulf War, but in fact Iraq had wasted much of its potential on the earlier conflict with Iran—and on mismanagement, corruption, and inefficiency. With its good agricultural land and its enormous oil income, Iraq should be one of the economic success stories of the entire realm. Instead, its failed leadership has made it one of the world's tragedies.

Syria

If there is one issue that can galvanize virtually the entire realm (other than Islam and oil), it is Israel. During the Gulf War, Iraq sought to draw Israel into the conflict by aiming missiles at it. Getting Israel into the fray would have been easier if Iraq had what Syria does: a common border with the Jewish state. Iraq, which once tried to build nu-

clear weapons with which to threaten Israel, is the only Middle Eastern country without territorial contact with the Israelis.

Syria has historic proof of its proximity: it lost a piece of territory, the Golan Heights, to Israel in 1967. Thirty years later, negotiators are trying to find a way to arrange the return of all or part of this area to Syria, but there are many problems. Jewish settlers have occupied parts of the Golan Heights; opposition to their return is strong in Israel, a democratic society where voters can oust representatives whose policies they do not approve. On the Syrian side there is the problem of continuity. Syria is not a democracy; since 1963 it has been a republic under a military regime. Moreover, while Syria's population is about 75 percent Sunni Muslim, the ruling elite comes from a smaller sect based there, the Alawites. Thus, although there is a governmental structure and two vice presidents are ready to succeed to the country's highest office, there is no certainty that such a succession will ensure continued stability.

Like Lebanon and Israel, Syria has a Mediterranean coastline where unirrigated agriculture is possible. Behind this densely populated coastal zone, Syria has a much larger interior than its neighbors, but areas of productive capacity are quite dispersed. Damascus, in the southwest corner of the country, was built on an oasis and is considered to be the world's oldest continuously inhabited city. Now the capital of Syria, its population exceeds 2 million.

The far northwest is anchored by Aleppo, the focus of cotton- and wheat-growing areas in the shadow of the Turkish border. Here the Orontes River is the chief source of irrigation water, but in the eastern part of the country the Euphrates valley is the crucial lifeline. It is in Syria's interest to develop its eastern provinces, and recent discoveries of oil there will speed that process. Syria is self-sufficient in staple grains and earns substantial revenues from its cotton exports, but oil earns most of its income. In addition, Syria is a country where opportunities for the expansion of agriculture still exist. Their realization would improve the cohesion of the state and bring its separate subregions into a tighter spatial framework.

Jordan

None of this can be said for Jordan (4.4 million), the desert kingdom that lies east of Israel and south of Syria. It, too, was a product of the Ottoman collapse, but it suffered heavily when Israel was created, more than any other Arab state. In the first place, Jordan's trade used to flow through Haifa, now an Israeli port, so that Jordan has to depend on destabilized Lebanon's harbors or the tedious route via the Gulf of Aqaba in the far south. Second, Jordan's final independence in 1946 was achieved with a total population of only about 400,000, including nomads, peasants, villagers, and a few urban dwellers. Then, with the partition of Palestine and the creation of Israel in 1948, Jordan received more than half a million Arab refugees. Soon it also found itself responsible for another half million Palestinians who, although living on the western side of the Jordan River, were incorporated into the state. Thus refugees outnumbered residents by more than two to one, and internal political problems were added to external ones—not to mention the economic difficulties of beginning national life as a very poor country.

Nonetheless, Jordan has survived with U.S., British, and other aid, but its problems have hardly lessened. Many Jordanian residents still have only minimum commitment to the country, do not consider themselves its citizens, and give little support to the hard-pressed monarchy. Dissatisfied groups constantly threaten to drag the country into another conflict with Israel. The 1967 war was disastrous for Jordan, which lost the West Bank (its claim was formally surrendered to the Palestine Liberation Organization in 1988) as well as its sector of Jerusalem (the kingdom's second largest city). Where hope for progress might lie—for example, in the development of the Jordan River valley—political conflicts intrude. The capital city, Amman, reflects the limitations and poverty of the country. Without oil, without much farmland, without unity or strength, and overwhelmed with refugees, Jordan presents one of the bleaker pictures in the Middle East.

Lebanon

Lebanon, Israel's northern coastal neighbor on the Mediterranean Sea, is one of the exceptions to the rule that the Middle East is the world of Islam: one-fourth of the population of 3.8 million adheres to the Christian rather than Muslim faith. Less than one-eighth the territorial size of Jordan (and only half the size of Israel), Lebanon has a long history of trade and commerce, beginning with the Phoenicians of old who were based here. Lebanon must import much of its staple food, wheat. The coastal belt below the mountains, though intensively cultivated, normally cannot produce enough grain to feed the entire population.

But normality has not prevailed in Lebanon during recent times. The country fell apart in 1975 when a civil war broke out between Muslims and Christians. This was a conflict with many causes. Lebanon for several decades had functioned with a political system that divided power between these two leading communities, but the basis for that system had become outdated. In the 1930s, Muslims and Christians in Lebanon were at approximate parity; but over the ensuing decades, the Muslim population increased at a much faster rate than the more urbanized and generally wealthier Christians (many of whom have emigrated from Lebanon in recent years). The Muslims' displeasure with

an outdated political arrangement (developed during the French occupation) was expressed during several outbreaks of rebellion prior to full-scale civil war. By then, Lebanon had also become a base for over 300,000 Palestinian refugees. These people, many of them living in squalid camps, were never satisfied with Lebanon's moderate posture toward Israel. When the first fighting between Muslims and Christians broke out in the northern coastal city of Tripoli, the Palestinians joined the conflict on the Muslim side. In the process, Lebanon was wrecked.

Beirut, the capital, once a city of great architectural beauty and often described as the Paris of the Middle East, was heavily damaged as Christian, Sunni Muslim, Shi'ite Muslim, Druze, and Palestinian militias—as well as the Lebanese and Syrian armies—fought for control. By the end of the 1980s, Beirut approached total destruction, and only the poorest 150,000 war-ravaged residents remained of the 1.5 million who had lived there as recently as the late 1970s. Since 1990, the conflict has abated. Beirut today has embarked on a long rebuilding process, and its population (now 2.1 million) is rising once more.

As the Muslims' strength intensified, the Christians concentrated in an area along the coast between Beirut and Tripoli (Fig. 6-12). In the meantime, Israel had taken control over a security zone in the south, and various competing Muslim factions occupied and controlled parts of the country. In 1976, the Syrians involved themselves in the situation, trying to pacify Lebanon through a military presence. Eventually the Syrians, too, fell victim to Lebanon's fractious character. By backing various client factions, Syria helped reduce the violence outside Beirut but then found itself inextricably enmeshed in the country's troubles.

In the mid-1990s, an exhausted Lebanon finally seemed to be moving toward some form of internal accommodation, but old enmities still threatened the fragile peace. Whether Lebanon will survive as a political entity remains uncertain.

Israel

Israel lies at the very heart of the Arab world (Fig. 6-8). Its neighbors are Lebanon and Syria to the north and northeast, Jordan to the east, and Egypt to the southwest—all in some measure still resentful of the creation of the Jewish state in their midst. For the past half-century since 1948, when Israel was created as a homeland for the Jewish people on the recommendation of a United Nations commission, the Arab-Israeli conflict has overshadowed all else in the Middle East.

Indirectly, Israel was the product of the collapse of the Ottoman Empire. Britain gained control over the Mandate of Palestine, and British policy supported the aspirations of European Jews for a homeland in the Middle East, em-

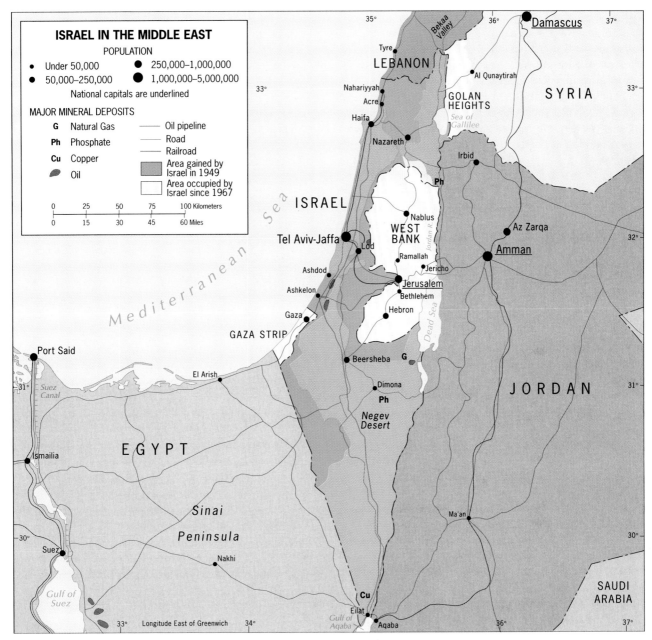

FIGURE 6-13

bodied in the concept of Zionism. In 1946, the British granted independence to the territory lying east of the Jordan River, and "Transjordan" (now the state of Jordan) came into being. Shortly afterward, the territory west of the Jordan River was partitioned by the United Nations, and the Jewish people got slightly more than half of it—including, of course, some land that had been occupied by Arabs. Jews actually owned only about 8 percent of Palestine's land, but they constituted more than one-third of the area's population.

As soon as the Jewish people declared the independent state of Israel on May 14, 1948, the new country was at-

tacked by its Arab neighbors, who rejected the scheme. In the ensuing battle, Israel not only held its own but also gained some crucial territory in its central and northern areas as well as in the Negev Desert to the south (Fig. 6-13). At the end of this first Arab-Israeli war, in 1949, the Jewish population controlled 80 percent of what had been Palestine west of the Jordan River. As Figure 6-13 shows, Israel now owned not only the territory allocated to it by the United Nations but additional areas facing Egypt, Lebanon, and the West Bank (the area west of the Jordan River shown on the map). During the conflict, troops from Transjordan (the independent Arab kingdom east of the Jordan

From the Field Notes

"Cultural contrast pervades Jerusalem's urban landscape; social tension is palpable. From almost any high vantage point you can see the places of worship and the holy ground that means so much to Jew, Christian, and Muslim: synagogues, churches, and mosques, walled cemeteries, sacred shrines, historic sites, all juxtaposed as in this panorama. In the distance, in another direction, you can see the Jordan Valley; Jerusalem is where the West Bank meets Israel. When the Israeli government decided to make Jerusalem its political headquarters, it was a move designed to proclaim the nation's dominance here: Jerusalem would not become an international city. This makes Jerusalem a recent example of a *forward* capital, the vanguard of the state in potentially contested territory."

River) invaded this West Bank area, and in 1950 the king formally annexed it to his country, which he renamed Jordan.

This early conflict proved to be only the first in a series of wars between Israel and its Arab neighbors. In 1967, a week-long military conflict resulted in a major Israel victory: Israel took the Golan Heights from Syria, the West Bank from Jordan, and the Sinai Peninsula (up to the Suez Canal itself) from Egypt. In 1973, another brief war led to Israel's withdrawal from the Suez Canal to truce lines in the Sinai Peninsula, and later all of the conquered Sinai Peninsula was returned to Egypt.

Since then, a fragile peace has been sustained, and in the 1990s some progress has been made on key issues involving the relationships between Israel and its Arab neighbors and between Israel and the leadership of the Palestinians who have been under Israeli control in the Gaza Strip and in the West Bank. Agreements normalizing relationships with Egypt and Jordan were major steps on this difficult road, and the establishment of the Palestinian Authority (PA) to govern Gaza and Jericho (a city on the West Bank east of Jerusalem) was another achievement. Many obstacles still stand in the way, however, including the following:

1. *The Golan Heights.* The return of the Golan Heights to Syria may be a precondition for normalization of relations with Syria, but it may not be possible given the political climate in Israel itself.

2. *The Security Zone.* Israel has taken control over a strip of territory in Lebanon immediately to the north of the Israeli border, to inhibit terrorist incursions. Israel patrols this zone, but Arab governments decry its existence.

3. *Jerusalem.* Holy city for Jews, Christians, and Muslims, Jerusalem was to become an international city under the U.N. blueprint for Palestine. It fell into Israeli hands when Jewish forces captured the West Bank in the 1967 war, and Israel now regards it as its capital. Most countries, however, have kept their embassies in Tel Aviv.

4. *The West Bank.* Even after its capture by Israel in 1967, the West Bank might have become a Palestinian homeland (and possibly a state) in the future, but Jewish immigration made such concessions difficult. In 1977 there were only 5,000 Jewish residents on the West Bank; by 1996 there were over 130,000, making up nearly 10 percent of the population and creating a seemingly inextricable jigsaw of Jewish and Arab settlements (Fig. 6-14).

5. *The Palestinians.* Although Palestinian issues tend to focus on Gaza and the West Bank, and on Arabs who continue to call themselves Palestinians in countries that neighbor Israel (principally Jordan), nearly 1 million Palestinian Arabs continue to live in Israel proper (see box titled ''The Palestinian Dilemma'').

6. *Arab/Islamic Disruption.* Extremist organizations such as Hamas and Hezbollah have attempted to derail the peace process through acts of terrorism and intimidation. The PA opposes such acts, but also raises opposition to further negotiation and ''concessions'' among the Israeli public.

With the help of massive foreign aid (mainly from the United States), large remittances by Jews living in other parts of the world, and the energies of its settlers, Israel

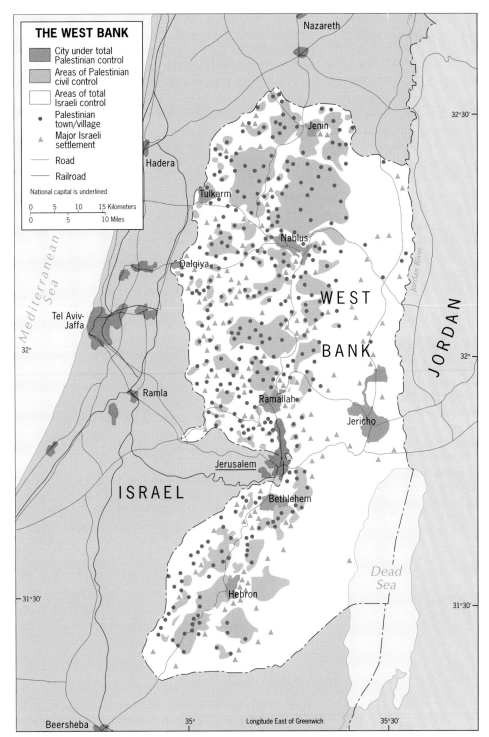

FIGURE 6-14

has become a high-income society in the midst of comparative poverty. (Israel's per capita GNP is more than 10 times that of Jordan and 20 times that of Egypt.) Irrigated areas have been expanded, dryland cultivation has succeeded with technological advances disseminated around the world, oil is bought from Egypt under the normaliza-

tion agreement, and an iron and steel plant at Tel Aviv is maintained for strategic reasons. Israel's population of nearly 6 million is highly urbanized, with just over 90 percent living in such cities as Tel Aviv, Jaffa, Haifa, and Jerusalem. The core area, including Tel Aviv and Haifa and the coastal zone between them, contains over three-quarters of the country's population, by far the most modernized cluster in any non-oil-producing country anywhere in the realm.

As the peace process advances, new obstacles constantly arise. One involves water. Over 30 percent of Israel's water supply comes from aquifers that lie under the West Bank. Of this, about one-fifth is allocated to the 1.4 million Palestinians living on the West Bank, and a smaller quantity goes to the 130,000 Jewish settlers. The rest is consumed in Israel proper, mainly Tel Aviv and Jerusalem. When the next stage in the transfer of land from Israeli to PA control occurs, some of the most important pumping stations will come under Palestinian jurisdiction. The Israelis do not want to change existing patterns and systems of water use, and they want the Palestinians to promise that no new wells will be drilled that would divert water from critical aquifers. Modern societies like Israel consume large

amounts of water; Israelis on average use four times as much water as Palestinians. The Palestinians argue that they should not have to support this huge disproportion. Israelis insist that without a separate agreement on water use, autonomy prospects for West Bank towns will diminish.

But Israel's future depends on a satisfactory settlement with its Palestinian minority, as well as normalized relations with moderate and secular Arab states. In the heart of the realm, Israel is at the focus of an animosity that has a 13-century history. Now, in the age of nuclear weapons and long-range missiles, its negotiations are a race against time.

❖ THE ARABIAN PENINSULA

About the regional identity of the Arabian Peninsula there can be no doubt: south of Jordan and Iraq, the entire peninsula is encircled by water. This is a region of old-style emirates and sheikdoms made wealthy by oil; here, too, lies the source of Islam.

As a region, the Arabian Peninsula is environmentally dominated by a desert habitat and politically dominated by the Kingdom of Saudi Arabia (Fig. 6-15). With its huge territory of 830,000 square miles (2,150,000 sq km), Saudi Arabia is the realm's fourth biggest state; only Kazakstan, Sudan, and Algeria are somewhat larger. On the peninsula, Saudi Arabia's neighbors (moving clockwise from the head of the Persian Gulf) are Kuwait, Bahrain, Qatar, the United Arab Emirates, the Sultanate of Oman, and the new Republic of Yemen (created in 1990 through the unification of former North Yemen and South Yemen). Together, these countries on the eastern fringes of the peninsula contain about 18 million inhabitants; the largest by far is Yemen, with 14 million. The interior boundaries of these states, however, are still inadequately defined, here in one of the world's last remaining frontier-dominated zones.

THE PALESTINIAN DILEMMA

■ ■ ■

Ever since the creation of Israel in 1948 in what had been the British Mandate of Palestine, Arabs who called Palestine their homeland have lived as refugees in neighboring countries. A large number have been assimilated into the societies of Israel's neighbors, but many still live in refugee camps. Israel is now a half-century old, and the great majority of Palestinians were born after the partition.

The Palestinians call themselves a nation without a state (much as the Jews did before Israel was founded), although they and their descendants make up the majority of Jordan's population today. They demand that their grievances be heard and that progress be made toward a territorial solution in the form of a Palestinian state. The first steps have been taken as part of the Arab-Israeli peace process, and negotiations continue.

Current estimates of Palestinian populations in the realm are as follows:

Israel and Occupied Territories		3,275,000
Israel	975,000	
West Bank	1,450,000	
Gaza Strip	850,000	
Jordan		2,300,000
Lebanon		450,000
Syria		400,000
Saudi Arabia		300,000
Iraq		80,000
Egypt		65,000
Kuwait		32,000
Libya		28,000
Other Arab States		520,000
TOTAL		**7,450,000**

■ ■ ■

Saudi Arabia

Saudi Arabia itself has only 20 million inhabitants in its vast territory, but the kingdom's importance is reflected in Figure 6-7: the Arabian Peninsula contains the Earth's largest concentration of known petroleum reserves. Saudi Arabia occupies most of this area and by some estimates may possess as much as one-quarter of all the world's remaining oil. These reserves lie in the eastern part of the country, particularly along the Persian Gulf coast and in the Rub al Khali (Empty Quarter) to the south.

The national state that is Saudi Arabia was only consolidated as recently as the 1920s through the organizational abilities of King Ibn Saud. At the time, it was a mere shadow of its former greatness as the source of Islam and the heart of the Arab World. Apart from some permanent settlements along the coasts and in scattered oases, there was little to stabilize the country; most of it is desert, with annual rainfall almost everywhere under 4 inches (10 cm). The land surface rises generally from east to west, so that the Red Sea is fringed by mountains that reach nearly

10,000 feet (3,000 m). Here the rainfall is slightly higher, and there are some farms (coffee is a cash crop). These mountains also contain known deposits of gold, silver, and other metals, and the Saudis hope to diversify their exports by adding minerals from the west to the oil from the east.

Figure 6-15 reveals that most economic activities in Saudi Arabia are concentrated in a wide belt across the

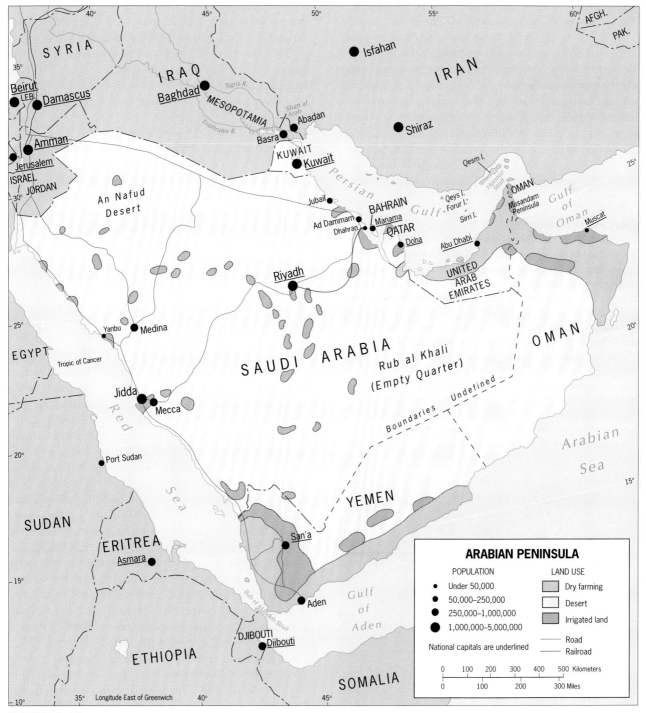

FIGURE 6-15

''waist'' of the peninsula, from the boom town of Dhahran on the Persian Gulf through the national capital of Riyadh in the interior to the Mecca-Medina area near the Red Sea. A fully effective internal transportation and communications network, one of the world's most modern, has recently been completed. But in the more remote zones of the interior, Bedouin nomads still ply their ancient caravan routes across the vast deserts. For several decades, Saudi Arabia's aristocratic royal families were virtually the sole beneficiaries of their country's incredible wealth, and there was hardly any impact on the lives of villagers and nomads.

That is now changing, however. Agriculture in particular is receiving major government investments because the Saudis want to prevent the food weapon from being used against them (as they themselves have occasionally wielded the oil weapon). As a result, widespread well drilling has significantly enhanced water supplies to support crops, and surpluses have been quickly produced. These achievements have been enormously expensive, however, for they necessitate the pumping of water from aquifers far below the desert surface as well as the construction of extensive center-pivot irrigation systems. Despite these efforts, there is now a growing realization that the Saudi Arabian goal of attaining self-sufficiency in food production is elusive. To begin with, the underground water supplies are proving to be a one-time-only, nonrenewable resource. Desalinization plants serve the coastal cities and towns and some interior places as well, but even an oil-rich kingdom cannot afford the price of raising its crops with desalinized water. Even more importantly, the Saudi population is growing at a much faster pace (3.2 percent, thereby doubling in just 22 years) than the rate at which the domestic food supply can be increased.

The country's rulers also have instituted reforms in housing, medical care, and education, and since 1970 they have spent hundreds of billions of dollars on national development programs. The collapse of oil prices during the 1980s slowed the progress of these programs, but overall living standards have improved significantly. Industrialization, too, has been stimulated. The new planned city of Jubail, north of Dhahran on the Persian Gulf, has become an industrial center with state-of-the-art petrochemical and metal fabrication plants. Similarly, Yanbu, north of Jidda on the Red Sea coast, lies at the end of a trans-Arabian oil pipeline, an incipient industrial outpost for the west. Still, as we noted earlier in this chapter, much of Saudi Arabia remains tradition-bound, and in the villages of the interior, the age of oil and development seems remote.

From the Field Notes
"The port of Mutrah, Oman, like the capital of Muscat nearby, lies wedged between water and rock, the former encroaching by erosion, the latter crumbling as a result of plate movement. From across the bay one can see how limited Mutrah's living space is, and one of the dangers here is the frequent falling and downhill sliding of large pieces of rock. It took about five hours to walk from Mutrah to Muscat; it was extremely hot under the desert sun but the cultural landscape was fascinating. Oil also drives Oman's economy, but here you do not find the total transformation seen in Kuwait or Dubai. Townscapes (as in Mutrah) retain their Arab–Islamic qualities; modern highways, hotels, and residential areas have been built, but not at the cost of the older and the traditional. Oman's authoritarian government is slowly opening the country to the outside world after long-term isolation."

On the Periphery

Five of Saudi Arabia's six neighbors on the Arabian Peninsula face the Persian and Oman Gulfs (Fig. 6-15) and are monarchies in the Islamic tradition. All five also derive substantial revenues from oil. Their populations range from 0.5 to 2.8 million; these are not strong or (OPEC aside) influential states. They do, however, display considerable geographic diversity. *Kuwait*, at the head of the Persian Gulf, comes close to cutting Iraq off from the open sea, an issue that was not settled by the Gulf War. *Bahrain* is an island state, a tiny territory, with dwindling oil reserves. Bahrain's approximately 630,000 people are 50 percent Shi'ite and only 35 percent Sunni. Neighboring *Qatar* consists of a peninsula jutting out into the Persian Gulf, a featureless, sandy wasteland made habitable by oil (now declining in importance) and natural gas (rising simultaneously). *The United Arab Emirates (UAE)*, a federation of seven emirates, faces the Persian Gulf between Qatar and Oman. The sheik is an absolute monarch in each of the emirates, and the seven sheiks form the Supreme Council of Rulers. In terms of oil revenues, however, there is no equality: two emirates—Abu Dhabi and Dubai—have most of the reserves. The eastern corner of the Arabian Peninsula is occupied by the Sultanate of *Oman*, another absolute monarchy centered on the capital, Muscat. Figure

6.1

6-15 shows that Oman consists of two parts: the large eastern corner of the peninsula and a small but critical cape to the north, the Musandam Peninsula, that protrudes into the Persian Gulf to form a narrow choke point—the Hormuz Strait (Iran lies on the opposite shore). Tankers that leave the other states must negotiate this narrow channel at slow speed, and during politically tense times they have had to be protected by warships. Several small islands near the Strait, owned by the UAE, are claimed by Iran, a potential source of dispute.

This leaves what is, in many ways, Saudi Arabia's most substantial and potentially most difficult neighbor: *Yemen.* Our chart (see Appendix A) states that Yemen has an area of 203,900 square miles, but that is an estimate: the boundary between Yemen and Saudi Arabia was never properly defined. In the early 1990s, following the unification of the two Yemens, exploration for oil was begun in the promising borderland. Saudi Arabia immediately demanded that it be stopped until the boundary had been defined and delimited. (The Saudis' relationship with the Yemenis, never very friendly, deteriorated after Yemen chose Iraq's side during the 1991 Gulf War, and Saudi Arabia expelled some 1 million Yemenis working in the kingdom.)

The size of Yemen's population also is uncertain and is variously reported between 13 and 16 million. Significantly, nearly half the people are Shi'ites. San'a, formerly the capital of North Yemen, was chosen as the headquarters of the unified country, and Aden is the only port of consequence. The new Yemen was born as a multiparty, secular, democratic state, the only one in the region. But its economy is by far the weakest, with very limited oil production.

Yemen occupies a strategic part of the peninsula. Between Yemen and Djibouti in Africa's Horn, the mouth of the Red Sea narrows to the confined Bab el Mandeb Strait, another choke point on Arabia's periphery. The immediate prospects for instability, however, are on land, not at sea. Royal, Sunni, oil-rich, modernizing Saudi Arabia stands in stark contrast to democratic, strongly Shi'ite, agricultural, poor, underdeveloped Yemen. Such a contrast, against a backdrop of territorial disagreement, spells trouble.

❖ THE EMPIRE STATES

North of the eastern Mediterranean Sea, the Middle East, and the Persian Gulf lies a tier of states that connects the turbulent present to a powerful, imperial past. Five states constitute this region, where Arab ethnicity gives way but Islamic culture continues. Two of these states, Turkey and Iran, are modern (or rather, current) manifestations of a greater history. Three others, Afghanistan, Azerbaijan, and Cyprus, have seen imperial power descend on them (Fig. 6-16).

So many things change along the border between the Middle East and these Empire States that we seem to be entering a different realm here. The physical geography strengthens this impression: we leave the rocky and sandy expanses of desert and enter the higher relief of mountains and plateaus. Gone is the Arabic language we heard from Morocco to Oman and from Saudi Arabia to the Horn of Africa. Gone, too, is the Arab nationalism that so strongly influences political geography from the Maghreb to the Middle East. But here to stay is the overarching imprint of Islam, in cities, villages, and countryside, not only in the cultural landscapes but also in ways of life and views of community and world. And when we get beyond the plateaus of Turkey and the mountains of Iran, we find that this part of the realm, too, has its deserts, oases, caravans, and clustered populations. We have crossed a regional boundary, but we have not left the realm.

Turkey

Before we focus on Turkey, center of the once-vast Ottoman Empire that brought Islam to the doorstep of Vienna and to the shores of the Arabian Peninsula, let us consider who the Turks are and why Turkey's influence may again expand far beyond its present borders. Historical geographers report that it is the Chinese who, perhaps as early as the third century, gave the name *Tukiu* to the nomadic peoples then living far from modern Turkey, in the steppes and forests of Siberia. By the sixth century, these nomadic Turks had created an empire stretching from Mongolia to the Black Sea. Over time they migrated into what is today northern Iran and into the domain of the collapsing Byzantine Empire, Turkey. In the process, the spread their *Turkic* language far and wide. But despite this wide diffusion, basic Turkish remains a *lingua franca* for a vast area extending from western China to southern Bulgaria.

The Turks were thinly spread, and their loosely knit states rose and fell. The Uzbeks, led by Tamerlane, with Samarqand as his capital city, conquered territory extending from the Volga to Pakistan during the second half of the fourteenth century. The Kazaks and Uygurs also founded large empires, but the Turkish peoples had little cultural continuity—until the arrival of Islam. The Arab-propelled Islamic wave touched the Turkish peoples before the end of the seventh century and diffused throughout their domain over the next several hundred years. Its effect was to make the far-flung Turkish peoples part of a multicultural, multinational, but uniform religious civilization. Turkish lands now had linkage, from the Aegean Sea in the west to western China in the east. Their renewed vigor sent them on imperial conquests in various directions, and thus the Islamic wave penetrated India from Turkestan and Europe from Turkey.

Earlier in this chapter, we noted the rise and decline of

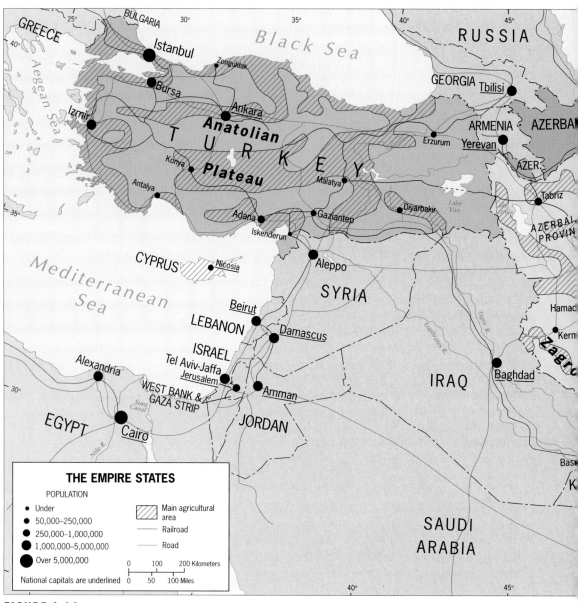

FIGURE 6-16

the Turkish Ottoman Empire. At the beginning of the twentieth century, the country we now know as Turkey lay at the center of this decaying and corrupt empire, ripe for revolution and renewal. This occurred in the 1920s and thrust into national and international prominence a leader who became known as the father of modern Turkey: Mustafa Kemal, known after 1933 as Atatürk.

The ancient capital of Turkey was Constantinople (now Istanbul), located on the strategic straits connecting the Black and Mediterranean seas. But the struggle for Turkey's survival had been waged from the heart of the country, the Anatolian Plateau, and it was here that Atatürk decided to place the new seat of government. Ankara, the new capital, possessed certain advantages: it would remind

the Turks that they were (as Atatürk always said) Anatolians; it lay nearer the center of the country than Istanbul; and it could therefore act as a stronger unifier. Istanbul, located on the western side of the Bosporus waterway, lies on the threshold of Europe, with the minarets and mosques of this largest and most varied Turkish city rising above a somewhat Eastern European townscape.

Although Atatürk moved the capital eastward and inward, his orientation was westward and outward. To implement his plans for Turkey's modernization, he initiated reforms in almost every sphere of life within the country. Islam, formerly the state religion, lost its official status. The state took over most of the religious schools that had controlled education. The Roman alphabet replaced the Ara-

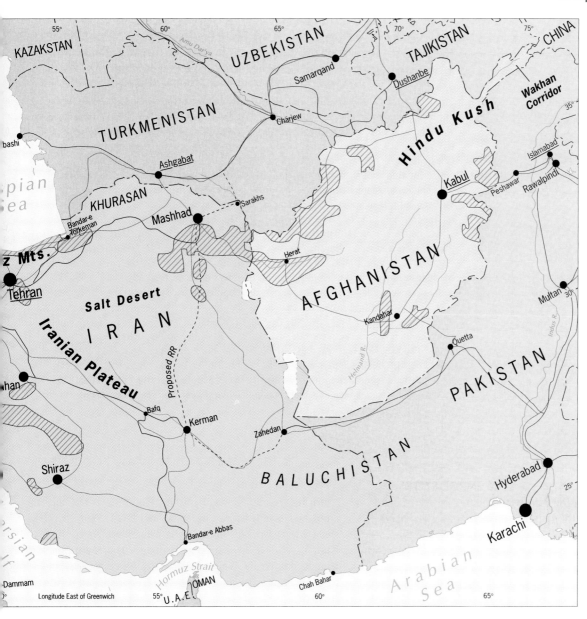

bic. Islamic law was replaced by a modified Western code. Symbols of old—growing beards, wearing the fez—were prohibited. Monogamy was made law, and the emancipation of women began. The new government emphasized Turkey's separateness from the Arab World, and ever since it has remained aloof from the affairs that engage other Islamic states.

Atatürk's government naturally faced opposition from religious leaders, notably in the rural areas. Turkey was, and remains, largely an agricultural country, with tens of millions of villagers living in remote farming areas, where resistance to change, especially rapid change, was to be expected. But there were other issues to confront as well. One of these continues to agitate Turkey today: its history

of mistreatment of minorities. Soon after the outbreak of World War I, the (pre-Atatürk) regime decided to expel all the Armenians living in its national territory, most of them in the northeast. Nearly 2 million Turkish Armenians were uprooted and brutally forced out; as many as 600,000 were killed in a campaign that still arouses anti-Turkish emotions among Armenians today. Many fled to Russia, where the Soviet communist regime later established an Armenian republic—now Turkey's neighbor.

In modern times Turkey has faced another minority problem, evident from Figure 6-12: its large and regionally concentrated Kurdish population. About one-fifth of Turkey's population of 63 million is Kurdish, and successive Turkish governments have mishandled relationships with

■ AMONG THE REALM'S GREAT CITIES . . .

Istanbul

From both shores of the Sea of Marmara, northward along the narrow Bosporus toward the Black Sea, sprawls the fabulous city of Istanbul, known for centuries as Constantinople, headquarters of the Byzantine Empire, capital of the Ottoman Empire, and, until Atatürk moved the seat of government to Ankara in 1923, capital of the modern Republic of Turkey as well.

Istanbul's site and situation are incomparable. Situated where Europe meets Asia and where the Black Sea joins the Mediterranean, the city was built on the requisite seven hills (as Rome's successor), rising over a deep waterway that enters the Bosporus from the west, the famous Golden Horn. A sequence of empires and religions endowed the city with a host of architectural marvels that give it, when approached from the water, an almost surreal appearance—and what is seen today is a mere remnant of history's accruement, survivals of earthquakes, fires, and combat.

Turkey's political capital may have moved to Ankara, but Istanbul remains its cultural and commercial headquarters. It also is the country's leading urban magnet, luring millions from the poverty-stricken countryside. In the heart of the city known as Stamboul and in the "foreign" area north of the Golden Horn called Beyoglu, modern buildings vie for space and harbor views, blocking the vistas that long made Istanbul's cultural landscape unique. On the outskirts, shantytowns emerge so rapidly that Istanbul's population is doubling every 13 years at current rates, having topped 8.5 million in 1995.

Istanbul's infrastructure is crumbling under this unprecedented influx, its growth is virtually without planning controls, and its legacy of two millennia of cultural landscapes is under threat.

---

this nation, even to the point of prohibiting the use of Kurdish speech and music in public places at one time (see box titled "A Future Kurdistan?", p. 302). Seeking to suppress a small but violent extremist group among the Kurds, the Turks violated the human rights of the entire minority, an issue that not only roils national politics but also affects Turkey's place in the community of nations. Turkish aspirations to join the European Union are imperiled by human rights questions arising from the Kurdish situation.

Turkey is a mountainous country of generally moderate relief and, as Figure I-7 indicates, considerable environmental diversity ranging from steppe to highland. On the dry Anatolian Plateau, villages are small and subsistence farmers grow cereals and raise livestock. Coastal plains are of limited size, but they are productive and densely populated. Textiles (from home-grown cotton) and farm products dominate the export economy, but Turkey also has substantial mineral reserves, some oil in the southeast, massive dam-building projects under way on the Tigris and Euphrates rivers, and a small steel industry based on domestic raw materials. Normally self-sufficient in staples, Turkey would seem to have a bright future.

In the late 1990s, Turkey faced several serious challenges. Potentially the most consequential was the rise of Islamic fundamentalism, which in part was the result of the

Kurdish problem. Several million Kurds have left their embattled southeastern homeland and have come to the shantytowns around Istanbul (where an estimated 3 million Kurds live on the outskirts) and Ankara. There, a militant Islamic group, the Welfare Party, is providing bread and basic accommodation to homeless and unemployed Kurds, recruiting them in large numbers to its fundamentalist cause. When the economy goes through a downturn, as it did during the 1990s, the Welfare Party benefits by appealing to those affected most. As a result, the party has won local elections in Istanbul and Ankara, vowing to reverse the emancipation of women, resegregating public transport, even attacking Western ballet performances as "degenerate" art.

Coupled with Europe's reluctance to take the first steps that would lead to Turkey's admission to the European Union, these developments cast an ominous shadow over Turkey's future and, by extension, over the future of the area at the heart of which it lies. Turkey has signaled its interest in joining the European Union, but the human rights issue and Greece's objections diminish its prospects. Ultimately the potential advantages of Turkey's membership in the European economic framework may override these obstacles; for one, Turkey's admission would prove that Europe is not an exclusive "Christian club." Islamic fundamentalists in Turkey hope that the drive for admission will be stalled long enough for them to take power in Ankara. In any case, the delays, in their eyes, constitute proof of Europe's anti-Islamic stance. If Europe closes the door against Turkey, Islamic extremists will benefit enormously. Should Turkey experience what happened in neighboring Iran during the 1970s and more recently in Algeria, the consequences would be disastrous.

Compared to such prospects, Turkey's other problems seem minor: the question of still-divided Cyprus (see box titled "The Problem of Cyprus"), relations with neighboring Iraq and Syria over dam building and oil pipelines, a quarrel with Syria over the Hatay boundary, and a dispute with Greece over maritime boundaries off the west coast. Turkey's relative location gives it an importance far greater than its dimensions, population, economy, and social fabric alone. Here turbulent Eastern Europe meets the world of Islam, here lies the choke point between the Black Sea and the Mediterranean, here is the gateway to Transcaucasia and Turkestan. In this world of growing ethnic and cultural consciousness, Turkey lies at the most crucial of crossroads.

Iran

Iran, Turkey's neighbor to the east, also has a history of imperial conquest. In 1971, the then-reigning shah and his family celebrated the 2,500th anniversary of Persia's first monarchy with unmatched scenes of royal splendor. But before the end of that decade, revolution engulfed Iran, and Shi'ite fundamentalists drove the shah from power. The monarchy was replaced by an Islamic republic, and a frightful wave of retribution followed.

As Figure 6-16 shows, Iran also occupies a critical area in this turbulent realm. It controls the entire corridor between the Caspian Sea and the Persian Gulf. To the west it adjoins Turkey and Iraq, both historic enemies. To the north (west of the Caspian Sea) Iran borders Azerbaijan and Armenia, where once again Muslim confronts Christian. To the east Iran meets Pakistan and Afghanistan, and east of the Caspian Sea lies volatile Turkmenistan.

In many places, Iran seems to spill over into its neighbors, a legacy of its expansionist days. As the map shows, Azerbaijan is the name of an Iranian province as well as an independent republic, and would be a part of Iran today

THE PROBLEM OF CYPRUS

■ ■ ■

In the northeastern corner of the Mediterranean Sea (see Fig. 6-16), much farther from Greece than from Turkey or Syria, lies Cyprus (population 770,000). Since ancient times this island has been predominantly Greek in its population, but in 1571 it was conquered by the Turks, under whose control it remained until 1878. Following the decay of the Ottoman state, the British soon took control of the island. By the time Britain was prepared to offer independence to much of its empire in the years following World War II, it had a problem in Cyprus because the Greek majority among the (then) 600,000 population preferred *enosis*—union with Greece. It is not difficult to understand that the 20 percent of the Cypriot population that was Turkish wanted no such union.

By 1955, the dispute had reached the stage of violence because differences between Greeks and Turks are deep, bitter, and intense. It had become impossible to find a solution to a problem in which the residents of this island country thought of themselves as Greeks or Turks first rather than Cypriots. Nevertheless, in 1960 Cyprus was granted independence under a complicated constitution designed to permit majority rule but with a guarantee of minority rights.

The fragile order finally broke down in 1974. As a civil war engulfed the island, Turkish armed forces intervened (about 30,000 soldiers were still present in the mid-1990s), and a major redistribution of population occurred. The northern 40 percent of Cyprus became the stronghold of more than 100,000 Turkish Cypriots; only a few thousand Greeks remained there. The rest of the island—south of the U.N.-patrolled "Green Line" that was demarcated across the country to divide the two ethnic communities—became the domain of the Greek majority, now including about 200,000 in refugee camps.

In 1983, again bringing the stalemated conflict to the attention of the world, the Turkish community (today comprising one-fifth of the population) seceded by declaring itself the new independent Turkish Republic of Northern Cyprus. The situation is especially difficult because of the many factions involved. The Turkish Cypriots, many of whom have expressed a desire for reintegration, are discrete from the Turkish nation and express a strong attachment to their island homeland. The Greek Cypriot majority is ideologically divided; the crisis of 1974 began not as a Greek-Turkish conflict but as a result of a coup by Greek Cypriot soldiers against the Greek-dominated government of the island. The Greek majority always was riven by factionalism.

Today, two *de facto* states exist, although the international community recognizes the government on the Greek side as the legitimate government of the whole island. With a population of about 600,000 and a relatively prosperous economy based on agriculture and tourism, the (official) Republic of Cyprus has emerged as a high-income economy. The Turkish Republic of Northern Cyprus, on the other hand, has needed support from Turkey, and its per capita GNP is less than half of that shown in Appendix A. Time is deepening, not softening, the contrasts between north and south.

■ ■ ■

were it not for Russian-Soviet intervention (see box titled "Cross-Border Azerbaijan"). Again, there is an Iranian Baluchistan and a Pakistani Baluchistan; the border marks a retreat enforced by British colonial power. The war of the 1980s with Iraq started as a territorial conflict, Iran having been perceived as encroaching on Iraq near the head of the Persian Gulf.

Iran, as Figure 6-16 demonstrates, is a country of mountains and deserts. The heart of the country is an upland, the Iranian Plateau, that lies surrounded by even higher mountains including the Zagros in the west, the Elburz in the north along the Caspian Sea coast, and the mountains of Khurasan to the northeast. The Iranian Plateau therefore is actually a huge highland basin marked by salt flats and wide expanses of sand and rock. The highlands wrest some moisture from the air, but elsewhere only oases break the arid monotony—oases that for countless centuries have been stops on the area's caravan routes.

In western Iran, neighboring Pakistan, and Afghanistan, people still move with their camels, goats, and other livestock along routes that are almost as old as the human history of this realm. Usually they follow a seasonal and annual cycle, visiting the same pastures year after year, pitching their tents near the same stream. It is a lifestyle especially associated with this realm: **nomadism**. In Iran as elsewhere, nomads are not aimless wanderers across boundless plains. They know their terrain intimately, and they carefully judge the time to linger and the moment to depart based on long experience along the route. Nor are nomads simply scavengers of the land. A nomadic community has its division of labor in which some are skilled at crafts and make leather and metal objects for sale or trade when contact occurs; others are herders; still others are farmers who may grow a crop when the group stops long enough during the summer at a suitable site. Caravans of nomads still travel the trails of Southwest Asia and North Africa, but their numbers are dwindling. Their lands are being penetrated by roads, the borders they used to cross freely are being sealed, and their environments are changing as desertification claims the steppe.

In ancient times, Persepolis in southern Iran was the focus of a powerful Persian kingdom, a city dependent on *qanats*, underground tunnels carrying water from moist mountain slopes to dry flatland sites many miles away. Today, Iran's population of 65 million is 57 percent urban, and the capital, Tehran, lies far to the north, on the southern slopes of the Elburz Mountains. This mushrooming metropolis, at the heart of modern Iran's core area, still depends in part on the same kinds of *qanats* that sustained Persepolis more than 2,000 years ago. As such, Tehran symbolizes the internal contradictions of Iran: a country in which modernization has taken hold in the cities, but in which little has changed in the vast countryside, where the *mullahs* led their peasant followers in a revolution that overthrew a monarchy and installed a theocracy.

CROSS-BORDER AZERBAIJAN
■ ■ ■

As Figure 6-16 shows, the boundaries of Iran over much of their length lie along coastlines or across deserts. But in the west and northwest, the situation is different. In the west, in the border area with Iraq and Turkey, lies Iran's part of "Kurdistan" (Fig. 6-12). And in the northwest lies the Iranian province of Azerbaijan. This also is the name of a new country across the border to the north: former Soviet Azerbaijan (which is part of the Transcaucasian Transition Zone, and is discussed on pp. 149–152).

The Azeris (short for Azerbaijanis) on both sides of the border are of the same ancestry: they are a Turkish people divided between the (then) Russian and Persian empires by a treaty signed in 1828. Since then, Iran's Azeris have been assimilated into the Persian (later Iranian) state, but Soviet Azerbaijan became an independent republic in 1991 as a result of the collapse of the U.S.S.R. When the Soviet Azeris mounted their campaign to oust the Soviets, thousands of Iranian Azeris came to the border to shout support and to smuggle arms across. The two peoples had not forgotten their common heritage.

Independent Azerbaijan has a population of 7.5 million. Estimates of the Azeri population in Iran vary but are as high as 15 million (with about 4 million living near the Azerbaijan border), making this a substantial nation. Obviously, the potential for irredentism exists: Azerbaijan, like Iran, is dominantly Shi'ite, its religion steadily reviving after the Soviet period. Iran may face an unfamiliar challenge here: an independent, greater Azerbaijan may appeal to Azeris as much as an autonomous Kurdistan interests the nearby Kurds.

■ ■ ■

The last shah of Iran, Muhammad Reza Pahlavi, hoped to be the Atatürk of his country, a modernizer and reformer who used Iran's enormous income from oil to introduce industrialization, update agriculture, improve medical services, emancipate women—and to build a huge security apparatus to contain his opposition and armed forces to establish his regional power base. In the end, Pahlavi turned out to be no Atatürk, although he ruled in very different times. His policies ran counter to the Islamic traditions of the majority of his people. Among the Muslim leaders who opposed him was an exiled *ayatollah* (Allah-ordained) named Khomeini. In time, this fanatical religious figure became the symbol of the Islamic revolution that exploded around his return to Iran from exile in 1979 and continued to run its course beyond his death 10 years later.

As Figure 6-7 shows, Iran has a large share of the oil riches in this part of the world. About 90 percent of the country's income is derived from petroleum and petroleum products. The reserves lie in a zone along the southwestern periphery of Iran's territory, and Abadan became its "oil capital" near the head of the Persian Gulf. But Iran is a

Iran's oil comes from reserves that lie beneath the shallow waters of the Persian Gulf as well as the mainland. Export is facilitated by an extensive system of pipelines that links the producing fields to export terminals, where ocean-going tankers are filled. A close look at the map of the head of the Persian Gulf shows, on the Iranian side of the median line, several small islands that serve as offshore oil-export bases. Among these the most important is Khark Island (Jazireh-ye-Khark), connected by pipelines to the mainland. Tankers can be seen in the upper right, directly opposite the cluster of holding containers.

large and populous country, and the wealth generated by oil could not bring about the transformation the last shah intended, a transformation that might have staved off the revolution. Modernization remained but a veneer: in the villages away from Tehran's polluted air, the holy men continued to dominate the lives of ordinary Iranians. As elsewhere in the Muslim world, urbanites, villagers, and nomads remained enmeshed in a web of production and profiteering, serfdom, and indebtedness that has always characterized traditional society here. The revolution swept this system away, but in the end it did not improve the lot of Iran's millions. A devastating war with Iraq (1980–1990), into which Iran ruthlessly poured hundreds of thousands of its young men, sapped the coffers as well as the energies of the state. When it was over, Iran was left poorer, weaker, and aimless, its revolution spent on unproductive pursuits.

In the mid-1990s, Iran's government is reaching outward in new ways. Still avowedly fundamentalist and devotedly Shi'ite, the modern Iranians realize that their country's future is being built outside its borders—in Transcaucasia, in Turkestan, and in Afghanistan. Iran's old imperial rival, secular Turkey, is everywhere in Iran's hinterlands, promoting its brand of Islam and its views on statehood—and extending bank credits and loans. And so Iran must compete: to strengthen its ties with Shi'ites outside its borders, to promote its model of an Islamic republic, to organize mosques, and to rally fundamentalists. Even as its representatives disperse into Iran's neighboring lands, Iran is purchasing weapons from former Soviet republics and is resurrecting its old ties with Moscow. The revolution may have ended a monarchy, but it has not extinguished all ties to an imperial past.

Afghanistan

The easternmost country in this region is Afghanistan, a landlocked, mountainous, divided, embattled product of foreign rivalries. Afghanistan exists because the British and the Russians, competing in this area during the nineteenth century, agreed to create it as a cushion, or **buffer state**, between them. This is how northern Afghanistan acquired the narrow corridor of land leading from the main territory eastward, the Wakhan Corridor (Fig. 6-16).

The Wakhan Corridor in effect makes China one of Afghanistan's neighbors. The country's more important adjoining states are Pakistan to the east and south, Iran to the west, and, following the collapse of the Soviet Union, Turkmenistan, Uzbekistan, and Tajikistan to the north. With so many neighbors, Afghanistan is fragmented ethnically and culturally. Its population of about 20 million has no majority; the largest minority are the Pushtuns (or Pathans), whose homeland lies in the east and southeast along the border with Pakistan. In the north, the Tajiks have a territorial base contiguous to neighboring Tajikistan, and the Uzbeks are concentrated farther westward. Turkmen, Hazara, and smaller minority groups create a complicated ethnic mosaic in a country that is also divided by physical geography. Texas-sized Afghanistan is a desert plateau in the west and south, and the center and northeast are dominated by the soaring ranges of the Hindu Kush.

There is little to unite the peoples of Afghanistan and much to divide them. Sunni Islam prevails here, but the faith has not been an intercultural bond. During the 1970s, the Soviet Union began to see an opportunity in the country's weakness, and in 1979 the Soviets invaded Afghanistan to prop up a puppet government. This *did* create a common cause, and while millions of refugees streamed into Pakistan and Iran, anti-Soviet rebels, with Western weapons and other assistance, mounted a fierce resistance. In 1989, the Soviets conceded defeat and withdrew, and in 1992 the government in Kabul (the capital) collapsed and was replaced by a revolutionary council representing some (but not all) the factions. The situation began to resemble that of the pre-Soviet past: a feudal pattern with a weak and ineffectual government in Kabul.

Afghanistan thus remains one of the realm's weakest and poorest countries. Urbanization still is below 20 percent, circulation is minimal, agricultural and pastoral subsistence remain the dominant livelihoods, and there is little national integration. Fruits, grown in the fertile lowlands adjacent to the Hindu Kush in the north, and carpets are the main exports. A small oil and natural gas industry was emerging in the north, but the Soviets, as they departed, capped the wells.

Afghanistan does have one coveted asset, one it cannot sell—its relative location. The country lies at the juncture of powerful Islamic states and near others. Its weakness may attract intervention by neighbors that already have stakes in the country, ethnic bridgeheads the Soviets could not claim. Pakistan may have an interest in redefining its boundary with Afghanistan, and Tajikistan may have similar objectives. Will Afghanistan become the Lebanon of this region?

❖ TURKESTAN

The old regional term *Turkestan* has reappeared on the map of the Islamic world. In the five former Soviet republics of Central Asia—now independent countries in which Islam is reviving itself even as politics and economics are changing—a momentous transformation is taking place. The configuration of this region is not yet final, but its general contents are clear: it will comprise most or all of five states. As shown in Figure 6-17, these are: (1) *Kazakstan*, territorially larger than the other four combined but situated astride an ethnic transition zone between Russia and Turkestan; (2) *Uzbekistan*, the most populous state, located at the heart of the region; (3) *Turkmenistan*, large but sparsely peopled and fronting the Caspian Sea; (4) *Tajikistan*, its population spilling over into adjoining Afghanistan and into China; and (5) *Kyrgyzstan*, land of the great Tian Shan range.

This is Turkestan today, but the region so defined does not incorporate all of what once was the Turkic domain. Turkic peoples inhabit all of northwestern China (turn to Fig. 9-8 for evidence), and the resurrection of an Islamic Turkestan will challenge the Chinese on their western frontier.

In Turkestan, the Soviet communist regime long suppressed Islam, but the faith was not extinguished. It is now being revived (as is Christianity in Russia itself), and a new wave of Islamic diffusion is spreading over the region. In previous editions of this book, politico-geographical realities forced us to classify this region as part of the former Soviet Union; today Turkestan is being reconnected to the realm of Islam.

Kazakstan

In one respect, Kazakstan resembles Chad and Sudan in the African Transition Zone: part of it lies in the realm of Islam, and part of it lies in a neighboring realm. As we noted in Chapter 2, the neighboring realm is Russia. Northern and northeastern Kazakstan have been Russified; the south remains Turkestan.

In southern Kazakstan, the physical and cultural geography resume their Dry-World, Islamic-world characteristics, two of the defining qualities of the North Africa/Southwest Asia realm. From the Caspian Sea in the west to the Chinese border in the east, Kazakstan is a land of vast desert and steppe; scattered, water-dependent population clusters; few permanent surface communications; and nomadic herding. The largest settled population concentrations lie in the better-watered east, where the capital, Almaty also is located.

The Kazaks are united by their common descent, by their Turkic language, and by Sunni Islam. But southern

FIGURE 6-17

Kazakstan is by no means a culturally homogeneous area. The Kazaks themselves constitute a minority of an estimated 40 percent of the republic's total population of 17 million (Russians, concentrated in the north, make up about 38 percent). In southern Kazakstan, Kazaks are in the majority, but they share their land with numerous minorities, large and small. The final Soviet census of 1989 indicates that nearly 100 minorities can be identified in all of Kazakstan, many of them in the Kazak-dominated south.

In Chapter 2 we mentioned the grand Soviet planning projects in Turkestan, notably those centered on the Aral Sea area. As Figure 6-17 shows, the Aral Sea lies on the border between Kazakstan and Uzbekistan, and here the Soviets brought millions of acres of desert land into production through massive irrigation. However, while the cotton fields expanded, natural drainage into the Aral Sea was disrupted, pesticides polluted the groundwater, and an ecological and human disaster occurred. Deprived of replenishment, the Aral Sea has been destroyed; illnesses caused by chemical contamination have devastated the people of the area.

The Soviets also plundered Kazakstan's considerable mineral and fuel resources. The city of Qaraghandy (Karaganda) was founded to facilitate the mining of a huge

Plants grow, flowers bloom, sand shifts, and boats rust where water once stood. The name Aral Sea has become synonymous with environmental disaster, a once-stable, fish-rich, high-volume water body destroyed by stream diversion in its catchment area for irrigation purposes. Soviet planners wanted to turn the heart of Central Asia into a vast cotton-producing zone, and by capturing the streams that fed the Aral Sea they succeeded, but at a terrible ecological (and medical) price. This landscape reflects the cost of unconstrained environmental devastation.

coalfield; today, that city looks to Moscow, not to Mecca, for its future. In the west, Kazakstan possesses major oil supplies, and some estimates suggest that the reserves in the Tengiz Basin at the northeastern end of the Caspian Sea (now being developed under contract with a major U.S. oil company) may be among the largest in the world. There also are large refineries and, near Atyrau (Guryev), a nuclear power station. All this gives Kazakstan bargaining power during the transition now under way, but the Kazaks are a loosely knit nation, and national unity will be both necessary and difficult to achieve.

Kazakstan's political and cultural mixes remain explosive, touching off both anti-Russian and interethnic strife. In another familiar pattern, there are latent territorial disputes: Kazakstan's new government has made known its dissatisfaction with parts of its boundaries with Uzbekistan and Turkmenistan. The government has also announced plans to shift the Kazak capital northward from peripheral Almaty in the southeast to Aqmola in the heart of the Russian Transition Zone—which would make the latter a *forward capital* in an area that looms prominently in Kazakstan's future.

Uzbekistan

The former Soviet republic of Uzbekistan lies at the heart of Turkestan, a populous, California-sized country with 24 million inhabitants, nearly 75 percent Uzbeks. To a greater degree even than Kazakstan, Uzbekistan was blighted by Soviet economic policy. To make Uzbekistan the world's third largest cotton producer, vast farmlands in the arid Aral Sea area were opened up for irrigation. Now locally grown foods are unsafe, drinking water is polluted, infant mortality has soared, and cancer rates are high.

At the other (eastern) end of Uzbekistan lies the Farghona (Fergana) valley, densely populated and poverty-ridden. Here, too, lies the capital, Toshkent (Tashkent), on the very doorstep of Kazakstan. Between these extremities, Uzbekistan typifies Turkestan: expanses of dry, open desert and steppe broken only by oases and ribbons of water. Uzbekistan is therefore a disjointed country, but the Uzbeks are the largest national group in Turkestan. Converted many centuries ago to Sunni Islam, they ruled Central Asia until their khanates of Khiva and Bukhara were absorbed into the Soviet Union in 1924. Russians never came here in great numbers, at least not compared to Kazakstan: they constitute only about 7 percent of the population today, a proportion that is declining.

As Figure 6-17 shows, the far northwestern part of Uzbekistan is named Qoraqalpoghistan. The Soviets established an autonomous Socialist Soviet Republic here which they called the Karakalpak ASSR, justified by the ethnic and cultural distinctiveness of the Qoraqalpog (Karakalpak) people who inhabit this area. The status of this territory, now part of Uzbekistan, is uncertain; Uzbeks form a minority here.

The Uzbeks also form sizeable minorities in neighboring republics: nearly 25 percent in Tajikistan, 13 percent in Kyrgyzstan, and about 10 percent in Turkmenistan. The government-controlled press in Toshkent has published editorials about a "greater Uzbekistan" in the region's future.

Uzbekistan borders every other state in this region and Afghanistan as well; it occupies a central and pivotal position in central Asia. Harshly exploited by Moscow during Soviet times for cotton and mineral production (gold and uranium), Uzbekistan after independence has seen troubled times. Relations with its neighbors are tense; irredentism toward Uzbek minorities in neighboring republics is rising; friction with internal non-Uzbeks occurs; Muslim fundamentalism is growing. Especially worrisome is the nature of Uzbekistan's post-Soviet government, which has been repressive and undemocratic, and has allowed the economy to collapse. Instability in Uzbekistan would affect the entire region.

Turkmenistan

In the southwest of Turkestan lies the desert republic of Turkmenistan, inhabited by just 4.7 million people. Extending eastward from the shores of the Caspian Sea to the boundary with Afghanistan, and bordering Iran for over 700 miles (1,100 km), this area was part of the old Muslim Turkestan before it became a Soviet republic in 1925. As large as Nevada and Utah combined, Turkmenistan was the frontier domain of many nomadic peoples when Soviet efforts to modernize the area began. Although 72 percent of the republic's inhabitants are Turkmen, many ethnic divisions still fragment the predominantly Muslim population. In addition, there are Russians (9 percent, mostly in the towns), Uzbeks (another 9 percent), Kazaks, Ukrainians, Tatars, and Armenians.

Soviet efforts to stabilize the population and to make it more sedentary centered on a massive project: the Garagum (Kara Kum) Canal, begun in the 1950s to bring water from mountains to the east into the heart of the desert. The canal eventually will reach the Caspian Sea; in 1996, it was more than 700 miles (1,100 km) long, and about 3 million acres had been brought under cultivation. But by diverting water from the Amu-Darya River (Fig. 6-17), the engineers who planned the Garagum Canal contributed greatly to the drying up of the Aral Sea.

Farmers in the irrigated lands grow cotton, corn, vegetables, and fruits, but many Turkmen still herd sheep and sell Astrakhan fur for export. Turkmenistan's real hope, however, lies in its sector of the Caspian Basin, where oil and natural gas abound. Already, oil and natural gas rank at the top of the country's exports by value, but major new developments are under way involving foreign companies. A key problem has to do with egress: how to send the

products to markets overseas. Iran, always interested in building influence in Turkestan, has offered to support construction of a pipeline from western Turkmenistan to its Persian Gulf port of Abadan. Other options would involve shipping the oil from the Turkmenistan port of Türkmenbashi (Balkhan) to Baki (Baku) in Azerbaijan, and thence by pipeline across the Transcaucusus and Turkey to a Mediterranean port (Iskenderun), or via Russia to a Black Sea port (Novorossiysk). Turkmenistan is *not* endowed with a favorable location!

Tajikistan

In the rugged, mountainous terrain of southeastern Turkestan, the political geography is as complex as it is in Transcaucasia. Tajikistan, Kyrgyzstan, and eastern Uzbekistan look like pieces of a jigsaw puzzle, their territories interlocked and, in Tajikistan's case, functionally fragmented (Fig. 6-17).

Iowa-sized Tajikistan borders Afghanistan and China. Its eastern reaches are dominated by the gigantic Pamirs, where high glacier-sustaining mountains feed rivers such as the Amu-Darya, source of irrigation water for the neighboring desert republics of Uzbekistan and Turkmenistan. The tallest peaks in this range, both over 23,000 feet (7,000 m), tower above some of the world's most spectacular highland terrain.

The Tajiks are a separate people in the region; they are of Persian (Iranian), not Turkic, origin and speak a Persian (and thus Indo-European) language. They constitute 62 percent of the population of 6 million, but a larger part of the Tajik nation inhabits neighboring Afghanistan (there also is a Tajik minority across the border to the east in China). The Tajiks share their land with Uzbeks in the west and northwest (24 percent), Russians (a shrinking 7 percent), Armenians, and others. Most Tajiks, despite their Persian affinities, are Sunni Muslims, not Shi'ites.

The capital of Tajikistan, Dushanbe, lies in the far west. When the Tadzhik Soviet Socialist Republic (as the communists spelled it) was established in 1929, this was an era of seminomadic herding and handicraft industries. Soviet rule and economic planning brought large industrial enterprises, mining, and irrigated agriculture, but most Tajiks remained farmers and herders.

Comparisons with Transcaucasia are appropriate. As a detailed relief map will show, Tajikistan is a country of remote and isolated areas not well connected to the center, with difficult mountain routes to outlying parts.

Along with the other four Central Asian states, Tajikistan declared its independence as the Soviet Union began to unravel in 1991. But the transition almost immediately deteriorated into rioting, bloodshed, and civil war. As Moscow's grip relaxed, fundamentalist Muslims rebelled against the entrenched communist leadership and seized the upper hand through much of 1992. But Tajikistan's newly independent neighbors and Russia, alarmed by the prospect of the forging of a Tajik Islamic republic with strengthening ties to nearby Iran and adjacent Afghanistan, gave support to the old rulers to counterattack the rebels. This effort soon succeeded, destroying much of Tajikistan's economy in the process.

In 1996, Tajikistan had an elected government but still confronted violent opposition. The government since 1994 has been making efforts to strengthen its ties with Russia, even designating the Russian ruble as Tajikistan's currency and appealing to Moscow for total economic integration. This might not be an attractive prospect for a Russia that is all too well aware of the pitfalls facing it in the Transcaucasus.

Kyrgyzstan

The great mountain ranges of the Tian Shan straddle the center of Kyrgyzstan, a republic whose outline on the map seems to relate neither to topography nor to culture. Once part of the Turkic Empire, this territory fell to Russian expansionism during the nineteenth century when it was a remote frontier land. Once under Soviet control, it acquired its present boundaries and gained identity as a republic.

Kyrgyzstan's boundaries were defined on the basis of nationality as interpreted by Soviet planners in the 1920s, but today the indigenous inhabitants, the Kyrgyz, represent barely 50 percent of the population of 4.6 million. Russians constitute more than 20 percent, Uzbeks about 13 percent. Ukrainians, Tatars, and other minorities also inhabit pockets of mountain-divided, Nebraska-sized Kyrgyzstan. Like their neighbors to the west and south, most Kyrgyz are Sunni Muslims.

The capital, Bishkek, lies on the north slope, very near the border with Kazakstan (Fig. 6-17). Surface communications between northern Kyrgyzstan and neighboring Kazakstan are better than those within Kyrgyzstan itself. The Tian Shan remains a formidable barrier, although a tunnel is under construction. In the meantime, towns in southwestern Kyrgyzstan are linked by road (and some by rail) to Uzbekistan, from where many immigrants have come to work in factories and to farm.

Pastoralism is the republic's mainstay. In addition to sheep and cattle, the Kyrgyz raise yaks for meat and milk; the yak can survive on high-altitude pastures unsuitable for other livestock. Irrigated lowlands adjacent to the mountains yield wheat, fruits, and vegetables. Most industries are processing and packaging plants for locally produced textiles and foods and turn out consumer goods. The economy, in general terms, remains typical of a low-income, disadvantaged country. What potential there is centers on oil and natural gas, textile manufacturing, and (possibly in the future) tourism. Remote and disjointed, Kyrgyzstan has little national identity. In the aftermath of the Soviet Empire, it remains an internally fragmented, isolated, poverty-afflicted entity in a region with an uncertain destiny.

The Future: The Realignment of Turkestan

For more than a century, Russian tsars and Soviet dictators kept Central Asia's Muslim peoples under tight control. The Soviets laid out a new boundary framework in what was Turkestan; they revolutionized local economies; they collectivized domestic farming and vastly expanded it in the fragile ecology of the desert of Kazakstan and Uzbekistan. In their final decade, they sought to expand their sphere of influence in the region by intervening in Afghanistan, a military and political adventure that ended in disaster and made refugees out of millions of Afghans.

When independence came to Turkestan's republics, it was welcomed by several sectors of the population: by former Soviets who anticipated that they would seize control in their respective countries, by Islamic groups hoping to regain the right to worship and spread the faith, by long-suppressed minorities that envisaged a better future, by intellectuals who hoped for democracy and greater freedom. These expectations were not always compatible. The Soviets had suppressed intraregional rivalries and ethnic disputes; their departure lifted those controls and stirred conflict. Furthermore, the end of Soviet rule led to migrations and associated instability. Peoples who had been exiled to Soviet Central Asia during the brutal communist regime saw their opportunity to return. Even as some moved out, others moved in. Thousands of Armenians were given refuge when they were threatened in Azerbaijan, and they settled in Turkmenistan.

For what may lie ahead, let us also look beyond the region's borders. The word ''Turk'' appears here in many forms: in the name of the region (Turkestan), in the name of one of the republics (Turkmenistan), in the Turkic languages spoken here. In the days of the Ottoman Empire's vast wealth and power, Turkey was the source and the beacon for this sprawling domain. Today, Turkish interest in Central Asia is reawakening. Turkey is also Southwest Asia's beacon of another kind: it is a secular state in a Muslim-dominated realm, a state that adheres to the principle that religion and government—mosque and state—must be separate. In Central Asia, Soviet domination suppressed but did not extinguish Islam. The kind of Islamic revival that will take place concerns Turkey greatly. If it takes a course toward Turkey's model, then Turkey's leadership may again be felt in an area from which it has been long excluded.

But the Turks are not the only potential contestants for such involvement. Geographically better situated are the Iranians, who gave strong irredentist support to the Azerbaijanis when they rose against Soviet domination in 1990. The Iranians hold the card of Islamic fundamentalism, and in Turkestan fundamentalism is on the rise. True, the Iranians are Shi'ite Muslims and most of Turkestan's Muslims are Sunnis, but the next decade of Islamic revival will present a window of opportunity for conversion. The former shah of Iran had dreams of a greater Iranian political empire. The current Islamic leaders of Iran, now an Islamic state by law, have visions of a greater religious empire. Turkestan to the north—and Transcaucasia—present an opportunity the shah never had.

And there is a third potential competitor: Pakistan. The Islamic state of Pakistan was severely affected by the Soviet intervention in Afghanistan: several million refugees, many of them Tajiks, came across the border into Pakistan and changed the political geography of its northwestern border zone. Now Pakistan adjoins a weakened, divided Afghanistan whose large Tajik population has affinities with the Tajiks of Turkestan. Should the region become destabilized again, Pakistan—whose military power and perhaps even nuclear capability are making it a major regional force—may find itself encouraged, if not compelled, to assert itself in a region that has long buffeted it.

Still another country has a major stake in the course of events in Turkestan: India. With a Muslim minority of almost 110 million within its borders, India has a vital interest in the region's future. India faces Islam's upsurge with trepidation but not with inaction. Visit the cities of Turkestan, and Indian entrepreneurs are everywhere, establishing linkages that will, in time to come, bear political as well as economic fruit. As one of these businesspeople stated to this chapter's author: "We Indians do not want to become the Israelis of South Asia. We see Central Asia's Islamic revival as a challenge, sure, but also as an opportunity to improve India's future."

And so events in Turkestan again confirm the interconnectedness of our world. Russians try to redirect their influence on the region; Islamic proselytizers from Iran, bankers from Turkey, politicians from Germany, economists from the United States, merchants from Pakistan, and managers from India are now engaged in a region that was long closed to the outside world. So we are witnessing the beginnings of a major realignment, the further expansion and integration of an already strengthening Muslim world that will stretch not only from Morocco to Indonesia but also from Kazakstan to Somalia. It is a vital lesson in regional geography: realms and regions are anything but static.

■ PRONUNCIATION GUIDE ■

Abadan (ahba-DAHN)
Abu Dhabi (ah-boo-DAH-bee)
Addis Ababa (adda-SAB-uh-buh)
Aden (AID-nn)
Afars (AH-farz)
Afghanistan (aff-GHANN-uh-stan)
Ahaggar (uh-HAG-er)
Al Andalus (ahl ahnda-LOOSE)
Alawite (AH-lah-wyte)
Aleppo (uh-LEP-poh)
Alexandria (ah-leg-ZAN-dree-uh)
Algeria (al-JEERY-uh)
Algiers (al-JEERZ)
Ali (ah-LEE)
Allah (AHL-ah)
Almaty (AHL-mah-tee)
Amhara (am-HAH-ruh)
Amharic (am-HAH-rick)
Amman (uh-MAHN)
Amu-Darya (uh-moo-DAHR-yuh)
Anatolia (anna-TOH-lee-uh)
Andalucía (ahn-duh-loo-SEE-uh)
Ankara (ANG-kuh-ruh)
Antalya (ant-ull-YAH)
Aozou (OO-zoo)
Aqaba (AH-kuh-buh)

Aqmola (AHK-moh-luh)
Aquifer (ACK-kwuh-fer)
Aral (ARREL)
Aramco (uh-RAM-koh)
Armenia (ar-MEENY-uh)
Ashgabat (ASH-kuh-bahd)
Ashkelon (OSH-kuh-lahn)
Ashkenazim (osh-kuh-NAH-zim)
Astrakhan (ASTRA-kahn)
Aswan (as-SWAHN)
Atatürk (ATTA-tyoork)
Atheism (AYTHEE-izm)
Atyrau (AH-tee-rau)
Ayatollah (eye-uh-TOH-luh)
Azeri (ah-ZERRY)
Azerbaijan (ah-zer-bye-JAHN)
Baathist (BATH-ist)
Bab el Mandeb (bab ull MAN-dub)
Babylon (BABBA-lon)
Bahrain (bah-RAIN)
Baki (Baku) (bah-KOO)
Balkhan (bahl-KAHN)
Baluchistan (buh-loo-chih-STAHN)
Basra (BAHZ-ruh)
Bedouin (BEH-doo-in)
Beirut (bay-ROOT)

Bekaa (buh-KAH)
Benghazi (ben-GAH-zee)
Berber (BERR-berr)
Beyoglu (bay-uh-GLOO)
Bishkek (BISH-kek)
Bosnia (BOZZ-nee-uh)
Bosporus (BAHSS-puh-russ)
Brunei (broo-NYE)
Buddhist (BOOD-ist)
Bukhara (boo-KAHR-ruh)
Burkina Faso (ber-keena FAHSSO)
Byzantine (BIZZ-un-teen)
Cairo (KYE-roh)
Caliph (KAY-liff)
Caliphate (KALLA-fate)
Castile (kass-STEEL)
Caucasus (KAW-kuh-zuss)
Chechen (CHEH-chen)
Constantinople (kon-stant-uh-NOPLE)
Córdoba (KORR-duh-buh)
Cyprus (SYE-pruss)
Cyrenaica (sear-uh-NAY-icka)
Damascus (duh-MASK-uss)
Damietta (dam-ee-ETTA)
Dhahran (dah-RAHN)
Dhow (DAU)

Diyarbakir (dih-yahr-buh-KEER)
Djezira-al-Maghreb (juh-ZEER-uh
    ahl-mahg-GRAHB)
Djibouti (juh-BOODY)
Druze (DROOZE)
Dubai (doo-BYE)
Dushanbe (doo-SHAHM-buh)
Eilat (AY-laht)
Elam (EE-lum)
Elburz (el-BOORZ)
Emirate (EMMA-rate)
Enosis (ee-NOH-sis)
Eritrea (erra-TRAY-uh)
Ethiopia (eeth-ee-OH-pea-uh)
Euphrates (yoo-FRATE-eeze)
Façade (fuh-SAHD)
Farghona (far-GOH-nuh)
Fatah (fuh-TAH)
Fellaheen (fella-HEEN)
Fergana (fare-GAHN-uh)
Fezzan (fuh-ZANN)
Gadhafi, Muammar (guh-DAHFI,
    MOO-uh-mar)
Gambia (GAM-bee-uh)
Ganges (GAN-jeeze)
Garagum (gah-rah-GOOM)
Gaza (GAH-zuh)
Gezira (juh-ZEER-uh)
Gibraltar (jih-BRAWL-tuh)
Giralda (hih-RAHL-duh)
Giza (GHEE-zuh)
Golan (goh-LAHN)
Gorno-Badakhshan
    (gore-noh-bah-dahk-SHAHN)
Guryev (GHOOR-yeff)
Hägerstrand, Torsten (HAYGER-strand,
    TOR-stun)
Haifa (HYE-fah)

Hamas (hah-MAHSS)
Hamitic (ham-MITTICK)
Hatay (hah-TYE)
Hazara (huh-ZAHR-ah)
Herodotus (heh-RODDA-tuss)
Hezbollah (HIZZ-boh-luh)
Hierarchical (hire-ARK-uh-kull)
Hindu Kush (hin-doo KOOSH)*
Homo sapiens (hoh-moh SAY-pea-enz)
Hormuz (hoar-MOOZE)
Husayn (hoo-SINE)
Hussein, Saddam (hoo-SAIN,
    suh-DAHM)
Ibn Saud (ib'n sah-OOD)
Imam (ih-MAHM)
Ingush (in-GOOSH)
Interdigitate (intuh-DID-juh-tate)
Intifada (intee-FAH-duh)
Iran (ih-RAN/ih-RAHN)
Iranian (ih-RAIN-ee-un)
Iraq (ih-RAK/ih-RAHK)
Iraqi (ih-RAKKY)
Irredentism (irra-DENT-tism)
Isfahan (iz-fuh-HAHN)
Iskenderun (iz-ken-duh-ROON)
Islam (iss-LAHM)
Ismaili (izz-MYE-lee)
Israel (IZ-rail)
Issas (ee-SAHZ)
Istanbul (iss-tum-BOOL)
Jabal (JAB-ull)
Jazirah (juh-ZEER-uh)
Jericho (JEH-ruh-koh)
Jerusalem (juh-ROO-suh-lum)
Jubail (joo-BILE)
Judaic (joo-DAY-ick)
Judaism (JOODY-ism)
Kabul (KAH-bull)

Kara Kum (kahr-ruh KOOM)
Karaganda (karra-gun-DAH)
Karakalpak (karra-kal-PAK)
Kazak (KUZZ-uck)
Kazakstan (kuzz-uck-STAHN/
    KUZZ-uck-stahn)
Kemal, Mustafa (keh-MAHL,
    moo-stah-FAH)
Khark (KARG)
Khartoum (kar-TOOM)
Khiva (KEE-vuh)
Khomeini (hoh-MAY-nee)
Khurasan (koor-uh-SAHN)
Kibbutz (kih-BOOTS)*
Kirkuk (keer-KOOK)*
Koran (kaw-RAHN)
Kosovo (KOSS-oh-voh)
Krasnovodsk (kruzz-noh-VAUGHTSK)
Kufra (KOO-fruh)
Kurd (KERD)
Kurdistan (KERD-duh-stahn)
Kuwait (koo-WAIT)
Kyrgyz (KEER-geeze)
Kyrgyzstan (KEER-geeze-stahn)
Levant (luh-VAHNT)
Lingua franca (LEEN-gwuh
    FRUNK-uh)
Luxor (LUK-soar)
Maghreb (mahg-GRAHB)
Malaysia (muh-LAY-zhuh)
Mali (MAH-lee)
Marmara (MAH-muh-ruh)
Mauritania (maw-ruh-TANEY-uh)
Mecca (MEH-kuh)
Medina (muh-DEENA)
Mesopotamia (messo-puh-TAY-mee-uh)
Mesquita (meh-SKEE-tuh)
Minaret (MINNA-ret)

* Double ''o'' pronounced as in *book*.

Monogamy (muh-NOG-ah-mee)
Mosque (MOSK)
Mubarak, Hosni (moo-BAH-rahk, HOZZ-nee)
Muhammad (moo-HAH-mid)
Mullah (MOOL-ah)*
Muscat (MUH-skaht)
Musandam (muh-SAND-um)
Muslim (MUZZ-lim)
Mutrah (MUTT-truh)
Nasser, Gamal Abdel (NASS-er, guh-MAHL AB-dul)
Negev (NEH-ghev)
Niger (nee-ZHAIR)
Nigeria (nye-JEERY-uh)
Novorossiysk (noh-voh-ruh-SEESK)
Ogaden (oh-gah-DEN)
Oman (oh-MAHN)
Oromo (AW-ruh-moh)
Orontes (aw-RAHN-teeze)
Pahlavi, Shah Muhammad Reza (puh-LAH-vee, shah mooh-HAH-mid RAY-zuh)
Pakhtuns (puck-TOONZ)
Pakistan (PAH-kih-stahn)
Palestine (PAL-uh-stine)
Palestinian (pal-uh-STINNY-un)
Pamir (pah-MEER)
Pathan (puh-TAHN)
Persepolis (per-SEPP-uh-luss)
Persia (PER-zhuh)
Pharaonic (fair-ray-ONNICK)
Phoenicians (fuh-NEE-shunz)
Pleistocene (PLY-stoh-seen)
Polygamy (puh-LIG-ah-mee)
Polytheism (polly-THEE-izm)
Port Said (port-sah-EED)
Pushtun (PAH-shtoon)

Qanat (KAH-naht)
Qaraghandy (kah-rah-GONDY)
Qatar (KOTTER)
Qattarah (kuh-TAR-ruh)
Qoraqalpog (kora-kal-PAHG)
Ramadan (rahma-DAHN)
Riverine (RIVER-reen)
Riyadh (ree-AHD)
Rub al Khali (rube ahl KAH-lee)
Rumailah (roo-MYE-luh)
Sahara (suh-HARRA)
Sahel (suh-HELL)
Samarqand (sahm-ahr-KAHND)
San'a (suh-NAH)
Saudi Arabia (SAU-dee uh-RAY-bee-uh)
Sedentary (SEDDEN-terry)
Semitic (seh-MITTICK)
Senegal (sen-ih-GAWL)
Sephardic (suh-FAHRD-ick)
Sharia (SHAH-ree-uh)
Shatt al Arab (shot ahl uh-RAHB)
Sheikdom (SHAKE-dum)
Shia (SHEE-uh)
Shi'ism (SHEE-izm)
Shi'ite (SHEE-ite)
Shiraz (shih-RAHZ)
Sidra (SIDD-ruh)
Sinai (SYE-nye)
Somalia (suh-MAHL-yuh)
Stamboul (STAHM-bool)
Sudan (soo-DAN)
Sudanese (soo-duh-NEEZE)
Suez (SOO-ez)
Suleyman (SOO-lay-mahn)
Sumer (SOO-mer)
Sunni (SOO-nee)
Syr-Darya (seer-DAHR-yuh)

Syria (SEARY-uh)
Tajik (TUDGE-ick)
Tajikistan (tah-JEEK-ih-stahn)
Tana (TAHNA)
Tatar (TAHT-uh)
Tehran (tay-uh-RAHN)
Tel Aviv-Jaffa (tella-VEEVE-JOFF-uh)
Tengiz (TEN-ghiz)
Thebes (THEEBZ)
Theocracy (thee-OCK-ruh-see)
Tian Shan (tyahn-SHAHN)
Tigris (TYE-gruss)
Toledo (toh-LAY-doh)
Toshkent (tahsh-KENT)
Transcaucasia (tranz-kaw-KAY-zhuh)
Tripoli (TRIPPA-lee)
Tripolitania (trip-olla-TANEY-yuh)
Tuareg (TWAH-regg)
Tukiu (too-KYOO)
Tunis (TOO-niss)
Tunisia (too-NEE-zhuh)
Turkestan (TERK-uh-stahn)
Türkmenbashi (tyoork-men-BAH-shee)
Turkmenistan (terk-MEN-uh-stahn)
Ural-Altaic (YOOR-ull-al-TAY-ick)
Uygur (WEE-ghoor)
Uzbek (OOZE-beck)
Uzbekistan (ooze-BECK-ih-stahn)
Wakhan (wah-KAHN)
Xian (shee-AHN)
Yanbu (YAN-boo)
Yemen (YEMMON)
Yugoslavia (yoo-goh-SLAH-vee-uh)
Zagros (ZAH-gruss)
Ziggurat (ZIG-goo-raht)
Zionism (ZYE-un-izm)
Zoroaster (zorro-AST-tuh)
Zoroastrian (zorro-ASTREE-un)

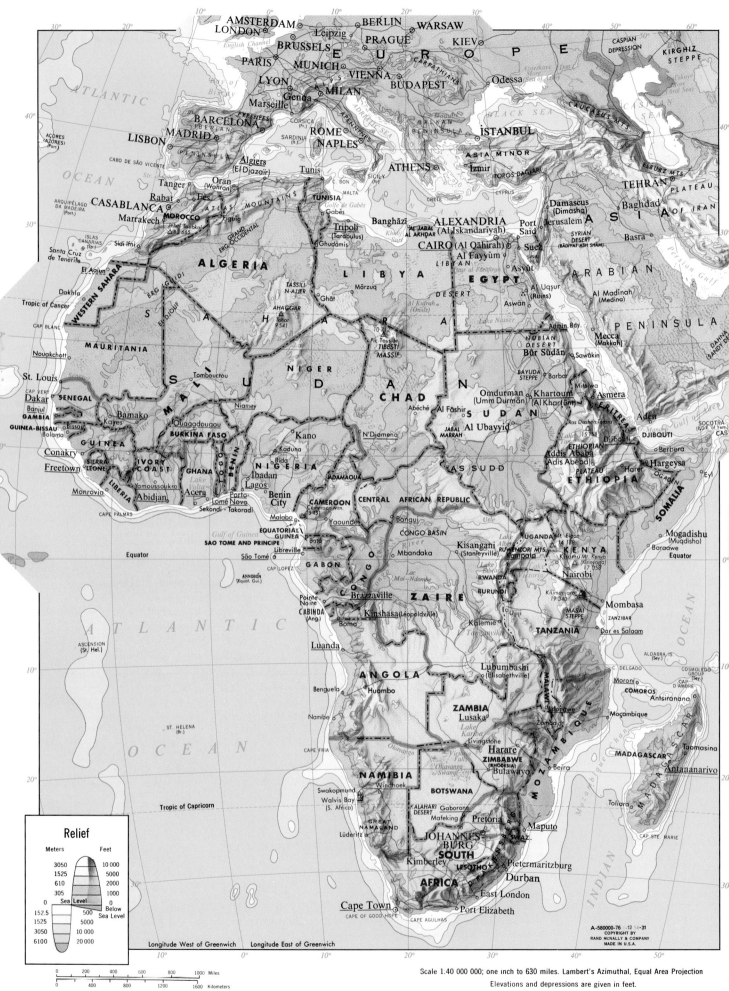

CHAPTER 7

Subsaharan Africa: Realm of Reversals

The African continent occupies a special place in the world. Beneath its surface lie some of the oldest parts of the planet's crust. Today Africa forms the heart of the Earth's landmasses, a relative location that may yet become one of its greatest assets. And current archeological research indicates that our species, *Homo sapiens*, emerged in Africa. We may have spread around the globe, but at the source we are all Africans.

For millions of years, Africa was the stage for the great drama of **human evolution**. The first tools ever made were used by our Hominid ancestors in eastern Africa. The origins of language probably lie in Africa. The first settled human communities plausibly formed in Africa. Our first artistic expression likely took place somewhere in Africa.

More recently, North Africa's Nile Valley was the scene of one of the earliest durable and creative civilizations, a culture hearth whose innovations radiated in all directions, Subsaharan Africa included. Ancient Egypt was to Africa what, thousands of years later, ancient Greece was to Europe: a source of knowledge and ideas. When West Africa's earliest states formed, their rulers modeled the political system on the Egyptian example.

In those times, more than 2,000 years ago, the African continent was a single geographic realm. The partition of Africa, through the Islamization and Arabization of the north, the European colonization, and the desiccation of

IDEAS & CONCEPTS	
Human evolution	Colonialism
Continental drift	Linguistic geography
Medical geography	Plural society
Endemism	Refugees
Epidemic	Periodic market
Pandemic	Landlocked state
Green Revolution	Exclave
Regional complementarity	

REGIONS	
West Africa	East Africa
Equatorial Africa	Southern Africa

the Sahara, came later. Today, Africa is a continent of two geographic realms: the north, a part of the greater Islamic realm that connects it to Southwest Asia and beyond; and the rest—Subsaharan Africa—an Africa defined by languages, lifestyles, and cultural landscapes.

DEFINING THE REALM

AFRICA'S PHYSIOGRAPHY

Before we investigate Subsaharan Africa's human geography, we should take note of the continent's unique physical geography. To begin with, Africa has an unusual location. No other landmass is so squarely positioned astride the equator, reaching almost as far to the north as to the south. This location has much to do with Africa's vegetation, soils, agricultural potential, and population distribution. The continent lies entirely in tropical and subtropical latitudes.

Africa also is a very large landmass, accounting for about one-fifth of the Earth's entire land surface. The north coast of Tunisia lies 4,800 miles (7,700 km) from the south coast of South Africa. Coastal Senegal, on the extreme western *Bulge* of Africa, lies 4,500 miles (7,200 km) from the tip of the *Horn* in easternmost Somalia. These distances have environmental implications. Much of Africa is far from maritime sources of moisture. In addition, as Figure I-7 shows, large parts of the landmass lie in latitudes where global atmospheric circulation systems produce arid conditions. The Sahara in the north and the Kalahari in the south form part of this globe-girdling desert zone. Water supply is one of Africa's great problems; damaging droughts occur frequently.

Mountains

When we examine Africa's topographic map, however, it becomes clear that the adjective "unique" is no exaggeration to describe the continent's physiography. Take, for example, the distribution of the world's linear mountain ranges. Every major landmass has at least one mountainous backbone: South America's Andes, North America's Rocky Mountains, Europe's Alps, Asia's Himalayas. Yet Africa, covering one-fifth of the land surface of the Earth, has nothing comparable. The Atlas Mountains of the far north occupy a mere corner of the landmass, and the Cape Ranges of the far south are not of continental dimensions. And where Africa does have high mountains, as in Ethiopia and South Africa, these are really deeply eroded plateaus— or, as in East Africa, high, snowcapped volcanoes. Missing in Africa are those elongated, parallel ranges of the Andes or Alps.

THE MAJOR GEOGRAPHIC QUALITIES OF SUBSAHARAN AFRICA

■ ■ ■

1. Physiographically, Africa is a plateau continent without a linear mountain backbone, with a set of great lakes, with variable rainfall, generally low-fertility soils, and mainly savanna and steppe vegetation.

2. Dozens of nations, hundreds of ethnic groups, and numerous smaller entities make up Subsaharan Africa's culturally rich and varied population.

3. The majority of Subsaharan Africa's peoples remain dependent on farming for their livelihood.

4. Health and nutritional conditions in Subsaharan Africa are in need of improvement as the incidence of disease remains high and diets are often unbalanced.

5. Africa's boundary framework is a legacy of the colonial period; many boundaries were drawn without adequate knowledge of or regard for the human and physical geography they divided.

6. The realm is rich in raw materials vital to industrialized countries, but much of Subsaharan Africa's population has little access to the goods and services of the world economy.

7. Patterns of raw-material exploitation and export routes set up during the colonial period still prevail in most of Subsaharan Africa. Interregional connections are poor.

8. During the Cold War, great-power competition magnified conflicts in several Subsaharan African countries, the aftermath of which will be felt for generations.

9. The rate of natural increase in Subsaharan Africa's population remains the highest in the world despite difficult environments and poor health conditions.

10. Severe dislocation affects many Subsaharan African countries, from Liberia to Rwanda. This realm has the largest refugee population in the world today.

11. Government mismanagement and the failure of leadership afflict the economies of many Subsaharan African countries.

■ ■ ■

Medical Geography

In April 1995, news reports around the world described a gruesome event: at a place called Kikwit in the Equatorial African country of Zaïre, an outbreak of Ebola fever was killing dozens. No one knew how the Ebola virus struck or why it happened here, in this remote town. What *was* clear was that the mortality rate was about 90 percent of those infected, and that death came, almost inevitably, after four or five days of horrible suffering.

It was not the first time that the Ebola virus, the most aggressive of all viruses, had struck in Africa. The first outbreaks occurred in 1976 in southern Sudan and northern Zaïre (the virus is named after a river there), killing hundreds. The virulence of these outbreaks gave rise to doomsday scenarios in which viruses of African origins would sweep the world and decimate the global population.

Frightening as these apparently random outbreaks of Ebola fever are, they have been confined to remote and comparatively isolated locales. The death toll has run in the hundreds, but Africa copes with much worse. Sleeping sickness, for example, whose origins are quite well known, kills over 200,000 people in Zaïre every year. That amounts to some 550 people a day, but these victims command no headlines. There is no risk that sleeping sickness will diffuse to New York or Paris. The field of medical geography puts reality and speculation in perspective.

Medical geography is the study of people's health in spatial context. Many diseases that afflict human populations have their origins in the environment. Diseases have source (core) areas; they spread (diffuse) through populations along identifiable routes; and they affect large clusters of populations (regions) when at their widest distribution. As usual, the map is medical geography's powerful ally. As long ago as 1854, a British physician-geographer mapped the home addresses of people who became ill with cholera in a district of London (this map is reproduced in Appendix B). Soon he noted a strong

concentration of those cases around a public water pump. The map pointed to the link between cholera and contaminated water, and the defeat of cholera could begin.

Today, medical geographers use computer maps to analyze the causes, trace the routes of diffusion, and record the impact of diseases. Tropical Africa, the source of many illnesses, is the focus of much of their work. Environmental conditions, including the heat and humidity that favor the carriers (*vectors*) of infectious diseases such as flies, mosquitoes, fleas, and snails, as well as cultural traditions that facilitate transmission, such as sexual habits, food preparation, and personal hygiene, all can be mapped. And comparing maps can lead to evidence that can help in combating the scourge.

In Africa today, hundreds of millions of people carry one or more diseases, often without knowing exactly what ails them. When a disease is carried by a large number of infected people (*hosts*) in a kind of equilibrium, without causing a rapid and widespread death toll, that disease is *endemic* to the population. In the United States, many people carry venereal diseases, whose incidence still is rising; this is another example of **endemism**. But make no mistake: the people affected may not die suddenly or dramatically, yet their health does deteriorate, energy levels are lowered, and the quality of life is reduced. In tropical Africa, various forms of hepatitis, venereal diseases, and hookworm are but a few of the threats to public health in this category.

The death-dealing outbreak of Ebola fever at Kikwit had a different pattern. When a high percentage of people are afflicted and a substantial number of victims die, the disease has **epidemic** proportions. Epidemics can have local dimensions, as in the case of the Ebola outbreak, but they also spread regionally. Sleeping sickness in tropical Africa, vectored by the tsetse fly, has regional epidemic proportions and claims countless thousands of victims

every year. But the regional extent of the epidemic is limited by the range of the vector. Where there are no tsetse flies, there is no sleeping sickness.

When a disease breaks out in one area and eventually spreads worldwide, it is **pandemic**. Various strains of influenza have started in China and diffused around the world, carried by their infected hosts on airplanes and ships. To spread the "flu," no intermediate vector is needed: a handshake, a kiss, even a sneeze can transmit it. And influenza pandemics still claim their victims by the millions around the world.

The most recent pandemic of African origin involves AIDS (see box titled, "AIDS and Subsaharan Africa" on p. 335). The disease probably originated long before it was recognized for what it is, and after some 20 years of dissemination it still continues to spread. Blood- and semen-carried, the HIV virus has diffused to virtually all corners of the world, but no geographic realm is as severely afflicted as Africa. The geographic source of the disease still is uncertain, although the evidence points to the forested eastern interior of Zaïre. Today, the incidence of AIDS in tropical Africa is so high that it has begun to suppress overall population growth rates in some African countries.

Medical geographers contribute to research on diseases and their diffusion; they also address problems of health care delivery. Even in high-income countries, the location of hospitals and clinics may not be satisfactory for the population in greatest need of medical help. In lower-income countries such as those of tropical Africa, limited facilities must be located so as to maximize their effectiveness. This requires knowledge not only of medicine but also of cultural conditions, people's daily movements, transport systems, and related circumstances. Once again, the alliance between geographic and health fields yields practical results.

▪ ▪ ▪

Lakes and Rift Valleys

This discovery should stimulate us to look more closely at the physiographic map of Africa (Fig. 7-1). What else is unusual about Africa's landscapes? As the map shows, Africa has a set of great lakes, concentrated in the east-central portion of the continent.

With the single exception of Lake Victoria, these lakes are remarkably elongated, from Lake Malawi in the south to Lake Turkana in the north. What causes this elongation and the persistent north-south alignment that can be observed in these lakes? The lakes occupy portions of deep trenches cutting through the East African Plateau, trenches that can be seen to extend well beyond the lakes themselves. Northeast of Lake Turkana, such a trench cuts the Ethiopian Highlands into two sections, and the entire Red Sea looks much like a northward continuation of it. On both sides of Lake Victoria, smaller lakes lie in similar trenches, of which the western one runs into Lake Tanganyika and the eastern one extends completely across Kenya, Tanzania, and Malawi (Fig. 7-1).

The technical term for these trenches is *rift valleys*. As the name implies, they are formed when huge parallel cracks or faults appear in the Earth's crust and the strips of crust between them sink or are pushed down to form great linear valleys. Altogether, these rift valleys stretch more than 6,000 miles (9,600 km) from the north end of the Red Sea to Swaziland in Southern Africa. In general, the rifts from Lake Turkana southward are between 20 and 60 miles (30 and 90 km) wide, and the walls, sometimes sheer and sometimes step-like, are well defined.

River Courses

Next, our attention is drawn to Africa's unusual river systems (Fig. 7-1). Africa has several great rivers, with the Nile and Zaïre (or Congo) ranking among the most noteworthy in the world. The *Niger* rises in the far west of Africa, on the slopes of the Futa Jallon Highlands, but first flows inland toward the Sahara. Then, after forming an interior delta, it suddenly elbows southeastward, leaves the desert area, and plunges over falls as it cuts through the plateau area of Nigeria, creating another large delta at its mouth. The *Zaïre (Congo)* River begins as the Lualaba River on the Zaïre-Zambia boundary; for some distance, it actually flows northeast before turning north, then west and southwest, finally cutting through the Crystal Mountains to reach the ocean. Note that the upper courses of these first two rivers appear quite unrelated to the continent's coasts where they eventually exit. In the case of the *Zambezi* River, whose headwaters lie in Angola and northwestern Zambia, the situation is the same; the river first flows south, toward the inland delta known as the Okovango Swamp,

and then turns northeast and southeast, eventually to reach its delta south of Lake Malawi. Finally, there is the famed erratic course of the *Nile* River, which braids into numerous channels in the Sudd area of southern Sudan and, in its middle course, actually reverses direction and flows southward before resuming its flow toward the Mediterranean delta in Egypt. With so many peculiarities among Africa's river courses, could it be that all have been affected by the same event at some time in the continent's history? Perhaps—but first let us look further at the map.

Plateaus and Escarpments

All continents have low-lying areas—witness the Gulf-Atlantic Coastal Plain of North America or the riverine lowlands of Eurasia and Australia. But as the map shows, coastal lowlands are few and of limited extent in Africa. In fact, it is reasonable to call Africa a *plateau continent*; except for coastal Moçambique and Somalia and along the northern and western coasts, almost the entire continent lies above 1,000 feet (300 m) in elevation, and fully half of it is over 2,500 feet (800 m) high. Even the Zaïre (Congo) Basin, Equatorial Africa's huge tropical lowland, lies well over 1,000 feet above sea level in contrast to the much lower-lying Amazon Basin across the Atlantic.

Although Africa is mostly plateau, this does not mean that the surface is completely flat and unbroken. In the first place, the rivers have been eroding the surface for millions of years and have made some fairly good cuts in it. For example, Victoria Falls on the Zambezi is 1 mile (1,600 m) wide and over 300 feet (90 m) high. Volcanoes and other types of mountains, some of them erosional leftovers, stand well above the landscape in many areas. The Sahara, where the Ahaggar and Tibesti Mountains both reach about 10,000 feet (3,000 m) in elevation, is no exception. In several places the plateau has sagged under the weight of accumulating sediments. In the Zaïre Basin, for example, rivers transported sand and sediment downstream for tens of millions of years and dropped their erosional loads into what was once a giant lake the size of an interior sea. Today the lake is gone, but the thick sediments that press this portion of the African surface into a giant basin are proof that it was there. And this was not the only inland sea. To the south, the Kalahari Basin was filling with sediments that now constitute that desert's sand; far to the north, in the Sahara, three similar basins lie centered on Sudan, Chad, and what today is Mali (the Djouf Basin).

The margins of Africa's plateau are of significance, too. Much of the continent, because it is a plateau, is surrounded by an escarpment. In Southern Africa, where this feature is especially pronounced, the *Great Escarpment* (as it is called there) marks the plateau's edge along many hundreds of miles; here the land drops precipitously from more

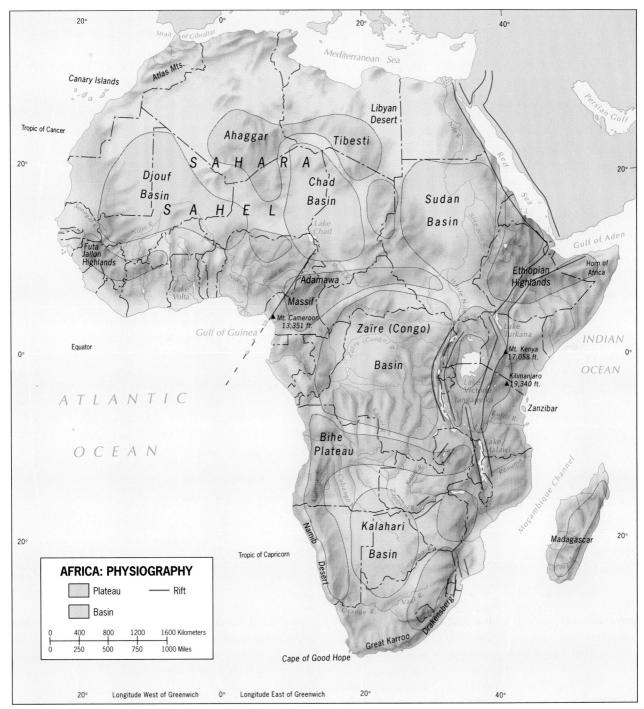

FIGURE 7-1

than 5,000 feet (1,500 m) in elevation to a narrow, hilly coastal belt. From Zaïre to Swaziland and intermittently on or near most of the African coastline, a scarp bounds the interior upland. Such escarpments are found in other parts of the world, too: in Brazil at the eastern margins of the Brazilian Highlands, and in India at the western edge of its Deccan Plateau. But Africa, even for its size, has a disproportionately large share of this topographic phenomenon.

Continental Drift and Africa

Africa's remarkable and unusual physiography was one of the pieces of evidence that the geographer Alfred Wegener used, early in this century, to construct his hypothesis of **continental drift**. According to this idea (first introduced on p. 9), all the landmasses on Earth were assembled into one giant continent named *Pangaea*; the southern continents constituted *Gondwana*, the southern part of this su-

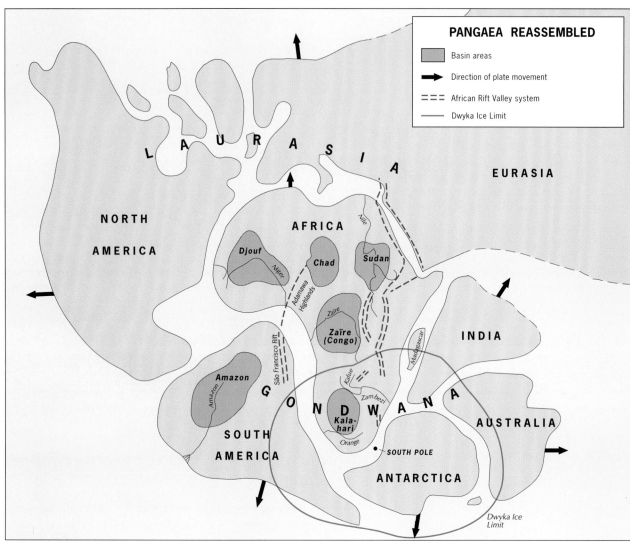

FIGURE 7-2

percontinent (Fig. 7-2). After a long (geologic) period of unity, this huge landmass began to break up more than 200 million years ago. Africa, which lay at the heart of Gondwana, attained the approximate shape we see today when North and South America, Antarctica, Australia, and India drifted radially away. Later, geophysicists gave the name *plate tectonics* to this process, and it is now understood that the breakup of Pangaea was only the most recent in a series of continental collisions and separations that spans much of the Earth's history (see Fig. I-3).

Africa's situation at the heart of the supercontinent explains much of what we see in its landscapes today. The Great Escarpment is a relic of the giant faults (fractures) that formed when the neighboring landmasses split off. The rift valleys are only the most recent evidence of the pulling forces that affect the African Plate; the Red Sea is an advanced stage of such rifts, and we may expect East Africa to separate from the rest of Africa much as Arabia did

earlier (and Madagascar before that). Africa's rivers once filled lakes within the continent, but they did not reach the distant ocean. Today, the lakes are drained by rivers that eroded inland when the oceans began to wash Africa's new shorelines.

And why does Africa not have those mountain chains that led us to look at the map more closely? The answer may lie in the direction and distance of plate motion. South America moved far westward, its plate colliding with the plate under the waters of the Pacific Ocean; the Andes crumpled up, accordion-like, in this collision. India moved northeastward, wedging into the Asian landmass; similarly, the Himalayas rose upward. But Africa moved comparatively little, being more affected by pulling tensional forces that create rifts than by the converging pressures of plate collision that generate the uplifting of mountains. Decidedly, the physiographic map reveals more than just the location of rivers, plains, and lakes.

NATURAL ENVIRONMENTS

African environments tend to be dominated by high temperatures, although plateau elevations moderate the equatorial heat in East Africa and the Horn. As noted earlier, drought is the other dominant factor in Africa. During the 1980s and 1990s, much of Southern Africa suffered under severe drought conditions, and many countries needed grain imports far beyond their normal requirements. Rapid population growth and dwindling farm output spell trouble for these low-income economies.

Climate

As Figure I-7 (p. 17) shows, Africa's climatic regions are distributed almost symmetrically about the equator. Since the equator virtually bisects the continent, the atmospheric conditions that affect the African surface tend to be similar in both halves. The hot, rainy climate of the Zaïre (Congo) Basin merges gradually, both northward and southward, into climates with distinct winter-dry seasons. Outside this great equatorial basin, therefore, annual rainfall becomes less abundant *and* less dependable. The dry season becomes longer and longer until desert conditions take over. Compare Figures I-7 and I-6 to get an idea of the steep decline in precipitation from more than 80 inches (200 cm) in equatorial Zaïre to a mere 4 inches in interior Algeria (north) and coastal Namibia (south). Note especially how dry the Horn and parts of East Africa are.

Vegetation

In common with other world areas where humid equatorial conditions prevail, the Zaïre (Congo) Basin and parts of West Africa support extensive tropical rainforest. But note, in Figure I-8, how comparatively small this rainforest zone is. And it is shrinking rapidly under population pressure and deforestation. Beyond the rainforest, where precipitation declines and dry seasons make their appearance, the rainforest gives way to the tree-studded grasslands of the savanna, a sometimes park-like vegetation that has long been the principal range for the continent's great remaining herds of wildlife. Moving poleward, the savanna yields to the drier tropical steppe, the sparse grasslands that are so easily overgrazed by livestock, risking desertification because beyond the steppe lies the desert. A close look at Figure I-7 reveals that the Sahara in the north, and the Kalahari in the south, in turn give way to steppe and, on the coast, the dry-summer subtropical climate and vegetation known as Mediterranean appears.

Soils

Tropical soils are not especially fertile, and whereas Africa has areas of good soil and adequate moisture, the overall situation is not favorable. Soils that develop under rainforest conditions are excessively leached and support the forest only because the forest supports itself, using nutrients from its own decaying biomass. Cut the forest down, and the soil soon ceases to support crops. Nor are savanna soils, in general, capable of intensive cultivation. Where conditions are good, as in the Ethiopian Highlands, on the volcanic soils of Mount Kilimanjaro, in the moister areas of higher-latitude South Africa, in parts of coastal West Africa, and in upland areas of East Africa, farm yields are high. But these are small areas compared to Africa's vastness, and in reality Subsaharan Africa has nothing to compare to the vast alluvial valleys of India and China that support hundreds of millions of people through their prodigious production of rice and wheat. Africa's staple grains are corn (maize) and millet, crops far less capable of providing the enormous per-acre yields than those of Asia.

ENVIRONMENT, PEOPLE, AND HEALTH

By global standards, Africa south of the Sahara is not a densely peopled realm. As Figure I-9 shows, large population clusters are few. The major concentrations lie in Nigeria in West Africa, around Lake Victoria in East Africa, and in several smaller areas of Southern Africa.

What Figure I-9 does not reveal, however, is the overall health and well-being of the population. Tragically, in Subsaharan Africa the people suffer from human dislocation on a massive scale, local and even regional famines still threaten, diets are not well balanced, infant mortality remains high, and life expectancies are lower than in any other world geographic realm. Why is so much of Africa afflicted this way?

The realm's physical geography has much to do with it. Tropical areas are breeding grounds for organisms that carry disease, such as flies, mosquitoes, fleas, even snails. Most Africans (more than 70 percent) live in rural areas, where they depend on river or well water that may be contaminated. Africa's climates and soils over large areas limit the range of foods that can be produced locally, and people generally do not have the money to spend on meats or other protein sources that might balance their diets. Poor nutrition weakens the body and makes it susceptible to illness; many Africans spend their entire lives in a state of poor health. Children who do not get enough protein in their diets, for instance, develop such food-deficiency disorders as kwashiorkor and marasmus; they may survive, but their resistance to diseases of older age will be reduced.

Tropical Diseases

In terms of casualties over the centuries, *malaria* heads any list of diseases that threaten Africans. A killer of as many as 1 million children per year, malaria in recent years has again been on the rise, showing resistance to the pesticides and drugs that once seemed to promise its demise. The vector (carrier) of malaria, a mosquito, prevails in almost all of tropical Africa. African children who survive childhood are likely to suffer from malaria to some degree, with a debilitating effect.

Another virulent malady is *African sleeping sickness*, transmitted by the tsetse fly and now affecting most of tropical Africa (Fig. 7-3). The fly infects not only wild animals, which constitute the great *reservoir* of this disease, but also people and their livestock. This disease appears to have originated in a West African source area around the beginning of the fifteenth century (Fig. 7-4). Since then, it has diffused far and wide, and has inhibited the development of livestock herds where meat would have provided a crucial balance to seriously protein-deficient diets. It channeled the migrations of cattle herders through fly-free corridors in East Africa (Fig. 7-3), destroying herds that moved into infested zones. Most of all, it ravaged the human population, depriving it not only of potential livelihoods but also of its health.

Still another serious and widespread disease in rural areas is *yellow fever*, which is also transmitted by a mosquito. African monkeys form the reservoir for this scourge, which reached pandemic (global) proportions until early in the twentieth century. Its defeat in the Americas made possible the construction of the Panama Canal, but in Africa, where it had existed longer and was endemic, it has persisted. An epidemic (regional outbreak) in Senegal in the 1960s claimed more than 20,000 victims, but yellow fever's greatest impact is on infants and children. The risk in the cities now is small, but in the countryside millions are infected who, if they survive, acquire a certain level of immunity.

To this depressing list must be added *schistosomiasis* (also called bilharzia), which is transmitted by snails. The parasites enter via body openings when people swim or wash in slow-moving water infested by the snails. Many development projects in Africa that dammed fast-moving streams or channeled irrigation water through farmlands inadvertently improved the habitat of the snail vector, spreading schistosomiasis among populations not previously affected. Internal bleeding, loss of energy, and pain result, although schistosomiasis is not itself fatal. It is endemic today to more than 200 million people worldwide, most of them residing in Africa.

These are among the major continentwide diseases; there are many others of regional and local distribution. The dreaded *river blindness*, caused by a parasitic worm that is transmitted by a small fly, is endemic to the savanna belt south of the Sahara from Senegal all the way across to Kenya. In northern Ghana alone, it blinds a large percentage of the adult villagers.

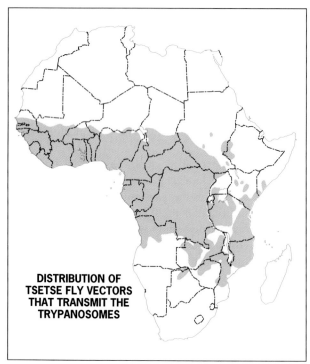

DISTRIBUTION OF TSETSE FLY VECTORS THAT TRANSMIT THE TRYPANOSOMES

FIGURE 7-3

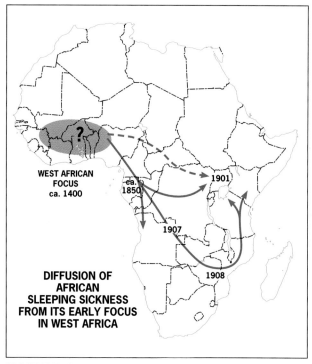

DIFFUSION OF AFRICAN SLEEPING SICKNESS FROM ITS EARLY FOCUS IN WEST AFRICA

FIGURE 7-4

In the 1970s, Africa was struck by still another scourge, AIDS, an epidemic that seems to have begun in Equatorial Africa and which, by the 1990s, had attained pandemic proportions (see box entitled "AIDS and Subsaharan Africa"). This is a nonvectored disease whose ultimate source may lie in the monkey population of the Zaïrian rainforest. The virus is carried in blood and semen, so that intimate body contact involves high risk of transmission. Although AIDS has spread worldwide, no realm on Earth is as severely afflicted by it as Subsaharan Africa, where more than 12 million, and possibly as many as 20 million, are now infected.

African peoples continue to be ravaged by diseases, even those that have been defeated or been driven back in other areas of the world. Poverty is a powerful barrier to improved health conditions, and the low-income countries of tropical Africa do not have the resources to provide needed medical services, to immunize children, and to reach families in remote rural areas. Subsaharan Africa's medical geography records an urgent need for remedial action.

THE PREDOMINANCE OF AGRICULTURE

The great majority of Africans live today as their predecessors lived—by subsistence farming, herding, or both. Not that the African environment is particularly easy for the cultivator who tries to grow food crops or raise cattle:

AIDS AND SUBSAHARAN AFRICA

■ ■ ■

AIDS—Acquired Immune Deficiency Syndrome—became a pandemic in the early 1990s. Since 1990, its impact on Africa has been devastating, afflicting millions and creating prospects of negative population growth in some areas.

Persons infected by HIV (human immunodeficiency virus) do not immediately or even soon display visible symptoms of AIDS. In the early stages, only a blood test will reveal infection, and then only by indicating that the body is mobilizing antibodies to fight HIV. People can carry the virus for years without being aware of it; during that period, they can unwittingly transmit it to others. Official reports of actual cases of AIDS thus lag far behind the reservoir of those infected. In the United States, for instance, the Centers for Disease Control in Atlanta had recorded about 496,000 cases as of mid-1996, but estimates of the number infected approached 2 million.

These U.S. data, however, pale before those from Africa. Official statistics for African countries still give no indication of the magnitude of the AIDS epidemic there, and for obvious reasons. The medical system, already overwhelmed by the long-prevailing maladies of tropical Africa, cannot cope with this new onslaught. Many of those ill with AIDS live in remote villages or in the vast shantytowns of Nairobi, Kinshasa, and other large cities, and they do not see a doctor. Yet staggering evidence of the impact of AIDS is everywhere, and surveys by the World Health Organization have begun to unveil the real magnitude of the disease in Subsaharan Africa: 20 million HIV cases as of 1996.

Most disconcerting are the trends. Because AIDS is a blood-borne disease and is transmitted most efficiently through (often undetected) blood contact, sexual practices in a society can presage routes of spatial diffusion. Patronage of prostitutes along the East African highway linking Kampala, Uganda, and Kenya's Nairobi and Mombasa, for example, is known to be one such diffusion pathway. In their 1991 book, *The Geography of AIDS*, medical geographers Gary Shannon, Gerald Pyle, and Rashid Bashshur reported a series of tests on female prostitutes in Nairobi. In 1981, 4 percent tested positive for the virus; in 1984, 61 percent; and in 1986, 85 percent. In Blantyre, Malawi's largest city, pregnant women in 1984 showed an infection rate of 2 percent; in 1990, the rate was 22 percent. About one-third of the babies born to infected women are themselves infected with the AIDS virus. Geographer Peter Gould, in his 1993 book, *The Slow Plague*, calls Africa "a continent in catastrophe," and describes the ways African governments conceal the real dimensions of the tragedy for various reasons: a sense of shame, concern that the real figures will drive investors and tourists away, and an arrogant power on the part of military rulers to deny reality. The situation, he states, is far worse in Africa than official U.N. data suggest; many doctors in Africa estimate that AIDS cases are 80 to 90 percent underreported.

If so, one in 23 Africans in the so-called AIDS Belt (from East Africa across Zaïre to the countries of West Africa) may now be infected. The urban centers are the great reservoirs, but AIDS diffuses outward along truck routes and rural roads, and the villagers are no safer than the city dwellers. In Uganda, Gould calculated in 1993, 1.3 million of the (then) population of 18 million were infected, or more than 7 percent of the entire citizenry. (A comparable rate in the United States would mean 20 million cases.)

The international medical battle against AIDS has not produced a solution, and the present state of medical knowledge holds out no long-term hope for those infected. Thus AIDS, as Africa moves into the twenty-first century, will have a devastating impact on societies and communities from Dakar to Dar es Salaam and from Djibouti to Durban.

■ ■ ■

tropical soils are notoriously unproductive, and much of Africa suffers from excessive drought.

Most African families still depend on subsistence farming for their living. Along the coast (especially the west coast) and the major rivers, some communities depend primarily on fishing. Otherwise, the principal mode of life is farming, with grain crops dominant in the drier areas and root crops in the more humid zones. In the methods of farming, the sharing of the work between men and women, the value and prestige attached to herd animals, and in other cultural aspects, subsistence farming provides the opportunity to gain insight into the Africa of the past. Colonialism changed the subsistence form of livelihood only indirectly. Tens of thousands of villages all across Africa were never fully brought into the economic orbit of the European invaders, and life in these settlements went on more or less unchanged.

Africa's pastoralists more often than not mix farming with their livestock-raising pursuits, and few of them are exclusively herders. In Africa south of the Sahara, there are two belts of herding activity. One extends laterally along the West African savanna-steppe transition zone and connects with the East African area where the Maasai drive their herds; the other centers on the plateau of South Africa. Especially in East and Southern Africa, cattle are less important as a source of food than they are as a measure of their owners' wealth and prestige in the community. Hence African cattle-owners in these areas have always been more interested in the size of their herds than in the quality of the animals. The staple foods here are the grain crops of corn, millet, and sorghum (together with some concentrations of rice), which overlap with the herding areas. Probably a majority of Africa's cattle-owners are sedentary farmers, although some pastoralists—such as the Maasai—engage in a more or less systematic cycle of movement, following the rains and seeking pastures for their livestock.

Cattle, of course, are not the only livestock in Africa. Chickens are nearly ubiquitous, and there are millions of

Shifting agriculture in Africa's tropical forest zones still forms the mainstay for millions. Here in the Ituri forest area in Zaïre, trees have been felled, the land cleared, and between stumps and rotting trunks the people farm cassava, yams, and bananas. The soil will sustain these crops for some years, but then yields will diminish, and the land will be abandoned, so that another plot must be cleared. Women do the traditional farming in Africa; here three women from the village of Ngodi are weeding the field while two youngsters look on.

goats everywhere—in the forest, in the savanna, in the steppe, even in the desert. The goats always seem to survive, and no African village would be complete without them. Where conditions are favorable, goats multiply swiftly, and then they denude the countryside and promote soil erosion. Thus goats constitute both an asset to their individual owners and a serious liability to the state.

A gradual change in the Africans' involvement in the cash economy has taken place during the past three decades. Excluded during the colonial period from most activities that could produce income for themselves and their families, many Africans have chosen, since independence, to make changes that promise to bring in money. Farmers have introduced cash crops onto their small plots, at times replacing the subsistence crops altogether. These developments notwithstanding, Africa continues to fall further behind in meeting its food needs. Across the continent, agricultural output is declining. Per capita food production in Subsaharan Africa during the mid-1990s was more than 10 percent below what it was a decade earlier. Even though total production is higher today, with 100 million more mouths to feed in 1997 than in 1987, all food increases are more than offset by the rapid rate of population growth (see box titled "A Green Revolution for Africa?").

AFRICA'S HISTORICAL GEOGRAPHY

Africa is the cradle of humanity. Archeological research has chronicled 7 million years of transition from Australopithecenes to Hominids to *Homo sapiens*. It is therefore something of an irony that comparatively little is known about Subsaharan Africa from 5,000 to 500 years ago— that is, prior to the onset of European colonialism. This is partly due to the colonial period itself, during which African history was neglected, many African traditions and artifacts were destroyed, and many misconceptions about African cultures and institutions arose and became entrenched. It is also a result of the absence of a written history over most of Africa south of the Sahara until the sixteenth century—and over a large part of it until much later than that. The best records are those of the savanna belt immediately south of the Sahara, where contact with North African peoples was greatest and where Islam achieved a major penetration.

The absence of a written record does not mean, as some scholars have suggested, that Africa does not have a history as such prior to the coming of Islam and Christianity. Nor does it mean that there were no rules of social behavior, no codes of law, no organized economies. Modern historians, encouraged by the intense interest shown by Africans generally, are now trying to reconstruct the African past,

A GREEN REVOLUTION FOR AFRICA?

■ ■ ■

The gap between world population and food production has been narrowed through the **Green Revolution**, the development of more productive, higher-yielding types of grains. Where people depend mainly on rice and wheat for their staples, the Green Revolution pushed back the specter of hunger. The Green Revolution has had less impact in Subsaharan Africa, however. In part, this relates to the realm's high rate of population growth (which is substantially higher than that of India or China). Other reasons have to do with Africa's staples. Rice and wheat support only a small part of Africa's population. Corn (maize) supports many more, along with sorghum, millet, and other grains. In moister areas, root crops such as the yam and cassava as well as the plantain (similar to the banana) supply the bulk of calories. These crops were not priorities in Green Revolution research.

Lately there have been a few signs of hope. Even as terrible famines struck several parts of the continent and people went hungry in places as widely scattered as Moçambique and Mali, scientists worked to accomplish two goals: first, to develop strains of corn and other crops that would be more resistant to Africa's virulent crop diseases, and second, to increase the productivity of those strains. But these efforts faced serious problems. An average African acre of corn, for example, yields only about 0.5 tons of corn, where the world average is 1.3 tons. Now a virus-resistant corn variety has been developed that also yields better than the old, and it has been introduced throughout Subsaharan Africa. In Nigeria, where it was first distributed, production rose significantly, contributing to a near doubling of farm production in that country since 1980 according to U.N. data. Hardier types of root crops, such as yams and cassava, also raised yields significantly.

But Africa needs more even than this if the cycle of food deficiency is to be reversed. It has been estimated that, for the realm as a whole, food production has fallen about 1 percent annually even as population grew by 3 percent. A lack of capital, inefficient farming methods, inadequate equipment, soil exhaustion, male dominance, apathy, and a series of devastating droughts contributed to this decline. The seemingly endless series of civil conflicts (Uganda, Ethiopia, Liberia, Angola, Rwanda) also reduced farm output. The Green Revolution may narrow the gap between production and need, but the battle for food sufficiency in Africa is far from won.

■ ■ ■

not only from the meager written record but also from folklore, poetry, art objects, buildings, and other such sources. Much has been lost forever, though. Almost nothing is known of the farming peoples who built well-laid terraces on the hillsides of northeastern Nigeria and East Africa or of the communities that laid irrigation canals and constructed stone-lined wells in Kenya; and very little is known about the people who, perhaps a thousand years ago, built the great walls of Zimbabwe. Porcelain and coins from China, beads from India, and other goods from distant sources have been found in Zimbabwe and other points in East and Southern Africa, but the trade routes within Africa itself—let alone the products that circulated on them and the people who handled them—still remain the subject of guesswork.

African Genesis

Africa on the eve of the colonial period was in many ways a continent in transition. For several centuries, the habitat in and near one of the continent's most culturally and economically productive areas—West Africa—had been changing. For 2,000 years, probably more, Africa had been innovating as well as adopting ideas. In West Africa, cities were developing on an impressive scale; in central and Southern Africa, peoples were moving, readjusting, sometimes struggling with each other for territorial supremacy. The Romans had penetrated to southern Sudan, North African peoples were trading with West Africans, and Arab *dhows* were sailing the waters along the eastern coasts, bringing Asian goods in exchange for gold, copper, and a comparatively small number of slaves.

Consider the environmental situation in West Africa as it relates to the past. As Figures I-6 through I-8 indicate, the environmental regions in this part of the continent exhibit a decidedly east-west orientation. The isohyets (lines of equal rainfall totals) run parallel to the southern coast (Fig. I-6); the climatic regions, now positioned somewhat differently from where they were two millennia ago, still trend strongly east-west (Fig. I-7); the vegetation map, though generalized, also reflects this situation (Fig. I-8), with a coastal forest belt yielding to savanna (tall grass with scattered trees in the south; shorter grass in the north), which gives way, in turn, to steppe and desert.

Early Trade

Essentially, then, the situation in West Africa was such that over a north-south span of a few hundred miles, there was an enormous contrast in environments, economic opportunities, modes of life, and products. The peoples of the tropical forest produced and needed goods that were quite different from the products and requirements of the peoples of the dry, distant north. As an example, salt is a prized commodity in the forest, where the humidity precludes its formation, but salt is in plentiful supply in the desert and steppe. Thus the desert peoples could sell salt to the forest peoples, but what could be offered in exchange? Ivory and spices could be sent north, along with dried foods. Thus there was a degree of **regional complementarity** between the peoples of the forest and the peoples of the drylands. And the savanna peoples—those located in between— were beneficiaries of this situation, for they found them-

selves in a position to channel and handle the trade (an activity that is always economically profitable).

The markets in which these goods were exchanged prospered and grew, and a number of true cities arose in the savanna belt of West Africa. One of these old cities, now an epitome of isolation, was once a thriving center of commerce and learning and one of the leading urban places in the world—Timbuktu. Others, predecessors as well as successors of Timbuktu, have declined, some of them into oblivion. Still other savanna cities, such as Kano in the northern part of Nigeria, continue to have considerable importance.

Early States

States of impressive strength and amazing durability arose in the West African cultural hearth (see Fig. 6-2). The oldest state about which anything is known is Ghana. Ancient Ghana was located to the northwest of the coastal country that has taken its name in the postcolonial period. It covered parts of present-day Mali and Mauritania along with some adjacent territory. Ghana lay astride the upper Niger River and included gold-rich streams flowing off the Futa Jallon Highlands, where the Niger has its origins. For a thousand years, perhaps longer, old Ghana managed to weld various groups of people into a stable state. The country had a large capital city complete with markets, suburbs for foreign merchants, religious shrines, and, some miles from the city center, a fortified royal retreat. There were systems of tax collection for the citizens and extraction of tribute from subjugated peoples on the periphery of the territory; tolls were levied on goods entering the Ghanaian domain; and an army maintained control. Muslims from the northern drylands invaded Ghana about A.D. 1062, when the state may already have been in decline. Even so, Ghana continued to show its strength: the capital was protected for no less than 14 years. However, the invaders had ruined the farmlands, and the trade links with the north were destroyed. Ghana could not survive, and it finally broke apart into a number of smaller units.

In the centuries that followed, the focus of politico-territorial organization in the West African culture hearth shifted almost continuously eastward—first to ancient Ghana's successor state of Mali, which was centered on Timbuktu and the middle Niger River valley, and then to the state of Songhai, whose focus was Gao, also a city on the Niger and one that still exists today. One possible explanation for this eastward movement may lie in the increasing influence of Islam. Ghana had been a pagan state, but Mali and its successors were Muslim and sent huge, rich pilgrimages to Mecca along the savanna corridor south of the desert. Indeed, hundreds of thousands of citizens of the modern Republic of Sudan trace their ancestry to the lands now within northern Nigeria, their ancestors having settled there while journeying to or from Mecca.

Unmistakably, the West African savanna region was the scene of momentous cultural, political, and economic developments for many centuries, but it was not alone in its progress in Africa. In what is today southwestern Nigeria, a number of urban farming communities became established, the farmers having concentrated in these walled and fortified places for reasons of protection and defense. Surrounding each "city of farmers" were intensely cultivated lands that could sustain thousands of people clustered in the towns. In the arts, too, Nigeria produced some great achievements, and the bronzes of Benin are true masterworks. In the region of the Zaïre River mouth, a large state named Kongo existed for centuries. In East Africa, trade with China, India, Indonesia, and the Arab domain brought crops, customs, and merchandise from these distant regions. In Ethiopia and Uganda, populous kingdoms emerged. Nevertheless, much of what Africa was in those earlier centuries remains to be reconstructed. Yet with all this external contact, it was clearly not isolated.

The Colonial Transformation

The period of European involvement in Subsaharan Africa began in the fifteenth century. This period was to interrupt the path of indigenous African development and irrevers-

ibly alter the entire cultural, economic, political, and social makeup of the continent. It started quietly enough, with Portuguese ships groping their way along the west coast and rounding the Cape of Good Hope not long before the opening of the sixteenth century. Their goal was to find a sea route to the spices and riches of the Orient. Soon, other European countries were sending their vessels to African waters, and a string of coastal stations and forts sprang up. In West Africa, the nearest part of the continent to European spheres in Middle and South America, the initial impact was strongest. At their coastal control points, the Europeans traded with African middlemen for the slaves who were wanted on New World plantations, for the gold that had been flowing northward across the desert, and for ivory and spices.

Suddenly, the centers of activity lay not with the cities of the savanna but in the foreign stations on the Atlantic coast. As the interior declined, the coastal peoples thrived. Small forest states rose to power and gained unprecedented wealth, transferring and selling slaves captured in the interior to the European traders on the coast. Dahomey (now called Benin) and Benin (now part of neighboring Nigeria) were states built on the slave trade. When the practice of slavery eventually came under attack in Europe, those who had inherited the power and riches it had brought opposed abolition vigorously in both continents.

Although slavery was not new to West Africa, the kind

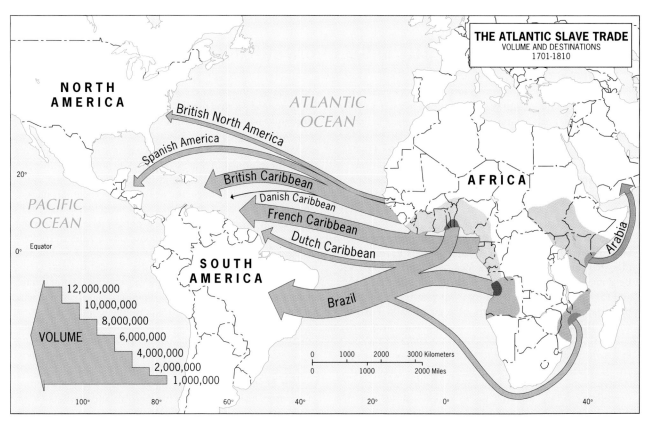

FIGURE 7-5

of slave raiding and trading introduced by the Europeans certainly was. In the savanna, kings, chiefs, and prominent families took slaves, but their number was small and their status was unlike anything that lay in store for those to be carried in bondage across the Atlantic. In fact, large-scale slave trading had been introduced in East Africa long before the Europeans brought it to West Africa. African middlemen from the coast raided the interior for able-bodied men and women and marched them in chains to the Arab markets on the coast (Zanzibar was a notorious one). There, packed in specially built *dhows*, they were carried off to Arabia, Persia, and India. When the European slave trade took hold in West Africa, however, its volume was far greater. Europeans, Arabs, and collaborating Africans combined to ravage the continent, forcing perhaps as many as 30 million persons away from their homelands in bondage (Fig. 7-5). Families were destroyed, as were whole villages and cultures; those affected suffered (if they survived their exile) a degree of human misery for which there is no measure.

The European presence on the West African coast brought about a complete reorientation of trade routes, for it initiated the decline of the interior savanna states and strengthened the coastal forest states. Moreover, it ravaged the population of the interior through its insatiable demand for slaves. But it did not lead to any major European thrust toward the interior, nor did it produce colonies overnight.

The African middlemen were well organized and strong, and they managed to maintain a standoff with their European competitors, not just for a few decades but for centuries. Although the European interests made their initial appearance in the fifteenth century, West Africa was not carved up among them until nearly 400 years later, and in many areas not until after the beginning of the twentieth century.

In all of Subsaharan Africa, the only area where European penetration was both early and substantial was at its southernmost end: at the Cape of Good Hope. There the Dutch founded Cape Town, a way-station to their developing empire in the East Indies. Soon, they entered and settled in the town's hinterland and migrated deeper into the interior. They brought thousands of slaves from Southeast Asia to the Cape, and intermarriage was common. Today, people of mixed ancestry form the largest sector of Cape Town's population. Later the British took control not only of the Cape but also of the expanding frontier, and here they transported tens of thousands of South Asians to work on their plantations. Multiracial South Africa was in the making.

Elsewhere in the realm, the European presence remained confined almost entirely to the coastal trading stations, whose economic influence was very strong. No real frontiers of penetration developed. Individual travelers, missionaries, explorers, and traders went into the interior,

THE BERLIN CONFERENCE

In November 1884, the imperial chancellor and architect of the German Empire, Otto von Bismarck, convened a conference of 14 powerful states (including the United States) to settle the political partitioning of Africa. Bismarck wanted not only to expand German spheres of influence in Africa but also to play off Germany's colonial rivals against one another to the Germans' advantage. The major colonial contestants in Africa were (1) the British, who held beachheads along the West, South, and East African coasts; (2) the French, whose main sphere of activity was in the area of the Senegal River and north of the Zaïre (Congo) Basin; (3) the Portuguese, who now desired to extend their coastal stations in Angola and Moçambique deep into the interior; (4) King Leopold II of Belgium, who was amassing a personal domain in the Congo; and (5) Germany itself, active in areas where the designs of other colonial powers might be obstructed, as in Togo (between British holdings), Cameroon (a wedge into French spheres), South West Africa (taken from under British noses in a swift strategic move), and East Africa (where the effect was to break the British design for a Cape-to-Cairo axis).

When the conference convened in Berlin, more than 80 percent of Africa was still under traditional African rule.

Nonetheless, the colonial powers' representatives drew their boundary lines across the entire map. These lines were drawn through known as well as unknown regions, pieces of territory were haggled over, boundaries were erased and redrawn, and sections of African real estate were exchanged in response to urgings from European governments. In the process, African peoples were divided, unified regions were ripped apart, hostile societies were thrown together, hinterlands were disrupted, and migration routes were closed off. All of this was not felt immediately, of course, but these were some of the effects when the colonial powers began to consolidate their holdings and the boundaries on paper became barriers on the African landscape (Fig. 7-6).

The Berlin Conference was Africa's undoing in more ways than one. The colonial powers superimposed their domains on the African continent. By the time independence returned to Africa after 1950, the realm had acquired a legacy of political fragmentation that could neither be eliminated nor made to operate satisfactorily. The African politico-geographical map is thus a permanent liability that resulted from three months of ignorant, greedy acquisitiveness during the period of Europe's insatiable search for minerals and markets.

but nowhere else in Africa south of the Sahara was there an invasion of white settlers comparable to Southern Africa's.

Colonization

After more than four centuries of contact, the European powers finally laid claim to virtually all of Africa during the second half of the nineteenth century. Parts of the continent had been ''explored,'' but now representatives of European governments and rulers arrived to create or expand African spheres of influence for their patrons. Soon there was intense competition. Spheres of influence began to crowd each other. It was time for negotiation, and in late 1884 a conference was convened in Berlin to sort things out. This conference laid the groundwork for the now-familiar politico-geographical map of Africa (see box titled ''The Berlin Conference'').

Figure 7-6 shows the result. The French dominated most of West Africa, and the British East and Southern Africa. The Belgians acquired the vast Congo. The Germans held four colonies, one in each of the realm's regions. The Portuguese held a small dependency in West Africa and two large colonies in Southern Africa (see the map dated 1910).

From the Field Notes

"From the top of the tallest building in the city, on Liberty Square, Dakar looks very much like a French-Mediterranean city with its red-tiled roofs, whitewashed buildings, and palm-lined avenues. Indeed, this was the French colonial intent: to create cities in the image of those of France, dominant centers the way Paris dominates France itself. Dakar was the colonial headquarters not only for Senegal, of which it is now the capital, but of the entire former French West African empire. But when the administrative divisions of former French West Africa became independent states, Dakar lost much of its functional hinterland."

The European colonial powers shared one objective in their African colonies: exploitation. But in the way they governed their dependencies, they reflected their differences as well. Some colonial powers were themselves democracies (Great Britain and France); others were dictatorships (Portugal, Spain). The British established a system of *indirect rule* over much of their domain, leaving indigenous power structures in place and making local rulers representatives of the Crown. This was unthinkable in the Portuguese colonies, where harsh, direct control was the rule. The French sought to create culturally assimilated elites that would represent French ideals in the colonies. The Belgians mirrored their divisions at home: in their sprawling Congo, the mining companies, government administrators, and the Roman Catholic Church each pursued their sometimes competing interests.

Colonialism transformed Africa, but in its post-Berlin form it lasted less than a century. In Ghana, for example, the Asante Kingdom still was fighting the British in the early years of the twentieth century; by 1957, Ghana was independent again. In a few years, much of Subsaharan Africa will have been independent for half a century, and the colonial period becomes an interlude rather than a paramount chapter in modern African history.

Legacies

Nevertheless, some of the legacies of the colonial period will remain on the map for centuries to come. The boundary framework has been modified in a few places, but it is a politico-geographical axiom that boundaries, once established, tend to become entrenched and immovable. The transport systems were laid out to facilitate the movement of raw materials from interior sources to coastal outlets. Internal circulation was but a secondary objective, and in

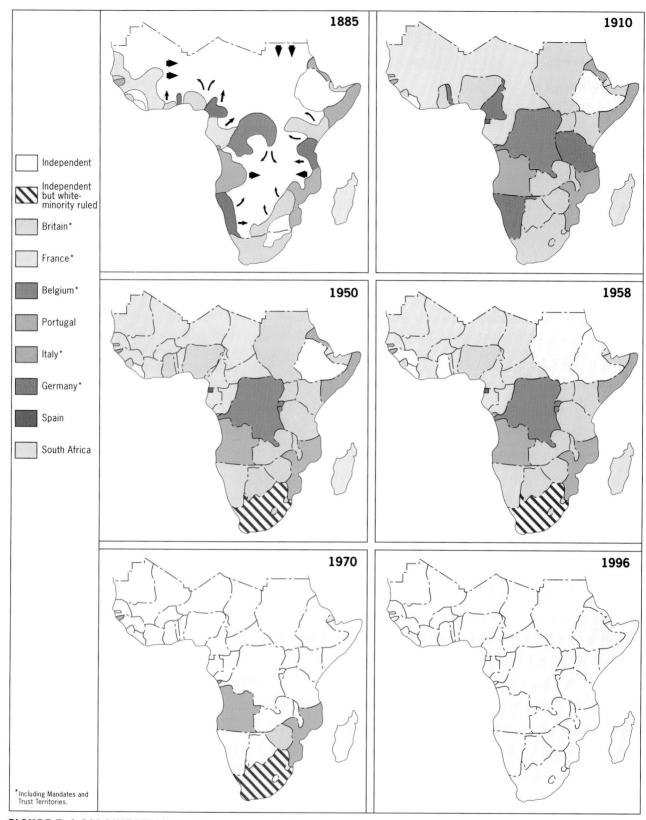

Independent

Independent but white-minority ruled

Britain*

France*

Belgium*

Portugal

Italy*

Germany*

Spain

South Africa

*Including Mandates and Trust Territories.

FIGURE 7-6 COLONIZATION AND DECOLONIZATION, 1885–1996

general African countries still are not well connected to each other. Many of Subsaharan Africa's cities were founded by the colonizers or built on the sites of small towns or villages. Some African states have repositioned their capitals, but the urban system overall is a colonial legacy that, like the political boundaries, is deeply entrenched on the African map.

In many countries, the elites that gained advantage and prominence during the colonial period also have retained their preeminence. This has led to authoritarianism in some of the postcolonial African states (for details, see the regional discussion below), and to violent reaction and civil conflict in others. Military takeovers of national governments have been a by-product of decolonization, and hopes for democracy in the struggle against colonialism have all too often been dashed. Yet some African countries have overcome the odds and have achieved representative government; among these are Tanzania, Zambia, Namibia, and Senegal.

Although not strictly a colonial legacy, the impact of the Cold War (1945–1990) on African states should be noted. In three countries—Ethiopia, Somalia, and Angola—civil wars that would perhaps have been inevitable in any case were magnified by great-power competition, weapons, military advisers, and, in the case of Angola, foreign armed forces. In other countries, ideological adherence to foreign dogmas led to political and social experiments (such as one-party Marxist regimes and costly farm collectivization) that cost Africa dearly. Today, debt-ridden Africa is again being told what to do, this time by foreign financial institutions. The cycle of poverty that followed the period of colonialism exacts a high price of African societies.

CULTURAL PATTERNS

We may tend to think of Africa in terms of its prominent countries and famous cities, its development problems and political dilemmas, but Africans themselves have another perspective. The colonial period created states and capitals, introduced a foreign language to serve as the *lingua franca*, and brought railroads and roads. The colonizers stimulated labor movements to the mines they opened, and they stopped other migrations that had been part of African life for many centuries. But they did not change the ways of life of the vast majority of the people. More than 70 percent of the realm's population still live in, and work near, Africa's hundreds of thousands of villages. They speak one of more than a thousand languages in use in the realm. The villagers' concerns are local; they focus on subsistence, health, and safety. They worry that the conflicts of others will engulf them, as has happened to millions in Ethiopia, Sudan, Liberia, Rwanda, Moçambique, and Angola in just

the past decade. Africa's largest peoples are major nations, such as the Yoruba of Nigeria and the Zulu of South Africa; Africa's smallest groups number a few thousand. As a geographic realm, Subsaharan Africa has the most intricate cultural mosaic on Earth.

African Languages

Earlier we noted that the northern boundary of the Subsaharan Africa geographic realm lies along the southern edge of the Sahara, a wide transition zone that extends from Senegal eastward to Ethiopia. Figure 7-7 shows how the Afro-Asiatic (2) languages (such as Arabic) give way to the family of languages called Niger-Congo (3) on this map. Crossing this linguistic frontier is a vivid geographic experience. Quite suddenly, the Arabic that can be used from Morocco to Sudan ceases to be useful; a desert-margin language, Hausa, replaces it. But where Hausa fades out, the intricate linguistic jigsaw of Subsaharan Africa takes over—dozens of languages in a single country, hundreds in a single region.

Africa's **linguistic geography** is one of its most intriguing aspects. Here, in an area about one-seventh of the inhabited world, is spoken one-third of all living languages by a population numbering less than one-tenth of all humankind. This is one of Subsaharan Africa's most distinctive regional properties, part of the richness of its mosaic of cultures—and one of its problems in modern times. In a sense, Figure 7-7 is misleading, because the vast extent of the Niger-Congo family (3) suggests a simplicity that does not exist. In Nigeria alone, more than 250 languages are in use, and most of them are mutually unintelligible; thus all school instruction in Nigeria takes place in the colonial *lingua franca*, English.

Nor are the Niger-Congo languages the only language family extant in Subsaharan Africa. In the area of northern Zaïre and southern Sudan, Sudanic languages (5) are in use; they are as different from the Niger-Congo languages as Europe's Germanic languages are from the Slavic. In the southwest, the Khoisan language family (6) represents the oldest surviving African languages, and at the Cape, Afrikaans (1) is a derivative of Dutch spoken by Europeans and Eurafricans. And the island of Madagascar is not part of Subsaharan Africa's linguistic geography at all. First settled by immigrants from Southeast Asia, this is an outlier of the Malay-Polynesian family of languages (7).

Religion in Africa

A map of African traditional belief systems would be almost as complicated as the map of languages. African societies, long before the universal religions reached them,

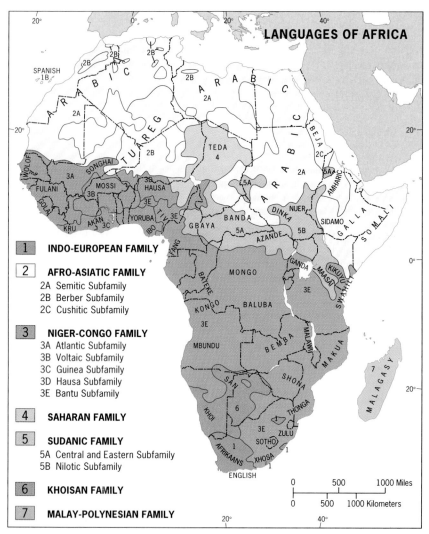

LANGUAGES OF AFRICA

1 **INDO-EUROPEAN FAMILY**

2 **AFRO-ASIATIC FAMILY**
  2A Semitic Subfamily
  2B Berber Subfamily
  2C Cushitic Subfamily

3 **NIGER-CONGO FAMILY**
  3A Atlantic Subfamily
  3B Voltaic Subfamily
  3C Guinea Subfamily
  3D Hausa Subfamily
  3E Bantu Subfamily

4 **SAHARAN FAMILY**

5 **SUDANIC FAMILY**
  5A Central and Eastern Subfamily
  5B Nilotic Subfamily

6 **KHOISAN FAMILY**

7 **MALAY-POLYNESIAN FAMILY**

FIGURE 7-7

had their own ways of worshipping gods and ancestors.

The first great wave of proselytism to reach Subsaharan Africa was Islamic, and Figure 6-1 shows how far it reached. Note the extent to which the religious boundary between Islam and non-Islam in Africa coincides with the linguistic boundary shown in Figure 7-7; here is further justification for the geographic-realm boundary we have established. Islam crossed the desert and converted the rulers of the savanna states, and it moved southward along the Indian Ocean coast to reach present-day Kenya and Tanzania.

But Islam's push was slowed by the European colonization of Subsaharan Africa and the attendant spread of the Christian religions. Christian missionaries fanned out across the realm, converting villagers in large numbers. Soon, while northern Nigeria was Muslim, southern Nigeria became Christian. So it was—and is—in Sudan and in Chad. The African Transition Zone (discussed on pp. 299–300) is a challenge to the countries that lie astride it.

Islam has been a more pervasive force, where it prevails, than Christianity in its arena. From Mauritania to Somalia, the population is virtually 100 percent Islamized, and Islam dominates life. In Subsaharan Africa, millions who profess themselves to be Christians have actually welded their traditional beliefs to Christian doctrine in a way the *mullahs* would never tolerate. This fundamental difference between Islamic dogma and Christian-animist accommodation creates a regionalism that affects all African countries in which both religions have substantial numbers of adherents.

Nations and Peoples

Given the complexity of Africa's ethnolinguistic map and the manner in which the realm's political boundaries were drawn, it is not surprising that only a few small countries could be designated as nation-states (see box p. 340). Perhaps Lesotho and Swaziland might qualify, but these are

The survival of Nigeria as a unified state is an African success story. The Nigerians have overcome strong centrifugal forces in a multiethnic country that is dominantly Muslim in the north, Christian in the south. Here, a mosque in Kano, a major city of the north, attracts a multitude of the faithful for Friday (sabbath) prayers. But now, some radical Muslim clerics are calling for an Islamic Republic in Nigeria. Can this country avoid the fate of Sudan?

ministates in any case. All the larger countries have **plural societies**.

But make no mistake: Africa's population includes some very large nations as well as numerous smaller groups. These large nations may not be as well known as, say, the Japanese or the French because they do not have their "own" countries. But they have learned to live in some form of harmony with their neighbors within the state which, as the former Yugoslavia reminds us, is a substantial achievement.

Three of Africa's larger nations form part of the Nigerian state: the Hausa-Fulani (30 million) in the north, the Yoruba (22 million) in the southwest, and the Ibo (20 million) in the southeast. Each could easily form the heart of a separate nation-state, but despite a civil war the Nigerian state has survived with three core areas and a contentious political system long dominated by the military. In Kenya, the powerful Kikuyu number more than 6 million; in multicultural South Africa, the largest single nation is the Zulu nation of about 8.5 million. In Zimbabwe, the Shona nation also numbers about 8 million. In Ethiopia, the peoples known collectively as the Oromo constitute about 40 percent of the total population, a nation of some 25 million; their Amhara-Tigrean neighbors, also in Ethiopia, number nearly 20 million.

If you compare these numbers to, say, the listing of European populations in Appendix A, you will note that many African nations are larger than the populations of several states such as Greece, Portugal, or Sweden. To refer to these African nations as "tribes" obviously is inappropriate (many anthropologists disavow the use of the term *tribe* altogether, whatever the size or status of the population). On the other hand, they are set in a mosaic of numerous smaller peoples and groups, many numbering in the thousands. Zaïre, with a population of nearly 47 million in 1997, has no dominant nation; it has four larger peoples and more than 200 smaller groups.

Ethnic conflict has ravaged parts of Africa during the postcolonial period, but external forces have contributed to this state of affairs. In Liberia, the seemingly endless cycle of violence started with an uprising against the long-entrenched American-Liberian community, which held power for 150 years until the military intervened in 1980. In Sudan, the southern provinces were united with the Islamic north by colonial edict; they have been fighting off Islamization ever since. In Moçambique, it was South African involvement that worsened a power struggle and turned it into an ethnic conflict. In Angola, Cuban troops under the communist aegis fought on one side of a post-independence ethnic conflict; South African forces aided the other side. In Rwanda and Burundi, old adversaries—the Hutu and Tutsi—saw their differences enlarged under Belgian rule and their spatial options limited by a political boundary that was essentially a reward for Belgian action against the Germans during World War I.

Not all ethnic strife can be attributed to the colonial past or to outside intervention, of course. Dreadful events in Uganda, the Central African Republic, and Burkina Faso (among others) had indigenous causes. Today, Subsaharan Africa, with less than one-tenth of the world's population, contains about 40 percent of the world's **refugees**. It is a measure of the challenge of ethnic diversity, cultural complexity, and economic deprivation confronting Africa as it confronts no other part of the world.

REGIONS OF THE REALM

On the face of it, Africa would seem to be so massive, so compact, and so unbroken that any attempt to justify a contemporary regional breakdown is doomed to failure. No deeply penetrating bays or seas create peninsular fragments as in Europe. No major islands (other than Madagascar) provide the broad regional contrasts we see in Middle America. Nor does Africa really taper southward to the peninsular proportions of South America. And Africa is not cut by an Andean or a Himalayan mountain barrier. Given Africa's colonial fragmentation and cultural mosaic, is regionalization possible? Indeed it is.

Maps of environmental distributions, ethnic patterns, cultural landscapes, historic culture hearths, and colonial frameworks yield the four-region structure shown in Figure 7-8:

1. **West Africa** includes the countries of the western coast and Sahara margin from Senegal and Mauritania in the west to Nigeria and Niger (and part of Chad) in the east.

2. **Equatorial Africa** centers on the giant state of Zaïre and includes Congo, Gabon, Cameroon, and the Central African Republic, a part of Chad, and southern Sudan.

3. **East Africa** also lies astride the equator, but its environments are moderated by elevation. Kenya and Tanzania are the coastal states; Uganda, Rwanda, and Burundi are landlocked. Highland Ethiopia also forms part of this region.

4. **Southern Africa** extends from the southern borders of Zaïre and Tanzania to the continent's southernmost cape. Ten countries, including Angola and Zimbabwe, form part of this region, whose giant is South Africa.

The island of *Madagascar*, in the Indian Ocean opposite Moçambique, cannot be incorporated into either East or Southern Africa, for geographic reasons to be discussed later.

Three of Africa's four regions are dominated, in one way or another, by a giant state. In West Africa, the leader is Nigeria by virtue of its huge population of over 100 million. In Equatorial Africa, Zaïre dominates by reason of its enormous territory, larger than the rest of the region combined. In Southern Africa, the giant is South Africa, not because of its size, but because of its economic power and influence.

Major Cities of the Realm

Cities	Population* (in millions)
Abidjan, Ivory Coast	3.2
Accra, Ghana	1.9
Addis Ababa, Ethiopia	2.4
Cape Town, South Africa	2.9
Dakar, Senegal	2.2
Dar es Salaam, Tanzania	1.6
Durban, South Africa	1.2
Harare, Zimbabwe	1.2
Ibadan, Nigeria	1.6
Johannesburg, South Africa	2.0
Kinshasa, Zaïre	4.7
Lagos, Nigeria	11.9
Lusaka, Zambia	1.5
Mombasa, Kenya	0.5
Nairobi, Kenya	2.4

*Based on 1997 estimates.

❖ WEST AFRICA

West Africa occupies most of Africa's Bulge, extending south from the margins of the Sahara to the Gulf of Guinea coast, and from Lake Chad west to Senegal (Fig. 7-9). Politically, the broadest definition of this region includes all those states that lie to the south of Morocco, Algeria, and Libya and to the west of Chad (itself sometimes included) and Cameroon. Within West Africa, rough division is sometimes made between the very large, mostly steppe and desert states that extend across the southern Sahara (Chad could also be included here), and the smaller, better-watered coastal states.

Apart from once-Portuguese Guinea-Bissau and long-independent Liberia, West Africa comprises four former British and nine former French dependencies. The British-influenced countries (Nigeria, Ghana, Sierra Leone, and Gambia) lie separated from one another, whereas Francophone West Africa is contiguous. As Figure 7-9 shows, political boundaries extend from the coast into the interior so that from Mauritania to Nigeria, the West African habitat is parceled out among parallel, coast-oriented states. Across these boundaries, especially across those between former British and former French territories, there is very little interaction. For example, in terms of value, Nigeria's

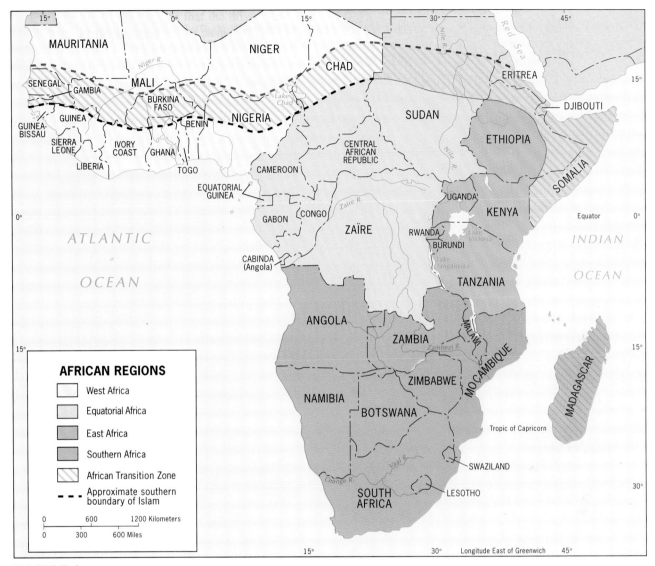

FIGURE 7-8

trade with Britain is about 100 times as great as its trade with nearby Ghana. The countries of West Africa are not interdependent economically, and their incomes are to a large extent derived from the sale of their products on the international market.

Given these cross-currents of subdivision within West Africa, what are the justifications for the concept of a single West African region? First, there is the remarkable cultural and historical momentum of this part of the realm. The colonial interlude failed to extinguish West African vitality, expressed not only by the old states and empires of the savanna and the cities of the forest but also by the vigor and entrepreneurship, the achievements in sculpture, music, and dance, of peoples from Senegal to Nigeria's southeastern Iboland. Second, West Africa contains a set of parallel east-west ecological belts, clearly reflected in Figures I-6 to I-8, whose role in the development of the region is pervasive. As the transport-route pattern on the map of

West Africa indicates, overland connections within each of these belts, from country to country, are quite poor; no coastal or interior railroad was ever built to connect this tier of countries. Yet, spatial interaction is stronger across these belts, and some north-south economic exchange does take place, notably in the coastal consumption of meat from cattle raised in the northern savannas. And third, West Africa received an early and crucial imprint from European colonialism, which—with its maritime commerce and slave trade—transformed the region from one end to the other. This impact was felt all the way to the heart of the Sahara, and it set the stage for the reorientation of the whole area, from which emerged the present patchwork of states.

The effects of the slave trade notwithstanding, West Africa today is Subsaharan Africa's most populous region (Fig. I-9). In these terms, Nigeria (whose census results are in doubt, but with perhaps 110 million people) is Africa's

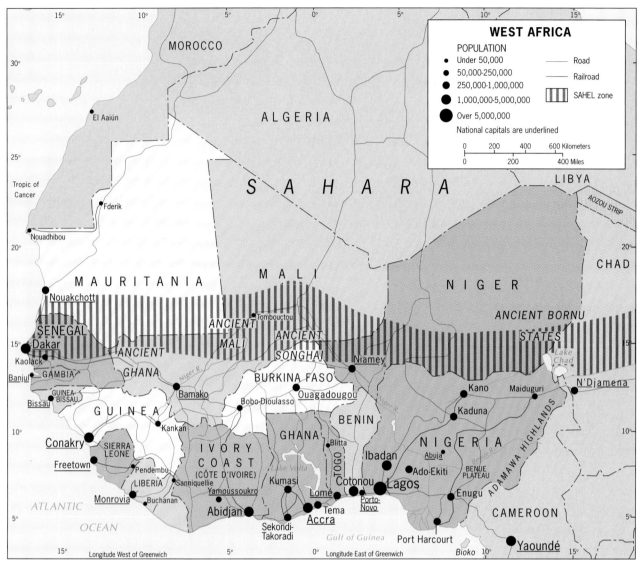

FIGURE 7-9

largest state; Ghana (19 million) ranks high as well. As Figure I-9 shows, West Africa also claims regional identity in that it constitutes one of Africa's major population clusters. The southern half of the region, understandably, is home to the majority of the people. Mauritania, Mali, and Niger include too much of the unproductive Sahel's steppe and the arid Sahara to sustain populations comparable to those of Nigeria, Ghana, or Ivory Coast.

The peoples along the coast reflect the modern era introduced by the colonial powers: they prospered in their newfound roles as middlemen in the coastward trade. Later, they were in a position to experience the changes of the colonial period; in education, religion, urbanization, agriculture, politics, health, and many other endeavors, they adopted new ways. The peoples of the interior, in contrast, retained their ties with a very different era in African history. Distant and aloof from the main theater of

European colonial activity and often drawn into the Islamic orbit, they experienced significantly less change. But the map reminds us that Africa's boundaries were not drawn to accommodate such differences. Both Nigeria and Ghana possess population clusters representing the interior as well as the coastal peoples, and in both countries the wide gap between north and south has produced political problems.

Nigeria: West Africa's Cornerstone

When Nigeria achieved full independence in 1960, it was endowed with a federal political structure that consisted of three regions based on the three major population clusters within its borders—two in the south and one in the north.

▪ AMONG THE REALM'S GREAT CITIES . . .

Lagos

In a realm that is only 27 percent urbanized, Lagos, former capital of federal Nigeria, is the exception: a teeming metropolis of 12 million sometimes called the Calcutta of Africa.

Lagos evolved over the past three centuries from a Yoruba fishing village, Portuguese slaving center, and British colonial headquarters into Nigeria's largest city, major port, leading industrial center, and first capital. Situated on the country's southwestern coast, its site consists of a group of low-lying islands and sand spits between the swampy shoreline and Lagos Lagoon. The center of the city still lies on Lagos Island, where the high-rises adjoining the Marina overlook Lagos Harbor and, across the water, Apapa Wharf and the Apapa industrial area. The city expanded southward onto Ikoyi Island and Victoria Island, but after the 1970s most urban sprawl took place to the north, on the western side of Lagos Lagoon.

Lagos's cityscape is a mixture of modern high-rises, dilapidated residential areas, and squalid slums. From the top of a high-rise one sees a seemingly endless vista of rusting corrugated roofs, the houses built of cement or mud in irregular blocks separated by narrow alleys. On the outskirts lie the shantytowns of the less fortunate, where shelters are made of plywood and cardboard and even the most basic facilities are lacking.

By world standards, Lagos ranks among the most severely polluted, congested, and disorderly cities. Mismanagement and corruption by officials are endemic. Laws, rules, and regulations, from zoning to traffic, are flouted. The international airport is notorious for its inadequate security and for extortion by immigration and customs officers. In many ways, Lagos is a city out of control.

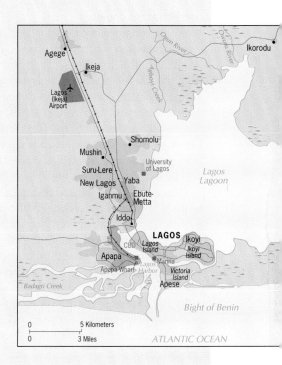

Around the Yoruba core in the southwest lay the Western Region. The Yoruba are a people with a long history of urbanization, but they are also farmers. In the old days, they protected themselves in walled cities around which they practiced intensive agriculture. The colonial period brought coastal trade, increased urbanization, cash crops (the mainstay, cocoa, was introduced from Fernando Póo [now Bioko] in the 1870s), and, eventually, a measure of security against encroachment from the north. Lagos, the country's first federal capital and now a teeming metropolis of almost 12 million people, grew up around port facilities on the region's south coast. A new, more centrally located capital is today being completed at Abuja (Fig. 7-9). Ibadan, also one of Subsaharan Africa's largest cities, evolved from a Yoruba settlement founded in the late eighteenth century. At independence, Nigeria's Western Region, more than any other part of Nigeria, had been transformed by the colonial experience.

The so-called Eastern Region, in the southeast quadrant of Nigeria, lay to the east of the Niger River and south of the Benue River. This was (and is) the stronghold of the Ibo nation, which now numbers about 20 million. At independence, Iboland was a world apart from the Yoruba core area: less urbanized, more densely peopled, less affected by the changes colonialism had brought. Over the

years, many Ibo left their area to seek work in the west, in the Muslim north, even outside Nigeria itself. No one knew that major oil reserves lay beneath the delta of the Niger, which would transform the Nigerian economy.

Nigeria's third federal region at independence was not only the largest but also the most populous: the Northern Region, extending across the full width of the country from east to west and from the northern border southward to (and even beyond) the Niger and Benue rivers. It was a fateful marriage of two southern, Christianized, coastal regions with one northern, Islamized, landlocked region where indirect rule, a feudal political system, conservative Islamic traditionalism, and resistance to change were the formative forces. But in Nigeria (unlike other coastal West African states to the west), the north had the advantage of numbers: the Hausa-Fulani population cluster is the largest in the country. Look again at Figure 7-9 and note that Nigeria extends farther northward than any other coastal West African country up to the turn of the coast at Senegal.

Nigeria's three-region federation did not last long. In 1967, the Eastern Region tried to secede as a separate political entity to be called Biafra, but this effort failed after a costly civil war lasting more than three years. Ever since, Nigeria's federal system has been modified to adjust to changing political circumstances. Today there are 30 States as well as the Federal Capital Territory of Abuja. But in effect, Nigeria has not functioned as a federation for many years. The military takeover of government in 1983 has led to a series of dictatorships that enriched the rulers and impoverished the country.

Oil exploitation in the Niger Delta brought Nigeria a brief period of prosperity. By the early 1980s, more than 90 percent of the country's export revenues were derived from the sale of petroleum and petroleum products. But mismanagement, corruption, poor planning, and (later) the worldwide decline in oil prices left Nigeria poorer than it was before the boom began. The 1983 version of Figure I-11 would have shown Nigeria as a lower-middle-income economy, and there was talk of a Nigerian "takeoff." Today, Nigeria is one of the world's poorest economies.

During the 1990s, Nigeria became a tragic, oft-cited example of a promising economy destroyed by military misrule. In 1993, the military regime permitted an election but then refused to accept the voters' choice for president; when the winner, Moshood Abiola, declared himself President he was imprisoned by the latest military dictator, General Abacha.

In 1995, a long-brewing conflict over land rights and environmental damage between radical leaders of the Ogoni people, who inhabit oil-rich areas of the Niger Delta,

and Abacha's regime had disastrous consequences. After a quick trial, and despite worldwide calls for clemency, nine Ogoni leaders, including an internationally known activist, were executed. As President-elect Abiola languished in jail, Nigeria took apartheid-era South Africa's place as the pariah of the continent.

All this background information is important, because Nigeria is not only West Africa's cornerstone; it is also one of Africa's most important states. About one-fifth of the entire population of the Subsaharan Africa geographic realm lives in Nigeria. In recent years, the relationship between the Muslim north and the non-Muslim south has been worsened by the country's economic and political downturns. Religious conflict, attended by large numbers of casualties and the burning of mosques and churches, has begun to afflict a society that once was able to restore peace after a major civil war. Nigerian society is being torn by centrifugal forces; the breakdown of the Nigerian state would be a calamity of immeasurable proportions.

Coast and Interior

Nigeria is one of 17 states (counting Chad and Cape Verde, the latter not shown in Figure 7-9) that constitute West Africa. Four of these states, comprising a huge territory but a small population, are landlocked: Mali, Burkina Faso, Niger, and Chad. Figure I-7 shows clearly how the natural environments of these four interior states are dominated by steppe and desert conditions. Figure I-9 reveals the way population is concentrated in the steppe zone and along the ribbon of water provided by the Niger River.

But even the coastal states do not escape the dominance of the desert over so much of West Africa. Mauritania lies

From the Field Notes

"This was to be a reprise of a trip taken nearly 20 years earlier: the train ride from Dakar north to St. Louis and back, a pleasant, instructive experience in 1960. But even before departure it was clear that Senegal's railway infrastructure had suffered. The coaches were in need of repair and, as it turned out, the track was unreliable. Various delays turned a ride of a few hours into a day-long adventure. Senegal, like so many other African countries, lacked the funds to maintain a railroad that had been built, in the first place, to serve colonial ends."

almost entirely under desert conditions; Senegal, as Figure 7-9 shows, is a Sahel country; and not only northern Nigeria but also northern Benin, Togo, and Ghana have interior steppe zones. The loss of pastures to desertification is a constant worry for the livestock herders there.

West Africa's states share the effects of the environmental zonation depicted in Figures I-6 to I-8, but they also have distinct regional geographies. Benin, Nigeria's neighbor, has a growing cultural and economic link with the Brazilian State of Bahia, where many of its people were taken in bondage and where elements of West African culture have survived. Ghana, once known as the Gold Coast, was the first West African state to achieve independence (1957), with a sound economy based on cocoa exports. Two grandiose postindependence schemes can be seen on its map: the port of Tema, which was to serve a vast West African hinterland, and Lake Volta, the region's largest dam project. Neither fulfilled expectations, and Ghana's economy collapsed. In the mid-1990s, a period of stable and democratic government (following a military regime) produced some recovery.

Ivory Coast (officially *Côte d'Ivoire*) translated three decades of autocratic but stable rule into economic progress that gave the country lower-middle-income status based mainly on cocoa and coffee sales. Continued French involvement in the country's affairs contributed to this prosperity; the capital, Abidjan, reflected its comparative well-being. But in a familiar pattern, Ivory Coast's president-for-life first engineered the transfer of the capital to his home village, Yamoussoukro, and then spent tens of millions of dollars building a Roman Catholic basilica there to rival that of St. Peter's in Rome. It was dedicated just as the country's economy was slowing and its social conditions worsened. By the mid-1990s, Ivory Coast had slipped back to low-income status (Fig. I-11).

Part of Ivory Coast's problem lay next door: in Liberia. In 1990, Liberia's political order (a military dictatorship) collapsed in the chaos of a ruinous civil war that pitted several ethnic groups against each other. Rubber plantations and iron mines ceased to function as the violence drove hundreds of thousands of refugees to neighboring countries—some 300,000 into Ivory Coast and 700,000 into Guinea and Sierra Leone. An estimated 230,000 people, almost 10 percent of the entire population, died in this catastrophic conflict, which has destabilized all three of Liberia's neighbors.

Compared to what has happened in Liberia and its neighbors, Senegal's situation, its obvious environmental problems notwithstanding, seems favorable. Its capital, Dakar, was the headquarters of the vast French West African colonial empire. Democratic government has existed here for nearly four decades. Plans to unite Senegal with Gambia have had to be abandoned, and a small separatist movement has centered in northern Senegal, but none of this has threatened democracy here. Senegal's greatest risks are en-

vironmental: it suffered severely during the 1970s drought that made the word "Sahel" synonymous with disaster, and as an agricultural economy (phosphate mining and fishing provide the only alternatives) it can never be secure. But as Appendix A shows, Senegal's per capita GNP is now the highest in the region, having surpassed declining Ivory Coast. All this is the case despite local droughts, reduced market prices for its exports, and a high rate of population growth—the story of many an African country.

Periodic Markets

As the foregoing discussion suggests, the great majority of the people of West Africa are not involved in the production of exports for world markets but subsist on what they can grow and raise—and trade. Their local transactions take place at small markets in villages. These village markets are not open every day but operate at regular intervals. In this way, several villages in an area get their turn to attract the day's trade and exchange, and each benefits from its participation in the wider network of interactions. People come to these **periodic markets** on foot, by bicycle, on the backs of their animals, or by whatever other means available. Periodic markets are not exclusively a West African phenomenon. They also occur in interior Southeast Asia, in China, and in Middle and South America as well as in other parts of Africa. The intervals between market days vary. In much of West Africa, village markets tend to be held on every fourth day, although some areas have two-day, three-day, or eight-day cycles.

Periodic markets, then, form an interlocking network of exchange places that serve rural areas where there are few or no roads. As each market in the network gets its turn, it will be near enough to one portion of the area so that the people who live in the vicinity can walk to it, carrying what they wish to sell or trade. In this way, small amounts of produce filter through the market chain to a larger regional market, where shipments are collected for interregional or perhaps even international trade.

What is traded, of course, depends on where the market is located. A visit to a market in West Africa's forest zone will produce very different impressions from a similar visit in the savanna zone. In the savanna, sorghum, millet, and shea-butter (an edible oil drawn from the shea-nut) predominate, and you will see some Islamic influences on the local scene. In the southern forest zone, such products as yams, cassava, corn, and palm oil change hands. Here, too, one is more likely to see some imported manufactured goods passing through the market chain, especially near the relatively prosperous areas of cocoa, coffee, rubber, and palm oil production. In general, however, the quantities of trade are very small—a bowl of sorghum, a bundle of firewood—and their value is low. These markets serve people who mostly live at or near the subsistence level.

❖ EQUATORIAL AFRICA

The Equatorial African region adjoins West Africa along the crests of a mountain range, the Adamawa Highlands, and across the heart of a steppe, in Chad. As a geographic region, Equatorial Africa is the low-lying part of Africa's equatorial zone and consists of Zaïre, Congo, the Central African Republic, Gabon, Cameroon, Equatorial Guinea, the southern part of Chad, and the southern subregion of Sudan (Fig. 7-10).

Zaïre

The territorial giant of Equatorial Africa is Zaïre. This vast country contains a wide range of resources, the region's largest cities, and its greatest potential for development. But it has been hampered by a lack of national cohesion, by corrupt and autocratic rule, by infrastructure problems, and by falling prices for its exports.

For so large a country, Zaïre's population of 47 million seems modest. The country's core area lies around the capital, Kinshasa, and along its narrow corridor to the port of Matadi near the Atlantic Ocean. But its economic focus has long been in the distant southeast, around the city of Lubumbashi, where most of the valuable mineral resources (chiefly copper) lie. The great Zaïre River system would seem to create a natural transport network, but a series of rapids interrupts this route, and several transshipments are necessary. Under normal circumstances, it is easier to send exports through Angola than through Zaïre itself. Internal circulation in Zaïre is plagued by breakdowns and delays because the huge forested basin that forms the center of the country remains a formidable obstacle even to modern equipment. During the independence period, Zaïre's infrastructure has virtually disintegrated. Trains no longer run, bridges have collapsed, and roads are impassable. Bands of armed thugs demand "tolls" of the traveler along stretches of serviceable roads.

As Figure 7-10 shows, Zaïre has no less than 10 neighbors (counting Cabinda), which underscores this country's potential importance in this region and in Africa at large. By providing refuge and permitting weapons traffic, it affected the civil war in Angola. When Burundi and, more recently, Rwanda experienced civil strife, hundreds of thousands of refugees crossed into Zaïre. A stable and viable Zaïre could form the foundation for the entire region.

Today, Zaïre is an African tragedy, a country often cited by development experts as proof that even a rich raw-material base and ample opportunities are not enough to ensure growth. In truth, the absence of centripetal forces in a vast land with more than 200 cultural groups and severe transport impediments would test a stable and representative government. As it is, self-preservation and enrichment have driven what leadership Zaïre has had.

Lowland Neighbors

Only two of the seven remaining countries of Equatorial Africa are landlocked: the Central African Republic, one of the continent's least-developed states, and Chad, of which the forested southern sector may be included in this region. Southern Sudan, which physiographically and culturally is a part of Equatorial Africa, is politically connected to that country's north, which has an outlet on the Red Sea. The other four countries (Congo, Gabon, Equatorial Guinea, and Cameroon) have coastlines on the Atlantic Ocean.

Congo, lying west of Zaïre directly across the Zaïre (Congo) River, has few natural resources but does enjoy some locational advantages. Brazzaville, its capital, was French Equatorial Africa's headquarters, and Pointe Noire was its major port. This gave Congo a transit function that brought in revenues, but unfortunately the country itself has little to trade. Its western neighbor, Gabon, has brighter prospects; its diversifying economic geography includes oil reserves, manganese, uranium, and iron ores as well as a productive (but destructive) logging industry. But the agricultural sector is weak, and large investments in transport systems were necessary (even though Gabon is only one-ninth the size of Zaïre). As Figure I-11 shows, Gabon's exports have made it an upper-middle-income economy, the first country in this region to accomplish this, largely through revenues from oil and logging. How long this commodity-based boom can last is a troubling question.

Oil also raised the economic prospects neighboring Camreroon, but Cameroon is not as well endowed as Gabon. When dwindling output combined with lower prices buffeted Cameroon's economy in the 1980s, this country had what Nigeria did not: a strong agricultural sector. Forested in the south and open savanna in the north, Cameroon's physiography also displays east-west contrasts based on elevation, relief, and proximity to the coast. As a result, the country is not only self-sufficient in food but exports such products as coffee, cocoa, cotton, and palm oil. Timber exports, too, helped weather the storm. Spatially, it is western Cameroon where development has forged ahead: here lie the capital, Yaoundé, and the port of Douala.

As a region, Equatorial Africa remains depressingly poor. But in its coastal Francophone states, at least, there is hope for the future.

❖ EAST AFRICA

East of the row of Great Lakes that marks the eastern border of Zaïre (Lakes Albert, Edward, Kivu, and Tanganyika), Africa takes on quite a different character. The land rises from the basin of Zaïre to the plateau of East Africa.

FIGURE 7-10

Here the rainforest disappears and is replaced by savanna. Great volcanoes rise above a rift-valley-dissected highland. At the heart of the region lies Lake Victoria. In the north the surface rises above 10,000 feet (3,300 m), and so deep are the trenches cut by faults and rivers there that the land was called, appropriately, Abyssinia.

Five countries, as well as the highland part of Ethiopia, form this East African region: Tanzania, Kenya, Uganda, Rwanda, and Burundi. Here the Bantu peoples that form the great majority of the population met Nilotic peoples from the north, including the Maasai. In the hills of Rwanda and Burundi, a stratified society developed in which the minority Tutsi, in their cattle-owning kingdoms, dominated the Hutu peasantry. The coast was the scene of many historic events: the arrival of Islam, the visit of Ming Dynasty Chinese fleets, the quest for power by the Turks, the Arab slave trade, the European colonial competition. Here developed the East African *lingua franca*, Swahili.

Kenya

Kenya is neither the largest nor the most populous country in East Africa, but over the past half-century it has been the dominant state in the region. Its skyscrapered capital, Nairobi, is the region's largest city; its port, Mombasa, is the region's busiest. During the 1950s, the Kikuyu nation led a vicious anticolonial rebellion that hastened the departure of the British from the region.

After independence, Kenya chose a capitalist path of development, aligning itself with Western interests. Without major known mineral deposits, Kenya depended on coffee and tea exports and on a tourist industry based on its magnificent national parks (Fig. 7-11). Tourism became the country's largest single earner of foreign exchange, and Kenya prospered, apparently proving the wisdom of its capitalist course.

But serious problems arose. Kenya during the 1980s had the highest rate of population growth in the world, and population pressure on farmlands and on the fringes of the wildlife reserves mounted. Poaching took on damaging proportions, and the numbers of visitors began to decline. Corruption in government siphoned off funds that should have been invested. Democratic principles were violated, and relationships with Western allies were strained. When the AIDS epidemic struck Kenya, it was another setback to a country that, in the early 1970s, appeared headed for an economic takeoff.

Today, Kenya's prospects are uncertain. Geography, history, and politics have placed the Kikuyu (22 percent of the population of 30 million) in a position of power. But there are a number of other major peoples (see Fig. 7-11) and several smaller ones. The Luhya, Luo, Kalenjin, and Kamba together constitute about 50 percent of the population, and on the territorial margins of the country there are peoples such as the Maasai, Turkana, Boran, and Galla. Whether Kenya can achieve democratic government and defeat corruption is an open question.

Tanzania

Tanzania (a name derived from *Tan*ganyika plus *Zan*zibar) is the largest East African country in terms of territory and population (31 million). Its total area exceeds that of the other four countries combined. Tanzania has been described as a country without a core, because its clusters of population and its zones of productive capacity lie dispersed—mostly on its margins on the east coast (where the capital, Dar es Salaam, is located), near the shores of Lake Victoria in the northwest, near Lake Tanganyika in the far west, and near Lake Malawi in the interior south. This is in sharp contrast to Kenya, which has a well-defined core area in the Kenya Highlands (centered on its capital in the heart of the country). Moreover, Tanzania is a country of many peoples, none with a numerical advantage to enable domination of the state. About 100 ethnic groups exist; one-third of the population, mainly those on the coast, adhere to Islam.

In contrast to Kenya, Tanzania after independence em-

From the Field Notes

"Nairobi's CBD remains one of tropical Africa's most modern, a world apart from the country of which it is the capital. Walking the streets here you rub elbows with one of the world's most varied urban populations including Africans, Asians, Europeans, Arabs, and many others. Prosperous as central Nairobi looks, the city has not escaped the consequences of rapid urbanization. Beyond a middle zone of comparatively well-off neighborhoods, and interspersed with these in some areas, lie some of Africa's most poverty-afflicted shantytowns. Poverty breeds desperation that can lead to crime, and street crime is a serious problem now in burgeoning Nairobi."

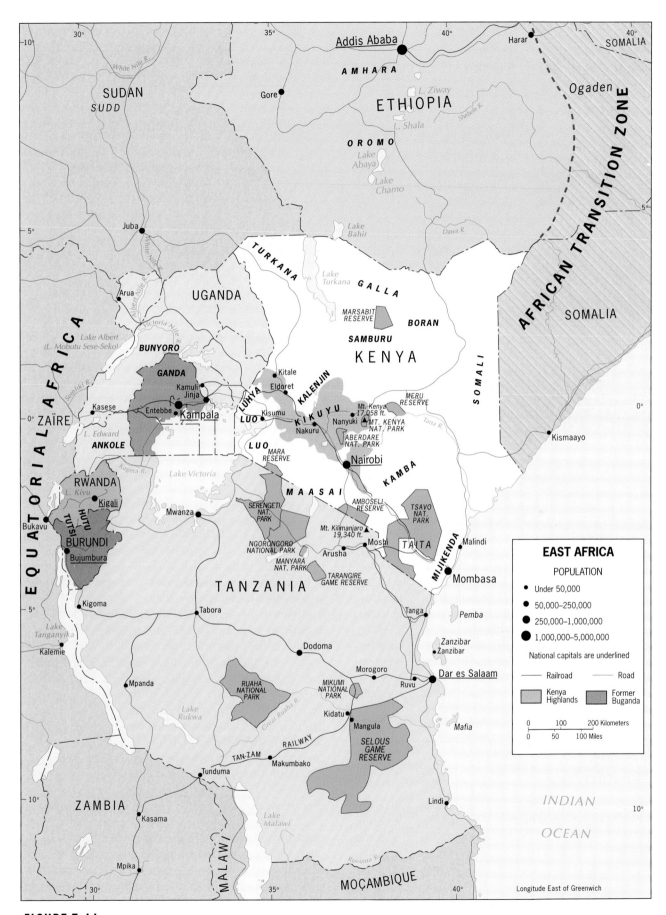

EAST AFRICA
POPULATION

- • Under 50,000
- ● 50,000–250,000
- ⬤ 250,000–1,000,000
- ⬤ 1,000,000–5,000,000

National capitals are underlined

Railroad	Road
Kenya Highlands	Former Buganda

0   100   200 Kilometers
0   50   100 Miles

Longitude East of Greenwich

FIGURE 7-11

■ AMONG THE REALM'S GREAT CITIES...

Nairobi

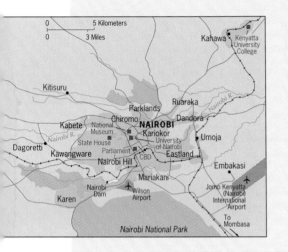

Nairobi is the quintessential colonial legacy: there was no African settlement on this site when, in 1899, the railroad being laid by the British from the port of Mombasa to the shores of Lake Victoria reached it. There was, however, something even more important: water. The fresh stream that crossed the railway line was known to the Maasai cattle herders as Enkare Nairobi (Cold Water).

Railroad building continued toward the interior, but Nairobi grew. Indian traders set up shop. The British established their administrative headquarters here. When Kenya became independent in 1963, Nairobi naturally was the national capital.

Nairobi owes its primacy to its governmental functions, which ensured its priority through the colonial and independence periods, and to its favorable situation. To the north and northwest lie the Kenya Highlands, the country's leading agricultural area and the historic base of the dominant nation in Kenya, the Kikuyu. Beyond the rift valley to the west lie the productive lands of the Luo in the Lake Victoria Basin. To the east, elevations drop rapidly from Nairobi's 5,000 feet (1,660 m), so that highland environs change to tropical savanna that, in turn, yields to steppe. Figure I-7 reveals this transition.

A moderate climate, a modern city center, several major visitor attractions (including Nairobi National Park, on the city's doorstep), and a state-of-the-art airport have boosted Nairobi's fortunes as a major tourist destination. In recent years, wildlife destruction, security concerns, and political conditions have damaged the industry.

Nairobi is Kenya's principal commercial, industrial, and educational center. But its growth (to 2.4 million today) has come at a price: its modern CBD stands in stark contrast to the squalor in the shantytowns that house the countless migrants attracted by its apparent opportunities.

---

barked on a socialist course toward development, including a massive farm collectivization program that was imposed without adequate planning. Communist China helped Tanzania construct a railroad, the Tan-Zam Railway, from Dar es Salaam to Zambia; but the project failed. So did an effort to move the capital from colonial Dar es Salaam to Dodoma in the interior (Fig. 7-11). The country's limited tourist infrastructure was allowed to degenerate, giving Kenya virtually the entire market in this arena.

In the late 1980s, Tanzania reversed course and announced a market-oriented economic recovery program. As Appendix A shows, however, agriculture-dependent Tanzania in the late 1990s remained one of the poorest countries in the world.

Uganda

Uganda contained the most important African political entity in the region when the British entered the scene during the second half of the nineteenth century. This was the Kingdom of Buganda (shown in dark brown in Fig. 7-11), which faced the north shore of Lake Victoria, had an impressive capital at Kampala, and was stable—as well as ideally suited for indirect rule over a large hinterland. The British established their headquarters at nearby Entebbe on the lake (thus adding to the status of the kingdom) and proceeded to organize their Uganda Protectorate in accordance with the principles of indirect rule. The Baganda (the people of Buganda) became the dominant people in Uganda, and when the British departed, they bequeathed to Uganda a complicated federal system that was designed to perpetuate Baganda supremacy.

Although a **landlocked state** dependent on Kenya for an outlet to the ocean, Uganda at independence (1962) had better economic prospects than many other African countries. It was the largest producer of coffee in the British Commonwealth. It had major cotton exports and also exported tea, sugar, and other farm products. Copper was mined in the southwest, and an Asian immigrant population of about 75,000 played a leading role in the country's commerce. Nevertheless, political disaster overtook the country. Resentment at Baganda overlordship fueled revolutionary change, and a brutal dictator, Idi Amin, took control in 1971. He ousted the Asians, exterminated his opposition, and destroyed the economy. Eventually, in 1979, a military campaign supported by neighboring Tanzania drove Amin from power, but by then Uganda lay in ruins. Recovery has been slow, complicated by the high cost of the AIDS epidemic, which struck Uganda with particular severity. The departure of the British colonizers set in motion a sequence of calamities from which the country may never fully recover.

Rwanda and Burundi

Immediately to the southwest of Uganda, and in the northwestern corner of what would seem to be Tanzania's territory, lie two territorially small countries, Rwanda and Burundi (Fig. 7-11). Indeed, these were part of Tanzania when it was German East Africa. But Belgian forces penetrated here from the (then) Belgian Congo, pushed the Germans back, and were awarded these territories after the end of World War I. What the Belgians got was a populous area. They used it as a labor pool for the mines of their neighboring Congolese domain.

In terms of population, Rwanda (8 million) and Burundi (7 million) are not small by African standards. More significantly, the population is stratified into three layers. Historically, the first group, the minority Tutsi, have been dominant, owning land and cattle and setting up kingdoms to perpetuate their power. The Hutu always were the peasants, often tenants on Tutsi land, numerically in the majority (85 to 90 percent of the population) but weaker politically. The third group, the Twa (pygmy) population, makes up just 1 to 2 percent and has been lost in the crises that have engulfed these lands.

Ever since the Belgians withdrew, Tutsis and Hutus in both Burundi and Rwanda have engaged in a series of contests for power marked by such carnage that all else in Africa—even Liberia and Sudan—has paled in comparison. In 1994, as many as 500,000 Rwandans, including a majority of the country's Tutsi, lost their lives in an orchestrated campaign by Hutu militias against Tutsis and ''moderate'' or collaborationist Hutus. More than 2 million refugees fled into Zaïre and Tanzania, where many succumbed to disease in makeshift camps. It will not be the last in this series of cultural conflicts, in which the basis for hostility is not ethnicity but status in society.

Highland Ethiopia

As Figure 7-11 shows, the East African region also encompasses the highland zone of Ethiopia, including the capital, Addis Ababa, the source of the Blue Nile, Lake Tana, and the Amharic core area that was the base of the empire that survived the colonial onslaught unconquered and only lost its independence briefly during the 1930s and 1940s. Ethiopia, mountain fortress of the Coptic Christians who held their own here, eventually became a colonizer itself. Its forces came down the slopes of the highlands and conquered much of the Islamic part of Africa's Horn, including present-day Eritrea and the Ogaden area, a Somali territory. (Geographically, these are parts of what we have mapped as the African Transition Zone in Fig. 6-11.)

Physiographically and culturally, a case can be made that highland Ethiopia is part of an East African region. But because Ethiopia was not colonized, and because Ethiopia's natural outlets are to the Red Sea, not southward to Mombasa, effective interconnections between former British East Africa and highland Ethiopia never developed. Now, however, things may change. Eritrea has become an independent state, and now-landlocked Ethiopia's natural outlet, the port of Djibouti, also is under a foreign flag. For this reason alone, Ethiopia may look southward. Add to this the Islamic-Somali front, faced by both Ethiopia and Kenya, and the East Africanization of Ethiopia may well accelerate.

From the Field Notes

"About 10 miles (16 km) from the Kenyan town of Meru the landscape showed signs of severe erosion. We stopped to talk with the people of these homesteads, and asked them about their crops. Yes, they knew that farming on slopes as steep as these would lead to 'gullying,' but they saw no alternative. You get a crop one or two years, and that's better than nothing, they said. Some neighbors whose village had lost most of its land had gone to the city, we were told, and now the place where they lived was like a desert."

❖ SOUTHERN AFRICA

Southern Africa, as a geographic region, consists of all the countries and territories lying south of Equatorial Africa's Zaïre and East Africa's Tanzania (Fig. 7-12). Thus defined, the region extends from Angola and Moçambique (on the Atlantic and Indian Ocean coasts, respectively) to the turbulent Republic of South Africa, and includes a half-dozen landlocked states. Also marking the northern limit of the region are Zambia and Malawi. Zambia is nearly cut in half by a long land extension from Zaïre, and Malawi penetrates deeply into Moçambique. The colonial boundary

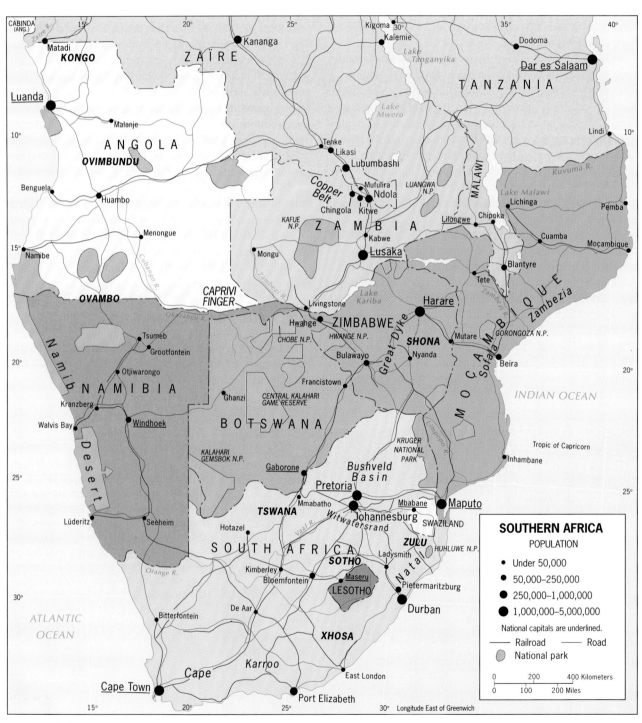

FIGURE 7-12

framework, here as elsewhere, produced many liabilities.

Southern Africa constitutes a geographic region in physiographic as well as human terms. Its northern zone marks the southern limit of the Zaïre Basin in a broad upland that stretches across Angola and into Zambia (the tan corridor extending eastward from the Bihe Plateau in Fig. 7-1). Lake Malawi is the southernmost of the East African rift-valley lakes; Southern Africa has none of East Africa's volcanic and earthquake activity. Most of the region is plateau country, and the Great Escarpment is much in evidence here. There are two pivotal river systems: the Zambezi (which forms the border between Zambia and Zimbabwe) and the Orange-Vaal (South African rivers that combine to demarcate southern Namibia from South Africa).

The social, economic, and political geographies of Southern Africa also confirm the regional definition. Landlocked Zambia, whose economic core area is the Copperbelt northwest of Ndola, always looked southward for its outlets to the sea, its electric power, and its fuels. Malawi's core area lies in the south of the country, and its outlets, too, are southward. Moçambique, with its lengthy Indian Ocean coastline, has served as an exit not only for Malawi but also for Zimbabwe (through the port of Beira) and South Africa (through Maputo) as well. Offshore Madagascar, however, remains a separate entity (see box titled "Distinctive Madagascar").

Southern Africa is the continent's richest region in material terms. A great zone of mineral deposits extends through the heart of the region from Zambia's Copperbelt through Zimbabwe's Great Dyke and South Africa's Bushveld Basin and Witwatersrand to the goldfields and diamond mines of the Orange Free State and northern Cape Province in the heart of South Africa. Ever since colonial exploitation of these minerals began, large numbers of migrant laborers have come to work in the mines. The range and volume of minerals mined in this belt are enormous, from the copper of Zambia and the chrome and asbestos of Zimbabwe to the gold, chromium, diamonds, platinum, coal, and iron ore of South Africa. However, not all of Southern Africa's mineral deposits lie in this central backbone. There is coal in western Zimbabwe at Hwange and in central Moçambique near Tete. In Angola, petroleum from oilfields along the north coast heads the export list; diamonds are mined in the northeast, manganese and iron on the central plateau. Namibia produces copper, lead, and zinc from a major mining complex anchored to the town of Tsumeb in the north, and diamond deposits lie in the beaches along the Atlantic Ocean in the south.

Southern Africa's mineral wealth is matched by its agricultural diversity. Vineyards drape the slopes of South Africa's Cape Ranges; tea plantations hug the eastern escarpment slopes of Zimbabwe. Before its civil war destroyed the economy, Angola was one of the world's leading coffee producers. South Africa's relatively high latitudes, and its range of altitudes, create environments for apple orchards, citrus groves, banana plantations, pineapple farms, and many other crops. Farther north, tobacco has long been one of Zimbabwe's leading products. And while corn (maize) is the staple grain for most of Southern Africa's farmers, wheat and other cereals also are grown. Even the pastoral industry is varied, with large herds of beef cattle on the grassy *highveld*, tens of millions of sheep on the southeast's pastures, and dairy farms around the big cities.

Despite this considerable wealth and potential, the countries of Southern Africa have not prospered. As Figure I-11 shows, several remain mired in the low-income category (Moçambique, with a per capita GNP of only $80, is one of the world's poorest states), and only South Africa is in the upper-middle-income rank. Civil wars, political instability, poor management, corruption, and environmental problems have combined to inhibit economic growth. Nevertheless, the situation is superior to that in any of the other three regions; more countries have risen above the low-income level than anywhere else in Subsaharan Africa. In the late 1990s, the end of several wars and the resurgence of South Africa gave hope that this region might finally burgeon, perhaps to lead the entire realm to a higher economic and social plane.

The Northern Tier

In the four countries that extend across the northern part of the region—Angola, Zambia, Malawi, and Moçambique—problems abound. Angola (12 million), formerly a Portuguese dependency, is one of Africa's richest countries in terms of raw materials and agricultural potential. It had a thriving economy at independence but was engulfed by civil war that was worsened by outside involvement. The government, centered in the north and supported by northern peoples, chose a communist course. When a rebel movement emerged in the south with support by the (then) South African regime, Angola's leaders called on Moscow for help. Thousands of Cuban troops arrived in the capital, Luanda, escalating the war but securing the state.

The impact of all this unrest on the Angolan economy was devastating. Farms lay abandoned, fields were mined, railroads destroyed, ports damaged. Casualties among the civilian population were high. But the government could pursue the war for a reason shown in Figure 7-12. Note that a small part of Angola lies separated as an **exclave** across the Zaïre (Congo) River. This exclave, Cabinda, contains oil reserves that produced revenues for the Luanda government during its war.

After several failures, the civil war was settled through U.N. mediation, and in the late-1990s Angola seemed ready to rebound. But its infrastructure lay in ruins, and it will take decades to overcome the damage and to realize the country's potential.

DISTINCTIVE MADAGASCAR

Maps showing the geographic regions of the Subsaharan Africa realm often leave Madagascar unassigned. And for good reason: Madagascar differs strongly from Southern Africa, the region to which it is nearest—just 250 miles (400 km) away. It also differs from East Africa, from which it has received some of its cultural infusions. Madagascar is the world's fourth largest island, a huge block of Africa that separated from the main landmass 160 million years ago. About 2,000 years ago, the first settlers arrived—not from Africa (although perhaps via Africa) but from Southeast Asia. Malay communities strengthened in the interior highlands of the island, which resembles Africa in a prominent eastern escarpment and a central plateau (Fig. 7-13). Here was formed a powerful kingdom, the empire of the Merina. Its language, Malagasy, of Malay-Polynesian origin, became the indigenous tongue of the entire island (Fig. 7-7).

The Malay and Indonesian immigrants brought Africans to the island as wives and slaves, and from this forced immigration evolved the African component in Madagascar's population of 16 million. In all, nearly 20 discrete ethnic groups coexist, among which the Merina (4 million) and Betsimisaraka (2 million) are the most numerous. Like mainland Africa, Madagascar experienced colonial invasion and competition. Portuguese, British, and French colonists appeared after 1500, but the Merina were well organized and long resisted the colonial conquest. Eventually Madagascar became part of France's empire, and French became the *lingua franca* of the educated elite.

Logically, Madagascar's staple food is rice, not corn. There are some minerals, including chromite, iron ore, and bauxite, but the economy is weak, damaged by long-term political turmoil and burdened by rapid population growth. The infrastructure crumbled; in 1990, the "main road" from the capital to the nearest port (Fig. 7-13) was a potholed 150 miles that took 10 hours for a truck to navigate.

Meanwhile, Madagascar's unique flora and fauna retreated before the human onslaught. Madagascar's long-term isolation kept evolution here on a separate track so distinct that the island is a discrete zoogeographic realm. Primates living here are found nowhere else; 33 varieties of lemurs are unique to Madagascar. Many species of birds, amphibians, and reptiles are exclusive to this island. Their home, the rainforest, covered 65,000 square miles in 1950, but today only about one-third of it is left. Slash-and-burn agriculture is destroying it; severe droughts in the 1980s and 1990s intensified the practice. Obviously, Madagascar should be a global conservation priority, but funds are limited and the needs are enormous. Malnutrition and poverty are powerful forces when survival is at stake for villages and families.

Madagascar's cultural landscape retains its Southeast Asian imprints, in the towns as well as the paddies. The capital, Antananarivo, is the country's primate city, its architecture

and atmosphere combining traces of Asia as well as Africa. Poverty dominates the townscape here, too, and there is little to attract in-migrants (Madagascar is only 22 percent urbanized). But perhaps the most ominous statistic is Madagascar's population doubling time of just 22 years. In this respect Madagascar resembles Africa, not Southeast Asia.

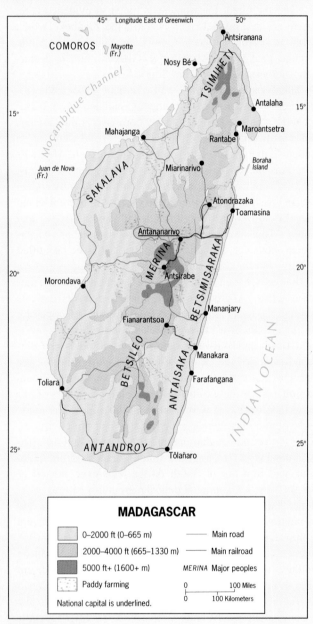

FIGURE 7-13

From the Field Notes
"The building of the Kariba Dam was one of the largest engineering projects ever attempted in Africa, and seeing the site in 1955 confirmed its dimensions. This photograph, taken from the Zambian (north) side, shows how the main structure was built within coffer dams that were later destroyed. The Zambezi River, once contained, backed up into one of the world's largest artificial lakes—but the promise of this megaproject never was fully realized. Political, ecological, and structural problems limited the impact of the Kariba project."

On the opposite coast, Moçambique (19 million) was even less fortunate. Also a former Portuguese domain that chose a Marxist course for development, Moçambique had far fewer resources than Angola. Before independence, its chief sources of income came from its cashew and coconut plantations and from its relative location. As the map shows, Moçambique's port of Beira is ideally situated to handle the external trade of Zimbabwe and southern Malawi, and its capital, Maputo, is the closest port to the great mining and industrial complex centered on the Witwatersrand in South Africa. In better times, goods streamed through these ports, and when a hydroelectric plant was built on the Zambezi River at Cabora Bassa, prospects seemed fair. Then, however, the economy collapsed under bad management, a rebel movement aided for a time by South African interests destroyed the social order, famines broke out, and Moçambique descended into chaos. More than a million refugees streamed into Malawi, the hydroelectric project was damaged, the port facilities at Beira ceased functioning, and Maputo's transit role was no more.

In the late 1990s, some semblance of stability had returned to Moçambique. The government abandoned its Marxist principles, the rebel movement laid down its arms, refugees returned, and the new South Africa was offering agricultural and technical help rather than arms to insurgents. But it will take generations for the country to climb from the depths of impoverishment to which it has been condemned.

Between Angola and Moçambique lie three landlocked states that formed part of the British colonial empire: Zambia, Malawi, and Zimbabwe. Zambia (9.6 million) and Malawi (10.3 million) are less developed than Zimbabwe, although Zambia contains the mineral-rich Copperbelt. The decline of world prices for its minerals, as well as the prob-

lems and costs associated with their long-distance transportation, have hurt the economy of Zambia. Malawi's economic geography is almost completely agricultural, with a variety of crops that include tea, cotton, tobacco, and peanuts. This has helped cushion the economy against market swings and has also involved a larger percentage of the labor force in productive work. Still, more workers labor in the mines of other countries than are gainfully employed within Malawi itself.

Southern States

Six countries constitute Africa's southernmost tier of states and form a distinct subregion within Southern Africa: Zimbabwe, Namibia, Botswana, Swaziland, Lesotho, and the Republic of South Africa. As Figure 7-12 shows, four of the six are landlocked. Botswana occupies the heart of the Kalahari Desert and surrounding steppe; only 1.6 million people inhabit this Texas-sized country. Lesotho (2.1 million) is encircled by South African territory, and Swaziland (1 million), ancestral home of the Swazi nation, is largely enclosed by the Republic as well. All three of these countries depend heavily on the income earned by workers who are employed by South African mines, factories, and farms.

Apart from South Africa, the region's giant, the most important southern state undoubtedly is Zimbabwe (12 million), landlocked but well endowed with mineral and agricultural resources. Zimbabwe (the country is named after a complex of stone ruins located in its interior) is mostly an elevated plateau between the Zambezi and Limpopo rivers, with the desert to the west and the Great Escarpment to the east. Its core area is defined by the mineral-rich Great Dyke and its environs, extending southwest from the vicinity of the capital, Harare, to the country's second city, Bulawayo. Copper, asbestos, and chromium (of which Zimbabwe is one of the world's leading sources) are among major mineral exports, but Zimbabwe is not just an ore-exporting country. Farms produce tobacco, tea, sugar, cotton, and other crops (corn is the staple). A bitter civil war preceded independence, but many whites stayed on their farms after the majority took power, encouraged by the government to do so. Disputes over land policy did arise later, but Zimbabwe's political transition did not derail the economy as it did in so many other countries.

The political compromise, the mixed economy, and an avoidance of involvement in neighbors' conflicts have served Zimbabwe well. But in the period after 1983, Zimbabwe faced a problem no enlightened policies could solve: a disastrous drought, the worst in its recorded history. Livestock died in large numbers, farms lay abandoned, and, inevitably, social and political tensions rose.

By the late 1990s, the environmental situation had improved, although its aftermath will affect Zimbabwe for many years to come. At the same time, another obstacle to economic progress was growing: the widespread corruption in a government that, by 1996, had been in power for 16 years. Infrastructure modernization was hampered by corporate payoffs (''commissions'') and legislative obstruction of private attempts to compete with operations in which government members had personal interest. The cumulative cost of this to Zimbabwe's future is devastating.

The western tip of Zimbabwe just barely touches the northeastern ''Caprivi Finger'' of Namibia, the region's youngest independent state. Once a German colony named South West Africa, the country was endowed with that narrow strip of land to connect it to the Zambezi River by the mapmakers in Berlin. When South Africa took control of the colony after World War I, the Caprivi Strip took on strategic importance; in the 1980s, there were rumors of missile (even nuclear) tests in the area. But Namibia achieved independence in 1990, and South Africa publicly abandoned its nuclear program. In 1994, the last territorial vestige of South Africa was erased when the port of Walvis Bay, a South African exclave, was transferred to Namibia.

Namibia is named after a desert, and appropriately so. Only about 1.6 million people, the majority concentrated in the moister northern border areas, inhabit this huge country which is about as large as Texas and Oklahoma combined. The capital, Windhoek, is centrally situated and is thus far from this northern cluster. That is because the major economic activities—mining in the Tsumeb area and ranching across the steppe country of the south—happen elsewhere, not in the subsistence-dominated north.

But that is where the voters are, and Namibia will undoubtedly undergo structural readjustment. Much of the productive capacity of the country remains in foreign hands, and food must be imported to supplement what little can be produced locally. For Namibia, the long haul toward true sovereignty has just begun.

South Africa

The Republic of South Africa is the giant of Southern Africa, an African country at the center of world attention, a bright ray of hope not only for Africa but for all humanity.

Long in the grip of one of the world's most notorious racial policies (*apartheid* and its derivative, ''separate development''), South Africa today is shedding its past and building a new future. That virtually all parties to the earlier debacle are now working cooperatively to restructure the country under a new flag, a new national anthem, and a new leadership is one of the great events of the twentieth century. Now, in the late 1990s, South Africa is poised to take its long-awaited role as the economic engine for the region—and perhaps beyond.

South Africa stretches from the warm subtropics in the north to Antarctic-chilled waters in the south. With a land area in excess of 470,000 square miles (1.2 million sq km) and a heterogeneous population of 45 million, South Africa is the dominant state in Southern Africa. It contains the bulk of the region's minerals, the majority of its good farmlands, its largest cities, best ports, most productive factories, and most developed transport networks. Mineral exports from Zambia and Zimbabwe move through South African ports. Workers from as far away as Malawi and as nearby as Lesotho work in South Africa's mines, factories, and fields.

Historical Geography

South Africa's location has much to do with its human geography. On the continent, peoples migrated southward, first the Khoisan-speakers and then the Bantu peoples, into the South African cul-de-sac. On the oceans, the Europeans arrived to claim the Cape as one of the most strategic places on Earth, the gateway from the Atlantic to the Indian Ocean, a way-station on the route to Asia's riches. The Dutch East India Company founded Cape Town as early as 1652, and soon the Hollanders began to bring Southeast Asians to the Cape to serve as domestics and laborers. By the time the period of Dutch dominance had run its course, about 150 years later, Cape Town had a substantial population of mixed ancestry, the source of today's so-called *Coloured* sector of the country's citizenry.

After the Dutch came the British, and they, too, altered the demographic mosaic by bringing tens of thousands of indentured laborers from their South Asian domain to work on the sugar plantations of Natal. Most of these laborers elected to remain in South Africa after their period of indenture was over, and today South Africa counts about 1.1

million *Indians* among its people. Most are still concentrated in Natal and prominently so in the city of Durban.

As we noted earlier, South Africa was an arena of European penetration and occupation long before the colonial "scramble for Africa" gained momentum. The Dutch, their control over the Cape ended by the British, trekked into the South African interior and, on the high plateau they called the *highveld*, founded republics to stake their claims. When diamonds and gold were found in their territories, the British lost no time in challenging the *Boers* (descendants of the Dutch settlers) for these prizes. Before the end of the nineteenth century, the two European invaders were locked in the Boer War (1899–1902). The British won that war, and British capitalists took control of the economic as well as the political life of South Africa. But the Boers negotiated an arrangement that gave them considerable power, and eventually they translated that power into hegemony. Having long since shed their European links, they now called themselves *Afrikaners*, their word for Africans, and proceeded to erect the system known as *apartheid*.

These foreign immigrations and struggles took place on lands that had already been entered—and fought over—by Africans themselves. When the Europeans reached the Cape, Bantu nations were overpowering the weaker Khoi and San peoples, driving them into less hospitable territory or forcing them to work in bondage. One great contest was taking place in the east and southeast, below the Great Escarpment. The Xhosa nation was moving toward the Cape along this natural corridor. Behind them, in Natal (Fig. 7-12), the Zulu Empire during the nineteenth century expanded rapidly to become the region's most powerful entity. On the highveld, the North and South Sotho, the Tswana, and other peoples could not stem the tide of European aggrandizement, but their numbers ensured survival (unlike many of Australia's Aboriginal groups).

In the process, South Africa became Africa's most plural and most heterogeneous society. People had converged on the country from Western Europe, Southeast Asia, South Asia, and other parts of Africa itself. Toward the end of the twentieth century, Africans outnumbered non-Africans by about 7 to 2 (Table 7-1).

Social Geography

Heterogeneity also marks the spatial demography of South Africa. Despite centuries of migration and (at the Cape) intermarriage, labor movement to the mines, farms, and factories, as well as massive urbanization, regionalism marks the human mosaic. The Zulu nation still is largely concentrated in the province the Europeans called Natal. The Xhosa still cluster in the Eastern Cape, from the city of East London to the Natal border and below the Great Escarpment. The Tswana still occupy ancestral lands along the Botswana border. Cape Town still is the core area of the Coloured population; Durban still has the strongest In-

7.1

TABLE 7-1
Demographic Data for South Africa

Population Groups	Estimates, 1997 (in millions)
African nations	35.0
Zulu	8.5
Xhosa	7.0
Sotho (N and S)	6.3
Tswana	3.3
Others (6)	9.9
Europeans	5.1
Afrikaners	2.9
English-speakers	1.9
Others	0.3
Mixed (Coloureds)	3.5
African/Europeans	3.2
Malayan	0.3
South Asian	1.1
Hindus	0.8
Muslims	0.3
TOTAL	44.7

dian imprint on its cityscape. Travel through South Africa, and you will recognize the diversity of rural cultural landscapes as they change from Swazi to Ndebele to Venda.

This persistent regionalism induced the Afrikaner-led white government of the *apartheid* (racial separation) period, after 1949, to extend its policies and lock the country in a grand design called *separate development*. By creating "homelands" that would later become "independent" republics, the white minority could assign all Africans to their national source area, making them foreigners in white South Africa while they worked there. This was a form of what is now called ethnic cleansing, and between 1960 and 1980 as many as 3.5 million black Africans were forced from their homes and relocated in their assigned abodes. The social cost of this program was enormous and will be felt well into the twenty-first century. Under the provisions of separate development, nearly 80 percent of South Africa was designated as white-owned, and barely more than 15 percent was to accommodate the African homelands. There, overcrowding led to excessive pressure on the land, accelerating soil erosion and general environmental deterioration. Crowding also produced conflict among homeland residents, but more serious strife pitted those who were willing to cooperate with the white regime (for example, by serving in government-approved homeland administrations) against those who opposed *apartheid* and separate development by all possible means.

For all the misery and dislocation it caused, separate development could not stem the tide of urbanization in South Africa. Millions of workers, jobseekers, and illegal migrants created vast shantytown rings around South Africa's cities. The existing "townships" or "locations," as the older African settlements were called, that once were the poorest parts of the cities, now took on a semblance of well-being by comparison. Black townships such as Soweto (for *So*uth *We*stern *To*wnships) in metropolitan Johannesburg were approved under *apartheid* laws, even acquiring some services. In the shantytowns, no legalities protected the residents, and the government's efforts to eradicate them never stopped. Bulldozers would sweep away whole settlements overnight, their residents loaded on trucks and sent to the homelands.

Still the number of both township residents and squatters continued to grow, and eventually the townships became the hubs of opposition to *apartheid*. Violent uprisings and crippling strikes proved the growing power of the majority and the inability of the state to control it. By the mid-1980s, international attention was focused on South Africa, economic sanctions were imposed, and the end of the *apartheid* era was in sight. The South African president, P. W. Botha, began a series of discussions with the country's most famous political prisoner, Nelson Mandela. The long-banned African National Congress (ANC) was permitted to engage in political activity again, and exiled leaders began to return home. When, in February 1990, Mandela walked out of prison after 28 years behind bars, he and the last Afrikaner president, F. W. de Klerk, began a process of negotiation and accommodation that would lead to the transfer of power from the white regime to a gov-

ernment elected by all the voters. In April 1994, a new era began in South Africa. Nelson Mandela, of distinguished Xhosa ancestry, had become president of an ANC-dominated government at Cape Town.

Political Reorganization

The route from prison gate to presidency was not an easy one either for Mandela or for the South Africa he hoped to lead; and again geography had a big role. De Klerk led the white minority, and Mandela the black majority, but there was a third force: the largest single nation in South Africa, the Zulu, whose leader, Chief Mangosuthu Buthelezi, saw a conspiracy between the two principals that would damage Zulu interests. Buthelezi's political movement, Inkatha, demanded equity in the new South Africa and promised to scuttle the conferences working toward the compromise South Africa needed. Inkatha's people put action with their words. In the townships of the major cities and in the hostels of the mining companies, Inkatha ''warriors'' killed thousands in a campaign of violence that threatened to devolve into a civil war. In the province of Natal, Buthelezi and his followers talked of secession and independence if the new and democratic South Africa was not to their liking.

In the end, geography came to South Africa's rescue. What the country needed was an election result that would

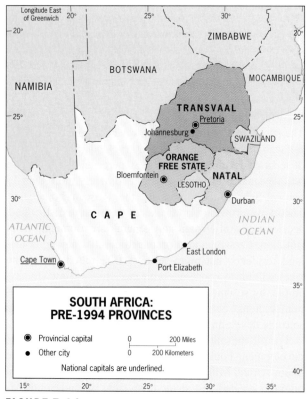

FIGURE 7-14

FIGURE 7-15

give the ANC its majority, but not a majority so large that it could overwhelm minority interests. To gain Inkatha's cooperation, an Inkatha victory in its domain would be crucial. To give substance to the promise of democracy in South Africa, a victory by parties other than the ANC in one other province would be desirable.

To accomplish this (not to say to rig the election), the conference that prepared for the power transfer redrew the map, and thus the electoral map, of South Africa. Ever since 1910, South Africa had had four provinces: the Cape, Natal, the Orange Free State (north of the great Orange River), and the Transvaal (''across the Vaal,'' the major Orange tributary). These four provinces essentially represented the European occupation of the country: the Cape and Natal were British strongholds, and the Orange Free State and Transvaal were Boer republics until their defeat in the Boer War (Fig. 7-14). The new map (Figure 7-15) created nine provinces, leaving Natal intact (but calling it Kwazulu-Natal) and also keeping the Orange Free State's boundaries unchanged. Three new provinces and part of a fourth were carved out of the Cape: the Western Cape, essentially Cape Town and its hinterland, where the electorate is mainly Coloured and white; the Eastern Cape, dominated by the Xhosa nation; the Northern Cape, by far the least populous, mainly rural province; and the Northwest, consisting of part Cape and part Transvaal, where the Tswana nation is concentrated. The remainder of the Transvaal became three provinces as well: Gauteng, the heart of the country's core area centered on Johannesburg; Northern Province, home to many Afrikaner opponents to the course of events; and Mpumalanga, a mix of wealthy white farmers and crowded Swazi and other African areas.

In the 1994 elections, the ANC won seven of the nine provinces. It lost in Kwazulu-Natal, where an alliance of Zulu, Asian, and white voters prevailed over the ANC, and in the Western Cape, where white and Coloured voters combined to beat ANC candidates. Nor did the ANC gain the 70 percent majority it would have needed to govern without regard to the opposition. What was accomplished in South Africa, by all parties, in the aftermath of more than four decades of oppression and mismanagement, was an example to a world quick to reach for weapons when political goals cannot be achieved.

South Africa's difficult transition to post-*apartheid* normalcy and stability, however, is not yet completed. President Mandela provided the leadership the country needed, and his name now stands with other great African statesmen who led their countries through decolonization and served as their first presidents. But South Africa now has to confront a post-Mandela era. In other African states, the succession from founder-of-the-nation to inheritor of the presidency generally has not gone well, but in South Africa the structure for succession was enshrined by the planning conference.

Another long-term problem centers on the relationship between the central government and Kwazulu-Natal. Buthelezi, the Zulu leader, agreed to serve in the Mandela administration, but friction continued. In 1995, when the government moved to assume responsibility for the payment of traditional chiefs from the provinces, Buthelezi objected and revived the issue of secession.

Still another challenge may come from the disaffected conservative whites, most of them farmers and many of them concentrated in the Northern Province, who were overtaken by the rapid sequence of events and whose future in the remote countryside may not be any easier than that of white farmers in neighboring Zimbabwe. The reform of land ownership, changed labor relations, security concerns, and economic policies all will affect this substantial minority, which owns large tracts of good farmland. During

From the Field Notes

"Flying over the Witwatersrand south of Johannesburg, you get a sense of the dimensions of the mining operations that have continued here for more than a century. The 'mine dumps,' pulverized rock left behind after gold extraction, are as large as entire suburbs, their walls towering over those of the houses nearby. So much rock has been removed from beneath the 'Rand' that settling produces major Earth tremors in the area."

■ AMONG THE REALM'S GREAT CITIES . . .

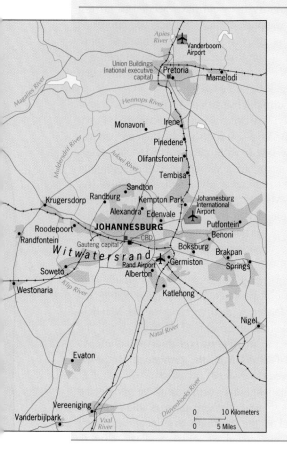

Johannesburg

Subsaharan Africa displays only one incipient conurbation, and Johannesburg, South Africa lies at the heart of it. Little more than a century ago, Johannesburg was a small (though rapidly growing) mining town based on the newly discovered gold reserves of the Witwatersrand. Today Johannesburg forms the focus of a megalopolis of nearly 7 million, extending from Pretoria to the north to Vereeniging to the south, and from Springs in the east to Krugersdrop in the west. Metropolitan Johannesburg itself had a 1997 population of 2 million, second only to Cape Town.

Johannesburg's skyline is the most impressive in all of Africa, a forest of skyscrapers reflecting the wealth generated here over the past hundred years. Look southward from a high vantage point, and you see the huge mounds of yellowish-white slag from the mines of the "Rand," the so-called mine dumps, partly overgrown today, interspersed with suburbs

and townships. In a very general way, Johannesburg developed as a white city in the north and a black city in the south. Soweto, the black township, lies to the south; Houghton and other spacious, upper-class suburbs, once exclusively white residential areas, lie to the north.

Johannesburg has neither the scenery of Cape Town nor the climate and beaches of Durban. The city is a mile above sea level, and its thin air often is polluted, as a result of smog created by automobiles, factories, mine-dump dust, and countless cooking fires in the townships and shantytowns that ring the metropolis.

Over the past century, the Johannesburg area has produced nearly one-half of all the world's gold by value. But today, Johannesburg lies at the heart of an industrial, commercial, and financial complex whose name on the new map of South Africa is Gauteng.

a pre-transition referendum, they voted overwhelmingly against then-President de Klerk's policy of cooperation toward majority government.

Economic Geography

Undoubtedly the most serious problems for the new South Africa are economic. Ever since diamonds were discovered at Kimberley in the 1860s, South Africa has been synonymous with minerals. The Kimberley finds, made in a remote corner of what was then the Orange Free State (the British soon annexed it to the Cape), set into motion a new economic geography. Rail lines were laid from the coast to the "diamond capital" even as fortune seekers, capitalists, and tens of thousands of African workers, many from as far afield as Lesotho, streamed to the site. One of the capitalists was Cecil Rhodes, now of scholarship fame, who used his fortune to help British advancement through Southern Africa.

Just 25 years later, prospectors found what was to be, for some time to come, the world's greatest goldfield on a ridge called the Witwatersrand. This time the site lay in the

so-called South African Republic (the Transvaal), and again the Boers were unable to hold the prize. Johannesburg became the gold capital of the world, and a new and even larger stream of foreigners arrived, along with a huge influx of African workers. Cheap labor enlarged the profits. Johannesburg grew explosively, satellite towns developed, black townships mushroomed. The Boer War was only an interlude here on the mineral-rich Witwatersrand.

During the twentieth century, South Africa proved to be even richer than had been foreseen. Additional goldfields were discovered in the Orange Free State. Coal was found in abundance, and so was iron ore; this gave rise to a major iron and steel industry. Other metallic minerals, including chromium and platinum, yielded large revenues on world markets. Asbestos, manganese, copper, nickel, antimony, and tin were mined and sold; a thriving metallurgical industry developed in South Africa itself. Capital flowed into the country, white immigration grew, farms and ranches were laid out, and markets mushroomed.

South Africa's cities grew apace. Johannesburg was no longer just a mining town; it became an industrial complex and a financial center as well. The old Boer capital, Pre-

From the Field Notes

"Looking down on this enormous railroad complex, we were reminded of the fact that almost an entire continent was turned into a wellspring of raw materials carried from interior to coast and shipped to Europe and other parts of the world. This complex lies near Witbank in the eastern Rand, a huge inventory of freight trains ready to transport ores from the plateau to Durban and Maputo. But at least South Africa acquired a true transport network in the process, ensuring regional interconnections; in most African countries, railroads serve almost entirely to link resources to coastal outlets."

toria, became the country's administrative center during *apartheid*'s days, just 30 miles north of the Witwatersrand. In the Orange Free State, major industrial growth (including oil-from-coal technology) matched the expansion of mining. While the core area developed megalopolitan characteristics, coastal cities expanded as well. Durban's port served not only the Witwatersrand but a wider regional hinterland as well. Cape Town was on its way to becoming South Africa's largest city, with its port, industries, and productive agricultural hinterland giving it primacy over a large area.

Apartheid ruined all these prospects and exposed the economy's weaknesses. The Afrikaner government made huge investments in its separate development policy, which went counter to many fundamental principles of economic geography. International sanctions against *apartheid* severely damaged the economy, labor unrest weakened it further, and lower prices for those products that did get sold hurt even more. In the black townships, the school systems fell apart under the cry of "liberation before education," creating a huge undereducated (later unemployable) mass of young people.

For South Africa, later is now, and the resulting problems are here. In 1997, South Africa's economy still depended heavily on the export of metals and minerals, but mineworkers demanding higher wages cut into the profits. Manufactures figured hardly at all in the export picture. Unemployment stood at high levels even as employed workers sought to raise their wages. In the cities, the whites fled to the suburbs, transforming the downtowns into commuter workplaces and the streets into bazaars. The political crime of the pre-transfer period was replaced by a mounting incidence of violent robberies. Most ominous of all is the revolution of rising expectations, a phenomenon experienced by many African countries after decolonization. Land pressure and housing needs cannot be accommodated overnight, but the newly empowered majority will expect its government to act expeditiously. Joblessness, housing shortages, and land pressure can form a potent mix to destabilize a society, and the South African government faces a daunting challenge.

On maps of development indices, South Africa is portrayed as an upper-middle-income economy, but averages mean very little in this country of strong core-periphery contrasts. In its great cities, industrial complexes, mechanized farms, and huge ranches, South Africa resembles a high-income economy, much like Australia or Canada. But outside the primary core area (centered on Johannesburg) and beyond the secondary cores and their linking corridors lies quite a different South Africa, where conditions are more like those of rural Zambia or Zimbabwe. In terms of such indices as life expectancy, infant and child mortality, overall health, nutrition, education, and many others, a wide range marks the population sectors listed in Table 7-1. South Africa has been described as a microcosm of the world, exhibiting in a single state not only a diversity of cultures but also a wide range of human conditions. If South Africa can keep on course, the one-time pariah of *apartheid* will become a guidepost to a better world.

■ PRONUNCIATION GUIDE ■

Abacha (ah-BAH-chuh)
Abidjan (abb-ih-JAHN)
Abiola, Moshood (abbey-OH-luh,
    moh-SHOOT
Abuja (uh-BOO-juh)
Abyssinia (abb-ih-SINNY-uh)
Accra (uh-KRAH)
Adamawa (add-uh-MAH-wuh)
Afrikaans (uff-rih-KAHNZ)
Afrikaner (uff-rih-KAHN-nuh)
Ahaggar (uh-HAH-gahr)
Amhara (am-HAH-ruh)
Amharic (am-HAH-rik)
Amin, Idi (uh-MEEN, iddy)
Angola (ang-GOH-luh)
Antananarivo (anta-nana-REE-voh)
Apapa (uh-PAHP-uh)
Apartheid (APART-hate)
Asante (ah-SAHN-tay)
Australopithecene (aw-strallo-
    PITH-uh-seen)
Autocracy (aw-TOCK-ruh-see)
Baganda (bah-GAHN-duh)
Bahia (bah-EE-yah)
Bantu (ban-TOO)
Bantustan (BAN-too-stahn)
Beira (BAY-ruh)
Benin (beh-NEEN)
Benue (BANE-way)
Betsimisaraka (bet-simmy-
    sah-RAH-kah)
Biafra (bee-AH-fruh)
Bihe (bee-HAY)
Bilharzia (bill-HARZEE-uh)
Bioko (bee-OH-koh)
Blantyre (BLAN-tyer)
Boer (BOOR)
Bophuthatswana (boh-pooh-
    taht-SWAHNA)
Boran (boar-RAN)

Botha (BAW-tuh)
Botswana (bah-TSWAHN-uh)
Brazzaville (BRAHZ-uh-veel)
Buganda (boo-GAHN-duh)
Bulawayo (boo-luh-WAY-oh)
Burkina Faso (berr-keena FAHSSO)
Burundi (buh-ROON-dee)
Buthelezi, Mangosuthu
    (boo-teh-LAY-zee, mungo-SOO-too)
Cabinda (kuh-BIN-duh)
Cabora Bassa (kuh-boar-rah BAHSSA)
Caprivi (kuh-PREE-vee)
Chingola (ching-GOH-luh)
Ciskei (SISS-skye)
Côte D'Ivoire (COAT deev-WAH)
Dahomey (dah-HOH-mee)
Dakar (duh-KAHR)
Dar es Salaam (dahr ess suh-LAHM)
Deccan (DECKEN)
Dhow (DAU)
Djibouti (juh-BOODY)
Djouf (JOOF)
Dodoma (DOH-duh-mah)
Douala (doo-AHLA)
Drakensberg (DRAHK-unz-berg)
Durban (DER-bun)
Ebola (ee-BOH-luh)
Endemism (en-DEM-izm)
Enkare (en-KAH-ray)
Entebbe (en-TEBBA)
Eritrea (erra-TRAY-uh)
Ethiopia (eeth-ee-OH-pea-uh)
Fernando Póo (fer-nahn-doh POH)
Fulani (foo-LAH-nee)
Futa Jallon (food-uh juh-LOAN)
Gabon (gah-BAW)
Galla (GAH-lah)
Gambia (GAM-bee-uh)
Gao (GAU)
Gauteng (GAU-teng)

Ghana (GAH-nuh)
Ghanaian (gah-NAY-un)
Gondwana (gond-WON-uh)
Guinea (GHINNY)
Guinea-Bissau (ghinny-bih-SAU)
Harare (huh-RAH-ray)
Hausa (HOW-sah)
Hegemony (heh-JEH-muh-nee)
Hominid (HOM-ih-nid)
Homo sapiens (hoh-moh SAY-pea-enz)
Houghton (HOH-t'n)
Hutu (HOO-too)
Hwange (WAHNG-ghee)
Ibadan (ee-BAHD-N)
Ibo [land] (EE-boh [land])
Ikoyi (ee-KOY-yee)
Inkatha (in-KAH-tah)
Ituri (ih-TOORY)
Johannesburg (joh-HANNIS-berg)
Kalahari (kallah-HAH-ree)
Kalenjin (kuh-LEN-jin)
Kamba (KAHM-bah)
Kampala (kahm-PAH-luh)
Kano (KAH-noh)
Kariba (kuh-REE-buh)
Katanga (kuh-TAHNG-guh)
Kenya (KEN-yuh)
Khoikhoi (KHOY-khoy)
Khoisan (khoy-SAHN)
Kibo (KEE-boh)
Kigali (kih-GAH-lee)
Kikuyu (kee-KOO-yoo)
Kikwit (KIK-wit)
Kilimanjaro (kil-uh-mun-JAH-roh)
Kinshasa (kin-SHAH-suh)
Kivu (KEE-voo)
Kruger (KROO-guh)
Krugersdorp (KROO-guzz-dorp)
Kwashiorkor (kwuh-shee-OAR-koar)
Kwazulu (kwah-ZOO-loo)

Lagos (LAY-gohss)
Lemur (LEE-mer)
Lesotho (leh-SOO-too)
Liberia (lye-BEERY-uh)
Limpopo (lim-POH-poh)
Lingua franca (LEAN-gwuh
   FRUNK-uh)
Lualaba (loo-uh-LAH-buh)
Luanda (loo-AN-duh)
Lubumbashi (loo-boom-BAH-shee)
Luhya (LOO-yuh)
Luo (LOO-oh)
Maasai (muh-SYE)
Madagascar (madda-GAS-kuh)
Maize (MAYZ)
Malagasy (malla-GASSY)
Malan, Daniel François (muh-LAHN,
   DAN-yell frawn-SWAH)
Malawi (muh-LAH-wee)
Malay (muh-LAY)
Mali (MAH-lee)
Mandela (man-DELLA)
Maputo (mah-POOH-toh)
Marasmus (muh-RAZZ-muss)
Matadi (muh-TAH-dee)
Mauritania (maw-ruh-TAY-nee-uh)
Merina (meh-REE-nuh)
Meru (MAY-roo)
Moçambique (moh-sum-BEEK)
Mombasa (mahm-BAHSSA)
Mpumalanga (mm-pooma-LAHNG-guh)
Nairobi (nye-ROH-bee)
Namib (nah-MEEB)
Namibia (nuh-MIBBY-uh)
Natal (nuh-TAHL)
Ndebele (en-duh-BEH-leh)
Ndola (en-DOH-luh)
Ngodi (eng-GOH-dee)
Niger [Country] (nee-ZHAIR)

Niger [River] (NYE-jer)
Nigeria (nye-JEERY-uh)
Nilotic (nye-LODDIK)
Nkrumah, Kwame (en-KROO-muh,
   KWAH-mee)
Nyasaland (nye-ASSA-land)
Ochre (OH-ker)
Ogaden (oh-gah-DEN)
Ogoni (oh-GOH-nee)
Okovango (oh-kuh-VAHNG-goh)
Olduvai (OLE-duh-way)
Oromo (AW-ruh-moh)
Pangaea (pan-GAY-uh)
Pariah (puh-RYE-uh)
Pointe-Noire (pwahnt-nuh-WAHR)
Polygamy (puh-LIG-uh-mee)
Pretoria (prih-TOR-ree-uh)
Proselytism (PRAH-sell-eh-tizm)
Pygmy (PIG-mee)
Rhodesia (roh-DEE-zhuh)
Rinderpest (RIN-duh-pest)
Rio Muni (ree-oh-MOOH-nee)
Rwanda (roo-AHN-duh)
Sahara (suh-HARRA)
Sahel (suh-HELL)
St. Louis (sah-loo-EE)
San (SAHN)
Schistosomiasis (shistoh-soh-
   MYE-uh-siss)
Senegal (sen-ih-GAWL)
Shari (SHAH-ree)
Shona (SHOH-nuh)
Sisal (SYE-sull)
Somalia (suh-MAHL-yuh)
Songhai (SAWNG-hye)
Sotho (SOO-too)
Soudan (soo-DAH)
Soweto (suh-WETTO)
Sudan (soo-DAN)

Sudd (SOOD)*
Swahili (swah-HEE-lee)
Swazi [land] (SWAH-zee [land])
Tanganyika (tan-gun-YEEKA)
Tanzania (tan-zuh-NEE-uh)
Tema (TAY-muh)
Tete (TATE-uh)
Tigrean (tih-GRAY-un)
Timbuktu (tim-buck-TOO)
Tsetse (TSETT-see)
Transkei (TRUN-skye)
Transvaal (TRUNZ-vahl)
Tsumeb (SOO-meb)
Tswana (TSWAHN-uh)
Tunisia (too-NEE-zhuh)
Turkana (ter-KANNA)
Tutsi (TOOTSIE)
Twa (TOO-wah)
Uganda (yoo-GAHN-duh/yoo-GANDA)
Vaal (VAHL)
Veld (VELT)
Vereeniging (fuh-REEN-ih-king)
Voortrekker (FOR-trecker)
Walvis (WAHL-vuss)
Wegener (VAY-ghenner)
Windhoek (VINT-hook)
Witbank (WHIT-bank)
Witwatersrand (WITT-waw-terz-rand)
Xhosa (SHAW-suh)
Yamoussoukro (yahm-uh-SOO-kroh)
Yaoundé (yown-DAY)
Yoruba (YAH-rooba)
Zaïre (zah-EAR)
Zambezi (zam-BEEZY)
Zambia (ZAM-bee-uh)
Zanzibar (ZANN-zih-bar)
Zimbabwe (zim-BAHB-way)
Zulu (ZOO-loo)

*Double ''o'' pronounced as in ''book.''

CHAPTER 8

South Asia: Resurgent Regionalism

From Iberia to Arabia and from Malaysia to Korea, Eurasia is a landmass ringed by peninsulas. Among these, none can compare to the great triangle of India that juts deep into the Indian Ocean—a vast, varied, volatile subcontinent.

India, the world's second most populous country (after China), lies at the heart of the South Asian geographic realm. With a population that will exceed 1 billion before the end of the twentieth century, India dominates—but this is a realm of giants. To India's east lies Bangladesh, with some 125 million inhabitants; to its west lies Pakistan, home to nearly 140 million. Only Nepal, in the mountainous north, and Sri Lanka, in the insular south, have modest populations of roughly 20 million each.

This huge population cluster lies in a comparatively well-defined physical realm. To the north, South Asia is separated from the rest of the continent by the Earth's greatest mountain range, the Himalayas, and its extensions east and westward. To the east, mountains and dense forests separate South Asia from neighboring Southeast Asia. Only the west affords avenues of entry, none of them easy. From there, across deserts and through mountain passes, came influences that repeatedly heightened the complexity of South Asian society. Today, South Asia is a Babel of languages, a Jerusalem of religions, a Lebanon of politics—and yet its conflicts have remained localized. Sur-

IDEAS & CONCEPTS	
Physiographic realm	Social stratification
Wet monsoon	Forward capital
Population geography	Irredentism
Population distribution	Federal system
Population density	Centrifugal forces
Doubling time	Centripetal forces
Population movement	Caste system
Population explosion	Intervening opportunity
Demographic transition	Natural hazards
Population structure	Insurgent state

REGIONS	
Pakistan	Southern Islands
India	Northern Mountains
Bangladesh	

prisingly, only one major political change has occurred over the past half-century of postcolonial independence. But how long will this record stand?

DEFINING THE REALM

To the north, east, and south, the South Asian realm is demarcated by mountains, forests, and coastlines, and as such constitutes one of the world's best-defined **physiographic realms**. As maps of religions and languages (see Figs. 6-1 and 8-4) underscore, South Asia is also clearly

defined in terms of cultural criteria. The main problem of regional definition lies in the northwest, where physical and cultural boundaries are less clear.

As defined here, South Asia consists of five regions: (1) India; (2) the west, Pakistan; (3) the east, centered on

■■■■■■■■■■■■ **FOCUS ON A SYSTEMATIC FIELD** ■■■■■■■■■■■

Population Geography

The population issue forms a backdrop to virtually any geographic topic as it relates to India. Economic prospects, health conditions, urbanization, education—all must be seen in context of India's huge population, its growth rate, distribution, movements, regional densities, and structure.

This is the stuff of **population geography**, the spatial view of demography. As always, the map is a powerful device to help us discern problems and detect solutions. To begin, look again at the South Asian part of the world population map (Fig. I-9 on p. 23). **Population distribution** can be mapped in various ways; Figure I-9 is a *dot map* in which each individual dot represents a certain number of people. For South Asia, it reveals an important reality: hundreds of millions of people continue to depend directly on the ribbons of life in this realm, the great rivers. From the population map you can almost trace the courses of the Indus (Pakistan), the Ganges (India), and the Brahmaputra (Bangladesh).

Dot maps are less useful in assessing **population density**. From Figure I-9 we get a general idea of the emptiness of southwestern Pakistan and the lower density (compared to the Ganges Basin) of parts of western and central India, but density—the number of people per unit area in a country, province, or some other territory—is better illustrated by other methods. Population density is often reported by country, as is done in Appendix A, but averages tend to deceive. The total population (967 million in India) is divided by its number of square miles, resulting in a density of 781 people per square mile in 1997. This is India's *arithmetic density*; yet the map shows us how uneven the distribution is. So that average is rather meaningless unless you relate it to other factors. One way to arrive at a more meaningful index is to calculate the number of people in a country per unit area of farmland, the *physiologic density*.

In India in 1997, this index was 1,300 people per square mile (the U.S. figure is just 169).

A key measure of any population is its **rate of natural increase**, and this gives great concern for India's future. Although India's rate has declined over the past several decades, it remains at 1.9 percent per year. With a base approaching 1 billion people, India adds nearly 20 million to its numbers annually. Again, the geographic perspective reminds us that the situation is far worse, in general, in eastern India than in western India. But no matter how it is viewed, India's rate of natural increase needs to be slowed. One way to view it is in terms of its **doubling time**. For every rate of increase, there is a period during which the population will double. For India in the late 1990s, that doubling time is just 36 years. Barring a substantial slowing of the rate of increase, India's 967 million of 1997 will become more than 1.9 billion by the year 2033, an almost unimaginable situation.

There is a difference between a population's rate of natural increase and its overall growth rate, and this has to do with **population movement**, another hot topic in population geography. The population of the United States, for example, increases naturally at only 0.7 percent (doubling time: 105 years), but its overall growth rate is estimated to be 1.6 percent. The reason: substantial legal and illegal immigration, adding millions to the total every year. The countries of South Asia experience more emigration than immigration, and the figures reported for their natural increase are more nearly representative of overall growth.

The population of the world as a whole still continues to grow at about 1.5 percent annually (doubling time: 45 years), a **population explosion** to which South Asia contributes enormously. In 1820, world population was about 1 billion; in 1930, it was just 2 billion. By 1975, it had

doubled again to 4 billion, and we are just about to pass the 6 billion mark now. The population spiral continues.

Will this cycle go on indefinitely? It had better not—and here geography again comes into play. The bulk of global population growth is occurring in the lower-income economies. But in the high-income economies (other than the oil-based economies of North Africa and Southwest Asia), population growth has leveled off. These countries have gone through a so-called **demographic transition**. As their economic development and associated industrialization and urbanization proceeded, high birth rates declined, and the gap between annual births and deaths narrowed to the point that some of these high-income countries have no population growth at all (see Appendix A; some Western European countries actually show *negative* population growth!). So there may be hope that the poorer countries of today, too, will grow economically and experience their own demographic transition. To be sure, conditions in many of the lower-income countries are not really comparable to Western Europe, but on the other hand, who would have predicted, a century ago, that Japan would be on the way to zero population growth today?

Population structure also matters in population geography. The age-sex structure of populations differs significantly. The high-income economies with their static populations now have large and growing numbers of oldsters who must be taken care of. The lower-income economies with their fast-growing populations have large numbers of youngsters in the lowest age groups, youngsters who must be nourished, educated, and employed.

Studies in population geography often reveal what basic statistics conceal: regional disparities *within* countries. As this chapter shows, when it comes to population geography, there is more than one India.

■ ■ ■

THE MAJOR GEOGRAPHIC QUALITIES OF SOUTH ASIA

1. South Asia is well-defined physiographically, extending from the southern slopes of the Himalayas to offshore Sri Lanka and the Maldive Islands.

2. Two river systems, the Ganges-Brahmaputra and the Indus, form crucial lifelines for hundreds of millions of people in this realm. The annual wet monsoons are a critical environmental element.

3. India lies at the heart of the world's second largest population cluster, which by 2025 will be the first.

4. No part of the world faces demographic problems with dimensions and urgency comparable to those of South Asia.

5. All the states of South Asia have low-income economies. Food shortages occur; nutritional imbalance prevails.

6. Agriculture in South Asia in general is comparatively inefficient and less productive than in other parts of Asia.

7. The great majority of South Asia's peoples live in villages and subsist directly on the land.

8. Strong cultural regionalism marks South Asia. The Hindu religion dominates life in India; Pakistan is an Islamic state; Buddhism thrives in Sri Lanka.

9. The South Asian realm's politico-geographical framework results from the European colonial period, but important modifications took place after the European withdrawal.

10. India constitutes the world's largest and most complex federal state.

Bangladesh; (4) the southern island areas of Sri Lanka and the Maldives; and (5) the northern mountain territories from Kashmir to Nepal and Bhutan. The largest region, India, divides into several subregions, which are discussed later in this chapter.

The inclusion of Pakistan in the South Asian (rather than the Islamic North Africa/Southwest Asia) realm can be debated. The political boundary between India and Pakistan forms a strong cultural divide as well: Pakistan is an Islamic state, whereas India is a secular state in which the Hindu religion and its ways of life dominate. On this basis it could be argued that Pakistan is not part of the South Asian realm, but consider the following: ethnic continuity links Pakistan more strongly to India than to Iran or Afghanistan. Historic migrations and diffusions into India have advanced across Pakistan. During the period of British colonial rule, Pakistan was part of Britain's South Asian empire, and English remains the universal language in both countries. And of India's 967 million people, about 110 million are Muslims. Another 100 million Muslims live in Bangladesh, so the Islamic population of the rest of South Asia far exceeds that of Pakistan. The cultural divide between the east and the west in South Asia, therefore, is not sufficient to justify the realm's partition. Pakistan continues to form the western flank of South Asia.

PHYSIOGRAPHIC REGIONS

Before we look into the complex and fascinating cultural geography of South Asia, it is useful to gain some insight into the physical stage of this populous realm. South Asia is a realm of immense physiographic variety, of snow-capped peaks and forest-clad slopes, of vast deserts and broad river basins, of high plateaus and spectacular shores. The collision of two of the Earth's great tectonic plates created the world's highest mountain ranges, their icy crests yielding meltwaters for great rivers below. The workings of the Earth's atmosphere put South Asia in the path of tropical cyclones and annual monsoons. This is a realm of almost infinite variety, a world unto itself.

In very general terms, it is possible to recognize three rather clearly defined physiographic zones in South Asia: the northern mountains, the southern peninsular plateaus, and, between these two, a belt of river lowlands. Superimposed on this configuration is an east-west precipitation gradient from wet (Bangladesh) to dry (western Pakistan) that is clearly visible in Figures I-6 and I-7, broken only by the strip of high moisture along India's Malabar Coast (Fig. 8-1).

Northern Mountains

The northern mountains extend from the Hindu Kush and Karakoram ranges in the northwest through the Himalayas in the center (Mount Everest, the world's tallest peak, lies in Nepal) to the ranges of Bhutan and the Indian State of Arunachal Pradesh in the east. Dry and barren in the west on the Afghanistan border, the ranges become green and tree-studded in Kashmir, forested in the lower sections of Nepal, and even more densely vegetated in Arunachal Pradesh. Transitional foothills, with many deeply eroded valleys cut by rushing meltwaters, lead to the river basins below.

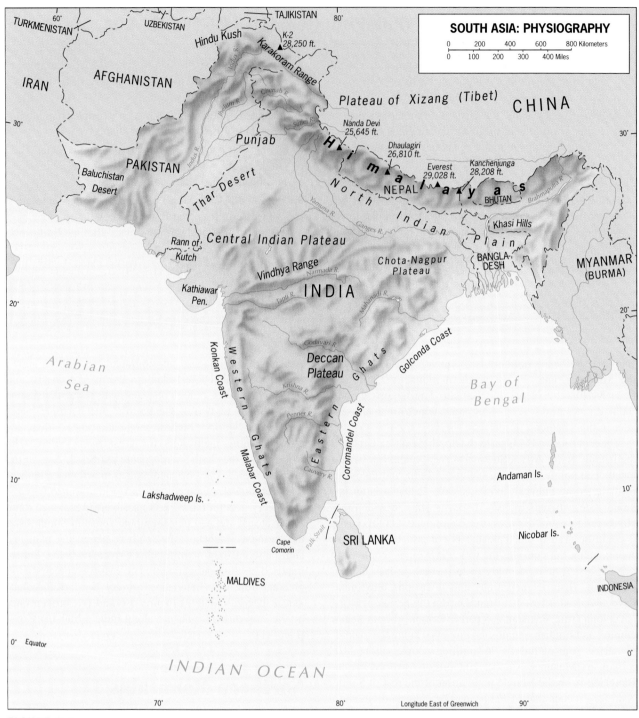

FIGURE 8-1

River Lowlands

The belt of river lowlands extends eastward from Pakistan's lower Indus valley (the area known as Sind) through the wide plain of the Ganges valley of India and on across the great double delta of the Ganges and Brahmaputra in Bangladesh (Fig. 8-1). In the east, this physiographic region often is called the North Indian Plain. To the west lies the lowland of the Indus River, which rises in Tibet, crosses Kashmir, and then bends southward to receive its major tributaries from the Punjab (''Land of Five Rivers'').

The dynamics of daily summer seabreezes are well known. The hot sun heats up the land, the air over this surface rises, and cool air from over the water blows in across the beach, replacing it. Teeming beaches and high-cost seafront apartments attest to the efficacy of this natural air conditioning.

Not only local beaches but, under certain circumstances, whole regions are affected by this kind of circulation. When an entire landmass heats up, a huge low-pressure system forms over it. This system can draw vast volumes of air from over the oceans onto the land. It is not just a local, daily phenomenon. It takes months to develop, but once it is in place, it persists for months as well. When the inflow of moist oceanic air starts over South Asia, the **wet monsoon** has arrived. It may rain for 60 days or more. The countryside turns green. The paddies fill. Another dry season's dust and dirt are washed away. The region is reborn.

Not all continents or coasts experience monsoons. A particular combination of topographic and atmospheric circumstances is needed. Figure 8-2 shows how it works in South Asia. It is sometime in June. For months, a low-pressure system has been building over northern India. Now, this cell has become so powerful that it draws air from over a large area of the warm, tropical Indian Ocean toward the interior. Some of that moisture-laden air is forced upward against the Western Ghats ①, cooling as it rises and condensing large amounts of rainfall. Other streams of air flow across the Bay of Bengal and get caught up in the convection over northeastern India and Bangladesh ②. A much larger area now is inundated by seemingly endless rain, including the whole North Indian Plain. The mountain wall of the Himalayas stops the air from spreading and the rain from dissipating ③. The air moves westward, drying out as it flows toward Pakistan ④.

After persisting for many weeks, the system finally breaks down and the monsoon gives way to periodic rains and, eventually, another dry season. Then the anxious wait begins for the next year's monsoon, for without it, India would face disaster. And in recent years, the monsoon has shown signs of irregularity and, as in 1987, partial failure. In India, life hangs by a meteorological thread.

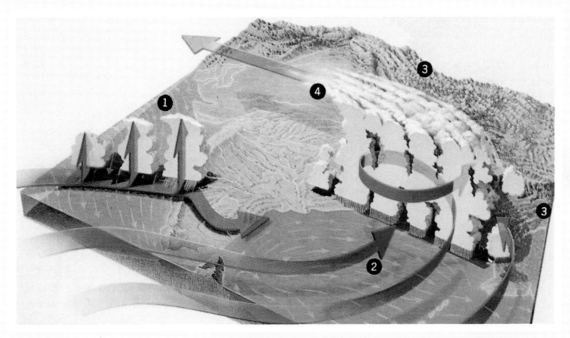

FIGURE 8-2

Southern Plateaus

Peninsular India is mostly plateau country, dominated by the massive Deccan, a tableland built of basalt that poured out when India separated from Africa during the breakup of Gondwana. The Deccan (meaning "South") is tilted to the east, so that its highest areas are in the west and the major rivers flow into the Bay of Bengal. North of the Deccan lie two other plateaus, the Central Indian Plateau to the west and the Chota-Nagpur Plateau to the east (Fig. 8-1). On the map, note the Eastern and Western Ghats ("hills") that descend from plateau elevations to the narrow coastal plains below. Onshore winds of the annual monsoon bring ample rain to the Western Ghats (see box titled "South Asia's Life-Giving Monsoon"). As a result, here lies one of India's most productive farming areas and one of southern India's largest population concentrations.

THE HUMAN SEQUENCE

South Asia is a realm of great river basins. Between the mountains of the north and the uplands of the south lie the broad, populous valleys of the Brahmaputra, the Ganges,

INDIA

■■■

Our use of the name *India* for the heart of the South Asian realm derives from the Sanskrit word *sindhu*, used to identify the ancient civilizations in the Indus valley. This word became *sinthos* in Greek descriptions of the area and then *sindus* in Latin. Corrupted to *indus*, which means "river," it was first applied to the region that now forms the heart of Pakistan. Subsequently, it was again modified to *India* to refer generally to the land of river basins and clustered peoples from the Indus in the west to the lower Brahmaputra in the east.

■■■

and the Indus. In one of these, the Indus (see box titled "India"), lies evidence of the realm's oldest civilization, contemporary to and interacting with ancient Mesopotamia. Unfortunately, much of the earliest record of this civilization lies buried beneath the present water table in the Indus valley, but those archeological sites that have yielded evidence indicate that here was a quite sophisticated culture with large, well-organized cities. As in Mesopotamia and the Nile valley, considerable advances were made in the technology of irrigation, and the civilization was based on the productivity of the irrigated soils in the Indus lowlands (Fig. 8-3).

Development of the River Lowlands

The Indus valley civilization was to India what the lower Nile culture hearth was to Africa: ideas and innovations diffused from there eastward and southward through the many different communities and societies that coexisted in

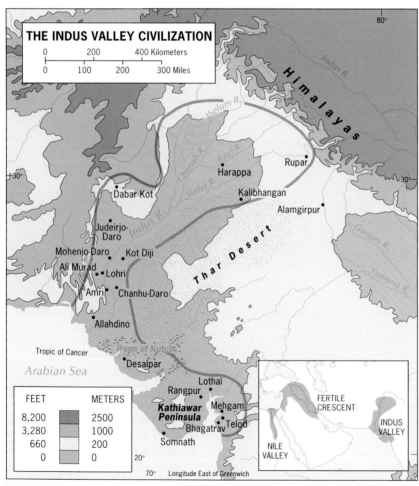

FIGURE 8-3

the peninsula (see Fig. 6-2). But far more than the ancient Egyptians did in Africa, the Indus culture brought coherence to India, and over an even longer period of diffusion. By the time the Indus civilization, anchored by the cities of Harappa and Mohenjo-Daro, began to decline about 4,000 years ago, many Indian cultures shared the legacy of the Harappans.

Now a new force entered the scene: the *Indo-Europeans* (or *Aryans*), who, from about 3500 B.C. onward, invaded the Indus valley from the direction of Iran, adopted many of the innovations of the Indus civilization, and pushed their frontier of settlement eastward beyond the valley into the Ganges lowland, where they founded an urban culture of their own. Pushing southward into the peninsula, they conquered and absorbed the tribes they found there. Their language, *Sanskrit*, began to differentiate into the linguistic complex of modern-day India (Fig. 8-4).

In the centuries following this invasion, Indian culture went through a period of growth and development. From a formless collection of isolated tribes and their villages, regional organization emerged. Towns developed, arts and crafts blossomed, and trade with Southwest Asia increased.

Most important of all, *Hinduism* emerged from the beliefs and practices brought to India by the Indo-Europeans, and a whole new way of life, based on this faith, evolved. A multilayered **social stratification** developed, controlled and administered by powerful priests. Ruling Brahmans stood at the head of a complex, bureaucratic hierarchy—a *caste system*—in which soldiers, artists, merchants, peasants, and all others had their place. But all was not peaceful in the India of, say, 3,000 years ago. Aggressive and expansionist kingdoms arose, competing and struggling with each other for greater power and influence.

It was in one of these kingdoms, located in northeastern India, that Prince Siddhartha, better known as Buddha, was born. His birth in the sixth century B.C. (more than 2,500 years ago) was unremarkable, but his actions were unique: he voluntarily gave up his princely position to seek salvation and enlightenment through religious meditation. His teachings demanded the rejection of earthly desires and prescribed a reverence for all forms of life. He walked the length and breadth of India and attracted a substantial following, but his teachings did not have a major impact on Hindu-dominated society during his lifetime. That impact

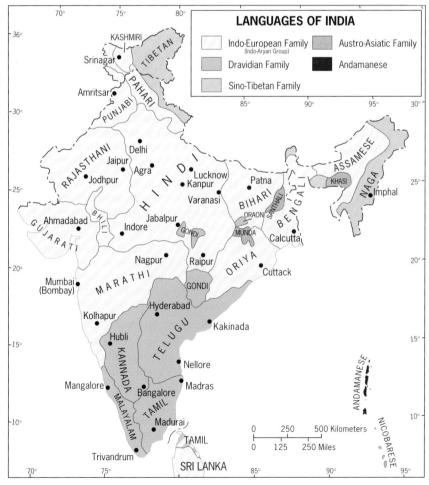

FIGURE 8-4

came later, when the ruler of a powerful Indian state decided to make Buddhism the state religion.

Before this occurred, however, South Asia was buffeted by yet another series of outside penetrations from the west and northwest. First the Persians pushed into the Indus Basin, and next the Greeks, under Alexander the Great, invaded not only the Indus valley but also the very heartland of India, the Gangetic Plain, late in the fourth century B.C.

The Southern Peninsula

While all this was happening in northern India, the peninsular south lay comparatively isolated, protected by distance from the arena of cultural innovation, infusion, and conflict. Southern India had been occupied by ancient peoples long before the Indus and Ganges civilizations arose—peoples whose historic linkages are not clear. Their physical appearance suggests connections to Africans and indigenous Australians; their languages, too, are distinctive and not related to those of the Indo-European north. As Figure 8-4 indicates, the south is a distinct subregion of India: both the peoples and their languages are known as *Dravidian*. There has, of course, been considerable intermixture with the north, but southern India's regional discreteness still prevails. The four major Dravidian languages—Telugu, Tamil, Kanarese (Kannada), and Malayalam—all have long literary histories. Today, Telugu and Tamil are spoken by nearly one-fifth of India's 967 million citizens.

Aśoka's Mauryan Empire

When the Greeks withdrew from the Ganges Basin and the Hindu heartland was once again free, a powerful empire arose there—the first true empire in the realm. This, the Mauryan Empire, extended its influence over India as far west as the Indus Valley (thus incorporating the populous Punjab) and as far east as Bengal (the double delta of the Ganges and Brahmaputra); it reached as far south as the modern city of Bangalore.

The Mauryan Empire was led by a series of capable rulers who achieved considerable stability over a vast domain. Undoubtedly the greatest of these leaders was Aśoka, who reigned for nearly 40 years during the middle part of the third century B.C. Aśoka was a believer in Buddhism, and it was he who elevated this religion from obscurity to regional and ultimately global importance.

In accordance with Buddha's teachings, Aśoka reordered his government's priorities from conquest and expansion to a Buddhist-inspired search for stability and peace. He sent missionaries to the outside world to carry Buddha's teachings to distant peoples, thereby also contributing to the diffusion of Indian culture. As a result, Buddhism became permanently established as the dominant religion in Sri Lanka (formerly Ceylon), and it established footholds as far afield as Southeast Asia and Mediterranean Europe. Ironically, Buddhism thrived in these remote places even as it declined in India itself. With Aśoka's death, the faith lost its strongest supporter.

The Mauryan Empire represented the greatest of political and cultural achievements of India in its day, and when it collapsed, late in the second century A.D., India fragmented into a mosaic of states. Once again, India lay open to infusions from the west and northwest, and across present-day Pakistan they came: Persians, Afghans, Turks, and others driven from their homelands or attracted by the lands of the Ganges.

The Power of Islam

In the late tenth century, the wave of Islam came rolling like a giant tide across the subcontinent, spreading from Persia in the west and Afghanistan in the northwest. Of course, the Indus valley lay directly in the path of this Islamic advance, and virtually everyone was converted. Next the Muslims penetrated the Punjab, the subregion that lies astride the present Pakistan-India border, and there perhaps as many as two-thirds of the inhabitants became converts. Then Islam crossed the bottleneck where Delhi is situated and diffused eastward and southeast into the Gangetic Plain and the subregion known as *Hindustan*—India's evolving core area. Here Islam's proselytizers had less success, persuading perhaps one in eight Indians to become Muslims. In the meantime, Islam arrived at the Ganges delta by boat, and present-day Bangladesh became overwhelmingly Islamic. (To the south of the Ganges heartland, however, Islam's diffusion wave lost its energy: Dravidian India never came under Muslim influence.)

Islam's vigorous, often violent onslaught brought far-reaching changes to Indian society. As in West Africa, Islam often was superimposed by political control: the rulers of states were converted, and their subjects followed. By the early fourteenth century, a sultanate centered at Delhi controlled more of the subcontinent than even the Mauryan Empire had earlier. Later, the Islamic Mogul Empire (the similarity to the word *Mongol* is no coincidence) constituted the largest political entity ever to unify the realm in precolonial times. To many Hindus of lower caste, Islam formed a welcome alternative to the rigid socioreligious hierarchy in which they were trapped at the bottom. Thus Islam was the faith of the ruling elites and of the disadvantaged, a powerful cultural force in the heartland of Hinduism.

Just as Islam weakened in southern Europe, so its force became spent in vast and populous India. For all the Muslims' power, they never managed to convert even a majority of South Asians to their faith. They dominated the northwest corner of the realm (present-day Pakistan), where Lahore became one of Islam's greatest cities. But in all of what is today India, less than 15 percent of the population became and remained Muslim. And throughout the period of Islamic intervention, the struggle for cultural supremacy went on. Placid Hinduism and aggressive Islam did not easily coexist.

The European Intrusion

Into this turbulent complexion of religious, political, and linguistic disunity still another element began its intrusion after 1500: European powers in search of raw materials, markets, and political influence. The Europeans profited from the Hindu-Muslim contest, and they were able to exploit local rivalries, jealousies, and animosities. British merchants gained control over the trade with Europe in spices, cotton, and silk goods, ousting their competitors—the French, Dutch, and Portuguese. The British East India Company's ships also took over the intra-Asian sea trade between India and Southeast Asia, long in the hands of Arab, Indonesian, Chinese, and Indian merchants. In effect, the East India Company became India's colonial administration.

As time went on, however, the East India Company faced problems it could not solve. Its commercial activities remained profitable, but it became entangled in a widening effort to maintain political control over an expanding Indian domain. The Company proved to be an ineffective governing agent at a time when the increasing Westernization of India brought to the fore new and intense frictions. Christian missionaries were challenging Hindu beliefs, and many Hindus believed that the British were out to destroy the caste system. Changes also came in public education, and the role and status of women began to improve. Aristocracies saw their positions threatened as Indian landowners had their estates expropriated. Finally, in 1857 the three-month Sepoy Rebellion broke out and changed the entire situation. It took a major military effort to put it down, and from that time the East India Company ceased to function as the government of India. Administration was turned over to the British government, the Company was abolished, and India formally became a British colony—a status it held for the next 90 years until this *raj* (rule) ended in 1947.

Colonial Transformation

Four centuries of European intervention in South Asia greatly changed the realm's cultural, economic, and political directions. Certainly, the British made positive contributions to Indian life, but colonialism also brought serious negative consequences. In this respect there are important differences between the South Asian case and that of Sub-

saharan Africa. When the Europeans came to India, they found a considerable amount of industry, especially in metal goods and textiles, and an active trade with both Southwest and Southeast Asia in which Indian merchants played a leading role. The British intercepted this trade, changing the whole pattern of Indian commerce.

India now ceased to be South Asia's manufacturer, and soon the country was exporting raw materials and importing manufactured goods—from Europe, of course. India's handicraft industries declined; after the first stimulus, the export trade in agricultural raw materials also suffered as other parts of the world were colonized and linked in trade to Europe. Thus the majority of India's people (who were farmers then as now) suffered an economic setback as a result of the manipulations of colonialism. Although in total volume of trade the colonial period brought considerable increases, the composition of the trade India now supported by no means brought a better life for its people.

Neither did the British manage to accomplish what the Mauryans and the Moguls had tried to do: unify the subcontinent and minimize its internal cultural and political divisions. When the Crown took over from the East India Company in 1857, about 750,000 square miles (nearly 2 million sq km) of Indian territory were still outside the British sphere of influence. Slowly, the British extended their control over this huge unconsolidated area, including several pockets of territory already surrounded but never integrated into the previous corporate administration. Moreover, the British government found itself obligated to support a long list of treaties that the Company's administrators had made with numerous Indian princes, regional governors, and feudal rulers.

These treaties guaranteed various degrees of autonomy for literally hundreds of political entities in India, ranging in size from a few acres to Hyderabad's more than 80,000 square miles (200,000 sq km). The British Crown saw no alternative but to honor these guarantees, and India was carved up into an administrative framework under which there were more than 600 ''sovereign'' territories in the subcontinent. These ''Native States'' had British advisors; the large British provinces such as Punjab, Bengal, and Assam had British governors or commissioners who reported to the viceroy of India, who in turn reported to Parliament and the monarch in London. In all, this near-chaotic amalgam of modern colonial control and traditional feudalism reflected and in some ways deepened the regional and local disunities of the Indian subcontinent. Although certain parts of India quickly adopted and promoted the positive contributions of the colonial era, other areas rejected and repelled them, thereby adding yet another element of division to an increasingly complicated human spatial mosaic.

Colonialism did produce assets for India. The country was bequeathed one of the best railroad and highway transport networks of the colonial domain. British engineers laid

Flight was one response to the 1947 partition of what had been British India, resulting in one of the greatest mass population transfers in human history. Here, two trainloads of eastbound Hindu refugees fleeing (then) West Pakistan arrive at the station in Amritsar, the first city inside India.

out irrigation canals through which millions of acres of land were brought into cultivation. Settlements that had been founded by Britain developed into major cities and bustling ports, led by Bombay, Calcutta, and Madras. These three cities are still three of India's largest urban centers, and their cityscapes bear the unmistakable imprint of colonialism. Modern industrialization, too, was brought to India by the British on a limited scale. In education, an effort was made to combine English and Indian traditions; the Westernization of India's elite was supported through the education of numerous Indians in Britain. Modern practices of medicine were also introduced. Moreover, the British administration tried to eliminate features of Indian culture that were deemed undesirable by any standards—such as the burning alive of widows on the funeral pyres of their husbands, female infanticide, child marriage, and the caste system. Obviously, the task was far too great to be achieved successfully in barely three generations of rule, but post-independence India itself has continued these efforts where necessary.

Partition

Even before the British government decided to yield to Indian demands for independence, it was clear that British India would not survive the coming of self-rule as a single political entity. As early as the 1930s, the idea of a separate Pakistan was being promoted by Muslim activists, who circulated pamphlets arguing that British India's Muslims were a nation distinct from the Hindus and that a separate state consisting of Sind, Punjab, Baluchistan, Kashmir, and a portion of Afghanistan should be created from the British South Asian Empire in this area. The first formal demand for such partitioning was made in 1940, and, as later elections proved, the idea had almost universal support among the realm's Muslims.

As the colony moved toward independence, a political crisis developed: India's majority Congress Party would not even consider partition, and the minority Muslims refused to participate in any future unitary government. But partition would be no simple matter. True, Muslims were in the majority in the western and eastern sectors of British India, but there were many Islamic clusters scattered through the realm (Fig. 8-5). Any new boundaries between Hindus and Muslims to create an Islamic Pakistan and a Hindu India would have to be drawn right through areas where both sides coexisted. People by the millions would be displaced.

Nor were Hindus and Muslims the only people affected by partition. The Punjab area, for example, was home to millions of Sikhs, whose leaders were fiercely anti-Muslim. But a Hindu-Muslim border based only on those two groups would leave the Sikhs in Pakistan. Even before independence day, August 15, 1947, Sikh leaders talked of revolt, and there were some riots. But no one could have foreseen the dreadful killings and mass migrations that followed the creation of the boundary and the formation of independent Pakistan and India. Just how many people felt compelled to participate in the ensuing migrations will never be known; 15 million is the most common estimate. It was human suffering on an incomprehensible scale.

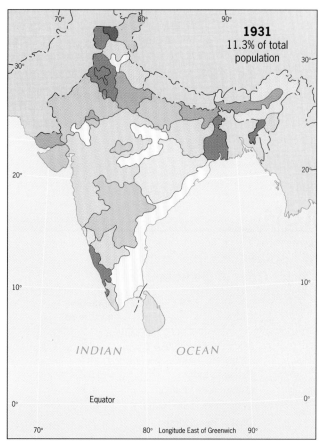

1931
11.3% of total
population

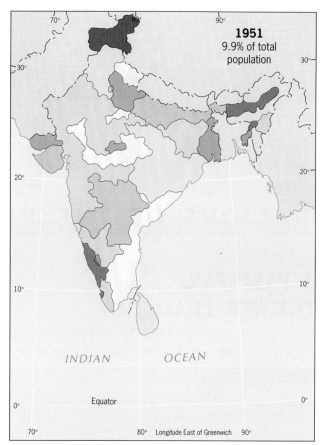

1951
9.9% of total
population

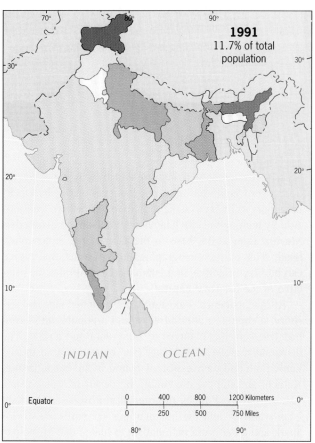

1991
11.7% of total
population

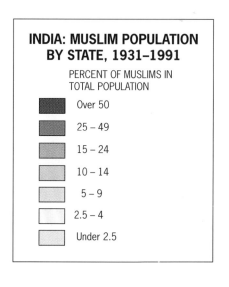

**INDIA: MUSLIM POPULATION
BY STATE, 1931–1991**

PERCENT OF MUSLIMS IN
TOTAL POPULATION

Over 50

25 – 49

15 – 24

10 – 14

5 – 9

2.5 – 4

Under 2.5

FIGURE 8-5

Even so large a flow of cross-border refugees, however, hardly began to ''purify'' India of Muslims. After the initial mass exchanges, there still were tens of millions of Muslims in India (Fig. 8-5). Today, the Muslim minority in Hindu-dominated India is almost as large as the whole Islamic population of Pakistan, having more than tripled since the late 1940s. It is the world's largest minority, far more than a mere remnant of the days when Islam ruled the realm. This force will play a growing role in the India of the future.

REGIONS OF THE REALM

❖ PAKISTAN: THE NEW CHALLENGE

If Egypt is the gift of the Nile, then Pakistan is the gift of the Indus. The Indus River and its major tributary, the Sutlej, sustain the ribbons of life that form the heart of this populous country (Fig. 8-6).

Territorially, Pakistan is not very large by Asian standards; its area is about the same as that of Texas plus Louisiana. But Pakistan's population of 137 million makes this one of the world's 10 most populous states. Among Muslim countries (officially, it is known as the Islamic Republic of Pakistan) only Indonesia is larger, but Indonesia's Islam is much less pervasive than Pakistan's. Pakistan is an active participant in the worldwide resurgence of Islamic fervor; Indonesia is not—at least not yet.

Pakistan lies like a giant wedge between Iran and Afghanistan to the west and India to the east. This wedge extends from the shores of the Arabian Sea in the south, where it is widest, to the snowcapped mountains of the north, where it reaches a point. Here in the far north of Pakistan, the boundary framework is complex and jurisdictions are uncertain. India, China, Pakistan, and, until recently, the Soviet Union have all vied for control over parts of this mountainous northern frontier. The Soviets, by virtue of their temporary rule over Afghanistan, controlled the Wakhan Corridor, (the long narrow segment of land labeled ''Afghanistan'' in the inset of Fig. 8-6). The Chinese have claimed areas to the east (inset map). But the major territorial conflict has been between Pakistan and India over the areas of Jammu and Kashmir. This issue has plagued relations between these two countries for decades (see box titled ''The Problem of Kashmir'').

When Pakistan became an independent state following the partition of British India (1947), its capital was Karachi on the south coast, near the western end of the Indus delta. As the map shows, however, the present capital is Islamabad, near the larger city of Rawalpindi in the north, not far from Kashmir. By moving the capital from the ''safe''

Major Cities of the Realm

Cities	Population* (in millions)
Ahmadabad, India	3.9
Bangalore, India	5.2
Calcutta, India	12.2
Colombo, Sri Lanka	0.8
Delhi-New Delhi, India	10.8
Dhaka, Bangladesh	9.0
Hyderabad, India	6.0
Karachi, Pakistan	11.0
Kathmandu, Nepal	0.5
Lahore, Pakistan	5.7
Madras, India	6.3
Mumbai (Bombay), India	16.6
Varanasi, India	1.2

*Based on 1997 estimates.

coast to the embattled interior and by placing it on the doorstep of contested territory, Pakistan made clear to its neighbors its intent to stake a claim to its northern frontiers. And by naming the city Islamabad, Pakistan proclaimed its Muslim foundation, here in the face of the Hindu challenge. This politico-geographical use of a national capital can be very effective, and Islamabad exemplifies the principle of the **forward capital**.

At independence, Pakistan had a bounded national territory, a capital, a cultural core, and a population—but it had few centripetal forces to bind state and nation. The disparate regions of Pakistan (see below) shared the Islamic faith and an aversion of Hindu India, but little else. Karachi and the coastal south, the desert of Baluchistan, Lahore and the Punjab, the rugged northwest on Afghanistan's border, and the mountainous far north are worlds apart, and a Pakistani nationalism to match that of India at independence did not exist. Successive Pakistani governments, civilian as well as military, turned to Islam to pro-

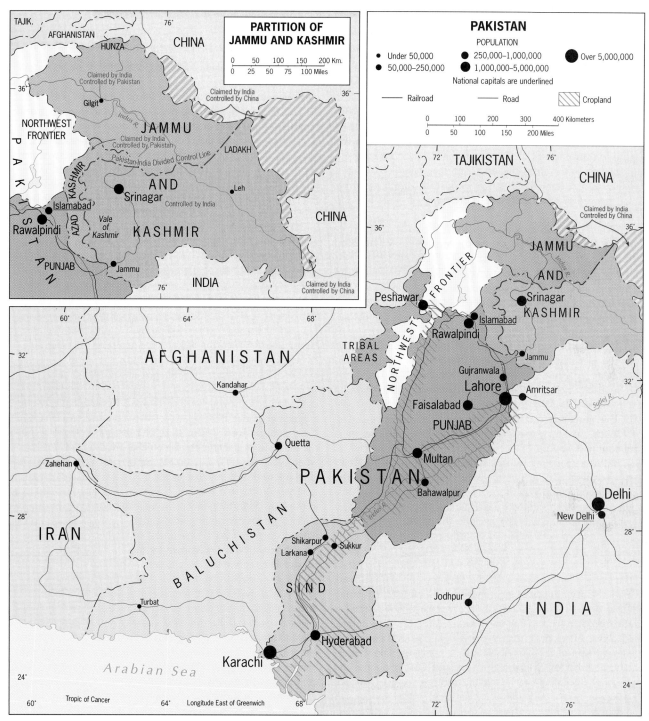

FIGURE 8-6

vide the common bond that history and geography had conspired to deny the nation. In the process, Pakistan became one of the world's most theocratic states; its common law, based on the English model, was gradually transformed into a Koranic system with Islamic *sharia* courts and associated punishments.

Despite the Islamization of Pakistan's plural society,

this remains a strongly regionalized country in which Urdu is the official language and English still serves as the *lingua franca* of the elite. Yet several other major languages prevail in diverse parts, and lifeways vary from nomadism in Baluchistan to irrigation farming in the Punjab to pastoralism in the northern highlands. We look next at Pakistan's varied regional framework.

THE PROBLEM OF KASHMIR

■ ■ ■

Kashmir is a territory of high mountains surrounded by Pakistan, India, China, and, along several miles in the far north, Afghanistan (Fig. 8-6). Although known simply as Kashmir, the area actually consists of several political divisions, including the State properly referred to as *Jammu and Kashmir* (one of the 562 Indian States at the time of independence) and the administrative areas of Gilgit in the northwest and Ladakh (including Baltistan) in the east. The main conflict between India and Pakistan over the final disposition of this territory has focused on the southwest, where Jammu and Kashmir are located.

When partition took place in 1947, the existing States of British India were asked to decide whether they would go with India or Pakistan. In most of the States, this issue was settled by the local authority, but in Kashmir there was an unusual situation. There were about 5 million inhabitants in the territory at that time, nearly half of them concentrated in the basin known as the Vale of Kashmir (where the capital, Srinagar, is located). Another 45 percent of the people were concentrated in Jammu, which leads down the foothill slopes of the Himalayas to the edge of the Punjab. The small remainder of the population was scattered through the mountains, including Pakhtuns in Gilgit and other parts of the northwest. Of these population groups, the people of the mountain-encircled Vale of Kashmir are almost all Muslims, whereas the majority of Jammu's population is Hindu.

But the important feature of the State of Jammu and Kashmir was that its rulers were Hindu, not Muslim, although the overall population was more than 75 percent Muslim. Thus the rulers were faced with a difficult decision in 1947—to go with Pakistan and thereby exclude themselves from Hindu India, or to go with India and thereby incur the wrath of the majority of the people. Hence, the maharajah of Kashmir sought to remain outside both Pakistan and India and to retain the status of an autonomous unit. This decision was followed, after the partitioning of India and Pakistan, by a Muslim uprising against Hindu rule in Kashmir. The maharajah asked for the help of India, and Pakistan's forces came to the aid of the Muslims. After more than a year's fighting and through the intervention of the United Nations, a cease-fire line was established that left Srinagar, the Vale, and most of Jammu and Kashmir (including nearly four-fifths of the territory's population) in Indian hands (Fig. 8-6, inset map). In due course, this line began to appear on maps as the final boundary settlement,

and Indian governments have proposed that it be so recognized.

Why should two countries, whose interests would be served by peaceful cooperation, allow a distant mountainland to trouble their relationship to the point of war? There is no single answer to this question, but there are several areas of concern for both sides. In the first place, Pakistan is wary of any situation whereby India would control vital irrigation waters needed in Pakistan. As the map shows, the Indus River, the country's lifeline, crosses Kashmir. Moreover, other tributary streams of the Indus originate in Kashmir, and it was in the Punjab that Pakistan learned the lessons of dealing with India for water supplies. Second, the situation in Kashmir is analogous to the one that led to the partition of the whole subcontinent: Muslims are under Hindu domination. The majority of Kashmir's people are Muslims, so Pakistan argues that free choice would deliver Kashmir to the Islamic Republic. A free plebiscite is what Pakistanis have sought—and the Indians have thwarted. Furthermore, Kashmir's connections with Pakistan prior to partition were much stronger than those between Kashmir and India, although India has invested heavily in improving its links to Jammu and Kashmir since the military stalemate. In recent years, it did seem more likely that the cease-fire line (now called the Divided Control Line) would indeed become a stable boundary between India and Pakistan. (The incorporation of the State of Jammu and Kashmir into the Indian federal union was accomplished as far back as 1975, when India was able to reach an agreement with the State's chief minister and political leader.)

The drift toward stability, however, was sharply reversed in the late 1980s as a new crisis engulfed Kashmir. Extremist Muslim groups demanding independence escalated a long-running insurgency into a new phase of violence in 1988. This brought on a swift crackdown by the Indian military, whose harsh tactics prompted the citizenry to far more strongly support the separatists. In the early 1990s, Pakistan charged the Indians with widespread human rights violations, and the Indians accused the Pakistanis of inciting secession and supplying arms. Between 1990 and 1995, nearly 10,000 people were killed in the sporadic fighting that continues to afflict this scenic area. In 1995, Muslim extremists began a campaign that involved capturing and killing foreign visitors, a move that was condemned by representatives of both sides. Nevertheless, the memory of three wars between India and Pakistan defied all attempts at compromise.

■ ■ ■

Subregions

Punjab

Pakistan's core area is the Punjab, the Muslim heartland across which the postcolonial boundary between Pakistan and India was superimposed. (As a result, India also has a

region called Punjab, sometimes spelled *Panjab* there). Pakistan's Punjab is home to nearly 60 percent of the country's population. In the triangle formed by the Indus River and its tributary, the Sutlej, live some 80 million people. Punjabi is the language here, and intensive farming of wheat is the mainstay.

Lahore, center of historic importance, religious significance, and architectural magnificence, is the functional focus of Pakistan's core area, the Punjab. Among its most impressive mosques is the Badshahi Mosque, whose walled courtyard on holy days accommodates more than a half million worshippers. Centrally located, this mosque (like others in the Islamic world) serves as a social gathering place as well; the courtyard is normally empty, but on this hot summer day cotton canopies shelter visitors from the sun.

Three cities anchor this core area: Lahore, the outstanding center of Islamic culture in the realm, Faisalabad, and Multan. Lahore, now home to about 6 million people, lies very close to the India-Pakistan border. The city received hundreds of thousands of refugees and continued to grow rapidly following the partition. Founded about 2,000 years ago, Lahore was situated favorably to become a great Muslim center during the Mogul period, when the Punjab was a corridor into India. As a center of royalty, Lahore was adorned with numerous magnificent buildings, including a great fort, several palaces, and many mosques displaying superb stonework and marble embellishments. The site of an old university and many magnificent gardens, Lahore is the cultural focus of Islam. Upon partition, Lahore lost all of its eastern hinterland, but its new role in modern Pakistan sustained its growth.

Sind

South of the Punjab lies Sind, centered on the chaotic port city of Karachi on the coast and on Hyderabad upriver. As Figure 8-6 suggests, Sind has three dominant landscapes: rocky hills to the west, desert to the east, and between these the fertile, irrigated alluvial valley of the Indus River. Large irrigation systems were built here during colonial times, and improved seed strains have made Sind a Pakistani breadbasket for wheat and rice. Commercially, cotton is king here, supporting major textile industries in the cities and towns.

A flood of immigrants from India moved into Sind during and after partition, and relations between immigrants and locals have not always been good. Karachi has become the focus of conflict between immigrant Muhajirs and their descendants and other groups in this poverty-stricken,

crime-ridden city. Karachi grew explosively during and after the refugee influx, and the city's infrastructure crumbled even as millions were added to its population. The social order broke down, law enforcement was unable to cope, and in the mushrooming shantytowns, gang battles over turf spilled over into the built-up areas. Large, multicultural cities can put severe strain on low-income economies when conditions spin out of control.

Baluchistan

A trip westward from Sind and Karachi into the rocky and sandy deserts makes one wonder whether this is the same country. Baluchistan's arid landscapes extend far into neighboring Iran, and the sparse population of this remote and forbidding territory still moves across the borders into and out of Iran along ancient caravan routes. Pakistan is an Islamic country, but about 20 percent of the people adhere to Shi'ite principles. Shi'ism arrived in Pakistan along these desert routes from its Iranian stronghold.

Baluchistan may figure more importantly in the economic future of Pakistan than it does today. The geology indicates considerable mineral potential, and some oil and natural gas have been found here.

Northwest Frontier

The fourth subregion of Pakistan is appropriately called the *Northwest Frontier*. This is the area that faces turbulent and disintegrating Afghanistan, and it has been the scene of massive refugee influxes, local separatist movements, and social dislocation. The Northwest Frontier is dominated by mountain ranges, mountain-encircled basins, and strategic passes. The most famous pass of all is the Khyber

Pass to Afghanistan. The Turks invaded the upper Indus valley by this route; later the Moguls streamed through it on their way to India; and in recent decades millions of war-weary emigrants have walked this road to safety in Pakistan.

The largest city in the Northwest Frontier province is Peshawar, but after Baluchistan this is Pakistan's least urbanized area. Peshawar lies in an alluvium-filled, fertile valley where fields of wheat and corn cover the countryside. But with its large refugee population (which, during the 1980s, reached in excess of 4 million), the Northwest Frontier has faced difficult times, only partly eased by United Nations assistance.

The ethnic and cultural backgrounds of the population in the Northwest Frontier vary. Even before the refugee influx from Afghanistan, a large part of the population consisted of Pathans (also called Pakhtuns, Pashtuns, or Pushtuns), who were closely related to the Pathans of central Afghanistan. At times, the Pathans of Afghanistan urged their kinspeople in Pakistan to demand greater autonomy, if not outright independence—a practice we know as **irredentism**. When the Soviets invaded Afghanistan in 1979 and war broke out, several million Pathans moved from Afghanistan to the Northwest Frontier. Pakistan not only managed to accommodate this refugee population but also countered the secession movement by improving the integration of the Northwest Frontier province with the rest of the country through road building, economic aid programs, and enhanced educational opportunities.

In the late 1990s, Pakistan's Northwest Frontier remained a sensitive zone. It is only 170 miles (270 km) from Peshawar to Afghanistan's capital, Kabul, and conditions in Afghanistan have always influenced the Northwest Frontier. When the Soviets withdrew from Afghanistan during the late 1980s, Afghan refugees began to return home—but soon, Afghanistan's diverse ethnic-cultural groups were in conflict again, this time over control of the capital and the government. A new outflow of refugees started, and the prospect of a collapsing Afghanistan, right on Pakistan's border, arose. Today, Pakistan's relationship with, and role in, a future Afghanistan is a crucial question for the entire region.

Livelihoods

For all its size and growing regional influence, Pakistan remains a low-income economy. In the late 1990s, urbanization was just above 30 percent. Population growth was high at 2.9 percent, implying a doubling time of only 24 years. Subsistence farming of food crops still occupied the great majority of the people. Life expectancy for a child born today is just 61 years. Illiteracy still afflicts 65 percent of the population over 15 years of age.

Nevertheless, Pakistan has made significant economic progress during its half-century of independence. Irrigation has expanded enormously; land reform has progressed; and the cotton-based textile industry has generated important revenues. Despite a very limited mineral resource base, some manufacturing growth has taken place, including a steel mill at Port Qasim near Karachi.

To the rest of the world, Pakistan sells cotton textiles, carpets and tapestries, leather goods—and rice. Despite its large and growing population, Pakistan in the 1990s was able to export rice, which is a measure of the success of the Green Revolution (see p. 337) as well as its agricultural expansion program. In recent years, Pakistan also has become the largest producer of heroin in South and Southwest Asia, and it produces opium and hashish for the international drug trade despite government efforts to eradicate the fields. The legitimate economy has been growing in recent years at 5 to 6 percent annually, and as Appendix A shows, Pakistan has the highest per capita GNP of all the states of mainland South Asia. With a lower rate of population increase, Pakistan's prospects would improve.

Emerging Regional Power

A country's underdevelopment is not necessarily a barrier against its emergence as a power to be reckoned with in international affairs. When Pakistan became independent in 1947, the country was weak, disorganized, and divided. Just 50 years later, Pakistan is a major military force and, probably, a nuclear power.

To say that Pakistan in 1947 was a divided country is no exaggeration. In fact, upon independence, present-day Pakistan was united with present-day Bangladesh, and the two countries were called West Pakistan and East Pakistan respectively. The basis for this scheme was Islam: in Bangladesh, too, Islam is the state religion. Between the two Islamic wings of Pakistan lay Hindu India. But there was little else to unify the easterners and westerners, and their union lasted less than 25 years. In 1971, a costly war of secession led to independence for East Pakistan, which then took the name of *Bangladesh*; at the same time, West Pakistan became *Pakistan*.

To Pakistan, the loss of Bangladesh was no disaster: in virtually every respect, Bangladesh was (and remains) even more severely impoverished than Pakistan (see appropriate data in Appendix A). Pakistan's challenges lay closer to home—in Kashmir, in the Northwest Frontier, and, most important, to the east in India. India had supported Bangladesh in its campaign for independence and did not resume diplomatic relations with Pakistan until five years later. Ever since, the political relationship between India and Pakistan has been tense. Apart from the conflict in Kashmir, their other differences are numerous. India remained a democracy while Pakistan became a military dictatorship; India's treatment of its Muslim minorities frequently riled Pakistan; India developed a close relationship with the same Soviet Union seen as a threat by Islamabad. During the period of the Cold War, India tilted toward Moscow, while Pakistan, despite frequent disputes, was favored by Washington. In 1979, the United States cut off aid to Pakistan because Washington believed the Pakistanis were secretly preparing to build nuclear weapons. But when the Soviet invasion of Afghanistan later that year put Moscow's armed forces on the doorstep of northwest Pakistan, the United States sold F-16 fighter planes to the Pakistani regime.

Mutual suspicion raised the military stakes in both Pakistan and India, and it is likely that both countries now possess nuclear arms. In the meantime, Pakistan now finds itself not only near disintegrating Afghanistan but also close to the scene of one of the great geopolitical, economic, and ideological transformations of the twentieth century: Turkestan. No longer merely a newly decolonized, economically disadvantaged country trying to survive, Pakistan today is a powerful bulwark of traditional Sunni Islam, militant in its regional aspirations and situated in a critical part of our changing world.

❖ INDIA: FIFTY YEARS OF FEDERATION

Nearly three-quarters of the great land triangle of South Asia is occupied by a single country—India, the world's most populous democracy and, in terms of human numbers, the world's largest federation. Consider this: India has nearly as many inhabitants as live in all the countries of Subsaharan Africa plus North Africa/Southwest Asia *combined*—74 of them. At present rates of population growth, India not only will outnumber these two realms at some time during the next century but will overtake even China. If India holds together as a single state, it is likely to become the world's most populous country, bar none.

That India has endured as a unified country is one of the politico-geographical miracles of the twentieth century. India is a cultural mosaic of immense ethnic, religious, linguistic, and economic diversity and contrast; it is a state of many nations. The period of British colonialism gave India the underpinnings of unity: a single capital, an inter-regional transport network, a *lingua franca*, a civil service. And upon independence in 1947, India adopted a **federal system** of government, giving regions and peoples some autonomy and identity, and allowing others to aspire to such status. Unlike Africa, where federal systems failed and where military dictatorships replaced them (notably in Nigeria), India, for a half-century now, has remained essentially democratic and has retained its federal framework.

This does not mean that India has not suffered from *centrifugal forces*. Border conflicts, frontier wars, civil strife, and international clashes (most recently involving nearby Sri Lanka) have buffeted and at times threatened India's stability. But India has prevailed where others in the postcolonial world have failed.

Political Spatial Organization

The politico-geographical map of India shows a federation divided into 25 States and 7 Union Territories (UTs) (Fig. 8-7). Delhi/New Delhi, the Union Territory that contains the capital, is the only Union Territory that has substantially more than 1 million inhabitants; the other six UTs are small territorially as well as demographically.

Several Indian States have territories and populations that are themselves larger than many countries and nations of the world. As Figure 8-7 shows, the (areally) largest States lie on the great southward-pointing peninsula. Uttar Pradesh (159 million inhabitants)* and Bihar (98 million) constitute much of the Ganges River Basin and are the core area of modern India (see box titled ''Solace and Sickness from the Holy Ganges''). Maharashtra (90 million), anchored by the great coastal city of Bombay (renamed *Mumbai* in 1996), also has a population larger than that of most of the world's countries. West Bengal, the State that adjoins Bangladesh, has 78 million residents, 12.2 million of whom live in its urban focus, Calcutta.

These are staggering numbers, and they do not decline much toward the south. Southern India consists of four States linked by a discrete history and by their distinct Dravidian languages. Facing the Bay of Bengal are Andhra Pradesh (76 million) and Tamil Nadu (62 million), both part of the hinterland of the megacity of Madras, located on the coast very near their joint border. Facing the Arabian Sea are Karnataka (50 million) and Kerala (31 million). Kerala, often at odds with the federal government in New Delhi, long has had the highest literacy rate in India and one of the lowest rates of population growth. Strong local government and strictly enforced policies are credited with these achievements here. ''It's a matter of geography,'' explained a teacher in the Kerala city of Cochin. ''We are here about as far away as you can get from the capital, and we make our own rules.''

As Figure 8-7 shows, India's smaller States lie mainly in the northeast, on the far side of Bangladesh, and in the northwest, toward Jammu and Kashmir. North of Delhi, India lies flanked by China and Pakistan, and physical as well as cultural landscapes change from the flatlands of the Ganges to the hills and mountains of spurs of the Himalayas. In the State of Himachal Pradesh, forests cover the hillslopes and living space is reduced by relief; only 5.7

*All State population data are estimates for 1997, extrapolated from India's last (1991) census.

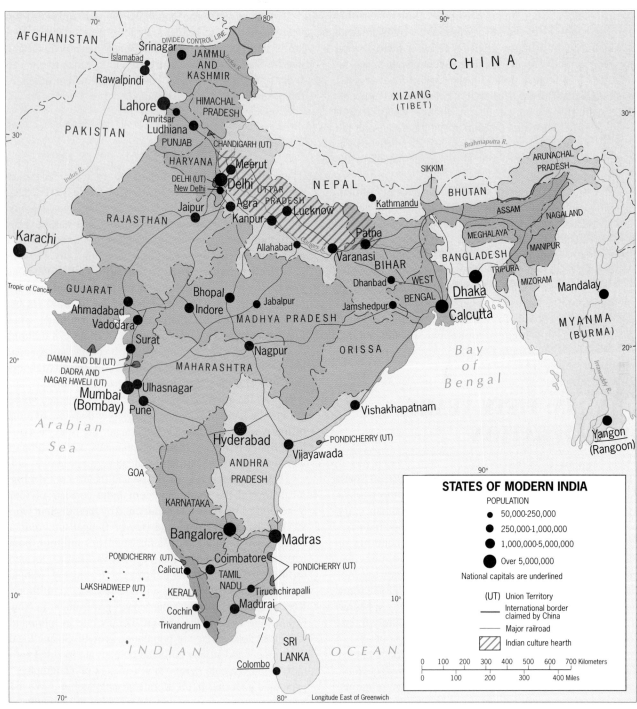

FIGURE 8-7

million people live here, many in small, comparatively isolated clusters. Before independence and political consolidation, the colonial government called this area the "Hill States."

But the map becomes even more complex in the distant northeast, beyond the narrow corridor between Bhutan and Bangladesh. The dominant State here is Assam, famed for its tea plantations and important to India because of its oil

and gas production, amounting to more than 40 percent of the domestic total. Assam attained full Statehood in 1972, just after Bangladesh became independent from Pakistan. The great majority of its 25 million residents live in the lowland of the Brahmaputra River, upstream from Bangladesh; illegal migration from Bangladesh into Assam has at times strained relations between India and its Muslim neighbor.

SOLACE AND SICKNESS FROM THE HOLY GANGES

Stand on the banks of the Ganges River in Varanasi, Hinduism's holiest city, and you will see people bathing in the holy water, drinking it, and praying as they stand in it—while the city's sewage flows into it nearby, and the partially cremated corpses of people and animals float past. It is one of the world's most compelling—and disturbing—sights.

The Ganges (*Ganga*, as the Indians call it) is Hinduism's holy river. Its ceaseless flow and spiritual healing power are earthly manifestations of the almighty. Therefore, tradition has it, the river's water is immaculate, and no amount of human (or other) waste can really pollute it. On the contrary: just touching the water can wash away a believer's sins.

At Varanasi, Allahabad, and other cities and towns along the Ganges, the river banks are lined with Hindu temples, decaying ornate palaces, and dozens of wide stone staircases called *ghats*. These stepped platforms lead down to the water, enabling thousands of bathers to enter the river. They come from the city and from afar, many of them pilgrims in need of the healing and spiritual powers of the water. It is estimated that more than a million people enter the river somewhere along its 1,600-mile (2,800-km) course every day. During religious festivals, the number may be 10 times as large.

By any standards, the Ganges is one of the world's most severely polluted streams, and thousands among those who enter it become ill with diarrhea or other diseases; many die. In 1986, then-Prime Minister Rajiv Gandhi launched a major scheme to reduce the level of pollution in the river, a decade-long construction program of sewage treatment plants and other facilities. In the mid-1980s, Gandhi was told, nearly 400 million gallons of sewage and other wastes were being disgorged into the Ganges every day. The plan called for the construction of nearly 40 sewage treatment plants in riverfront cities and towns.

Many Hindus, however, did not support this costly program to clean up the Ganges. To them, the holy river's spiritual purity is all that matters. Getting physically ill is merely incidental to the spiritual healing power a drop of the *Ganga*'s water contains.

The stone steps beside the Ganges in Varanasi, India. This is Hinduism's holiest city, and millions descend these *ghats* every year.

In the Brahmaputra valley, Assam resembles the India of the Ganges. But in almost all directions from Assam, things change. To the north, in sparsely populated Arunachal Pradesh (1 million), we are in the Himalayan offshoots again. To the east, in Nagaland 1.6 million), Manipur (2.1 million), and Mizoram (845,000), lie the forested and terraced hillslopes that separate India from Myanmar (Burma). This is an area of numerous ethnic groups (more than a dozen in Nagaland alone) and of frequent rebellion against Delhi's government. And to the south, the States of Meghalaya (2.1 million) and Tripura (3.3 million), hilly and still wooded, border the teeming floodplains of Bangladesh. Here in the country's northeast, where peoples are always restive and where population growth still soars, India faces one of its strongest regional challenges.

India's Changing Map

The present map of India's States and UTs (Fig. 8-7) is not the one with which India was born as a sovereign state in 1947, and it is sure to change again in the future. After independence, the government first had to contend with several hundred "princely states," fiefdoms whose rights had been protected during the colonial period. These were absorbed into the States, and the privileged "princely orders" were phased out by 1972.

Next, the Indian government reorganized the country on the basis of its major regional languages (see Fig. 8-4). Hindi, spoken by more than one-third of the population, was designated the country's official language, but 13 other major languages also were given national status by the Indian constitution, including the four Dravidian languages of the south. English, it was anticipated, would become India's common language, its *lingua franca* at government, administrative, and business levels. Indeed, English not only remained the language of national administration but also became the chief medium of commerce in growing urban India. English was the key to better jobs, financial success, and personal advancement, and the language constituted a common ground in higher education.

The newly devised framework based on the major regional languages, however, proved to be unsatisfactory to many communities in India. In the first place, many more languages are in use than the 14 that had been officially recognized. Demands for the establishment of additional States soon arose. In 1960, the State of Bombay was divided into two language-based States, Gujarat and Maharashtra. Other changes were made to accommodate linguistic pressures.

A second problem involved the smaller ethnic groups of the northeast. The Naga, a group of peoples whose domain had been incorporated into Assam State, rebelled soon after India's independence. A protracted war brought federal troops into the area; following a truce and a decade of negotiations, Nagaland was proclaimed a State in 1961. This led the way for other politico-geographical changes in India's problematic northeastern wing.

Still another dilemma involves India's Sikh population. The Sikhs (the word means "disciples") adhere to a religion that was created about five centuries ago to unite warring Hindus and Muslims into a single faith. This faith's principles rejected negative aspects of Hinduism *and* Islam, and it gained millions of followers in the Punjab and

adjacent areas. During the colonial period, many Sikhs supported British administration of India, and by doing so they won the respect and trust of the British, who employed tens of thousands of Sikhs as soldiers and policemen. By 1947, there was a large Sikh middle class in the Punjab. When independence came, many left their rural homes and moved to the cities to enter urban professions. Today, they still exert a strong influence over Indian affairs, far in excess of the 2 percent of the population they constitute.

After independence, the Sikhs demanded that the original Indian State of Panjab (Punjab) be divided into a Sikh-dominated west and a Hindu-majority east. The government agreed, so that Punjab as now constituted (Fig. 8-7) is India's Sikh stronghold, whereas neighboring Haryana State is mainly Hindu. Unfortunately, this redelimitation did not defuse all the pressures in the area. A militant Sikh minority demanded that an even more autonomous Sikh State be created in the Punjab, to be called *Khalistan*. Gradually this minority gained strength, and when the radical Sikh leaders held the State in a grip of violence and intimidation, the federal government sent troops to contain them. This led to a disastrous confrontation in 1984 at the Golden Temple in the capital, Amritsar, Sikhism's holiest shrine. In its aftermath, two of then-Prime Minister Indira Gandhi's Sikh bodyguards assassinated her, and the crisis deepened. To this day, relationships between Sikhs and non-Sikhs are tense; cycles of violence persist, affecting life in the otherwise comparatively prosperous Punjab.

These ethnic, cultural, and regional problems are but a sample of the stresses on India's federal framework. There is no Muslim State in India, but India has about 110 million Muslims within its borders—the largest cultural minority in the world. As Figure 8-5 shows, the percentage of Muslims is highest in remote Jammu and Kashmir, but it also is substantial in such widely dispersed States as Kerala, Assam, and Uttar Pradesh. It should also be noted that the Muslim population in the 1990s (approximately 11.7 percent) constitutes a larger percentage than it did after partition (9.9 percent). Moreover, as a sector of India's population, the Muslim minority today is among the most rapidly growing. It also is strongly urbanized. Nearly half the population of India's largest city, Mumbai (Bombay), is Muslim.

Centrifugal Forces: From India to Hindustan?

In Chapter 1 we introduced the concept of **centrifugal** and **centripetal forces** (see box p. 56), respectively, the dividing and unifying forces that continuously affect all states. No country in the world exhibits greater cultural diversity than India, and variety in India comes on a scale unmatched anywhere else on Earth. Such diversity spells strong centrifugal forces, although, as we will see, India also has powerful consolidating bonds.

Among the centrifugal forces, Hinduism's stratification of society into castes remains a pervasive reality. Under Hindu dogma, castes are fixed layers in society whose ranks are based on ancestries, family ties, and occupations. The **caste system** may have its origins in the early social divisions into priests and warriors, merchants and farmers, craftspeople and servants; it may also have a racial basis, for the Sanskrit term for caste is color. Over the centuries, its complexity grew until India came to possess several thousand castes, some with a few hundred members, others containing millions. Thus, in city as well as village, communities were segregated according to caste, ranging from the highest (priests, princes) to the lowest (the untouchables).

A person was born into a caste based on his or her actions in a previous existence. Hence, it would not be appropriate to counter such ordained caste assignments by permitting movement (or even contact) from a lower caste to a higher one. Persons of a particular caste could perform only certain jobs, wear only certain clothes, worship only in prescribed ways at particular places. They or their children could not eat with, play with, or even walk with people of a higher social status. The untouchables occupying the lowest tier were the most debased, wretched members of this rigidly structured social system. Although the British ended the worst excesses of the caste system, and postcolonial Indian leaders—including Mohandas (Mahatma) Gandhi (the great spiritual leader who sparked the independence movement) and Jawaharlal Nehru (the first prime minister)—have worked to modify it, centuries of class consciousness are not wiped out in a few decades. In traditional India, caste provided stability and continuity; in modernizing India, it constitutes an often painful and difficult legacy.

Today, it is possible to discern a geography of caste—a degree of spatial variation in its severity. Cultural geographers estimate that about 15 percent of all Indians are of lower caste, about 40 percent of backward caste (one important rank above the lower caste), and some 18 percent of upper caste (at the top of which are the *Brahmans*, men in the priesthood). (The caste system does not extend to the Muslims, Sikhs, and other non-Hindus in India, which is why the percentages above do not total 100.) Not only the colonial government but also successive Indian governments have tried to help the lowest castes. This has had more effect in the urban areas than in the rural parts of India. In the isolated villages of the countryside, the untouchables often are made to sit on the floor of their classroom (if they go to school at all); they are not allowed to draw water from the village well because they might pollute it; and they must take off their shoes, if they wear any, when they pass higher-caste houses. But in the cities, untouchables have reserved places in the schools, a fixed percentage of government jobs at State as well as federal levels, and a quota of seats in national and State legislatures.

Much of this was the result of efforts by Mohandas Gandhi, who took a special interest in the fate of the untouchables (*harijans*) in Indian society.

The caste system remains a powerful centrifugal force, not only because it fragments society but also because efforts to weaken it often result in further division (see box titled "Jharkand: A New State in the Making?"). Gandhi himself was killed, only a few months after independence, by a Hindu fanatic who opposed his work for the least fortunate in Indian society. Today, India is being swept by a wave of Hindu fundamentalism that is caused, at least in part, by continuing efforts to help the poorest. Higher castes see themselves as disadvantaged, and they take refuge in a "return" to fundamental Hindu values.

The radicalization of Hinduism, and the infusion of Hindu nationalism into India's politics, loom today as twin threats to the country's unity. Hindu nationalist political parties are polarizing the electorate at the State as well as federal level; in Maharashtra, radical Hindu political leaders managed to rename Bombay, the capital, giving this great international city an old Hindu toponym, Mumbai. If Indian politics fragment along religious lines, the miracle of Indian unity may come to an end.

Centripetal Forces

In the face of all these divisive forces, what bonds have kept India unified for so long? Without question, the dominant binding force in India is the cultural strength of Hinduism, its sacred writings, holy rivers, and general influence over Indian life. For the great majority, Hinduism is a way of life as much as it is a faith, and its diffusion over virtually the entire country (Muslim, Sikh, and Christian minorities notwithstanding) brings with it a national coherence that constitutes a powerful antidote to regional divisiveness. Over the long term, the key ingredients of this Hinduism, however, have been its gentility and introspection, radical outbursts notwithstanding. Now the specter of Hindu fanaticism threatens this vital bond.

Another centripetal force lies in India's democratic institutions. In a country as culturally diverse and as populous as India, reliance on democratic institutions has been a birthright ever since independence, and democracy's survival—raucous, sometimes corrupt, always free—has been a crucial unifier.

Furthermore, communications in much of India are better than they are in many other developing countries, and the continuous circulation of people, ideas, and goods helps bind the disparate state together. Before independence, opposition to British rule was a shared philosophy, a strong centripetal force. After independence, the preservation of the union was a common objective, and national planning made this possible.

India's capacity for accommodating major changes and its flexibility in the face of regional and local demands have

JHARKAND: A NEW STATE IN THE MAKING?

Where the Indian States of Bihar, West Bengal, Orissa, and Madhya Pradesh meet, in the zone west of Calcutta, lies one of India's poorest areas. This is tribal India, where meager subsistence farming supports millions of families, where schools are few, hospitals are virtually unknown, and serfdom is a lingering condition. The people here are at the bottom rung of the caste ladder, the poorest of the poor. They have had no say in State or federal affairs; industrialization has passed them by; planners have ignored them; land has been stolen from them.

Until now. In the 1980s, people in this four-State area mapped below (Fig. 8-8) began to organize themselves politically, and in Bihar their party won several seats in the legislature. Soon there were demands for a separate State whose tribal interests would be paramount and whose representatives would press for change at the federal level. This new State would be called *Jharkand*. In the 1990s, this intensifying demand was punctuated by road and railroad blockades, strikes, and bombings.

But neither the federal government nor the government of the most-affected State, Bihar, has been prepared to negotiate with the separatists. As Figure 8-13 shows, southern Bihar is part of India's major eastern industrial region. Although the tribespeople here have been little affected by industrialization, the revenues generated by this manufacturing activity are critical to Bihar's budget. Losing its south to a Jharkand State would not only put these industries under a rival State government but would also cause a crisis in (remaining) Bihar's finances. Once again, the lines between the establishment and the ethnic upstarts are drawn—and still another part of India's map hangs in the balance.

FIGURE 8-8

also served as a centripetal force. Boundaries have been shifted; internal political entities have been created, relocated, or otherwise modified; and secessionist demands have been handled with a mixture of federal power and cooperative negotiation. Indians in South Asia have accomplished what Europeans in Yugoslavia could not, and India's history of success is itself a centripetal force.

Finally, no discussion of India's binding forces would

be complete without mention of the country's strong leadership. Gandhi, Nehru, and their successors did much to unify India by the strength of their compelling personalities. For many years, leadership was a family affair: Nehru's daughter, Indira Gandhi, twice took decisive control (in 1966 and 1980) following episodes of governmental weakness, and *her* son, Rajiv Gandhi (who later was also assassinated), served as prime minister during the late 1980s. That era now appears to be over, and the crucial question of India's leadership again hangs in the balance.

The Population Dilemma

The human population of the realm's seven countries today totals nearly 1.3 billion—more than one-fifth of all humankind. India alone has 967 million inhabitants, second only to China among the countries of the world, and on course to overtake it. Such rapid population growth poses a threat to national development. India's federal government and the governments of its States have enacted legislation and implemented programs to reduce the rate of population growth. Although these initiatives have had some effect, India's population in the late 1990s was still growing at an annual rate of 1.9 percent (the realm as a whole was expanding by 2.1 percent).

Earlier, we noted the implications for India of population growth so rapid that the country's doubling time is only 36 years. Economic gains are overtaken by growing numbers; the risk of famine arises intermittently. In Appendix A, note that except for the tiny Maldives, the country with the slowest population growth also is the country with the highest per capita GNP—Sri Lanka.

India's Demographic Challenge

India's struggle to contain its population explosion will set the course for the realm as a whole. The question is whether India will go through its own version of the *demographic transition* mentioned in the box on population geography (p. 372). The model demographic transition carries a population from a first stage of high births and high death rates through two intermediate stages to a condition of low birth and low death rates and virtual stability (Fig. 8-9).

When the British ruled India during the nineteenth century, the country still was in the first stage, with high birth rates and high death rates; the high death rates were caused not only by a high incidence of infant and child mortality but also by famines and epidemics. As Figure 8-9 indicates, the population during stage 1 does not grow or decline much, but it is anything but stable.

But then India entered the second stage. Birth rates remained high, but death rates declined because medical services improved (soap came into widespread use), food

From the Field Notes
"Walking toward the memorial to Gandhi, we came upon a symbolic scene: a university class was being instructed in history beneath a statue to India's leader in the struggle for independence and first Prime Minister—Nehru. In a country as diverse as India, memories of great leaders are part of the glue of nationhood. The instructor told his students that Gandhi's ideals and Nehru's abilities were sadly lacking in present-day India."

distribution networks were made more effective, farm production expanded, and urbanization developed. In the 1920s, India's population still was growing at a rate of only 1.04 percent, but by the 1970s, that rate had shot up to 2.22 percent per year (Fig. 8-10). Note that India gained 28 million people during the decade of the 1920s, but a staggering 135 million during the 1970s.

Has India entered the third stage, when the death rate begins to level off and birth rates decline substantially, narrowing the gap and slowing the annual increase? The rate of increase suggests it: from 2.22 percent during the 1970s, it dropped to 2.11 percent in the 1980s and a projected 1.88 percent during the 1990s (Fig. 8-10). But India has another problem. During its population explosion, its numbers grew so large that even a declining rate of natural increase continues to add ever greater gains to its total. In Figure 8-10, note that while the decadal rate of increase dropped from 2.22 to 2.11 between 1970 and 1990, the millions added grew from 135 in the 1970s to 161 in the 1980s. At a still lower rate of 1.88 percent, the 1990s promise to add still more: 170 million (123 million have already been added between 1991 and 1997). If India has indeed entered the third stage of the demographic transition, its effects will not be felt for some time to come.

Some population geographers theorize that all national

THE MODEL DEMOGRAPHIC TRANSITION

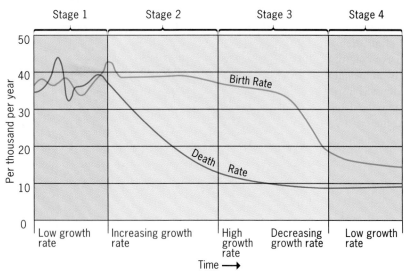

FIGURE 8-9

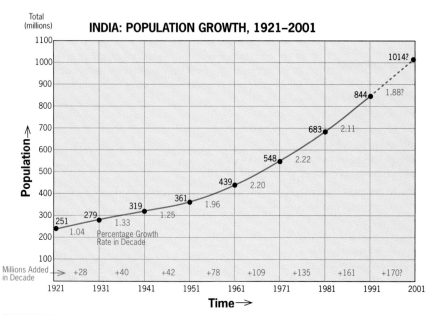

FIGURE 8-10

populations will eventually stabilize at some level, just as Europe's did, so that the fourth stage of the demographic transition will affect India and China as it has Western Europe and Japan. But even if such stabilization were just one doubling time away, India still would have an astronomical total of nearly 2 billion citizens.

Geography of Demography

Statistics for a country as large as India tend to lose their usefulness unless they are put in geographic context. In its demographic as in so many of its other aspects, there is not just one India but several regionally different and distinct Indias. Figure 8-11 takes population growth down to the State level and provides a comparison between the census periods 1971–1981 and 1981–1991.

During the period from 1971 to 1981, the highest growth rates were recorded in the States of the northeast and the northwest, and only three States had growth rates below 2 percent. But between 1981 and 1991, no fewer than eight States, including Goa, had growth rates below 2 percent. Comparing these two maps tells us that little has

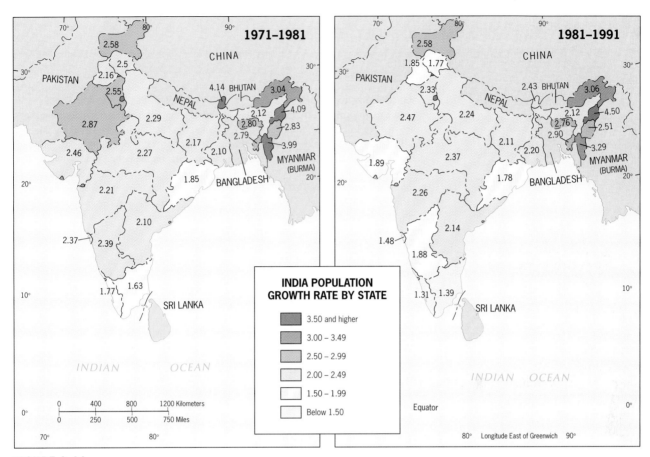

FIGURE 8-11

changed in India's heartland, where the populous States of Uttar Pradesh and Bihar show slight decreases but West Bengal and Madhya Pradesh display equally slight increases. The most important reductions in rates of natural increase are recorded in the northwest, west, and south. In the tip of the peninsula, Kerala and neighboring Tamil Nadu have growth rates even lower than that of the region's leading country, Sri Lanka.

Figure 8-11 suggests a broad contrast between eastern India and western India. We will soon see that this is not just a matter of demography.

Urbanization

When people move to the cities, they tend to have smaller families. Incentives that cause rural couples to have many children do not prevail in the cities, where living space is highly confined. Urbanization, therefore, contributes to reduced rates of natural increase in populations.

India is not yet a highly urbanized society, but we should not lose sight of the dimensions. Only 26 percent of India's population was urbanized in 1997, but that 26 percent represents more than 250 million people—nearly as many as the entire population of the United States.

And India's rate of urbanization is on the upswing. People by the hundreds of thousands are arriving in the already-teeming cities, swelling urban India by about 5 percent annually, more than twice as fast as the overall population. Not only do the cities attract as they do everywhere; many villagers are driven off the land by the desperate conditions in the countryside. As villagers manage to establish themselves in Calcutta or Madras or Mumbai, they help their relatives and friends to join them in squatter settlements that often are populated by newcomers from the same area, bringing their language and customs with them and cushioning the stress of the move.

As a result, India's cities are places of staggering social contrasts. Squatter shacks without any amenities at all crowd against the very walls of modern high-rise apartments and condominiums. Hundreds of thousands of homeless roam the streets and sleep in parks, under bridges, on sidewalks. As crowding intensifies, social stresses multiply. Disorder never seems far from the surface; sporadic rioting, often attributable to the actions of rootless urban youths unable to find employment, has become commonplace in India's cities.

India's modern urbanization has its roots in the colonial period, when the British selected Calcutta, Bombay, and

■ AMONG THE REALM'S GREAT CITIES . . .

Mumbai (Bombay)

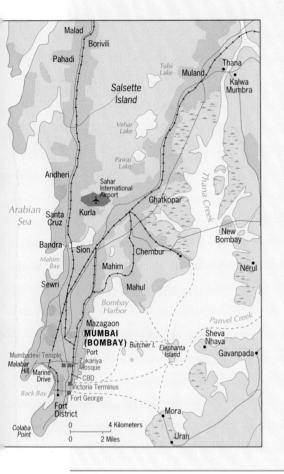

Another historic name is disappearing from the map: Bombay. In precolonial times, fishing folk living on the seven small islands at the entrance to this harbor named the place after their local Hindu goddess, *Mumbai*. The Portuguese, first to colonize it, called it *Bom Bahia*, ''Beautiful Bay.'' The British, who came next, corrupted both to Bombay, and so it remained for more than three centuries. Now local politics has taken its turn. Leaders of a Hindu nationalist party that control the government of the State of which Bombay is the capital, Maharashtra, passed legislation to change its name back to Mumbai. In January 1996, India's federal government approved this change.

Mumbai's 16.6 million people make this India's largest city. Maharashtra is India's economic powerhouse, the State that leads the country in virtually every respect. Locals dream of an Indian Ocean Rim of which Mumbai will be the anchor.

As such, Mumbai is a microcosm of India, a burgeoning, crowded, chaotic, fast-moving agglomertion of humanity. The Gothic-Victorian architecture of the city center is a legacy of the British colonial period. Shrines, mosques, temples, and churches evince the pervasive power of religion in this multicultural society. Street signs come in a bewildering variety of scripts and alphabets. Creaking double-decker buses compete with oxwagons and handcarts on the congested roadways. The throngs on the sidewalks—businesspeople, holy men, sari-clad women, beggars, clerks, homeless wanderers—spill over into the streets.

Mumbai is an urban agglomeration of city-sized neighborhoods, each with its own cultural landscape. The seven islands have been connected by bridges and causeways, and the resulting Fort District still is the center of the city, with many of its monuments and architectural landmarks. Marine Drive leads to the wealthy Malabar Hill area, across Back Bay. Northward lie the large Muslim districts, the Sikh neighborhoods, and other ethnic and cultural enclaves. Beyond are some of the world's largest squatter settlements, where desperate poverty prevails.

It is situation, not site, that gave Bombay primacy in South Asia. The opening of the Suez Canal in 1869 made Bombay the nearest Indian port to Europe. Today, Mumbai lies at the focus of India's fastest-growing economic zone—not yet a tiger, but on the move.

8.1 Madras as regional trading centers and as fortified ports. Madras was fortified as early as 1640; Bombay (1664) had the situational advantage of being the closest of all Indian ports to Britain; and Calcutta (1690) lay on the margin of India's largest population cluster and had the most productive hinterland, to which it was connected by the Ganges delta's countless channels. This natural transport network made the city an ideal colonial headquarters, but the population of Bengal was often rebellious. In 1912, the British moved their colonial government from Calcutta to the safer interior city of New Delhi, built adjacent to the old Mogul headquarters of Delhi.

Figure 8-7 displays the distribution of major urban centers in India. Except for Delhi–New Delhi, the largest cities have coastal locations, Calcutta dominating the east, Madras the south, and Mumbai the west. But considerable urbanization also has taken place in the interior, notably in the core area. The surface interconnections among India's cities still are inadequate (notably the road network), but an Indian urban system is emerging.

Economic Geography

If India has faced problems in its great effort to achieve political stability and national cohesion, these are more than matched by the difficulties that lie in the way of economic growth and development. The large-scale factories and power-driven machinery of the colonial powers wiped out a good part of India's indigenous industrial base. Indian trade routes were taken over. European innovations in health and medicine sent the rate of population growth soaring, without introducing solutions for the many problems this spawned. Surface communications were improved and food distribution systems became more efficient, but local and regional food shortages occurred (and

■ AMONG THE REALM'S GREAT CITIES . . .

Calcutta

The name *Calcutta* is synonymous with all that can go wrong in large cities: poverty, dislocation, disease, pollution, crime, corruption. To call a city the Calcutta of its region is to summarize urban catastrophe.

But what of the real Calcutta? The city got its dreadful reputation during colonial times, when plague, malaria, and other diseases claimed countless thousands in legendary epidemics. The British chose the site, 80 miles (130 km) up the Hooghly River from the Bay of Bengal and less than 30 feet (9 m) above sea level, not far from some unhealthy marshes but well placed for commerce and defense. When the British East Asia Company was granted freedom of trade in the populous hinterland, Calcutta's heyday began; when (in 1772) the British made it their colonial capital, the city prospered. The British sector was drained and raised, and so much wealth accumulated here that Calcutta became known as the "city of palaces." Outside the British town, rich

Indian merchants built magnificent mansions. Beyond lay neighborhoods often based on occupational caste, whose names are still on the map today (such as Kumartuli, the potters' district). Almost everywhere, on both banks of the Hooghly, lay the huts and hovels of the poorest of the poor. Searing social contrasts characterized Calcutta.

The twentieth century has not been kind to Calcutta. In 1912, the British moved their colonial capital to New Delhi. The 1947 partition that created Pakistan also created then-East Pakistan (now Bangladesh), cutting off a large part of Calcutta's hinterland and burdening the city with a flood of refugees. The Indian part of the city had arisen virtually without any urban planning, and the influx created almost unimaginable conditions. Today, Calcutta counts more than 12 million residents, including as many as 500,000 homeless. Beyond the façade of the downtown lies what may be the sickest city of all.

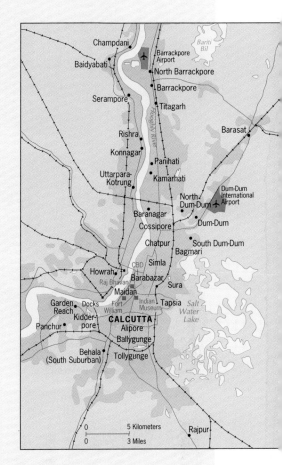

Delhi New and Old

Fly directly over the Delhi-New Delhi conurbation into the regional airport, and you may not see the place at all. A combination of smog and dust creates an atmospheric soup that can limit visibility to a few hundred feet for weeks on end. Relief comes when the rains arrive, but Delhi's climate is mostly dry. The tail-end of the wet monsoon reaches here during June or July, but altogether the city gets only about 25 inches (60 cm) of rain a year.

When the British colonial government decided to leave Calcutta and build a new capital city adjacent to Delhi, conditions were quite different. South of the old city lay a hill about 50 feet (15 m) above the surrounding countryside, on the right bank of the southward-flowing Yamuna tributary of the Ganges. Compared to Calcutta's hot, swampy environment, Delhi's was agreeable. In 1912, it was not yet a mega-city. Skies were mostly clear. Raisina Hill became the site of a New Delhi.

This was not the first time rulers chose

Delhi as the seat of empire. Ruins of numerous palaces mark the passing of powerful kingdoms. But none brought to the Delhi area the transformation the British did. In 1947, the Indian government decided to keep its government here. In 1970, the urban population exceeded 4 million. By 1997, it was 10.8 million.

Delhi's popularity as a seat of government has the same cause as its present expansion: the city has a fortuitous relative location. The regional topography creates a narrow corridor through which all land routes from northwest India to the North Indian Plain must pass, and Delhi lies in this gateway. Thus the twin cities not only contain the government functions; they also anchor the core area of this populous country.

Old Delhi once was a small, traditional, homogeneous town. Today Old and New Delhi form a multicultural, multifunctional urban giant.

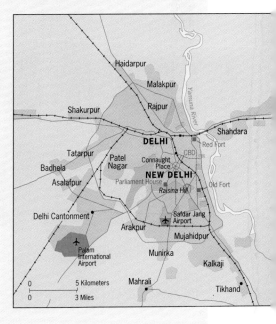

From the Field Notes

"Searing social contrasts abound in India's overcrowded cities. Even in Mumbai (Bombay), India's most prosperous large city, hundreds of thousands of people live like this, in the shadow of modern apartment buildings. Within seconds we were surrounded by a crowd of people asking for help of any kind, their ages ranging from the very young to the very old. Somehow this scene was more troubling here in well-off Mumbai than in Calcutta or Madras (the latter's modest dirt streets are shown at the left), but it typified India's urban problems everywhere."

still do) as droughts frequently caused crop failures. Today, nearly half of India's 950-million-plus people live in abject poverty, and the prospects of reducing that high level of human misery anytime soon are not encouraging. (Yet even with its modest annual per capita GNP [U.S. $290], the sheer *size* of India's population creates a very big overall economy—the world's sixth largest according to the latest rankings.)

Agriculture

India's underdevelopment is nowhere more apparent than in its agriculture. Traditional farming methods continue, and yields per acre and per worker remain low for virtually every crop grown under this low-technology system. Moreover, movement of agricultural commodities is hampered by the transportation inefficiencies of the traditional farming system. In 1995, only 41 percent of India's 600,000 villages were accessible by motorable road, and today animal-drawn carts still outnumber motor vehicles nationwide.

As the total population grows, the amount of cultivated land per person declines. Today, this *physiologic density* is 1,300 per square mile (500 per sq km). However, this is nowhere near as high as the physiologic density in neighboring Bangladesh, where the figure is more than twice as great (3,100 and 1,200, respectively). But India's farming is so inefficient that this comparison is deceptive. More than two-thirds of India's huge working population depends directly on the land for its livelihood, but the great majority of Indian farmers are poor and unable to improve their soils, equipment, or yields. Those areas in which India has made substantial progress toward the modernization of its agriculture (as in Punjab's wheat zone) remain islands in a sea of agrarian stagnation.

This stagnation has persisted in large measure because India, after independence, failed to implement a much-needed nationwide land reform program. In the early 1990s, about one-quarter of India's entire cultivated area was still owned by less than 5 percent of the country's farming families, and little land redistribution was taking place. Perhaps half of all rural families own either as little as an acre or no land at all. Independent India inherited inequities from the British colonial period, but the individual States of the federation would have had to cooperate in any national land reform program. As always, the large landowners retained considerable political influence, so the program never got off the ground.

To make matters worse, much of India's farmland is badly fragmented as a result of local rules of inheritance, thereby inhibiting cooperative farming, mechanization, shared irrigation, and other opportunities for progress. Not surprisingly, land consolidation efforts have had only limited success except in the States of Punjab, Haryana, and

Agriculture in India, overall, is among the least efficient in all of Asia. Poor organization, inadequate equipment, lack of capital, and nature's constraints combine to keep yields comparatively low.

Uttar Pradesh, where modernization has gone farthest. Certainly, official agricultural development policy, at the federal as well as State level, has also contributed to India's agricultural malaise and the uneven distribution of progress. Unclear priorities, poor coordination, inadequate information dissemination, and other failures have been reflected in the country's disappointing output.

It is instructive to compare Fig. 8-12, showing the distribution of crop regions and water supply systems in India, with Figure I-6, which shows mean annual precipitation in India and the world. In the comparatively dry northwest, notably in the Punjab and neighboring areas of the upper Ganges, wheat is the leading cereal crop. Here, India has made major gains in annual production through the introduction of high-yielding grain varieties developed under the banner of the Green Revolution, the international research program that has had a key role in overcoming the food crises of the 1960s. These "miracle crops" led to the expansion of cultivated areas, the construction of new irrigation systems, and the more intensive use of fertilizer (a mixed blessing, for fertilizers tend to be expensive and the "miracle" crops are more heavily dependent on them).

Toward the moister east, and especially in the wet-monsoon-drenched areas (Fig. 8-12), rice takes over as the dominant staple. About one-fourth of India's total farmland lies under rice cultivation, most of it in the States of Assam, West Bengal, Bihar, Orissa, and eastern Uttar Pradesh and along the Malabar-Konkan coastal strip facing the Arabian Sea. These areas receive over 40 inches (100 cm) of rainfall annually, and irrigation supplements precipitation where necessary.

The same paddyfields before and after the arrival of the monsoon rains in Goa State on the west-central Arabian Sea coast, revealing the stunning contrast in the agricultural landscape of the dry and wet seasons.

India has more land devoted to rice cultivation than any other country, but yields per acre remain among the world's lowest—despite the introduction of "miracle rice." Nevertheless, the gap between demand and supply has narrowed, and in the late 1980s India actually *exported* some grain to Africa as part of a worldwide effort to help refugees there. The situation remains precarious, however. As the population map (Fig. I-9) shows, there is a considerable degree of geographic covariation between India's rice-producing zones and its most densely populated areas. India is just one poor-harvest year away from another food crisis.

Figure 8-12 is a simplification of India's complex agricultural mosaic. It is intended to show the dominant crop in the country's agricultural regions without additional detail. Obviously, wheat is not grown throughout mountainous Kashmir, but where crops are grown, wheat is the most widespread. Again, rice dominates in the lighter-green areas, but not to the total exclusion of all other crops.

Subsistence continues to be the fate of many tens of millions of Indian villagers who cannot afford fertilizers, cannot cultivate the new and more productive strains of rice or wheat, and cannot escape the cycle of poverty. Perhaps as many as 170 million of these people do not even own a plot of land and must live as tenants, always uncertain of their fate. This is the enduring reality against which optimistic predictions of improved nutrition in India must be weighed. True, rice and wheat yields have increased at slightly more than the rate of population growth since the Green Revolution. But food security still is an elusive goal, and India continues to face the risks inherent in its burgeoning population's always-growing needs.

Industrialization

Notwithstanding the problems faced by its farmers, agriculture must be the foundation for development in India. Agriculture employs approximately two-thirds of the workers, generates most of the government's tax revenues, contributes many of its chief exports by value (cotton textiles, tea, fruits and vegetables, jute products, and leather goods all rank high), and produces most of the money the country can spend in other sectors of the economy. Add to this the compelling need to grow more and more food crops, and India's heavy investment in agriculture is understandable.

In 1947, India inherited the mere rudiments of an industrial framework. After more than a century of British control over the economy, only 2 percent of India's workers were engaged in industry, and manufacturing and mining combined produced only about 6 percent of the national income. Textile and food-processing industries dominated. Although India's first iron-making plant was opened in 1911 and the first steel mill began operating in 1921, the initial major stimulus for heavy industrialization came after the outbreak of World War II. Manufacturing was concentrated in the largest cities: Calcutta led, Mumbai (Bombay) was next, and Madras ranked third.

The geography of manufacturing today still reflects those beginnings, and industrialization in India has proceeded slowly, even after independence (Fig. 8-13). Calcutta now anchors India's eastern industrial region—the Bihar-Bengal District—where jute manufactures dominate, but cotton, engineering, and chemical industries also exist. On the nearby Chota-Nagpur Plateau to the west, coalmining and iron and steel manufacturing have developed.

On the opposite side of the subcontinent, two industrial

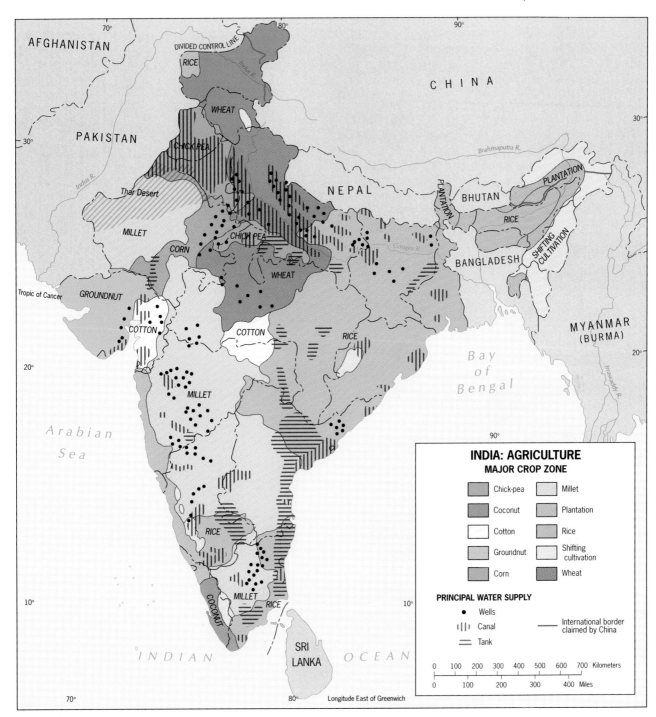

FIGURE 8-12

areas dominate the western manufacturing region: one is centered on Mumbai and the other on Ahmadabad. This region, lying in Maharashtra and Gujarat States, specializes in cotton and chemicals, with some engineering and food processing. Cotton textiles have long been an industrial mainstay in India, and this was one of the few industries to derive some benefit from the nineteenth-century economic order imposed by the British. With the local cotton harvest, the availability of cheap yarn, abundant and inexpensive labor, and the power supply from the Western Ghats' hydroelectric stations, the industry thrived and today outranks Britain itself in the volume of its exports.

Finally, the southern industrial region consists chiefly of a set of linear, city-linking corridors focused on Madras, specializing in textile production and light engineering activities. In the 1990s, all of India's manufacturing regions

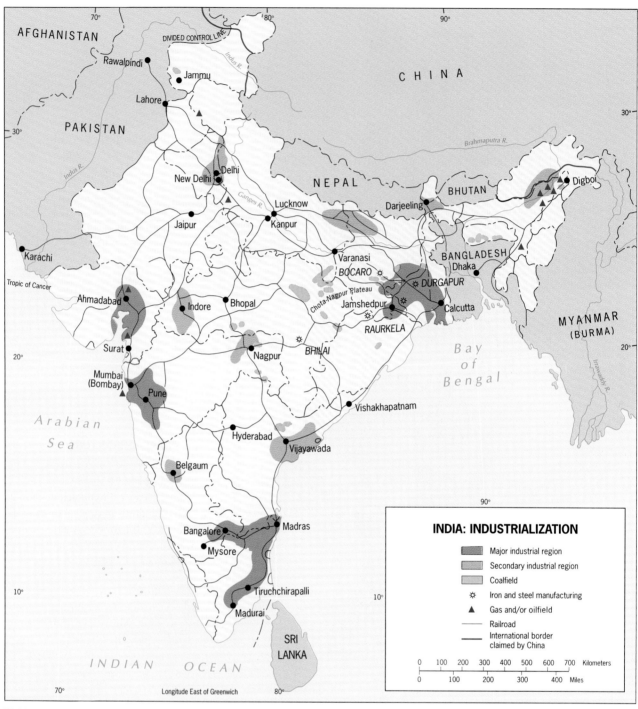

FIGURE 8-13

are increasing their output of ready-to-wear garments—another legacy of the early development of cotton textiles. Today, clothing has become India's second-leading export by value; the production of gems and jewelry, another growing specialization, ranks first.

Despite some imbalances and inefficiencies, India's industrial resource base is quite well endowed. Limited high-quality coal deposits are exploited in the Chota-Nagpur

area. In combination with large lower-grade coalfields elsewhere, the country's total output is high enough to rank it among the world's 10 leading coal producers. In the absence of known major petroleum reserves (some oil comes from Assam, Gujarat, Punjab, and offshore from Mumbai), India must spend heavily on energy every year. Major investments have been made in hydroelectric plants, especially multipurpose dams that provide electricity, enhance

From the Field Notes

"It was a scene reminiscent of Africa in some ways: here was a town of modest size and even more modest amenities, but on the coast lay a huge facility designed to unload iron ore from incoming trains and coastwise barges, and load oceangoing freighters in short order. This was Marmagao, about 25 miles (40 km) south of Goa's capital of Panaji; the large ship being loaded in the distance was from Japan. Now officially an Indian State, Goa leads India in several economic indices."

irrigation, and facilitate flood control. India's iron ores in Bihar (northwest of Calcutta) and Karnataka (in the heart of the Deccan) may rank among the largest in the world. Jamshedpur, located west of Calcutta in the eastern industrial region, has emerged as India's leading steelmaking and metals-fabrication center. Yet India still exports iron ore as a raw material to the higher-income industrialized countries, mainly Japan. For low-income, revenue-needy countries, entrenched practices are difficult to break.

India East and West

The most commonly cited, and most clearly evident, regional division of India is between north and south. The north is India's heartland, the south its Dravidian appendage; the north speaks Hindi as its *lingua franca*, the south prefers English over Hindi; the north is bustling, testy, the south seems slower, less agitated.

But there is another, as yet less obvious, but potentially more significant divide across India. In Figure 8-13, draw a line from Lucknow, on the Ganges River, south to Madurai, near the southern tip of the peninsula. To the west of this line, India is showing signs of economic progress, the kind of economic activity that has brought Pacific Rim countries such as Thailand and Indonesia a new life. To the east, India has more in common with less promising countries also facing the Bay of Bengal: Bangladesh and Myanmar (Burma).

As with other regional divides, there are exceptions to our east-west delineation. Indeed, our map seems to suggest that much of India's industrial strength lies in the east. But what the map cannot reveal is the profitability of those industries. True, the east is rich in iron and coal, but the heavy industries built by the state in the 1950s are now outdated, uncompetitive, and in decline. The hinterland of Calcutta now contains India's Rustbelt. The state keeps

many industries going, but at a high cost. Old industries, such as carpetmaking and cottonweaving, continue to use child labor in order to remain viable. The State of Bihar represents the stagnation that afflicts much of India east of our line: by several measures it is the poorest of the 25 States.

Compare this to western India. The State of Maharashtra, the hinterland of Mumbai, leads India in many categories, and Mumbai leads Maharashtra. Numerous smaller, private industries have emerged here, manufacturing goods ranging from umbrellas to satellite dishes and from toys to textiles. Across the Arabian Sea lie the oil-rich economies of the Arabian Peninsula. Hundreds of thousands of workers from western India have found jobs there, sending money back to families from Punjab to Kerala. More importantly, many have used their foreign incomes to establish service industries back home. Outward-looking western India, in contrast to the inward-looking east, has begun to establish other links with the outside world. The beaches of Goa, the small State immediately to the south of Maharashtra, lie in an advantageous situation with respect to the tourist markets of Europe. This is, in fact, a classic case of **intervening opportunity** because resorts have sprung up along Goa's coast, and European tourists who once went to the more distant Maldives and Seychelles are coming to Goa. Maharashtra's economic success also has spilled over into Gujarat to the north, and even landlocked Rajasthan (the next State to the north) is experiencing the beginnings of what, by Indian standards, is a boom.

With the boom, however, have come political problems. Not only is Maharashtra State an economic power; it also is the base of a strong Hindu nationalist political movement whose leaders object to foreign intrusions, and have stopped major development projects and other enterprises from going forward. A huge industrial scheme was halted about halfway through, and a fast-food operation was

8.2

deemed incompatible with local culture and closed. Such clashes between foreign interests and domestic traditions are not unique to India, of course, but the rising tide of Hindu fundamentalism has uncertain prospects and unsettles investors to whom India remains a high-risk calculation.

Nevertheless, India's east-west divide shows a growing contrast that puts the west far ahead. The hope is that Maharashtra's success will spread northward and southward along the Arabian Sea coast, and will ultimately diffuse eastward as well. But for this to happen, India's population spiral will have to be brought under control.

❖ BANGLADESH: PERSISTENT POVERTY

On the map of South Asia, Bangladesh looks like another State of India: the country occupies the area of the double delta of India's great Ganges and Brahmaputra rivers, and it lies almost completely surrounded by India on its landward side (Fig. 8-14). But Bangladesh is an independent country, born in 1971 following its brief war for independence against Pakistan, with a territory about the size of Wisconsin. Today it is one of the poorest and least developed countries on Earth, with a population of 125 million that is doubling in under 30 years.

Natural Hazards

In the spring of 1991, Bangladesh was struck by yet another in an endless series of natural disasters: a devastating hurricane (or cyclone, as these tropical storms are called in this part of the world) that killed perhaps as many as 150,000 people (the exact toll will never be known). The storm, on a curving northward path across the Bay of Bengal, pushed a surging wall of water nearly 20 feet high across the islands and flatlands of the delta and swept most of the southeastern port city of Chittagong off the map. The storm surge forced its way well inland along the winding channels of the Ganges-Brahmaputra delta, causing death and destruction even far from the exposed coastlands along the bay. When the waters receded, the bodies of countless people and animals were carried out to sea, later to wash up on the beaches. It was a catastrophe of unimaginable proportions—but it was not, by far, the worst calamity that Bangladesh has suffered. During the twentieth century, 8 of the 10 costliest natural disasters in the entire world have struck this single country. What is it that makes Bangladesh so vulnerable to these **natural hazards**?

Look at Figure 8-14 again. The land of Bangladesh lies just barely above sea level; the deltaic plain of the Ganges-Brahmaputra is a labyrinth of stream channels. Only in the extreme east and southeast do these flatlands yield to hills and mountains. The delta's alluvial soils are extremely fertile, and every available patch of it is under crops: rice and wheat for subsistence, jute and tea for cash. The rivers' annual floods renew the farmlands' fertility by bringing silt; at the seaward margins of the delta, the silt piles up to form new islands. Even as this new land builds up, people move in to farm it. The crush of ever more mouths to feed compels this migration.

Now consider the inverted-funnel shape of the Bay of Bengal (Fig. 8-1). Cyclones form often in this warm-water, humid-air environment (in contrast to the Arabian Sea, with its drier air and desert coasts, on the western side of India). When cyclones form in the Bay of Bengal, they often move northward along a rightward curve. As they do so, the water that piles up ahead of the storm has no place to go: the bay becomes ever narrower. And so, time and again, storm surges rise across the delta, sweeping people, livestock, and crops from the land. After the storm abates, the returning outrush of water causes added devastation as the delta's normally placid channels become troughs of raging torrents.

Unlike the comparatively wealthy Dutch, the Bangladeshis cannot combat their environmental enemy. Flood and storm warning systems are insufficient; escape plans and routes are inadequate. A program was recently begun to construct concrete, storm-proof shelters on pillars, to which trapped villagers might flee. Few were saved by the available shelters when the 1991 cyclone struck. Even this is too costly for Bangladesh to bear.

Stability and Subsistence

Bangladesh's economic condition is reflected by its GNP per capita (a mere U.S. $220) and by its level of urbanization (only 17 percent in the late 1990s). This is a land of subsistence farmers, with one of the highest physiologic densities in the world: 3,100 per square mile (1,200 per sq km). But there *are* bright spots: higher-yielding varieties of rice have helped close the gap between supply and demand, and wheat now plays a larger role on the farmlands, a rotation that improves food security. But diets remain unbalanced; overall, nutrition is barely adequate.

The other bright spot is political: despite occasional election-time skirmishes and notwithstanding some latent border disputes with India, the country has been relatively stable over the past decade. This stability has been crucial to its survival, because countries with fragile subsistence economies suffer disproportionately when political struggles cause dislocation. (Southern Africa's Moçambique, which is even poorer than Bangladesh, is a case in point.)

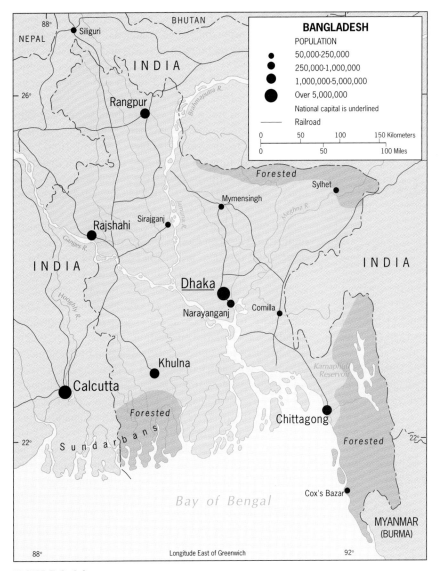

FIGURE 8-14

Nonetheless, the situation in Bangladesh *is* fragile. The country's infrastructure was badly damaged during the 1971 war for independence, and the communications system has never been repaired. Moreover, corruption in government and a recurrent insurgency problem in the forested, mountainous southeast pose additional challenges for the years to come.

Dhaka, the centrally situated capital, and the devastated port of Chittagong are the only urban centers of consequence in this dominantly rural country. Another measure of Bangladesh's economic misfortune lies in its transport system: there still is not a single road bridge across the Ganges River anywhere in the country, and there is only one such railway crossing. The Brahmaputra River has not been bridged at all. When you travel from Dhaka to any town some distance away, you must be prepared for crawling road traffic and time-consuming ferry transfers. Much

of the country can be reached only by boat, thousands of which ply Bangladesh's many waterways.

The Indian State of West Bengal adjoins Bangladesh to the west, and this part of India resembles Bangladesh (*Bangla* means Bengal) in physical as well as human ways. West Bengal, too, occupies part of the delta of the Ganges River. And like most Bangladeshis, the people here are Bengali. It was religion that put the boundary on the map in 1947: more than 80 percent of the people of Bangladesh adhere to Islam. But here in India's east, the cultural divide is much less sharp than it is vis-à-vis Pakistan in the west. More than 15 percent of the Bangladeshis are Hindu, while West Bengal contains a large Muslim minority.

Even more so than in much of India, population growth is Bangladesh's prime challenge. But let us not lose sight of the global context. A child born in Bangladesh will consume, during an equivalent lifetime, only 3 percent of the total consumption of a child born in the United States (by "consumption" we mean food, energy, minerals, and all other natural resources). Put another way, a single American child will consume what about 33 Bangladeshi children do. True, Bangladesh faces a population dilemma. But populous Bangladesh strains the world far less than a high-income country of the same dimensions. In Bangladesh, survival is the leading industry; all else is luxury.

❖ THE SOUTHERN ISLANDS

Sri Lanka: South Asian Tragedy

Sri Lanka (known as Ceylon prior to 1972), the compact, pear-shaped island located just 22 miles (35 km) across the Palk Strait from the southern tip of the Indian peninsula, is the fourth independent state to have emerged from the British sphere of influence in South Asia (Fig. 8-15). Sovereign since 1948, Sri Lanka has had to cope with political as well as economic problems, some of them quite similar to those facing India and Pakistan, and others quite different.

There were good reasons to create a separate independence for Sri Lanka. This is neither a Hindu nor a Muslim country; the majority—some 70 percent—of its 19 million people are Buddhists. Furthermore, unlike India or Pakistan, Sri Lanka is a plantation country (a legacy of the European period), with export agriculture still the mainstay of the external economy.

The majority of Sri Lanka's people are not Dravidian but are of Aryan origin with a historical link to ancient northern India. After the fifth century B.C., their ancestors began to migrate to Ceylon, a relocation that took several centuries to complete and brought to this southern island the advanced culture of the northwestern portion of the subcontinent. Part of that culture was the Buddhist religion; another component was a knowledge of irrigation techniques. Today, the descendants of these early invaders, the *Sinhalese*, speak a language (Sinhala) belonging to the Indo-European linguistic family of northern India.

The Dravidians from southern India never came in sufficient numbers to challenge the Sinhalese. They introduced the Hindu way of life, brought the Tamil language to northern Sri Lanka, and eventually came to constitute a substantial minority (now 18 percent) of the country's population. Their numbers were markedly strengthened during the second half of the nineteenth century when the British brought hundreds of thousands of Tamils from the adjacent mainland to Ceylon to work on the plantations that were being laid out. Sri Lanka once sought the repatriation of this ethnic element in its population, and an agreement to

that effect was even signed with India. After independence, successive Sri Lankan governments marginalized the Tamil minority (although Tamil was made a "national language" in 1978). The result was a costly armed rebellion in the far north that evolved into a full-scale separatist movement.

Sri Lanka is not a large island (about the size of West Virginia), but it is quite mountainous. The highest uplands, in the south, reach over 8,000 feet (2,500 m), from where steep, thickly forested slopes lead down to an encircling lowland. The north, including the Jaffna Peninsula, is entirely low-lying. The rivers from the uplands feed the paddies. Rice, not wheat, is the staple crop.

The moist southwest has long been the leading agricultural zone. The plantations introduced by the European colonizers still function, producing coconuts in the lowlands, rubber at intermediate elevations, and tea (for which Sri Lanka is famous) in the highlands. Tea still accounts for one-fourth of the country's exports by value. Rice is another story. In the 1960s, a combination of family-planning policies and malaria eradication, coupled with the repopulation of lowlands and the expansion of rice growing, produced self-sufficiency for Sri Lanka. But, as in India, Sri Lankan paddy farming is not known for its efficiency, population growth has quickened again, and today Sri Lanka must import rice to meet domestic demand.

Colombo is the focus for what little industry has developed in Sri Lanka, most of it in food processing and otherwise largely dependent on the small local market. Colombo's cityscape mirrors the fate of the country since independence; an early period of optimism and modernization was interrupted by the civil war involving the Tamil northerners. In the mid-1990s, the prospect of a settlement contributed to some renewed investment; two tall office towers were under construction in 1995, the first major change in the city's skyline for many years. But hopes for a permanent truce were dashed as the war heated up again. A sizeable Tamil minority clusters in Colombo's Pettah District, and violence has intermittently struck the city. The tourist industry, once a major revenue earner, has been devastated.

8.3

Fragmentation?

A look at the map suggests that Sri Lanka, off the coast of India, might share some situational advantages with Taiwan, off the coast of China. But Sri Lanka is no economic tiger on an Indian Ocean Rim. Certainly Sri Lanka has opportunities (and despite the war its per capita GNP still is the highest in the region except the tiny Maldives). The tragedy of Sri Lanka is that what happened could have been averted by more enlightened political leadership. The Tamils of the north and east had long given warning that they were unable to achieve equal rights in education, employment, land ownership, and linguistic and political repre-

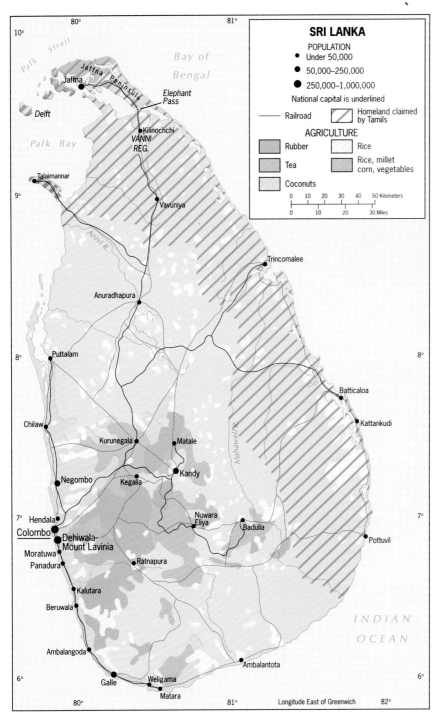

FIGURE 8-15

sentation. When the Sri Lankan government failed to accommodate Tamil demands, an insurrection broke out, and Tamil leaders called for a Cyprus-like partition of the island. Tamil forces (the Tamil Tigers) managed to secure the Jaffna Peninsula and some adjacent areas, and demanded a greater sovereign state, *Eelam*, in their domain (see the striped area in Fig. 8-15).

The sequence of events is depressingly familiar in this devolutionary world. Just as the Turks of northern Cyprus received support from nearby Turkey, so Sri Lanka's Tamils got help from India's more than 50 million Tamils. Sri Lanka asked the government of India to help curb the Tamil uprising, which embroiled India in a foreign conflict with dangerous domestic implications. In 1991, the leading Indian politician campaigning for the office of prime minister was assassinated near the Tamil city of Madras.

From the Field Notes

"The long civil war involving Tamil demands for a separate state has severely damaged Sri Lanka's economy. Its impact can be seen in the capital, Colombo, where the tourist industry collapsed and new investments dwindled. In the 1990s, with hopes for a resolution of the crisis rising, some new building finally began in Colombo: the twin towers seen rising above the townscape were built by Singaporean investors anticipating a resumption of economic progress in this embattled country. But even the capture by government forces of the Tamil stronghold of Jaffna did not lower the tension in the capital. We saw armed military checkpoints everywhere." (Note: Shortly after this photo was taken, a bomb exploded in the area shown, killing more than 80 people, injuring over 1000, and proving that Tamil extremists were unwilling to negotiate a solution to Sri Lanka's crisis).

The Insurgent State Phenomenon

When the Tamil rebels in the north managed to oust the government's forces and create a Tamil-controlled area in Sri Lanka, the process was familiar to political scientists who study such events. It involves three stages they identify as (1) *contention*, the period of initial rebellion; (2) *equilibrium*, when the rebels gain control over some territory; and (3) *counteroffensive*, the full-scale conflict that will decide the future.

These stages were translated into spatial terms by the geographer Robert McColl, who suggested that the equilibrium stage marks the emergence of an **insurgent state**, in effect a state within a state. Such an insurgent state has an informally bounded territory with a distinct core area, systems of government and administration that operate from headquarters based in this core, and social services such as schools and hospitals that substitute for those of the formal state. In Sri Lanka, the Jaffna Peninsula became

the core of an insurgent state, with Tamil Eelam headquarters based in the town of Jaffna. The stage of equilibrium prevailed for several years as Tamil rebels controlled not only the Jaffna Peninsula but also the large Vanni territory in the northern part of the island.

The counteroffensive stage was reached in 1995, when a massive government force attacked the Jaffna Peninsula and drove the Tamil Tigers out of Jaffna town. But the fall of Jaffna alone was not enough to destroy the insurgent state, and in 1996 the conflict continued sporadically even as the Colombo government searched for ways to accommodate Tamil demands within limits acceptable to the Sinhalese majority. Meanwhile, the town of Kilinochchi became the new Tamil rebel capital, and with the population of the Vanni region swollen by over 200,000 refugees from the Jaffna conflict, the future appeared anything but stable.

The insurgent-state model certainly has had its manifestations in recent decades: in Cuba, Nicaragua, Peru, Cyprus, Bosnia, and elsewhere. In some instances, the third stage did lead to partition, notably in Cyprus (although the "Turkish Republic of Northern Cyprus" [see box p. 315] has not achieved international recognition). In retrospect, one thing is clear: wherever the insurgent state arose, it damaged the state and its economy, even where it ultimately failed.

The Maldives: A Sri Lankan Progeny?

Imagine a country of more than a thousand tiny islands, whose combined area is just 115 square miles (less than 300 sq km), located more than 400 miles (650 km) from the nearest continent, where the highest elevation is barely more than 6 feet (2 m) above sea level in a region where tropical storms prevail. Add a population of about 280,000, and we have described the Republic of the Maldives, the southernmost political entity of the South Asian realm (Fig. 8-1).

As you sail toward the capital, the island of Male, the Maldives look like large lily pads on the ocean. Small boats ply the waters, linking the inhabited islands (about 200 have settlements) to each other and to the capital. Male's townscape is dominated by a large mosque, and a few buildings rise above two stories. Here, buildings have replaced the ubiquitous palm trees that almost completely cover the other islands.

The inhabitants of the Maldives speak a language that is related to an old form of Sinhalese, and it is believed that the first settlers came from Sri Lanka. If they were Buddhists, however, their successors were converted to Islam, and for centuries during the colonial period, when Portuguese and Dutch and British ruled here, the Maldives were a sultanate. Three years after the Maldives became independent, the sultanate was abolished, and the country

8.4

has been a republic since 1968. Small as this country may be, it has not escaped the fractiousness that seems to mark so much of South Asia. Resentment between the out-islanders, who see themselves as disadvantaged compared to the residents of wealthier Male, has led to occasional strife.

The Maldives' magnificent, palm-fringed beaches lure visitors from colder climes, and tourism has become the major industry, although fishing remains important as well. The people here have a special reason to worry about predictions of global warming and sea-level rise: if these predictions come true, their entire country will disappear.

❖ THE MOUNTAINOUS NORTH

South Asia, as we noted earlier, is one of the world's most clearly defined geographic realms in physical as well as cultural terms. Walls of mountains stand between India and China—mountains that defy penetration even in this age of modern highways. And those mountains are more than barriers: South Asia's great life-giving rivers rise here, their courses fed by melting snow, sustaining tens of millions in the valleys and plains far below. Control over those source areas has caused centuries of conflict, and the results are etched on the political map.

As the map shows, a tier of landlocked countries and territories lies across this mountainous northern zone. From Afghanistan in the west and through Jammu and Kashmir and Nepal to Bhutan in the east, these isolated, remote, vulnerable entities are products of a long and complicated frontier history. Their vulnerability is underscored by the recent misfortunes of one of them, Afghanistan, and the disappearance (as a separate country) of another, Sikkim. The latter, wedged between Nepal and Bhutan, was absorbed by India in 1975 and made one of its 25 States. The kingdoms of Nepal and Bhutan, however, retain their independence.

Nepal

Nepal, the size of Illinois and containing a population of 24 million, lies directly northeast of India's Hindu coreland. It is a country of three geographic zones (Fig. 8-16): a southern, subtropical, fertile lowland called the Terai; a central belt of Himalayan foothills with swiftly flowing streams and deep valleys; and the spectacular high Himalayas themselves (topped by Mount Everest) in the north. The capital, Kathmandu, lies in the east-central part of the country in an open valley of the central hill zone.

Nepal is a materially poor but culturally rich country.

The Nepalese are a people of many sources, including India, Tibet, and interior Asia; about 90 percent are Hindu, but Nepal's Hinduism is a unique blend of Hindu and Buddhist ideals. Thousands of temples and pagodas ranging from the simple to the ornate grace the cultural landscape, especially in the valley of Kathmandu, the country's core area. Although over a dozen languages are spoken, 90 percent of the people also speak Nepali, a language related to Indian Hindi.

Nepal's problems are those of underdevelopment and centrifugal political forces. Living space is limited, population pressure is high and rising steadily, and environmental degradation is a consequence. Deforestation is particularly severe—over one-third of Nepal's alpine woodlands have been cut over into wastelands since the 1960s. The growing population of subsistence farmers has been forced to expand into higher-altitude wilderness zones for sufficient crop-raising space (on steep terraces) and to obtain the firewood that supplies most of Nepal's energy needs. But soil quality is poor throughout the uplands, new farms are soon abandoned after a few seasons of declining productivity, and the land denudation process intensifies. Moreover, the steep slopes and the awesome power of the wet-monsoon rains accelerate soil erosion in treeless areas, and so much silt is now transported out of the Himalayas that, according to some researchers, the process is heightening the risk of flooding in the crowded lower Ganges and Brahmaputra basins. With about half its farmland already abandoned to erosion and with 95 percent of its population engaged in subsistence agriculture (rice, corn, wheat, and millet), Nepal today faces a serious ecological crisis.

As the data in Appendix A underscore, Nepal is a severely underdeveloped country; its per capita GNP (U.S. $160) is the lowest in the entire realm of South Asia, lower even than that of Bangladesh. The country's infrastructure is weak, and regionalism is strong. In terms of political geography, support for the old monarchy in the core area was not enough to forestall a nationwide demand for more democracy that, during the late 1980s, created a period of costly disruption. In 1991, democratic elections ushered in a new era.

But the end of absolute monarchy did not solve Nepal's economic woes. Nepal needs integration and improved communications; the southern Terai zone, with its tropical lowlands resembling neighboring India, is a world apart from the hills of the central zone. And the peoples of the west have origins and traditions quite different from those in the east. There is always a fear of domination by the giant to the south, but even relations with neighboring Bhutan have been problematic, especially because Nepal now has representative government whereas Bhutan continues to be an absolute monarchy. Landlocked, regionally fragmented, economically deteriorating, and culturally splintered, Nepal faces the future with many liabilities and few assets. Survival as a coherent state is its greatest challenge.

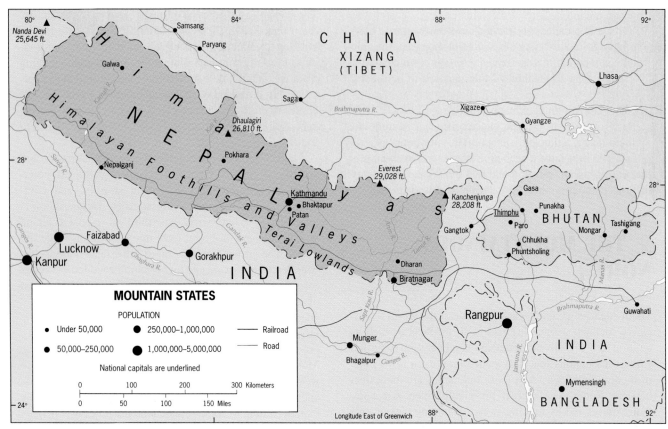

FIGURE 8-16

Bhutan

East of Nepal lies the Kingdom of Bhutan, territorially only one-third as large and with just 880,000 inhabitants. As Figure 8-16 shows, an extension of India's territory lies between Nepal and Bhutan (the State of Sikkim, referred to earlier), but more than one-quarter of Bhutan's population is of Nepalese origin. Following Nepal's overthrow of the monarchy, some Nepalese leaders began to express concern over the fate of Nepalese citizens of Bhutan, and relations between the two countries worsened. Many Nepalese people from Bhutan moved as refugees to Nepal.

Not only was Bhutan the sole remaining absolute monarchy in the realm, but here, Buddhism is the state religion—although most Nepalese in Bhutan are Hindus. Bhutan's official language is related to Tibetan, and the country is more isolated and northward-looking even than Nepal. The capital, Thimphu, has a mere 35,000 residents and is located in the west-central part of the country.

Bhutan had a special position in South Asia during the period of British colonial control and was allowed to remain a separate entity. After India became independent in 1947, a treaty between the two was drawn up in 1949 that permitted Bhutan to continue to control its internal affairs but that required its rulers to consult India on matters of foreign relations.

This would appear to have been a prescription for eventual annexation by India, but Bhutan has survived—in part because of its protective remoteness and in part because it has been self-sufficient in food. Bhutan has a subsistence economy, but it has sustained the country; the monarchy has treated it like the mountain fortress it is. Certainly there are economic possibilities. Forestry, hydroelectric power generation, and tourism have potential; moreover, the country is known to possess deposits of limestone, coal, slate, dolomite, lead, graphite, copper, talc, marble, gypsum, graphite, mica, pyrites, beryl, and tufo. But development has been minimal, and for visitors this is a country where time seems to have stood still. Nevertheless, the clock is ticking, and even Bhutan will not escape the interactions that have forged this volatile South Asian geographic realm.

Figure I-11 reminds us of a troubling reality: South Asia is the only geographic realm on Earth that consists entirely of countries in the lowest-income category. It also contributes by far the largest share to the ongoing global population spiral. It is a realm where Islam meets infidel and where contentious neighbors have nuclear weapons. The wealthier world cannot ignore the issues and problems arising in this walled-off, peninsular corner of the planet.

Afghanistan (aff-GHANN-uh-stan)
Agra (AHG-ruh)
Ahmadabad (AH-muh-duh-bahd)
Allahabad (ALLA-huh-bahd)
Amritsar (um-RIT-sahr)
Andhra Pradesh (ahn-druh pruh-DESH)
Anuradhapura (unna-rahd-uh-POOR-uh)
Arunachal Pradesh (AHRA-NAHTCH-
    ull pruh-DESH)
Aryan (AHR-yun)
Aśoka (uh-SHOH-kuh)
Assam (uh-SAHM)
Badshahi [Mosque] (bud-SHAH-hee
    [MOSK])
Baluchistan (buh-loo-chih-STAHN)
Bangalore (BANG-guh-loar)
Bangladesh[i] (bang-gluh-DESH [ee])
Bengal[i] (beng-GAHL [ee])
Beryl (BERREL)
Bharatiya Janata (buh-RUH-tee-uh
    juh-NAH-tuh)
Bhopal (boh-PAHL)
Bhutan (boo-TAHN)
Bihar (bih-HAHR)
Bom Bahia (bom bah-EE-yah)
Bombay (bom-BAY)
Brahmaputra (brahm-uh-POOH-truh)
Buddhism (BOOD-izm)
Calcutta (kal-KUTT-uh)
Caste (CAST)
Ceylon (seh-LONN)
Chittagong (CHITT-uh-gahng)
Chota Nagpur (choat-uh-NAHG-poor)
Cochin (koh-CHIN)
Colombo (kuh-LUM-boh)
Coromandel (kor-uh-MANDLE)
Deccan (DECKEN)
Delhi (DELLY)
Deltaic (dell-TAY-ick)
Dhaka (DAHK-uh)
Dravidian (druh-VIDDY-un)
Eelam (EE-lum)
Eurasia (yoo-RAY-zhuh)
Faisalabad (fye-SAHL-ah-bahd)
Gandhi (GONDY)
Ganga (GUNG-guh)
Ganges/Gangetic (GAN-jeez/
    gan-JETTICK)
Ghats (GAHTSS)
Gilgit (GILL-gutt)
Goa (GO-uh)
Godavari (guh-DAH-vuh-ree)
Gondwana (gond-WON-uh)
Gujarat (goo-juh-RAHT)
Gurkha (GHOOR-kuh)
Harappa (huh-RAP-uh)
Harijan (hah-ree-JAHN)

Haryana (hah-ree-AHNA)
Himachal Pradesh (huh-MAHTCH-ull
    pruh-DESH)
Himalayas (him-AHL-yuzz/
    himma-LAY-uzz)
Hindu Kush (hin-doo KOOSH)*
Hindustan (hin-doo-STAHN)
Hooghly (HOO-glee)
Hyderabad (HIDE-uh-ruh-bahd)
Iran (ih-RAN/ih-RAHN)
Islamabad (iss-LAHM-uh-bahd)
Jaffna (JAHF-nuh)
Jammu (JUH-mooh)
Jamshedpur (JAHM-shed-poor)
Jharkand (JAR-kahnd)
Kabul (KAH-bull)
Kanarese (KAHN-uh-reece)
Karachi (kuh-RAH-chee)
Karakoram (kahra-KOR-rum)
Karnataka (kahr-NAHT-uh-kuh)
Kashmir (KASH-meer)
Kathmandu (kat-man-DOOH)
Kerala (KEH-ruh-luh)
Khalistan (kahl-ee-STAHN)
Khyber (KYE-burr)
Kilinochchi (kih-luh-NOCK-chee)
Konkan (KAHNG-kun)
Kumartuli (koo-MAR-too-lee)
Ladakh (luh-DAHK)
Lahore (luh-HOAR)
Lingua franca (LEAN-gwuh
    FRUNK-uh)
Littoral (LIT-uh-rull)
Lucknow (LUCK-nau)
Madhya Pradesh (mahd-yuh
    pruh-DESH)
Madras (muh-DRAHSS)
Madurai (mahd-uh-RYE)
Maharajah (mah-hah-RAH-juh)
Maharashtra (mah-huh-RAH-shtra)
Malabar (MAL-uh-bahr)
Malayalam (mal-uh-YAH-lum)
Malaysia (muh-LAY-zhuh)
Maldives (MALL-deeves)
Male (MAH-lee)
Manipur (man-uh-POOR)
Marmagao (marma-GAU)
Mauryan (MAW-ree-un)
Meghalaya (may-guh-LAY-uh)
Meghna (MAIG-nuh)
Mizoram (mih-ZOR-rum)
Mohenjo Daro (moh-hen-joh-DAHRO)
Muhajir (MOO-hah-jeer)
Multan (mool-TAHN)
Mumbai (MOOM-bye)
Myanmar (mee-ahn-MAH)
Naga[land] (NAHGA-[land])

Nagpur (NAHG-poor)
Nehru, Jawaharlal (NAY-roo
    juh-WAH-hur-lahl)
Nepal (nuh-PAHL)
Orissa (aw-RISSA)
Pagoda (puh-GOH-duh)
Pakhtuns (puck-TOONZ)
Pakistan (PAH-kih-stahn)
Palk (PAWK)
Panaji (pah-NAH-jee)
Pashtuns [see Pushtuns]
Pathans (puh-TAHNZ)
Peshawar (puh-SHAH-wahr)
Pettah (PET-uh)
Progeny (PRAH-juh-nee)
Punjab (pun-JAHB)
Pushtuns (PAH-shtoonz)
Qasim (kah-SEEM)
Raisina (rye-SEENA)
Raj (RAHDGE)
Rajasthan (RAH-juh-stahn)
Rajiv (ruh-ZHEEV)
Rawalpindi (rah-wull-PIN-dee)
Sepoy (SEE-poy)
Seychelles (say-SHELLZ)
Sharia (SHAH-ree-uh)
Shi'a (SHEE-uh)
Shi'ism (SHEE-izm)
Shi'ite (SHEE-ite)
Siddhartha (sid-DAHR-tuh)
Sikh (SEEK)
Sikhism (SEEK-izm)
Sikkim (SICK-um)
Sinhala (sin-HAHLA)
Sinhalese (sin-hah-LEEZE)
Sri Lanka (sree-LAHNG-kuh)
Srinagar (srih-NUG-arr)
Sunni (SOO-nee)
Sutlej (SUTT-ledge)
Taj Mahal (TAHJ muh-HAHL)
Tamil [Nadu] (TAMMLE [NAH-doo])
Tapti (TAHP-tee)
Telugu (TELLOO-goo)
Terai (teh-RYE)
Thimphu (thim-POOH)
Tibet (tuh-BETT)
Tripura (TRIP-uh-ruh)
Tufo (TOO-foh)
Turkestan (TER-kuh-stahn)
Urdu (OOR-doo)
Uttar Pradesh (ootar-pruh-DESH)
Vanni (VAH-nee)
Varanasi (vuh-RAHN-uh-see)
Vindhya (VIN-dyuh)
Wakhan (wah-KAHN)
Yamuna (YAH-muh-nuh)

---

*Second double ''o'' pronounced as in ''book''

411

Scale 1:16 000 000; one inch to 250 miles. Polyconic Projection
Elevations and depressions are given in feet

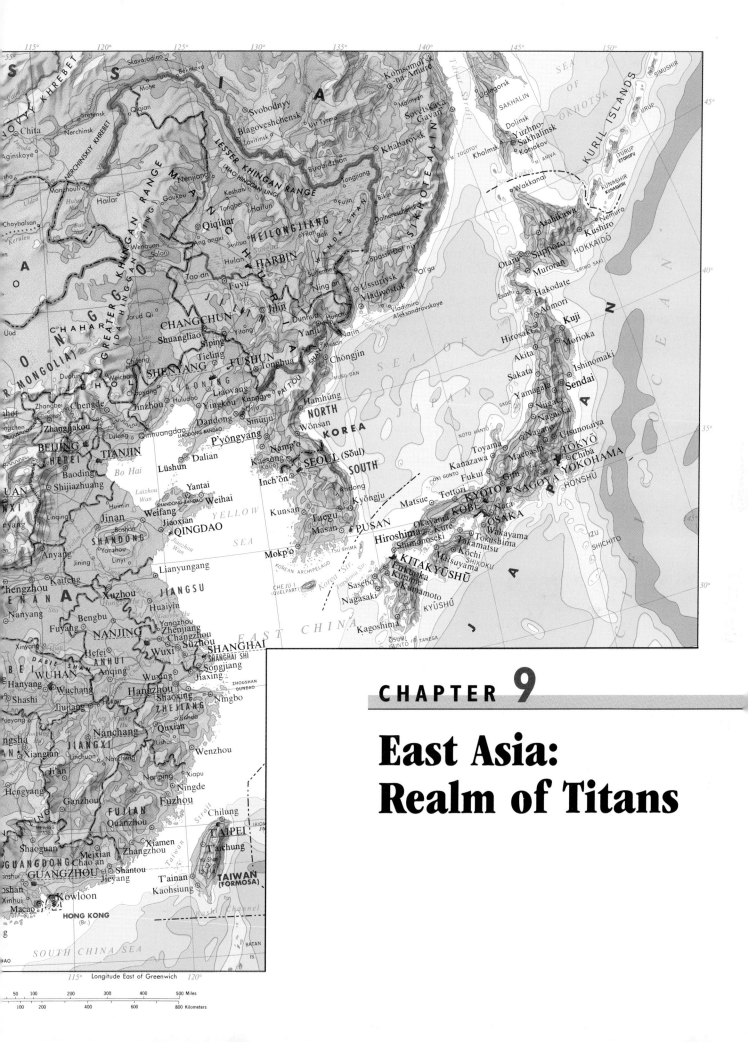

CHAPTER 9

East Asia:
Realm of Titans

East Asia is a geographic realm like no other. At its heart lies the world's most populous country. On its periphery lies one of the globe's most powerful national economies. Along its coastline, on its peninsulas, and on its islands an economic miracle is transforming cities and countrysides. Its interior contains the world's highest mountains and vast deserts. It is a storehouse of raw materials. The basins of its great rivers produce food that can sustain more than a billion people.

The East Asian geographic realm consists of six political entities: China, Mongolia, North Korea, South Korea, Japan, and Taiwan. Note that we refer here to "political entities" rather than "states." In changing East Asia, the distinction is significant. Taiwan, officially called by its government the Republic of China, functions as a state but is regarded by mainland China (the People's Republic of China) as a temporarily wayward province. North Korea is not a full member of the United Nations, and the division of the Korean Peninsula may be temporary. A seventh East Asian political entity, Hong Kong (Xianggang), was scheduled for reunification with China in 1997; and an eighth, Portuguese Macau (Macao), will be absorbed (also by China) in 1999.

As here defined, East Asia lies between the vast expanses of Russia to the north and the populous countries of South and Southeast Asia to the south. This geographic realm extends from the deserts of Central Asia to the Pa-

cific islands of Japan and Taiwan. Environmental diversity is one of its hallmarks.

East Asia also is the hub of the evolving regional phenomenon called the *Pacific Rim*. From Japan to Taiwan and from South Korea to Hong Kong (Xianggang), the Pacific frontage of East Asia is being transformed. Sky-

IDEAS & CONCEPTS

Development	Areal functional organization
Extraterritoriality	Regional complementarity
Hegemony	State capitalism
Core area	Special Economic Zone (SEZ)
Buffer state	Economic tiger
Jakota Triangle	Open Cities
Modernization	Open Coastal Areas
Relative location	

REGIONS

China Proper	Mongolia
Xizang (Tibet)	Jakota Triangle
Xinjiang	

THE MAJOR GEOGRAPHIC QUALITIES OF EAST ASIA

■ ■ ■

1. East Asia is the world's most populous geographic realm and contains the world's most populous nation.

2. East Asia is the leading stage for the emergence of the western Pacific Rim, which extends from eastern Australia in the south to Japan in the north.

3. The Jakota Triangle (Japan–Korea–Taiwan) lies at the vanguard of Pacific Rim development and presages the kind of economic growth that may lie ahead for other parts of East and Southeast Asia.

4. The Pacific Rim in East Asia is the western anchor of a greater Pacific Rim phenomenon that discontinuously encircles the Pacific Ocean from southern New Zealand clockwise to central Chile.

5. From China's western Islamic frontier to Taiwan's burgeoning cities, political and economic developments are transforming traditional cultural landscapes.

6. Intensifying regional disparities mark large regions of the East Asian realm.

7. East Asia was the scene of a successful campaign of colonial expansion by a non-European power, Japan; the impact of Japan's hegemony still lingers in its former empire.

8. Population in the East Asian geographic realm remains concentrated in its eastern regions, which reflects the historical-geographical importance of the lower basins of the realm's great rivers.

9. The political geography of East Asia displays potential instability and change in numerous locales, including Japan's Kurile Islands, divided Korea, secessionist Taiwan, reabsorbed Xianggang (Hong Kong), and colonial Tibet.

10. East Asia may witness the rise of the world's next superpower as China's economic and military strength and influence grow—and if China avoids the devolutionary forces that fractured the Soviet Union.

■ ■ ■

■ ■ ■ ■ ■ ■ ■ ■ ■ ■ ■ **FOCUS ON A SYSTEMATIC FIELD** ■ ■ ■ ■ ■ ■ ■ ■ ■ ■ ■

The Geography of Development

This chapter is not only about the countries of East Asia: it also focuses on the regional phenomenon we call the Pacific Rim. The Pacific Rim is the scene of rapid, sometimes dramatic economic development. Japan lay at the vanguard of it; China is at the heart of it today. The eastern zone of East Asia is synonymous with development.

Scholars from many academic disciplines study **development** systematically. In the Introduction we took a geographic perspective, displaying on a map the global distribution of countries based on their national incomes (Fig. I-11, pp. 28–29). We noted that the regional distribution of development is changing, so that it is no longer appropriate to distinguish between developed and underdeveloped realms. East Asia is a case in point. Japan is one of the most developed countries on Earth, and Mongolia is one of the least developed. Virtually all researchers still distinguish between developed countries (DCs) and underdeveloped countries (UDCs), but even at this larger scale the distinction can be misleading. Along its Pacific Rim, China in places is taking on the characteristics of a developed country. But in its interior, China displays symptoms of severe underdevelopment.

Development is measured by many yardsticks, including (1) the gross national product (GNP) per person, which is determined by adding up all incomes of all citizens in a given year and dividing that sum by the total population; (2) the occupational structure of the labor force, that is, how many people work on farms as opposed to factories; (3) the consumption of energy and energy resources per person; (4) the quantity of transportation and communication facilities per person; (5) the amount of manufactured metals a country requires during a given year, converted to an index per person; (6) the productivity per worker in the labor force;

and (7) a set of rates such as literacy, nutrition, and savings. These data do not, of course, tell us *why* countries exhibit the level of development they do, and that question often has a geographic answer. The geography of development leads us into questions of raw-material distribution, environmental conditions, cultural traditions (including elitism, patronage, and corruption), and the lingering effects of colonialism.

Because scholars from so many disciplines do research on development problems, there may be more cross-disciplinary interaction in this arena than in most others, with good results. Geographers tend to look at development spatially, others structurally. A global model of the development process, still much discussed, was formulated by the economist Walt Rostow in the 1960s. His model suggested that all developing countries follow an essentially similar path through five interrelated growth stages. In the earliest of these stages, a *traditional society* engages mainly in subsistence farming, is locked in a rigid social structure, and resists technological change. When Stage 2—*preconditions for takeoff*—is reached, a progressive group of leaders moves the country toward greater flexibility, openness, and diversification. Old ways are abandoned by substantial numbers of people; birth rates decline; more products than farm-related goods are made and sold; transportation improves. A sense of national unity and purpose may emerge to help nurture this progress. If that happens, Stage 3—*takeoff*—is reached.

Now the country experiences something akin to an industrial revolution, and sustained growth takes hold. Industrial urbanization proceeds, and technological and mass-production breakthroughs occur. If the economy continues to develop, it enters Stage 4—*drive to maturity*. Technologies diffuse nationwide, sophisticated

industrial specialization occurs, international trade expands. Some countries reach a still more advanced Stage 5, that of *high mass consumption*, marked by high incomes, the robust production of consumer goods and services, and a majority of workers in the tertiary and quaternary sectors. In the 1990s, we might add a further stage to Rostow's model: the *postindustrial* stage of ultrasophisticated high technology (see Chapter 3).

On today's world map (refer again to Fig. I-11), the low-income economies of Stages 1 and 2 remain heavily concentrated in tropical Africa and southern and eastern Asia. Note that China, for all its explosive Pacific Rim growth, still is a low-income economy. The high-income DCs associated with Stages 4 and 5 lie mainly in Western Europe and North America (the oil-derived incomes of some Persian Gulf states create a special case: income with limited real development).

Any assessment of the stage of development of a country involves difficult judgments. For instance, by some measures countries in the "takeoff" stage include Brazil, Thailand, and Turkey. Others, however, are stagnating or sliding backward. One example is Nigeria, which during its oil boom of the 1970s seemed poised for takeoff; but now it has many of the characteristics of a Stage 2, not a Stage 3, society.

Despite Pacific Rim developments and other signs of progress in a few places (most notably Indonesia), the prognosis for many UDCs in the late 1990s is not good. They have been thrust into a global system of exchange and capital flow over which they have little or no control. Their borrowing creates debts that rise every year. In Africa and Asia, we are reminded that the geography of development is the geography of inequality.

■ ■ ■

scrapers tower over the city of Haikou, capital of the once-dormant island of Hainan. Luxury automobiles from Europe and America ply the streets of Dalian, long the drab port for China's Northeast. A forest of construction cranes marks the emergence of Pudong, a huge industrial zone where Shanghai stakes its entry into the Pacific Rim's burgeoning economy. Millions of people are on the move, abandoning their farms and villages to seek work in such projects.

Japan was the leader in East Asia's Pacific Rim development. Long before this regional term even came into general use, Japan had built a giant economy with global connections, and Japan stood alone as a highly developed country in an Asia plagued by stagnation. While China lay isolated and South Korea struggled in the aftermath of its terrible war against communism (1950–1953), Japan built

a society unlike any other in eastern Eurasia, and Japan merited recognition as a discrete geographic realm. But Pacific Rim developments are diminishing the contrasts between Japan and its East Asian neighbors. South Korea today challenges Japan on international markets, selling goods ranging from electronics to automobiles. Taiwan's economy prospers. In southeastern China, Guangdong Province is an economic juggernaut that includes Shenzhen, the world's fastest-growing city. Japan's own investments in Pacific Rim economies have helped not only to lower the contrasts, but also to enhance the linkages between Tokyo and the mainland. In short, Japan has become, again, a region within the East Asian realm.

No matter how powerful Japan's economy became during the second half of the twentieth century, the colossus of East Asia remains China. One of the world's oldest con-

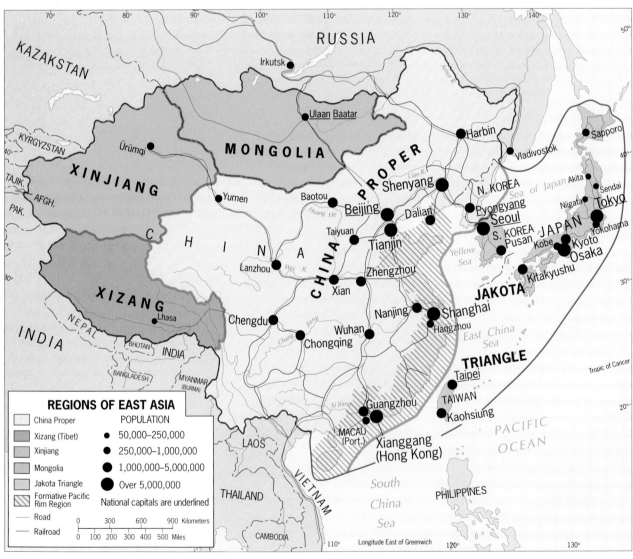

FIGURE 9-1

tinuous civilizations, China was a major culture hearth when Japan remained an isolated frontier inhabited by Ainu communities. Migrations that originated in China advanced through the Korean Peninsula and reached the islands. Philosophies, religions, and cultural traditions (including, for example, urban design and architecture) diffused from China to Korea and Japan, and to other locales on the margins of the Chinese hearth. Over the past millennium, China repeatedly expanded to acquire empires, only to collapse in chaos when its center failed to maintain control. European colonial powers took such opportunities to partition China among themselves. The Russians and the Japanese pushed their empires deep into China. But always China recovered, and today China may be poised to expand its own sphere of influence over a wider range.

REGIONS OF THE EAST ASIAN REALM

East Asia, therefore, presents us with an opportunity to illustrate the changeable nature of regional geography. Our regional delimitation takes account of current reality, and it predicts the way the framework may change. It is anything but static.

Toward the turn of the twenty-first century, we identify five geographic regions in the East Asian realm. These are:

1. *China Proper*

   Almost any map of China's human geography—population distribution, urban centers, surface communications, agriculture—emphasizes the strong con-

centration of Chinese presence and activity in the country's eastern sector. This is the ''real'' China, where its great cities, populous farmlands, and new Pacific Rim growth poles are located. Long ago, scholars called this China Proper, and it is a good regional designation. (Fig. 9-1)

2. *Xizang (Tibet)*

   The tall mountains and high plateaus of Xizang, still widely known as Tibet, form a stark contrast to teeming China Proper. Here, right next to one of the world's most populous regions, lies one of the emptiest.

3. *Xinjiang*

   The vast desert basins and mountain rims of Xinjiang form the third East Asian region. Again physical as well as human geographic criteria come into play: here China meets Islamic Central Asia in the form of Kazakstan, Kyrgyzstan, and Tajikistan, and Xinjiang takes on the properties of a frontier between rimland Asia and inner Asia.

4. *Mongolia*

   East Asia's fourth region is constituted by the desert state of Mongolia. As we note later, Mongolia has the properties of a buffer state between China and Russia. With well under 3 million inhabitants in an area more than twice as large as Texas, Mongolia stands in stark contrast to populous China to the south.

5. *The Jakota Triangle*

   Japan anchors East Asia's fifth region, forming, with South Korea and Taiwan, the *Jakota Triangle*, the first full-scale regional manifestation of the new Pacific Rim era into which this realm is moving. This region is likely to expand, perhaps eventually to incorporate North Korea (following unification) and, possibly, a contiguous coastal Pacific Rim zone in China.

❖ CHINA PROPER

It is often said and written these days that the disintegration of the Soviet Union marked the collapse of the world's last remaining great empire. Russian imperialism forged a vast colonial empire inherited by Moscow's communist despots, but now the pieces of that empire are independent states. The age of empires, say some scholars, is over.

From a geographic viewpoint, that verdict is premature. By many measures, an empire far more populous than Russia's ever was still dominates the eastern quadrant of Asia.

It is a realm without democracy or multiparty elections. It is a power core in control of territories that are colonies in all but name. It is a land of numerous disadvantaged minorities. It is a domain that still lays claim to territories beyond its borders. It is a regime that threatens neighbors. It is China, the last of the twentieth century's devolving empires.

China has not managed to shut out the winds of political and economic change that are sweeping the world. China's east is affected by the momentous changes marking the western Pacific Rim, and China lies exposed to the post-colonial transformation of Turkestan (see Chapter 6). And in 1989 the very heart of China was riven by student-led, labor-supported, pro-democracy movements that crested tragically at central Beijing's Tiananmen Square in a bloody suppression. Ever since, human rights issues have roiled China's international relations. None of this, however, has impelled China's communist rulers to do what Soviet leaders did. The Communist party's monopoly over the politics of China remains as strong as ever. The world may have turned its back on communist dogma, but not China. And China, we should remember, contains more than one-fifth of all humankind. Those who now write about a postcommunist world are—again—too hasty.

From the Field Notes

"Pacific Rim China's southernmost outpost is Hainan Island, and here the transformation brought by economic growth is dramatic; the entire capital of Haikou resembles one huge construction site. Everywhere on the city's outskirts, billboards proclaim what the future will look like: villages will be swept away, farms plowed under, and high-rise developments will replace them. This advertisement stood above a grocery store on a main street (the store itself already modernized)."

A LAND OF CONTRADICTIONS

China is nonetheless changing in fundamental ways. In the political sphere, authoritarian communism persists, but in the economic arena, China has opened its doors to capitalist enterprise. China's rulers are determined to keep political control, thereby to avoid the turmoil that has plagued devolving empires from the British to the Russian. But they also are determined not to miss the opportunities of market economics and to share in the economic opportunities arising on the Pacific Rim.

In the mid-1990s, while North American and European economies stagnated, China's economy was growing at an annual rate of 9 percent. Geographically, this growth is transforming the country's structure. Zones along China's Pacific coast are booming, their urban landscapes transformed by new skyscrapers and modern factories. But China's vast interior is comparatively unchanged, still awaiting the impact of the new economic policies. In time, China will have to face up to its fundamental political-economic contradiction. When that moment arrives, communist dogma will confront capitalist energy, and China will face

the social consequences of unequal incomes and the spatial problems of regional inequalities. It will take more than a military clampdown on protesters in city squares to resolve China's paradox.

Whatever happens, its impact will be felt globally, and it will be critical for us in the Western world to understand China as best we can. Many years ago, two prominent geographers, Halford Mackinder and Nicholas Spykman, debated the geopolitical prospects of Eurasia's power cores. Mackinder, nearly a century ago, assessed the distribution of resources and populations, environmental conditions and natural advantages, and concluded that the "heartland" power, Russia or its successor, would become the continent's dominant force. Spykman interpreted the data differently. The rim of Asia, he argued, contained the demographic, environmental, locational, and resource potentials to produce even greater power. Spykman prophetically used the term "rimland" as early as 1944. Today, Asia's rimlands are emerging, and the former Soviet heartland is in decline.

Where will China's ascent lead? When an economy grows at an annual rate of 9 percent, the total national wealth doubles every eight years; hence, China is gaining rapidly in its quest for development. Japan, China's neighbor, twice in recent history rose from nowhere to global power: once during the second half of the nineteenth century, when it emerged in a few decades as a military power that conquered a colonial empire and attacked the United States; and again when it arose from the ashes of World War II to become a mighty economic force. China now is poised to take, in the words of its communist godfather Mao Zedong, a "great leap forward." Yet this great leap will occur because China's rulers, in the post-Mao period, have been able to loosen the reins on the economy. If China's new-age communists can hold the country together, if they can resist political reform while channeling economic change, the future will witness China's turn as a challenger for the role of global superpower—if not world domination.

In all this, the Pacific Ocean will become the world's crucial arena, just as the Atlantic was a century ago. Across it, and along its margins, the United States and China will trade and compete. The nineteenth century saw the rise of European colonial empires that spanned the world. The twentieth century witnessed Nazi Germany's quest for world domination and the Soviet Union's emergence as a global superpower. The twenty-first century will see the ascent of China—if the last empire lasts.

CHINESE PERSPECTIVES

When we in the Western world chronicle the rise of civilization, we tend to focus on the historical geography of Southwest Asia, the Mediterranean, and Western Europe. Ancient Greece and Rome were the crucibles of culture; Mediterranean and Atlantic waters were the avenues of its diffusion. China lay remote, so we believe, barely connected to this Western realm of achievement and progress. When an Italian adventurer named Marco Polo visited China during the thirteenth century and described the marvels he had seen, his work did little to change European minds. Europe was and would always be the center of civilization.

The Chinese, naturally, take quite a different view. Events on the western edge of the great Eurasian landmass were deemed irrelevant to theirs, the most advanced and refined culture on Earth. Roman emperors were rumored to be powerful, and Rome surely was a great city, but nothing could match the omnipotence of China's rulers. Certainly the Chinese city of Xian far eclipsed Rome as a center of sophistication. Chinese civilization existed long before ancient Greece and Rome emerged, and it was still there long after they collapsed. China, the Chinese teach themselves, is eternal. It was, and always will be, the center of the civilized world.

We should remember this when we study China's regional geography, because 4,000 years of Chinese culture and perception will not change overnight—not even in a generation. Time and again, China overcame the invasions

NAMES AND PLACES

■ ■ ■

In 1958, the government of the People's Republic of China adopted the so-called *pinyin* system of standard Chinese, which replaced the Wade-Giles system used since colonial times. This was not done to teach foreigners how to spell and pronounce Chinese names and words but to establish a standard form of the Chinese language throughout China. The pinyin system is based on the pronunciation of Chinese characters in Northern Mandarin, the Chinese spoken in the region of the capital and in the north in general.

The new linguistic standard caught on rather slowly outside China, but today it is in general use. The old name of the capital, Peking, has become Beijing. Canton is now Guangzhou. The Yangtze Kiang (River) is now the Chang Jiang. Tientsin, Beijing's port, is now Tianjin. A few of the old names persist, however. The Chinese call their colony Xizang, but many maps still carry the name Tibet.

Personal names, too, were affected by pinyin usage. China's long-time ruler, Mao Tse-tung, now is called Mao Zedong. His eventual successor, Teng Hsiao-ping, is more simply Deng Xiaoping. And remember: when the Chinese write their names, they use the last name first. Xiaoping, therefore, is Mr. Deng's first name. To the Chinese, it is Clinton Bill and Dole Bob, not the other way around!

■ ■ ■

and depredations of foreign intruders, and afterward the Chinese would close off their vast country against the outside world. Barely two decades ago, in the early 1970s, there were just a few *dozen* foreigners in the entire country with its (then) nearly 1 billion inhabitants. The institutionalization of communism required this, and, following a quarrel, even the Soviet advisers had been thrown out. But then China's rulers decided that an opening to the Western world would be advantageous, and U.S. President Richard Nixon was invited to visit Beijing. That historic occasion, in 1972, ended this latest period of isolation—as always, on China's terms. Since then, China has been open to tourists and businesses, teachers and investors. Tens of thousands of Chinese students were sent to study at American and other Western institutions. Long-suppressed ideas flowed into China, and the pro-democracy movement arose. China's rulers knew that their violent repression of this movement would anger the world, but this did not matter: foreign condemnation was deemed irrelevant. Foreigners in China had done much worse. And Westerners had no business interfering in China's domestic affairs.

ETERNAL CHINA

Most of China's 1.3 billion people are heirs to what may indeed be the world's oldest continuous national culture and civilization. The present capital, Beijing, lies near the nucleus of ancient China. Near the city lies a cave in which were found fossils of "Peking Man," a representative of *Homo erectus*, possibly an ancestor of modern humans, who lived here half a million years ago. Anthropologists are not yet in agreement on human origins, but one theory holds that modern humans evolved from several regional predecessors, including an East Asian species of *Homo erectus*. If this turns out to be the case, modern Chinese people may be able to trace their ancestries deep into the Pleistocene.

In any case, Chinese culture and civilization have ancient roots, traceable at least 4,000 years. Again, we in the Western world look to Mesopotamia and Southwest Asia generally for clues to the origin of states, but the process may have started even earlier in China. In the Lower Basin of the Huang He (Yellow River), near the confluence of the Huang and Wei rivers, early Chinese societies developed state-like organizations. At times, a single state seems to have dominated; at other times, there was fragmentation and competition. China's early culture hearth expanded and contracted as its environmental and other fortunes fluctuated. But the thread of culture never broke, and over more than 40 centuries the Chinese created a society with strong traditions, values, and philosophies. Unlike Mesopotamia, this was an isolated society, confident in its strength, continuity, and ultimate superiority. It could reject and repel

foreign influences or sometimes absorb and assimilate them. Yet through most of its long period of gestation, China lay remote, protected by distance and physiography from outside influences.

RELATIVE LOCATION FACTORS

The Chinese themselves, throughout their nation's history, have contributed to the isolation of their country from foreign influences. As we noted, the most recent episode of exclusion and closure happened just a few decades ago. It is one of China's recurrent traditions, spawned by China's relative location and Asia's physiography. Many a writer has commented on China's "splendid isolation—an isolation made possible by geography.

A physiographic map of East Asia offers an explanation (Fig. 9-2). To China's north lie the rugged mountain ranges of eastern Russia and the vast Gobi Desert. To the northwest, beyond Xinjiang, the mountains open—but onto the huge, dry Kirghiz Steppe. To the west and southwest lie the legendary Tian Shan and Pamirs, snowcapped ranges that might as well be thousand-foot-high walls. To the south, as a result of China's colonization of Tibet (Xizang), China's empire is bounded by the incomparable Himalayas, the effective boundary between the East Asian and South Asian realms. And high relief also separates China from much of Southeast Asia. On the physiographic map, China looks like a mountain-, desert-, and forest-encircled fortress.

Equally telling is the factor of distance. Until recently, China always lay far from the modern source areas of industrial innovation and change. True, China—as the Chinese emphasize—was itself such a hearth, but China's contributions to the outside world were very limited. China did interact to some extent with Korea, Japan, Taiwan, and parts of Southeast Asia, and millions of Chinese did emigrate to neighboring areas. But compare this to the impact of the Arabs, who ranged far and wide and who brought their knowledge, religion, and political influence to areas from Mediterranean Europe to Bangladesh and from West Africa to Indonesia. Later, when Europe became the center of change, China found itself farther removed, by land or sea, than almost any other part of the world.

Today, modern communications notwithstanding, China still is distant from anywhere else on Earth. Going by rail from Beijing to Moscow, the heart of China's Eurasian neighbor, is a tedious journey of several days. Direct overland connections with India are practically nonexistent. Communications with Southeast Asian countries, though improving, remain tenuous.

The most significant changes in China's relative loca-

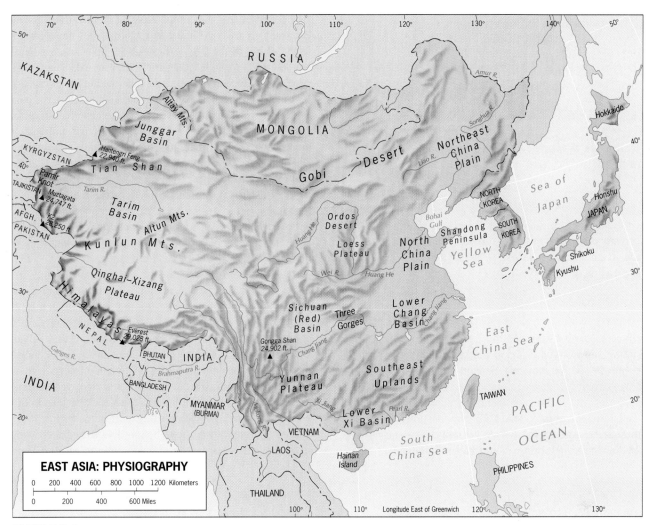

FIGURE 9-2

tion and external connections are occurring on the Pacific Rim, although political impediments still exist. China's co-communist alliance with North Korea has weakened as a result of North Korea's lack of enthusiasm for China's economic reforms. In contrast, links with South Korea are strengthening after a long period of isolation. And Japanese business and investment are making huge inroads in China. For the first time in history, China not only lies spatially near a world-class hearth of technological innovation and financial strength; it also has opened its doors to Japan's projects and ideas.

Further changes are in progress. China's 1997 takeover of formerly British Hong Kong (Xianggang) will soon be followed by its reacquisition of the nearby Portuguese dependency of Macau (Macao). This leaves Taiwan as the unresolved issue: although China has recognized the independence of devolved imperial fragments from Estonia to Tajikistan, it will not acknowledge Taiwan's sovereignty. This is the kind of decision that imperial rulers accustomed to ignoring world opinion feel they can afford to make.

EXTENT AND ENVIRONMENT

China's total area is only very slightly larger than that of the United States including Alaska: each has about 3.7 million square miles (9.5 million sq km). As Figure 9-3 reveals, the longitudinal extent of China and the 48 contiguous U.S. States also is quite similar. Latitudinally, however, China is considerably wider. Miami, near the southern limit of the United States, lies halfway between Shanghai and Guangzhou. China's area extends well into the tropics. Southern China thus takes on characteristics of tropical Asia. In the northeast, too, China incorporates much of what would in North America be Quebec and Ontario. Westward, China's land area becomes narrower and similarities increase. Of course, China has no west coast!

Now compare the climate maps of China and the United States in Figure 9-4 (which are enlargements of the appropriate portions of the world climate map in Figure I-7).

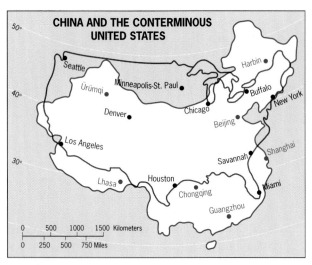

FIGURE 9-3

Note especially the comparative location of the U.S. and Chinese *Cfa* areas in Figure 9-4. China's lies much farther to the south. In the United States, the *Cfa* climate extends beyond 40° North latitude, but in China, very cold and generally winter-dry *D* climates take over at the latitude of Virginia. Beijing has a warm summer but a bitterly cold and long winter. Northeast China, in the general latitudinal range of Canada's lower Quebec and Newfoundland, is much more severe than its North American equivalent. Harsh environments prevail over vast regions of China, but, as we will see later, nature makes up for this in spectacular fashion. From the climatic zone marked *H* (for highlands) in the west come great life-giving rivers whose wide basins contain enormous expanses of fertile soils. Without these, China would not have a population more than four times that of the United States.

If you were to travel in China, environments and distances would at times seem quite familiar. The distance from Shanghai to the capital, Beijing, is not much greater than that from Washington, D.C. to Chicago. Flying cross-country would take about the same amount of time, given similar aircraft. But be prepared: China is not yet a country of modern transport facilities. Efficient airports, cross-country superhighways, or bullet trains have yet to make their impact. Still, return visitors are struck by the continuous improvement in China's transport infrastructure. Several competing airlines have bought and leased a growing fleet of planes. A new superhighway links the port of Tian-

Note that both China and the United States have a large southeastern climatic region marked *Cfa* (that is, humid, temperate, warm-summer), flanked in China by a zone of *Cwa* (where winters become drier). Westward in both countries, the *C* climates yield to colder, drier climes. In the United States, moderate *C* climates develop again along the Pacific coast. China, however, stays dry and cold as well as high in elevation at equivalent longitudes.

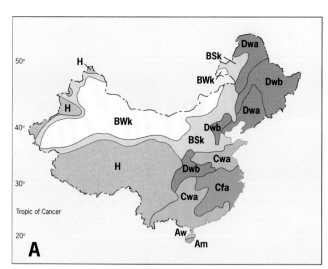

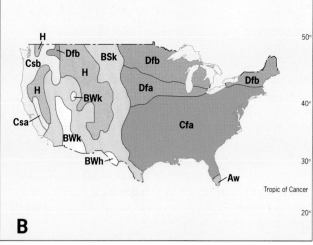

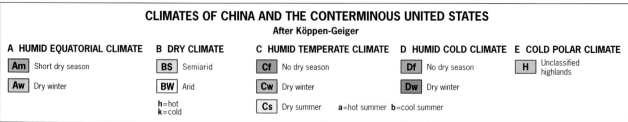

CLIMATES OF CHINA AND THE CONTERMINOUS UNITED STATES
After Köppen-Geiger

A HUMID EQUATORIAL CLIMATE
- **Am** Short dry season
- **Aw** Dry winter

B DRY CLIMATE
- **BS** Semiarid
- **BW** Arid
- **h**=hot
- **k**=cold

C HUMID TEMPERATE CLIMATE
- **Cf** No dry season
- **Cw** Dry winter
- **Cs** Dry summer
- **a**=hot summer **b**=cool summer

D HUMID COLD CLIMATE
- **Df** No dry season
- **Dw** Dry winter

E COLD POLAR CLIMATE
- **H** Unclassified highlands

FIGURE 9-4

jin with Beijing, and another (a toll road!) connects Shenzhen to Guangzhou. Developing, modernizing China is on the move.

EVOLUTION OF THE CHINESE STATE

As the physiographic map (Fig. 9-2) indicates, China is a vast and varied physical stage. From interior mountains, major rivers flow across wide, fertile plains: the Liao in the northeast, the Huang and Chang in the middle, and the Xi in the south. The North China Plain, centered on Beijing, produces wheat; the Lower Chang Basin, in Shanghai's hinterland, grows rice. In the narrower Xi Basin (the Xi becomes the famous Pearl River near its mouth), China is warm enough to double-crop rice. In the interior, the fertile and productive Sichuan Basin contains one of the largest inland population agglomerations on Earth. Westward lie the great mountains and plateaus of Xizang (Tibet) and the mountains and desert basins of Xinjiang. As the climate map reminds us, this is forbidding country.

The great and lengthy drama of China's evolution has taken place in the eastern sector of this huge and diverse physical stage. It has been suggested that the early Chinese, perhaps as long as 4,000 years ago, received stimuli from the river-based civilizations of Southwest and South Asia, but the evidence is thin. The Chinese are proven innovators, and Chinese cultural individuality was established very long ago.

China's history and historical geography are chronicled as *dynasties*. From very early on, rulers of the same family succeeded each other, often over periods lasting several centuries, until the lineage broke down. Each dynasty left its imprint on Chinese culture and on the map of China's evolving empire. Thus we refer to "Han," "Ming," or "Manchu" China, using dynastic names to identify significant phases in Chinese national evolution.

The oldest dynasty of which much is known is the Shang (sometimes called Yin), which formed when China was taking shape in the area of the confluence of the Huang and Wei rivers (Fig. 9-5). This dynasty lasted from about 1770 B.C. until around 1120 B.C. Walled cities were built during the Shang period, and the Bronze Age commenced during Shang rule. For more than a thousand years after the beginning of the Shang Dynasty, North China was the center of development in this part of Asia. The Zhou Dynasty (roughly 1120–221 B.C.) sustained and consolidated what had begun during the Shang period. The building of defensive walls against northern invaders began during this dynasty.

Eventually, agricultural techniques and population numbers combined to press settlement in the obvious direction—southward—where the best opportunities for further expansion lay. During the brief Qin (Chin) Dynasty (221–207 B.C.), the lands of the Chang Jiang were opened up, and settlement spread as far south as the Lower Xi Basin. In the north, the Qin rulers connected existing defensive walls into a continuous structure and built a system of fortified watchtowers to improve communication.

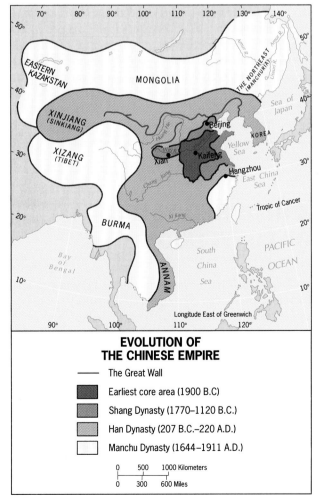

**EVOLUTION OF
THE CHINESE EMPIRE**

——— The Great Wall

▮ Earliest core area (1900 B.C)

▮ Shang Dynasty (1770–1120 B.C.)

▮ Han Dynasty (207 B.C.–220 A.D.)

▢ Manchu Dynasty (1644–1911 A.D.)

```
0    500    1000 Kilometers
0    300    600 Miles
```

FIGURE 9-5

Han China

A pivotal period in the historical geography of China now lay ahead: the Han Dynasty (207 B.C.–A.D. 220). The Han rulers brought unity and stability to China, and they enlarged the Chinese sphere of influence to include Korea, the Northeast, Mongolia, Xinjiang, and, in Southeast Asia, Annam (located in what is now central Vietnam). Thus the Chinese rulers established control over the nomads of the surrounding steppes, deserts, and mountains, and were able to protect the Silk Route across Xinjiang, the link between Xian and Rome.

The Han period was a formative time in the evolution of China (Fig. 9-5). China's military power grew stronger than ever before. The old feudal order broke down, and systems of land ownership changed, allowing individuals to own property. The Silk Route prospered. Great achievements were made in the arts, architecture, and the sciences. Many of China's durable social norms were established during this period, and to this day the people of China Proper call themselves the *People of Han*.

Han China was the Roman Empire of East Asia, and like its European counterpart, greatness was followed by decline and disarray. Not until the Tang Dynasty (618–907) did stability return.

Insiders and Outsiders

Following the Tang period, when China once again was a great national state, came the Sung Dynasty (960–1279). Now Chinese culture reached unprecedented heights. Nomadic attackers from the north caused frequent problems (to the point that the Sung rulers in 1127 abandoned their capital, Kaifeng, on the Huang River and moved it to Hangzhou in the southeast), but China thrived nevertheless. In many ways, China during the Sung Dynasty was the world's most advanced state. It had several cities with more than 1 million inhabitants; paper money was in use; many schools were built; commerce expanded; the arts prospered; and the philosophies of Kongfuzi (Confucius), for many centuries China's guide, were modernized, printed, and mass-distributed for the first time.

The Sung Dynasty was overthrown by invading outsiders, the Mongols. Led by Kublai Khan, the Mongols took control of China in 1279, making the country another part of their vast Eurasian empire. In Chinese historical geography, Mongol rule is called the Yuan Dynasty, but it lasted less than a century (to 1368), and the Mongols had little effect on Chinese culture. Indeed, it was the Mongols who were acculturated as they adopted may of the norms of Chinese civilization!

When a local ruler overthrew the last of the Yuan overlords, there was little to indicate that China was about to embark on one of its greatest periods, perhaps the greatest since the Hans. The Ming Dynasty (1368–1644) reconsolidated China from the Great Walls of the north (which the Ming rulers strengthened and expanded) to Annam in the south, and from Korea in the northeast to Burma in the southwest. Beijing was the Ming capital, and the Ming rulers established a tradition of autocratic government that concentrated power at the top and controlled the provinces tightly. Enforced stability was rewarded with success in many spheres. Chinese fleets plied the Indian Ocean and reached East Africa. Links with Europe were established.

Great achievements in architecture, the arts (notably in blue and white porcelain), literature, and education refined Chinese culture. The civil service was expanded and made more efficient. China was at the height of its powers.

But autocratic rule breeds reaction, and eventually it came. In 1644, a rebel leader laid siege to the capital, and the Ming commander there made the mistake of asking the Manchus, whose base was in the northeast beyond the Great Wall, for help. The Manchus came, but they did not rescue the Ming Dynasty. They occupied Beijing and took control of China, establishing the last great dynasty, the Manchu (or Qing). Once again, China was ruled by outsiders, but the Manchus were no Mongols. Building on what had been achieved during the Ming period, the Manchu nurtured Chinese civilization and culture, extended China's presence across a domain that was the largest ever (Fig. 9-5), and developed China's first national economy.

The Manchu Dynasty (1644–1911) saw China's population grow from an estimated 150 million to 450 million, not only through natural increase but also through the incorporation of peoples in Mongolia, Turkestan, Tibet, and Indochina. With this increase came problems of administration and food provision. Despite efforts to control the flows of China's great rivers and to expand farmlands, deadly famines claimed millions of victims. And the Manchu also had the misfortune of being in charge when the European powers as well as Japan's armies penetrated China. Chaos descended on the People of Han, and in 1911 the last of the dynasty's emperors, a boy of 12 in the corrupt imperial household in Beijing's Forbidden City, was forced to abdicate. It was the end of four millennia of continuity in China's governmental tradition, and the beginning of a century of upheaval.

A HUNDRED YEARS OF CONVULSION

As we try to better understand China's present political, economic, and social geography, it is important not to lose sight of the China that endured for thousands of years under a sequence of imperial dynasties as a single entity. Larger and more populous than Europe, China had its divisive, feudal periods, but always China came together again, ruled—mostly dictatorially—from a strong center as a unitary state. China was, and remains, a predominantly rural society, a powerless and subservient population controlled by an often ruthless bureaucracy. Retributions, famines, epidemics, and occasional uprisings took heavy tolls in the countryside. But in the cities, Chinese culture flourished, subsidized by the taxes and tribute extracted from the hinterlands.

China long withstood the European colonialists in East

Asia with a self-assured superiority based on the strength of its culture and the continuity of the state. There was no market for the British East India Company's rough textiles in a country long used to finely fabricated silks and cottons. There was little interest in the toys and trinkets the Europeans produced in the hope of bartering for Chinese tea and pottery. And even when Europe's sailing ships made way for steam-driven vessels and newer and better factory-made textiles were offered in trade for China's tea and silk, China continued to reject the European imports that were still, initially at least, too expensive and too inferior to compete with China's handmade goods. Long after India had fallen into the grip of mercantilism and economic imperialism, China was able to maintain its established order. This was no surprise to the Chinese. After all, they had held a position of undisputed superiority among the countries of eastern Asia as long as could be remembered, and they had dealt with overland and seaborne invaders before.

A (Lost) War on Drugs

The nineteenth century finally shattered the self-assured isolationism of China as it proved the superiority of the new Europe. On two fronts, economic and political, the European powers destroyed China's invincibility. In the economic sphere, they succeeded in lowering the cost and improving the quality of manufactured goods, especially textiles, and the handicraft industry of China began to collapse in the face of unbeatable competition. In the political sphere, the demands of the British merchants and the growing English presence in China led to conflicts. In the early part of the century, the central issue was the importation into China of opium, a dangerous and addictive intoxicant. Opium was destroying the very fabric of Chinese culture, weakening the society, and rendering China an easy prey for colonial profiteers. As the Manchu government moved to stamp out the opium trade in 1839, armed hostilities broke out, and soon the Chinese found themselves losing a war on their own territory. The First Opium War (1839–1842) ended in disaster: China's rulers were forced to yield to British demands, and the breakdown of Chinese sovereignty was under way.

British forces penetrated up the Chang Jiang and controlled several areas south of that river (Fig. 9-6); Beijing hurriedly sought a peace treaty. As a result, leases and concessions were granted to foreign merchants. Hong Kong Island was ceded to the British, and five ports, including Guangzhou (Canton) and Shanghai, were opened to foreign commerce. No longer did the British have to accept a status that was inferior to the Chinese in order to do business; henceforth, negotiations would be pursued on equal terms. Opium now flooded into China, and its impact on Chinese society became even more devastating. Fifteen years after the First Opium War, the Chinese tried again to stem the

From the Field Notes
"To pass through the gate from Tiananmen Square and enter the Manchu Emperors' Forbidden City was a riveting experience. Look back now toward the gate, and you see throngs of Chinese visiting what less than a century ago was the exclusive domain of the Qing Dynasty's absolute rulers—so absolute that unauthorized entry was punished by instant execution. The vast, walled complex at the heart of Beijing not only served as the imperial palace; it was a repository of an immense collection of Chinese technological and artistic achievements. Some of this is still here, but most of all you are awed by the lingering atmosphere, the sense of what happened here in these exquisitely constructed buildings that epitomized what high Chinese culture could accomplish. Inevitably you think of the misery of millions whose taxes and tribute paid for what stands here."

disastrous narcotic tide, and again they were bested by the foreigners who had attached themselves to their country. Now the cultivation of the opium poppy in China itself was legalized. Chinese society was disintegrating, and the scourge of this drug abuse was not defeated until the revival of Chinese power in the twentieth century.

But before China could reassert itself, much of what remained of China's independence was steadily eroded away (see box titled "Extraterritoriality"). The Germans obtained a lease on Qingdao on the Shandong Peninsula in 1898, and in the same year the French acquired a sphere of influence in the far south at Zhanjiang (Fig. 9-6). The Portuguese took Macau; the Russians obtained a lease on Liaodong in the Northeast as well as railway concessions there; even Japan got into the act by annexing the Ryukyu Islands and, more importantly, Formosa (Taiwan) in 1895.

After millennia of cultural integrity, economic security, and political continuity, the Chinese world lay open to the aggressions of foreigners whose innovative capacities China had denied to the end. But now ships flying Euro-

FIGURE 9-6

pean flags lay in the ports of China's coasts and rivers; the smokestacks of foreign factories rose above the landscapes of its great cities. The Japanese were in Korea, which had nominally been a Chinese vassal, and the Russians had entered China's Northeast. The foreign invaders even took to fighting among themselves, as did Japan and Russia in ''Manchuria'' (as the foreigners now called the Northeast) in 1904 (see box titled ''Manchuria: The Northeast'').

Rise of a New China

In the meantime, organized opposition to the foreign presence in China was gathering strength, and the twentieth century opened with a large-scale revolt against all outside elements. Bands of revolutionaries roamed the cities as well as the countryside, attacking not only the hated foreigners but also Chinese citizens who had adopted Western

EXTRATERRITORIALITY

■ ■ ■

A sign of China's weakening during the second half of the nineteenth century was the application in its cities of a European doctrine of international law—**extraterritoriality**. This principle denotes a situation in which foreign states or international organizations and their representatives are immune from the jurisdiction of the country in which they are present. This, of course, constitutes an erosion of the sovereignty of the state hosting these foreign elements, especially when the practice goes beyond the customary immunity of embassies and persons in diplomatic service.

In China, extraterritoriality reached unprecedented proportions. The best residential suburbs of the large cities, for example, were declared to be "extraterritorial" parts of foreign countries and were made inaccessible to Chinese citizens. Sha Mian Island in the Pearl River in Guangzhou was a favorite extraterritorial enclave of that city. A sign at the only bridge to the island stated "No Dogs or Chinese." In this way, the Chinese found themselves unable to enter their own public parks and many buildings without permission from foreigners. Christian missionaries fanned out into China, their bases fortified with extraterritorial security. To the Chinese, this involved a loss of face that contributed to a bitter opposition to the presence of all foreigners—a resentment that finally exploded in the Boxer Rebellion of 1900.

Opponents of the present Chinese regime's economic policies, which give major economic privileges and exemptions to foreign firms active in certain coastal cities, have argued that this practice is reviving extraterritoriality in a modern guise. This is a sensitive issue in a China that has not forgotten the indignities of the colonial era.

■ ■ ■

cultural traits. Known as the Boxer Rebellion (after a loose translation of the Chinese name for these revolutionary groups), the 1900 uprising was put down with much bloodshed. Simultaneously, another revolutionary movement was gaining support, aimed against the Manchu leadership itself. In 1911, the emperor's garrisons were attacked all over China, and in a few months the 267-year-old dynasty was overthrown. Indirectly, it too was yet another casualty of the foreign intrusion, and it left China divided and disorganized.

The end of the Manchu era and the proclamation of a republican government in China did little to improve the country's overall position. The Japanese captured Germany's holdings on the Shandong Peninsula, including the city of Qingdao, during World War I. When the victorious European powers met at Versailles in 1919 to divide the territorial spoils, they affirmed Japan's rights in the area. This led to another significant demonstration of Chinese reassertion as nationwide protests and boycotts of Japanese goods were organized in what became known as the May

Fourth Movement. One participant in these demonstrations was a charismatic young man named Mao Zedong.

Nonetheless, China after World War I remained a badly divided country. By the early 1920s, there were two governments—one in Beijing and another in the southern city of Guangzhou (Canton), where the famous Chinese revolutionary, Sun Yat-sen, was the central figure. Neither government could pretend to control much of China. The Northeast was in complete chaos, petty states were emerging all over the central part of the country, and the Guangzhou "Parliament" controlled only a part of Guangdong Province in the Southeast. Yet it was just at this time that the power groups that were ultimately to struggle for supremacy in China were formed. While Sun Yat-sen was trying to form a viable Nationalist government in Guangzhou, the Chinese Communist Party was formed by a group of intellectuals in Shanghai. Several of these intellectuals had been leaders in the May Fourth Movement, and in the early 1920s they received help from the Communist Party of the Soviet Union. Mao Zedong was already a prominent figure in these events.

Initially, there was cooperation between the new Communist Party and the Nationalists led by Sun Yat-sen. The Nationalists were stronger and better organized, and they hoped to use the communists in their antiforeign (especially anti-British) campaigns. By 1927, the foreigners were on the run; the Nationalist forces entered cities and looted and robbed at will while aliens were evacuated or, failing that, sometimes killed. As the Nationalists continued their drive northward and success was clearly in the offing, internal dissension arose. Soon, the Nationalists were as busy purging the communists as they were pursuing foreigners. The central figure to emerge in this period was Chiang Kai-shek. Sun Yat-sen died in 1925, and when the Nationalists established their capital at Nanjing (Nanking) in 1928, Chiang was the country's leader.

MANCHURIA: THE NORTHEAST

■ ■ ■

Three provinces, Liaoning, Jilin, and Heilongjiang, constitute China's Northeast, a region bounded by Russia to the north and east, Mongolia to the west, and North Korea to the southeast (Fig. 9-10). This was the home of the conquering Manchus and, later, the scene of foreign domination. Russians and Japanese struggled for control over the area; ultimately, Japan created a dependency here in the 1930s that it called Manchukuo. More recently, the region came to be called Manchuria, but this is not an accepted regional appellation in China. Chinese geographers refer to the region encompassed by these three northeastern provinces as simply the Northeast—a practice followed in this chapter.

■ ■ ■

Three-Way Struggle

The post–Manchu period of strife and division in China was quite similar to other times when, following a lengthy period of comparative stability under dynastic rule, the country fragmented into rival factions. In the first years of the Nanjing government's **hegemony**, the campaign against the communists intensified and many thousands were killed. Chiang's armies drove them ever deeper into the interior. (Mao himself escaped the purges only because he was in a remote rural area at the time.) For a while, it seemed that Nanjing's armies would break the back of the communist movement in China.

The Long March

A core area of communist peasant forces survived in the zone where the provinces of Jiangxi and Hunan adjoin in southeastern China, and these forces defied Chiang's attempts to destroy them. Their situation grew steadily worse, however, and in 1933 the Nationalist armies were on the verge of encircling this last eastern communist stronghold. The communists decided to avoid inevitable strangulation by leaving. Nearly 100,000 people—armed soldiers, peasants, local leaders—gathered near Ruijin and started to walk westward in 1934. This was a momentous event in modern China, and among the leaders of the column were Mao Zedong and Zhou Enlai. The Nationalists rained attack after attack on the marchers, but they never succeeded in wiping them out completely; as the communists marched, they were joined by new sympathizers.

The Long March (see the route in Fig. 9-6), as this drama has come to be called, first took the communists to Yunnan Province, where they turned north to enter western Sichuan. They then traversed Gansu Province and eventually reached their goal, the mountainous interior near Yanan in Shaanxi Province. The Long March covered nearly 6,000 miles (10,000 km) of China's more difficult terrain, and the Nationalists' continuous attacks killed an estimated 75,000 of the original participants. Only about 20,000 survived the epic migration, but among them were Mao and Zhou, who were convinced that a new China would arise from the peasantry of the rural interior to overcome the urban easterners whose armies could not eliminate them.

The Japanese

While the Nanjing government was pursuing the communists, foreign interests made use of the situation to further their own objectives in China. The Soviet Union held a sphere of influence in Mongolia and was on the verge of annexing a piece of Xinjiang. Japan was dominant in the Northeast, where it had control over ports and railroads.

The Nanjing government tried to resist the expansion of Japan's sphere of influence. The effort failed, and the Japanese set up a puppet state in the region; they appointed a ruler and called their new possession Manchukuo.

The inevitable full-scale war between the Chinese and the Japanese broke out in 1937. For a while, the Chinese communists and the Nationalists stopped fighting each other in order to concentrate on the war against Japan, but soon their factional war erupted again. Now Chinese communists were fighting Chinese Nationalists while both fought the Japanese. In the process, China broke up into three regions: the Japanese sphere in the north and east (Fig. 9-6), the Nationalists' domain centered on their capital of Chongqing in the Sichuan Basin, and the communist zone in the interior west. The Japanese, by pursuing and engaging Chiang's Nationalist forces, gave the communists an opportunity to build their strength and foster their reputation and prestige in China's western areas.

The Japanese campaign in China was accompanied by unspeakable atrocities. Millions of Chinese citizens were shot, burned, drowned, subjected to gruesome chemical and biological experiments, and otherwise victimized. When China's economic reforms of the 1980s and 1990s led to a renewed Japanese presence in China, the Chinese public and its leaders called for Japanese acknowledgment and apology for these wartime abuses. In Japan, this pitted apologists against strident nationalists, causing a political crisis. In 1992, Emperor Akihito visited China and referred to the war but stopped short of a formal apology. On this issue, the book is not yet closed.

Communist China Arises

After Japan was defeated by the US.-led Western powers in 1945, the civil war in China quickly resumed. The United States, hoping for a stable and friendly government in China, sought to mediate the conflict but did so while recognizing the Nationalist faction as the legitimate government. Furthermore, the United States aided the Nationalists militarily, destroying any chance of genuine and impartial mediation. By 1948, it was clear that Mao Zedong's well-organized militias would defeat Chiang Kai-shek. Chiang kept moving his capital—back to Guangzhou, seat of Sun Yat-sen's first Nationalist government, then back to Chongqing. Late in 1949, following a series of disastrous defeats in which hundreds of thousands of Nationalist forces were killed, the remnants of Chiang's faction gathered Chinese treasures and valuables and fled to the island of Taiwan. There, they took control of the government and proclaimed their own Republic of China.

Meanwhile, in Beijing, standing in front of the assembled masses at the Gate of Heavenly Peace on Tiananmen Square, Mao Zedong, on October 1, 1949, proclaimed the birth of the People's Republic of China.

CHINA'S HUMAN GEOGRAPHY

After a half-century of communist rule, China is a society transformed. It has been said that the year 1949 actually marked the beginning of a new dynasty not so different from the old, an autocratic system that dictated from the top. Mao Zedong, in that view, simply bore the mantle of his dynastic predecessors. Only the family lineage had fallen away; no communist ''comrades'' would succeed each other.

And certainly some of China's old traditions continued during the communist era, but in many other ways China became a totally overhauled society. Benevolent or otherwise, the dynastic rulers of old China headed a country in which—for all its splendor, strength, and cultural richness—the fate of landless people and of serfs often was

undescribably miserable; in which floods, famines, and diseases could decimate the populations of entire regions without any help from the state; in which local lords could (and often did) repress the people with impunity; in which children were sold and brides were bought. The European intrusion made things even worse, bringing slums, starvation, and deprivation to millions who had moved to the cities.

The communist regime, dictatorial though it was, attacked China's weaknesses on many fronts, mobilizing virtually every able-bodied citizen in the process. Land was taken from the wealthy. Farms were collectivized. Dams and levees were built by the hands of thousands. The threat of hunger for millions receded. Health conditions improved. Child labor was reduced. Mao's long tenure (1949–1976) and apparent omnipotence may remind us of the dynastic rulers' frequent longevity and absolutism, but

POLITICAL DIVISIONS OF CHINA

— – — International boundary
– – – – Province boundary
National capital is underlined

| 0 | 200 | 400 | 600 | 800 | 1000 Kilometers |
| 0 | | 200 | | 400 | 600 Miles |

FIGURE 9-7

the new China he left behind was vastly different from the old.

In several areas, however, Mao's strict adherence to Marxist dogma had problematic consequences for China. One of these consequences is familiar to anyone who has studied communist planned economies anywhere: the control by government over all productive capacity, not only agriculture but also industry. China's industrialization program bequeathed the country thousands of inefficient, costly, uncompetitive, state-owned manufacturing plants. Another of Mao's dictums had to do with population. Like the Soviets (and influenced by a horde of Soviet communist advisers and planners), Mao refused to impose or even recommend any form of population policy. As a result, China's population grew explosively during his rule. After Mao's death in 1976, reform-minded communists imposed a one-child policy on China's families and used tough tactics, including forced abortions, to sustain that policy. And, as we will see later, they also adopted a program of limited market activity accompanied by the closure of some of the state-controlled factories. This led the growth of China's Pacific Rim industrial complexes, but it also created economic havoc in, among other places, the industrial Northeast, now China's "Rustbelt." Today, a debate rages among China's leaders as to how far privatization of state industries should go.

The Political Map

Before we proceed to investigate the emerging human geography of China, we should acquaint ourselves with the country's political and administrative framework (Fig. 9-7). For administrative purposes, China is divided into 3 central-government-controlled municipalities, 5 autonomous regions, and 22 provinces.

The three central-government-administered municipalities are the capital, Beijing, its nearby port city, Tianjin, and China's largest metropolis, Shanghai. These are, respectively, the foci of the core areas of China's two most populous and important regions, and direct control over them provides the regime with added security.

The five autonomous regions were established in recognition of the non-Han minorities living there. Some laws that apply to Han Chinese do not apply to certain minorities. As we saw in the case of the former Soviet Union, however, demographic changes and population movements affect such regions, and the policies of the 1940s may not work in the 1990s. Han Chinese immigrants now outnumber several minorities in their own regions. The five autonomous regions (A.R.s) are: (1) Nei Mongol A.R. (Inner Mongolia); (2) Ningxia Hui A.R. (adjacent to Inner Mongolia); (3) Xinjiang Uygur A.R. (China's northwest corner); (4) Guangxi Zhuang A.R. (far south, bordering Vietnam); and (5) Xizang A.R. (Tibet).

TABLE 9-1
China: Population by Major Administrative Divisions*

Provinces	(1990 Census)	1997 Estimate
Anhui	56,180,813	60,104,986
Fujian	30,048,224	33,374,185
Gansu	22,371,141	24,039,444
Guangdong	62,829,236	70,965,672
Guizhou	32,391,066	34,897,450
Hainan	6,557,482	7,049,871
Hebei	61,082,439	67,113,782
Heilongjiang	35,214,873	36,943,443
Henan	85,509,535	91,257,554
Hubei	53,969,210	57,897,111
Hunan	60,659,754	64,993,563
Jiangsu	67,056,519	72,001,347
Jiangxi	37,710,281	40,593,334
Jilin	24,658,721	26,014,949
Liaoning	39,459,697	41,821,931
Qinghai	4,456,946	4,934,755
Shaanxi	32,882,403	34,634,722
Shandong	84,392,827	89,012,765
Shanxi	28,759,014	30,872,563
Sichuan	107,218,173	114,634,114
Yunnan	36,972,610	39,334,923
Zhejiang	41,445,930	44,242,589

Autonomous Regions	(1990 Census)	1997 Estimate
Guangxi Zhuang	42,245,765	45,087,991
Nei Mongol	21,456,798	24,783,331
Ningxia Hui	4,655,451	4,973,348
Xinjiang Uygur	15,155,778	16,889,890
Xizang	2,196,010	2,285,332

*Population data for the three centrally-administered municipalities are shown in the table on p. 434.

China's 22 provinces, like U.S. States, tend to be smallest in the east and larger toward the west. Territorially smallest are the three easternmost provinces on China's coastal bulge: Zhejiang, Jiangsu, and Fujian. The two largest are Qinghai, flanked by Tibet, and Sichuan, China's Midwest.

As with all large countries, some provinces are more important in the national picture than others. The province of Hebei nearly surrounds Beijing and occupies much of the core of the country. The province of Shaanxi is centered on the great ancient city of Xian. In the southeast, momentous economic developments are taking place in the province of Guangdong, whose urban focus is Guangzhou. When, in the pages that follow, we refer to a particular province or region, Figure 9-7 is a useful locational guide.

YOU'D BETTER HAVE ONE CHILD ONLY

From the Field Notes
"We arrived in the Sichuan city of Chengdu on June 17, 1981, and found the campaign to limit families to one offspring in full gear. Over the main square of the city towered a huge billboard with a message that could be interpreted as threatening: YOU'D BETTER HAVE ONE CHILD ONLY! The message was printed in English as well as Chinese; but very few of the citizens of Chengdu spoke the former. It seemed as though English was being used here to convey not only a demographic but also a modernizing message."

With regard to population, many Chinese provinces have more inhabitants than most of the world's countries. According to the latest projections based on the country's 1990 census, China's most populous province, Sichuan, has more than 114 million residents; Shandong and Henan provinces, covering much of the North China Plain, each contain about 90 million. By contrast, the Xizang Autonomous Region (Tibet) has fewer than 2.3 million inhabitants (Table 9-1).

Population Challenges

These enormous numbers should be viewed in context of China's population growth rate. As we have observed previously, high rates of population growth inhibit the national economic development prospects of emerging countries. China faces a special problem. Even a modest rate of population growth adds millions to its citizenry every year.

Mao Zedong refused to institute policies that would reduce the rate of population growth because he believed (encouraged in this belief by Soviet advisers) that this would play into the hands of his capitalist enemies. Fast-growing populations, he said, constituted the only asset many poor, communist countries had. If the capitalist world wanted to reduce growth rates, it must be an anti-communist plot.

But there were leaders in China who realized that their country's progress would be stymied by rapid population growth, and who tried to stem the tide through local propaganda and education. Some of these leaders lost their positions as ''revisionists,'' but their point was not lost on China's leaders of the future. After Mao's death, China embarked on a vigorous population-control program. In the

early 1970s, the annual rate of natural increase was about 3 percent; by the mid-1980s, it was down to 1.2 percent. Families were ordered to have one child only and were penalized for having more—by losing tax advantages, educational opportunities, even housing privileges.

Although it may be said that the one-child policy gained general acceptance, the program (vigorously policed by members of the Communist Party in towns, villages, and hamlets) also had serious negative consequences. The number of abortions soared, sometimes in the third trimester and coerced by party enforcers. Female infanticide also rose as families wanted their one child to be a boy. Families tried to hide their second and third children by sending the ''illegal'' child to family members in other villages; upon discovery by party members, some parents had their houses burned down. When reports of such dreadful byproducts of the population-control campaign leaked to the outside world, the policy was relaxed somewhat during the late 1980s. Certain farming and fishing families were allowed to have a second child to fulfill future labor needs, and rural families could request permission to have a second child if their first baby was a girl. Ethnic minority groups were mostly exempted.

Partly because of disobedience, and partly because of the slightly relaxed implementation of the one-child policy, China's growth rate has inched upward again and is now about 1.2 percent, which means that China adds more than 15 million people every year. The objective of China's planners was to reach 1.2 billion people by the year 2000, but it now appears that the turn-of-the-century total will be close to 1.3 billion. That means an additional *100 million* people to be housed, educated, treated, and, eventually, employed. Against this background, China's economic surge looks rather less promising.

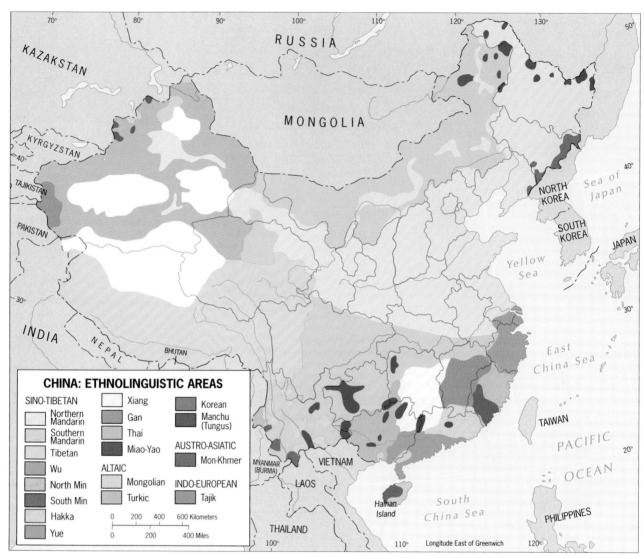

FIGURE 9-8

The Minorities

We asserted earlier that China remains an empire, its government in control of territories and peoples that are in effect colonized. The ethnolinguistic map (Fig. 9-8) appears to confirm this proposition.

The map should be seen in context, however. It does *not* show majority populations, but instead reveals where minorities are concentrated. For example, the Mongolian population in China is shown to be clustered along the southeastern border of Mongolia in the Autonomous Region called Nei Mongol (see Fig. 9-7). But even in that A.R., the Han Chinese, not the Mongols, are now in the majority.

Nevertheless, the map gives definition to the term *China Proper* as the home of the People of Han, the ethnic (Mandarin-speaking) Chinese depicted in tan and light orange

in Figure 9-8. From the far Northeast to the border with Vietnam and from the Pacific coast to the margins of Xinjiang, this Chinese majority dominates. When you compare this map to that of population distribution (to be discussed shortly), it will be clear that the minorities constitute only a small percentage of the country's total. The Han Chinese form the largest and densest clusters.

In any case, China controls non-Chinese areas that are vast, if not populous. The Tibetan group numbers under 3 million, but it extends over all of settled Xizang. Turkic peoples inhabit large areas of Xinjiang. Thai, Vietnamese, and Korean minorities also occupy areas on the margins of Han China. As we will see later, the Southeast Asian minorities in China have participated strongly in the Pacific Rim developments on their doorstep. Hundreds of thousands have migrated from their Autonomous Regions to the economic opportunities along the coast.

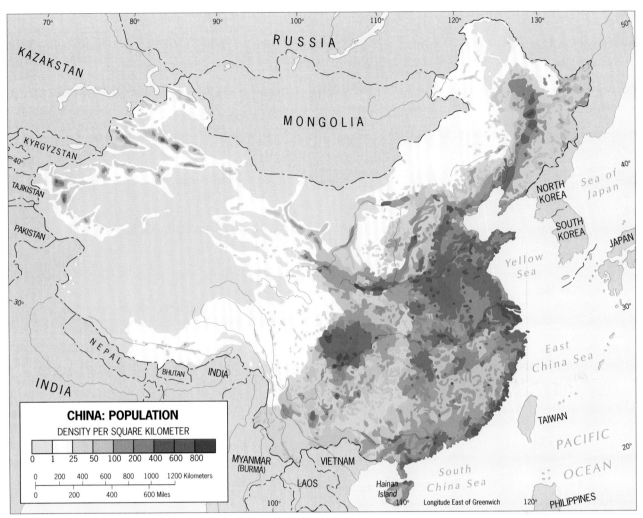

FIGURE 9-9

Numerically, Chinese dominate in China to a far greater degree than Russians dominated their Soviet Empire. But territorially, China's minorities extend over a proportionately larger area. The Ming and Manchu rulers bequeathed the People of Han an empire.

PEOPLE AND PLACES OF CHINA PROPER

A map of China's population distribution (Fig. 9-9) reveals the continuing relationship between the physical stage and its human occupants. In technologically advanced countries, we have noted, people shake off their dependence on what the land can provide; they cluster in cities and in other areas of economic opportunity. This has the effect of depopulating rural areas that may once have been densely inhabited. In China, that stage has not yet been reached. True, there are large cities here, but as in India the great majority of the people (over 70 percent) still live on—and from—the land. Thus the map of population distribution reflects the livability and productivity of China's basins, lowlands, and plains. Compare Figures 9-4A and 9-9, and China's continuing dependence on soil, water, and warmth will be evident.

The population map also suggests that in certain areas limitations of the environment are being overcome. Industrialization in the Northeast, irrigation in the Inner Mongolia Autonomous Region, and oil-well drilling in Xinjiang enabled millions of Chinese to migrate from China Proper into these frontier zones, where they now outnumber the indigenous minorities.

Nevertheless, physiography and demography remain closely related in China. To get a sense of this relationship, it is useful to compare Figures 9-2 (physiography) and 9-9 (population) as our discussion proceeds. On the population

From the Field Notes

"Our first field experience in one of China's Autonomous Regions, the Guangxi-Zhuang A.R., designated for non-Han minorities, had mixed results. Land degradation here was more advanced than in any other part of China visited; desertification seemed to be in progress in many areas. The cause: overuse of the land, and the collapse of what appeared to have been sound terracing systems. A Chinese colleague told us that China's rules for population control *and* land use were relaxed in these Autonomous Regions, often leading to ecological damage."

map, the darker the color, the denser the population: in places it is possible to follow the courses of major rivers through this key. Look, for example at China's Northeast. A ribbon of population follows the Liao and the Songhua River basins. Also note the huge, nearly circular population concentration on the western edge of the red zone. That is the Sichuan Basin, in the upper course of the Chang Jiang.

Four major river basins contain more than three-quarters of China's nearly 1.3 billion people:

1. The Liao-Songhua Basin or the Northeast China Plain
2. The Lower Huang He (Yellow River) Basin called the North China Plain
3. The Upper and Lower Basins of the Chang Jiang (Yangzi [Yangtze] River)
4. The Basins of the Xi (West) and Pearl River

Not all the people living in these river plains are farmers, of course. China's great cities also have developed in these populous areas, from the industrial centers of the Northeast to the Pacific Rim upstarts of the South. Both Beijing and Shanghai lie in river basins. And, as the map shows, the hilly areas between the river basins are not exactly sparsely populated either. Note that the hill country south of the Chang Basin (opposite Taiwan) still has a density of 130-260 people per square mile (50-100 per sq km).

Major Cities of the Realm

Cities	Population* (in millions)
Beijing, China	13.3
Chongqing, China	3.8
Guangzhou, China	4.4
Nanjing, China	3.2
Osaka, Japan	10.6
Pusan, South Korea	4.2
Seoul, South Korea	12.0
Shanghai, China	16.2
Shenyang, China	5.7
Shenzhen, China	3.3
Taipei, Taiwan	3.7
Tianjin, China	11.6
Tokyo, Japan	27.3
Wuhan, China	4.7
Xianggang (Hong Kong), China	5.6
Xian, China	3.6

*Based on 1997 estimates.

Northeast China Plain

The Northeast China Plain is the heartland of China's Northeast, ancestral home of the Manchus who founded China's last dynasty, battleground between Japanese and Russian invaders, Japanese colony, heartland of communist industrial development, and now, losing ground to the industries of the Pacific Rim. As the administrative map (Fig. 9-7) shows, there are three provinces here: Liaoning in the south, facing the Yellow Sea; Jilin in the center; and Heilongjiang, by far the largest, in the north.

When you are familiar with the vast farmlands of the North China Plain, the southernmost part of the Northeast China Plain looks so similar that it seems to be a continuation. Near the coast, farms look as they do near the mouth of the great Huang He; to the east, the Liaodong Peninsula looks very much like the Shandong Peninsula across the water. That impression soon disappears, however. Travel northward, and the Northeast China Plain reveals the effects of cold climate and thin soils. Farmlands are patchy; smokestacks rise, it seems, everywhere. This is industrial, not primarily farming, country. But the landscape looks like a Rustbelt: equipment often is old and outdated, transport systems are inadequate, buildings are in disrepair. Stream, land, and air pollution is dreadful. The Northeast in some ways resembles industrial zones in former communist Eastern Europe.

The Northeast has seen many ups and downs. Although Japanese colonialism was ruthless and exploitive, Japan did build railroads, roads, bridges, factories, and other components of the regional infrastructure. (At one time,

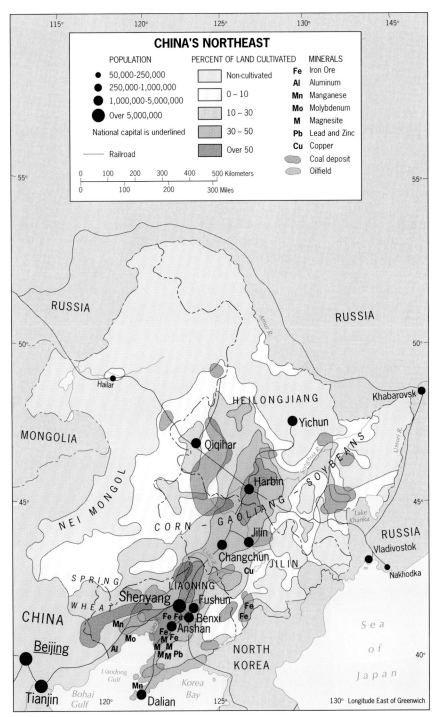

CHINA'S NORTHEAST

POPULATION
- 50,000-250,000
- 250,000-1,000,000
- 1,000,000-5,000,000
- Over 5,000,000

National capital is underlined

— Railroad

PERCENT OF LAND CULTIVATED
- Non-cultivated
- 0 – 10
- 10 – 30
- 30 – 50
- Over 50

MINERALS
- **Fe** Iron Ore
- **Al** Aluminum
- **Mn** Manganese
- **Mo** Molybdenum
- **M** Magnesite
- **Pb** Lead and Zinc
- **Cu** Copper
- Coal deposit
- Oilfield

FIGURE 9-10

half of the entire railroad mileage in China was in the Northeast.) After the Japanese were ousted the Soviets entered the area and looted it of machinery, equipment, and other goods. During the late 1940s, the Northeast was a ravaged frontier. But then the communists took power, and they made the industrial development of the Northeast a priority. From the 1950s until the 1970s, the Northeast led the nation in manufacturing growth. The population of the

region, just a few million in the 1940s, mushroomed to more than 100 million. Towns and cities grew by leaps and bounds.

All this was based on the Northeast's considerable mineral wealth (Fig. 9-10). Iron ore deposits and coalfields lie concentrated in the Liao Basin; aluminum ore, feroalloys, lead, zinc, and other metals are plentiful, too. As the map shows, the region also is well endowed with oil reserves;

the largest of these lie in the Daqing Reserve between Harbin and Qiqihar. The communist planners made Changchun capital of Jilin Province, the Northeast's automobile manufacturing center. Shenyang, capital of Liaoning Province, became the leading steelmaking complex. Fushun and Anshan were assigned other functions, and the communists encouraged Han Chinese to migrate to the Northeast to find employment there. To attract them, the government built apartments, schools, hospitals, recreational facilities, even old-age homes, all to speed the industrial development of the region.

For a time, it worked. During the 1970s, the Northeast produced fully one-quarter of the entire country's industrial output. Stimulated by the needs of the growing cities and towns, farms expanded and crops diversified. The Songhua River Basin in the hinterland of Harbin, capital of Heilongjiang Province, became a productive agricultural zone despite the environmental limitations here. The Northeast was a shining example of what communist planning could accomplish.

Then came the economic restructuring of the 1980s. Under the new, market-driven economic order, the very state companies that had led China's industrial resurgence were unable to adapt. Suddenly they were aging, inefficient, obsolete. Official policy encouraged restructuring and privatization, but the huge state factories could not (and much of their labor force would not) comply. Today, the Northeast's contribution to China's industrial output is a mere 12 percent as the production of factories in the Southeast, from Shanghai to Guangdong, rises annually.

Economic geographers, however, will not write off the prospects of this northeastern corner of China; this area, as we noted, has had its downturns before. The Northeast still contains a storehouse of resources, including extensive stands of oak and other hardwood forests on the mountain slopes that encircle the Northeast China Plain. The locational advantages of the major centers are undiminished. Harbin, for example, lies at the convergence of five railroads and is situated at the head of navigation on the Songhua River, link to the northeast corner of the region and, beyond the Amur River, to Russia's Khabarovsk.

When the Koreas reunite, producing what may become the western Pacific Rim's second Japan, and when the Russian Far East takes off (Vladivostok and Nakhodka are the natural ports for the upper Northeast), this northern frontier of China may yet reverse its decline and enter still another era of economic success.

North China Plain

Take the new superhighway between the port of Tianjin and the capital, Beijing, or the train from Beijing southward to Anyang, and you see a physical landscape almost entirely devoid of relief and a cultural landscape that repeats itself endlessly. Villages, many showing evidence of the collectivization program of the early communist era, pass at such regular intervals that the cultural landscape seems to be controlled by some kind of geometry. Carefully diked canals stretch in all directions. Rows of trees mark fields and farms. Depending on the season, the air is clear, as it is during the moist summer when the wheat crop covers the countryside like a carpet, or choked with dust, which can happen almost any other time of the year. The soil here is a fertile mixture of alluvium and loess (deposits of very fine, windborne silt or dust of glacial origin), and it is so fine that the slightest breeze stirs it up. In the early spring the dust hangs like a fog over the landscape, and everything is covered with a fine powder. Add this to the urban pollution of the cities, and these become as unhealthy as any in the world.

For all we see, there are geographic explanations. The geometry of the settlement pattern results from the flatness of the uniformly fertile surface and the distance villagers could walk to neighboring villages and back during daylight. The canals form part of a vast system of irrigation channels designed to control the waters of the Huang River. The rows of trees serve as windbreaks. And while wheat dominates among the crops of the North China Plain, you will also see millet, sorghum, increasing acreages of soybeans, cotton for the textile industry, tobacco (the Chinese are heavy smokers), and fruits and vegetables for the urban markets.

The North China Plain is one of the world's most heavily populated agricultural areas. Figure 9-9 shows that most of it has a density of more than 1,000 people per square mile (400 per sq km), and in some parts the density is twice as high. Here, the ultimate hope of the Beijing government lay less in land redistribution than in raising yields through improved fertilization, expanded irrigation facilities, and the more intensive use of labor. A series of dams on the Huang River now somewhat reduces the flood danger, but outside the irrigated areas the ever-present problem of rainfall variability and drought persists. The North China Plain has not produced any substantial food surplus even under normal circumstances; thus when the weather turns unfavorable, the situation soon becomes precarious. The specter of famine may have receded, but the food situation is still uncertain in this very critical part of China Proper.

The North China Plain forms an excellent example of the concept of a national **core area**. Not only is it a densely populated, highly productive agricultural zone, but it also is the site of the capital and other major cities, a substantial industrial complex, and several ports, among which Tianjin ranks as one of China's largest (Fig. 9-11). Tianjin, on the Bohai Gulf, is linked by rail and highway (less than a two-hour drive) to Beijing. Like many of China's harbors, that of Tianjin's river port is not particularly good; but Tianjin is well situated to serve not only the northern sector of the North China Plain and the capital, but also the Upper

Huang Basin and Inner Mongolia beyond (see the atlas map at the beginning of this chapter). Tianjin, again like several other Chinese ports, had its modern start as a treaty port, but the city's major growth awaited communist rule. For many decades it had remained a center for light industry and a flood-prone harbor, but after 1949 a new artificial port was constructed and flood canals were dug. Also, Tianjin was chosen as a site for major industrial development, and large investments were made in the chemical industry (in which Tianjin still leads China), in iron and steel production, in heavy machinery manufacturing, and in textiles. Today, with a population exceeding 11 million,

FIGURE 9-11

■ **AMONG THE REALM'S GREAT CITIES...**

Beijing

Beijing (13.3 million), capital of China, lies at the northern apex of the roughly triangular North China Plain, just over 100

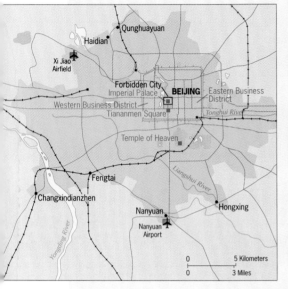

miles (160 km) from its port, Tianjin, on the Bohai Gulf. Urban sprawl has reached the hills and mountains that bound the Plain to the north, a defensive barrier fortified by the builders of the Great Wall. You can reach the Great Wall from central Beijing by road in little more than an hour.

Although settlement on the site of Beijing began thousands of years ago, the city's rise to national prominence began during the Mongol (Yuan) Dynasty, more than seven centuries ago. The Mongols preferred a capital close to their historic heartland, and they endowed the city with walls and palaces. Following the Mongols, later rulers at times moved their capital southward, but the government always came back to Beijing. From the third Ming emperor onward, Beijing was China's focus; it was ideally situated for the Manchus when they took over. During the twentieth century, China's Nationalists

again chose a southern capital, but when the communists prevailed in 1949 they reestablished Beijing as their (and China's) headquarters.

Ruthless destruction of historic monuments, from the Mongols to the communists (and now during China's modernization), has diminished, but not destroyed, Beijing's heritage. From the Forbidden City of the Manchu emperors to the fifteenth-century Temple of Heaven, the capital remains an open-air museum and the cultural focus of China. Successive emperors and aristocrats, moreover, bequeathed the city numerous parks and other recreational spaces, lacking in other cities.

Today Beijing is being transformed as a result of China's new economic policies. A forest of high-rises towers above the traditional cityscape, avenues have been widened, and a beltway is in the making. A new era dawns in this historic capital.

Tianjin is China's third largest metropolis and the center of one of its leading industrial complexes.

Northeast of Tianjin lies the smaller city of Tangshan. In 1976, a devastating earthquake, with its epicenter between Tangshan and Tianjin, killed an estimated 700,000 people, probably the worst natural disaster of this century. So inept was the relief effort mounted by the government in nearby Beijing that widespread resentment may have contributed to the end of the Maoist period of communist rule in China.

Beijing, unlike Tianjin, is China's political, cultural, and educational center. Industrial development here has not matched Tianjin's. The communist administration did, however, greatly expand the municipal area of Beijing, which (as we noted), is not controlled by the province of Hebei but is directly under the central government's authority. In one direction, Beijing was enlarged all the way to the Great Wall—30 miles (50 km) to the north—so that the ''urban'' area includes hundreds of thousands of farmers. Not surprisingly, an enormous total population has been circumscribed, enough to rank Beijing, with its 13.3 million inhabitants, as the seventh largest metropolitan concentration on Earth.

Inner Mongolia

Northwest of the core area as defined by the North China Plain, along the border with the state of Mongolia, lies the Nei Mongol Autonomous Region. Originally established to protect the rights of the approximately 5 million Mongols who live outside the Mongolian state, Nei Mongol (Inner Mongolia) has been the scene of massive immigration by Han Chinese, so that today Chinese outnumber Mongols here by nearly four to one. Near the Mongolian border, Mongols still traverse the steppes with their tents and herds, but elsewhere irrigation and industry have created an essentially Chinese landscape here. The A.R.'s capital, Hohhot, has been eclipsed by Baotou on the Huang He, which supports a ribbon of farm settlements as it crosses the dry land here on the margins of the Gobi and Ordos deserts. With about 25 million inhabitants, Inner Mongolia still cannot compare to the huge numbers that crowd the North China Plain, but recent mineral discoveries have boosted industry in Baotou and livestock herding continues to expand. Nei Mongol may retain its special administrative status, but it is a functional part of Han China in all but name.

■ AMONG THE REALM'S GREAT CITIES...

Shanghai

Sail into the mouth of the great Chang Jiang, and there is little to prepare you for your encounter with China's largest city. For that, you turn left into the narrow Huangpu River, and for the next several hours you will be spellbound. To starboard lies a fleet of Chinese warships. On the port side you pass oil refineries, factories, and neighborhoods. Soon, rusty tankers and freighters line both sides of the stream. High-rise tenements tower behind the cluttered, chaotic waterfront where cranes, sheds, boatyards, piles of rusting scrap iron, mounds of coal, and stacks of cargo vie for space. In the river, your boat competes with ferries, barge trains, cargo ships. Large vessels, anchored in midstream, are being offloaded by dozens of lighters tied up to them in rows. The air is acrid with pollution. The noise—bells, horns, whistles—is deafening.

What strikes you is the vastness and the sameness of Shanghai's cityscape, until you pass beneath the first of two gigantic suspension bridges. Suddenly, everything changes. To the left, or east, the space-age Oriental Pearl Television Tower that looks like a rocket on its launchpad rises above a row of modern, glass-and-chrome skyscrapers that make the Huangpu look like Hong Kong's Victoria

Harbor. To the right stands a row of Victorian buildings, monuments to the British colonialists who made Shanghai a treaty port and started the city on its way to greatness. Everywhere, construction cranes rise above the cityscape. Shanghai looks like one vast construction zone—complete with a 1510-foot office tower (called the World Financial Center) that will become the world's tallest building when it opens before 2000.

Shanghai is an Open Coastal City, where China invites foreign investment by offering privileges to companies. The government is spending heavily to make Shanghai a tiger on the Pacific Rim, and the scene on the east side of the Huangpu may be unmatched in East Asia. There, an immense business and industrial complex, named *Pudong*, has arisen since 1990. Its growth rivals that of Shenzhen. Its future, say locals, will eclipse Xianggang (Hong Kong).

Meanwhile, Shanghai's past is being obliterated. Whole neighborhoods in the central city are falling to the wreckers' bulldozers, their occupants sent far away to those tenements we saw along the river. Even as citizens are ousted, jobseekers from the hinterland jam the streets. Nearly 3 million converged on Shanghai when

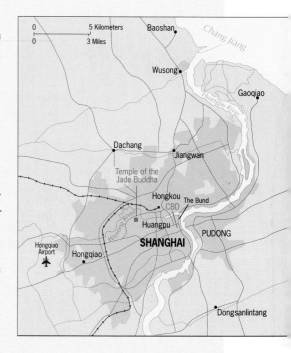

word of the Pudong project spread. About 1 million found work. With 16.2 million residents today, Shanghai remains China's largest city, and by century's end it may be the world's fifth largest. But Shanghai may have traded its soul for a stock market.

Basins of the Chang Jiang

Whereas the North China Plain is one contiguous, agricultural-urban-industrial region that defines China's core area, the Chang Jiang (or Yangzi [Yangtze] River) and its tributaries flow across several basins in their upper, middle, and lower courses, and unlike the shallow, muddy, loess-carrying Huang He—understandably called the *Yellow* River flowing into the *Yellow* Sea—the Chang Jiang is China's most navigable waterway (Fig. 9-11). Oceangoing ships can sail over 600 miles (1,000 km) up the river through Nanjing to the Wuhan conurbation (Wuhan is short for Wuchang, Hanyang, and Hankou, three coalescing cities). Several of the Chang's tributaries also are navigable, and a total of more than 18,500 miles (30,000 km) of water transport routes operate in the entire Chang drainage basin. Therefore, the Chang Jiang is one of China's leading transport corridors. With its tributaries it handles

the trade of a vast area, including nearly all of middle China and sizeable parts of the north and south. The North China Plain may be the core area of China, but the Lower Chang Basin in many ways is its heart.

9.2

Early in Chinese history, when the Chang's basin was being opened up and rice and wheat cultivation began, a canal was built to link this granary to the northern core of old China. Over 1,000 miles (1,600 km) in length, this was the longest artificial waterway in the world, but during the nineteenth century it fell into disrepair. Known as the *Grand Canal*, it was dredged and rebuilt during the period when the Nationalists held control over eastern China. After 1949, the communist regime continued this restoration effort, and much of the canal is now once again open to barge traffic, supplementing the huge fleet of vessels that hauls domestic interregional trade along the east coast (Fig. 9-11).

As Figure 9-12 shows, the Lower Chang Basin is an

FIGURE 9-12

area of rice as well as wheat farming, further proof of its pivotal situation between south and north in the heart of China Proper. Shanghai lies at the entrance to this productive region on a small tributary of the Chang River, the Huangpu (inset map, Fig. 9-11). The city has an immediate hinterland of some 20,000 square miles (50,000 sq km)—smaller than West Virginia—containing more than 50 million people. About two-thirds of these are farmers who produce food, silk filaments, and cotton for the city's industries.

Travel upriver along the Chang Jiang, and you are met by an unending stream of vessels large and small, including numerous "barge trains," as many as six or more barges pulled by a single tug in an effort to save fuel (time is of less concern). In the middle course of the Chang, beyond Yichang, you reach a place where the Chang River cuts through a series of narrow valleys and gorges. The Gezhouba Dam here has already been completed, controlling floods and producing electricity. But now Chinese planners are embarked on what will become the largest engineering

From the Field Notes

"On the right bank of the Huangpu, across from Shanghai's famous waterfront Bund and its Victorian buildings, one of the world's greatest development projects (Pudong) is under way. An entire section of the city has been leveled to make way for an industrial zone that will, China's planners say, eclipse even Xianggang (Hong Kong) and Shenzhen. Billboards and unfinished structures now dot a townscape that will soon transform eastern China's industrial map. But, as we learned during a visit in 1995, a serious problem looms. The weight of all the new construction, plus the increased drawing of water from beneath the city, is causing a measurable subsidence of the ground—in a part of the Chang Jiang delta where the surface already is barely above sea level."

project of its kind in the world, the *New China Dam*. The Three Gorges segment of the Chang valley (Fig. 9-11) is a series of canyons of majestic proportions, which has inspired Chinese painters and poets for many centuries. The new dam, more than 600 feet (180 m) high, will submerge scenic cliffs, displace more than 1 million people, and destroy natural environments. A debate over the project has raged in China for years, and Chinese environmentalists continue to protest the project even as scientists are gathering what they can of the geology and botany before the gorges are inundated. In 1996 it appeared that nothing would stop this gigantic scheme, which eventually will provide electricity for a vast region of central China.

Still farther up the Chang River we reach the city of Chongqing, a river port that, before the dam projects got under way, could be reached by boats of up to 1,000 tons

all the way from the coast. But even without this link, the river remains important here as a transport route, because here lies the Sichuan Basin, the upper basin of the Chang, with almost 115 million inhabitants in 1997. Sichuan's comparative remoteness always has made this a problem province for China's rulers, imperial or communist. The capital, Chengdu, was one of China's most active centers of dissent during the 1989 pro-democracy movement. Sichuan's fertile, well-watered soils yield an amazing variety of crops. In addition to grains (rice, wheat, corn), there are soybeans, tea, sweet potatoes, sugarcane, and many fruits and vegetables. In many ways the Sichuan Basin is a country unto itself. The people here are Han Chinese, but cultural landscapes reveal the area's proximity to Southeast Asia. Along the great riverine axis of the Chang, China changes character time and again.

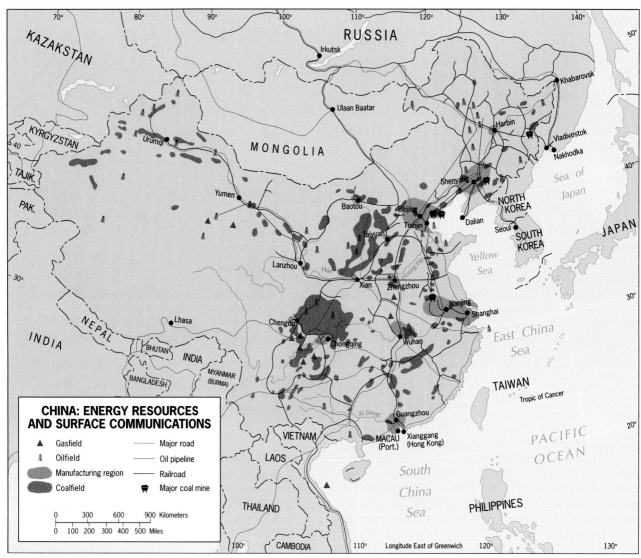

FIGURE 9-13

Basins of the Xi (West) and Pearl River

In fact, the Xi River *is* the Pearl River, a colonial name the Xi continues to carry in its lowest course. As the population map (Fig. 9-9) reveals, however, the Xi River and its basin are no match for the Huang or the Chang. Local relief is higher, farmlands are more confined, and water supply is a problem, especially in interior areas. On the other hand, the long growing season of the warm, subtropical climate permits the double-cropping of rice.

Again unlike the Huang and the Chang, which rise in the interior mountains of Xizang (Tibet), the Xi is a relatively shorter stream that has its source on the Yunnan Plateau. It does, however, also have important cities at and near its mouth: Guangzhou, the capital of Guangdong Province, and, at its estuary, Xianggang (Hong Kong).

South China, for which the Xi (the name means *West*) serves as the major artery, is an important subregion of China for reasons other than agriculture. This area continues to be the focus of China's Pacific Rim development, and here a new China is being born. This is so despite a relative paucity of natural resources (Fig. 9-13). The map shows the location of China's major industrial complexes; none lies farther from energy (or other) resources than the South. Note that the Sichuan Basin, centered on Chengdu, and Shaanxi Province (anchored by Xian) are both well endowed with coal reserves. In the southern part of Sichuan, moreover, lie important reserves of natural gas.

CHINA PROPER'S CONTINUING DEVELOPMENT

As Figure 9-13 shows, energy resources are widely dispersed throughout eastern and northern China, and additional discoveries are likely. The continental shelf off the China coast is especially promising. Pipelines connect oilfields in the Northeast to the industrial region of Beijing–Tianjin, and the development of pipeline links is continuing in Xinjiang as well. Unlike nearby Japan, China has its own substantial energy resources.

China's railroad network (also shown in Fig. 9-13) is inadequate. Sun Yat-sen had envisioned a China whose distant regions would be connected by a well-planned rail system, but his design could not be executed because the country's instability and lack of capital held it back. Indeed, much of China's network, notably in the Northeast, was laid out and built under Japanese and (later) Soviet engineers. A national scheme did not evolve prior to the communist period; the first railroad bridges across the Chang River (at Wuhan and Chongqing) were not built until the 1950s. Double-tracking is proceeding, but the system is constantly strained by the country's growing needs. China does not yet have a web of superhighways with trucks competing against the railroads for cargoes. Road travel is slow and inefficient, with the roads clogged with vehicles ranging from bicycles and horse-drawn wagons to cars and buses. Expanding and improving China's rail network remains a national priority in this developing country with its substantial distances, far-flung resources, and huge population.

China Proper, as we have seen, in many ways is the dominant region of the East Asian realm. All the other regions—Xizang (Tibet), Xinjiang, Mongolia, and Japan–Korea–Taiwan—lie on its periphery. And, as we have noted, Pacific Rim developments are creating a new region on China's own coastal margin. Before focusing on China's Pacific Rim transformation, we first examine the remaining four regions of East Asia.

❖ XIZANG (TIBET)

As the climatic and physiographic maps demonstrate, harsh physical environments dominate this colony of China. It is a vast, sparsely peopled area designated as an Autonomous Region, but in fact Tibet is an occupied society. In terms of its human geography, we should take note of two areas.

The first of these is the core area of Tibetan culture, between the Himalayas to the south and the Transhimalayas lying not far to the north. In this area, some valleys lie below 7,000 feet (2,100 m); the climate is comparatively mild, and some cultivation is possible. Here lies Tibet's main population cluster, including the crossroads capital of Lhasa. The Chinese government has made investments to develop these valleys, which contain excellent sites for hydroelectric power projects (some have been put to use in a few light industries) and several promising mineral deposits. The second area of interest is the Qaidam Basin in the north. This basin lies thousands of feet below the surrounding Kunlun and Altun Mountains and has always contained a concentration of nomadic pastoralists. Recently, however, exploration has revealed the presence of oilfields and coal reserves below the surface of the Qaidam Basin, and the development of these resources is now under way.

As Figure 9-5 reminds us, Tibet came under Chinese domination during the Manchu (Qing) Dynasty in 1720, but the territory regained its separate status in the late nineteenth century. China's communist regime regained control following the invasion of 1950; in 1959, China crushed a Tibetan uprising after Tibetan villagers tried to resist the Chinese presence. Tibetan society had been organized around the fortress-like monasteries of Buddhist monks who paid allegiance to their supreme leader, the Dalai Lama. The Chinese wanted to modernize this feudal system, but the Tibetans clung to their traditions. In 1959, they proved no match for the Chinese armed forces; the Dalai Lama was ousted, and the monasteries were emptied. The Chinese destroyed much of Tibet's cultural heritage, looting the region of its religious treasures and works of art. Their harsh rule took a severe toll on Tibetan society, but after Mao's death in 1976 the Chinese relaxed their tight control. Although amends were made (religious treasures were returned to Xizang, monastery reconstruction was permitted, and Buddhist religious life resumed), pro-independence rioting has been frequent since 1987, and the Chinese have again tightened their grip. Since its annexation in 1965, Xizang has been administered as an Autonomous Region. Although it is large in territorial size, Tibet's population remains slightly more than 2 million.

Visiting Chinese-occupied Tibet is a depressing experience for the neutral observer. Tibetan domestic architecture has beauty and expressiveness, but the Chinese have built ugly, gray structures to house the occupiers—often directly in front of Tibetan monasteries and shrines. This juxtaposition of Chinese banality and Tibetan civility seems calculated to perpetuate hostility, an anachronism in this supposedly postcolonial world. During the rule of Mao Zedong, the Chinese officially regarded Tibet as a territorial cushion against India (then allied with the Soviet Union). The demise of the Soviet Empire has not softened Beijing's stand.

❖ XINJIANG

As Figure 9-7 shows, an elongated Chinese province, Gansu, borders Inner Mongolia to the west and south. A look at surface communications indicates why Gansu is there: it contains the vital Hexi Corridor, a road and rail route that links Xian (and thus Han China via the provincial capital, Lanzhou) to Xinjiang, China's potentially vital northwestern corner.

Xinjiang is China's largest single administrative area, an Autonomous Region covering more than one-sixth of the entire country. It borders Russia, Kazakstan, Kyrgyzstan, Tajikistan, the Wakhan Corridor of Afghanistan, Pakistan, and Jammu and Kashmir (where China has a boundary dispute with India). As the map of ethnic minorities (Fig. 9-8) reveals, this is a region of many and diverse peoples.

Since 1949, the Chinese have made a major effort to develop the limited agricultural potential of the Tarim Basin in southwestern Xinjiang (Fig. 9-2). Canals and *qanats* were built, oases enlarged, and the acreage of productive farmland has at least quadrupled from what it was 45 years ago. The long-neglected northern rim of the Tarim Basin, in particular, has been brought into the sphere of development, and fields of cotton and wheat now attest to the success of the program.

The Junggar Basin in the far north (Fig. 9-2), although it contains only about one-third of the Xinjiang Uygur Autonomous Region's approximately 17 million inhabitants, has a number of assets of importance to China. First, it has long been the site of strategic east-west routes. Second, the main westward rail link toward Kazakstan runs from Xian in eastern China through Yumen in Gansu and Ürümqi in Junggar. Third, the Junggar Basin is proving to contain sizeable oilfields (as is the Tarim Basin), notably around the town of Karamay near the border with Kazakstan. Pipelines have been laid all the way east to Yumen and Lanzhou, where refineries have been built. Thus it is not altogether surprising that the Autonomous Region's capital of Ürümqi lies in the less populous but strategically important northern part of Xinjiang.

Xinjiang's population is now more than one-third Chinese (up from 5 percent in 1949) as a result of Beijing's determination to integrate this distant region into the national framework. However, the ethnic Chinese proportion stopped growing here during the mid-1980s and is now stable. The majority of the people in Xinjiang are Muslim Uygurs, Kazaks, and Kyrgyz, with cultural affinities across the border to all five of the republics of Turkestan (see Fig. 6-17). The Uygurs, most numerous among these peoples (about 50 percent of the population), have for centuries been concentrated in the oases of the Tarim Basin, and the mobile Kazaks and Kyrgyz circulate along nomadic routes that cross international borders.

We also should contemplate the potential strategic significance of Xinjiang. China's space program and nuclear weapons development program long have been based here, and this region of China lies closest to the Russian heartland. The prospect of a confrontation between China and Russia seems remote today, but the two giants have historic, territorial, and ideological differences. Should a power struggle develop, Xinjiang would be one of China's greatest assets. More immediately, however, the challenge may be one of continued control and stability. The transformation of Turkestan (see Chapter 6) already is affecting this remote Chinese frontier, where notions of independence and Islamic revival have long been suppressed.

❖ MONGOLIA

Between China's Inner Mongolia and Xinjiang to the south, and Russia's Eastern Frontier region to the north, lies a vast, landlocked, isolated country called Mongolia (Fig. 9-1). With only 2.3 million inhabitants in an area larger than Alaska, Mongolia is a steppe- and desert-dominated vacuum between two of the world's most powerful countries. Mongolia's historical geography is tied to Russia and the former Soviet Union as well as to China. Mongolia was part of the great Mongol Empire, source of those great waves of horsemen who rode into the Slavic heartland seven centuries ago. The Mongols also invaded China, establishing the short-lived Yuan Dynasty (1279–1368). Eventually, however, the Mongol domain fell under Chinese control. From the late 1600s until 1911, Mongolia was part of the Chinese Empire.

The Mongolians seized independence in 1911 while China was in revolutionary chaos. In those days, the area of present-day Mongolia was known as Outer Mongolia; the ill-defined region to its southeast and east was called Inner Mongolia. As the Chinese tried once again to assert their hegemony over Outer Mongolia, the Russian Revolution broke out. This gave the (Outer) Mongolians another chance to hold off the Chinese. In the early 1920s, the country became a People's Republic on the Soviet model, with Soviet protection. Inner Mongolia remained a Chinese-dominated frontier and became the Nei Mongol Autonomous Region under Beijing. The border between the two Mongolias is still heavily militarized.

Today, with its vast deserts, grassy plains, forest-clad mountains, and fish-filled lakes, Mongolia lies uneasily between larger, unstable neighbors, neither of which can permit its incorporation into the other. Despite its historic associations, ethnic affinities, and cultural involvement with China, Mongolia's development for some 70 years was guided by the Soviets. The country was made a member of COMECON (the Council for Mutual Economic Assistance, a now-defunct, Soviet-led economic union constituted by its then Eastern European satellite countries); the Soviet Communist Party controlled its political life; and Soviet armed forces were stationed within its borders. The capital, Ulaan Baatar, the country's largest city, lies close to the Lake Baykal subregion of Russia's Eastern Frontier. China seems far away from here, across the forbidding Gobi Desert.

But Mongolia's cultural and economic landscapes are East Asian, despite the Soviet/Russian veneer, and the winds of political change have reached here too. The Mongolian alphabet, abolished under Soviet educational regulations, has recently been reinstated. Public demonstrations in support of political reforms led to democratic elections. Mongolian nationalism is on the rise, but what are its options? Russia's eastern zone is not immune to the instability that now marks its western and southern periphery; and China remains a communist state politically.

Mongolia today is a landlocked, isolated, vulnerable **buffer state**, wedged between larger, more populous, more powerful neighbors. Most of its people still ride their horses and herd their livestock across the Wyoming-like countryside. Below the surface, however, lie valuable minerals. In the north, Mongolian territory leaves only a narrow corridor of Russian soil between it and Lake Baykal. Will Mongolia escape the fate of those other inner-Asian buffers, Tibet and Afghanistan?

❖ THE JAKOTA TRIANGLE

The past still hangs heavily over the China of Sichuan and Shaanxi, Xizang and Xinjiang. We turn now to the region that reveals the future of East Asia, where Pacific Rim economic development has transformed community and society: Japan, Korea (South Korea at present, but probably a reunited Korea in the future), and Taiwan. We call this region the *Jakota Triangle* (Fig. 9-14). It is a region of great cities, prodigious consumption of raw materials, state-of-the-art industrialization, voluminous exports, global linkages, trade surpluses, and rapid development. It also is a region of social problems, political uncertainties, and vulnerabilities.

JAPAN

When we assess the prospects of China assuming a position of superpower on the world stage, it is well to remember what happened in Japan in the nineteenth century. In 1868, a group of reform-minded modernizers seized power from the old guard, and by the end of the century Japan was a military as well as an economic force. From the factories in and around Tokyo and from urban-industrial complexes elsewhere poured forth a stream of weapons and equipment the Japanese used to embark on a campaign of colonial expansion. By the mid-1930s, Japan lay at the center of an empire that included all of the Korean Peninsula, the whole of China's Northeast (which the Japanese called Manchukuo), the Ryukyu Islands, and Taiwan, as well as the southern half of Sakhalin Island (called Karafuto). Not even a disastrous earthquake, which destroyed much of Tokyo in 1923, could slow the Japanese drive.

The Second World War saw Japan expand its domain farther than the architects of the 1868 "modernization" could have anticipated. By early December 1941, Japan had conquered large parts of China Proper, all of French

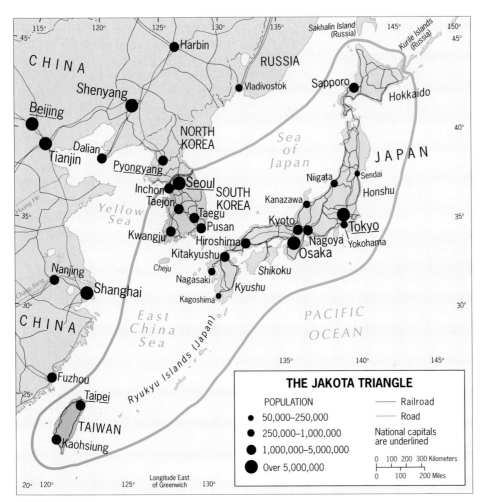

FIGURE 9-14

Indochina to the south, and most of the small islands in the western Pacific. Then, on December 7, 1941, Japanese-built aircraft carriers moved Tokyo's warplanes within striking range of Hawaii, and the surprise attack on Pearl Harbor underscored Japan's confidence in its war machine. Soon the Japanese overran the Philippines, the (then) Netherlands East Indies, Thailand, and British Burma and Malaya, and drove a wide corridor through the heart of China to the border with Vietnam.

A few years later, Japan's expansionist era was over. Its armies had been driven from virtually all its possessions, and when American nuclear bombs devastated two Japanese cities in 1945, the country lay in ruins. But once again Japan, aided this time by an enlightened U.S. postwar administration, surmounted disaster.

Japan's economic recovery and its rise to the status of world economic superpower has been the story of the second half of the twentieth century. Japan lost the war and its empire, but it scored many economic victories in a new global arena. Japan became an industrial giant, a technological pacesetter, a fully urbanized society, a political

power, and a thriving affluent nation. No city in the world today is without Japanese cars in its streets; few photography stores lack Japanese cameras and film; laboratories the world over use Japanese optical equipment. From microwave ovens to VCRs, from oceangoing ships to camcorders, Japanese goods flood the world's markets.

Japan's brief colonial adventure helped lay the groundwork for other economic successes along the western Pacific Rim. The Japanese ruthlessly exploited Korean and Formosan (Taiwanese) natural and human resources, but they also installed a new economic order there. After World War II, this infrastructure facilitated the economic transition—and soon made both Taiwan and South Korea competitors on world markets.

Spatial Limitations

In discussing the economic geography of Europe, we noted how the comparatively small island of Britain became the crucible of an Industrial Revolution that gave it a huge head start and advantage over the mainland, across which the Industrial Revolution spread decades later. Britain was not large, but its insular location, its local raw materials, the skills of its engineers, the presence of a large labor force, and the nature of its social organization combined to endow it not only with industrial strength but also with a vast overseas empire.

After the Japanese modernizers took over control of the country in 1868—an event known as the *Meiji Restoration* (the return of "enlightened rule" centered on the Emperor Meiji)—they turned to Britain for guidance in their efforts to reform their nation and its economy. The British, in the decades that followed, made many contributions. They advised the Japanese on the layout of cities and the construc-

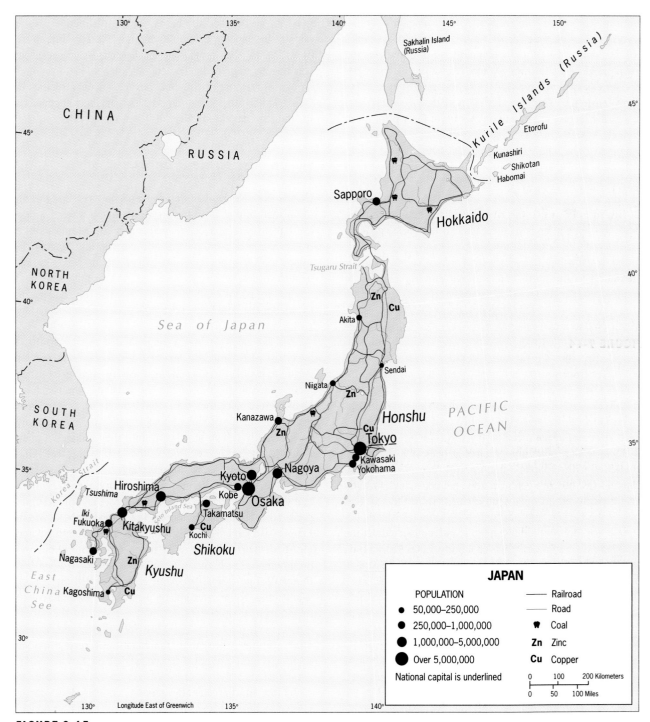

FIGURE 9-15

tion of a railroad network, on the location of industrial plants and the organization of education. The British influence still is visible in the Japanese cultural landscape today: the Japanese, like the British, drive on the left side of the road. Consider the consequences of this in the U.S. effort to open the Japanese market to U.S. automobiles!

The Japanese reformers of the late nineteenth century undoubtedly saw many geographic similarities between Britain and Japan. At that time, most of what mattered in Japan was concentrated on the country's largest island, Honshu (literally, "mainland"). The ancient capital, Kyoto, lay in the interior, but the modernizers wanted a coastal, outward-looking headquarters. So they chose the town of Edo, on a large bay where Honshu's eastern coastline turns sharply (Fig. 9-15). They renamed the place Tokyo ("eastern capital"), and little more than a century later it was the largest urban agglomeration in the world. From Honshu's coasts, it was no great distance to the shores of mainland Asia, where sources of raw materials and potential markets for Japanese products lay. The notion of a greater Japanese empire followed naturally from the British example.

But in other ways, the British and Japanese archipelagoes, at opposite ends of the Eurasian landmass, differed considerably. In terms of total area, Japan is larger. In addition to Honshu, Japan has three other large islands—Hokkaido to the north and Shikoku and Kyushu to the south—plus numerous small islands and islets, for a total land area of about 146,000 square miles (377,000 sq km). Much of this area is mountainous and steep-sloped, geologically young, earthquake-prone, and studded with volcanoes. Britain has lower relief, is older geologically, does not suffer from severe earthquakes, and has no active volcanoes. And in terms of the raw materials for industry, Britain was much better endowed than Japan. Self-sufficiency in iron ore and high-quality coal gave Britain the head start that lasted a century.

Japan's high-relief topography has been an ever-present challenge. All of Japan's major cities, except the ancient capital of Kyoto, are perched along the coast, and virtually all lie partly on artificial land claimed from the sea. Sail into Kobe harbor, and you will pass artificial islands designed for high-volume shipping and connected to the mainland by automatic space-age trains. Enter Tokyo Bay, and the refineries and factories to your east and west stand on huge expanses of landfill that have pushed the bay's shoreline outward. With 125 million people, the vast majority (77 percent) living in towns and cities, Japan uses its habitable living space very intensively—and expands it wherever possible.

As Figure 9-16 shows, farmland in Japan is both limited and regionally fragmented. Urban sprawl has invaded much cultivable land. In the hinterland of Tokyo lies the Kanto Plain; around Nagoya, the Nobi Plain; and surrounding Osaka, the Kansai District—each a major farming zone under relentless urban pressure. All three of these plains

lie within Japan's fragmented but well-defined core area (delimited by the red line on the map), the heart of Japan's prodigious manufacturing complex.

For nearly a half-century, Japan's seemingly insurmountable economic lead, its almost unique (for so large a population) ethnic homogeneity, its insular situation, and its cultural uniformity combined to justify its designation as a discrete geographic realm, not a region of a greater East Asia. But times are moving rapidly on the rim of the western Pacific, and change—both within Japan and beyond it—has eroded the bases for this distinction. In the mid-1990s, Japan's economic engine faltered while those of its Pacific Rim neighbors accelerated, its social fabric was torn by unprecedented acts of violence, its foreign relations were soured by disputes over wartime atrocities in China and Korea. Japan stood alone in the vanguard of Pacific Rim developments, but it is no longer discrete.

Early Directions

On the map, the Korean Peninsula seems to extend like an unfinished bridge from the Asian mainland toward southern Japan. Two Japanese-owned islands, Tsushima and Iki, in the Korea Strait come close to completing the connection (Fig. 9-15). In fact, Korea *was* linked to Japan in prehistoric times, when sea levels were lower. Peoples and cultural infusions reached Japan from Asia time and again, even after the land bridge between them was inundated by rising waters.

The Japanese islands were inhabited even before the ancestors of the modern Japanese arrived from Asia. The Ainu, a people of Caucasian ancestry, had established themselves on all four of the major islands thousands of years before. But the Ainu could not hold their ground against the better organized and armed invaders, and they were steadily driven northward until they retained a mere foothold on Hokkaido. Today, the last vestiges of Ainu ancestry and culture are disappearing.

By the sixteenth century, a highly individualistic culture had evolved in Japan, characterized by social practices and personal motivations based on the Shinto belief system. People worshiped many gods, venerated ancestors, and glorified their emperor as a divine figure, infallible and omnipotent. The emperor's military ruler, or *shogun*, led the campaign against the Ainu and controlled the nation. The capital, Kyoto, became an assemblage of hundreds of magnificent temples, shrines, and gardens. In their local architecture, modes of dress, festivals, theater, music, and other cultural expressions, the Japanese forged a unique society. Handicraft industries, based on local raw materials, abounded.

In the mid-nineteenth century, nonetheless, Japan hardly seemed destined to become Asia's leading power. For 300 years the country had been closed to outside influences, and Japanese society was stagnant and tradition-

FIGURE 9-16

bound. When the European colonizers first appeared on Asia's shores, their merchants and missionaries were tolerated, even welcomed, in Japan. But as the European presence in East Asia grew stronger, the Japanese began to shut their doors. Toward the end of the sixteenth century, the emperor decreed that all foreign traders and missionaries should be expelled. Christianity (especially Roman Catholicism) had gained a considerable foothold in Japan, but now this was feared as a prelude to colonial conquest. Early in the seventeenth century, the *shogun* launched a massive and bloody campaign that practically stamped out Christianity.

From the Field Notes

"Japan may have modernized, but here in Kyoto you can contemplate what the Japan before Tokyo was like. Japanese tradition and concepts of beauty flourish here amid the urban sprawl: magnificent temples, tranquil gardens, graceful arches over busy streets. During the Second World War, U.S. bombs devastated many Japanese cities, but American strategists knew that this once-powerful and ever-spiritual capital of the country must be saved. As a result, Kyoto's heritage is a direct link to Japan's past, when *Shogun* rulers and Shinto priests created here the pinnacle of Japanese culture in the form of castles and temples. This is just one of about 1600 temple entrances, an elaborate wooden gate flanked by stone lions and equipped with a large bronze bell that allows visitors to announce themselves to the gods."

Determined not to suffer the same fate as the Philippines, which had by then fallen to Spain, the Japanese henceforth allowed only the most minimal contact with Europeans. A few Dutch traders, confined to a little island near the city of Nagasaki, were for many decades the sole representatives of Europe in all of Japan. And thus Japan entered a long period of isolation, lasting past the middle of the nineteenth century.

Japan could maintain its aloofness while other areas were falling victim to the colonial tide because of its strong central authority and well-organized military, because of the natural protection provided by its islands, and because

of its remoteness along East Asia's difficult northern coast. Also, Japan's isolation was far less splendid than that of China, whose exquisite silks, prized teas, and skillfully made wares attracted traders and usurpers.

When Japan finally came face to face with the new weaponry of its old adversaries, it had no answers. In the 1850s, the steel-hulled "black ships" of the American fleet sailed into Japanese harbors and the Americans extracted one-sided trade agreements. Soon the British, French, and Dutch were also on the scene, seeking similar treaties. When there was local resistance to their show of strength, the Americans quickly demonstrated their superiority by shelling parts of the Japanese coast. By the late 1860s, even as Japan's modernizers were about to overturn the old order, no doubt remained that Japan's protracted isolation had come to an end.

A Japanese Colonial Empire

When the architects of the Meiji Restoration confronted the challenge to build a Japan capable of competing against powerful adversaries in a changing world, they took stock of the country's assets and liabilities. Material assets, they found, were limited. To achieve industrialization, coal and iron ore were needed. In terms of coal, there was enough to support initial industrialization. Coalfields in Hokkaido and Kyushu were located near the coast; since the new industries also were on the coast, cheap water transportation was possible. As a result, the shores of the (Seto) Inland Sea (Fig. 9-15) became the sites of many factories. But there was little iron ore, certainly nothing like what was available domestically in Britain, and not enough to sustain the massive industrialization the reformers had in mind. This commodity would have to be purchased overseas and imported.

On the positive side, manufacturing—light manufacturing of the handicraft type—already was widespread in Japan. In cottage industries and in community workshops, the Japanese produced textiles, porcelain, wood products, and metal goods. The small ore deposits in the country were enough to supply these local industries. Power came from human arms and legs and from wheels driven by water; the chief source of fuel was charcoal. Importantly, Japan did have an industrial tradition and an experienced labor force that possessed appropriate manufacturing skills. All this was not enough to lead directly to industrial modernization, but it did hold promise for capital formation. The community and home workshops were integrated into larger units, hydroelectric plants were built, and some thermal (coal-fired) power stations were constructed in critical areas.

Now for the first time, Japanese goods (albeit of the light manufacture variety) began to compete with Western products on international markets. The Japanese planners resisted any infusion of Western capital, however. While Western support would have accelerated the industrialization process, it would have cost Japan its economic autonomy. Instead, farmers were more heavily taxed, and the money thus earned was poured into the industrialization effort.

Another advantage for Japan's planners lay in the country's military tradition. Although the *shogun*'s forces had been unable to repel the invasions of the 1850s, this was the result of outdated equipment, not lack of manpower or discipline. So while the economic transformation gathered momentum, military forces, too, were modernized. Barely more than a decade after the Meiji Restoration, Japan laid claim to its first Pacific prize: the Ryukyu Islands (1879). A Japanese colonial empire was in the making.

When Japan gained control over its first major East Asian colonies, Taiwan and Korea, its domestic raw-material problems were essentially solved, and a huge labor pool fell under its sway as well. High-grade coal, high-quality iron ore, and other resources were shipped to the factories of Japan; and in Taiwan as well as in Korea, the Japanese built additional manufacturing plants to augment production. This, in turn, provided the equipment to sustain the subsequent drive into China and Southeast Asia.

Modernization

The reformers, who in 1868 set Japan on a new course, probably did not anticipate that three generations later their country would lie at the heart of a major empire sustained by massive military might. They set into motion a process of **modernization**, but they managed to build on, not replace, Japanese cultural traditions. We in the Western world tend to equate modernization with Westernization: urbanization, the spread of transport and communications facilities, the establishment of a market (money) economy, the breakdown of local traditional communities, the proliferation of formal schooling, the acceptance and adoption of foreign innovations. In the non-Western world, the process often is viewed differently. There, "modernization" is seen as an outgrowth of colonialism, the perpetuation of a system of wealth accumulation introduced by foreigners driven by greed. The local elites who replaced the colonizers in the newly independent states, in that view, only continue the disruption of traditional societies, not their true modernization. Traditional societies, they argue, can be modernized without being Westernized.

In this context, Japan's modernization is unique in many ways. Having long resisted foreign intrusion, the Japanese did not achieve the transformation of their society by importing a Trojan horse; it was done by Japanese planners, building on the existing Japanese infrastructure, to fulfill Japanese objectives. Certainly Japan imported foreign technologies and adopted innovations from the British and others. But the Japan that was built, a unique combination of modern and traditional elements, was basically an indigenous achievement.

Relative Location

Japan's changing fortunes over the past century reveal the influence of relative location in the country's development. Until well after the Meiji Restoration took place, Britain, on the other side of the Eurasian landmass, lay at the center of a global empire. The colonization and Europeanization of the world were in full swing. The United States was still a developing country, and the Pacific Ocean was an avenue for European imperial competition. Japan, even while it was conquering and consolidating its first East Asian colonies (the Ryukyus, Taiwan, Korea), lay remote from the mainstream of global change.

Then Japan became embroiled in the Second World War and dealt severe blows to the European colonial armies in Asia. The Europeans never recovered: the French lost Indochina, and the Dutch were forced to abandon their East Indies (now Indonesia). When World War II ended, Japan was a defeated and devastated country, but the Japanese had done much to diminish the European presence in the Pacific Basin. Moreover, the global situation had changed dramatically. The United States, Japan's trans-Pacific neighbor, had become the world's most powerful and wealthiest country, whereas Britain and its global empire were fading. Suddenly Japan was no longer remote from the mainstream of global action: now the Pacific was becoming the avenue to the world's richest markets. Japan's **relative location**—its situation relative to the economic and political foci of the world—had changed. Therein lay much of the opportunity the Japanese seized after the postwar rebuilding of their country.

▶ Japan's Spatial Organization

Imagine this: 125 million people crowded into a territory the size of the State of Montana (population 900,000), most of it mountainous, subject to frequent earthquakes and volcanism, with no domestic oilfields, very little coal and few raw materials for industry, and not much level land for farming. If Japan today were an underdeveloped country in need of food relief and foreign aid, explanations would abound: overpopulation; inefficient farming; energy shortages.

True, only an estimated 18 percent of Japan's national territory is designated as habitable. And Japan's large population is crowded into some of the world's largest cities. Moreover, Japan's agriculture *is* not especially efficient. But Japan defeated the odds by calling on old Japanese virtues: organizational efficacy, massive productivity, dedication to quality, and adherence to common goals. Even before the Meiji Restoration, Japan was a tightly organized, populous country with as many as 30 million citizens. Historians suggest that Edo, even before it was selected as the capital and renamed Tokyo, may have been the world's largest urban center. The Japanese were no strangers to urban life; they knew manufacturing, and they prized social and economic order.

Areal Functional Organization

All this proved invaluable to the modernizers when they set Japan on its new course. The new industrial growth of the country could be based on the urban and manufacturing development that was already taking place. As we noted, Japan does not possess major domestic raw-material sources, so no substantial internal reorganization was necessary. However, some cities did possess advantages over others in terms of their sites and their situations relative to those limited local resources and, more importantly, relative to external sources of raw materials. As Japan's regional organization took shape, a hierarchy of cities developed; Tokyo took and kept the lead, but other cities also grew rapidly as industrial centers.

This process was governed by a geographic principle that Allen Philbrick called **areal functional organization**, a set of five interrelated tenets that help explain the evolution of regional organization, not only in Japan but throughout the world. Human activity, Philbrick reasoned, has spatial focus. It is concentrated in some locale, whether a farm or factory or store. Every one of these establishments occupies a particular location; no two of them can occupy exactly the same spot on the Earth's surface (even in high-rises, there is a vertical form of absolute location). Nor can any human activity proceed in total isolation, so that interconnections develop among these various establishments. This system of interconnections grows more complex as human capacities and demands expand. Each system (for example, farmers sending crops to market and buying equipment at service centers) forms a unit of areal functional organization. Recall that in the Introduction we referred to functional regions as systems of spatial organization; Philbrick's units of areal functional organization can be mapped as regions. These regions evolve as a result of what he called ''creative imagination'' as people apply their total cultural experience as well as their technological knowhow when they organize and rearrange their living space. Finally, Philbrick suggests, it is possible to recognize levels of development in areal functional organization, a ranking of places and regions based on the type, extent, and intensity of exchange.

In the broadest sense, we can rank regions of human organization into three categories: subsistence, transitional, and exchange. Japan's areal organization reflects the third category. Within each of these categories, individual places can also be ranked on the basis of the number and kinds of activities they generate. Even in regions where subsistence activities dominate, some villages have more interconnections than others. The map of Japan (Fig. 9-16) showing its resources, urban settlements, and surface communications can tell us a great deal about the kind of economy the country has. It looks just like maps of other parts of the world where an exchange type of areal organization has developed—a hierarchy of urban centers ranging from the largest cities to the tiniest hamlets, a dense network of railways and roads connecting these places, and productive agricultural areas near and between the urban centers.

In Japan's case, however, the map shows us something else: it reflects the country's external orientation, its dependence on foreign trade. All primary and secondary regions lie on the coast. Of all the cities in the million-size class, only Kyoto lies in the interior (Fig. 9-15). If we were to deduce that Kyoto does not match Tokyo–Yokohama, Osaka–Kobe, or Nagoya in terms of industrial development, that would be correct—the old capital remains a center of small-scale light manufacturing. Actually, Kyoto's ancient character has been deliberately preserved, and large-scale industries have been discouraged. With its old temples and shrines, its magnificent gardens, and its many workshop and cottage industries, Kyoto remains a link with Japan's premodern past.

Leading Economic Regions

As Figure 9-16 shows, Japan's dominant region of urbanization and industry (along with very productive agriculture) is the Kanto Plain, which contains about one-third of the Japanese population and is focused on the Tokyo–Yokohama–Kawasaki metropolitan area. This gigantic cluster of cities and suburbs (the world's largest urban agglomeration), interspersed with intensively cultivated farmlands, forms the eastern anchor of the country's elongated and fragmented core area. Besides its flatness, the Kanto Plain possesses other advantages: its fine natural harbor at Yokohama, its relatively mild and moist climate, and its central location with respect to the country as a whole. (The region's only disadvantage is its vulnerability to earthquakes—see box titled ''When the Big One Strikes.'') It has also benefited enormously from Tokyo's designation as the modern capital, which coincided with Japan's embarkation on its planned course of economic development. Many industries and businesses chose Tokyo as their headquarters in view of the advantages of proximity to the government's decision makers.

The Tokyo–Yokohama–Kawasaki conurbation has become Japan's leading manufacturing complex, producing more than 20 percent of the country's annual output. The raw materials for all this industry, however, come from far away. For example, the Tokyo area is among the chief steel producers in Japan, using iron ores from the Philip-

From the Field Notes

"Standing on the second level of the Tokyo Tower (Tokyo's answer to the Eiffel Tower) you can see part of the vast cityscape and, in the distance, Tokyo Bay. Under those placid waters lurk mortal danger: there, three tectonic plates are colliding, a process that has already yielded devastating earthquakes and death-dealing tsunamis (seismic sea waves). Ominously, Tokyo is structurally more vulnerable today than it was in 1923, when an earthquake, resulting firestorm, and ensuing tsunami killed some 140,000 people; the city is also far more populous now."

pines, Malaysia, Australia, India, and even Africa; most of the coal is imported from Australia and North America, and the petroleum from Southwest Asia and Indonesia. The Kanto Plain cannot produce nearly enough food for its massive resident population. Imports must come from Canada, the United States, and Australia as well as from other areas in Japan. Thus Tokyo depends completely on its external trading ties for food, raw materials, and markets for its wide variety of products, which run the gamut from children's toys to high-precision optical equipment to the world's largest oceangoing ships.

The second-ranking economic region in Japan's core area is the Osaka–Kobe–Kyoto triangle—also known as the Kansai District—located at the eastern end of the Inland Sea. The position of the Osaka–Kobe conurbation, with respect to the old Manchurian Empire (Manchukuo) created by Japan, was advantageous. Situated at the head of the Inland Sea, Osaka was the major Japanese base for the China trade and the exploitation of Manchuria, but it suffered when the empire was destroyed and its trade connections with China were lost after World War II. Kobe (like Yokohama, which is Japan's chief shipbuilding center) has remained one of the country's busiest ports, handling Inland Sea traffic as well as extensive overseas linkages. Kyoto, as we noted, remains much as it was before Japan's great leap forward, a city of small workshop industries. The Kansai District is also an important farming area. Rice, of course, is the most intensively grown crop in the warm, moist lowlands, but this agricultural zone is not as large as the one on the Kanto Plain. Here is another huge concentration of people that requires a large infusion of foodstuffs every year.

The Kanto Plain and Kansai District, as Figure 9-16 indicates, are the two leading primary regions within Japan's core. Between them lies the Nobi Plain (also called the Chubu District), focused on the industrial metropolis of Nagoya, Japan's leading textile producer. The map indicates some of the Nagoya area's advantages and liabilities. The Nobi Plain is larger than the lowlands of the Kansai District; thus its agricultural productivity is greater, although not as great as that of the Kanto Plain. But Nagoya has neither Tokyo's centrality nor Osaka's position on the Inland Sea: its connections to Tokyo are via the tenuous Sen-en Coastal Strip. Its westward connections are somewhat better, and there are signs that the Nagoya area is coalescing with the Osaka–Kobe conurbation. Still another disadvantage of Nagoya lies in the quality of its port, which is not nearly as good as Tokyo's Yokohama or Osaka's Kobe and has been plagued by silting problems.

Westward from the three regions just discussed—which together constitute what is often called the *Tokaido* megalopolis—extends the Inland Sea, along whose shores the remainder of Japan's core area is continuing to develop. The most impressive growth has occurred around the western entry to the Inland Sea (the Strait of Shimonoseki), where Kitakyushu—a conurbation of five cities on northern Kyushu—constitutes the fourth Japanese manufacturing complex and primary economic region. Honshu and Kyushu are connected by road and railway tunnels, but the northern Kyushu area does not have an urban-industrial equivalent on the Honshu side of the strait. The Kitakyushu conurbation includes Yawata, site of northwest Kyushu's (rapidly declining) coal mines. It was on the basis of this coal that the first steel plant in Japan was built there; for many years it was the country's largest. The advantages of transportation here at the western end of the Inland Sea are obvious: no place in Japan is better situated to do business with Korea and China. As relations with mainland Asia expand, this area will reap many of the benefits. Elsewhere on the Inland Sea coast, the Hiroshima–Kure urban area has a manufacturing base that includes heavy industry. And on the coast of the Korea Strait, Fukuoka and Nagasaki are

WHEN THE BIG ONE STRIKES

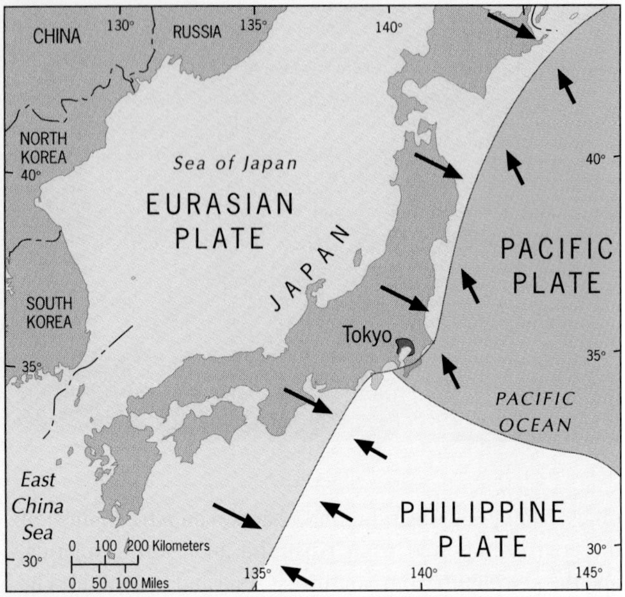

The Tokyo–Yokohama–Kawasaki urban area is the biggest metropolis on Earth. But Tokyo is more than a large city: it constitutes the most densely concentrated financial and industrial complex in the world. In 1996, 9 of the world's 12 largest banks were headquartered in Tokyo. Two-thirds of Japan's businesses worth more than $50 million are also clustered here, many with vast overseas holdings. More than half of Japan's huge industrial profits (averaging about U.S. $100 billion annually since the late 1980s) are generated in the factories of this gigantic metropolitan agglomeration.

But Tokyo has a worrisome environmental history because three active tectonic plates are converging here (Fig. 9-17). All Japanese know about the "70-year" rule: over the past three-and-a-half centuries, the Tokyo area has been struck by major earthquakes roughly every 70 years—in 1633, 1703, 1782, 1853, and 1923. The Great Kanto Earthquake of 1923 set off a firestorm that swept over the city and killed an estimated 140,000 people. Moreover, Tokyo Bay virtually emptied of water; then a *tsunami* (seismic sea wave) roared back in, sweeping homes, factories, and all else before it. That Japan could overcome this disaster was evidence of the strength of its ongoing economic miracle.

Today, Tokyo is more than a national capital. It is a global financial and manufacturing center in which so much of the world's wealth and productive capacity are concentrated that an earthquake comparable to the one of 1923 would have a calamitous effect worldwide. Ominously, the Tokyo of the 1990s is a much more vulnerable place than the Tokyo of the 1920s. True, building regulations are stricter and civilian preparedness is better. But whole expanses of industries have been built on landfill that will liquefy; the city is honeycombed by underground gas lines that will rupture and stoke countless fires; congestion in the area's maze of narrow streets will hamper rescue operations; and many older high-rise buildings do not have the structural integrity that has lately emboldened builders to build skyscrapers of 50 stories and more. Add to

FIGURE 9-17

this the burgeoning population of the Kanto Plain—approaching 30 million on what may well be the most dangerous 4 percent of Japan's territory—and we realize that the next big earthquake in the Tokyo area will not be a remote, local news story.

The Japanese would need cash to rebuild, which would force them to sell many of their worldwide holdings. This would quickly precipitate a global financial crisis. It is a measure of the interconnectedness of our world that we should all hope that nature will break its "70-year" rule in the 1990s and prolong Tokyo's stability.

---

the principal centers—the former an industrial city, the latter a center of large shipyards.

Only one major Japanese manufacturing complex lies outside the belt extending from Tokyo in the east to Kitakyushu in the west—the secondary region centered on Toyama on the Sea of Japan. The advantage here is cheap power from nearby hydroelectric stations, and the cluster of industries reflects it: paper manufacturing, chemical factories, and textile plants have located here. Of course, Figure 9-16 also gives an inadequate picture of the variety and range of industries that exist throughout Japan, many of them oriented to local (and not insignificant) markets. Thousands of manufacturing plants operate in cities and towns other than those shown on this map, even on the

cold northern island of Hokkaido, connected to Honshu by the Seikan rail tunnel, the world's longest, beneath the treacherous Tsugaru Strait.

The map of Japan's areal organization shows the country's core area to be dominated by four primary regions; each of them is primary because they duplicate to some degree the contents of the others. Each contains iron and steel plants, each is served by one major port, and each lies in or near a large, productive farming area. What the map does not show is that each also has its own external connections for the overseas acquisition of raw materials and the sale of finished products. These linkages may even be stronger than those among the four internal regions of the core area. Only in the case of Kyushu and its coal have

■ AMONG THE REALM'S GREAT CITIES . . .

Tokyo

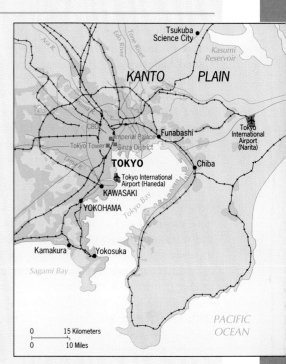

Many an urban agglomeration is named after the city that lies at its heart, and so it is with the largest of all: Tokyo (27.3 million). Even its longer name—Tokyo–Yokohama–Kawasaki—does not begin to describe the congregation of cities and towns that form this crowded metropolis that encircles the head of Tokyo Bay and continues to grow, outward and upward.

Near the waterfront, some of Tokyo's neighborhoods are laid out in a grid pattern. But the urban area has sprawled over hills and valleys, and much of it is a maze of narrow, winding streets and alleys. Circulation is slow and traffic jams are legendary. The train and subway systems, however, are models of efficiency—although at rush hours you must get used to the *shirioshi* pushing you into the cars to get the doors closed.

At the heart of Tokyo lies the Imperial Palace and its moats and private parks. Across the street, buildings retain a respectful low profile, but farther away Tokyo's skyscrapers seem to ignore the peril of earthquakes. Nearby lies one of the world's most famous avenues, the Ginza, lined by department stores and luxury shops. In the distance you can see an edifice that looks like the Eiffel Tower, only taller: this is the Tokyo Tower, a multipurpose structure designed to test lighter Japanese steel, transmit television and radio signals, detect Earth tremors, monitor air pollution, and attract tourists.

Tokyo is the epitome of modernization, but Buddhist temples, Shinto shrines, historic bridges, and serene gardens still grace this burgeoning urban complex, a cultural landscape that reflects Japan's successful marriage of the modern and the traditional.

---

domestic raw materials played much of a role in shaping the nature and location of heavy manufacturing, and these resources are all but depleted today. In the structuring of the country's areal functional organization, therefore, more than just the contents of Japan itself is involved. In this respect Japan is not unique: all countries that have exchange-type organizations must, to some degree, adjust their spatial forms and functions to the external interconnections required for progress. But it would be difficult to find a country in which this is truer than in Japan.

Food: Production and Price

Japan's economic modernization so occupies center stage that it is easy to forget the country's considerable achievements in the venerable field of agriculture. Japan's planners, who today are no less interested in closing the food gap than in expanding industries, have created extensive networks of experiment stations to promote mechanization, optimal seed selection and fertilizer use, and information services to distribute to farmers as rapidly as possible knowledge that is useful for enhancing crop yields. Although this program has been very successful, Japan faces the unalterable reality of its stubborn topography: it simply does not have sufficient land to farm. Populous Japan has one of the highest physiologic population densities—6,170 per square mile (2,380 per sq km)—on Earth (this concept is discussed on p. 372).

The Japanese go to extraordinary lengths to maximize their limited farming opportunities, making huge investments in research toward higher-yielding rice varieties, mechanization, irrigation, terracing, fertilization, and other practices. More than 90 percent of farmland is assigned to food crops. Hilly terrain is given over to cash crops such as tea and grapes (Japan has a thriving wine industry). Vegetable gardens ring the outskirts of all the cities.

But all this costs money, and Japan's food prices are high. Millions of Japanese have moved from the farms to the cities, depopulating the countryside where farmers are needed. Japan's political system is designed to give underpopulated rural areas disproportionate influence, and prices of produce are kept artificially high, all to induce farmers to stay on the land. It has not worked. Between 1920 and 1995, full-time farmers declined from 50 to 4 percent of the labor force. Yet the policies persist. In the early 1990s, American farmers proved that they could provide rice to Japanese consumers at *one-sixth* of the price they were paying for home-grown rice. But the government resisted pressure to open its markets to imports.

With their national diet so rich in such starchy foods as rice, wheat, barley, and potatoes, the Japanese need protein to balance it. Fortunately, this can be secured in sufficient quantities, not by making foreign purchases but by harvesting it from rich fishing grounds near the Japanese islands. With customary thoroughness, the Japanese have developed a fishing industry that is larger than that of the United States or any of the long-time fishing nations of northwestern Europe, and it now supplies the domestic market with a second staple after rice. Although mention of the Japanese fishing industry brings to mind a fleet of ships scouring the oceans and seas far from Japan, the fact is that most of this huge catch (about one-seventh of the world's annual total) comes from waters within a few dozen miles of Japan itself. Where the warm Kuroshio and Tsushima currents meet colder water off Japan's coasts, a rich fishing ground exists that yields sardines, herring, tuna, and mackerel in the warmer waters and cod, halibut, and salmon in the seas to the north. Along Japan's coasts there are about 4,000 fishing villages, and tens of thousands of small boats ply the waters offshore to bring home catches that are distributed both locally and in city markets. The Japanese also practice aquaculture—the "farming" of fresh-water fish in artificial ponds and flooded paddy (rice) fields, seaweeds in some aquariums, and oysters, prawns, and shrimp in shallow bays. They are even experimenting with the cultivation of algae for their food potential.

When you walk the streets of Japan's cities, you will notice something that Japanese vital statistics are reporting: the Japanese are getting taller and heavier. Coupled with this change is evidence that heart disease and cancer are rising as causes of mortality. The reason seems to lie in the changing diets of many Japanese, especially the younger people. When rice and fish were the staples for virtually all, and calories available were limited but adequate, the Japanese as a nation were among the world's healthiest and longest-lived. But as fast-food establishments diffused throughout Japan, and Western (especially American) tastes for red meat and fried food spread among children and young adults, their combined impact on the population was soon evident. Coupled with the heavy cigarette smoking that prevails in all the Asian countries of the Pacific Rim, this is changing the region's medical geography.

Japan's Future

In the late 1990s, Japan remains the economic giant on the western Pacific Rim, an industrial powerhouse whose products dominate world markets and whose investments span the globe. Nevertheless, Japan faces problems. Competition from other Pacific Rim countries, particularly South Korea and Taiwan, is undercutting Japanese products on world markets, much as Japan's own products once undercut more expensive Western goods. Many of Japan's overseas investments have lost much of their value. And always, Japan's dependence on foreign oil is a potential

THE WAGES OF WAR

■ ■ ■

Japan and the former Soviet Union never signed a peace treaty to end their World War II conflict. The reason? There are actually four reasons, and all of them are on the map just to the northeast of Japan's northernmost large island, Hokkaido (Fig. 9-15). Their names are Habomai, Shikotan, Kunashiri, and Etorofu. The Japanese call these rocky specks of the Kurile Island chain their "Northern Territories." The Soviets occupied them late in the war, but they never gave them back to Japan. Now they are part of Russia, and the Russians have not turned them over either.

The islands themselves are no great prize. During the Second World War, the Japanese brought 40,000 forced laborers, most of them Koreans, to the islands to mine the minerals found there. When the Red Army overran them in 1945, the Japanese were ordered out, and most of the Koreans fled. Today the population of about 50,000 is mostly Russian, many of them members of the military based on the islands, and their families. At their closest point the islands are only 3 miles (5 km) from Japanese soil, a constant and visible reminder of Japan's defeat and the loss of Japanese land. Moreover, their territorial waters bring Russia even closer, so that the islands' geostrategic importance far exceeds their economic potential.

Attempts to settle the issue have failed. In 1956, Moscow offered to return the tiniest two, Shikotan and Habomai, but the Japanese refused, demanding all four islands back. In 1989, then-Soviet President Gorbachev visited Tokyo in the hope of securing an agreement. The Japanese, it was widely reported, offered an aid-and-development package worth U.S. $26 billion to develop Russia's eastern zone—its Pacific Rim and the vast resources of the eastern Siberian interior. This would have begun the transformation of Russia's Far East, stimulated the ports of Nakhodka and Vladivostok, and made Russia a participant in the spectacular growth of the western Pacific Rim.

But it was not to be. Subsequently, Russian President Boris Yeltsin also was unable to come to terms with Japan on this issue, facing opposition from the islands' inhabitants and from his own government in Moscow. And so World War II, a half-century after its conclusion, continues to cast a shadow over this northernmost segment of the western Pacific Rim.

■ ■ ■

weakness: rising costs and supply interruptions would have grave consequences for the country's energy-dependent economy.

Nonetheless, Japan still has enormous further potential, much of it arising from its relative location. As we noted in Chapter 2, the Russian Far East, lying opposite northern Honshu and Hokkaido, is a storehouse of raw materials awaiting exploitation, and the Japanese are ideally located to help achieve this (a major obstacle, however, has been a group of small islands northeast of Hokkaido—see box titled ''The Wages of War''). Moreover, Japan's technological prowess is finding an ever more sizeable market in developing China, and again the Japanese are situated on the doorstep of growing opportunity.

All this should be viewed against the backdrop of a changing social and cultural Japan. The population of 125 million is aging, and demographers report that it will soon stabilize and then begin a slow decline. That is why, as we noted earlier, Japan some years ago began to face a labor shortage, a new experience for this burgeoning economy. But more recently, Japan confronted still another dilemma: its economic growth slowed, factories (including automobile plants) had to be closed, and the labor shortage was replaced by unemployment. As the value of investments declined, confidence in the system plummeted. All this came at a time when the growing number of elderly people was increasing the state's costs for health care, retirement, and other services for the aged.

Meanwhile, modernization put additional strains on society. The Japanese have a strong tradition of family cohesion and veneration of its oldest members, but this custom is breaking down. Younger people want more privacy and self-determination, and they are less willing than their parents and grandparents to accept overcrowded housing and limited comforts. Yet in Japan, for all its material wealth, family homes tend to be small, cramped, often flimsily built, and sometimes without basic amenities considered indispensable in other developed countries. With their high incomes, the Japanese have come to rank among the world's most-traveled tourists, and, having seen how Americans and Europeans of similar income levels live, they have returned dissatisfied.

This dissatisfaction also spills over into the workplace and into the schools. Japan's work ethic is world renowned, but it is sustained at a high cost to the individual and often involves enormous demands of time and dedication to the corporate good. In education the requirements are perhaps the most rigorous anywhere, and the competition is fierce. Now some leaders are beginning to question the appropriateness of such standards and regulations in the new, economic-world-power Japan.

In recent years the country has displayed other, unprecedented symptoms of trouble. In 1995, members of a secret cult began a campaign of terror designed to kill large numbers of citizens and to disrupt the social order. Although the death toll was limited, the campaign spread fear throughout Japan. The dimensions of the cult's network, the intensity of its hatred for the established order, and the apparent failure of the police to provide protection shook Japanese society. All this also heightened the self-doubt that seems now to pervade the country.

Despite such signs of weakness, Japan remains an economic and social model for the countries of the Pacific Rim—a model to be emulated but also challenged. Within the Jakota Triangle, workers in Taiwan are paid less than

in Japan, and in South Korea they are paid less than in Taiwan, on average. And so their cheaper products compete with those of the Japanese. But from other places (China, Malaysia, Thailand) come cheaper goods still.

Japan stands at a crossroads: of great opportunity on East Asia's threshold, but of growing rivalry from its Pacific Rim neighbors to the south.

KOREA

On the Asian mainland, directly across the Sea of Japan, lies the peninsula of Korea (Fig. 9-18), a territory about the size of the State of Idaho, much of it mountainous and rugged, and containing a population of 70 million. Unlike Japan, however, Korea has long been a divided country, and the Koreans a divided nation. For uncounted centuries Korea has been a pawn in the struggles of more powerful neighbors. It has been a dependency of China and a colony of Japan. When it was freed from Japan's oppressive rule at the end of World War II (1945), the victorious Allied powers divided Korea for administrative purposes. That division gave North Korea (north of the 38th parallel) to the forces of the Soviet Union, and South Korea to those of the United States. In effect, Korea traded one master for two new ones. The country was not reunited for the rest of the century because North Korea immediately fell under the communist ideological sphere and evolved as a dictatorship in the familiar (but in this case extreme) pattern. South Korea, with massive American aid, became part of East Asia's capitalist perimeter. Once again, it was the will of external powers that prevailed over the desires of the Korean people.

In 1950, North Korea sought to reunite the country by force and invaded South Korea across the 38th parallel. This attack drew a United Nations military response led by the United States. Thus began the devastating Korean War (1950–1953) in which North Korea's forces pushed far to the south only to be driven back across its own half of Korea almost to the Chinese border. Then China's Red Army entered the war and drove the U.N. troops southward again. A cease-fire was arranged in 1953, but not before the people and the land had been ravaged in a way that was unprecedented even in Korea's violent past. The cease-fire line shown in Figure 9-18 became a heavily fortified *de facto* boundary. For more than four decades there has been virtually no contact of any kind across this border, which divides economies and families alike.

As a result, the Jakota Triangle, as a region of East Asia, must include South Korea but exclude the North. North Korea remains a subregion of the region dominated by China: the state of its development, the nature of its economy, the character of its political system, and its limited external connections combine to resemble the norms of

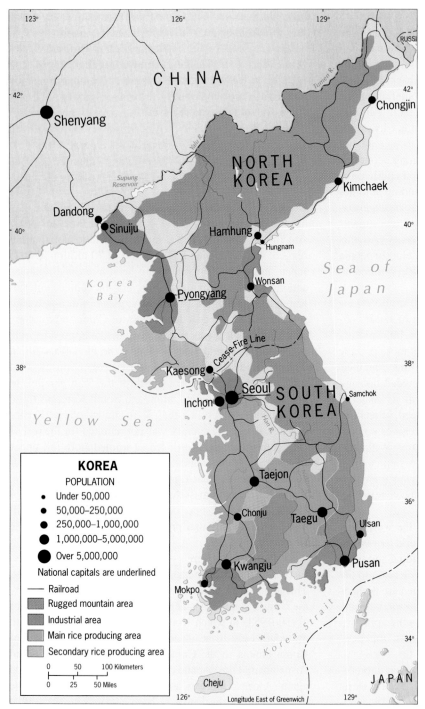

FIGURE 9-18

interior China rather than the Pacific Rim on which it lies. Indeed, North Korea is more reminiscent of China during its Maoist isolation. You will not find any North Korean products in stores around the world, or North Korean students at Western universities, or North Korean representatives at international economic conferences. The few travelers who have managed to enter North Korea report on the amazing contrasts between North and South: the regimented townscape of the North's capital, Pyongyang, compared to Seoul, the South's bursting-at-the-seams, chaotic, economic powerhouse; the North's antiquated state enterprises and the South's modern factories; the North's changeless rural areas and the South's intensive, increasingly mechanized agriculture.

During the mid-1990s, the specter of renewed conflict on the Korean Peninsula emerged, involving Western concerns that the North Koreans were constructing facilities that would be capable of producing nuclear bombs. In the past, the North Korean communist regime was accused of engaging in terrorist acts that included the bombing of a political gathering in which a number of South Korean officials were killed and the destruction of an airliner in midflight. Fears that nuclear weapons in the hands of such people would threaten not only South Korea but also Japan and the peace in East Asia led to threats from the West and, eventually, negotiations. These negotiations produced the first glimpse of a future normalization of relations between North and South.

Existing conditions dictate, however, that we focus here on South Korea, the country that, with 45 percent of Korea's land area but with two-thirds of the Korean population, has emerged as one of the *economic tigers* on the Pacific Rim. There is no telling what Korea might have achieved as an undivided country, because North and South Korea are in a situation that geographers call **regional complementarity**. Such a condition arises when two adjacent regions complement each other in economic-geographic terms. In this case, North Korea has raw materials that the industries of South Korea need; South Korea produces food required by the North; North Korea produces chemical fertilizers needed on farms in the South.

Under the political circumstances of the past half-century, however, the two Koreas were cut off from each other, and they developed in opposite directions. North Korea, whose large coal and iron ore deposits attracted the Japanese and whose hydroelectric plants produce electricity of which part could be used in the South, carried on what limited trade it generated with the Chinese and the Soviets. South Korea's external trade links were with the United States, Japan, and Western Europe.

In the early postwar years, there were few indications that South Korea would emerge as a major economic force on the Pacific Rim and, indeed, on the world stage. Over 70 percent of all workers were farmers; agriculture was inefficient; the country was stagnant. But huge infusions of aid, first from the United States and later also from Japan, coupled with the reorganization of farming and the stimulation of industries (even if they lost money), produced a dramatic turnaround. Large feudal estates were parceled out to farm families in 3-hectare (7.5-acre) plots, and a program of massive fertilizer importation was begun. Production rose to meet domestic needs, and some years yielded surpluses.

The industrialization program was based on modest local raw materials, plentiful and capable labor, ready overseas (especially American) markets, and continuing foreign assistance. Despite recurrent political instability and social unrest, successive South Korean governments managed to sustain a rapid rate of economic growth that placed the country, by the late 1980s, among the world's top 10 trading powers. To get the job done, the government borrowed heavily from overseas, and it controlled banks and large industries. South Korea's growth resulted from **state capitalism** rather than free-enterprise capitalism, but it had impressive results. South Korea became the world's largest shipbuilding nation; its automobile industry grew rapidly; and iron and steel and chemical industries thrive. There has been less emphasis, however, on smaller, high-technology industries, and these are the strengths of other Pacific Rim economic tigers. So the future remains uncertain.

From the Field Notes

"Pusan, located at the southeast corner of the Korean Peninsula, is a major port, industrial center, commercial hub, and focus for a large agricultural region. Sail or fly into Pusan, however, and you get a geographic lesson on the problems a city's site can present during urban expansion. Pusan lies at the southern end of a rugged mountain chain (see Fig. 9-18) and the city itself sprawls up hillsides, into narrow valleys, and, in the port area, onto reclaimed land. Walking the central city, we found streets sometimes ending in stairways."

■ AMONG THE REALM'S GREAT CITIES...

Seoul

Seoul (12 million), located on the Han River, is ideally situated to be the capital of all of Korea, North and South (its name means capital in the Korean language). Indeed, it served as such from the late fourteenth century until the early twentieth, but events in this century changed its role. Today, the city lies in the northwest corner of South Korea, for which it serves as capital; not far to the north lies the tense demilitarized zone (DMZ) that contains the cease-fire line with North Korea. That line cuts across the mouth of the Han, depriving Seoul of its river traffic. Its ocean port, Inchon, has emerged as a result.

Seoul's undisciplined growth, attended by a series of recent accidents including the failure of a major bridge over the Han and the collapse of a six-story department store, reflects the unbridled expansion of the South Korean economy as a whole, as well as the political struggles that carried the country from autocracy to democracy. Central Seoul lies in a basin surrounded by hills to an elevation of about 1,000 feet (330 m), and the city has sprawled outward in all directions, even toward the DMZ. An urban plan designed in the early 1960s was overwhelmed by the immigrant flow.

During the period of Japanese colonial control, Seoul's surface links to other parts of the Korean Peninsula were improved, and this infrastructure played a role in the city's later success. Seoul is not only the capital but also the leading industrial center of South Korea, exporting huge volumes of textiles, clothing, footwear, and (increasingly) electronic goods. South Korea is aptly called an economic tiger on the Pacific Rim, and Seoul is its heart.

Today, South Korea is prospering, and its economic prowess can be seen on the map (Fig. 9-18). The capital, Seoul, with 12 million inhabitants, is the world's eleventh largest metropolis and the anchor of a huge industrial complex facing the Yellow Sea at the waist of the Korean Peninsula. Hundreds of thousands of farm families migrated to the Seoul area after the end of the Korean War (today only about 25 percent of South Koreans remain on the land). They also moved to Pusan, the nucleus of the country's second largest manufacturing zone, located on the Korea Strait opposite the western tip of Honshu. And the government-supported, urban-industrial drive here continues. Just 25 years ago, Ulsan City, 40 miles (60 km) north of Pusan along the coast, was a fishing center with perhaps 50,000 inhabitants; today its population exceeds 700,000, nearly half of them the families of workers in the Hyundai automobile factories and the local shipyards. The third industrial area shown in Figure 9-18, focused on the city of Kwangju near the southwestern tip of the peninsula, does not yet match the Seoul and Pusan complexes, but it has an advantageous relative location and is bound to develop further.

South Korea's success inevitably has brought pressure from its competitors, including its capitalist allies. Cheap South Korean textiles undercut the Japanese textile industry, and Japanese textile makers demanded protection. Low-priced Korean automobiles captured a share of the American and European markets, and Japanese as well as American automakers wanted to limit the Koreans' access to these markets. But South Korea diversified its production and continued its climb on the ladder already topped by Japan.

TAIWAN

The third component of the region we have identified as the Jakota Triangle is another economic success story on the Pacific Rim: Taiwan. Along with the tiny islands of Kinmen, Matsu, and Penghu, Taiwan politically was a part of China—indeed, a full-fledged province of it—for centuries, until the Japanese colonized the territory in 1895. Thus began more than a hundred years of turmoil from which Taiwan has emerged an economic powerhouse, but also as a political problem.

When China's aging Manchu Dynasty was overthrown following the rebellion of 1911, a Nationalist government in 1912 proclaimed the Republic of China (ROC), led by Sun Yat-sen. Naturally, this government wanted to oust all colonists, not only from Europe but also from Japan, and not only from the mainland, but also from Taiwan. On Taiwan, however, the Japanese held on, using the island as a source of food and raw materials as well as a market for

9.4

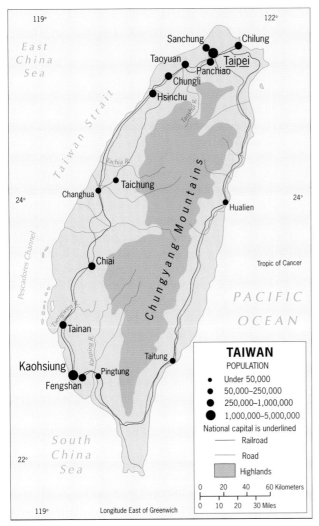

FIGURE 9-19

reconstruction of the war-damaged infrastructure. In the international arena, the ROC represented the country; in the still-young United Nations, the Nationalists of Taiwan occupied China's seat.

On the huge mainland, however, it was the communists in Beijing who ruled, and called their country the People's Republic of China (PRC). Thus was born the two-China dilemma which still confronts the international community. When the ROC was ousted from the United Nations to make way for the PRC, legitimacy also transferred.

All this might have remained an internal dispute but for two major developments: Taiwan's rise as an *economic tiger* on the Pacific Rim, coupled with remarkable strides toward democracy made in the ROC, and the emergence on the world political and economic stage of communist-ruled China following decades of isolation. Today, the power of the PRC is growing rapidly, and Taiwan's heirs to the ROC face an uncertain future; but it is the ROC, not the PRC, that has managed to combine economic success with democratization.

Taiwan, as Figure 9-19 shows, is not a large island. It is smaller than Switzerland but has a population much larger (22 million), most of it concentrated in an arc lining the western and northern coasts. The eastern half of the island is dominated by the Chungyang Mountains, an area of high elevations (some over 10,000 feet [3,000 m]), steep slopes, and dense forests. Westward, these mountains yield to a zone of hilly topography, and, facing the Taiwan Strait, a substantial coastal plain. Streams from the mountains irrigate the paddy fields, and farm production has more than doubled since 1950 even as hundreds of thousands of farmers left the fields to seek work in Taiwan's mushrooming industries.

Today, the lowland urban-industrial corridor of western Taiwan is anchored by the capital, Taipei (Taibei), at the island's northern end and rapidly growing Kaohsiung (Gaoxiong) in the far south.* Chilung (Jilong), Taipei's outport, was developed by the Japanese to export nearby coal, but now the raw materials flow the other way. Taiwan imports raw cotton for its textile industry, bauxite (for aluminum) from Indonesia, oil from Brunei, and iron ore from Africa. Taiwan has a developing iron and steel industry, nuclear power plants, shipyards, a large chemical industry, and modern transport networks. Increasingly, however, Taiwan's export products are those of high-technology industries: personal computers, telecommunications equipment, precision electronic instruments. Taiwan has enormous brainpower, and many foreign firms join in the R&D (research and development) carried on in such places as Hsinchu (Xinzhu), where the government has helped establish a high-tech research facility.

With less than one-fiftieth of the population of mainland

*The Taiwanese have retained the old Wade-Giles spelling of place names; China now uses the pinyin system. Names in parentheses are written according to the pinyin system.

Japanese products. To ensure all this, the Japanese launched a prodigious development program involving road and railroad construction, irrigation projects, hydroelectric schemes, mines (mainly for coal), and factories. Farmlands were expanded and farming methods improved. While mainland China was engulfed in conflict, Taiwan remained under Japanese control. Not until 1945, when Japan was defeated in World War II, did Taiwan become an administrative part of China again. But only briefly. As we noted earlier, the Nationalists, under the leadership of Chiang Kai-shek, now faced the rising tide of communism on the mainland, and the ROC government fought a losing war. In 1949, the flag of the Republic of China was taken to a last stronghold: offshore Taiwan. There, Chiang and his followers, with their military might, weapons, and wealth taken from the mainland, established what they (and the world, led by the United States) proclaimed as the legitimate government of all China. On the island, the Nationalists, helped by the United States, began a massive

China, Taiwan in the late 1990s still ranked ahead of its giant neighbor as a trading nation. A substantial portion of this trade is with the People's Republic of China; despite more than half a century of animosity, ties between the Taiwanese and the mainlanders remain strong. Ever since the 1940s, the British-colonial status of Hong Kong (Xianggang) facilitated this trade; by channeling ROC-PRC trade through Hong Kong, it could be construed as indirect trade. Hong Kong's reversion to China in 1997 presents both partners with a dilemma.

Both China and Taiwan are trying to adjust to the new era by using a principle that seems to hark back to extra-territoriality: the creation of "open" economic zones where the general rules of the state do not prevail. On the atlas map of East Asia, note two cities: Xiamen, on the coast of mainland China, and Kaohsiung, on the coast of Taiwan, across the Taiwan Strait from Xiamen. Both places have been designated as "open" economic zones (we will discuss the details later). This device may help sustain the massive China-Taiwan trade, which is in their mutual interest, without violating the rules against direct exchange.

In some ways, Taiwan's emergence as an economic tiger on the Pacific Rim is even more spectacular than Japan's. True, Taiwan received much Western assistance, but it got out of debt faster than could have been expected. Today, per-capita income is well above U.S. $10,000 per year, which is higher than in many European countries. Taiwan's trade surplus has yielded tens of billions of dollars in reserves, used now to further its own development and to invest overseas. Taiwan, for example, is the largest foreign investor in the rebuilding of Vietnam.

Taiwan's greatest problem, therefore, is political, not economic. The Nationalists who made the island their base in 1949 established a Republic of China that was authoritarian, but the ROC has evolved into a democratic state. This was not an easy process, but when Taiwan's own democracy movement culminated in 1901, it produced *not* a massacre, but free elections for a new parliament. Taiwan's leaders now proclaim that the direct and popular election of their president, in March, 1996, was the first

such vote not just in Taiwan or the ROC, but in the five-thousand-year history of the state. Taiwan thus presents itself as an alternate model to the communist-ruled PRC, proving to the world that democracy is compatible with Chinese culture and tradition.

Nevertheless, Taiwan is not recognized by the international community as an independent state, and a substantial segment of the Taiwanese electorate itself opposes any move toward outright sovereignty. Beijing dictates that Taiwan is off limits to U.S. cabinet members. In 1995, when the United States permitted the ROC's president to visit the American university of which he is a graduate, so that he might attend a class reunion, relations between Washington and Beijing plummeted.

The political geography of Taiwan, therefore, is at odds with its economic geography. The ROC stood firm against communism in the 1960s and 1970s; but its long-term ally, the United States, must publicly commit itself to a one-China policy, the one being the PRC, the communist giant. Even as goods and capital flow between Taiwan and the mainland, Beijing engages in threatening acts (such as nearby missile tests) when democracy in the ROC takes another step forward.

Today, Taiwan's political options are as limited as its economic horizons are endless. The communist Chinese have made it clear that a move toward independence, based on the likely outcome of any referendum, would lead to intervention. Though militarily well prepared, Taiwan could not long hold off the PRC's enormous military forces. In any case, independence for Taiwan has philosophical disadvantages for the leaders of the ROC, as it would diminish their claim to legitimacy as the alternative model for a reunified China. In practical terms, conflict on the Pacific Rim would be disastrous for both sides, whatever course it were to take.

And so Taiwan's future hangs in the balance, a wayward province with national aspirations, a quasi-state with global connections, a test of Chinese capacity for compromise and accommodation. In this era of devolutionary forces, the fate of Taiwan will be a harbinger of the world order in the twenty-first century.

THE PACIFIC RIM IN CHINA: EMERGING REGION?

We now return to what may already be a region in the making. It was noted earlier that many of the geographic generalizations we make about mainland China (its low level of urbanization, the inefficiency of its state-owned enterprises, the isolation of its rural areas, the changeless

poverty of remote villages) no longer apply to a vibrant, burgeoning zone along the country's Pacific coast. From Dalian on the doorstep of the Northeast to Hainan Island off China's southernmost shore, that zone is in a process of total transformation. It is an uneven, sometimes chaotic

KONGFUZI (CONFUCIUS)

■ ■ ■

Confucius (*Kongfuzi* or *Kongzi* in pinyin) was China's most influential philosopher and teacher. His ideas dominated Chinese life and thought for over 20 centuries.

Kongfuzi was born in 551 B.C. and died in 479 B.C. He was appalled at the suffering of ordinary people during the Zhou Dynasty, and he urged the poor to assert themselves and demand explanations for their harsh treatment by the feudal lords. He tutored the indigent as well as the privileged, giving the poor an education that had hitherto been denied them and ending the aristocracy's exclusive access to the knowledge that constituted power.

Kongfuzi's revolutionary ideas extended to the rulers as well as the ruled. He abhorred supernatural mysticism and cast doubt on the divine ancestries of China's aristocratic rulers. Human virtues, not godly connections, should determine a person's place in society. He proposed that the dynastic rulers turn over the reins of state to ministers chosen for their competence and merit. This was another Kongzi heresy, but in time that idea came to be accepted and practiced.

His Earthly philosophies notwithstanding, Kongfuzi took on the mantle of a spiritual leader after his death. His thoughts, distilled from the mass of philosophical writing (including Daoism) that poured forth during his lifetime, became the guiding principles of the formative Han Dynasty. The state, he said, should not exist just for the power and pleasure of the elite; it should be a cooperative system for the well-being and happiness of the people.

As time went on, a mass of writings evolved, much of which Kongfuzi never wrote. At the heart of this body of literature lay the *Confucian Classics*, 13 texts that became the basis for education in China for 2,000 years. From government to morality and law to religion, the *Classics* were Chinese civilization's guide. The entire national system of education (including the state examinations through which everyone, poor or privileged, could achieve entry into the arena of political power) was based on the *Classics*. Kongfuzi was a champion of the family as the foundation of Chinese culture, and the *Classics* prescribe a respect for the aged that was a hallmark of Chinese society.

But Kongfuzi's philosophies also were conservative and unchanging, and when the colonial powers penetrated China, Kongzi's *Classics* came face to face with practical Western education. For the first time, some Chinese leaders began to call for reform and modernization, especially of teaching. Kongzi principles, they said, could guide an isolated China, but they were found wanting in the new age of competition. But the Manchu rulers resisted this, and the Nationalists tried only to combine Kongzi and Western knowledge into a neo-Kongzi philosophy, an unworkable plan.

The communists who took power in 1949 attacked Kongzi thought on all fronts. The *Classics* were abandoned, indoctrination pervaded education, and even the family was, for a time, viewed as a thing of the past. But the communists miscalculated. It proved impossible to eradicate two millennia of cultural conditioning in a few decades. The fading spirit of Kongfuzi will haunt physical and mental landscapes in China for generations to come.

■ ■ ■

process that may eventually produce a Japan-like economic geography here. That day is a long way off, but of this there is no doubt: China's Pacific Rim is generating what will become a new region, possibly an extension of the Jakota Triangle. Already, parts of China's Pacific Rim differ as dramatically from interior China as South Korea does from the North. By all definitions, China still is an underdeveloped country, but not along its Pacific Rim. There, China has taken off.

REORGANIZATION UNDER COMMUNISM

Fifty years ago, when the communist era was about to begin in China, there was nothing to foreshadow these developments. The country had been wracked by war. Its economy lay in ruins. Famines threatened. Population growth soared. The Nationalists had fled and taken over Taiwan.

In the 1950s, the victorious communist regime launched massive programs of reconstruction and reform, with the Soviet model as its guide. Land was expropriated, and farming was collectivized, being later forced into the *commune* scheme. This draconian program, which assigned men and women to segregated "production teams," destroyed families, in effect orphaned millions of children, and regimented people's lives at the behest of the Communist Party. While the specter of famine receded, the impact of social dislocation was profound.

China's industries, too, were reorganized into state-owned, communal enterprises. Based on the Soviet model, huge investments were made in the establishment of heavy industries, and the Northeast became a vast complex of state-owned factories.

China's communist boss, Mao Zedong, had further plans for his country. He wanted to rid China of its age-old Kongzi (Confucian) traditions and to substitute the principles of communism as the philosophical basis for the new order (see box titled "Kongfuzi [Confucius]"). Meanwhile, he and his inner circle quarreled over communist doctrine. Mao regarded the Soviet Union as weak in its application of communist principles, and so the ''revisionist'' Soviet advisers were forced to leave the country. When Mao suspected that many of his own subordinates harbored revisionist views, he launched the so-called *Great Proletarian Cultural Revolution*, a campaign for a return to strict and orthodox communism. Schools were closed because they ostensibly promoted ''bourgeois'' ideas. Millions of young people found themselves at loose ends, ready to be recruited into the Red Guards, a paramilitary organization that was organized to combat the revisionists. In the power struggle that ensued, thousands of Communist Party leaders and other officials lost their positions to people more loyal to Mao. Soon, the Red Guards were fighting cadres organized to support threatened or ousted local leaders. Eventually, even more devoutly pro-Mao ''brigades'' challenged the Red Guards themselves. From 1966 onward, China once again was in chaos, and as under the emperors, the victims of the rulers in the capital were the people in the provinces. We should not gloss over the Cultural Revolution and its toll. Perhaps as many as 20 million Chinese died in another of communism's dreadful impacts on those under its heel.

No one would have predicted in the mid-1970s that, 20 years later, China would have a new economic geography forged by market principles rather than communist dogma, with stock markets in its major cities. Mao's death in 1976 left China on the verge of civil war, with a power struggle being waged in the capital between a clique of orthodox Maoists and a group who called themselves ''pragmatic moderates.'' When the moderates won the contest, a man named Deng Xiaoping took power, and his name will be forever etched on China's cultural landscape. Deng moved the regime toward the notion that communist political rule could be wedded to capitalist economic practices. He also favored opening China to foreign science and technology and allowed tens of thousands of Chinese students to attend foreign universities in order to bring back the skills China needed. All this set in motion the sequence of events that was to transform Pacific Rim China in a few short years and send ripples of change deep into China's interior.

GEOGRAPHY OF CONTRADICTIONS

China's communist era has witnessed the metamorphosis of the world's most populous nation. Despite the dislocations of collectivization, communization, the Cultural Revolution, power struggles, policy reversals, and civil conflicts, China as a communist state managed to stave off famine and hunger, control its population growth, improve its overall infrastructure and economy, strengthen its military capacity, and enhance its position in the world. Now,

China is absorbing one of the leading economic tigers on the Pacific Rim (Hong Kong), has rapidly growing trade with neighbors near and far, and doubles its national wealth every eight years at present economic growth rates. And still its government espouses communist dogma.

How was it possible to manipulate the coexistence of communist politics and market economics? That was the key question confronting Deng Xiaoping and his comrades when they took power in 1979. In terms of ideology, this objective would seem unattainable; an economic ''open door'' policy would surely lead to rising pressures for political democracy.

But Deng thought otherwise. If China's economic experiments could be spatially separated from the bulk of the country, the political impact would be kept at bay. At the outset, the new economic policies would apply mainly to China's bridgehead on the Pacific Rim, leaving most of the vast country comparatively unaffected. Accordingly, the government introduced a complicated system of *Special Economic Zones*, so-called *Open Cities*, and *Open Coastal Areas*, that would attract technologies and investments from abroad and transform the economic geography of eastern China (Fig. 9-20).

In these economic zones, investors are offered many incentives. Taxes are low. Import and export regulations are eased. Land leases are simplified. The hiring of labor under contract is allowed. Products made in the economic zones may be sold on foreign markets and, under some restrictions, in China as well. And profits made here may be sent back to the investors' home countries.

ECONOMIC ZONES

As Figure 9-20 shows, five **Special Economic Zones (SEZs)** were established:

1. *Shenzhen*, adjacent to booming Xianggang (Hong Kong)
2. *Zhuhai*, on the estuary of the Pearl River below Guangzhou and just above the soon-to-be former Portuguese colony of Macau
3. *Shantou*, directly opposite the southern end of productive Taiwan
4. *Xiamen*, also across from southern Taiwan
5. *Hainan Island*, China's southernmost province, located across the Gulf of Tonkin from northern Vietnam

Three of the five SEZs lie in Guangdong Province, the crucible of Chinese activity on the western Pacific Rim (Fig. 9-20). The last to be authorized was Hainan Island, now known not only for its free-enterprise economy but

FIGURE 9-20

also for its high-volume smuggling operations. But, in fact, four of the SEZs initially were less successful than the Deng regime had anticipated. Only Shenzhen took off in spectacular fashion because of its situation just across the porous border from Xianggang.

Shenzhen and Xianggang (Hong Kong)

As Figure 9-20 shows, the province of Guangdong lies at the heart of China's Pacific Rim development. This province (population 71 million) is the semicircular hinterland of a great, growing megalopolis consisting of four urban areas: Hong Kong (its Chinese name, *Xianggang*,* restored upon China's 1997 takeover); Shenzhen, the cornerstone of Deng's new economic policy; Macau, the Portuguese

9.1

colony also reverting to Chinese rule; and Zhuhai, the Special Economic Zone adjacent to Macau. These four economic hubs cluster around the Pearl River Estuary, once China's economic back door, now its front gate. As the map suggests, this conurbation is growing toward a fifth metropolis, the capital of the province, Guangzhou (formerly Canton), about 75 miles away. The drive from Guangzhou's factories to Hong Kong's harbor used to take a full day of stop-and-go congestion, but a Hong Kong tycoon built a six-lane superhighway that opened in 1995 (toll: U.S. $14) and cut the time to two hours. Already, development is mushrooming at the interchanges. The emerging megalopolis will soon encompass Guangzhou and number more than 20 million inhabitants.

None of this would have happened were it not for Hong Kong's spectacular rise. Hong Kong deserves its recognition as an economic miracle, the quintessential **economic tiger** on the western Pacific Rim. Here, on the left bank of the estuary of South China's Pearl River, nearly 6 million people (97 percent of Chinese ancestry) crowded onto a

*In this section we continue to use the name "Hong Kong," but all maps use "Xianggang (Hong Kong)."

From the Field Notes

"In 1996, all eyes were focused on the eastern side of the Pearl River estuary and on the 1997 transfer of Hong Kong from British to Chinese control. But what was happening on the opposite side was by no means insignificant. As we stood on a prominent hill in Macau, the Portuguese dependency also being transferred to China in the late 1990s, we could see the massive development of another of China's Special Economic Zones: Zhuhai, across the river. Macau is no Hong Kong, and Zhuhai is no Shenzhen, but there are advantages of site and situation on the western side of the estuary that presage a narrowing of the gap as time goes on. Already, Macau itself is undergoing major reconstruction and expansion, including one of Asia's longest bridges to link its largest islands to the mainland."

fragmented, hilly patch of territory covering just 400 square miles (1,000 sq km) under the tropical sun have created an economy that is larger than those of more than a hundred countries.

Hong Kong consists of three parts, of which the island named Hong Kong (32 square miles/82 sq km) is one. Other islands, including Lan Tao, the site of a new, state-of-the-art airport, fringe the mainland parts of Hong Kong: the Kowloon Peninsula and the New Territories (Fig. 9-21). The capital, Victoria, lies on Hong Kong Island and overlooks Victoria Harbor, one of the busiest in the world. As these place names suggest, Hong Kong was a British colony for a long time. The islands and the Kowloon Peninsula were ceded permanently by China to Britain in 1841 and 1860, respectively, but the rest, the New Territories, were leased on a 99-year basis in 1898. That lease expired in 1997, and the British and Chinese agreed on the transfer

of authority over all three of the dependency's components from London to Beijing on July 1 of that year. East Asia's economic tigers seem to share uncertain futures.

With its excellent deep-water harbor, Hong Kong long has been the major *entrepôt* of the western Pacific between Shanghai and Southeast Asia's Singapore. The Chinese could well have recaptured the colony during the late 1940s, when the communist forces were victorious throughout China. But both Beijing and London saw advantages in maintaining the status quo. During the period of isolation and communist reorganization, Hong Kong provided the People's Republic of China with a convenient place of contact with the Western world without any need for long-distance entanglement. Even when hundreds of thousands of Chinese from Guangdong and Fujian provinces fled to Hong Kong, and later when the city was a place of rest and recuperation for American forces during

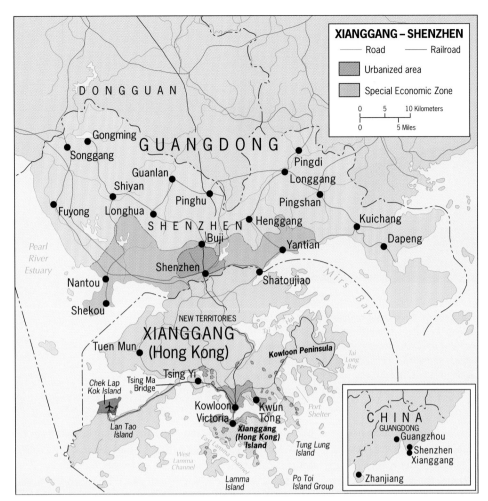

FIGURE 9-21

the Indochina War, the Chinese government continued to tolerate the British colonial presence. Indeed, the colony depends on China for vital supplies, including fresh water and food. It could easily have been choked into submission in a very short time.

Until the 1950s, Hong Kong was just another trading colony—a busy port but little more. Then the Korean War and the United Nations embargo on trade with China cut the colony's connections with its hinterland, and an economic reorientation was necessary. Hong Kong possessed no mineral resources, but it did have human resources in virtually unlimited numbers—people in often desperate need to earn a wage, no matter how small. All that was needed was a supply of raw materials, investment in equipment, and efficient organization. Within just a few years, a huge textile industry developed, along with many light manufacturing industries of other kinds. Products were made at very low cost; they found ready markets throughout the world. Accumulated capital was then used to establish factories making electrical equipment, appliances, and literally countless other consumer goods. As the flood of immigrants continued, labor costs remained low, and

products remained competitive. Today, textiles and fabrics still account for about 40 percent of the colony's total export revenues by value, but the range of products continues to expand. Simultaneously, Hong Kong became one of the world's leading financial centers, with a strong and broad-based stock market.

Hong Kong's consumer goods are exported to the United States, China, Germany, the United Kingdom, Japan, and many other countries. While Hong Kong had separate political status, its capitalists accounted for about 75 percent of all "foreign" investment in China itself, and about 25 percent of all of China's foreign trade passed through Hong Kong, including trade with Taiwan.

All this was in the balance as the date of China's takeover (July 1, 1997) approached and passed. During the negotiations with the British, Chinese leaders had committed themselves to allow Hong Kong to continue its capitalist ways for a minimum of 50 years. But as the date of transfer approached, political and social changes made by the British were deemed unacceptable in Beijing, and the terms of the transfer agreement appeared to be endangered. In the meantime, Hong Kong's wealthier people emigrated

From the Field Notes

"Nathan Road is Kowloon's main street, one of the most expensive shopping areas in Hong Kong, with real estate prices among the territory's highest and pedestrian traffic the largest. This makes the presence of this mosque, occupying an entire city block in this prime area, all the more remarkable, and forms a reminder of the worldwide reach and diffusion of Islam. Large posters facing Nathan Road invite passersby to visit the mosque and learn more about the faith. Saudi Arabian investments sustain the mosque."

in substantial numbers, the flow reaching 60,000 per year in the early 1990s. Many of the emigrants, however, kept a base in Hong Kong while establishing residencies in such places as Singapore, Canada, Australia, the United Kingdom, or the United States, and tens of thousands returned to Hong Kong after doing so, to await the fateful date. Should things go awry, they would have the option to leave for their foreign bases.

Immediately across the old colonial border from Hong Kong lies Shenzhen (Fig. 9-21), a name that has become synonymous with China's new economic policies and its Pacific Rim development. In 1980, Shenzhen was just a sleepy fishing and duck-farming village of about 20,000; in 1996 its population surpasses 3 million—perhaps the fastest-growing urban center anywhere on Earth. Visit Shenzhen today, and it is hard to believe you are in China: there you will find a raucous stock exchange, gleaming hotels, luxury stores, pet shops, fancy restaurants. Foreign symbols, from corporate logos to luxury automobiles, are everywhere.

The map reveals why Shenzhen has been the most successful of China's SEZs. Its proximity to Hong Kong, a genuine economic tiger on the Pacific Rim, at a time when border restrictions were relaxed, its access to port facilities, the availability of labor willing to work for wages even lower than Hong Kong's, and the advantages bestowed upon foreign firms through China's new economic policies combined to stimulate one of the greatest booms ever seen anywhere in the world. Many Hong Kong firms invested

heavily in Shenzhen, and other investments came from Japan, Taiwan, Singapore, the United States, and other sources. Many successful Chinese living in Southeast Asia also made use of the opportunity to reinvest in the motherland. In short, in just a few years Shenzhen evolved into a business and industrial giant complete with hundreds of skyscrapers, numerous factories, totally new transport systems, and a growing hinterland. Since 1980, when Shenzhen was designated an SEZ, exports have grown annually by 70 percent. In the mid-1990s, Shenzhen alone accounted for 15 percent of all of China's foreign trade.

But now Shenzhen faces a new era. Hong Kong no longer is the separate entity that helped give Shenzhen its early boost. Labor costs in Shenzhen, so low that they attracted major Hong Kong investment, are now higher than they are in its own hinterland, higher even than in Guangzhou. That means that investment may begin to leapfrog Shenzhen into the province's interior, setting up rival industries there. And there are signs of other problems. Real estate speculation has been out of control, becoming a major drain on productive investment. Corruption and organized crime far exceed the norms of Hong Kong. And the regime in Beijing appears to be reconsidering the privileges it gave to Shenzhen (and the other SEZs). These freedoms are deemed to favor the SEZs so strongly that the interior provinces suffer disproportionately, creating excessive regional disparities. This may be another way of expressing a more serious concern: that the Pearl Estuary megalopolis and its burgeoning economy, together with the whole province of Guangdong, will develop so fast that it will spin out of Beijing's control, so that the aging dogmatists in the capital will find themselves riding, rather than caging, this economic tiger.

Open Cities and Coastal Areas

China's new economic policy was not confined to the five SEZs listed above. To encourage investment and trade all along the Pacific coast, Beijing designated 14 coastal cities as **Open Cities** (Fig. 9-20). When most of these 14 ports experienced only limited growth, the program was scaled back to focus on only four (Dalian, Tianjin, Shanghai, and Guangzhou), and among these, Shanghai received the largest share of national investment.

Six **Open Coastal Areas** form yet another geographic layer of the new economic system. Again the objective was to attract foreign investment, although the economic incentives are less far-reaching. As Figure 9-20 shows, these Open Coastal Areas comprise parts of China's many Pacific coast deltas and peninsulas. It was and, on paper, it remains the intent of the Chinese regime to open virtually the entire Pacific littoral to market-driven economic activities, and if the program succeeds, a new geographic region will emerge here. To date, however, none of the other SEZs, Open Cities, or Open Coastal Areas has become a second Shenzhen, although spectacular growth is occurring in Haikou, the capital of Hainan, in Shanghai, and in Tianjin-Beijing. In the late 1990s, China's economic growth figures attested to the overall success of the new economic policy. Whether that is enough for the program to survive the end of the Deng era is an unanswered question.

URBAN CHINA

In our earlier discussion of China's subregions, we emphasized the rural character of Chinese society, and the villages and hamlets in which the great majority of the people live. In the present discussion our focus is on China's cities, old and new. Here we get a glimpse of the China of the future. China will, in time, become an urbanized society just as Japan and Western Europe did.

China's urban population is growing faster than the nation overall. In 1997, an estimated 350 million Chinese resided in cities and towns—a mere 28 percent of the population. Still, there are 50 million more Chinese urban residents than there are citizens of the United States and Canada combined.

China's cities are crowded, congested, polluted, massive, often bleak agglomerations of urban humanity. In some ways they resemble the cities of England during the early years of the Industrial Revolution. Soot-belching smokestacks rise like a forest from city blocks; factories and workshops large and small vie for space with apartment buildings, schools, and architectural treasures of the past. Soviet-style public buildings bear witness to the early communist period. No zoning regulations seem to exist, nor is there any protection for the urban environment. Turn a streetcorner, and the deafening roar of machines drowns out all else; walk along the street and as one roar fades, another ear-splitting racket replaces it. Across the street is a school, and one wonders how students can learn under such conditions day after day. Gaze upward and the sky is shrouded by an ever-present cloud of pollution so thick that the air is actually dangerous to breathe. Cross a bridge and look down: here is another vivid portrait of China's industrial march. The water below is little more than an industrial sewer, filled with waste and colored gray by tons of toxic chemicals.

By the standards of suburbanized Western cities, China's metropolises are bleak and severe places. In the streets, waves of bicycle riders vie with buses and trucks for space; humans carry and pull loads that seem to be beyond their capacity to do so. There are large public squares, but few leafy parks. The great majority of the people are dressed in workers' clothing. Luxuries are few—except in the bustling new centers on the Pacific coast.

And yet China's great cities are crucibles of the new China. Beijing's skyline reflects the post–Soviet age as modern buildings rise above the old. Shanghai's old colonial frontage is flanked by new high-rises, and the city's entire waterfront is under reconstruction. Guangzhou does not yet have suburbs on the American model, but its outskirts mark the impact of a new economic order. Cities farther west—Xian, Chengdu, Kunming—are less affected, but they, too, are beginning to change. Inevitably, the government's new economic policies will draw larger numbers of people from the rural areas to the cities. Efforts will be made to slow this flow, but China will urbanize just as Europe did two centuries ago.

As Figure 9-11 shows, most of China's large cities (like those in India) remain single centers, widely dispersed, interacting with and connected to large hinterlands. In the context of China's huge population, Shanghai, Beijing, and Guangzhou are not enormous in size. Japan has one-tenth of China's population, but metropolitan Tokyo has nearly twice as many inhabitants as China's largest city. Still, China too is showing signs of megalopolitan development, as in the case of Wuhan on the Chang Jiang and Shenzhen–Guangzhou–Zhuhai (plus Hong Kong and Macau as their colonial status ends) on the Pearl River Estuary. Urbanization is a mirror of development, and Figure 9-11 well reflects where China's action is. Already, the emerging Pacific Rim region is far more urbanized than China as a whole, providing further evidence of the regional disparities Pacific Rim development is generating. ◀▐

9.3

Xian

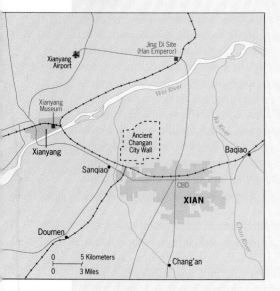

Xian (3.6 million) was the Rome of ancient China; it is the western gateway of modern China. Until the Mongols chose Beijing as their headquarters, Xian intermittently was China's capital for more than 900 years. China's ancient culture hearth lay in the upper Wei valley, and Xian was its focus. From here, Chinese power spread eastward and southward.

Xian's hinterland encompasses the Loess Plateau, a densely peopled physiographic region defined by the distribution of a fertile, wind-blown deposit of glacial origin, *loess*. Not only is loess extremely fertile: its prismatic structure gives it vertical stability. As a result, it can be hollowed out without collapsing, and many tens of thousands of people live underground. Strong earthquakes can collapse the tunnels, though, and have claimed numerous casualties.

Xian was the eastern terminus of the Silk Route, and great wealth accumulated here. Near the modern city, archeologists in the 1970s unearthed an emperor's burial ground, more than 2,000 years old, that contained an army of 6,000 life-sized horses and men, wagons and weapons, fashioned in terra cotta and lined up in formation, ready for action at the deceased ruler's command.

More recently in its varied history, Xian received the spreading waves of Buddhism and Islam (it remains a Muslim center), and played a key role during the struggle between the communists and the Nationalists. The Nationalists moved industries from the vulnerable east to Xian's safer interior location, and later the communists enlarged Xian's industrial base.

Today almost nothing remains of Xian's architectural history. The city is a large, bleak industrial complex with steel mills, textile factories, and power plants.

CHINA: A GLOBAL SUPERPOWER?

Two centuries ago, Napoleon remarked that China was a giant asleep; whoever would awaken the Chinese giant would be sorry. Today China is awake, aware of its power and its potential. As the twenty-first century looms, China at last is poised to take the world's center stage.

Who awakened China? This momentous development is yet another consequence of the Cold War between then-superpowers, the United States and the Soviet Union. The Cold War worsened many regional and even local conflicts as each superpower supported, and armed, the sides it favored. Most of those conflicts have come and gone. But when the Soviet communists became advisers and technicians to Mao's regime, the seeds of China's awakening were sown. Russia and China had long been adversarial neighbors in eastern Eurasia, and it was not long before Chinese ideologues and Soviet ''revisionists' were at loggerheads. Now China found itself at odds with an immediate neighbor in possession of a vast nuclear arsenal. Mao and his comrades decided that an opening to the capitalist West was in China's long-term interest, and in 1972 U.S.

President Nixon and his entourage arrived in Beijing at Mao's invitation. Before the decade was over, Mao had died, and the ensuing power struggle was won by those who favored the open-door policy. The rest, as they say, is geography. China's cultural and economic landscapes were transformed in two short decades. China not only was awake, it was astir.

Today, China is on the march. Beijing has taken the United Nations seat long occupied by Taiwan's mainlanders. Its leading trading partners are not neighboring Russia or India, but Japan and the United States. (Trade with the United States is boosted by China's Most-Favored Nation [MFN] status.) Japan is best situated to sell to China's more than 300 million city dwellers such products as air conditioners, electronic goods, and automobiles; the United States imports from China a wide range of goods including textiles and toys. China also is a major exporter of weapons. Its global trade network is expanding in ways unforeseen by Mao Zedong and his isolationist colleagues.

The Chinese rulers who took control in Beijing in the late 1970s, led by Deng Xiaoping, called themselves ''pragmatists.'' That appellation referred especially to their economic policies, but it is applicable to other practices as well. Self-interest, regardless of the views or actions of others in the world community, guides China's use of military power and strategic decision-making. In mid-1989, Chinese armed forces brutally destroyed a student-led democracy movement in the heart of Beijing and in other cities, with a huge loss of life. In the early 1990s, China established diplomatic relations with South Korea, facilitating economic ties but infuriating the communist leadership of North Korea. But shortly thereafter, when North Korea appeared to be trying to acquire the ability to build nuclear weapons and a U.N. coalition of countries tried to stop it, China took North Korea's side. When the United States in 1994 indicated that it might end China's MFN status because of human rights violations, the Chinese were prepared to pay that price—and the United States "decoupled" the two issues. When the Taiwanese leader was allowed to enter the United States for his university class reunion in 1995, China delayed the appointment of a U.S. ambassador to Beijing and sold Silkworm missiles to Iran over U.S. objections.

These might be mere incidents in the evolving relationship between China and the rest of the world, but there are more serious portents. Beijing's apparent unwillingness to move the Taiwan issue toward resolution may result in confrontation. China's actions in the South China Sea, to which it lays claim in contravention of United Nations rules, has serious implications; the Spratly Islands may become the focus of conflict. There, China has laid claim to, and built structures on, a reef that lies within the Philippines' Exclusive Economic Zone (a concept discussed in Chapter 12). Furthermore, China has actual and latent conflicts over boundaries and territories with several of its Eur-asian neighbors, including India, Kazakstan, and Russia.

What will China's position be in the world of the twenty-first century? China contains a wide range of natural resources, from fuels and metals to forests and fertile soils. Its human resources are enormous, and its peoples' skills and capacities have been demonstrated within China as well as outside its borders. Barring a Soviet-type organizational collapse, China will become the giant of East Asia, and not only in a demographic sense.

Thus China is likely to become more than an economic force of world proportions; China also appears on course to achieve global superpower stature. As long as it retains its autocratic form of government (which made possible the imposition of draconian population policies and comprehensive economic experiments without the inconvenience of electoral consultation), China will be able to practice the kind of state capitalism that—in another guise—made South Korea an economic power. And unlike Japan or South Korea, China has no constraints on its military power. China's armies served as security forces during the pro-democracy turbulence of 1989; the armed forces are enormous, and they are modernizing. China is a nuclear power adjoined by other nuclear powers, and its self-interest will compel continued preparedness.

The sequence of events set in motion during the 1970s already has had momentous consequences. Today, when we look ahead to a modernizing, mobilizing China in the coming century, it is not difficult to hear the echo of Napoleon's famous words.

■ PRONUNCIATION GUIDE ■

Ainu (EYE-noo)
Akihito (ah-kee-HEE-toh)
Algae (AL-jee)
Altay (AL-tye)
Altun (ahl-TUN)*
Amur (uh-MOOR)
Ancien regime (aw-see-EH ray-ZHEEM)
Anhui (ahn-HWAY)
Annam (uh-NAHM)
Anshan (ahn-SHAHN)
Anyang (ahn-YAHNG)
Assam (uh-SAHM)
Austral (AW-strull)
Baotou (bao-TOH)

*"U" or final "u" pronounced as in "put"

Baykal (bye-KAHL)
Beihai (bay-HYE)
Beijing (bay-ZHING)
Bohai (bwoh-HYE)
Brunei (broo-NYE)
Buddhist (BOOD-ist)
Bund (BUND)
Cadre (KAH-dray)
Canton (kan-TONN)
Chang Jiang (chung jee-AHNG)
Changchun (CHAHNG-CHOON)*
Chengdu (chung-DOO)
Chiang Kai-shek (jee-AHNG kye-SHECK)
Chilung [Jilong] (JEE-LOONG)
Chongqing (chong-CHING)
Chosen (CHOH-SEN)

Chubu (CHOO-BOO)
Chungyang (joong-YAHNG)
Confucius (kun-FEW-shuss)
Dalai Lama (dah-lye LAHMA)
Dalian (dah-lee-ENN)
Daoism (DAU-ism)
Daqing (dah-CHING)
Deng Xiaoping (DUNG shau-PING)
Dzungaria (joong-GAH-ree-uh)
Edo (EDD-oh)
Endoreic (en-doh-RAY-ick)
Etorofu (etta-ROH-foo)
Formosa (for-MOH-suh)
Fuji (FOODGY)
Fujian (foo-jee-ENN)
Fukuoka (foo-kuh-WOH-kuh)
Fushun (foo-SHUN)*

Fuzhou (foo-ZHOH)
Gansu (gahn-SOO)
Gaoliang (gow-lee-AHNG)
Gezhouba (guh-JOH-bah)
Ginza (GHIN-zuh)
Gobi (GOH-bee)
Guangdong (gwahng-DUNG)
Guangxi Zhuang (gwahng-shee JWAHNG)
Guangzhou (gwahng-JOH)
Giulin (gway-LIN)
Guizhou (gway-JOH)
Habomai (HAH-boh-mye)
Haikou (HYE-KOH)
Hainan (HYE-NAHN)
Han (HAHN)
Hangzhou (hahng-JOH)
Hankou (hahn-KOH)
Hanyang (hahn-YAHNG)
Harbin (HAR-bin)
Hebei (huh-BAY)
Hegemony (heh-JEH-muh-nee)
Heian (HAY-ahn)
Heilongjiang (hay-long-jee-AHNG)
Henan (heh-NAHN)
Hexi (huh-SHEE)
Himalayas (him-AHL-yuzz/ himma-LAY-uzz)
Hiroshima (hirra-SHEE-muh/ huh-ROH-shuh-muh)
Hohhot (huh-HOO-tuh)
Hokkaido (hoh-KYE-doh)
Homogeneity (hoh-moh-juh-NAY-eh-tee)
Hongkou (hong-KOH)
Honshu (HONN-shoo)
Hsinchu [Xinzhu] (shin-JOO)
Huang [He] (HWAHNG [huh])
Huangpu (hwahng-POO)

Hubei (hoo-BAY)
Hunan (hoo-NAHN)
Hyundai (HUN-dye)
Iki (EE-kee)
Inchon (int-CHON)
Islam (iss-LAHM)
Jakota (juh-KOH-duh)
Jiangsu (jee-ahng-SOO)
Jiangxi (jee-ahng-SHEE)
Jilin (jee-LIN)
Jilong (jee-LUNG)
Junggar (JOONG-gahr)
Kaifeng (kye-FUNG)
Kansai (KAHN-SYE)
Kanto (KAN-toh)
Kaohsiung [Gaoxiong] (gau-shee-AHNG)
Karafuto (kahra-FOO-toh)
Karamay (kah-RAH-may)
Kawasaki (kah-wah-SAH-kee)
Kazak (KUZZ-uck)
Kazakstan (KUZZ-uck-stahn)
Khabarovsk (kuh-BAHR-uffsk)
Kinki (kin-KEE)
Kinmen (kin-MEN)
Kirghiz (keer-GEEZE)
Kissinger (KISS-sin-jer)
Kitakyushu (kee-TAH-KYOO-shoo)
Kobe (KOH-bay)
Kongfuzi (kung-FOODZEE)
Kongzi (KUNG-dzee)
Korea (kuh-REE-uh)
Kowloon (kau-LOON)
Kublai Khan (koob-lye KAHN)
Kunashiri (koo-NAH-shuh-ree)
Kunlun (KOON-LOON)
Kunming (koon-MING)
Kure (KOOH-ray)
Kurile (CURE-reel)

Kuroshio (koo-roh-SHEE-oh)
Kwangju (GWONG-JOO)
Kyoto (kee-YOH-toh)
Kyrgyz (keer-GEEZE)
Kyrgyzstan (KEER-geeze-stahn)
Kyushu (kee-YOO-shoo)
Lan Tao (LAHN-DAU)
Lanzhou (lahn-JOH)
Lhasa (LAH-suh)
Lianyungang (lee-en-yoong-GAHNG)
Liao (lee-AU)
Liaodong (lee-au-DUNG)
Liaoning (lee-au-NING)
Littoral (LITT-uh-rull)
Loess (LERSS)
Macau [Macao] (muh-KAU)
Mackinder, Halford (muh-KIN-der, HAL-ferd)
Malaya (muh-LAY-uh)
Malaysia (muh-LAY-zhuh)
Manchu (man-CHOO)
Manchukuo (mahn-JOH-kwoh)
Manchuria (man-CHOORY-uh)
Mandarin (MAN-duh-rin)
Matsu (mah-TSOO)
Mao Zedong (MAU zee-DUNG)
Meiji (may-EE-jee)
Molybdenum (muh-LIB-dun-um)
Mongol (MUNG-goal)
Mongolia (mung-GOH-lee-uh)
Mosque (MOSK)
Myanmar (mee-ahn-MAH)
Nagasaki (nah-guh-SAHKEE)
Nagoya (nuh-GOYA)
Nakhodka (nuh-KAUGHT-kuh)
Nanjing (nahn-ZHING)
Nei Mongol (nay-MUNG-goal)
Ningbo (ning-BWOH)
Ningxia Hui (NING-shee-AH HWAY)

Nobi (NOH-bee)
Ordos (ORD-uss)
Osaka (oh-SAH-kuh)
Pamirs (pah-MEERZ)
Peking (pea-KING)
Penghu (pung-HOO)
Philippines (FILL-uh-peenz)
Pinyin (pin-YIN)
Pleistocene (PLY-stoh-seen)
Pudong (poo-DONG)
Pusan (POO-sahn)
Pyongyang (pea-AWHNG-yahng)
Qaidam (CHYE-DAHM)
Qanats (KAH-nahts)
Qin (CHIN)
Qing (CHING)
Qingdao (ching-DAU)
Qinghai (ching-HYE)
Qinhuangdao (chin-hwahng-DAU)
Qiqihar (chee-CHEE-har)
Quemoy (keh-MOY)
Rostow (ROSS-stoff)
Ruijin (rway-JEEN)
Ryukyu (ree-YOO-kyoo)
Sakhalin (SOCK-uh-leen)
Seikan (say-KAHN)
Sen-en (sen-NENN)
Seoul (SOAL)
Seto (SET-oh)
Shaanxi (shahn-SHEE)
Sha Mian (shah mee-AHN)
Shan (SHAHN)
Shandong (shahn-DUNG)
Shanghai (shang-HYE)
Shantou (SHAHN-TOH)
Shanxi (shahn-SHEE)
Shenyang (shun-YAHNG)

Shenzhen (shun-ZHEN)
Shi (SHEE)
Shikoku (shick-KOH-koo)
Shikotan (shee-koh-TAHN)
Shimonoseki (shim-uh-noh-SECKEE)
Shirioshi (shih-ree-OH-shee)
Shogun (SHOH-goon)
Sichuan (zeh-CHWAHN)
Singapore (SING-uh-poar)
Sinicization (sine-ih-sye-ZAY-shun)
Songhua (SUNG-hwah)
Spratly (SPRAT-lee)
Spykman (SPIKE-mun)
Sri Lanka (sree-LAHNG-kuh)
Sui (SWAY)
Sun Yat-sen (SOON yaht-SENN)
Sung (SOONG)
Taipei [Taibei] (tye-BAY)
Taiwan (tye-WAHN)
Tajikistan (tah-JEEK-ih-stahn)
Taklimakan (tahk-luh-muh-KAHN)
Tang (TAHNG)
Tangshan (tahng-SHAHN)
Tarim (TAH-REEM)
Thailand (TYE-land)
Tiananmen (TYAHN-un-men)
Tianjin (tyahn-JEEN)
Tian Shan (TYAHN SHAHN)
Tibet (tuh-BETT)
Tokaido (toh-KYE-doh)
Tokyo (TOH-kee-oh)
Tonkin (TAHN-KIN)
Toyama (toh-YAH-muh)
Tsugaru (tsoo-GAH-roo)
Tsukuba (tsoo-KOOB-uh)
Tsunami (tsoo-NAH-mee)
Tsushima (tsoo-SHEE-muh)

Turkestan (TERK-uh-stahn)
Ulaan Baatar (oo-lahn BAH-tor)
Ulsan (OOL-SAHN)
Ürümqi (oo-ROOM-chee)
Uygur (WEE-ghoor)
Versailles (vair-SYE)
Vietnam (vee-et-NAHM)
Vladivostok (vlad-uh-vuh-STAHK)
Wakhan (wah-KAHN)
Wei (WAY)
Wenzhou (whunn-JOH)
Wuchang (woo-CHAHNG)
Wuhan (woo-HAHN)
Wuzhou (woo-JOH)
Xi (SHEE)
Xiamen (shah-MEN)
Xian (shee-AHN)
Xianggang (see-AHNG-gahng)
Xi Jiang (SHEE jee-AHNG)
Xinjiang (shin-jee-AHNG)
Xizang (sheedz-AHNG)
Yanan (yen-AHN)
Yangshuo (YAHNG-SHWOH)
Yangzi (Yangtze) (YANG-dzee)
Yantai (yahn-TYE)
Yawata (yuh-WAH-tuh)
Yichang (yee-CHAHNG)
Yokohama (yoh-kuh-HAH-muh)
Yuan (YOO-ahn)
Yumen (YOO-mun)
Yunnan (yoon-NAHN)
Zhanjiang (JAHN-jee-AHNG)
Zhejiang (JEJ-ee-AHNG)
Zhou (JOH)
Zhou Enlai (JOH en-lye)
Zhu (JOO)
Zhuhai (joo-HYE)

Scale 1:16 000 000; one inch to 250 miles. Polyconic Project
Elevations and depressions are given in feet

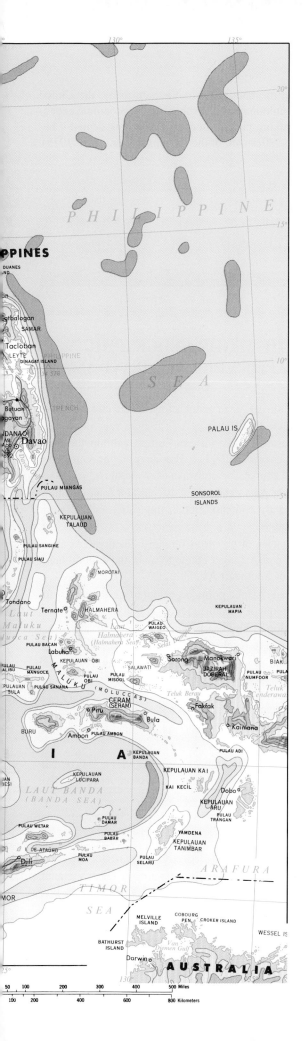

CHAPTER 10

Southeast Asia: Between the Giants

IDEAS & CONCEPTS

Buffer zone
Shatter belt
Political geography
Genetic boundary
   classification
State territorial morphology
   Compact states
   Protruded states

Elongated states
Fragmented states
Perforated states
Domino theory
Insurgent state model
Entrepôt
Translocation

❖ REGIONS

Mainland Region   Insular Region

southeast Asia . . . the very name roils American emotions. Here the United States owned its only major colony. Here American foreign policy suffered its most disastrous failure. Here the United States fought the only war it ever lost. Here Washington's worst Cold War fears failed to materialize. In Vietnam today, the United States has an opportunity to share in the economic expansion now taking place—an opportunity long denied it because of the aftermath of the Indochina War (1964–1975).

DEFINING THE REALM

Southeast Asia is a realm of peninsulas and islands, a corner of Asia bounded by India on the northwest and China on the northeast. Its western coasts are washed by the Indian Ocean, and to the east stretches the vast Pacific. From all these directions, Southeast Asia has been penetrated by outside forces. From India came traders; from China, settlers. From across the Indian Ocean came Arabs to engage in commerce and Europeans in pursuit of empires. And from across the Pacific came the Americans. Southeast Asia has been the scene of countless contests for power and primacy—the competitors have come from near and far.

Southeast Asia's geography in some ways is reminiscent of that of Eastern Europe. It is a mosaic of smaller countries on the periphery of one of the world's largest states. It has been a **buffer zone** between powerful adversaries. It is a **shatter belt** in which stresses and pressures from without and within have fractured the political geography. Like Eastern Europe, Southeast Asia is the scene of great cultural diversity. Again, as in Eastern Europe, the map has changed even in recent times. In 1965, Singapore broke away from Malaysia to become the realm's smallest political entity territorially, a contemporary city-state. Still more recently, a boundary disappeared from the map in 1976 when North and South Vietnam were united. And while the world's attention focused on the strife in former Yugoslavia, ethnic and cultural conflict claimed thousands of casualties and produced streams of refugees in Southeast Asia as well.

Because the politico-geographical map (Fig. 10-1) is so complicated, it should be studied attentively. One good way to strengthen your mental map of this realm is to follow the mainland coastline from west to east. The westernmost state in the realm is Myanmar (called Burma before 1989, and still referred to by that name), the only country in Southeast Asia that borders both India and China. Myanmar shares the "neck" of the Malay Peninsula with Thailand, heart of the *mainland* region. The south of the peninsula is part of Malaysia—except for Singapore, at the very tip of it. Facing the Gulf of Thailand is Cam-

THE MAJOR GEOGRAPHIC QUALITIES OF SOUTHEAST ASIA

■ ■ ■

1. The Southeast Asian realm is fragmented into numerous peninsulas and islands.

2. Southeast Asia, like Eastern Europe, exhibits the characteristics of a shatter belt. Pressures on this realm from external sources have always been strong.

3. Southeast Asia exhibits intense cultural fragmentation, reflected by complex ethnic, linguistic, and religious geographies.

4. The legacies of powerful foreign influences (Asian as well as non-Asian) continue to mark the cultural landscapes of Southeast Asia.

5. Southeast Asia's politico-geographical traditions involve frequent balkanization, instability, and conflict.

6. Population in Southeast Asia tends to be strongly clustered, even in rural areas.

7. Compared to neighboring regions, mainland Southeast Asia's physiologic population densities remain relatively low.

8. Rapid population growth has prevailed in the island regions of Southeast Asia, notably in the Philippines, during much of the twentieth century.

9. Intraregional communications in Southeast Asia remain inferior. External connections are often more effective than internal linkages.

■ ■ ■

bodia. Still moving generally eastward, we reach Vietnam, a strip of land that extends all the way to the Chinese border. And surrounded by its neighbors is landlocked Laos, remote and isolated. This leaves the islands that constitute *insular* Southeast Asia: the Philippines in the north and Indonesia in the south, and between them the offshore por-

Political Geography

Southeast Asia is a laboratory for the study of **political geography**. This systematic field, one of the oldest in geography, focuses on the spatial expressions of political behavior. Boundaries on land and on the oceans, the roles of capital cities, power relationships among states, administrative systems, voter behavior, conflicts over resources, and even matters involving outer space have politico-geographical dimensions. We have already been introduced to aspects of political geography when we studied the possible unification of Western Europe (*supranationalism*), the historic shattering of Eastern Europe (*balkanization*), the support Pakistan gave to Muslims in Kashmir (*irredentism*), and the competition for hegemony over parts of Eurasia (*geopolitics*).

The field of political geography grew from geographers' interest in the spatial nature of the national state. What are the ingredients of a nation-state? Why do some states survive over many centuries, while others (such as the Ottoman Empire, the Austro-Hungarian Empire, and now the Soviet Empire) collapse? About a century ago, Friedrich Ratzel (1844–1904) proposed a theory that likened the nation-state to a biological organism. Just as an organism is born, grows, matures, and eventually dies, Ratzel argued, states go through stages of birth (around a culture hearth or core area), expansion (perhaps by colonization), maturity (stability), and eventual collapse. Only the sporadic absorption of new land and people, he suggested, could stave off the state's decline. This was a blueprint for imperialism!

Later, political geographers realized that every state has ties that bind it (*centripetal forces*) and stresses that tend to break it apart (*centrifugal forces*). When

the centripetal or unifying forces are much stronger, the state succeeds; when the centrifugal or divisive forces prevail, the state fails (*devolution*). In Japan today, the centripetal forces are very strong; but in Afghanistan, centrifugal forces predominate. Identifying and measuring these forces, obvious as they may be, is one of the challenges of political geography.

We sometimes call countries *nations*, but many countries contain more than one nation, which is why it is better to call them *states*. It might be appropriate to call Finland a nation, but Nigeria (as we noted in Chapter 7) contains three nations and many minority peoples. States that are also pluralistic societies, such as Russia, exhibit powerful centrifugal forces that must be overcome. Even Canada confronts cultural division, spatially expressed, that may well lead to its fragmentation; thus Canada is not yet a nation.

The most basic device in political geography is the world political map (Fig. I-10). This map reveals the enormous range in the sizes of states; some are microstates or ministates, while others are giants. The map also suggests why the United Nations officially recognizes a group of Geographically Disadvantaged States (GDS): more than two dozen states have no *maritime boundaries* and are landlocked. As we noted in the case of Bolivia, this can have a disastrous effect on the fortunes of a country. Another aspect of the world map is the **state territorial morphology**, or physical shape of states. In this chapter we examine in some detail the shapes of Southeast Asian states, and discuss what effect their morphology may have had on their development.

States have capitals, core areas, administrative divisions, and boundaries. Boundaries are sensitive parts of the anatomy of a state: just as people are territorial about their individual properties, so nations and states are sensitive about their territories and limits. Boundaries, in effect, are contracts between neighboring states. The *definition* of a boundary is likely to be found in an elaborate treaty that verbally describes its precise location. Cartographers then perform the *delimitation* (official mapping) of what the treaty stipulates. Certain boundaries are actually placed on the ground as fences, walls, or other artificial barriers; this represents the *demarcation* of the boundary. The world map shows that some boundaries have a sinuous form, while others are straight lines. Boundaries can therefore be classified as *geometric* (straight-line or curved), *physiographic* (coinciding with rivers or mountain crests), or *anthropogeographic* (marking breaks or transitions in the cultural landscape).

Modern political geography also focuses on political behavior and the way this varies across the cultural landscape. In the United States, for example, voters in urban areas tend to be less conservative than those in rural areas (although suburbanization has blurred this contrast); there was a time when the "Solid South" was a Democratic stronghold. Explanations of such changes involve the analysis of such factors as voters' age, income, education, occupation, and much more. In Southeast Asia, as we will see, democratization has reached different stages in different countries. This, too, presents opportunities for the study of spatial variations in the political landscape, and for predictions of what may lie ahead.

---

tion of Malaysia, situated on the largely Indonesian island of Borneo. Completing the map is Brunei, the smallest country in the realm in terms of population but, as we will see, important in the regional picture.

These are the countries of a realm in which there is no

dominant state—no Brazil, no China, no India—although Indonesia is its population giant. Southeast Asia is a realm of mountain barriers, unproductive uplands, rugged coastlines, and far-flung islands, as well as fertile valleys and deltas, rich volcanic soils on terraced hillslopes, and pro-

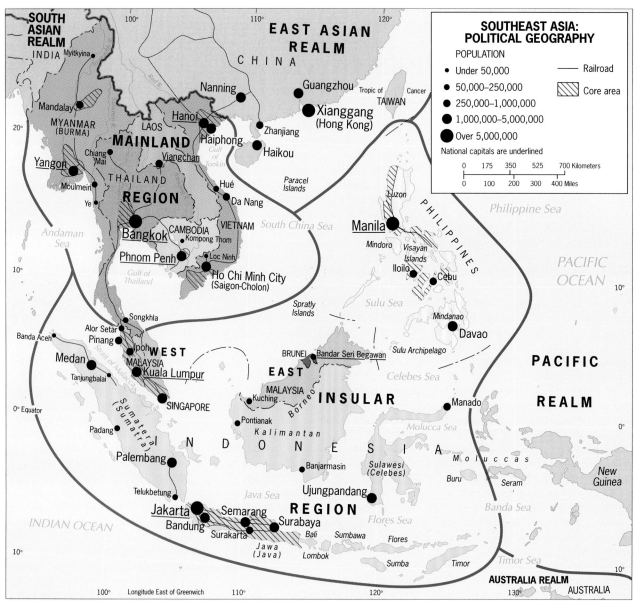

FIGURE 10-1

ductive plains watered by ample rainfall (Fig. 10-2). The realm contains some of the world's largest remaining stands of tropical rainforest. Its surviving wildlife includes endangered and threatened species such as the rhinoceros and one of our closest relatives, the orangutan. Towering over the landscape—often rising from the very shoreline— are hundreds of volcanoes, many of them active.

No single, dominant core of indigenous development emerged here. In the river basins and plains of the mainland, and on the islands offshore, a flowering of cultures produced a diversity of societies whose languages, religions, arts, music, foods, and other achievements formed an almost infinitely varied mosaic—but none of those cultures rose to imperial power. It was the European colonizers who

forged empires here, often by playing one local state off against another; the Europeans divided and ruled. Out of this foreign intervention came the modern map of Southeast Asia, as only Thailand survived the colonial era as an independent entity. Thailand was useful to two competing powers, the French to the east and the British to the west. It was a convenient buffer, and while the colonists carved pieces off Thailand's domain, the kingdom endured.

Indeed, the Europeans accomplished what local powers could not: the formation of comparatively large, multicultural states that encompassed diverse peoples and societies and welded them together. Were it not for the colonial intervention, it is unlikely that the 13,000 islands of far-flung Indonesia would today constitute the world's fourth

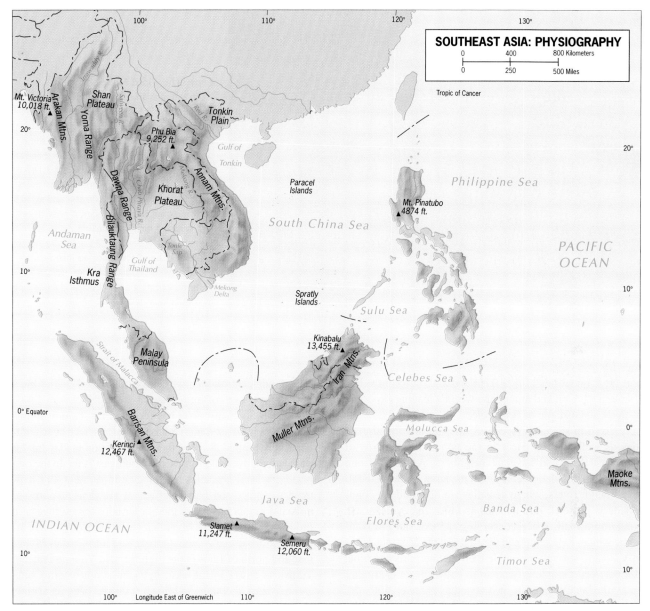

SOUTHEAST ASIA: PHYSIOGRAPHY

FIGURE 10-2

largest country in terms of population. Nor would the nine sultanates of Malaysia have been united, let alone with the peoples of northern Borneo across the South China Sea. For good or ill, the colonial intrusion consolidated a realm of few culture cores and numerous ministates into nine countries (the tenth, Singapore, was the only instance of postcolonial secession).

POPULATION GEOGRAPHY

Compared to the huge population numbers and densities in the habitable regions of South Asia and China, demo-graphic totals for the countries of Southeast Asia, with the sole exception of Indonesia, seem modest. Again, compar-isons with Europe come to mind. Three countries—Thai-land, the Philippines, and Vietnam—have populations be-tween 60 and 80 million. Laos, quite a large country territorially (comparable to the United Kingdom), had just 5.1 million inhabitants in 1997. Cambodia, substantially larger than Greece, had 11.2 million.

As Appendix A shows, arithmetic population densities are not especially high in Southeast Asia except in the ur-banized ministate of Singapore; physiologic densities like-wise are substantially lower than in eastern China or neigh-boring South Asian countries. When the political situation is stable, several Southeast Asian countries are able to ex-

From the Field Notes

"The Indonesian island of Bali's volcanic, well-watered soils and near-equatorial location, permit the production of as many as three crops of rice on the same paddy field every year. With over three million people on an island of just 2,147 square miles (5,561 sq km), Bali's farmers must produce a large harvest—and a growing one, since population growth on Bali remains near Indonesia's national average of 1.6 percent. As a result, much of Bali has been transformed into sets of terraces watered by elaborate irrigation systems, creating a rural cultural landscape of meticulous detail."

port large quantities of rice. Thailand and, in recent years, Vietnam have ranked among the world's leading exporters of this crucial staple of Asian diets.

Viewed in spatial perspective, one can discern similarities as well as differences in the population patterns of Southeast Asia and its giant neighbors. As in East and South Asia, population clusters in the large basins and deltas of major rivers, where densities are high and the countryside is parceled into paddies as far as the eye can see. Numerous small, space-conserving villages dot the landscape which, for example in the Red and the Mekong deltas in Vietnam, strongly resemble their counterparts in China. But between these low-lying, alluvial basins, in the higher areas, the landscape takes on a savanna character, with reddish, leached, less fertile soils and much sparser population. And in the islands of Indonesia and the Philippines, Southeast Asia has environments that are not present at all in India or China—well-watered volcanic soils supporting luxuriant natural vegetation and capable of sustaining intensive agriculture. Whole countrysides have been meticulously terraced to create paddies and to irrigate them.

People and Land

It is noteworthy that, of Southeast Asia's 500 million inhabitants, well over half (275 million, 55 percent) live on the islands of Indonesia and the Philippines, leaving the realm's seven mainland countries with just 45 percent of the population. When there is such high population pres-

sure in adjacent realms, why has Southeast Asia not been flooded by waves of immigrants?

In fact, Southeast Asia has received its share of immigrants, but it is true that overland invasions have been comparatively limited. Several factors have combined to hinder travel along overland routes into Southeast Asia. In the first place, physical obstacles inhibit such movement. In Chapter 8 we noted the barrier effect of the densely forested hills and mountains along the border between northeastern India and northwestern Myanmar (Burma). North of Myanmar lies forbidding Xizang (Tibet), and northeast of Myanmar and north of Laos is the high, rugged Yunnan Plateau. Transit is easier in the east, between southeastern China and northern Vietnam, and this has indeed been an avenue for contact and migration. As the ethnolinguistic map of China (Fig. 9-8) shows, migration from Southeast Asia into China also has occurred in this area.

Second—and the population distribution map (Fig. I-9) reflects this—the highly clustered agricultural opportunities in Southeast Asia lie separated by large stretches of less productive land, and these in turn are crossed by high-relief zones that hinder movement. The map of Southeast Asia reveals several concentrated areas of productive capacity but few corridors that would attract migrants.

As we noted earlier, immigrants to Southeast Asia over the past millennium have come mainly by sea, not by land. That is true not only for Arabs and Europeans but also for neighboring Chinese. Mostly the Chinese immigrants came not as farmers seeking land, but as traders and workers looking for opportunities in the cities and towns. "Chinatowns" form a vital part of most Southeast Asian cities today, notably the coastal ones.

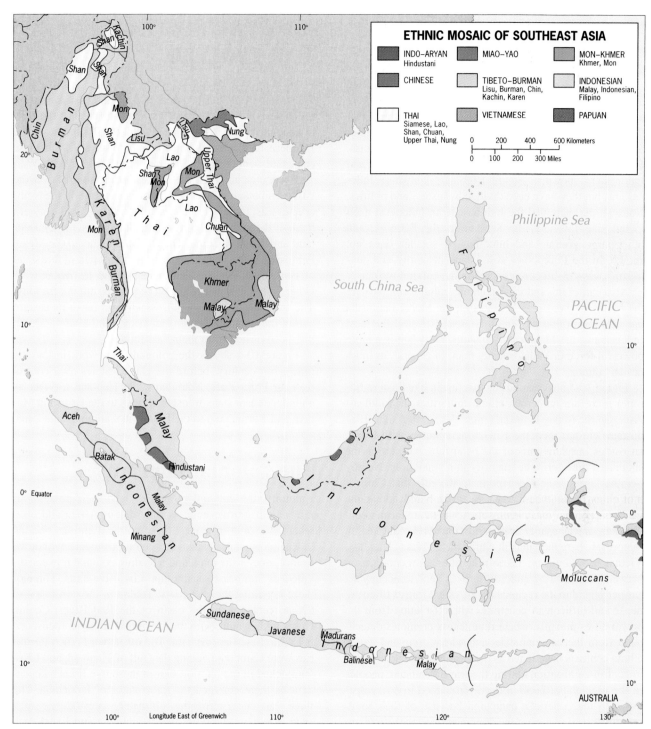

FIGURE 10-3

The Ethnic Mosaic

Southeast Asia's peoples come from a common stock just as (Caucasian) Europeans do, but this has not prevented the emergence of regionally or locally discrete ethnic or cultural groups. Figure 10-3 displays a generalized distribution of ethnolinguistic groups in the realm, but be aware

that this *is* a generalization. At the scale of this map, numerous small groups cannot be depicted.

The map shows the rough spatial coincidence, on the mainland, between major ethnic group and modern political state. The Burman dominate in the country once known as Burma (Myanmar); the Thai occupy the state once known as Siam (Thailand); the Khmer form the nation of Cambodia and extend northward into Laos; and the Vietnamese inhabit the long strip of territory facing the South China Sea.

Territorially, by far the largest population is classified in Figure 10-3 as Indonesian, the inhabitants of the great archipelago that extends from Sumatera* west of the Malay Peninsula to the Moluccas in the east, and from the lesser Sunda Islands in the south to the Philippines in the north. Collectively, all these peoples—the Filipinos, Malays, and Indonesians shown in Figure 10-3—are known as Indonesians, but they have been divided by history and politics. Note, on the map, that the Indonesians in Indonesia itself include the Javanese, the Sundanese, the Balinese, and other large groups, and hundreds of smaller ones not shown. In the Philippines, too, island insularity and contrasting ways of life are reflected in the cultural mosaic.

Also part of this Indonesian ethnic-cultural complex are the Malays, whose heartland lies on the Malay Peninsula but who form minorities in other areas as well. Like most Indonesians, the Malays are Muslims, but Islam is a more powerful force in Malay society than, in general, in Indonesian culture.

Figure 10-3 also reminds us that (again like Eastern Europe) there are numerous ethnic minorities in Southeast Asia. On the Malay Peninsula, note the South Asian (Hindustani) cluster; there are Hindu communities with Indian ancestries in many parts of the peninsula, but here in the southwest they form the majority in a small area. Nearby lies the largest single exception to the rule that Chinese minorities in Southeast Asia are urban-based: along the west coast of the Malay Peninsula, Chinese are farmers and villagers as well as city dwellers. In the north of the realm, minorities share the lands in which the Burman, Thai, and Vietnamese dominate. Those minorities, as a comparison between Figures 10-1 and 10-3 proves, tend to occupy areas peripheral to the regional cores, where forests often are dense and difficult to penetrate, where isolation from the power cores prevails, where remoteness promotes detachment from the national state, and where, time and again, ethnic conflicts have raged as the core-based majority tried to control the frontier. During the war in Vietnam, peoples in the highlands (the *Montagnards*) tended to be sympathetic to the American campaign because of their long-standing hostility toward the dominant (lowland) Vietnamese. Today, the Shan and the Karen in Myanmar (Burma; see map) are in a state of conflict with the regime based in the core area.

*As in Africa, names and spellings have changed with independence. In this chapter, the contemporary spellings will be used, except when reference is made to the colonial period. Thus Indonesia's four major islands are Jawa, Sumatera, Kalimantan (the Indonesian part of Borneo), and Sulawesi. The Dutch called them Java, Sumatra, Borneo, and Celebes, respectively.

HOW THE POLITICAL MAP EVOLVED

The leading colonial competitors in Southeast Asia were the Dutch, French, British, and Spanish (the last replaced by the Americans in their stronghold, the Philippines). The Japanese had colonial objectives here as well, but these came and went during the course of the Second World War.

The Dutch acquired the greatest prize: control over the vast archipelago now called Indonesia (formerly the Netherlands East Indies). France established itself on the eastern flank of the mainland, controlling all territory east of Thailand and south of China. The British conquered the Malay Peninsula, gained power over the northern part of the island of Borneo, and established themselves in Burma as well. Other colonial powers did gain footholds, but not for long. The exception was Portugal, which held on to its eastern half of the island of Timor (Indonesia) until after the Dutch had been ousted from their East Indies.

Figure 10-4 shows the colonial framework in the late nineteenth century, before the United States took control over the Philippines. Note that while Thailand survived as an independent state, it lost territories to the British in Malaya and Burma, and to the French in Cambodia and Laos.

The Colonial Imprint

Naturally, the colonial powers divided their possessions into administrative units as they did in Africa and elsewhere. In some instances, these political entities became independent states when the colonial powers withdrew. France, one of the mainland's leading colonial powers, divided its Southeast Asian empire into five units. Three of these lay along the east coast: Tonkin in the north next to China, centered on the basin of the Red River; Cochin China in the south, with the Mekong delta as its focus; and between these two, Annam. The other two French territories were Cambodia, facing the Gulf of Thailand, and Laos, landlocked in the interior. Out of these five French dependencies there emerged the three states of Indochina. The three east-coast territories ultimately became one such state, Vietnam; the other two (Cambodia and Laos) each achieved separate independence.

The British ruled two major entities in Southeast Asia (Burma and Malaya) in addition to a large part of northern Borneo and many small islands in the South China Sea. Burma was attached to Britain's Indian empire; it was governed from 1886 until 1937 from distant New Delhi. But when British India became independent in 1947 and split into several countries, Burma was not part of the grand design that created East and West Pakistan, Ceylon (now

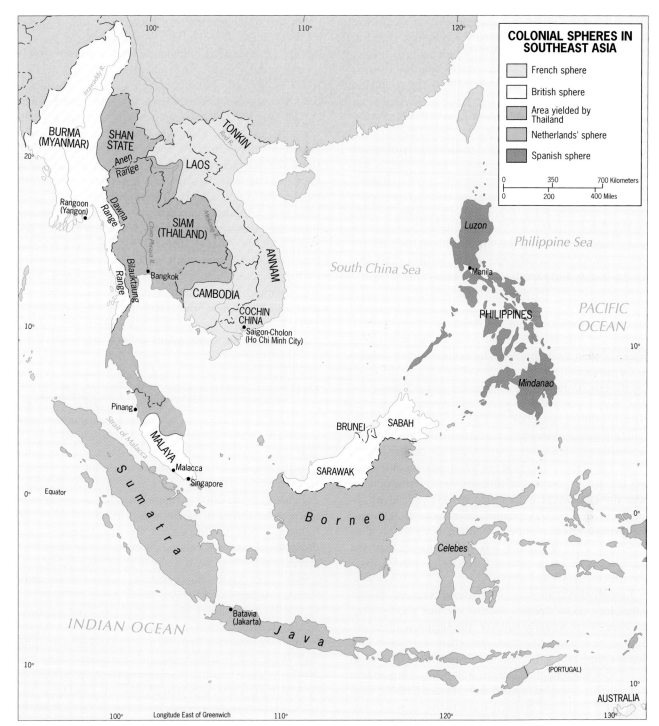

COLONIAL SPHERES IN SOUTHEAST ASIA

- French sphere
- British sphere
- Area yielded by Thailand
- Netherlands' sphere
- Spanish sphere

FIGURE 10-4

Sri Lanka), and India. Instead, Burma (now Myanmar) in 1948 was given the status of a sovereign republic.

In Malaya, the British developed a complicated system of colonies and protectorates that eventually gave rise to the equally complex, far-flung Malaysian federation. Included were the former Straits Settlements (Singapore was

one of these colonies), the nine protectorates on the Malay Peninsula (former sultanates of the Muslim era), the British dependencies of Sarawak and Sabah on the island of Borneo, and numerous islands in the Strait of Malacca and the South China Sea. The original federation of Malaysia was created in 1963 by the political unification of recently independent mainland Malaya, Singapore, and the former British dependencies on the largely Indonesian island of Borneo. Singapore, however, left the federation in 1965 to become a sovereign city-state, and the remaining units were later restructured into peninsular Malaysia and, on Borneo, Sarawak and Sabah. Thus the term *Malaya* properly refers to the geographic area of the Malay Peninsula, including Singapore and other nearby islands; the term *Malaysia* identifies the politico-geographical entity of which Kuala Lumpur is the capital city.

The Hollanders took control of the ''spice islands'' through their Dutch East India Company, and the wealth that was extracted from what is today Indonesia brought the Netherlands its Golden Age. From the mid-seventeenth to the late-eighteenth century, Holland could develop its East Indies sphere of influence almost without challenge, for the British and French were preoccupied with the Indian subcontinent. By playing the princes of Indonesia's states against one another in the search for economic concessions and political influence, by placing the Chinese in positions of responsibility, and by imposing systems of forced labor in areas directly under its control, the company had a ruinous effect on the Indonesian societies it subjugated. Java (Jawa), the most populous and productive island, became the focus of Dutch administration; from its capital at Batavia (now Jakarta), the company extended its sphere of influence into Sumatra (Sumatera), Dutch Borneo (Kalimantan), Celebes (Sulawesi), and the smaller islands of the East Indies. This was not accomplished overnight, and the struggle for territorial control was carried on long after the Dutch East India Company had yielded its administration to the Netherlands government. Dutch colonialism thus threw a girdle around Indonesia's more than 13,000 islands, paving the way for the creation of the realm's largest nation-state in terms of population (over 200 million today) and territory.

In the colonial tutelage of Southeast Asia, the Philippines, long under Spanish domination, had a unique experience. From as early as 1571, the islands north of Indonesia were under Spain's control (they were named for Spain's King Philip II). Spanish rule began at a time when Islam was reaching the southern Philippines via northern Borneo. The Spaniards spread their Roman Catholic faith with great zeal, and between them the soldiers and the priests consolidated Hispanic dominance over the mostly Malay population. Manila, a city founded in 1571, became a profitable way-station on the route between southern China and western Mexico (Acapulco usually was the trans-Pacific destination for the galleons leaving Manila's port). There was much profit to be made, but the indigenous people shared little in it. Great landholdings were awarded to loyal Spanish civil servants and to men of the church. Oppression eventually yielded revolution, and Spain was confronted with a major uprising when the Spanish-American War broke out elsewhere in 1898.

As part of the settlement of that war, the United States replaced Spain in Manila. That was not the end of the revolution, however. The Filipinos now took up arms against their new foreign ruler, and not until 1905, after terrible losses of life, did American forces manage to ''pacify'' their new dominion. Subsequently, U.S. administration in the Philippines was more progressive than Spain's had been. In 1934, the Philippine Independence Law was passed, providing for a 10-year transition to sovereignty. But before independence could be arranged, World War II intervened. In 1941, Japan invaded the islands, temporarily ousting the Americans; U.S. forces returned three years later and, with strong support from Filipino forces, defeated the Japanese in 1945. Now the agenda for independence could be resumed, and in 1946 the Sovereign Republic of the Philippines was proclaimed.

Today, all of Southeast Asia's states are independent, but centuries of colonial rule have left strong cultural imprints. In their urban landscapes, in their education systems, in their civil service, and in countless other ways, this realm still carries the marks of its colonial past.

Cultural-Geographic Legacies

The French, who ruled and exploited a crucial quadrant of Southeast Asia, had a name for their empire: *Indochina*. That name would be appropriate for the rest of the realm as well, because it suggests the two leading Asian influences that have affected it for the past 2,000 years. Periodic expansions of the Chinese Empire, notably along the eastern periphery, as well as the arrival of large numbers of Chinese settlers (see box titled ''Overseas Chinese''), infused the region with cultural norms from the north. The Indians came from the west by way of the sea, as their trading ships plied the coasts and settlers from India founded colonies on Southeast Asian shores in the Malay Peninsula, on the lower Mekong Plain, on Jawa and Bali, and on Borneo.

With the migrants from the Indian subcontinent came their faiths: first Hinduism and Buddhism, later Islam. The Muslim religion was also promoted by the growing number of Arab traders who appeared on the scene, and Islam became the dominant religion in Indonesia (where nearly 90 percent of the population adheres to it today). But in Myanmar, Thailand, and Cambodia, Buddhism remained supreme, and in all three countries the overwhelming majority of the people are now adherents. In culturally diverse Malaysia, the Malays are Muslims (to be a Malay *is* to be a Muslim), and almost all Chinese are Buddhists; but most Malaysians of Indian ancestry continue to follow the Hindu way of life. Although Southeast Asia has generated its own local cultural expressions, most of what remains in tangible form has resulted from the infusion of foreign elements. For instance, the main temple at Angkor Wat, constructed

OVERSEAS CHINESE

Chinese have been migrating into Southeast Asia for 2,000 years, but the largest movements occurred during the past six centuries (Fig. 10-5). The *precolonial period* (1370–1670) saw the arrival of small groups of Chinese who established neighborhoods in the larger towns. During the *early colonial period* (1670–1870), commercial opportunities attracted greater numbers of Chinese to the evolving colonial empires. The *late colonial period* (1870–1940) produced more than 20 million migrants as the Chinese were beckoned by expanding job as well as trade opportunities. Late during this period, however, colonial governments (as well as Thailand) imposed restrictions on Chinese immigration. The Chinese had become an influential class of middlemen, and both colonial rulers and native peoples feared potential Chinese "expansionism." During the *war and decolonization period* (1940–1970), Chinese suffered severe punishment, especially in Malaya, un-

der wartime Japanese rule, and later in Indonesia, during a murderous anticommunist campaign.

Not counting intermarriage, there may be as many as 30 million Chinese in Southeast Asia today, and many have become very successful businesspeople. They remember their roots in Guangdong or Fujian or other provinces of China, and China's recent reopening has rekindled long-dormant links to clan and village. China's overseas diaspora is wealthy far beyond their numbers, having taken control of large parts of countries' commercial and manufacturing sectors. Having always saved their money carefully, they are now in a position to invest in their ancestral homeland. Many are doing so, and the growing interconnections between China's overseas entrepreneurs and the old sources of emigration are helping forge the new China on the Pacific Rim.

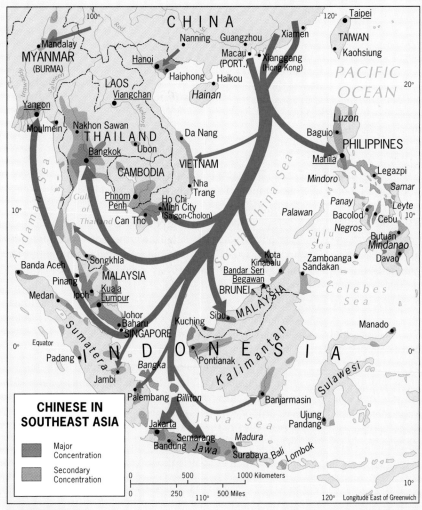

FIGURE 10-5

From the Field Notes

"Like most major Southeast Asian cities, Bangkok's urban area includes a large and prosperous Chinese sector. More than 12 percent of Thailand's population of 62 million is of Chinese ancestry, and the great majority of Chinese live in the cities. In Thailand, this large non-Thai population is well integrated into local society, and intermarriage is common. Still, Bangkok's 'Chinatown' is a distinct and discrete part of the great city. There is no mistaking Chinatown's limits: Thai commercial signs change to Chinese, goods offered for sale also change (Chinatown contains a large cluster of shops selling gold, for example), and the urban atmosphere, from street markets to bookshops, is dominantly Chinese. This is a boisterous, noisy, energetic part of multicultural Bangkok, a vivid reminder of the Chinese commercial success in Southeast Asia."

in Cambodia during the twelfth century, remains a monument to Indian architecture of that time.

The *Indo* part of Indochina, then, refers to the cultural imprints from South Asia: the Hindu presence, the importance of the Buddhist faith (which came to Southeast Asia via Sri Lanka [Ceylon] and its seafaring merchants), the influences of Indian architecture and art (especially sculpture), writing and literature, and social structures and patterns.

The Chinese role in Southeast Asia has been substantial as well. Chinese emperors coveted Southeast Asian lands, and China's power at times reached deeply into the realm. Social and political upheavals in China sent millions of Sinicized people southward. Chinese traders, pilgrims, sailors, fishermen, and others sailed from southeastern China to the coasts of Southeast Asia and established settlements there. Over time, those settlements attracted additional Chinese emigrants, and Chinese influence in the realm grew (Fig. 10-5). Not surprisingly, relations between the Chinese settlers and the earlier inhabitants of Southeast Asia have at times been strained, even violent. The Chinese presence in Southeast Asia is long-term, but the invasion has continued into modern times. The economic power of Chinese

minorities and their role in the political life of the area have led to conflicts.

The Chinese initially profited from the arrival of the Europeans, who stimulated the growth of agriculture, trade, and industries; here the Chinese found opportunities they did not have at home. They established rubber holdings, found jobs on the docks and in the mines, cleared the bush, and transported goods in their sampans. They brought with them skills that proved to be very useful, and as tailors, shoemakers, blacksmiths, and fishermen, they prospered. The Chinese also proved to be astute in business; soon, they not only dominated the region's retail trade but also held prominent positions in banking, industry, and shipping. Thus their importance has always been far out of proportion to their modest numbers in Southeast Asia. The Europeans used them for their own designs but found the Chinese to be stubborn competitors at times—so much so that eventually they tried to impose restrictions on Chinese immigration. The United States, when it took control of the Philippines, also sought to stop the influx of Chinese into those islands.

When the European colonial powers withdrew and Southeast Asia's independent states emerged, Chinese population sectors ranged from nearly 50 percent of the total in Malaysia (in 1963) to barely over 1 percent in Myanmar. In Singapore, Chinese today constitute 78 percent of the population of 3.1 million; when Singapore seceded from Malaysia in 1965, the Chinese component in Malaysia was reduced to about 35 percent. In Indonesia, the percentage of Chinese in the total population is not high (no more than 3 percent), but the Indonesian population is so large that even this small percentage indicates a Chinese sector of more than 6 million. In Thailand, on the other hand, many Chinese have married Thais, and the Chinese minority of about 12 percent has become a cornerstone of Thai society, dominant in trade and commerce.

In general, Southeast Asia's Chinese communities remained quite aloof and formed their own separate societies in the cities and towns. They kept their culture and language alive by maintaining social clubs, schools, and even residential suburbs that, in practice if not by law, were Chinese in character. There was a time when they were in the middle between the Europeans and Southeast Asians and when the hostility of the local people was directed toward white people as well as toward the Chinese. Since the withdrawal of the Europeans, however, the Chinese have become the main target of this antagonism, which remained strong because of the Chinese involvement in money lending, banking, and trade monopolies. Moreover, there is the specter of an imagined or real Chinese political imperialism along Southeast Asia's northern flanks.

The *china* in Indochina, therefore, represents a diversity of penetrations. The source of most of the old invasions was in southern China, and Chinese territorial consolidation provided the impetus for successive immigrations. Mongoloid racial features carried southward from East Asia mixed with the preexisting Malay stock to produce a transition from Chinese-like people in the northern mainland to darker-skinned Malay types in the distant Indonesian east. Although Indian cultural influences remained strong, Chinese modes of dress, plastic arts, types of houses and boats, and other cultural attributes were adopted throughout Southeast Asia. During the past century, and especially during the last half-century, renewed Chinese immigration brought skills and energies that propelled these minorities to positions of comparative wealth and influence.

The Boundaries

In general, the politico-geographical boundaries of Southeast Asia were better defined and delimited than was the case in Africa. Over a major part of their combined length, the political boundaries of mainland Southeast Asia lie across remote, relatively sparsely peopled, inland areas not fully integrated into the colonial spheres established by the Europeans. Even after independence, these areas retained

their frontier-like characteristics. In these border zones lie the roots of some of the region's persistent ethnic wars. These are fertile grounds, too, for revolutionary activities (and for other illicit activities as well—witness the huge opium poppy harvests of the notorious ''Golden Triangle'' where the borders of Myanmar, Laos, and Thailand converge).

In our Focus on Political Geography (p. 477), we saw how political boundaries can be categorized morphologically. Figure 10-3 shows that the boundary between Thailand and Myanmar, over long segments, is anthropogeographic, notably where the name *Karen*, for the Burman minority, appears on the map. Figure 10-2 as well as the atlas map on p. 474 reveal that a large part of the Vietnam–Laos boundary is physiographic-political, coinciding with the Annam (Annamese) Mountains.

Boundaries also can be classified genetically, that is, in the context of their evolution as it relates to the cultural landscapes they traverse. A leading political geographer, Richard Hartshorne (1899–1992), proposed a four-level **genetic boundary classification**. All four boundary types can be observed in Southeast Asia (although one technically lies outside the realm yet bounds a country within it).

Certain boundaries, Hartshorne reasoned, were defined and delimited before the present-day human landscape developed. In Figure 10-6 (upper left map), the boundary between Malaysia and Indonesia on the island of Borneo is an example of the first boundary type, the *antecedent*. Most of this border passes through sparsely inhabited tropical rainforest, and the break in settlement can even be detected on the small-scale world population map (Fig. I-9).

A second category of boundaries evolved as the cultural landscape of an area took shape, part of the ongoing process of accommodation. These *subsequent* boundaries are represented in Southeast Asia by the map in the upper right of Figure 10-6, which shows in some detail the border between Vietnam and China. This border is the result of a long process of adjustment and modification, the end of which may not yet have come.

The third category involves boundaries drawn forcibly across a unified or at least homogeneous cultural landscape. This was done by the colonial powers when they divided the island of New Guinea by delimiting a boundary in a nearly straight line (curved in only one place to accommodate a bend in the Fly River), as shown in the lower left map of Figure 10-6. The *superimposed* boundary they delimited gave the Netherlands the western half of New Guinea. When Indonesia became independent in 1949, the Dutch did not yield their part of New Guinea, which is peopled mostly by ethnic Papuans, not Indonesians. In 1962, the Indonesians invaded the territory by force of arms, and in 1969, the United Nations authorized its authority there. This made the colonial, superimposed boundary the eastern border of Indonesia and had the effect of

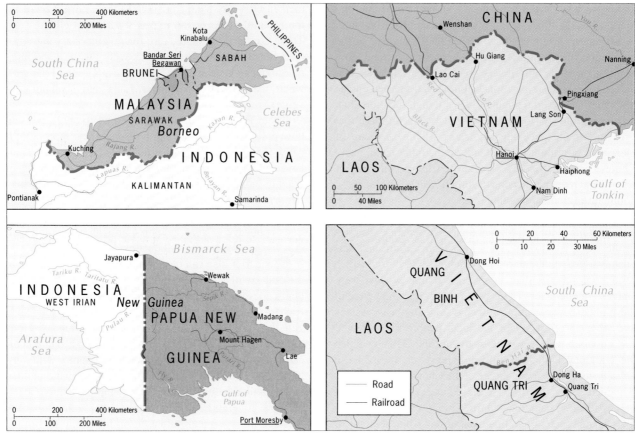

FIGURE 10-6

extending Indonesia from Southeast Asia into the Pacific Realm. Geographically, all of New Guinea forms part of the Pacific realm.

The fourth genetic boundary type is the so-called *relict* boundary—a border that has ceased to function, but whose imprints (and sometimes influence) are still evident in the cultural landscape. The boundary between the former North and South Vietnam (Fig. 10-6, lower right map) is a classic example: once demarcated militarily, it has had relict status since 1976 following the reunification of Vietnam in the aftermath of the Indochina War (1964–1975).

Southeast Asia's boundaries have colonial origins, but they have continued to influence the course of events in postcolonial times. Take one instance: the physiographic boundary that separates the main island of Singapore from the rest of the Malay Peninsula, the Johor Strait (see Fig. 10-10). That physiographic-political boundary facilitated, perhaps crucially, Singapore's secession from the state of Malaysia. Without it, Malaysia might have been persuaded to stop the separation process; at the very least, territorial issues would have arisen to slow the sequence of events.

As it was, no land boundary needed to be defined. The Johor Strait demarcated Singapore and left no question as to its limits.

State Morphology

Boundaries define and delimit states; they also create the mosaic of often interlocking territories that give individual countries their shape, also known as their *morphology*. Let no one doubt that the morphology of a state has an effect on its condition, even its survival in this world. Vietnam's extreme elongation has been a factor in its existence since time immemorial. Indonesia has tried to redress its fragmentation into thousands of islands by "translocating" Jawanese from the most populous island to many of the others in order to promote unity.

Political geographers identify five dominant state territorial configurations, all of which we have encountered in our world survey, but which we have not categorized until now. All but one of these shapes are represented in Southeast Asia. Figure 10-7 provides the terminology and examples from the realm:

- **Compact states** have territories shaped somewhere between round and rectangular, without major indentations. This encloses a maximum amount of territory within a minimum length of boundary. Southeast Asian example: Cambodia.

- **Protruded states** (sometimes called *extended*) have a substantial, usually compact territory from which extends a peninsular corridor that may be landlocked or coastal. Southeast Asian examples: Thailand, Myanmar.

- **Elongated states** (also called *attenuated*) have territorial dimensions in which the length is at least six times the average width, creating a state that lies astride environmental or cultural transitions. Southeast Asian example: Vietnam.

- **Fragmented states** consist of two or more territorial units separated by foreign territory or by water. Subtypes are mainland–mainland, mainland–island, and island–island. Southeast Asian examples: Malaysia, Indonesia, Philippines.

- **Perforated states** completely surround the territory of other states, so that they have a ''hole'' in them. No Southeast Asian example; the most illustrative current case is South Africa, perforated by Belgium-sized Lesotho.

In the discussion that follows, we will have frequent occasion to refer to this geographic property of Southeast Asia's states. For so comparatively small a realm, with so

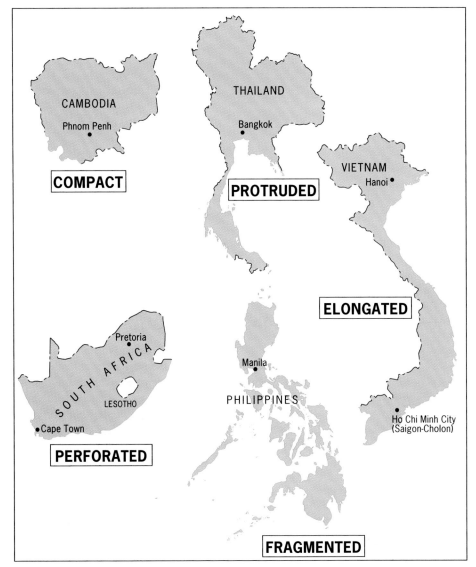

FIGURE 10-7

few states, Southeast Asia displays a considerable variety of state morphologies. When we link these features to other geographic aspects (such as relative location), we obtain useful insights into the regional framework.

One point of caution: states' morphologies do not *determine* their cohesion, viability, unity, or lack thereof; they can, however, *influence* these conditions. For example, Belgium is a compact state, but it is strongly divided cul-turally. Belgium's compactness has not ensured any national cohesion; but if Belgium's two principal regions, Flanders and Wallonia, had been separate islands in the North Sea, it is unlikely that a unified Belgium would have existed. In Belgium's case, compactness of shape helped overcome cultural disunity. In the pages that follow, we take note of the morphology of Southeast Asia's countries as we assess their geography.

REGIONS OF THE REALM

Southeast Asia's first-order regionalization must be based on its mainland–island fragmentation. But there are physiographic, historic, and cultural reasons to include the Malaysian (southern) part of the Malay Peninsula in the insular region (see Figs. I-12 and 10-1). Using the political framework as our grid, we see that the regions of Southeast Asia are constituted as follows:

- **Mainland Region**
      Vietnam, Cambodia, Laos, Thailand, Myanmar (Burma)
- **Insular Region**
      Malaysia, Singapore, Indonesia, Brunei, Philippines

Note, however, that the realm boundary excludes the Indonesian area of New Guinea, which is part of the Pacific geographic realm.

❖ MAINLAND SOUTHEAST ASIA

Five countries form the mainland region of Southeast Asia: two of them protruded, one compact, one elongated, and one landlocked. Two colonial powers, buffered by Thailand, shaped its modern historical geography. One religion, Buddhism, dominates cultural landscapes, but this is a multicultural, multiethnic region. Although one of the least urbanized regions in the world, it contains several major cities. And as Figure 10-1 shows, two countries (Vietnam and Myanmar) possess more than one core area each. We approach the region from the east.

Major Cities of the Realm

Cities	Population* (in millions)
Bangkok, Thailand	10.1
Hanoi, Vietnam	2.9
Ho Chi Minh City (Saigon), Vietnam	7.9
Jakarta, Indonesia	12.8
Kuala Lumpur, Malaysia	1.8
Manila, Philippines	10.1
Phnom Penh, Cambodia	1.5
Singapore, Singapore	3.1
Viangchan, Laos	0.6
Yangon, Myanmar (Burma)	4.2

*Based on 1997 estimates.

▮▶ Elongated Vietnam

Consider this: when the Indochina War ended in 1975, Vietnam had less than *half* the population it has today. In 1997, Vietnam's population approached 80 million, more than 60 percent of it under 21 years of age. For the great majority of Vietnamese, therefore, the terrible war of the 1960s and 1970s is history, not memory. What concerns most Vietnamese now is to overcome two decades of isolation, to reconnect to the world at large, and to join in the economic boom on the Pacific Rim. In 1995, the United States moved to normalize relations with Hanoi, opening an embassy and authorizing interactions long prohibited in the aftermath of the conflict.

10.3

Travel up the crowded road from the northern port of Haiphong to the capital, Hanoi, or sail up the river to the southern metropolis officially known as Ho Chi Minh

FIGURE 10-8

City—but called Saigon by almost everyone there—and you are quickly reminded of the cultural effects of Vietnam's elongation (Fig. 10-8). Vietnam was delimited by the French colonizers as a 1,200-mile (2,000-km) strip of land extending from the Chinese border in the north to the tip of the Mekong delta in the south. Substantially smaller than California, this coastal belt was the domain of the Vietnamese (Fig. 10-3). The French recognized that Vietnam, whose average width is under 150 miles (240 km), was no homogeneous colony, so they divided it into three

DOMINO THEORY

■ ■ ■

During the Indochina War (1964–1975), it was United States policy to contain communist expansion by supporting the efforts of the government of South Vietnam to defeat communist insurgents. Soon, the war engulfed North Vietnam as U.S. bombers attacked targets north of the border between North and South. In the later phases of the war, conflict spilled over into Laos and Cambodia. In addition, U.S. warplanes took off from bases in Thailand. Like dominoes, one country after another fell to the ravages of the war or was threatened.

Some scholars warned that this domino effect could eventually affect not only Thailand but also Malaysia, Indonesia, and Burma (today Myanmar): the whole Southeast Asia realm, they predicted, could be destabilized. But, as we know, that did not happen. The war remained confined to Indochina. And the domino "theory" seemed invalid.

But is the theory totally without merit? Unfortunately, some political geographers to this day make the mistake of defining this idea in terms of communist activity. Communist insurgency, though, is only one way a country may be destabilized (Peru came close to collapse). But right-wing rebellion (Nicaragua's Contras), ethnic conflict (Bosnia), religious extremism (Algeria), and even economic and environmental causes can create havoc in a country. Properly defined, the **domino theory** holds that destabilization from any cause in one country can result in the collapse of order in a neighboring country, starting a chain of events that can affect a series of contiguous states in turn.

In fact, any visitor to Laos and especially to Cambodia will see the disastrous long-term impact of the "Vietnam" War on these countries and societies; these dominoes certainly fell. Today, Indochina is relatively stable (although Cambodia suffers from sporadic strife). Now, the dominoes are falling in another shatter belt, Eastern Europe. Look at the map again: the struggle in former Yugoslavia has moved from Slovenia to Croatia, on to Bosnia and Serbia-Montenegro, and threatens to engulf Kosovo, Macedonia, Albania, and perhaps even Greece and Turkey. In the mid-1990s, it was U.S. and U.N. policy to attempt to contain ex-Yugoslavia's conflicts and prevent them from spreading to Kosovo and beyond. There may be something to the domino theory after all.

■ ■ ■

units: (1) Tonkin, land of the Red River delta and centered on Hanoi in the north; (2) Cochin China, region of the Mekong delta and centered on Saigon in the south; and (3) Annam, focused on the ancient city of Hué, in the middle (Fig. 10-4).

The Vietnamese (or *Annamese*, also *Annamites*, after their cultural heartland) speak the same language, although the northerners can quickly be distinguished from southerners by their accent. As elsewhere in their colonial empire, the French introduced their language as the *lingua franca* of Indochina, but their tenure was cut short by the Japanese, who invaded Vietnam in 1940. During the Japanese occupation, Vietnamese nationalism became a powerful force, and after the Japanese defeat in 1945, the French could not regain control. In 1954, the French suffered a disastrous final trouncing on the battlefield at Dien Bien Phu and were ousted.

But Vietnam did not become a unified state even after its forces routed the colonizers. Separate regimes took control: a communist one in Hanoi and a noncommunist version in Saigon. Vietnam's pronounced elongation had made things difficult for the French; now it played its role during the postcolonial period. Note, in Figure 10-8, that Vietnam is widest in the north and in the south, with a very narrow "waist" in its middle zone. North and South were worlds apart.

Many Americans still remember the way in which the United States became involved in the inevitable conflict between communists and noncommunists in Vietnam during the 1960s and 1970s. At first, American military advisers were sent to Saigon to help the shaky regime there cope with communist insurgents. When the tide turned against the South, military forces and equipment were committed to halt the further spread of communism (see box titled "Domino Theory"); at one time, more than half a million U.S. soldiers were in the country.

The conduct of the war, its mounting casualties, and its apparent futility created severe social tensions in the United States. If you had been a student during that period, your college experience would have been radically different from that of today. Protest rallies, "teach-ins," anti-draft demonstrations, "sit-ins," marches, strikes, and even hostage-taking disrupted campus life. Several student protesters were killed by the National Guard on an Ohio campus in 1970. The Indochina War threatened the stability of American society. It drove an American president (Lyndon Johnson) from office in 1968 and destroyed the electoral chances of his vice president (Hubert Humphrey), who would not disavow the U.S. role in the war.

In 1975, the Saigon government fell and the United States was ousted—just two decades after the French defeat at Dien Bien Phu. When the last helicopter left from the roof of the U.S. embassy in Saigon amid scenes of desperation and desertion, it marked the end of a sequence of events that closely conformed to the **insurgent state model** outlined on p. 408.

Vietnam Today

After the war's end, the Soviets, not the neighboring Chinese, became Vietnam's patrons. As Vietnam was finally unified in 1976 under a dogmatic communist regime, renewed conflict broke out, this time involving China (boundary issues in the north were the ostensible cause). But there was more to this situation. The Soviets exploited

■ AMONG THE REALM'S GREAT CITIES...

Saigon

Officially, it is known as Ho Chi Minh City, renamed in 1975 after the ouster of American forces, but only bureaucrats and atlases call it so. To the people here, it is still Saigon, city on the Saigon River, served by, it seems, a thousand companies using the name. Saigon Taxi, Saigon Hotel, Saigon Pharmacy—there is hardly a city block without the name Saigon on some billboard.

The phenomenon is significant. Hanoi's northerners ordained the change, but this still is South Vietnam, and many southerners see the North as peopled by slow, dogmatic communists obstructing the hopes and plans for their vibrant city. In truth, Saigon may be many times larger than Hanoi, but like the capital it suffers from decades of deterioration, a broken-down infrastructure, and undependable public services. When, in the early 1990s, businesspeople could find no modern accommodations or working telephones, an Australian firm built an eight-story float-

ing hotel with its own power plant, towed it up the Saigon River, and parked it on the downtown waterfront.

Therein lies Saigon's great advantage: the city can be reached by large ocean-going vessels, a three-hour sail up the meandering river. The Chinese developed its potential: in the 1960s, Chinese merchants controlled more than half of the South's exports and nearly all of its textile factories and foreign exchange. Saigon-Cholon (Cholon means ''great market'') was a Chinese success story, but after the war the Hanoi regime nationalized all Chinese-owned firms. Tens of thousands of middle-class Chinese left the city, its economy devastated.

Today, Saigon (population: 7.9 million) is fast recovering. A modern skyline is rising above the colonial one. A large Special Economic Zone is being created south of the city. Factories are being built. The Pacific Rim era is beginning. Will the South rise again?

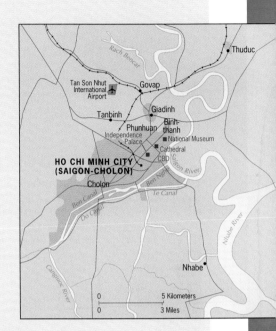

the historic distrust between Chinese and Vietnamese because Moscow was itself at odds with Beijing, and so life for Vietnam's Chinese minorities became much more difficult. Many Chinese (the number will never be known) joined in the tragic and disastrous exodus of *boat people* who sailed from Vietnam's coasts in small, often unseaworthy boats. Of the estimated 2 million of these refugees, more than half perished from storms, exposure, pirates, starvation, and sinkings. The great majority of those who survived were brought to the United States.

While accepting hundreds of thousands of Vietnamese refugees, the United States simultaneously placed a strong embargo on Vietnam, a policy that isolated the communist country and stifled its economy. When the Soviet Union collapsed in 1991, Vietnam was in a desperate position—with one saving grace. With two major river deltas and plenty of fertile farmland, Vietnam can produce large harvests of rice, enough to feed its own population *and* export to foreign markets. Therefore, Vietnam could at least survive its boycott.

At the same time, Vietnam could do little to maintain, let alone improve, what was left of its infrastructure. In 1996, makeshift, one-way bridges still slowed traffic on the congested road between the capital, Hanoi, and the port of

Haiphong, a 60-mile (100-km) journey that takes four hours on a good day (Fig. 10-8). Dilapidated buildings, potholed roads, torn-up sidewalks, and malfunctioning utility systems and telephone lines bedevil the capital, where the first signs of modernization are just appearing. With 2.9 million residents, Hanoi anchors the northern (Tonkin) core area of Vietnam, the basin of the Red River (its agricultural hinterland). In the paddies, irrigation water still is raised by bucket. On the roads, goods move by human- or animal-drawn cart. Little has yet changed here—except people's expectations.

10.1

Elongated Vietnam's southern headquarters, Saigon-Cholon (officially named Ho Chi Minh City), is far ahead of the capital. Nearly 8 million people inhabit this urban agglomeration on the Saigon River, which contains about 10 percent of the country's entire population—and a far larger share of its best-educated and capable people. Saigon-Cholon (Cholon is the city's Chinatown, much diminished after the war but now reviving) is changing rapidly as high-rises built by foreign investors tower over the still-colonial townscape, modern hotels are opening, and a large Special Economic Zone is being laid out just downstream from the port. Unlike Hanoi, Saigon can be reached by oceangoing vessels, and it lies north of, rather than within,

10.2

From the Field Notes
"The center of Vietnam's Ho Chi Minh City (Saigon) is changing. The French-era, red-tiled city hall that overlooks the main avenue of the CBD, Dong Khoi Street, now has a backdrop, a controversial Australian-built high-rise that, in the view of many, mars the view from street level. This, however, is only one of dozens of new buildings going up throughout downtown Saigon (and many older ones are under renovation). Saigon is the latest of the Asian Pacific Rim cities to see its skyline changed by the new economic realities."

the Mekong delta that is part of its hinterland. The city's streets are choked with bicycles, mopeds, handcarts, buses, taxis, and even a few private cars; consumer goods are everywhere, and commercial activity throbs. Saigon, too, has serious infrastructure problems, but it is recovering now from decades of neglect. One great need is a bridge to link the right bank of the Saigon River, on which downtown and most of Saigon lie, to the left bank, which remains merely a village.

A sense of distance and remoteness pervades north as well as south in Vietnam. To northerners, Saigon and the south are symbols still of the divisions that tore Vietnam

From the Field Notes
"After a four-hour, stop-and-go journey from the coast we arrived in Hanoi, Vietnam's capital. After experiencing Saigon's bustle and energy, we felt as though we were in a city where time had stood still. Here in the city center you could find little evidence of Vietnam's new economic era, no new, tall buildings as in Saigon, no significant signs of modernization. Hanoi's disadvantageous relative location, its weak surface links beyond its immediate hinterland, and the communist dogma of its administration combine to explain why Saigon is forging ahead while the capital, less than half Saigon's size, lags."

apart during the war. To southerners, the bureaucrats from the north who run much of Saigon are a burden, obstructing the rush toward modernization. This sense of alienation is reinforced by the actual as well as the functional distance from Saigon to Hanoi. Imagine a country where the road trip from the capital to the largest city can take as long as a week!

Hanoi, in imitation of Beijing, has chosen to retain its communist political system while opting for market economics. The new period is still young, but already it is possible to discern the elements of a new division in this attenuated state. The north is and will be the center of dogmatic communism; the south will become the focus of the economic transformation. Neither side is likely to be comfortable with the other—which would be nothing new in this narrow, double-core, two-river, two-city domain of the Vietnamese. ◀⎮

Compact Cambodia

Former French Indochina contained two additional entities—Cambodia and Laos. In Cambodia, the French possessed one of the greatest treasures of Hinduism: the city of Angkor, capital of the ancient Khmer Empire, and the temple complex known as Angkor Wat. The Khmer Empire prevailed here from the ninth to the fifteenth centuries, and Angkor Wat was built by King Suryavarman II to symbolize the universe in accordance with the precepts of Hindu cosmology. When the French took control of this area during the 1860s, Angkor lay in ruins, sacked by invading Vietnamese and Thai armies. The French created a protectorate, began the restoration of the shrines, established Cambodia's permanent boundaries, and restored the monarchy under their supervision.

Geographically, Cambodia enjoys several advantages, notably its compact shape. Compact states enclose a maximum of territory within a minimum of boundary and are without peninsulas, islands, or other remote extensions of the national spatial framework. Cambodia had the further advantage of strong ethnic and cultural homogeneity: 90 percent of its people are Khmers, with the rest equally divided between Vietnamese and Chinese.

As Figure 10-8 shows, Southeast Asia's greatest river, the Mekong, enters Cambodia from Laos and crosses it from north to south, creating a great bend before flowing into southern Vietnam. Phnom Penh (1.5 million), the country's present capital, lies on the river's western bank where it makes that bend. The ancient capital of Angkor lies in the northwest, not far from Tonle Sap, a major interior lake that drains into the Mekong.

Cambodia fell victim to the Indochina War in the most dreadful way, and neither its compactness nor its isolation could save it. The war first spilled over from Vietnam into its eastern border areas. Then, in 1970, the last king was deposed by military rulers; in 1975, that government was itself overthrown by communist revolutionaries, the so-called Khmer Rouge. These new rulers embarked on a course of terror and destruction in order to reconstruct *Kampuchea* (as they called Cambodia) as a rural society. They drove townspeople into the countryside where they had no place to live or work, emptied hospitals and sent the sick and dying into the streets, outlawed religion and family, and killed as many as 2 million Cambodians (out of a total of 8 million) in the process. In the late 1970s, Vietnam, victorious in its own war, invaded Cambodia to drive the Khmer Rouge away. But this led to new terror, and a stream of refugees crossed the border into Thailand.

Once a self-sufficient country that could feed others, Cambodia now must import food. The economy is dominantly agricultural, and most people grow rice and beans for subsistence. Its population, severely reduced by war, again surpassed 7 million around 1990 (it stands at 11.2 million today). But the country's future remains uncertain, with powerful adversaries on its borders and armed factions within them.

Landlocked Laos

North of Cambodia lies Southeast Asia's only landlocked country, Laos (Fig. 10-8). Interior and isolated, Laos changed little during 60 years of French colonial administration (1893–1953). Then, along with other French-ruled areas, Laos became an independent state. Soon its well-entrenched, traditional kingdom fell victim to rivalries between traditionalists and communists, and the old order collapsed.

Laos has no fewer than five neighbors, one of which is the East Asian giant, China. A long stretch of its western boundary is formed by the Mekong River, and the important sensitive border with Vietnam to the east lies in mountainous terrain. With 5.1 million people (about half of them ethnic Lao, related to the Thai of Thailand), Laos lies surrounded by comparatively powerful states. The country has no railroads, just a few miles of paved roads, and very little industry; it is only 19 percent urbanized (the capital, Viangchan, contains only about 600,000 people). Laos remains the region's poorest and most vulnerable entity.

I► Protruded Thailand

In virtually every way, Thailand is the leading state of the region under discussion. Its per capita GNP is higher than that of Vietnam, Cambodia, Laos, and Myanmar combined. Alone among its neighbors, Thailand has participated strongly in Pacific Rim economic development. Thailand's capital, Bangkok, is by far the largest urban center in the region, one of the world's most prominent primate cities. The country's population, about 62 million in 1997, is growing at the slowest rate in the entire realm except urbanized Singapore. Over the past decades, only political instability and uncertainty have inhibited economic progress. Thailand is a kingdom, but it is a constitutional monarchy; its moves toward stable democracy are thwarted by graft, corruption, and, sometimes, violent confrontation.

Thailand is the textbook example of a protruded state. From a relatively compact heartland, in which lie the core area, capital, and major areas of productive capacity, a 600-mile (1000-km) corridor of land, in places less than 20 miles wide, extends southward to the border with Malaysia (Fig. 10-9). The boundary that defines this protrusion runs down the length of the Malay Peninsula to the Kra Isthmus, where neighboring Myanmar peters out and Thailand fronts the Andaman Sea (an arm of the Indian Ocean) as well as the Gulf of Thailand. In the entire country, no place lies farther from the capital than the southern end of this tenuous protrusion.

As Figure 10-1 shows, Thailand occupies the heart of the mainland region of Southeast Asia. While Thailand has no Red, Mekong, or Irrawaddy delta, its central lowland is watered by a set of streams that flow off the northern highlands and the Khorat Plateau in the east. One of these streams, the Chao Phraya, is the Rhine of Thailand. From the head of the Gulf of Thailand to Nakhon Sawan, this river is a highway of traffic. Barge trains loaded with rice head for the coast, ferry boats head upstream, freighters transport tin and tungsten (of which Thailand is among the world's largest producers). Bangkok sprawls on both sides of the Chao Phraya, here flanked by skyscrapers, pagodas, factories, boatsheds, ferry landings, luxury hotels, and modest dwellings in crowded confusion. On the right bank of the Chao Phraya, Bangkok's west side, lie the city's remaining *klong* (canal) neighborhoods, where waterways and boats form the transport system. Bangkok still is known as the Venice of Asia, although many *klongs* have been filled in and paved over to serve as roadways.

Thailand's economic success is attributable to a number of circumstances, among which relative location and natural environment play important roles. At the head of the Gulf of Thailand, the country's core area opens to the South China Sea and hence the Pacific Ocean. The Gulf itself has long been a rich fishing ground and more recently a source of oil (the major refinery lies at the fast-developing

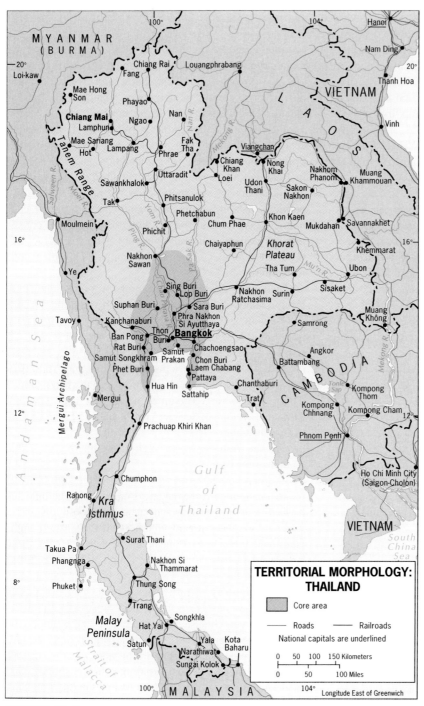

FIGURE 10-9

port area of Laem Chabang). But the key to Thailand's economic growth has been the Thai workforce, laboring under often dreadful conditions for very low wages to produce goods sold on foreign markets at low prices. This labor force attracted major overseas investment which, in turn, stimulated service industries. Today, working conditions and wages are improving, and Thailand has taken on

the trappings of a Pacific Rim economic tiger. Visit the port and industrial area of Laem Chabang, and you see huge holding lots full of ''Japanese'' cars made in Thailand for export to North America. Factories are being expanded or built in profusion, and the economic landscape reflects confidence in the country's future.

Thailand's natural environments and cultural land-

10.5

■ AMONG THE REALM'S GREAT CITIES...

Bangkok

Bangkok (10.1 million), mainland Southeast Asia's largest city, is a massive sprawling metropolis on the banks of the Chao Phraya River, an urban agglomeration without a center, an aggregation of neighborhoods ranging from the immaculate to the impoverished. In this city of great distances and high tropical temperatures, getting around is often difficult as roadways are choked with traffic. An (unfortunately diminishing) network of waterways affords the easiest way to travel, and life focuses on the busiest waterway of all, the Chao Phraya. Ferries and water taxis carry tens of thousands of commuters and shoppers across and along this bustling artery, flanked by a growing number of high-rise office buildings, luxury hotels, and ultramodern condominiums. Many of these modern structures reflect the Thais' fondness for domes, columns, and small-paned windows, creating a skyline unique in Asia.

Gold is Buddhist Thailand's symbol, adorning religious and nonreligious architecture alike. From a high vantage point in the city, you can see hundreds of golden spires, pagodas, and façades rising above the townscape. The Grand Palace, where royal, religious, and public buildings are crowded inside a white, crenellated wall more than a mile long, is a gold-laminated city within a city embellished by ornate gateways, monsters, dragons, and statuary. Across the mall in front of the Grand Palace lie government buildings sometimes targeted by rioters in Bangkok's volatile political atmosphere. Not far away are Chinatown and a myriad of markets; Bangkok has a throbbing commercial life.

Decentralized, dispersed Bangkok has major environmental problems, ranging from sinking ground (resulting in part from excessive water pumping) to some of the world's worst air pollution. Yet industries, port facilities, and residential

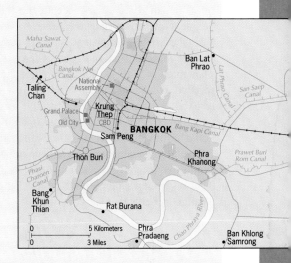

areas continue to expand as the city's hinterland grows and prospers with Pacific Rim investment. Bangkok is the pulse of Southeast Asia's mainland region.

From the Field Notes

"Thailand exhibits one of the world's most distinctive cultural landscapes, both in its urban and its rural areas. Graceful pagodas, stupas, and spirit houses of Buddhism adorn city and countryside alike. Gold-layered spires rise above, and beautify cities and towns that would otherwise be drab and featureless. The architecture of Buddhism has diffused into public architecture, so that many secular buildings, from businesses to skyscrapers, are embellished by something approaching a national style. The grandest expression of the form undoubtedly is a magnificent assemblage of structures within the walls of the Grand Palace in the capital, Bangkok, of which a small sample is shown here. Climb onto the roof of any tall building in the city's center, however, and the urban scene will display hundreds of such graceful structures, interspersed with the modern and the traditional townscape."

scapes attract millions of visitors annually. The magnificent architecture of Buddhism (the old name *Siam* was a Chinese word, meaning golden-yellow, to describe the gold-swathed temples and shrines seen everywhere) and the tropical coasts and beaches make Thailand a favorite destination for tourists from Germany to Japan. One of the world's most famous resorts is the island of Phuket, on the distant southwestern coast near the end of the Malay Peninsula protrusion.

Tourism is Thailand's leading source of foreign revenues, but there is a dark side to the industry. The Thai people's relaxed attitude toward sex has made Thai tourism substantially a sex industry, candidly advertised as such in foreign markets such as Japan, where "sex tours" attract planeloads of male participants. When the AIDS pandemic reached Thailand in the 1980s, it hardly changed the pattern, but by the early 1990s Thailand was, in the words of geographer Peter Gould, a catastrophe in the making that could ultimately exact a higher toll than the murderous Khmer Rouge campaign did in Cambodia. Although prostitution is officially illegal in Thailand, brothel-bars are a Thai institution, and cultural habits change but slowly. In Thailand, if change ever comes, it may be too late. ◀▮

Extended Myanmar

Myanmar (formerly Burma) shares with Thailand its protruded morphology, but little else. Long languishing under one of the world's most corrupt military regimes, Myanmar is one of the planet's poorest countries where, it seems, time has stood still for centuries. Whereas Thailand's peninsular protrusion is served by air, railroad, and road, there is not even an all-weather track to the southern end of Myanmar's. Whereas busy shipping lanes converge on Thailand's coastal core area, Myanmar's Irrawaddy delta is more reminiscent of Bangladesh.

Myanmar's territorial morphology is complicated by a shift in the Burmese core area that took place during colonial times. Prior to the colonial period, the focus of embryonic Burma lay in the so-called *dry zone* between the Arakan Mountains and the Shan Plateau (Fig. 10-2). The urban focus of the state was Mandalay (see the atlas map, p. 474), which had the advantage of central situation and relative proximity to the non-Burmese highlands all around. Then the British developed the agricultural potential of the Irrawaddy delta, and Rangoon (now called Yangon) became the hub of the colony. The old and the new core areas are linked by the Irrawaddy waterway, but the center of gravity has shifted to the south.

The political geography of Myanmar constitutes a particularly good example of the role and effect of state morphology on internal state structure. Not only is Myanmar a protruded state; in addition, Myanmar's core area is surrounded on the west, north, and east by a horseshoe of great mountains—mountains where many of the country's 11 minority peoples had their homelands before the British occupation. The colonial boundaries had the effect of incorporating these peoples into the domain of the Burman people, who constitute about two-thirds of the population (46.5 million today). When the British departed and Burma

From the Field Notes

"One of Southeast Asia's most spectacular Buddhist shrines is the golden Shwedogon Pagoda in the heart of Yangon, the capital of Myanmar (Burma). The golden dome (or *chedi*) is one of the finest in the entire realm, and its religious importance also is superior: here are preserved eight hairs of the Buddha. Vast amounts of gold have gone into the creation and preservation of the Shwedogon Pagoda; local rulers often gave the monks their weight in gold—or more. Today, the pagoda is a cornerstone of Buddhism, drawing millions of faithful to the site. It also is slowly becoming a tourist attraction as Myanmar's door to the outside world is slightly ajar and foreign visitors are arriving again."

became independent in 1948, the less numerous peoples traded one master for another. In 1976, nine of these indigenous peoples formed a union to demand the right to self-determination in their homelands.

As Figure 10-3 shows, the peripheral peoples of Myanmar occupy a significant part of the state. The Shan of the northest and far north, who are related to the neighboring Thai, constitute about 9 percent of the population, or more than 4 million. The Karen (7 percent, 3.2 million) live in the neck of Myanmar's protrusion and have proclaimed that they wish to create an autonomous territory within a federal Myanmar. Although the powerful military have dealt these aspirations a series of setbacks, centrifugal forces continue to bedevil the central regime. Its response has been to exert power by all available means rather than to accommodate these forces, even to the point of stifling political discourse (let alone opposition) among the Burman themselves.

This is all the more tragic since Myanmar has economic potential far superior to Bangladesh, its neighbor to the west, and to several other countries in its own region. It can feed itself, and it could export significant quantities of rice. Varied soils and diverse environments can yield a range of other crops. Tin and other metals have been found in the highlands. Today, Myanmar is exploiting the forests of the interior at an alarming rate; teak has risen to first place among exports by value.

Legitimate exports, that is. Myanmar also is the world's largest illicit producer of opium poppies and harvests other illegal crops for the drug trade as well. For a time, the regime in Yangon tried to control the trade, but its constant conflict with peoples outside the core area diverted its attentions and energies from that campaign.

For decades, the repressive policies of a brutal military regime have kept Myanmar in turmoil and all but halted its development. Infrastructure here is in worse shape even than in Vietnam (in 1993 there was *one* working telephone in Myanmar capable of making and receiving overseas calls in Yangon, the capital). While more than 80 percent of the adult population is literate, there are very few jobs to employ their talents. In the late 1990s there were signs of change, including the release of some political dissidents and the beginnings of tourism. But a change of heart by the regime will not soon undo the effects of policies that have pushed impoverished and exhausted Myanmar into the ranks of the world's poorest economies.

Of the five countries of Southeast Asia's mainland region, the leading one in almost every respect is the only one that was never colonized. While Thailand lost territory and subjects to the colonialists on all sides, the core of the kingdom persisted and forged its own brand of modernization in political as well as economic spheres.

❖ INSULAR SOUTHEAST ASIA

On the peninsulas and islands of Southeast Asia's southern and eastern periphery lie five of the realm's 10 states (Fig. 10-1). Few regions in the world contain so diverse a set of countries. Malaysia, the former British colony, consists of two major areas separated by hundreds of miles of South China Sea. The realm's southernmost state, Indonesia, sprawls across thousands of islands from Sumatera in the west to New Guinea in the east. North of the Indonesian archipelago lies the Philippines, a nation that once was a U.S. colony. These are three of the most severely fragmented states on Earth, and each has faced the challenges that such politico-spatial division brings. This insular region of Southeast Asia also contains two small but important sovereign entities: a city-state and a sultanate. The city-state is Singapore, once a part of Malaysia (and one instance in which internal centrifugal forces were too great to be overcome). The sultanate is Brunei, an oil-rich Muslim territory on the island of Borneo that seems transplanted from the Persian Gulf. Few parts of the world are more varied or interesting geographically.

▶ Mainland–Island Malaysia

The state of Malaysia represents one of the three types of fragmented states discussed earlier: the mainland–island type, in which one part of the national territory lies on a continent and the other on an island. Malaysia is a colonial political artifice that combines two quite disparate components into a single state: the southern end of the Malay Peninsula and the northern part of the island of Borneo. Respectively, these are known as West Malaysia and East Malaysia (Fig. 10-1).

The name *Malaysia* came into use in 1963, when the original Federation of Malaya, on the Malay Peninsula, was expanded to incorporate the areas of Sarawak and Sabah in Borneo. When the name Malaya is used, it refers to the peninsular part of the federation. Malaysia refers to the total entity.

The Malays of the peninsula, traditionally a rural people, displaced older aboriginal communities there and today make up about 57 percent of the country's population of 21 million. They possess a strong cultural identity expressed in adherence to the Muslim faith, a common language, and a sense of territoriality that arises from their perceived Malayan origins and their collective view of Chinese, Indian, European, and other foreign intruders.

The Chinese came to the Malay Peninsula and to northern Borneo in substantial numbers during the colonial period, and today they constitute nearly one-third of Malaysia's population (they are the largest single group in Sarawak). During the Second World War, when the Japanese ruled Malaya, they were ruthlessly persecuted; many were driven into the forested interior where they founded a communist-inspired resistance movement that long destabilized the area. Today, the commercially active, urbanized Chinese sector still is victimized by discrimination, now by the dominant Malays.

Hindu South Asians were in this area long before the Europeans, and for that matter the Arabs and Islam, arrived on these shores. Today they still form a substantial minority of over 9 percent of the population, clustered, like the Chinese, on the western side of the peninsula (Fig. 10-3).

The ethnic complexity of Sarawak and Sabah (East Malaysia) defies summation. Chinese are the single largest

PINANG: A FUTURE SINGAPORE?

■ ■ ■

In the northwest corner of Malaysia's Malay Peninsula, opposite the northern end of Indonesian Sumatera, where the Strait of Malacca opens into the Indian Ocean, lies an island called Pinang. A thriving outpost during the colonial period, Pinang (capital: Georgetown) was one of the so-called Straits Settlements. Singapore was another.

Today, Singapore no longer is a part of Malaysia, but Pinang's importance is rapidly growing. Now connected to the mainland by Asia's longest (9-mile) bridge, the 100-square-mile island has a population over 1 million. Chinese outnumber Malays by 60 to 32 percent, and Georgetown, the old colonial port, has become the focus of an expanding high-technology manufacturing complex for the international computer industry. The 65-story Tun Abdul Razak Complex, housing government offices, businesses, department stores, and recreational facilities, is just one landmark in an ultramodern skyline. From the top you can see modern industrial parks with neon-lit signs proclaiming the presence of such companies as Intel, SONY, Philips, Motorola, and Hitachi.

The rapid development of Pinang is a calculated strategy by the Malaysian government. The emphasis is on building a highly skilled labor force that is locally nurtured in technical schools and training facilities financed by the multinational corporations themselves. Malaysia makes infrastructure investments ranging from the new toll bridge to a major airport. Georgetown's colonial and multicultural townscape attracts a growing tourist trade, and new hotels and resorts are enlarging the service industry.

Malaysia hopes that Pinang will become the leading corner of a major *growth triangle* (a concept that is gaining popularity along the Pacific Rim), together with Indonesian Sumatera to the west and southern Thailand to the north. So far, the policy has been spectacularly successful, and Pinang is following in Singapore's economic footsteps. But Kuala Lumpur's politicians surely hope that Chinese-dominated Pinang will not follow Chinese-dominated Singapore's political course.

■ ■ ■

Towering buildings in city skylines always have symbolized economic power and national prestige. Earlier in the twentieth century, the world's tallest buildings and the most impressive skylines were synonymous with U.S. economic might. Now, other countries compete in the skyward race. These twin towers anchor the mushrooming skyline of Kuala Lumpur, the current capital of Malaysia; when they were completed in 1996, they constituted the tallest buildings in the world at 1,483 feet (450 m) including their decorative spires. The Petronas Towers, owned by Malaysia's national oil company, reflect the country's booming economy, but they will not be the world's tallest for long: Shanghai has plans for a higher building still. And Kuala Lumpur will cease to be Malaysia's capital. About 25 miles (40 km) south of the city, Malaysia is building an ultramodern, new capital, Putrajaya. A new airport and a huge dam project also are under way. Some economic geographers suggest that the cost of such megaprojects may form a risk for economies whose boom may not be long-term.

group in Sarawak, but each state counts more than two dozen ethnic identities. As a result, politics in East Malaysia are fractious and often combative, leading to repeated intervention by the federal government.

During colonial times, Britain focused its attention on the Malay Peninsula, creating a substantial economy there. The economic heartland developed on the western side of the peninsula, where the capital (Kuala Lumpur), the densest surface transport networks, the plantations, and most of the mines were concentrated (Fig. 10-1). The Strait of Malacca became one of the world's busiest and most strategic waterways, and Singapore, at the southern end of it, a

10.8

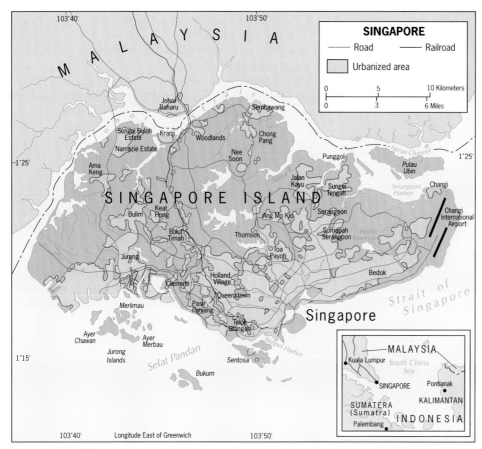

FIGURE 10-10

prized possession (Singapore separated from the Malaysian Federation in 1965).

Since 1980, Malaysia's economy has grown rapidly, and Malaysia has become an economic force on the Pacific Rim. Foreign manufacturers, attracted by the skills and modest wages of the Malaysian workforce, have established hundreds of factories here (see box titled "Pinang: A Future Singapore?"). While rubber, palm oil, and tin still rank high on the export list, electronics have made their appearance as well. One benefit from the merger with East Malaysia has been that sector's production of valuable timber and petroleum. Kuala Lumpur's colonial townscape has long since been overtaken by the symbols of modern economic times; in 1996, it acquired the world's tallest building (the twin Petronas towers [p. 500]) amid a skyline looking ever more like Singapore's. Malaysia's per capita GNP is third in the realm, after oil-rich Brunei and all-urban Singapore, and well ahead of Thailand's.

Malaysia has benefited from strong and stable (if somewhat authoritarian) leadership, from its colonial tutelage, from its varied resource base, and from its capable and diverse population—the skills of Malaysians often complementing those of the Chinese. The country's greatest challenge continues to lie in its ethnic and cultural heterogeneity, heightened by its territorial fragmentation.

Singapore

In 1965, a fateful event occurred in Southeast Asia: Singapore, crown jewel of British colonialism in Southeast Asia, seceded from the Malaysian Federation and became a sovereign state, albeit a ministate (Fig. 10-10). With its magnificent relative location, its human resources, and firm government, Singapore has overcome the limitations of space and the absence of raw materials to become one of the economic tigers on the Pacific Rim.

With a mere 240 square miles (600 sq km) of area, space is at a premium in Singapore, and this is a constant worry for the government. Singapore's only local spatial advantage over Hong Kong is that its small territory is less fragmented (there are just a few small islands in addition to the compact main island). With a population of 3.1 million and an expanding economy, Singapore must develop space-conserving, high-tech industries. In this effort it has been very successful, and the urban area's forest of high-rise buildings reflects the city-state's prosperity. You will find no slums in Singapore; there are dilapidated streets in the old Chinatown, but these are left to conserve a fragment of the city's past. Mostly, Singapore impresses with its newness and its modernity, and with the order of daily life.

As the map shows, Singapore lies where the Strait of

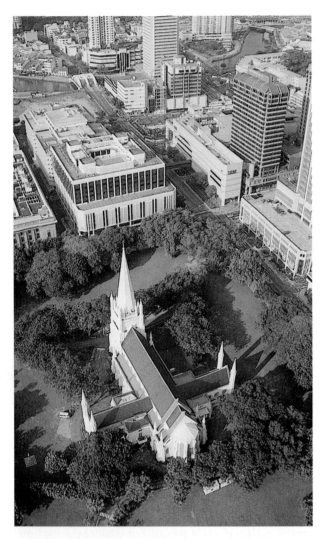

From the Field Notes

"Singapore is a multicultural country incorporating hundreds of religious shrines representing numerous religions and denominations. Saint Andrew's Cathedral reflects the wealth of religious institutions here: its spacious grounds occupy some of Singapore's most costly real estate between the commercial center and the hotel district. In addition to Christian churches, Singapore has several mosques including the magnificent Sultan Mosque in the Muslim district, many Hindu shrines in Little India and elsewhere in the city (among which the Chettiar Hindu Temple is the most elaborate), several Buddhist centers, and numerous Chinese temples. The authoritarian government of Singapore makes every effort to ensure religious freedom and coexistence in a country where religious opposites often find themselves in close proximity."

Malacca (leading westward to India) opens into the South China Sea and the waters of Indonesia (Fig. 10-10 inset). The port developed as part of the British Southeast Asian empire, and when independence came in 1963, it was made a part of the Malaysian Federation. However, two years later Singapore seceded from Malaysia and became a gen-

uine city-state on the Pacific Rim. No reunification, merger, or annexation looms here.

Benefiting from its relative location, the old port of Singapore had become one of the world's busiest (by number of ships served) even before independence. It thrived as an *entrepôt* between the Malay Peninsula, Southeast Asia, Japan, and other emerging economic powers on the Pacific Rim and beyond. Crude oil from Southwest Asia still is unloaded and refined at Singapore, then shipped to Asian destinations. Raw rubber from the adjacent peninsula and from Indonesia's island of Sumatera is shipped to Japan, the United States, China, and other countries. Timber from Malaysia, rice, spices, and other foodstuffs are processed and forwarded via Singapore. In return, automobiles, machinery, and equipment are imported into Southeast Asia through Singapore.

But that is the old pattern. Singapore's leaders want to redirect the city-state's economy and move toward high-tech industries for the future. In Singapore the government tightly controls business as well as other aspects of life. (Some newspapers and magazines have been banned for criticizing the regime, and there are even fines for such things as eating on the subway and failing to flush a public toilet.) Its overall success after secession has tended to keep the critics quiet: while GNP per capita from 1965 to 1993 increased more than tenfold—to over U.S. $19,000—that of neighboring Malaysia reached just U.S. $3,160. Among other things, Singapore became (and for many years remained) the world's largest producer of disk drives for small computers.

As multinational corporations settled in, Singapore's political stability on the volatile Pacific Rim was another advantage. Nonetheless, a problem emerged: unlike Taiwan, where brainpower is plentiful, Singapore's technical capacities were limited. In 1995, only 8 percent of the population was university educated. So while the government reduced the incentives for low-technology manufacturers to come here, and wages rose, the high-tech industries found themselves thinking about moving to nearby Malaysia and Thailand. That slowed Singapore's boom and challenged the government to find ways to stimulate a revival.

To accomplish this revival, Singapore has moved in several directions. First, it will focus on three growth areas: information technology, automation, and biotechnology. Second, there are notions of a "Growth Triangle" involving Singapore's developing neighbors, Malaysia and Indonesia; those two countries would supply the raw materials and cheap labor, and Singapore the capital and technical know-how. Third, Singapore has opened its doors to capitalists of Chinese ancestry who have been leaving Hong Kong and who wish to relocate their enterprises here. Singapore's population is 78 percent Chinese, 14 percent Malay, and 7 percent South Asian. The government is Chinese-dominated, and its policies have been to sustain Chinese control. Indeed, Singapore's combination of authori-

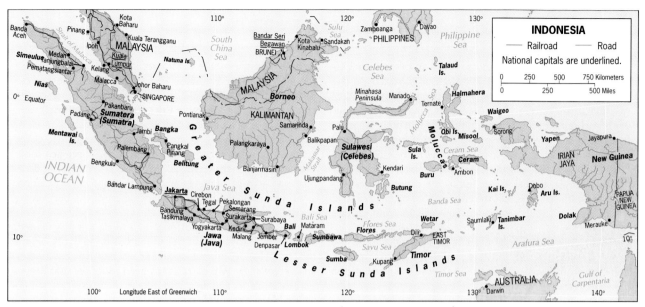

FIGURE 10-11

tarianism and economic success often is cited in China itself as proving that communism and market economics can coexist.

Singapore continues to build on its success. A second bridge across the Johor Strait to Malaysia has been constructed to accommodate the spiraling truck traffic between the two countries. An active land reclamation program already has added over 4 percent of territory to the crowded island; the next scheme is to connect the Jurong group of islands in the southwest (Fig. 10-10). Key to Singapore's progress has been its comparative lack of corruption, attracting many companies to establish themselves here as a base from which to do business in notoriously corrupt countries elsewhere in the realm. Still, Singapore is the smallest of the Pacific Rim's economic tigers, and it faces growing competition near and far. Over the long term its relative location, which yielded its early wealth, may turn out to be its most enduring guarantee of survival—if not prosperity.

Indonesia's Archipelago

The very term *archipelago* denotes fragmentation, and Indonesia is the world's most expansive archipelagic state. Spread across more than 13,000 islands, Indonesia's over 200 million people live separated and clustered—separated by water and clustered on islands large and small.

The map of Indonesia requires some attention (Fig. 10-11). Five large islands dominate the archipelago territorially, one of which, New Guinea in the east, is not part of the Indonesian culture sphere, although the western half of it is under Indonesian control. The other four major is-

lands are collectively known as the Greater Sunda Islands: Jawa (Java), smallest but by far the most populous and important; Sumatera (Sumatra) in the west, directly across the Strait of Malacca from Malaysia; Kalimantan (the Indonesian sector of large, compact, minicontinent Borneo); and wishbone-shaped, distended Sulawesi (Celebes) to the east. Extending eastward from Jawa are the Lesser Sunda Islands, including Bali and, near the eastern end, Timor. Another important island chain within Indonesia is the Moluccan Islands, between Sulawesi and New Guinea. The central water body of Indonesia is the Java Sea.

As the map shows, Indonesia does not control all of the archipelago. In addition to East Malaysia's territory on Borneo, there also is Brunei (see box titled ''Rich and Broken Brunei''). And a difficult politico-geographical situation has persisted in East Timor, the former Portuguese colony in the Lesser Sunda Islands, where Indonesia's military takeover in 1976 has met with armed resistance, a matter that still has not been finally resolved.

Indonesia is a Dutch colonial creation, and the Dutch chose Jawa as their colonial headquarters, making Batavia (now Jakarta) their capital. Today, Jawa remains the core of Indonesia. With about 120 million inhabitants, Jawa is one of the world's most densely peopled places and one of the most agriculturally productive. Sail into the port of Jakarta, and you pass hundreds of ships and boats at anchor, awaiting berths in one of the Pacific Rim's busiest harbors. Indeed, the whole city and its environs bear the mark of Pacific Rim economic impact. Indonesia offers foreign companies plentiful supplies of cheap labor willing to work for wages about one-third of what a Malaysian or a Thai worker commands, lower even than those of China. (In 1995, 30 cents per hour was the going rate in the textile

RICH AND BROKEN BRUNEI

■ ■ ■

Brunei is an anomaly in Southeast Asia—an oil-exporting Islamic sultanate far from the Persian Gulf. Located on the north coast of Borneo, sandwiched between Malaysian Sarawak and Sabah (see Fig. 10-1), the Brunei sultanate is a former British-protected remnant of a much larger Islamic kingdom that once controlled all of Borneo and areas beyond. Brunei achieved full independence in 1984. With a mere 2,225 square miles (5,700 sq km)—slightly larger than Delaware—and only 310,000 people, Brunei is dwarfed by the other political entities of Southeast Asia. But the discovery of oil in 1929 (and natural gas in 1965) heralded a new age for this remote territory.

Today, Brunei is one of the largest oil producers in the British Commonwealth, and recent offshore discoveries suggest that production will increase. As a result, the population is growing rapidly by immigration (70 percent of Brunei's residents are Malay, 19 percent Chinese), and the sultanate enjoys one of the highest standards of living in Southeast Asia

(gross national product per capita in 1994—the latest year for which data are available—was U.S. $14,240). Most of the people live near the oilfields in the western corner of the country and in the capital in the east—Bandar Seri Begawan. The evidence of a development boom can be seen in modern apartment houses, shopping centers, and hotels—a sharp contrast to many other towns on Borneo. There are some marked internal contrasts as well: Brunei's interior still remains an area of subsistence agriculture and rural isolation, virtually untouched by the modernization of the burgeoning coastal zone.

Brunei's territory is not only very small; it is fragmented. A sliver of Malaysian land separates the larger west, where the capital lies, from the east, and Brunei Bay provides a water link between the two parts. It has been proposed that Brunei purchase this separating corridor from the Malaysians, but no action has yet been taken.

■ ■ ■

factories around Jakarta; in faraway Sumatera, a factory worker may earn just $1 a day.) Only endemic corruption, bribery, cronyism, and bureaucratic obstruction seem to have impeded Indonesia's rise as a Pacific Rim economic force, although political uncertainties also matter. Indonesia's system of government is a mixture of democratic concept and authoritarian reality.

Nothing in Indonesia can compare to Jawa in terms of population, urbanization (still only 31 percent nationwide), communications, or productivity. As a cultural group, the Jawanese constitute about 60 percent of the entire population. Sumatera, much larger territorially than Jawa, has

10.10

the archipelago's second largest population, some 42 million, and ranks second also in economic terms with large rubber plantations in its eastern lowlands. Kalimantan, Indonesia's part of Borneo, is large but has just 11 million inhabitants, mostly in coastal towns and villages. And Sulawesi, east of Borneo, has an estimated 14 million.

Unity in Diversity

Viewed from this perspective, the persistence of Indonesia as a unified state is another politico-geographical wonder on a par with postcolonial India. Wide waters and high mountains have helped perpetuate cultural distinctions and differences; centrifugal political forces have been powerful and, on more than one occasion, have nearly pulled the country apart. But Indonesia's national integration appears to have strengthened, as the country's motto—*Unity in Diversity*—underscores. With more than 300 discrete ethnic clusters, over 250 individual languages, and just about

From the Field Notes

"In the urban mosaic of residential Jakarta, these are by no means the lowest-quality dwellings. This is part of the inner city of Jakarta, and people here have the good fortune of living in permanent abodes located not far from their place of work. The waterway, which leads to the river, serves as a sewer, and yet people wash their clothes in it; but these houses are not totally without amenities. Electricity has been brought in, and the satellite dish rising above the roofs proves that these Jawanese, while not living in luxury, are in touch with the outside world."

Jakarta

Jakarta, capital of Indonesia, sometimes is called the Calcutta of Southeast Asia. Stand on the elevated highway linking the port to the city center and see the villages built on top of garbage dumps by scavengers using what they find in the refuse, and the metaphor seems to fit. There is poverty here unlike that in any other city in the realm.

But there are other sides to Jakarta. Indonesia's economic progress has made its mark here, and the evidence is everywhere. Television antennas and satellite dishes rise like a forest from rusted, corrugated-iron rooftops. Cars (almost all, it seems, late-model), mopeds, and bicycles clog the streets, day and night. A meticulously manicured part of the city center contains a cluster of high-rise hotels, office buildings, and apartments. Billboards advertise planned communities on well-located, freshly cleared land.

Jakarta's population is a cross section of Indonesia's, and the silver domes of Islam rise above the cityscape alongside

Christian churches and Hindu temples. The city always was cosmopolitan, beginning as a conglomeration of villages at the mouth of the Ciliwung River under Islamic rule, becoming a Portuguese stronghold and later the capital of the Dutch East Indies (under the name Batavia). Advantageously situated on the northwest coast of Indonesia's most populous island, Jakarta bursts at the seams with growth. Sail into the port, and hundreds of vessels, carrying flags from Russia to Argentina, await berths. Travel to the outskirts, and huge shantytowns are being expanded by a constant stream of new arrivals. So vast is the human agglomeration—nobody knows how many millions have descended on this place (the official figure is 12.8 million)—that the majority live without adequate (or any) amenities.

A seemingly permanent dome of metallic-gray air, heavy with pollutants, rests over the city. The social price of economic progress is high, but Jakartans are willing to pay it.

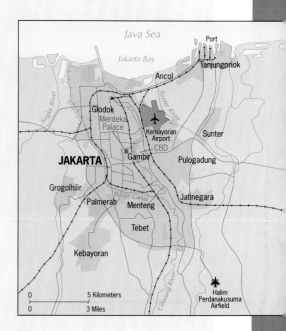

every religion practiced on Earth, Indonesian nationalism has faced enormous odds that have been overcome, to some extent, by development based on the archipelago's considerable resources. This includes sizeable petroleum reserves, large rubber and palm oil plantations (Sumatera shares the nearby Malay Peninsula's environments), extensive lumber resources, major tin deposits offshore from eastern Sumatera, and soils that produce tea, coffee, and other cash crops. However, Indonesia's large population continues to grow at a pace associated with a doubling time of 43 years—a longer-term threat to the country's future. Significantly, rice and wheat have already become prominent among annual imports.

What Indonesia has achieved is etched against the country's continuing cultural complexity. There are dozens of distinct aboriginal cultures; virtually every coastal community has its own roots and traditions. And the majority, the rice-growing Indonesians, include not only the numerous Jawanese—who are Muslims largely in name only and have their own cultural identity—but also the Sundanese (who constitute 15 percent of Indonesia's population), the Madurese (5 percent), and others. Perhaps the best impression of the cultural mosaic comes from the string of islands

that extends eastward from Jawa to Timor (Fig. 10-11). The rice-growers of Bali adhere to a modified version of Hinduism, giving the island a unique cultural atmosphere; the population of Lombok is mainly Muslim, with some Balinese Hinduism; Sumbawa is a Muslim community; Flores is mostly Roman Catholic. In western Timor, Protestant groups dominate; in the east, where the Portuguese ruled, Roman Catholicism prevails. Nevertheless, Indonesia nominally is the world's largest Muslim country: overall, 87 percent of the people adhere to Islam, and in the cities the silver spires of neighborhood mosques rise above the townscape. But Islam is not the issue it is in Malaysia, where observance generally is stricter and where minorities fear Islamization as Malay power grows.

To reduce population pressure in Jawa, to counter centrifugal forces arising from Indonesia's geography, to spread the official language (Bahasa Indonesia, a modified form of Malay), and to strengthen the core's influence over the outlying areas, the government has long pursued a policy known as **translocation**, the resettlement of Jawanese in other parts of the country. Indonesia's population still is growing at an annual rate of 1.6 percent, and population redistribution would reduce the disparities prevailing in the

10.11

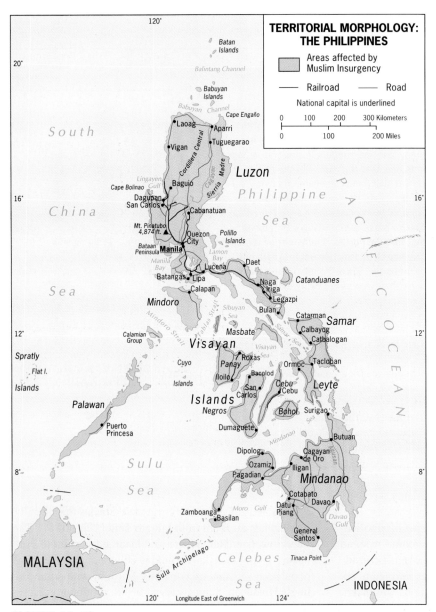

FIGURE 10-12

archipelago. But the policy has had mixed effects. It is seen locally as a policing action, and in the other islands, the Jawanese often are blamed for introducing methods and practices of administration and business that run counter to local traditions.

Indonesia, once one of the most harshly ruled colonies in the world, now is in the awkward position of being a colonial power itself. Not only has the conflict over East Timor attracted world attention, but Indonesia's control over West Irian (western New Guinea, also known as Irian Jaya) also has raised concern. The indigenous peoples of New Guinea are Papuans, not Indonesians, and the eastern half of New Guinea is an independent, Papuan state. As an Indonesian province, West Irian has 22 percent of Indonesia's land but just under 1 percent of its population, in-

cluding thousands of translocated Indonesians from Jawa, many of them concentrated in the provincial capital with the non-Papuan name of Jayapura. It is a world apart from Jawa, in effect a dependency of a country unable to bear the cost of its development. The time may come when Indonesia will rue its campaign to acquire West Irian, potential source of the kinds of centrifugal forces an archipelagic state must constantly confront. ◀▮

Fragmented Philippines

North of Indonesia, across the South China Sea from Vietnam, and south of Taiwan lies an archipelago of more than 7,000 islands (only about 460 of them larger than one square mile in area) inhabited by over 70 million people. Named after a sixteenth-century Spanish king, the Philippines was under Spanish colonial control for more than three centuries and under American tutelage for another 48 years until independence in 1946.

The inhabited islands can be seen as three groups: (1) Luzon, largest of all, and Mindoro in the north, (2) the Visayan group in the center, and (3) Mindanao, second largest, in the south (Fig. 10-12). Southwest of Mindanao lies a small group of islands, the Sulu Archipelago, where the Philippines have a large problem. A Muslim-based insurgency here, nearest to Indonesia, has kept this area in turmoil.

Few of the generalizations we have been able to make for Southeast Asia could apply in the Philippines without qualification. The country's location relative to the mainstream of change in this part of the world has had much to do with this situation. The islands, inhabited by peoples of Malay ancestry with Indonesian strains, shared with much of the rest of Southeast Asia an early period of Hindu cultural influence, which was strongest in the south and southwest and diminishing northward. Next came a Chinese invasion, felt more strongly on the largest island of Luzon in the northern part of the Philippine archipelago. Islam's arrival was delayed somewhat by the position of the Philippines well to the east of the mainland and to the north of the Indonesian islands. The few southern Muslim beachheads were soon overwhelmed by the Spanish invasion of the sixteenth century. Today the Philippines, adjacent to the world's largest Muslim state (Indonesia), is 84 percent Roman Catholic, 10 percent Protestant, and only 5 percent Muslim.

Out of the Philippine melting pot, where Mongoloid-Malay, Arab, Chinese, Japanese, Spanish, and American elements have met and mixed, has emerged the distinctive culture of the Filipino. It is not a homogeneous or a unified culture, but in Southeast Asia it is in many ways unique. One example of its absorptive qualities is demonstrated by the way the Chinese infusion has been accommodated: although the ''pure'' Chinese minority numbers less than 2 percent of the population (far lower than in most Southeast Asian countries), a much larger portion of the Philippine population carries a decidedly Chinese ethnic imprint. What has happened is that the Chinese have intermarried, producing a sort of Chinese-mestizo element that constitutes more than 10 percent of the total population. In another cultural sphere, the country's ethnic mixture and variety are paralleled by its great linguistic diversity. Nearly 90 Malay languages, major and minor, are spoken by the 71 million people of the Philippines; only about 1 percent still use Spanish. Visayan is the language most commonly spoken, and more than 50 percent of the population is able to use English. At independence in 1946, the largest of the Malay languages, Tagalog, or Pilipino, was adopted as the country's official language, and its general use is strongly promoted through the educational system. English is

The Philippines is an overwhelmingly Roman Catholic society in a realm dominated by Buddhism and Islam. About 5 percent of the population (thus amounting to more than 3.5 million people) are Muslims, and this Muslim minority is concentrated about as far from the capital and national core area as you can get: on the island of Mindanao and in the Sulu Archipelago, at the southern end of the country. Muslims here have staged uprisings against Manila. Cultural landscapes in such places as Zamboanga and Basilan (the port and its outrigger canoes shown here) are studded with mosques, in stark contrast to the Christian-dominated islands to the north.

learned as a subsidiary language and remains the chief *lingua franca*; an English-Tagalog hybrid (''Taglish'') is increasingly heard today, remarkably cutting across all levels of society.

The widespread use of English in the Philippines, of course, results from a half-century of American rule and influence, beginning in 1898 when the islands were ceded to the United States by Spain under the terms of the treaty that followed the Spanish-American War. The United States took over a country in open revolt against its former colonial master and proceeded to destroy the Filipino independence struggle, now directed against the new foreign rulers. It is a measure of the subsequent success of U.S. administration in the Philippines that this was the only dependency in Southeast Asia that sided against the Japanese during World War II in favor of the colonial power. United States rule had its good and bad features, but the Americans did initiate reforms that were long overdue, and they were already in the process of negotiating a future independence for the Philippines when the war intervened in 1941.

The Philippines' population, concentrated where the good farmlands lie in the plains, is densest in three general areas (Fig. I-9): (1) the northwestern and south-central part of Luzon, (2) the southeastern extension of Luzon, and (3) the islands of the Visayan Sea between Luzon and Min-

■ AMONG THE REALM'S GREAT CITIES . . .

Manila

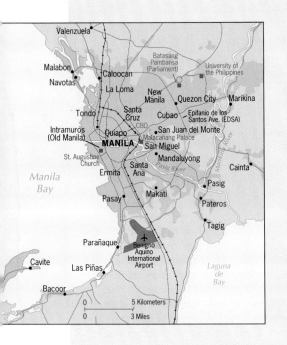

Manila, capital of the Philippines, was founded by Spanish invaders of Luzon Island more than four centuries ago. The colonists made a good choice in terms of site and situation. Manila sprawls at the mouth of the Pasig River where it enters one of Asia's finest natural harbors. To the north, east, and south a crescent of mountains encircles the city, which lies just 600 miles across the South China Sea from Hong Kong.

Manila, named after a flowering shrub in the local marshlands, is bisected by the Pasig River, which is bridged in numerous places. The old walled city, Intramuros, lies to the south. Despite heavy bombing during the Second World War, some of the colonial heritage survives in the form of churches, monasteries, and convents. St. Augustine Church, completed in 1599, is one of the city's landmarks.

The CBD of Manila lies on the north side of the Pasig River. Although Manila has a well-defined commercial center with several avenues of luxury shops and mod-ern buildings, the skyline does not reflect the high level of energy and activity common in Pacific Rim cities on the opposite side of the South China Sea. Neither is Manila a city of notable architectural achievements. Wide, long, and straight avenues flanked by palm, banyan, and acacia trees give it a look not unlike parts of San Juan, Puerto Rico.

In 1948, a newly built city immediately to the northeast of Manila was inaugurated as the *de jure* capital of the Philippines and called Quezon City. The new facilities were eventually to house all government offices, but many functions of the national government never made the move. In the meantime, Manila's growth overtook Quezon City, so that it became part of the Greater Manila metropolis (which today is home to 10.1 million). Although the proclamation of Quezon City as the Philippines' official capital was never rescinded, Manila remains the *de facto* capital of the country today.

---

danao. Luzon is the site of the capital, Manila-Quezon City (10.1 million, nearly one-seventh of the entire national population), a major metropolis facing the South China Sea. Alluvial as well as volcanic soils, together with ample moisture in this tropical environment, produce self-sufficiency in rice and other staples and make the Philippines a net exporter of farm products despite a high population growth rate of 2.1 percent in the late 1990s.

The Philippines seems to get little mention in discussions of developments on the Pacific Rim, and yet this country would seem to be well positioned to share in Pacific Rim economic growth. Governmental mismanagement and political instability have slowed the country's participation, but during the 1990s the situation improved. Despite a series of jarring events—the ouster of U.S. military bases, the damaging eruption of a volcano near the capital, the violent actions of Muslim insurgents, and a dispute over the nearby Spratly Islands in the South China Sea—the Philippines made substantial economic progress during the decade. Its electronics and textile industries (mostly in the Manila hinterland) expanded continuously, and more foreign investment arrived. The Philippines' economy continues to be dominated by agriculture, unemployment remains high, further land reform is badly needed, and social restructuring (reducing the controlling influence over national affairs by a comparatively small group of families) must occur; however, progress is being made. The country now is a lower-middle-income economy, and given a longer period of stability and success in reducing the population growth rate, it will rise to the next level and, finally, take its place among Pacific Rim growth poles.

In recent years, the population issue has divided this dominantly Roman Catholic society, with members of the government promoting family planning and members of the clergy opposing it. But behind this debate lies another of the Philippines' assets: in a realm of mostly undemocratic regimes, the Philippines has come out of its period of authoritarian rule as a rejuvenated if not yet robust democracy.

In the years ahead, Southeast Asia's states and economies will increasingly feel the effect of the rise to world-power status of their giant neighbor, China. Already, Vietnam has felt China's wrath over oil-drilling initiatives it has taken in the South China Sea; the Philippines has seen China set up structures on islands within its maritime jurisdiction (the eastern Spratly Islands; Fig. 10-12). Still another foreign intervention in the affairs of this realm may not be far in the future.

■ PRONUNCIATION GUIDE ■

Acacia (uh-KAY-shuh)
Acapulco (ah-kah-POOL-koh)
Andaman (ANN-duh-mun)
Angkor [Wat] (ANG-kor [WOT])
Annam (uh-NAHM)
Annam [ese]/[ite] (anna-[MEEZE]/
    [MITE])
Antecedent (an-tee-SEE-dunt)
Arakan (ah-ruh-KAHN)
Archipelago (ark-uh-PELL-uh-goh)
Attenuated (uh-TEN-yoo-aited)
Bahasa (bah-HAH-sah)
Bali [nese] (BAH-lee [NEEZE])
Banda (BAHN-duh)
Bandar Seri Begawan (BUN-dahr
    SERRY buh-GAH-wun)
Bangladesh (bang-gluh-DESH)
Basilan (bah-see-LAHN)
Bataks (buh-TAHKS)
Batavia (buh-TAY-vee-uh)
Borneo (BOAR-nee-oh)
Bosnia (BOZ-nee-uh)
Brunei (broo-NYE)
Buddha (BOOD-uh)
Buddhism (BOOD-izm)
Burmans (BURR-munz)
Cambodia (kam-BOH-dee-uh)
Celebes (SELL-uh-beeze)
Chao Phraya (CHOW pruh-yah)
Chedi (CHEDDY)
Chettiar (CHEDDY-ahr)
Chiang Mai (chee-AHNG mye)
Cholon (choh-LONN)
Ciliwung (SILL-uh-wong)
Cochin (KOH-chin)
Copra (KOH-pruh)
Croatia (kroh-AY-shuh)
Dien Bien Phu (d'yen-b'yen-FOOH)
Dong Khoi (dong-KOY)
Filipinos (filla-PEA-noze)
Flores (FLAW-rihss)
Fujian (foo-jee-ENN)
Guangdong (gwahng-DUNG)
Haiphong (hye-FONG)
Hanoi (han-NOY)
Hartshorne (HARTSS-horn)
Hegemony (heh-JEH-muh-nee)
Hindustani (hin-doo-STAHN-nee)
Ho Chi Minh [see Minh, Ho Chi]
Hué (HWAY)
Indonesia (indo-NEE-zhuh)
Intramuros (in-trah-MOOR-rohss)
Irian Jaya (IH-ree-ahn JYE-uh)
Irrawaddy (ih-ruh-WODDY)

Jakarta (juh-KAHR-tuh)
Java (JAH-vuh)
Jawa (JAH-vuh)
Jawanese (jah-vuh-NEEZE)
Jayapura (jye-uh-POOR-ruh)
Johor (juh-HOAR)
Jurong (juh-RONG)
Kachins (kuh-CHINZ)
Kalimantan (kalla-MAN-tan)
Kampuchea (kahm-pooh-CHEE-uh)
Karens (kuh-RENZ)
Kashmir (KASH-meer)
Kelang (kuh-LAHNG)
Khmer [Rouge] (kuh-MAIR [ROOZH])
Khorat (koh-RAHT)
Kompong Thom (kahm-pong TOM)
Kosovo (KAW-suh-voh)
Kra Isthmus (KRAH ISS-muss)
Kuala Lumpur (KWAHL-uh LOOM-
    poor)*
Kuwait (koo-WAIT)
Laem Chabang (lay-EMM
    chuh-BAHNG)
Lao (LAU)
Laos (LAUSS)
Laotian (lay-OH-shun)
Lesotho (leh-SOO-too)
Lingua franca (LEEN-gwuh
    FRUNK-uh)
Lombok (LAHM-bahk)
Luzon (loo-ZAHN)
Macedonia (massa-DOH-nee-uh)
Madurese (muh-dooh-REECE)
Malacca (muh-LAH-kuh)
Malay (muh-LAY)
Malaya (muh-LAY-uh)
Malaysia (muh-LAY-zhuh)
Mandalay (man-duh-LAY)
Mekong (MAY-kong)
Mestizo (meh-STEE-zoh)
Mindanao (min-duh-NOW)
Minh, Ho Chi (MINN, hoh chee)
Moluccans (muh-LUCK-unz)
Montagnards (MON-tun-yardz)
Moped (MOH-ped)
Mosque (MOSK)
Myanmar (mee-ahn-MAH)
Nakhon Sawan (nah-kawn-suh-WAHN)
New Guinea (noo-GHINNY)
Nicaragua (nick-uh-RAH-gwuh)

*Each "u" in Lumpur pronounced as in "look."

Orangutan (aw-RANG-gyoo-tan)
Papua [ns] (pahp-OO-uh[-unz])
Paracel (para-SELL)
Pasig (PAH-sig)
Petronas (peh-TROH-nuss)
Philippines (FILL-uh-peenz)
Phnom Penh (puh-NOM PEN)
Phuket (POO-KETT)
Pilipino (pill-uh-PEA-noh)
Pinang (puh-NANG)
Putrajaya (poo-truh-JYE-ru)
Quezon (KAY-soan)
Qinghai (ching-HYE)
Rangoon (rang-GOON)
Ratzel, Friedrich (RAHT-sull,
    FREED-rish)
Sabah (SAHB-ah)
Saigon (sye-GAHN)
Salween (SAL-ween)
Sarawak (suh-RAH-wahk)
Shan (SHAHN)
Shwedogon (SHWAY-duh-gonn)
Siam (sye-AMM)
Singapore (SING-uh-poar)
Sinicized (SYE-nuh-sized)
Spratly (SPRAT-lee)
Sri Lanka (sree-LAHNG-kuh)
Stupa (STOO-puh)
Sulawesi (soo-luh-WAY-see)
Sultan (sool-TAHN)
Sultanate (SULL-tuh-nut)
Sulu (SOO-loo)
Sumatera (suh-MAH-tuh-ruh)
Sumatra (suh-MAH-truh)
Sumbawa (soom-BAH-wuh)
Sunda [nese] (SOON-duh [NEEZE])
Suryavarman (soory-AHVA-mun)
Tagalog (tuh-GAH-log)
Thailand (TYE-land)
Tibet (tuh-BETT)
Timor (TEE-more)
Tonkin (TAHN-kin)
Tonle Sap (tahn-lay SAP)
Tun Abdul Razak (toon ahb-DOOL
    rah-ZAHK)
Viangchan (vyung-CHAHN)
Vietnam (vee-et-NAHM)
Vietnamese (vee-et-nuh-MEEZE)
Visayan (vuh-SYE-un)
Wallonia (wah-LOANY-uh)
Xizang (sheedz-AHNG)
Yangon (yahn-KOH)
Yunnan (yoon-NAHN)
Zamboanga (zahm-boh-AHNG-guh)

Relief

Meters		Feet
3050		10 000
1525		5000
610		2000
305		1000
152.5		500
0	Sea Level	0
152.5		500
1525		5000
3050		10 000
6100		20 000

Below Sea Level

A-590200-76 2-5 -15
COPYRIGHT BY
RAND McNALLY & COMPANY
MADE IN U.S.A.

Longitude 115° East of Greenwich

Scale 1:16 000 000; one inch to 250 miles. Lambert's Azimuthal, Equal Area Pro
Elevations and depressions are given in feet

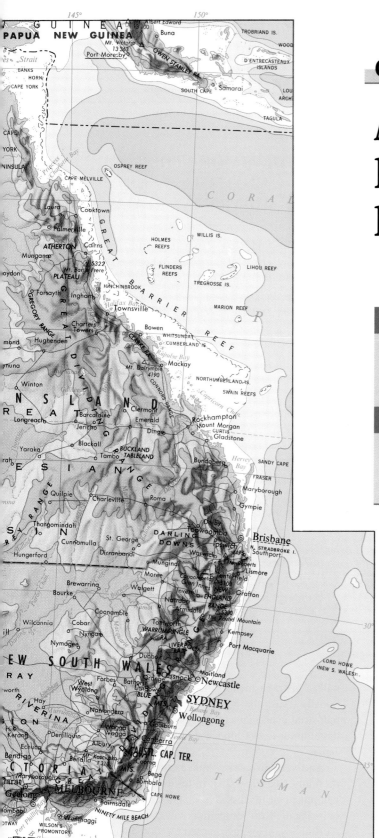

CHAPTER 11

Australia: Dilemmas Downunder

IDEAS & CONCEPTS

Biogeography
Ecosystem
Federalism

Import-substitution
industries
Aboriginal rights

REGIONS

Australia
  Core Area
  Outback

New Zealand

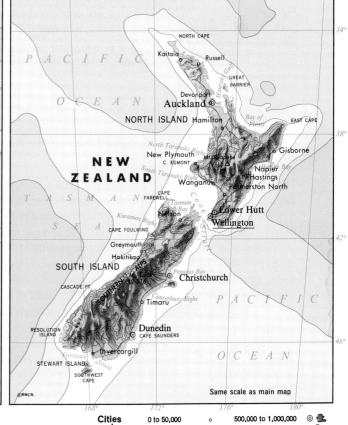

Cities
and
Towns

0 to 50,000   o   500,000 to 1,000,000 ◎

50,000 to 500,000 ⊙   1,000,000 and over

Same scale as main map

▶ THE AUSTRALIAN REALM

The Australian realm is geographically unique. It is the only geographic realm to lie entirely in the Southern Hemisphere. It is the only realm to have no land link of any kind to a neighboring realm, and is thus completely surrounded by ocean and sea. It is the world's least populous geographic realm. Appropriately, its name refers to its location (*Austral* means south)—a location far from the sources of its dominant cultural heritage, but close to its newfound economic partners on the western Pacific Rim.

Two countries constitute this Austral realm: Australia, in every way the dominant one, and New Zealand, physiographically more varied than its giant partner. Between them lies the Tasman Sea. To the west lies the Indian Ocean, to the east the Pacific, and to the south the frigid Southern Ocean.

This southern realm is at a crossroads. On the doorstep of populous Asia, its Anglo-European legacies are now infused by other cultural strains. Polynesian Maori in New Zealand and Aboriginal communities in Australia are demanding better terms of life. Pacific Rim markets are buying huge quantities of raw materials. Japanese and other Asian tourists fill hotels and resorts. Queensland's tropical ''Gold Coast'' resembles Honolulu's Waikiki. Visit Christchurch, a small city on the South Island of New Zealand, and you will find that it has become a favorite place for Japanese weddings. Wedding parties arrive by the planeload: it is cheaper to fly here than to marry in Japan. Tuxedo-rental shops and florists abound, all under Japanese signage. But what of the long-range future of this insular, remote realm?

❖ AUSTRALIA

Imagine a country on the Pacific Rim, 10 times the size of Texas, rich in mineral resources, well endowed with farmlands and vast pastures, major rivers, ample underground water sources, served by good natural harbors, and populated by about 18 million mostly well-educated people who have enjoyed stable government since the beginning of this century. Surely such a country would be a leading economic tiger in the region, a tough competitor for the others where population pressures, labor problems, scarce domestic resources, political uncertainties, and other obstacles must be overcome.

The answer is—no. Australia today is what some locals wryly call an NDC, a Newly Declining Country; a seller of raw materials, not finished ones; a purveyor of livestock, meat, and wheat on undependable world markets; a society deeply in debt; an economy with an uncertain future. Spa-

> ## THE MAJOR GEOGRAPHIC QUALITIES OF THE AUSTRALIAN REALM
>
> ■ ■ ■
>
> 1. Australia and New Zealand constitute a geographic realm by virtue of territorial dimension, relative location, and dominant cultural landscape.
>
> 2. Despite their inclusion in a single geographic realm, Australia and New Zealand differ physiographically. Australia is marked by a vast, dry, low-relief interior; New Zealand is mountainous.
>
> 3. Australia and New Zealand are marked by peripheral development—Australia because of its aridity, New Zealand because of its topography.
>
> 4. The populations of Australia and New Zealand are not only peripherally distributed but also highly clustered in urban centers.
>
> 5. The realm's human geography is changing—in Australia because of Aboriginal activism and Asian immigration, and in New Zealand because of Maori activism and Pacific-Islander immigration.
>
> 6. The economic geography of Australia and New Zealand is dominated by the export of livestock products (and in Australia by wheat production and mining).
>
> 7. Australia and New Zealand are being integrated into the economic framework of the western Pacific Rim, principally as suppliers of raw materials.
>
> ■ ■ ■

tially, Australia anchors the Pacific Rim in the south. Functionally, Australia has become an open-pit mine for the industrial economies to the north. Its manufactures are no threat to those of China, Taiwan, or South Korea, let alone Japan.

And yet Australia continues to rank as a high-income economy (see Fig. I-11). In terms of the indicators of development we discussed in Chapter 9, Australia is far ahead of its Pacific Rim competitors save Japan. In the late 1990s, Australians on average earned far more than Thais, Malaysians, Chinese, or Koreans. In terms of consumption of energy per capita, the number of automobiles and miles of roads, levels of health and literacy, Australia had all the properties of a developed country. Australian cities, where more than 85 percent of all Australians live, are not encircled by crowded shantytowns. The Australian countryside does not display a poverty-stricken peasantry.

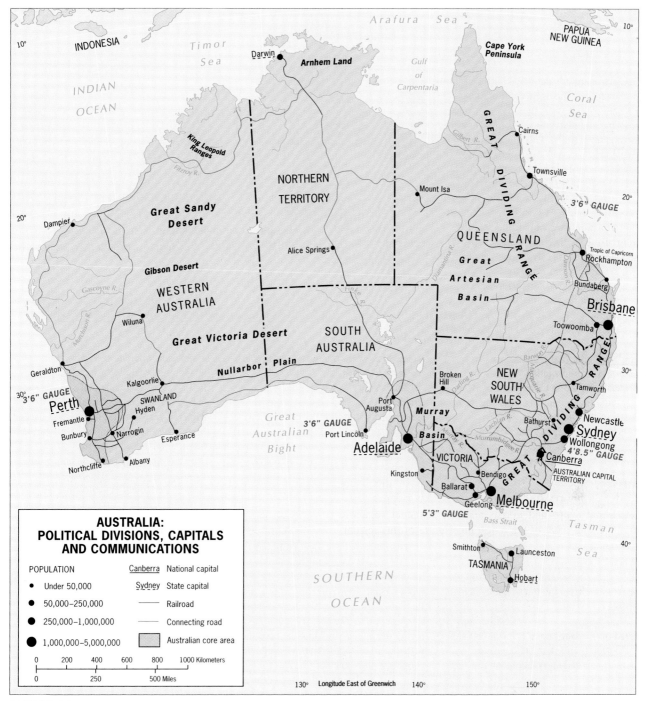

FIGURE 11-1

So why the concern? Because Australia's prospects are not nearly as good as they were in its comfortable past. Its national economy has been growing at rates far lower than those of most of its Pacific Rim neighbors. In 1995, the per capita GNPs of Hong Kong (Xianggang) and Singapore were higher than Australia's, and Japan's was nearly *twice* as high. Dependence on farm products and unfinished raw materials to earn external revenues is risky in a world where buyers may find cheaper alternatives. Furthermore, Australia's human geography is changing significantly, with uncertain outcomes. About 7 percent of the total population is Asian; one-quarter of all Australians today come from a non-English-speaking background. The challenges of multiculturalism are roiling Australian society.

Biogeography

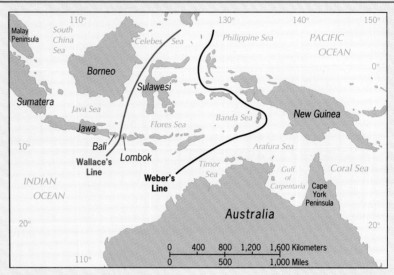

FIGURE 11-2

One of Australia's defining characteristics is its wildlife. This is the land of kangaroos and koalas, wallabies and wombats, possums and platypuses. These and numerous other *marsupials* (animals whose young are born very early in their development and then carried in an abdominal pouch) owe their survival to Australia's early isolation during the breakup of Gondwana (see Fig. 7-2). Before more advanced mammals could enter Australia and replace the marsupials, as happened in other parts of the world, the landmass was separated from Antarctica and South America, and today it contains the world's largest assemblage of marsupial fauna.

Australia's vegetation, too, has distinctive qualities, notably the hundreds of species of eucalyptus trees native to this geographic realm. Many other plants form part of Australia's unique flora, some with unusual adaptation to the high temperatures and low humidity that characterize much of the continent.

The study of fauna and flora in spatial perspective combines the disciplines of biology and geography in a field known as **biogeography**, and Australia is a giant laboratory for biogeographers. In the Introduction (pp. 17-20), we noted that several of the world's climatic zones are named after the vegetation that marks them: tropical savanna, steppe, tundra. When climate, soil, vegetation, and animal life reach a long-term, stable adjustment, vegetation forms the most visible element of this **ecosystem**.

Biogeographers are especially interested in the distribution of plant and animal species, and in the relationships between plant and animal communities and their natural environments. They seek explanations for the distributions the map reveals. For example, why are certain species found in certain areas but not in others that also appear suitable for them? And why do species exist cooperatively in some places but competitively in others? What is the effect of isolation (for example, on top of a butte or mesa, or on an island formed suddenly when a storm cuts off the tip of a peninsula) on the affected species? How are their interrelationships

affected? How do migration patterns change?

The founder of biogeography as a systematic field of study was Alexander von Humboldt (1769–1859), who traveled much of the world in search of plant and animal specimens; in South America's Andes he recognized the relationships among elevation, temperature, and vegetation. His name still is associated with the ocean current that moves past South America's west coast (also known as the Peru Current).

The division of biogeography into two fields, focused, respectively, on plants (phytogeography) and animals (zoogeography), came after von Humboldt's time. In 1876, Alfred Russel Wallace published a book entitled *The Geographical Distribution of Animals* in which he fired the first shot in a long debate: where does the zoogeographic boundary of Australia's fauna lie? Wallace's fieldwork in the area revealed that Australian forms exist not only in Australia itself but also in New Guinea and in some islands to the west. So Wallace proposed that the faunal boundary should lie between Borneo and Sulawesi, and just east of Bali (Fig. 11-2).

Wallace's Line soon was challenged by other researchers, who found species Wallace had missed and who visited islands Wallace had not. There was no question that Australia's zoogeographic realm

ended somewhere in the Indonesian archipelago, but where? Western Indonesia was the habitat of nonmarsupial animals such as tigers, rhinoceros, elephants, as well as primates; New Guinea clearly was part of the realm of the marsupials. How far had the more advanced mammals progressed eastward along the island stepping stones toward New Guinea? The zoogeographer Max Weber found evidence that led him to postulate his own *Weber's Line* which, as Figure 11-2 shows, lay very close to New Guinea.

Not all research in zoogeography or phytogeography deals with such large questions. Much of it focuses on the relationships between particular species and their *habitats*, that is, the environment they normally occupy and of which they form a part. Such environments change, and the changes can spell disaster for the species. In Australia, the arrival of the Aboriginal population (between 50,000 and 60,000 years ago) had little effect on the habitats of the extant fauna. But the invasion of the European colonizers and the introduction of their livestock led to the destruction of habitats and the extinction of many native species. Whether in East Africa or in Western Australia, the key to the conservation of what remains of natural flora and fauna lies in the knowledge embodied by the field of biogeography.

■ ■ ■

For most of the twentieth century, Australians could afford to ignore the obvious. Australia was a British progeny, a European outpost, and with nearby New Zealand it enjoyed standards and ways of living that were the envy of much of the world. With its wide range of environments, magnificent scenery, vast open spaces, and seemingly endless opportunities, Australia appeared invulnerable to the problems of the rest of the world. Europeans peopled the continent; the small remnant Aboriginal population (under 300,000 today) posed no threat. When the Japanese Empire expanded, Australia's remoteness saved the day. When immigration became an issue, Australia adopted an all-white admission policy (which was not officially terminated until 1976). Australia–New Zealand became an outlier of Europe where, it seemed, time had stopped, a geographic realm distinct from its Pacific surroundings and remote from its cultural sources.

Today that geographic realm still persists, but things are changing. It is estimated that at the beginning of this century, Australia's per capita GNP may have been the highest in the world. In 1995, it ranked twenty-second among the world's countries. Australia's share of world trade has fallen, over the past four decades, by more than half. Meanwhile, the Asian influx continues; in the largest city, Sydney, 1 in 8 residents has an Asian ancestry, a ratio that is projected to increase to 1 in 5 shortly after the turn of the century. By a wide margin, Japan has become Australia's leading trade partner, and Australian schools today teach Japanese to tens of thousands of children. Unmistakably, Australia is beginning to become part of the Asian world it adjoins.

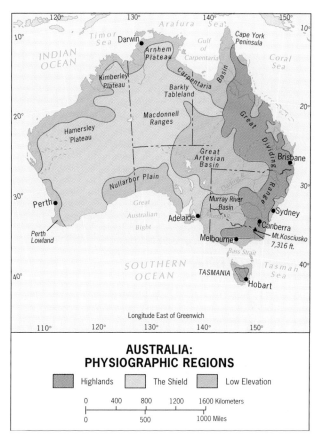

AUSTRALIA: PHYSIOGRAPHIC REGIONS

Highlands The Shield Low Elevation

0 400 800 1200 1600 Kilometers

0 500 1000 Miles

Longitude East of Greenwich

FIGURE 11-3

Australian Rimland

Australia is a large landmass, but the country's population is heavily concentrated in a core area that lies in the east and southeast, most of which faces the Pacific Ocean (here named the Tasman Sea between Australia and New Zealand). As Figure 11-1 shows, this crescent-like Australian heartland extends from north of the city of Brisbane to the vicinity of Adelaide and includes the largest city, Sydney, the capital, Canberra, and the second largest city, Melbourne. A secondary core area has developed in the far southwest, centered on Perth and its outpost, Fremantle.

The Outback

To better understand the evolution of this spatial arrangement, it helps to refer again to the map of world climates (Fig. I-7, p. 17). Note that the bulk of Australia's dry interior is desert (*BWh*) or steppe (*BSh/BSk*), inhospitable but containing many valuable mineral deposits. The core area described above coincides rather closely with the humid temperate climatic zone mapped as *Cfa/Cfb*, extending be-

tween the spine of the Great Dividing Range (Fig. 11-3) and the east coast, and from near the Tropic of Capricorn to Tasmania in the south. Around Adelaide and in the far southwest (including Perth), a dry-summer, Mediterranean regime (*Csa/Csb*) prevails. And the far north lies under tropical, wet-and-dry savanna (*Aw*) conditions, with a small monsoonal strip (*Am*) along the neck of the Cape York Peninsula.

Environmentally, Australia's most favored strips face the Pacific and Southern oceans, and they are not large. The country may be described as a coastal rimland with cities, towns, farms, and forested slopes giving way to a vast, arid interior Australians call the *Outback*. On the western flanks of the Great Dividing Range lie the extensive grassland pastures that catapulted Australia into its first commercial age—and on which still graze one of the largest sheep herds on Earth (over 160 million sheep, producing more than one-fifth of all the wool sold in the world). Where it is moister, to the north and east, cattle by the millions are ranched. This is frontier Australia, over which livestock have ranged for nearly two centuries.

Aboriginal Australians reached this landmass as long as 50,000 to 60,000 years ago, crossed the Bass Strait into Tasmania, and had developed a patchwork of indigenous cultures by the time Captain Arthur Phillip sailed into what

From the Field Notes

"The highway from Alice Springs to Yulara extends through the heart of Australia's great interior desert; any climatic map shows this is the epitome of *BW* (Köppen-Geiger) conditions. Yet this is not the sandy, barren waste the map might suggest. Indeed, the adaptive vegetation that covers the countryside makes this look like steppe country, or better. Nevertheless, this is meteorologically a true desert, with annual rainfall between 4 and 8 inches (10 and 20 cm)—most of it in sudden downpours that generate rapid runoff, cause dry stream beds to overflow, and is largely lost to evaporation. Near Yulara lie some of Aboriginal Australia's holiest sites, including Uluru (Ayers Rock) and the Olga Mountains."

is today Sydney Harbor (1788) to establish the beginnings of modern Australia. The Europeanization of Australia doomed the continent's Aboriginal societies. The first to suffer were those situated in the path of British settlement on the coasts, where penal colonies as well as free towns were founded. Distance protected the Aboriginal communities of the northern interior longer than elsewhere; in Tasmania, the indigenous Australians were exterminated in

a matter of decades after having lived there for perhaps 45,000 years.

Eventually, the major coastal settlements became the centers of separate, quite different colonies. There were seven of these, each with their own hinterlands, and by 1861 Australia was delimited by its now-familiar pattern of straight-line boundaries (Fig. 11-1). Sydney was the focus for New South Wales; Melbourne, Sydney's rival, anchored Victoria. Adelaide was the heart of South Australia, and Perth lay at the core of Western Australia. Brisbane was the nucleus of Queensland, and in Tasmania, the town of Hobart was the seat of government. The largest clusters of surviving Aboriginal people were in the so-called Northern Territory, with Darwin, on Australia's tropical north coast, its colonial city. Notwithstanding their shared cultural heritage, the Australian colonies were at odds not only with London over colonial policies but also with each other over economic and political issues. The building of an Australian nation during the late nineteenth century was a slow and difficult process.

A Federal State

On January 1, 1901, years of difficult negotiations finally produced the Australia we know today: the Commonwealth of Australia, consisting of six States and two Federal Territories (Table 11-1; Fig. 11-1). The special status of Federal Territory was assigned to the Northern Territory to protect the interests of the Aboriginal population there. A second Federal Territory was created in southern New South Wales for the establishment of a new federal capital (Canberra). Both Sydney and Melbourne had fiercely competed for this honor, but it was decided that a completely new seat of government, in a specially designated area, was the best solution. In 1927, the government buildings were ready for use and the administration moved into its new home, a city planned from scratch to serve a nation built from a group of quarrelsome colonies.

TABLE 11-1
States and Territories of Federal Australia, 1997

State	Area (Sq Mi)	Estimated Population (millions)	Capital	Estimated Population (millions)
New South Wales	309.5	6.2	Sydney	3.6
Queensland	666.9	3.1	Brisbane	1.5
South Australia	379.9	1.5	Adelaide	1.1
Tasmania	26.2	0.5	Hobart	0.2
Victoria	87.9	4.7	Melbourne	3.2
Western Australia	975.1	1.8	Perth	1.3
Territory				
Australian Capital Territory	0.9	0.3	Canberra	0.3
Northern Territory	519.8	0.2	Darwin	0.1

Today Canberra, Australia's only large city not situated on the coast, has a population of 342,000, less than one-tenth of Sydney's. As a growth pole, Canberra has been no great success. But as a symbol of Australian **federalism**, it has been a triumph. Australia's unification was made possible through the adoption of an idea with ancient Greek and Roman roots, one also familiar to Americans and Canadians: the notion of *association*. The term to describe it comes from the Latin *foederis*; in practice, it means alliance and coexistence, a union of consensus and common interest—a federation. It stands in contrast to the idea that states should be centralized, or *unitary*. For this, too, the ancient Romans had a term: *unitas*, meaning unity. Most European countries are unitary states, including the United Kingdom of Great Britain and Northern Ireland. Although the majority of Australians came from that tradition (a kingdom, no less), they managed to overcome their differences and establish a Commonwealth that was, in effect, a federation of states with different viewpoints, economies, and objectives, separated by vast distances along the rim of an island continent. And yet, the experiment succeeded: in a few years, Australia will celebrate 100 years of federation.

The Core: An Urban Culture

In this century of federal association, the Australians have developed an urban culture. For all those vast open spaces and romantic notions of frontier and Outback, fully 85 percent of all Australians live in cities and towns. On the map, Australia's areal functional organization is somewhat similar to Japan's: large cities lie along the coast, the centers of manufacturing complexes as well as the foci of agricultural areas. In Japan, mountainous topography contributed to this situation; in Australia, an arid interior. There, however, the similarity ends. Australia's territory is 20 times larger than that of Japan, and Japan's population is seven times that of Australia's. Japan's port cities are built to receive raw materials and to export finished products. Australia's cities forward minerals and farm products from the Outback to foreign markets and import manufactures from overseas. Distances in Australia are much greater, and spatial interaction (which tends to decrease with increasing distance) is less. In comparatively small, tightly organized Japan, you can travel from one end of the country to the other along highways, through tunnels, and over bridges with utmost speed and efficiency. In Australia, the overland trip from Sydney to Perth, or from Darwin to Adelaide, is time-consuming and slow. Nothing in Australia compares to Japan's high-speed bullet train.

For all its vastness and youth, Australia nonetheless developed a remarkable cultural identity, a sameness of urban and rural landscapes that persists from one end of the continent to the other. Sydney, often called the New York of Australia, lies on a spectacular estuarine site, its compact,

From the Field Notes
"The CBD of Sydney lies on the south side of its great harbor, and from almost any angle it presents an impressive skyline. This view is from the east and shows the entire CBD, including the observation tower (left) and the hotels at the ferry port (right). Although Sydney's downtown still dominates the vast urban area (population: 3.6 million, one-fifth of the entire country's), subsidiary business districts, marked by clusters of high-rise buildings, are emerging elsewhere in a pattern similar to other cities in the Western world. One of these lies directly across the harbor from the CBD, in part the product of difficult transport between the two sides."

high-rise central business district overlooking a port bustling with ferry and freighter traffic. Sydney is a vast, sprawling metropolis with multiple outlying centers studding its far-flung suburbs; brash modernity and reserved British ways blend here. Melbourne, sometimes regarded as the Boston of Australia, prides itself on its more interesting architecture and more cultured ways. Brisbane, the capital of Queensland, which also anchors Australia's Gold Coast and adjoins the Great Barrier Reef, is the Miami of Australia; unlike Miami, however, its residents can find nearby relief from the summer heat in the mountains of its immediate hinterland (as well as at its beaches). Perth, Australia's San Diego, is one of the world's most isolated cities, separated from its nearest Australian neighbor by two-thirds of a continent and from Southeast Asia and Africa by thousands of miles of ocean.

And yet, each of these cities—as well as the capitals of South Australia (Adelaide), Tasmania (Hobart), and, to a lesser extent, the Northern Territory (Darwin)—exhibits an Australian character of unmistakable quality. Life is orderly and unhurried. Streets are clean, slums are few, graffiti rarely seen. By American and even European standards, violent crime (though rising) is uncommon. Standards of public transportation, city schools, and health care provi-

11.3

Sydney

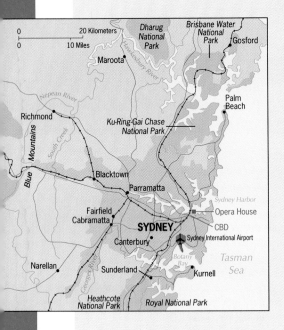

Just slightly over two centuries ago, Sydney was founded by Captain Arthur Phillip as a British outpost on one of the world's most magnificent natural harbors. The free town and penal colony that struggled to survive evolved into Australia's largest city. Today Sydney (3.6 million) is home to nearly one-fifth of the country's entire population. An early start, the safe harbor, fertile nearby farmlands, and productive pastures in its hinterland combined to propel Sydney's growth. Later, as road and railroad links made Sydney the focus of Australia's growing core area, industrial development and political power augmented Sydney's primacy.

With its incomparable setting and mild, sunny climate, its many city beaches, and its easy reach to the cool Blue Mountains of the Great Dividing Range, Sydney is one of the world's most liveable cities. Good public transportation (including an extensive cross-harbor ferry system from the doorstep of the waterfront CBD), fine cultural facilities headed by the multi-theatre Opera House, and numerous public parks and other recreational facilities make Sydney attractive to visitors as well. A booming tourist trade, much of it from Japan and other Asian countries, bolsters the city's economy. Sydney's hosting of the 2000 Olympic Games is certain to further enhance its global reputation.

Increasingly, Sydney also is a multi-cultural city. Its small Aboriginal sector is being overwhelmed by the arrival of tens of thousands of Asian immigrants. The Sydney suburb of Cabramatta has come to symbolize the impact: more than half of its 80,000 residents were born elsewhere, mostly in Vietnam, unemployment exceeds 25 percent, drug use is rampant, and crime and gang violence abound. The fact that tens of thousands of Asian immigrants have succeeded in establishing themselves in some profession is lost in the outcry over the behavior of a small minority.

But in that respect, Sydney is only coming of age. The end of Australia's isolation has brought Asia across the country's threshold, and once again it is the leading city that will show the way.

---

sion are high. Spacious parks, pleasing waterfronts, and plentiful sunshine make Australia's urban life more acceptable than almost anywhere else in the world. Critics of Australia's way of life say that this very pleasant state of affairs has persuaded Australians that hard work is not really necessary. But Australia's privileges are being eroded away, and standards of living are in danger of serious decline. The country's cultural geography evolved as that of a European outpost, prosperous and secure in its isolation. Now Australia must reinvent itself as a major link in an Australo-Asian chain, a Pacific partner in a transformed regional economic geography.

Economic Geography

Distance has been an ally as well as an enemy to Australia. Its remoteness helped save the day when Japan's empire expanded over the western Pacific. From the very beginning, however, goods imported from Britain (and later from the United States) were expensive, largely because of transport costs. This encouraged local entrepreneurs to set up their own industries in and near the developing cities—industries economic geographers call **import-substitution industries**.

When the prices of foreign goods became lower because transportation was more efficient and therefore cheaper, the local businesses demanded protection from the colonial governments. This was done by enacting high tariffs against imported goods. The local products now could continue to be made inefficiently because their market was guaranteed. Japan could not afford this—how could Australia? The answer can be seen on the map (Fig. 11-4). Even before federation in 1901, all the colonies could export valuable minerals whose earnings were used to shore up those inefficient, uncompetitive local industries. By the time the colonies unified, the income from the pastoral industries contributed as well. So the miners and the farmers paid for those imports Australians could not reproduce themselves, plus the products made in the cities. No wonder the cities grew: here were secure manufacturing jobs, jobs in state-run service enterprises, and jobs generated by a growing government bureaucracy. When we noted earlier that Australians once achieved the highest per capita GNP

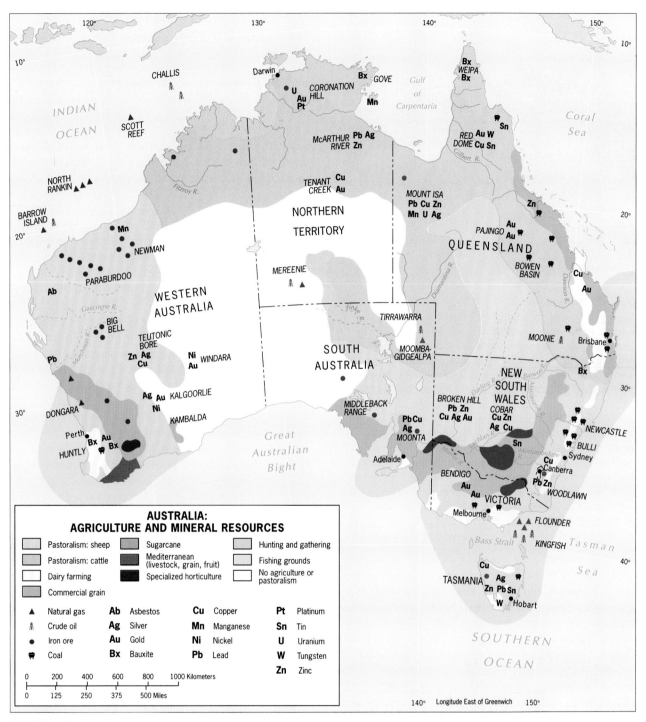

FIGURE 11-4

in the world, this was achieved in the mines and on the farms, not in the cities.

But the good times had to come to an end. The prices of farm products fluctuated, and international market competition increased. The cost of mining ores and minerals, transporting, and shipping them also rose. Australians like to drive (there are more road miles per person in Australia

than in the United States), and expensive petroleum imports were needed. Meanwhile, the government-protected industries had been further fortified by strong labor unions. Not surprisingly, as the economy declined the national debt rose, inflation grew, and unemployment crept upward. In 1995, only 18 percent by value of Australian exports were the kind of high-tech goods that have made East Asia's

From the Field Notes

"Darwin, on the north coast of the Northern Territory, is Australia's smallest and most remote subnational capital. The city, with a population approaching 100,000, hardly represents the Northern Territory in social terms: although it is a multicultural, multiethnic community, only about 3 percent of Darwinians are of Aboriginal descent. Rather, Darwin is an Australian outpost on northern waters sometimes crossed, these days, by Chinese and Southeast Asian would-be immigrants. There also is a sizeable Malay community, and Darwin boasts a mosque, Buddhist temple, synagogue, and Christian churches. With its excellent natural harbor, Darwin could play an important economic role in Australia in the future, given its Pacific Rim location. But assertive Aboriginal land claims in its hinterland could inhibit mining and other development. Virtually all the buildings seen in this photograph date from after the year 1974, when almost all of Darwin was destroyed, with great loss of life, by a devastating hurricane."

economic tigers so successful. That was double the 1985 figure, but still far below what the country needs to produce.

Agricultural Abundance

And yet, Australia has material assets of which other countries on the Pacific Rim can only dream. In terms of agriculture, the raising of sheep was the earliest of commercial ventures, but it was the technology of refrigeration that brought world markets within reach of Australian beef producers. Wool, meat, and wheat have long been the country's big three income earners; Figure 11-4 displays the vast pastures in the east, north, and west that constitute the ranges of Australia's huge herd. The zone of commercial grain farming forms a broad crescent extending from northeastern New South Wales through Victoria into South Australia, and covers a large area in the hinterland of Perth. Keep in mind the scale of this map: Australia is only

slightly smaller than the 48 contiguous states of the United States! Commercial grain farming in Australia is big business. As the climatic map would suggest, sugarcane grows along most of the warm, humid coastal strip of Queensland, and Mediterranean crops (including grapes for Australia's wines) cluster in the hinterlands of Adelaide and Perth. Mixed horticulture concentrates in the basin of the Murray River system, including rice, grapes, and citrus fruits, all under irrigation. And, as elsewhere in the world, dairying has developed near the large urban areas. With its considerable range of environments, Australia yields a diversity of crops.

Mineral Wealth

Australia's mineral resources, as Figure 11-4 shows, also are diverse. Major gold discoveries in Victoria and New South Wales produced a 10-year gold rush starting in 1851 and ushered in a new economic era. By the middle of that decade, Australia was producing 40 percent of the world's gold. Subsequently, the search for more gold led to the discoveries of other minerals. New finds are still being made today, and even oil and natural gas have been found both inland and offshore (see the symbols in Fig. 11-4 in the Bass Strait between Tasmania and the mainland, and off the northwestern coast of Western Australia). Coal is mined at numerous locations, notably in the east near Sydney and Brisbane but also in Western Australia and even in Tasmania; before coal prices fell, this was a very valuable export. In terms of metallic and nonmetallic minerals, major deposits abound—from the complex at Broken Hill and the mix of minerals at Mount Isa to the huge nickel deposits at Kalgoorlie and Kambalda, the copper of Tasmania, the tungsten and bauxite of northern Queensland, and the asbestos of Western Australia. A glance at the map reveals the wide distribution of iron ore (the red dots), and for this raw material as for many others, Japan has been Australia's best customer in recent years. In the late 1990s, Japan was buying more than one-third of all Australian mineral exports.

Manufacturing's Limits

Australian manufacturing, as we noted earlier, remains oriented to local domestic markets. Do not expect to find Australian automobiles, electronic equipment, or cameras challenging the Pacific Rim's economic tigers for a place on world markets—not yet, at any rate. Australian manufacturing is quite diversified, producing some machinery and equipment made of locally produced steel as well as textiles, chemicals, paper, and many other items. These industries cluster in and near the major urban areas where the markets are. The domestic market in Australia is not large, but despite declining real incomes, it remains a relatively affluent one. This makes it attractive to foreign producers, and Australia's shops are full of goods from Japan, South Korea, Taiwan, and Hong Kong. Indeed, despite its long-term protectionist practices, Australia still does not produce many goods that could be manufactured at home. Overall, the economy continues to display symptoms of a still-developing rather than a fully developed country.

Australia's Future

A new age is dawning in Australia, a country that just a decade ago (1988) celebrated its bicentennial—two centuries as a European outpost facing the Pacific Ocean. Now, as Australia enters its third century, its European bonds are weakening and its Asian ties are strengthening (see box titled "Republic of Australia?").

On the face of it, Australia and its northern neighbors on the Pacific Rim would seem to exhibit a geographic complementarity: Australia has excess food, metals, and minerals needed by Japan and the economic tigers, and Australia needs the cheap manufactures Asia produces. But it is not that simple. Australia still has tariff barriers against imported goods, and the Asian countries on the Pacific Rim maintain import barriers against processed foods and minerals, thus discouraging Australia from refining its exports and earning more from them. Asia's economic tigers, furthermore, are more interested in the potential of large markets (such as Indonesia, with 205 million people) than Australia, and its investors like the low wages of East and Southeast Asia. In any case, it is difficult for Australia to open its economy and to lower its protective tariffs without reciprocation from its Asian trading partners.

Nor has the United States been helpful to Australia's wheat farmers. U.S. government subsidies to American wheat farmers have long made it possible for U.S. farmers to sell their wheat at lower prices than Australian farmers can afford to sell it. In 1996, farm subsidies in the United States were under review—not because of long-term Australian appeals, but because of domestic political considerations. This further illustrates the risks inherent in an economy that depends substantially on revenues from agricultural exports.

Aboriginal Rights

11.2

In the late 1990s, Australia faces other challenges as well. The Aboriginal population of about 300,000 (including many of mixed ancestry) has been gaining influence in the country's affairs. In the 1980s, Aboriginal leaders began a campaign to obstruct exploration and mining on ancestral and sacred lands. They achieved a notable victory at Coronation Hill (Fig. 11-4) in the Northern Territory, where the government prohibited the development of platinum, gold, and palladium deposits because they were deemed to lie on an Aboriginal community's sacred land. In 1992,

REPUBLIC OF AUSTRALIA?

■ ■ ■

Australians in the 1990s were facing an emotional decision: whether to end their ties to the British Crown and make their country a republic. While the Australian economy deteriorated, politicians spent critical time arguing over an issue that, many agreed, would ultimately make little difference to the nation. But the question did raise fears in some quarters about Australia's future. State governments insisted that the matter must be put to a referendum—not nationally, but State by State. In Western Australia, public opinion seemed to favor the monarchy, and the suggestion was even raised that a "no" vote in that State might lead to secession. Other observers feared that republic status would eventually change the State system, marking the beginning of the end of a framework that has served Australia well. Thus even Australia is not immune to the politico-geographical stirrings that afflict today's world.

■ ■ ■

Australia's High Court ruled that a form of "native title" existed before white settlement and could still exist on any land where the government had not issued title.

This ruling, the so-called Mabo decision (after Eddie Mabo, a Torres Strait Aborigine), was followed in 1993 by passage in the Australian Senate of the Native Title bill. As a result, Aborigines and Torres Strait Islanders who long lived on "vacant" land could petition the courts for title to their habitat. Those who can prove that their land was taken away are entitled to compensation. And when a mining company's lease expires, Aborigines now have the right to claim the land back.

The Mabo ruling and the Native Title bill created considerable apprehension among miners, farmers, and even homeowners potentially affected by Aboriginal claims (activist Aboriginal groups claimed ownership to land in the heart of Sydney and parts of Brisbane). In fact, the two States most affected will be South Australia and Western Australia, where there are huge tracts of "Crown" land and Aboriginal peoples who can lay credible claim on them. There, State governments are now seeking to pass legislation that would void the effects of the Mabo ruling. The Mabo decision opened a new era in national affairs, and the outcome (and impact on the federal state) is yet uncertain.

It should be noted that only a few Aboriginal communities still pursue their original life of hunting and gathering and fishing in the remotest corners of the Northern Territory, northern Queensland, and northernmost Western Australia. The remainder are scattered across the continent, subsisting on reserves set aside for them by the government (large parts of the Northern Territory are so designated), working on cattle stations, or performing menial jobs in cities and towns. In comparatively affluent Australia, the

From the Field Notes

"The eucalyptus tree is native to Australia. It has been planted worldwide because of its rapid growth and utility as a building material. Australia has extensive forests of eucalyptus ("blue-gum") trees, home to wildlife such as the koala. But the oily eucalyptus is susceptible to fire, and burns fast once lightning or some other incident ignites it. One of Australia's constant environmental threats is the fast-moving eucalyptus fire; such wildfires have taken thousands of casualties. This forest was photographed days after a fire swept through it; some of the trees will recover, but all wildlife has been lost here, west of Newcastle, New South Wales."

Aboriginal peoples remain victims of poverty, disease, inadequate education, and even malnutrition. The **Aboriginal rights** question will loom large in Australia's future.

Environmental Degradation

Another growing issue involves environment and conservation. Australia's wealth was derived not only from its ores and minerals and from soils and pastures: its ecology, too, paid a heavy price (see Focus box p. 514). Great stands of magnificent forest were destroyed. In Western Australia, centuries-old trees were simply "ringed" and left to die so that the sun could penetrate through their leafless crowns to nurture the grass below. Then the sheep could be driven into these new pastures. In Tasmania, where Australia's native eucalyptus tree reaches its greatest dimensions (comparable to North American redwood stands), tens of thousands of acres of this irreplaceable treasure have been lost to chain saws and pulp mills. Australia's unique marsupial fauna have suffered the loss of numerous species, and many more are endangered or threatened. "Never have so few people wreaked so much havoc on the ecology of so large an area in so short a time," observed a geographer in Australia recently; but awareness of this environmental degradation is growing.

In Tasmania, the "Green" environmentalist political party has become a force in state affairs, and its activism has slowed deforestation, dam-building, and other "development" projects. Still, many Australians fear the environmental movement as an obstacle to economic growth at a time when the economy needs stimulation. This, too, is an issue for the future.

Immigration Issues

The population question has preoccupied Australia as long as Australia has existed—and even longer. Fifty years ago, when Australia had less than half the population it has today, 95 percent of the people were of European ancestry, and more than three-quarters of them came from the British Isles. Eugenic (race-specific) immigration policies until the 1970s maintained this situation. Today, the picture is dramatically different: of 18 million Australians, only about one-third have British-Irish origins, and Asian immigrants outnumber both European immigrants as well as the natural increase each year. During the early 1990s, nearly 150,000 legal immigrants arrived in Australia annually, the majority from Hong Kong, Vietnam, China, the Philippines, Malaysia, India, and Sri Lanka. (Immigration from Britain and Ireland continues at around 20,000 to 25,000 annually, and between 5,000 and 10,000 New Zealanders also emigrate to Australia each year.)

All this is happening at a time of economic difficulty, and the government is under pressure again to limit immigration. The quota was set at 80,000 for the next few years, but such restrictions touch sensitive nerves among many Australians. Some want to reduce the numbers still more: the Australian Nationalist Movement (headquartered in Perth) and the League of Rights, two extremist groups, oppose further immigration from Asia. So the population issue stresses Australia once again, and as the ethnic makeup of Australia changes, multiculturalism and its challenges will emerge as a national problem for a transforming society.

Major Cities of the Realm

Cities	Population* (in millions)
Adelaide, Australia	1.1
Auckland, New Zealand	1.0
Brisbane, Australia	1.5
Canberra, Australia	0.3
Melbourne, Australia	3.2
Perth, Australia	1.3
Sydney, Australia	3.6
Wellington, New Zealand	0.4

*Based on 1997 estimates.

In the meantime, thousands of Japanese tourists stream through Australia's many attractions. Bilingual signs, maps, and menus are everywhere; Japanese newspapers sit on city newsstands; Japanese is taught in hundreds of Australian schools. The two ends of *Austrasia* (the new regional term we introduced in the box on p. 36 to denote the western Pacific Rim) are sharply different in countless ways, but their economic meeting ground grows every year. East Asian freighters in Australian ports symbolize the interconnections and interactions that are forging a new geographic region, of which Australia will be an integral part.

❖ NEW ZEALAND

Fifteen hundred miles east-southeast of Australia, in the Pacific Ocean across the Tasman Sea, lies New Zealand. In an earlier age, New Zealand would have been part of the Pacific geographic realm because its population was Maori, a people with Polynesian roots. But New Zealand, like Australia, was invaded and occupied by Europeans. Today, the country's population of 3.6 million is 85 percent European, and the Maori form a minority of less than 400,000, with many of mixed Euro-Polynesian ancestry.

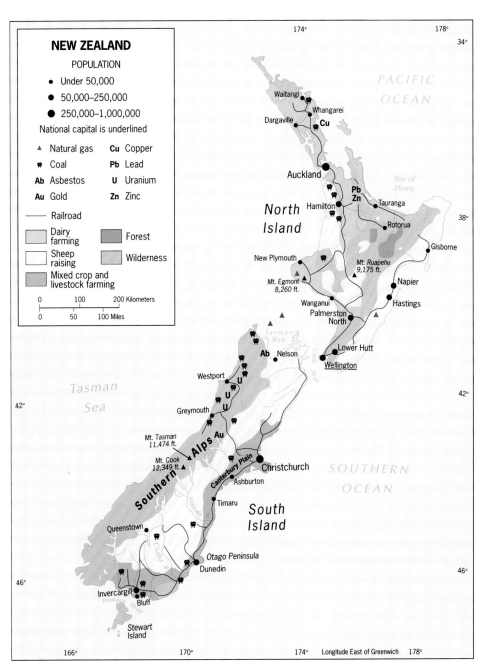

FIGURE 11-5

New Zealand consists of two large mountainous islands and numerous scattered smaller islands (Fig. 11-5). The two large islands, with the South Island somewhat larger than the North Island, look diminutive in the great Pacific Ocean, but together they are larger than Britain. In sharp contrast to Australia, the two islands are mainly mountainous or hilly, with several peaks rising far higher than any on the Australian landmass. The South Island has a spectacular snowcapped range appropriately called the Southern Alps, with peaks reaching beyond 11,700 feet (3,500 m). The smaller North Island has proportionately more land under low relief, but it also has an area of central highlands along whose lower slopes lie the pastures of New Zealand's chief dairying district. Hence, while Australia's land lies relatively low in elevation and exhibits much low relief, New Zealand's is on the average quite high and evinces mostly rugged relief.

Human Spatial Organization

Thus the most promising areas for habitation are the lower-lying slopes and lowland fringes on both islands. On the North Island, the largest urban area, Auckland, occupies a comparatively low-lying peninsula. On the South Island, the largest lowland is the agricultural Canterbury Plain,

From the Field Notes

"We headed inland from Christchurch across the wide, glacial-outwash Canterbury Plain toward the Southern Alps. Fields stood under cereals, vegetables, and fruits; soils obviously are fertile here, and the short but sunny growing season yields ample good harvests. Nearer the foothills lay pastures. Rows of trees served as windbreaks; strong winds blow from the mountains onto the Canterbury Plain when a low pressure system forms offshore, powerful enough to damage crops."

centered on Christchurch. What makes these lower areas so attractive, apart from their availability as cropland, is their magnificent pastures. Such is the range of soils and pasture plants that both summer and winter grazing can be carried on. Moreover, a wide variety of vegetables, cereals, and fruits can be produced in the Canterbury Plain, the chief farming region. About half of all New Zealand is pasture land, and much of the farming is done to supplement the pastoral industry in order to provide fodder when needed. Sheep and cattle dominate these livestock-raising activities, with wool, meat, and dairy products providing over one-third of the islands' export revenues.

Despite their contrasts in size, shape, physiography, and history, New Zealand and Australia have a great deal in common. Apart from their joint British heritage, they share a sizeable pastoral economy, a small local market, the problem of great distances to world markets, and a desire to stimulate (through protection) domestic manufacturing. The high degree of urbanization in New Zealand (85 percent of the total population) again resembles Australia: substantial employment in city-based industries, mostly the processing and packing of livestock and farm products, and government jobs.

More remote even than Australia, New Zealand also has been affected by developments on the western Pacific Rim. In the mid-1990s, Japan and other countries on the Pacific Rim together bought more of New Zealand's exports than did Australia or the United States (Britain takes only about 6 percent of New Zealand's exports). Australia and the United States still send New Zealand most of its imports (about 40 percent combined), but Japan's contribution is rising.

Spatially, New Zealand shares with Australia its pattern of peripheral development (Fig. I-9), imposed not by desert but by high rugged mountains. The country's major cities—Auckland and the capital of Wellington (together with its satellite of Hutt) on the North Island, and Christchurch and Dunedin on the South Island—are all located on the coast, and the entire railway and road system is peripheral in its configuration (Fig. 11-5). This is more pronounced on the South Island than in the north because the Southern Alps are New Zealand's most formidable barrier to surface communication.

The Maori Factor and the Future of New Zealand

One of New Zealand's historic sites is the wooden Treaty House at Waitangi, in the northern peninsula of the North Island. There, following a series of brutal conflicts, the British and the Maori signed the Treaty of Waitangi in 1840. The treaty granted the British sovereignty over New Zealand, but it also guaranteed Maori rights over tribal lands. February 6, the date of the signing of the treaty, is a public holiday in New Zealand called Waitangi Day. A ceremony at the Treaty House, attended by the prime minister and by Maori leaders, is a highlight of the day's events.

Visitors to the 1995 ceremony were in for a shock. Maoris disrupted the proceedings; the prime minister was jostled by members of the crowd, and an attempt was made to set fire to the Treaty House. For the first time in more than half a century, the ceremony had to be abandoned.

What precipitated this show of anger? In fact, it was only the latest in a series of events signaling a rise in Maori ethnic consciousness. Maori leaders insist that the terms of the Waitangi Treaty (partly abrogated in 1862) be enforced, that large tracts of land in urban as well as rural areas, amounting to more than half of the national territory, be awarded to their rightful Maori owners, and that Maori fishing rights be paramount over offshore waters.

Judicial rulings during the 1990s have begun to support the Maori position, and Maori success in one arena has led to new claims in others. During the present decade, the Maori question has become the leading national issue in New Zealand, and Maori activism has risen sharply. In this context, the events of February 6, 1995 were no surprise.

11.4   The Maori appear to have reached New Zealand during the tenth century A.D., and by the time the European colonists arrived they had had an enormous impact on the islands' ecosystems. One of the Maori's complaints is the persistently slow pace of integration of the Maori minority of under 400,000 into modern New Zealand society. In fact, the Maori have been joined by other Pacific is-

landers, so that Auckland today not only is New Zealand's largest city; it also may be the largest urban concentration of Polynesians anywhere, with over 100,000 Maori as well as Samoans, Cook Islanders, Tongans, and others making up one-sixth of the metropolitan population. Their neighborhoods give Auckland an ethnic patchwork that evinces the non-integration about which the Maori complain.

The Maori issue has other geographic overtones. Nearly all Maori live on the North Island (where three-quarters of all New Zealander reside), but Maori leaders are making huge claims, based on historic primacy, over much of the South Island. Remote and comparatively conservative, South Islanders have reacted to these claims in ways similar to the response in Western Australia to the Mabo decision concerning Aboriginal rights.

The Waitangi Treaty was intended to be the framework for a partnership between the British and the Maoris. If the current disputes can lead to a reconsideration and fulfillment of the treaty's terms in the context of modern New Zealand's human geography, then the country's future could be shaped by a harmonious, mainly bicultural society. If this fails, the specter of serious ethnic polarization looms against a demographic background that projects the rapidly growing Polynesian population of New Zealand as doubling to 25 percent of the national total by 2012.

Dominant cultural heritage and prevailing cultural landscape form two criteria on which the delimitation of the Australia–New Zealand geographic realm is based. But in both entities of the realm, the cultural mosaic is changing, and the convergence with neighboring realms proceeds.

◀▮

▪ PRONUNCIATION GUIDE ▪

Aboriginal (abb-uh-RIDGE-uh-null)
Adelaide (ADDLE-ade)
Auckland (AWK-lund)
Bali (BAH-lee)
Bauxite (BAWKS-ite)
Borneo (BOAR-nee-oh)
Brisbane (BRIZZ-bun)
Butte (BYOOT)
Cabramatta (kabbra-MADDA)
Canberra (KAN-burruh)
Canterbury (KAN-ter-berry)
Dunedin (duh-NEED-nn)
Eucalyptus (yoo-kuh-LIP-tuss)
Humboldt (HUMM-bolt)
Kakadu (KAH-kuh-doo)

Kalgoorlie (kal-GHOOR-lee)
Kambalda (kahm-BAHL-duh)
Koala (kuh-WAH-luh)
Köppen (KER-pun)
Mabo (MAY-boh)
Maori (MAU-ree/MAH-aw-ree)
Marsupial (mar-SOOPY-ull)
Melbourne (MEL-bun)
Mesa (MAY-suh)
Mount Isa (mount EYE-suh)
Mosque (MOSK)
Murray (MUH-ree)
New Guinea (noo-GHINNY)
Phytogeography (FYE-toh-jee-
    OG-gruh-fee)

Platypus (PLATT-uh-puss)
Samoan (suh-MOH-un)
Sulawesi (soo-luh-WAY-see)
Tasman (TAZZ-mun)
Tasmania (tazz-MAY-nee-uh)
Tongan (TONG-gun)
Torres (TOAR-russ)
Uluru (ooh-LOO-roo)
Waitangi (WYE-tonggy)
Wallaby (WALLA-bee)
Xianggang (see-AHNG-gahng)
Yulara (yoo-LAH-rah)
Zoogeography (ZOH-oh-jee-
    OG- gruh-fee)

ARCTIC OCEAN

NOVOSIBIRSKIYE O-VA

East Siberian Sea

VRANGELYA I.

ELLESMÈRE ISLAND

BROOKS RANGE

Arctic Circle

Beaufort Sea

VICTORIA ISLAND

BAFFIN ISLAND

VERKHOYANSKIYE KHREBET

ALASKA (U.S.A.)

ALASKA RANGE

Anchorage

Hudson Bay

RUSSIA

O. KAMCHATKA

Sea of Okhotsk

KORYAKSKY KHREBET

BERING SEA

ST. LAWRENCE I.

KODIAK ISLAND

CANADA

SAKHALIN I.

CHINA

HARBIN

Vladivostok

HOKKAIDO

KURILE ISLANDS

Petropavlovsk-Kamchatskiy

ALEUTIAN ISLANDS

QUEEN CHARLOTTE ISLANDS

Vancouver

Victoria

Seattle

ROCKY MOUNTAINS

MONTRÉAL

TORONTO

SHENYANG

NORTH KOREA

JAPAN

HONSHU

TOKYO

YOKOHAMA

PYONGYANG

SOUTH KOREA

SEOUL

KYUSHU

OSAKA

KITAKYUSHU

RYUKYU IS.

TAIPEI

TAIWAN

EAST CHINA SEA

BONIN IS. (JAPAN)

Tropic of Cancer

MIDWAY IS. (U.S.A.)

PACIFIC OCEAN

INTERNATIONAL DATE LINE

WAKE I. (U.S.A.)

HAWAIIAN IS. (U.S.A.)

Honolulu

CHICAGO

UNITED STATES

NEW YORK

St. Louis

Washington, D.C.

San Francisco

SIERRA NEVADA

COAST RANGES

LOS ANGELES

SAN DIEGO

New Orleans

Miami

HAVANA

CUBA

JAMAICA

MEXICO

SIERRA MADRE OCCIDENTAL

MEXICO CITY

BELIZE

Belmopan

GUATEMALA

HONDURAS

GUATEMALA CITY

San Salvador

EL SALVADOR

Managua

NICARAGUA

Tegucigalpa

San José

COSTA RICA

PANAMA

BOGO

QUITO

ECUADOR

LIMA

MANILA

PHILIPPINES

GUAM (U.S.A.)

NORTHERN MARIANA ISLANDS

MARSHALL ISLANDS

CAROLINE IS.

PALAU IS.

PALAU

FEDERATED STATES OF MICRONESIA

Majuro

HOWLAND I. (U.S.A.)

BAKER I. (U.S.A.)

KIRITIMATI

Equator

GALÁPAGOS IS. (Ecuador)

INDONESIA

NEW GUINEA

BISMARCK ARCH.

NAURU

KIRIBATI

PHILIPPINE SEA

ARAFURA SEA

PAPUA NEW GUINEA

Port Moresby

SOLOMON ISLANDS

Honiara

Funafuti

TUVALU

TOKELAU (N.Z.)

WESTERN SAMOA

MARQUESAS IS.

TIMOR SEA

VANUATU

NEW HEBRIDES

Port-Vila

WALLIS AND FUTUNA (Fr.)

Apia

AMERICAN SAMOA

COOK ISLANDS (N.Z.)

SOCIETY IS.

ÎLES TUAMOTU

NEW CALEDONIA (Fr.)

FIJI

Suva

TONGA

NIUE (N.Z.)

TAHITI

FRENCH POLYNESIA

LOYALTY IS.

Tropic of Capricorn

AUSTRALIA

GREAT DIVIDING RANGE

Brisbane

PITCAIRN (U.K.)

RAPA NUI (EASTER I.) (Chile)

Adelaide

Canberra

SYDNEY

MELBOURNE

Auckland

NEW ZEALAND

NORTH ISLAND

PACIFIC OCEAN

SANTIAGO

CHILE

TASMAN SEA

Wellington

SOUTH ISLAND

Christchurch

SOUTHERN OCEAN

TASMANIA

Hobart

Dunedin

Antarctic Circle

SOUTHERN OCEAN

ANTARCTICA

0	500	1000		2000 Miles
0		1000		2000 Kilometers

Scale 1:87,500,000; one inch to 1,381 miles
Apinus II projection
Elevations and depressions are given in feet

Relief

Meters

3050

1525

610

305

152.5

0   Sea Level

152.5   500

1525   5000

3050   10 000

6100   20 000

The Pacific Realm: Uncertain Futures

IDEAS & CONCEPTS

Marine geography
Territorial sea
Continental shelf
Exclusive Economic Zone
    (EEZ)
Maritime boundary
Choke point

Median line boundary
Claimant state
High seas
High islands
Low islands
Maritime environment

◆ REGIONS

Melanesia   Polynesia
Micronesia

DEFINING THE REALM

Between the Americas to the east and the western Pacific Rim to the west lies the vast Pacific Ocean, larger than all the world's land areas combined. In this greatest of all oceans lie tens of thousands of islands, some large (New Guinea is by far the largest), most small (many are uninhabited). Together, the land area of these islands is a mere 376,000 square miles, about the size of Texas plus New Mexico, and over 90 percent of this lies in New Guinea.*

The atlas map (p. 526) shows that the Pacific geographic realm—land *and* water—covers nearly an entire hemisphere of this world, the one commonly called the Sea Hemisphere (its opposite, the Land Hemisphere, is mapped in Fig. 1-1); this Sea Hemisphere meets the Russian and North American realms in the far north and merges into

the Southern Ocean in the south (see box titled ''Antarctic Realm?,'' p. 535). Despite the preponderance of water, this fragmented, culturally complex realm does possess regional identities. It includes the Hawaiian Islands, Tahiti, Tonga, and Samoa—fabled names in a world apart.

In terms of modern cultural and political geography, Indonesia and the Philippines are not part of the Pacific realm, although Indonesia's political system reaches into it; nor are Australia and New Zealand. Before the European invasion and colonization, Australia and New Zealand would have been included, Australia because of its Aboriginal population and New Zealand because its Maori population has Polynesian affinities. But black Australians and Maori New Zealanders have been engulfed by the Europeanization of their countries, and the regional geography of Australia and New Zealand today is decidedly not Pacific except in comparatively small locales. In New Guinea, on the other hand, Pacific peoples remain numerically and culturally the dominant element.

---

*The figures in Appendix A do not match these totals, because only the political entity of Papua New Guinea is listed, not the Indonesian part of the island (West Irian). Here, as Figure 10-1 showed, the political and the regional boundaries do not match.

From the Field Notes

"The port of Papeete on Tahiti forms a reminder that this is a last vestige of European colonialism in the Pacific. French Polynesia remains under the rule of Paris despite growing opposition here; when France again used Polynesia as a testing ground for nuclear weapons in 1995, violence erupted in Papeete and elsewhere. We saw much evidence of France's military presence (as in this photograph); but also evidence of France's substantial spending on its Pacific domain, as reflected by the oil storage and distribution facilities seen here. Those investments have restrained the independence movement; without them, this archipelago would be very poor in material terms."

THE MAJOR GEOGRAPHIC QUALITIES OF THE PACIFIC RIM

■ ■ ■

1. The Pacific realm's total area is the largest of all geographic realms. Its land area, however, is among the smallest, with the bulk of it lying on the island of New Guinea.

2. Papua New Guinea, with a population of 4.3 million, alone contains over three-fifths of the Pacific realm's population.

3. The highly fragmented Pacific realm consists of three regions: Melanesia (including New Guinea), Micronesia, and Polynesia.

4. The Pacific realm's islands and cultures may be divided into volcanic *high-island* cultures and coral *low-island* cultures.

5. In Polynesia, local culture is nearly everywhere severely strained by external influences. In Hawaii, as in New Zealand, indigenous culture has been submerged by Westernization.

6. Indigenous Polynesian culture exhibits a remarkable consistency and uniformity throughout the Polynesian region, its enormous dimensions and dispersal notwithstanding.

7. The Pacific realm is in politico-geographical transition as islands attain independence or redirect their political associations.

■ ■ ■

Politico-geographically, the Pacific realm is delimited into rectangular units marking the maritime boundaries of groups of islands (Fig. 12-1). The Pacific islands were colonized by the French, British, and Americans; an indigenous Polynesian kingdom in the Hawaiian Islands was annexed by the United States and is now the fiftieth State. Still, today, the map is an assemblage of independent and colonial territories. New Caledonia and French Polynesia remain under the control of Paris. Guam and American Samoa, the Line Islands, Wake Island, Midway Islands, and several smaller islands are under U.S. administration; in addition, the United States has special relationships with other territories, former dependencies now nominally independent. The British, through New Zealand, have responsibility for the Pitcairn group of islands, and New Zealand administers and supports the Cook, Tokelau, and Niue Islands. Easter Island, the storied speck of land in the southeastern Pacific, is part of Chile.

Other island groups have become independent states. The largest are Fiji, once a British dependency, the Solomon Islands (also formerly British), and Vanuatu (ruled jointly by France and Britain). Also on the current map, however, are such microstates as Tuvalu, Kiribati, Nauru, and Palau. Foreign aid is crucial to the survival of most of these countries. Tuvalu, for example, has a total area of about 10 square miles, a population of some 10,000, and a per capita GNP of about $600, derived from fishing, copra sales, and some tourism. But what really keeps Tuvalu going is an international trust fund set up by Australia, New Zealand, the United Kingdom, Japan, and South Korea. Annual grants from that fund, as well as money sent back to families by workers who have left for New Zealand and elsewhere, allow Tuvalu to survive.

MARITIME CLAIMS

What the Pacific island states lack in territory, they gain in marine domain (Fig. 12-2). When the UNCLOS convention codified the concept of an Exclusive Economic Zone (EEZ) for all coastal states (see Focus box p. 529), such

Marine Geography

When we look at a globe or an atlas map, land and sea meet along a well-defined border. The map may not reveal much detail; the scale may not allow the representation of all the bends in the coastline, or all the small islands offshore. If the map shows terrestrial relief, it also may show generalized seafloor topography, as does the map on p. 526.

Nation-states, however, do not end where the atlas map suggests they do. The territories of coastal states extend seaward on, in, and beneath the oceans. How far state jurisdiction extends, by what methods seaward boundaries are defined, delimited, and sometimes demarcated, and what rights and responsibilities states have in the remaining "high seas" are issues of maritime law—and of **marine geography**.

The twentieth century has witnessed what political geographers call the "scramble for the oceans." Actually, coastal (*littoral*) countries have laid claim to adjoining waters for centuries for purposes of fishing rights and self-defense. Those claims used to be relatively modest, as little as three nautical miles from shore (1 nautical mile = 1.15 statute miles). The strip of water so claimed became known as the **territorial sea**. All the laws and rights of states prevailing on their territory also applied over those territorial waters.

In the early years of the twentieth century, states' offshore claims began to widen, from 3 miles to 6, and then to 12. In the 1930s, the League of Nations saw a growing problem and called a conference to seek international agreement on a standard width and on a method to define it. Geographers gave testimony regarding the changing nature of coastlines, the way shorelines move with the ebb and flow of tides, the problems presented by fjorded coasts, and the difficulties caused by islands.

The conference could not reach agreement on the width of the territorial sea, but it did achieve one important aim: a universal method for delimiting it (whatever it would be) was approved. The so-called *envelope method*, using a drawing compass to draw an infinite number of arcs from the shoreline, has been in use ever since. The outermost intersections of all the arcs produce the seaward margin of the territorial sea.

After the Second World War, the scramble for the oceans gained momentum. A major stimulus came from the United States when, in late 1945, President Harry S. Truman issued a proclamation that claimed U.S. jurisdiction and control over the resources "in and on" the entire **continental shelf**, the shallow, sloping seafloor down to about 100 fathoms (600 feet). In some areas, the continental shelf off the eastern United States extends more than 200 miles seaward, and Washington did not want foreign countries drilling for oil just beyond the 3-mile territorial sea.

This unilateral claim set in motion a rush of other claims. In 1952, a group of South American countries, some with very little continental shelf to claim, issued the Declaration of Santiago, claiming exclusive fishing rights to a distance of 200 nautical miles (230 statute miles) off their coasts. Meanwhile, as part of the Cold War contest, the Soviet Union urged its allies to claim a 12-mile territorial sea.

Now the United Nations intervened, and a series of so-called UNCLOS (United Nations Conference on the Law of the Sea) meetings began. These addressed a multitude of issues. If a state wanted to close off a bay, what would be the maximum *baseline* length permitted to do so? When the joint land boundary of two states reaches the sea, what rules govern its seaward extension? If a state possesses an island outside its territorial sea, will that island have its own territorial waters?

Out of these UNCLOS conferences came a lengthy convention, now ratified by virtually all participants. Among its key provisions were the authorization of a 12-mile territorial sea for all countries, and the recognition of a 200-mile (230-statute-mile) **Exclusive Economic Zone (EEZ)** over which the coastal state would have exclusive rights. Resources in and under this EEZ (fish, oil, metals) belong to the coastal state. Since many states would not have the technological know-how to exploit these EEZ resources, it had the right to lease or sell these as it saw fit.

The implications of this convention in the Pacific realm are far-reaching. A ministate consisting of one small island acquired a vast EEZ (at least 166,000 square nautical miles). European colonial powers, still holding some minor Pacific possessions, saw their domains greatly expanded. Small, low-income countries could now bargain with large fishing nations over fishing rights in their EEZs. In the seas on the margins of the Pacific, such as the South China Sea, disputes broke out among coastal states over hitherto unclaimed groups of islands that partly extended into EEZs (such as the Spratly Islands, which lie partly in Philippine waters).

Both the 12-mile territorial sea and the 200-mile EEZ created **maritime boundary** problems. Many countries all over the world are separated by waters less than 24 miles wide, and in **choke points** such as the English Channel, Turkey's Bosporus, and the Straits of Malacca west of Southeast Asia's Malay Peninsula, it was necessary to draw **median line boundaries**. Neighboring countries could not always agree on these lines, even when the United Nations declared such waters to be "international" waters. And a host of countries lie closer than 400 nautical miles apart, requiring maritime-boundary delimitation to divide their EEZs. This, too, has produced numerous disputes. Practitioners of marine geography are having a busy time.

■ ■ ■

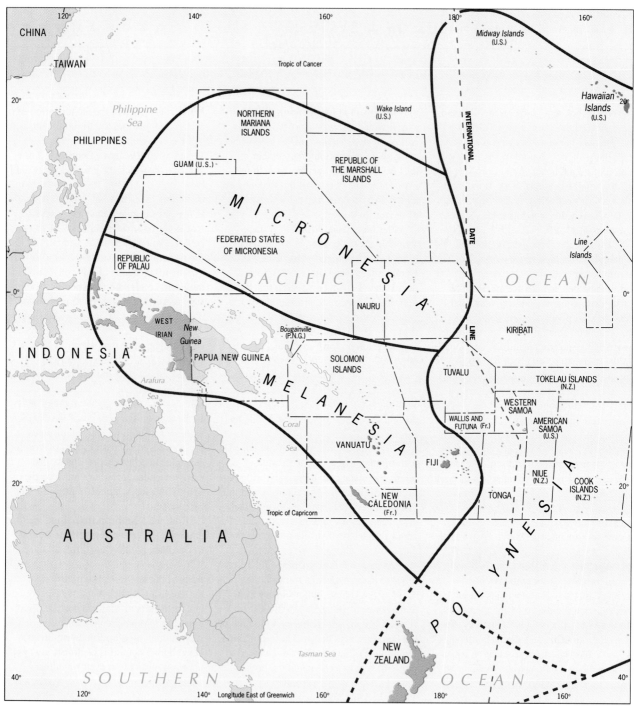

FIGURE 12-1

small Pacific island states as Kiribati and Vanuatu gained hundreds of thousands of square miles of maritime area, where they were able to sell fishing leases and, in some cases, mineral exploration rights to countries possessing the technology to exploit these opportunities. Before the EEZs were authorized, Pacific island states could claim only the 12-mile territorial sea from their coastlines, what-

ever the map of rectangular subdivisions of the Pacific Ocean (Fig. 12-1) showed. Those geometric boundaries were jurisdictional borders, not the limits of sovereign (maritime) domains. The EEZs, however, gave the Pacific ministates direct control over vast reaches of the ocean and the *subsoil*, as the ocean floor and what lies upon and beneath it are called in UNCLOS legalese.

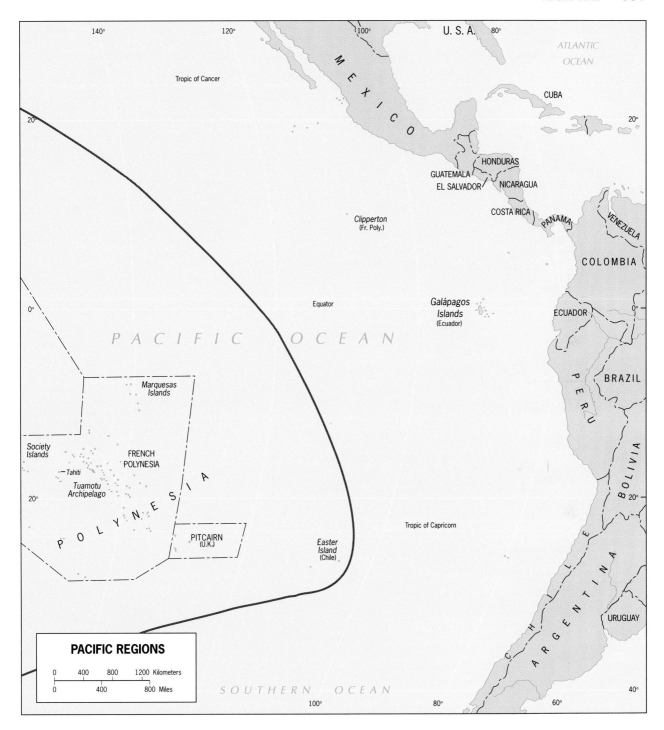

On the map, the Pacific-realm EEZs appear as circular areas around individual islands and groups of islands (Fig. 12-2). This map shows how severely the EEZs reduce the free and open **high seas** that once were a hallmark of the oceans. Whenever two islands or mainland–island territories converge within 400 nautical miles (460 statute miles), no high seas remain between them. Note the effect of this in the southwestern part of the Pacific realm and in neighboring Southeast Asia, where almost no high seas survive after delimitation of the EEZs. Only the middle part of the South China Sea remains open, but since it is surrounded by EEZ claims, this is little more now than a buffer between coastal states. Huge EEZ zones surround the Hawaiian Islands and New Zealand.

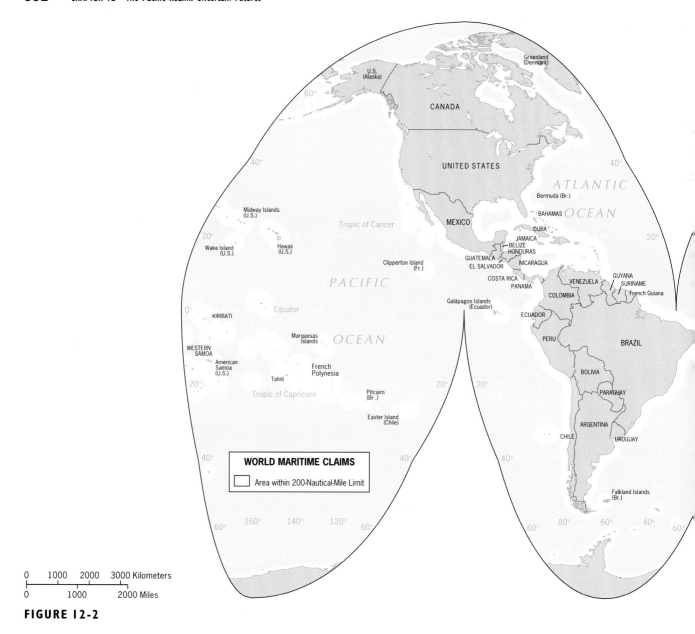

FIGURE 12-2

The delimitation of EEZs has not gone without incident. As the map shows, some EEZs must be established through median-line construction, and this has given rise to disputes among several Pacific and Pacific Rim states. Unpopulated, unclaimed islands suddenly take on great significance: a successful claim can expand a state's maritime jurisdiction by hundreds of thousands of square miles. Disputes involving the Philippines, Indonesia, and Malaysia over island claims make Figure 12-2, at least in part, tentative. And China has laid claim to the entire South China Sea on ''historic'' grounds, a claim that, it says, precedes the UNCLOS convention and its EEZ concept.

If, as some geographers predict, the twenty-first century will be the Pacific Century, the small island states of the Pacific Ocean and their maritime domains are likely to grow in importance. The French, who retained part of their Pacific colonial empire, not only gained enormously from the EEZ award but it also enabled their continued nuclear-weapons testing on a remote atoll in their archipelago. When the Philippine government asked the Americans to close their bases on Luzon Island not far from Manila, the strategic potential of nearby Pacific islands was underscored.

REGIONS OF THE PACIFIC REALM

In this realm as a whole, tourism may well be the largest overall revenue earner. Sail across the Pacific Ocean, and one spectacular vista follows another. Dormant and extinct volcanoes, sculpted by erosion into basalt spires draped by luxuriant tropical vegetation and encircled by reefs and lagoons, tower over azure waters. Low atolls with nearly snow-white beaches, crowned by stands of palm trees, seem to float in the water. Pacific islanders, where they have not been overtaken by foreign influences, appear to take life with enviable ease.

More serious investigations reveal that such Pacific sameness is more apparent than real. Even the Pacific realm, with its long sailing traditions, its still-diffusing populations, and its historic migrations, has a persistent regional framework. In Figure 12-1, three Pacific regions are outlined:

- **Melanesia**

  West Irian (Indonesia), Papua New Guinea, Solomon Islands, Vanuatu, New Caledonia, Fiji

From the Field Notes

"Arriving in the capital of New Caledonia, Nouméa, in February 1996, was an experience reminiscent of French Africa 40 years ago. The French tricolor was much in evidence, as were uniformed French soldiers. European French residents occupied hillside villas overlooking palm-lined beaches, giving the place a Mediterranean cultural landscape. And New Caledonia, like Africa, is a source of valuable minerals. It is the world's second-largest nickel producer, and from this vantage point you could see the huge treatment plants, complete with concentrate ready to be shipped (left, under conveyor). What you cannot see here is how southern New Caledonia has been ravaged by the mining operations, which have denuded whole mountainsides. Working in the mines and in this facilitiy are the local Kanaks, Melanesians who make up about 58 percent of the population of nearly 200,000. In 1986, violent clashes between Kanaks and French took many lives, and now a referendum on independence is scheduled for 1998. Already, northern parts of the island are out of French control. Will those French tricolors be lowered without war?"

- **Micronesia**

    Palau, Federated States of Micronesia, Northern Mariana Islands, Republic of the Marshall Islands, Nauru, western Kiribati, Guam

- **Polynesia**

    Hawaiian Islands, Western Samoa, American Samoa, Tuvalu, Tonga, eastern Kiribati, Cook and other New Zealand-administered islands, French Polynesia

Ethnic, linguistic, and physiographic criteria are among the bases for this regionalization of the Pacific realm, but we should not lose sight of the dimensions. Not only is the land area small, but also the population of this entire realm in 1997 was approximately 10 million, about the same as one very large city. Fewer people live in the Pacific realm than in another vast area of far-flung settlements, the oases of North Africa's Sahara.

Major Cities of the Realm

Cities	Population* (in millions)
Honolulu, Hawaii (U.S.)	0.9
Nouméa, New Caledonia	0.1
Port Moresby, Papua New Guinea	0.2
Suva, Fiji	0.2

*Based on 1997 estimates.

❖ MELANESIA

New Guinea lies at the western end of a Pacific region that extends eastward to Fiji and includes the Solomon Islands, Vanuatu, and New Caledonia (Fig. 12-1). These islands are inhabited by Melanesian peoples who have very dark skins and dark hair (*melas* means black); the region as a whole is called Melanesia. Some cultural geographers include the Papuan peoples of New Guinea in the Melanesian race, but others suggest that the Papuans are more closely related to the Aboriginal (indigenous) Australians. In any case, Melanesia is by far the most populous Pacific region (roughly 7 million in 1997). New Guinea alone has a population of almost 6 million (although statistics are unreliable), but this island is divided into two halves by a superimposed geometric boundary (Fig. 10-6) that separates West Irian, now an Indonesian province, from independent Papua New Guinea (P.N.G.)

With 4.3 million inhabitants today, P.N.G. became a sovereign state in 1975 after nearly a century of British and Australian administration. It is one of the world's poorest and least developed countries, with much of the mountainous interior—where the Papuan population is clustered in tiny villages—hardly touched by the changes that transformed neighboring Australia. The largest town and capital, Port Moresby, has more than 200,000 residents; only about one-sixth of the people of P.N.G. live in urban areas, but the current rate of urbanization is high. Although English is used by the educated minority, about 57 percent of the population remains illiterate, and over 700 languages are spoken by the Papuan and Melanesian communities. The Melanesians are concentrated in the northern and eastern coastal areas of the country, and here, as in the other islands of this region, they grow root crops and bananas for subsistence. However, recent discoveries of major mineral deposits (copper, gold, and oil) point to the country's bright development potential; there is also enormous scope for increasing the output of such profitable export crops as palm oil, coffee, and cocoa.

ANTARCTIC REALM?

▪ ▪ ▪

South of the Pacific geographic realm lies a vast expanse of water, the Southern Ocean, encircling the ice-covered continent of Antarctica. Persistent westerly winds drive frigid water, pack ice, and icebergs on an endless path around the polar landmass. The Southern Ocean is defined by this West Wind Drift, as it is often called. Where it meets the southernmost waters of the Indian, Atlantic, and Pacific oceans, major ecological changes—environmental, biological, chemical—occur along the Antarctic Convergence.

Do Antarctica and the Southern Ocean (which together cover not much less than one-sixth of the Earth's surface!) constitute a geographic realm? In physiographic terms yes, but not on the basis of the criteria we have used in this book. No functional regions have developed here as yet, no cities, no transport networks. This most remote of all the Earth's frontiers remains just that—a frontier. As such, it constitutes a physiographic region with several subregions, but it is not a full-fledged geographic realm.

Like virtually all frontiers, Antarctica has always attracted pioneers and explorers. Whale and seal hunters destroyed huge populations of Southern Ocean fauna during the eighteenth and nineteenth centuries, and explorers planted the flags of their countries on Antarctic shores. Between 1895 and 1914, the quest for the South Pole became an international obsession; Roald Amundsen, the Norwegian, reached it first in 1911. All this led to national claims in Antarctica during the interwar period (1918–1939). The geographic effect was the partitioning of Antarctica into pie-shaped sectors centered on the South Pole (Fig. 12-3). In the least frigid area of the continent, the Antarctic Peninsula, British, Argentinean, and Chilean claims overlapped—and still do. One sector,

Marie Byrd Land (shown in white on the map), was never formally claimed by any country.

Why should states be interested in territorial claims in so remote and difficult an area? Both land and sea contain raw materials that may some day become crucial: proteins in the waters, and fuels and minerals beneath the ice. Antarctica (5.5 million sq mi/14.2 million sq km) is almost twice as large as Australia, and the Southern Ocean is nearly as large as the North *and* South Atlantic. However distant actual exploitation may be, countries want to keep their stakes here.

But the **claimant states** (those with territorial claims) recognize the need for cooperation. During the late 1950s, they joined in the International Geophysical Year (IGY) that launched major research programs and established a number of permanent research stations throughout the continent. This spirit of cooperation led to the 1961 signing of the Antarctic Treaty, which ensures continued scientific collaboration, prohibits military activities, protects the environment, and holds national claims in abeyance. In 1991, when the treaty was extended under the terms of the Wellington Agreement, concerns were raised that it does not do enough to control future resource exploitation.

In an age of growing national self-interest and increasing raw-material consumption, the possibility exists that Antarctica and its offshore waters may yet become an arena for international rivalry. Until now, its remoteness and its forbidding environments have saved it from that fate. The entire world benefits from this, because evidence is mounting that Antarctica plays a critical role in the global environmental system, so that human modifications may have global (and unpredictable) consequences.

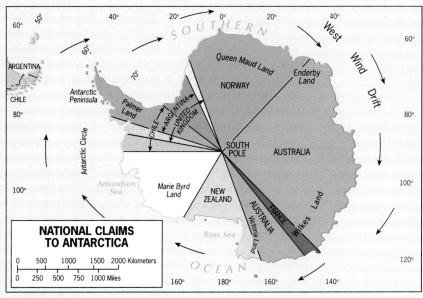

FIGURE 12-3

▪ ▪ ▪

From the Field Notes

"In the town of Lautoka, on Fiji's main island of Viti Levu, we found one answer to the omnipresent question: 'What can tropical islands produce in the way of renewable resources?' In western Viti Levu's fertile, volcanic, well-watered soils, pine trees grow fast. Large plantations now produce a nearly constant supply of soft wood converted into chips at the source and trucked to the port. 'The Japanese buy it as fast as we can produce it,' the man in charge of this portside facility told us. 'They compress this stuff into boards and use them as building materials.' Here is the Pacific Rim at work in the Pacific Ocean itself!"

Papua New Guinea has been beset by centrifugal forces since independence, and relations with neighboring Indonesia also have caused problems. The most serious challenge arose on the island of Bougainville (Fig. 12-1), where a separatist insurgency arose. Irredentism from the neighboring Solomon Islands, where the insurgents found support, caused tensions between the two neighboring Melanesian states. The rebels succeeded in capturing and closing Bougainville's productive copper mine, damaging P.N.G.'s economy. In late 1994, government forces recaptured the mine but were unable to stop the Bougainville Revolutionary Army from regrouping in the forested hill country. In terms of the insurgent state model (see p. 408), the situation in early 1996 remained one of contention, although a multinational peacekeeping force was in place.

Even as the Bougainville copper mine was taken back, disaster struck in the highlands of New Guinea in the form of a devastating explosion at the Porgera gold mine, one of the world's largest and a major source of income for P.N.G. Production ceased, and reconstruction was expected to take many months.

These incidents highlighted the vulnerability of Papua New Guinea in the postcolonial era. Indeed, the Bougainville issue is a direct legacy of colonialism. The map sug-

gests that Bougainville is the westernmost island in the Solomon archipelago, and at independence there was little enthusiasm in Bougainville for incorporation into Papua New Guinea. But the Australians, who had administered both territories, insisted that Bougainville join P.N.G., undoubtedly because of the revenues the copper mine would bring to Port Moresby's coffers. It was a forced marriage, and the result was an ongoing threat to stability in this area.

It is a measure of the cultural fragmentation of Melanesia that in the Solomon Islands, with a population of about 400,000, as many as 120 languages are in use, some spoken by just a few hundred villagers in isolated locales. Neighboring Vanuatu has less linguistic diversity, but its archipelago, inhabited by fewer than 200,000 people, consists of 80 islands. And New Caledonia, still under French rule (a referendum on independence is scheduled for 1998), also has a population of about 200,000, but Melanesians constitute only just over 40 percent of it. About 37 percent are of French ancestry, many of them descended from the inhabitants of the penal colony France established here. Others were attracted by the service industries associated with the mining of nickel; New Caledonia's nickel reserves are among the world's largest. Today, most of the French population lives in or near the capital city of Nouméa, on the southwestern side of the island. Most of the Melanesians live in villages, farming for subsistence. The two communities have not come to terms over the future, and when world prices for nickel recently declined and unemployment affected many Melanesian workers, relationships deteriorated.

On its eastern margins, Melanesia includes one of the Pacific realm's most interesting countries, Fiji. On two larger and over 100 smaller islands live nearly 800,000 Fijians, of whom 49 percent are Melanesians and 46 percent South Asians, the latter brought to Fiji from India during the period of British colonial occupation to work on the sugar plantations. When Fiji achieved independence in 1970, the native Fijians owned most of the land and held political control, while the Indians were concentrated in the towns (chiefly Suva, the capital) and dominated commercial life. It was a recipe for trouble, and it was not long in coming when, in a later election, the politically active Indians outvoted the Fijians for seats in the parliament. A military coup by the Fijian military was followed by a revision of the constitution, which now awards a majority of seats to ethnic Fijians. The military coup severely damaged the tourist industry, and coupled with the worst drought on record, the Fijian economy suffered at the worst possible time. This had the effect, however, of splintering the Fijian majority, and a coalition government is proving the constitution workable—and acceptable to the majority of Indian residents of the country. Undercurrents of cultural strain continue (most Melanesians are Christians whereas virtually all the Indians are Hindu), but Fiji has escaped the fate of Sri Lanka or Cyprus.

Melanesia, the most populous region in the Pacific realm, also is bedeviled by centrifugal forces of many kinds. No two countries present the same form of multiculturalism, each has its own challenges to confront, and some of these challenges spill over into neighboring (or more distant foreign) lands.

❖ MICRONESIA

North of Melanesia and east of the Philippines lie the islands that constitute the region known as Micronesia (Fig. /001012/1). In this case, the name (*micro* means small) refers to the size of the islands, not the physical appearance of the population. The 2,000-plus islands of Micronesia are not only tiny (many of them no larger than 1 square mile), but they are also much lower-lying, on an average, than those of Melanesia. There are some volcanic islands (**high islands**, as the people call them), but they are outnumbered by islands composed of coral, the **low islands** that barely lie above sea level. Guam, with 210 square miles (550 sq km), is Micronesia's largest island, and no island elevation anywhere in Micronesia reaches 3,300 feet (1,000 m).

The *high-island–low-island* dichotomy is useful not only in Micronesia, but also throughout the realm. Both the physiographies of these islands and the economies they support differ in crucial ways. High islands wrest substantial moisture from the ocean air; they tend to be well watered and have good volcanic soils. As a result, agricultural products show some diversity and life is reasonably secure. Populations tend to be larger on these high islands than on the low islands, where drought is the rule and fishing and the coconut palm are the mainstays of life. Small communities cluster on the low islands, and over time, many of these have died out. The major migrations, which sent fleets to populate islands from Hawaii to New Zealand, tended to originate in the high islands.

Until the mid-1980s, Micronesia was largely a United States Trust Territory (the last of the post–World War II trusteeships supervised by the United Nations), but that status has now changed. As Figure 12-1 shows, today Micronesia is divided into countries bearing the names of independent states. The Marshall Islands, where the United States tested nuclear weapons (giving prominence to the name Bikini), now is a republic in ''free association'' with the United States, having the same status as the Federated States of Micronesia and (since 1994) Palau. The Northern Mariana Islands are a commonwealth ''in political union'' with the United States. In effect, the United States provides billions of dollars in assistance to these countries, in return for which they commit themselves to avoid foreign policy actions that are contrary to U.S. interests. There are other conditions: Palau, for example, granted the United States

rights to existing military bases for 50 years following independence.

Also part of Micronesia are the U.S. territory of Guam, where independence is not in sight and where U.S. military installations and tourism provide the bulk of income, and the remarkable Republic of Nauru. With a population of just 10,000 and 8 square miles of land, this microstate got rich by selling its phosphate deposits to Australia and New Zealand, where they are used as fertilizer. Per capita incomes rose to $U.S. 10,000, making Nauru one of the Pacific's high-income societies. But the phosphate deposits are running out, and soon after the turn of the century Nauru will face a moment of truth.

From the Field Notes
"From a distance at sea, approaching Polynesia's Society Islands, Bora Bora presented one of the most magnificent scenes in the entire Pacific—an eroding, extinct volcano sculpted into sheer cliffs, draped in dense forest, lined by palm-fringed beaches. But a closer look produced a very different picture. Litter marred the beaches not owned by hotel chains, and worrisome signs of poverty afflicted the small settlements away from the beaten path."

Guam and Nauru are exceptions in this region of tiny islands. Most people here subsist on farming or fishing, and virtually all the countries need infusions of foreign aid to survive. The natural economic complementarity between the farming high-island cultures and the fishing low-island communities all too often is negated by distance, spatial as well as cultural. Life here may seem idyllic to the casual visitor, but for the Micronesians it often is a daily challenge.

❖ POLYNESIA

To the east of Micronesia and Melanesia lies the heart of the Pacific, enclosed by a great triangle stretching from the Hawaiian Islands to Chile's Easter Island to New Zealand. This is Polynesia (Fig. 12-1), a region of numerous islands (*poly* means many), ranging from volcanic mountains rising above the Pacific's waters (Muana Kea on Hawaii reaches nearly 13,800 feet [over 4,200 m]), clothed by luxuriant tropical forests and drenched by well over 100 inches of rainfall each year, to low coral atolls where a few palm trees form the only vegetation and where drought is a persistent problem. The Polynesians have somewhat lighter-colored skin and wavier hair than do the other peoples of the Pacific realm; they are often also described as having an excellent physique. Anthropologists differentiate between these original Polynesians and a second group, the Neo-Hawaiians, who are a blend of Polynesian, European, and Asian ancestries. In the U.S. State of Hawaii—actually an archipelago of more than 130 islands—Polynesian culture has not only been Europeanized but also Orientalized.

Its vastness and the diversity of its natural environments notwithstanding, Polynesia clearly constitutes a geographic region within the Pacific realm (population: 2 million). Polynesian culture, though spatially fragmented, exhibits a remarkable consistency and uniformity from one island to the next, from one end of this widely dispersed region to the other. This consistency is particularly expressed in vocabularies, technologies, housing, and art forms. The Polynesians are uniquely adapted to their **maritime environment**, and long before European sailing ships began to arrive in their waters, Polynesian seafarers had learned to navigate their wide expanses of ocean in huge double canoes as long as 150 feet (45 m). They traveled hundreds of miles to favorite fishing zones and engaged in interisland barter trade, using maps constructed from bamboo sticks and cowrie shells and navigating by the stars. However, modern descriptions of a Pacific Polynesian paradise of emerald seas, lush landscapes, and gentle people distort harsh realities. Polynesian society was forced to accommodate to much loss of life at sea when storms claimed their boats; families were ripped apart by accident as well as migration; hunger and starvation afflicted the inhabitants of smaller islands; and the island communities were often embroiled in violent conflicts and cruel retributions.

The political geography of Polynesia is complex. In 1959, the Hawaiian Islands became the fiftieth State to join the United States. The State's population is now 1.3 million, with over 80 percent living on the island of Oahu. There, the superimposition of cultures is illustrated by the panorama of Honolulu's skyscrapers against the famous extinct volcano at nearby Diamond Head. The Kingdom of Tonga became an independent country in 1970 after seven decades as a British protectorate; the British-administered

12.2

From the Field Notes
"In King Kamehameha's time, villages flanked this Oahu waterfront and canoes lay on the beach. Today a curtain of concrete lines the ocean here at Honolulu, and the people here are transients, tourists from Japan, North America, and other parts of the world. The Polynesians, whose last monarch reigned until the 1890s, have seen their Hawaiian archipelago transformed."

Ellice Islands were renamed Tuvalu, and along with the Gilbert Islands to the north (now renamed Kiribati), they received independence from Britain in 1978. Other islands continued under French control (including the Marquesas Islands and Tahiti), under New Zealand's administration (Rarotonga), and under British, U.S., and Chilean flags.

12.1

In the process of politico-geographical fragmentation, Polynesian culture has been dealt some severe blows. Land developers, hotel builders, and tourist dollars have set Tahiti on a course along which Hawaii has already traveled far. The Americanization of eastern Samoa has created a new society quite different from the old. Polynesia has lost much of its ancient cultural consistency; today, the region is a patchwork of new and old—the new often bleak and barren with the old under intensifying pressure.

The countries and cultures of the Pacific realm lie in an ocean on whose margins a great drama of economic and political transformation will play itself out during the twenty-first century. Already, the realm's own former margins—in Hawaii and in New Zealand—have been so recast by foreign intervention that little remains of the kingdoms and cultures that once prevailed. Now the Pacific world faces changes far greater even than those brought here by European colonizers. Once upon a time the shores and waters of the Mediterranean Sea formed an arena of regional transformation that changed the world. Then it was the Atlantic, avenue of the Industrial Revolution and stage of fateful war. Now it seems to be the turn of the Pacific as the world's largest country and next superpower (China) faces the richest and most powerful (the United States). Giants will jostle for advantage in the Pacific; how will the weak societies of the Pacific realm fare?

■ PRONUNCIATION GUIDE ■

Amundsen, Roald (AH-moon-sun, ROH-ahl)
Archipelago (ark-uh-PELL-uh-goh)
Atoll (AY-toal)
Bora Bora (boar-ruh-BOAR-ruh)
Bosporus (BAHSS-puh-russ)
Bougainville (BOO-gun-vil)
Copra (KOH-pruh)
Ellice (ELL-uss)
Fiji (FEE-jee)
Fijian (fuh-JEE-un)
Guam (GWAHM)
Hawaii (huh-WAH-ee)
Honolulu (honn-uh-LOO-loo)
Irian (IH-ree-ahn)
Irredentism (irruh-DEN-tism)
Kamehameha (kah-MAY-hah-may-hah)
Kanaks (KAH-nahkss)
Kiribati (KIH-ruh-bahss)
Lautoka (lau-TOH-kuh)

Littoral (LIT-oh-rull)
Luzon (loo-ZAHN)
Malay (muh-LAY)
Maori (MAU-ree/MAH-aw-ree)
Mariana (marry-ANNA)
Marquesas (mahr-KAY-suzz)
Mauna Kea (mau-nuh-KAY-uh)
Melanesia (mella-NEE-zhuh)
Micronesia (mye-kroh-NEE-zhuh)
Nauru (nah-OO-roo)
Nautical (NAW-dih-kull)
New Caledonia (noo-kalla-DOAN-yuh)
New Guinea (noo-GHINNY)
Niue (nee-OOH-ay)
Nouméa (noo-MAY-uh)
Oahu (uh-WAH-hoo)
Palau (puh-LAU)
Papeete (pahp-ee-ATE-tee)
Papua New Guinea
    (pahp-OO-uh noo-GHINNY)

Papuan (pahp-OO-un)
Pitcairn (PITT-kairn)
Polynesia (polla-NEE-zhuh)
Porgera (POAR-jeh-ruh)
Port Moresby (port MORZ-bee)
Rarotonga (rarra-TAHNG-guh)
Samoa (suh-MOH-uh)
Spratly (SPRAT-lee)
Sri Lanka (sree-LAHNG-kuh)
Suva (SOO-vuh)
Tahiti (tuh-HEET-tee)
Tokelau (TOH-kuh-lau)
Tonga (TAHNG-guh)
Tuvalu (too-VAHL-oo)
UNCLOS (UNN-klohss)
Vanuatu (vahn-uh-WAH-too)
Viti Levu (vee-tee-LEH-voo)
Waikiki (wye-kuh-KEE)

Area and Demographic Data for the World's States

(Smallest Microstates Omitted)

	Area (1000 sq mi)	Population (Millions)			1997 Population Density (per sq mi)	Annual Rate of Natural Incr (Percent)	Doubling Time (years)	Life Expectancy at Birth (years)	Percent Urban Population	1993 Gross Nat'l Product Per Capita ($US)
		1997	2004	2010						
World	51,510.8	5,878.5	6,425.6	7,023.6	114	1.5	45	66	43	4,500
Europe	2,261.0	581.9	587.3	592.8	257	0.2	332	73	72	11,870
Albania	11.1	3.6	3.8	4.1	322	1.8	39	72	37	340
Austria	32.4	8.1	8.2	8.3	250	0.1	533	77	54	23,120
Belarus	80.2	10.3	10.6	10.9	128	−0.2		69	68	2,840
Belgium	11.8	10.2	10.3	10.4	862	0.1	578	77	97	21,210
Bosnia	19.7	3.5	3.9	4.4	178	0.7	95	72	34	
Bulgaria	42.8	8.4	8.2	7.9	196	−0.3		71	67	1,160
Croatia	21.8	4.5	4.4	4.4	206	−0.1		70	54	
Czech Republic	30.6	10.4	10.4	10.5	339	0.0		73	75	2,730
Denmark	16.6	5.2	5.3	5.3	314	0.1	770	75	85	26,510
Estonia	17.4	1.5	1.4	1.4	84	−0.5		70	71	3,040
Finland	130.1	5.1	5.2	5.2	40	0.3	227	76	64	18,970
France	211.2	58.5	60.1	61.7	277	0.3	217	78	74	22,360
Germany	137.8	81.5	81.3	81.2	591	−0.1		76	85	23,560
Greece	50.9	10.5	10.3	10.2	206	0.0		77	63	7,390
Hungary	35.9	10.2	10.0	9.9	284	−0.3		69	63	3,330
Iceland	39.8	0.3	0.3	0.3	7	1.1	64	79	91	23,620
Ireland	27.1	3.6	3.6	3.5	134	0.5	139	75	57	12,580
Italy	116.3	57.7	57.1	56.5	496	0.0		77	68	19,620
Latvia	24.9	2.5	2.5	2.4	100	−0.5		68	69	2,030
Liechtenstein	0.1	0.1	0.1	0.1	523	0.6	108			
Lithuania	25.2	3.7	3.7	3.8	147	0.0		71	68	1,310
Luxembourg	1.0	0.4	0.4	0.4	411	0.4	193	76	86	35,850
Macedonia	9.9	2.2	2.2	2.3	218	0.8	85	72	58	780
Malta	0.3	0.4	0.4	0.4	1,257	0.7	102	75	85	
Moldova	13.0	4.4	4.6	4.8	337	0.4	193	68	47	1,180
Netherlands	15.9	15.6	16.2	16.9	980	0.4	182	77	89	20,710
Norway	125.2	4.4	4.5	4.7	35	0.3	224	77	73	26,340
Poland	120.7	38.8	39.5	40.2	321	0.2	301	72	62	2,270
Portugal	34.3	9.9	9.9	9.9	290	0.1	866	75	34	7,890
Romania	91.7	22.7	22.4	22.2	247	−0.1		70	55	1,120
Serbia	26.9	10.9	11.0	11.1	406	0.3	204	72	47	

	Area (1000 sq mi)	Population (Millions)			1997 Population Density (per sq mi)	Annual Rate of Natural Incr (Percent)	Doubling Time (years)	Life Expectancy at Birth (years)	Percent Urban Population	1993 Gross Nat'l Product Per Capita ($US)
		1997	2004	2010						
Slovakia	18.8	5.4	5.6	5.7	288	0.4	178	71	57	1,900
Slovenia	7.8	2.0	2.0	2.0	254	0.1	1,386	73	50	6,310
Spain	194.9	39.2	39.1	39.0	201	0.1	578	77	64	13,650
Sweden	173.7	8.9	9.0	9.2	51	0.1	990	78	83	24,830
Switzerland	15.9	7.1	7.3	7.6	446	0.3	224	78	68	36,410
Ukraine	233.1	51.6	52.3	53.0	222	−0.4		69	68	1,910
United Kingdom	94.2	59.0	60.0	61.0	626	0.2	385	76	92	17,970
Russia	6,592.8	145.7	147.6	149.5	22	−0.6		65	73	2,350
Armenia	11.5	3.8	4.0	4.2	331	0.8	83	71	68	660
Azerbaijan	33.4	7.5	8.2	9.0	225	1.6	43	71	54	730
Georgia	26.9	5.4	5.6	5.7	202	0.2	462	73	56	560
North America	7,509.9	296.8	314.9	334.2	40	0.7	105	76	75	24,340
Canada	3,831.0	30.0	31.7	33.6	8	0.7	102	78	77	20,670
United States	3,678.9	266.7	283.0	300.4	72	0.7	105	76	75	24,750
Middle America	1,055.0	168.9	186.7	206.1	125	2.3	30	71	65	3,090
Antigua and Barbuda	0.2	0.1	0.1	0.1	333	1.2	58	73	31	6,390
Bahamas	5.4	0.3	0.3	0.3	53	1.5	47	73	84	11,500
Barbados	0.2	0.3	0.3	0.3	1,329	0.7	98	76	38	6,240
Belize	8.9	0.2	0.3	0.3	26	3.3	21	68	48	2,440
Costa Rica	19.6	3.5	3.9	4.4	178	2.2	32	76	49	2,160
Cuba	44.2	11.3	11.8	12.3	256	0.7	102	75	74	
Dominica	0.3	0.1	0.1	0.1	243	1.3	55	77		2,680
Dominican Republic	18.8	8.2	8.9	9.7	434	2.1	32	70	61	1,080
El Salvador	8.7	6.2	6.9	7.6	710	2.6	27	68	46	1,320
Grenada	0.1	0.1	0.1	0.1	984	2.4	29	71		2,410
Guadeloupe	0.7	0.4	0.5	0.5	637	1.2	56	75	48	
Guatemala	42.0	11.3	13.4	15.8	269	3.1	22	65	38	1,110
Haiti	10.7	7.5	8.6	9.8	702	2.3	30	57	31	
Honduras	43.3	5.8	6.6	7.6	133	2.8	25	68	46	580
Jamaica	4.2	2.5	2.7	2.8	606	2.0	35	74	53	1,390
Martinique	0.4	0.4	0.4	0.4	969	1.1	62	76	81	
Mexico	761.6	97.8	107.3	117.7	128	2.2	34	72	71	3,750
Netherlands Antilles	0.4	0.2	0.2	0.2	510	1.3	55	76	92	
Nicaragua	50.2	4.7	5.6	6.7	93	2.7	26	65	62	360
Panama	29.8	2.7	3.0	3.3	92	2.1	33	72	54	2,580
Puerto Rico	3.4	3.7	3.9	4.1	1,103	1.0	67	74	73	7,020
Saint Lucia	0.2	0.1	0.2	0.2	748	2.0	34	72	48	3,040
St. Vincent and the Grenadines	0.2	0.1	0.1	0.1	601	1.8	38	73	25	2,130
Trinidad and Tobago	2.0	1.3	1.4	1.6	667	1.1	64	71	65	3,730
Virgin Islands	0.1	0.1	0.1	0.1	1,009	1.9	37	75		
South America	6,875.0	330.1	361.1	394.9	48	1.8	38	68	73	3,020
Argentina	1,068.3	35.5	38.0	40.8	33	1.3	55	71	87	7,290
Bolivia	424.2	7.8	8.9	10.2	18	2.6	27	60	58	770
Brazil	3,286.5	163.2	178.1	194.4	50	1.7	41	66	77	3,020

	Area (1000 sq mi)	Population (Millions)			1997 Population Density (per sq mi)	Annual Rate of Natural Incr (Percent)	Doubling Time (years)	Life Expectancy at Birth (years)	Percent Urban Population	1993 Gross Nat'l Product Per Capita ($US)
		1997	2004	2010						
Chile	292.3	14.7	15.9	17.3	50	1.7	41	72	85	3,070
Colombia	439.7	39.1	42.5	46.1	89	1.8	39	69	68	1,400
Ecuador	109.5	12.0	13.4	14.9	109	2.2	31	69	58	1,170
French Guiana	35.1	0.2	0.2	0.2	4	2.6	26	74		
Guyana	83.0	0.9	0.9	1.0	10	1.8	39	65	33	350
Paraguay	157.1	5.2	6.0	7.0	33	2.8	25	70	51	1,500
Peru	496.2	25.0	27.5	30.3	50	2.1	33	66	70	1,490
Suriname	63.0	0.4	0.5	0.5	7	2.0	36	70	49	1,210
Uruguay	68.0	3.2	3.3	3.5	47	0.7	102	73	90	3,910
Venezuela	352.1	23.0	25.7	28.7	65	2.6	27	72	84	2,840
North Africa/ Southwest Asia	7,783.6	483.9	560.0	648.0	62	2.5	28	63	46	
Afghanistan	250.0	19.5	24.6	31.1	78	2.8	24	43	18	
Algeria	919.6	29.8	33.6	38.0	32	2.4	29	67	50	1,650
Bahrain	0.3	0.6	0.7	0.8	2,111	2.5	28	74	88	7,870
Cyprus	3.6	0.8	0.8	0.8	210	0.9	76	77	68	10,380
Djibouti	8.9	0.6	0.7	0.8	68	2.2	32	48	77	780
Egypt	386.7	64.7	72.3	80.7	167	2.3	31	64	44	660
Eritrea	45.4	3.7	4.4	5.2	82	2.6	27	48	17	
Iran	636.3	64.8	73.6	83.7	102	2.9	24	67	57	2,230
Iraq	167.9	22.2	27.7	34.5	132	3.7	19	66	70	
Israel	8.0	5.7	6.3	6.9	712	1.5	47	77	90	13,760
Gaza	0.1	1.0	1.3	1.8	9,881	4.6	15	69	94	
West Bank	2.3	1.6	2.1	2.7	690	3.4	20	68		
Jordan	35.5	4.4	5.2	6.2	123	3.3	21	72	68	1,190
Kazakstan	1,049.2	17.2	17.8	18.4	16	0.9	74	69	57	1,540
Kuwait	6.9	1.6	2.0	2.5	228	2.2	31	75		23,350
Kyrgyzstan	76.6	4.6	5.1	5.6	60	1.8	38	68	36	830
Lebanon	4.0	3.8	4.4	5.0	962	2.0	34	75	86	
Libya	679.4	5.6	7.1	8.9	8	3.4	21	63	85	
Morocco	275.1	30.7	34.5	38.7	112	2.2	32	69	46	1,030
Oman	82.0	2.4	3.0	3.7	29	4.9	14	71	12	5,600
Qatar	4.3	0.5	0.6	0.6	126	1.8	39	73	91	15,140
Saudi Arabia	830.0	19.7	24.3	30.0	24	3.2	22	70	79	7,780
Somalia	246.2	9.8	11.9	14.5	40	3.2	22	47	24	
Sudan	976.5	29.8	35.2	41.5	30	3.0	23	55	27	
Syria	71.5	15.7	19.2	23.6	220	3.5	20	66	51	
Tajikistan	55.3	6.1	7.5	9.2	111	2.4	29	70	31	470
Tunisia	63.2	9.2	10.2	11.2	146	1.9	36	68	60	1,780
Turkey	301.4	63.3	70.8	79.2	210	1.6	44	67	51	2,120
Turkmenistan	188.5	4.7	5.3	5.9	25	2.5	28	66	45	1,380
United Arab Emirates	32.3	2.0	2.2	2.5	61	1.9	36	72	82	22,470
Uzbekistan	172.7	23.8	27.6	31.9	138	2.5	28	69	41	960
Yemen	203.9	14.2	17.6	21.9	69	3.6	19	52	25	
Subsaharan Africa	8,112.6	575.7	691.0	829.4	71	3.0	23	52	27	541
Angola	481.4	12.1	14.6	17.6	25	2.7	26	46	37	
Benin	43.5	5.7	6.9	8.3	132	3.1	22	48	30	420
Botswana	231.8	1.6	1.9	2.2	7	2.3	30	64	27	2,590

	Area (1000 sq mi)	Population (Millions)			1997 Population Density (per sq mi)	Annual Rate of Natural Incr (Percent)	Doubling Time (years)	Life Expectancy at Birth (years)	Percent Urban Population	1993 Gross Nat'l Product Per Capita ($US)
		1997	2004	2010						
Burkina Faso	105.9	11.0	12.6	14.5	104	2.8	24	45	15	300
Burundi	10.8	6.8	8.0	9.5	628	3.0	23	50	6	180
Cameroon	183.6	14.3	17.4	21.2	78	2.9	24	58	41	770
Cape Verde Islands	1.6	0.4	0.5	0.6	259	2.8	25	65	44	870
Central African Republic	240.5	3.3	3.6	3.9	14	2.0	34	41	39	390
Chad	495.8	6.7	7.9	9.3	13	2.6	27	48	22	200
Comoros Islands	0.7	0.6	0.7	0.9	840	3.6	20	58	29	520
Congo	132.1	2.6	2.9	3.2	20	2.3	31	46	58	920
Equatorial Guinea	10.8	0.4	0.5	0.6	41	2.6	27	53	37	360
Ethiopia	426.4	59.4	73.1	90.0	139	3.1	23	50	15	100
Gabon	103.4	1.4	1.6	1.9	13	2.2	32	54	73	4,050
Gambia	4.4	1.1	1.3	1.5	261	2.7	26	45	26	360
Ghana	92.1	18.5	22.2	26.6	201	3.0	23	56	36	430
Guinea	94.9	6.9	8.0	9.3	72	2.4	29	44	29	510
Guinea-Bissau	14.0	1.1	1.3	1.5	80	2.1	32	44	22	220
Ivory Coast	124.5	15.2	18.7	23.1	122	3.5	20	51	39	630
Kenya	225.0	30.1	36.2	43.6	134	3.3	21	56	27	270
Lesotho	11.7	2.1	2.5	3.0	182	1.9	36	61	22	660
Liberia	43.0	3.2	4.0	4.8	75	3.3	21	55	44	
Madagascar	226.7	15.7	19.1	23.3	69	3.2	22	57	22	240
Malawi	45.8	10.3	12.3	14.7	224	2.7	25	45	17	220
Mali	478.8	10.0	12.2	15.0	21	3.2	22	47	22	300
Mauritania	398.0	2.4	2.8	3.3	6	2.5	27	52	39	510
Mauritius	0.8	1.1	1.2	1.3	1,437	1.5	47	69	44	2,980
Moçambique	302.3	18.4	22.2	26.9	61	2.7	26	46	33	80
Namibia	318.3	1.6	1.9	2.2	5	2.7	26	59	32	1,660
Niger	489.2	9.8	12.0	14.8	20	3.4	21	47	15	270
Nigeria	356.7	107.6	132.0	162.0	302	3.1	22	56	16	310
Réunion	1.0	0.7	0.7	0.8	676	1.8	40	73	73	
Rwanda	10.2	8.2	9.2	10.4	799	2.3	30	46	5	200
São Tomé and Principe	0.4	0.1	0.2	0.2	368	2.6	27	64	46	330
Senegal	75.8	8.8	10.4	12.2	116	2.7	26	49	39	730
Sierra Leone	27.7	4.8	5.5	6.4	172	2.7	26	46	35	140
South Africa	471.4	44.7	50.7	57.5	95	2.3	30	66	63	2,900
Swaziland	6.7	1.0	1.3	1.6	154	3.2	22	57	30	1,050
Tanzania	364.9	30.2	35.9	42.8	83	3.0	23	49	21	100
Togo	21.9	4.7	5.9	7.4	216	3.6	19	58	30	330
Uganda	91.1	22.7	27.1	32.3	249	3.3	21	45	11	190
Zaïre	905.6	46.8	56.9	69.1	52	3.2	22	48	29	
Zambia	290.6	9.6	11.2	13.0	33	3.1	23	48	42	370
Zimbabwe	150.8	11.9	13.5	15.3	79	2.7	26	54	27	540
South Asia	1,701.1	1,272.2	1,420.4	1,586.0	748	2.1	34	60	25	300
Bangladesh	55.6	124.9	141.7	160.8	2,246	2.4	29	55	17	220
Bhutan	18.2	0.9	1.0	1.1	47	2.3	30	51	13	170
India	1,237.1	966.6	1,069.2	1,182.7	781	1.9	36	60	26	290
Maldives	0.1	0.3	0.3	0.4	2,797	3.6	19	65	26	820
Nepal	54.4	23.6	27.6	32.2	435	2.4	29	54	10	160

	Area (1000 sq mi)	Population (Millions)			1997 Population Density (per sq mi)	Annual Rate of Natural Incr (Percent)	Doubling Time (years)	Life Expectancy at Birth (years)	Percent Urban Population	1993 Gross Nat'l Product Per Capita ($US)
		1997	2004	2010						
Pakistan	310.4	137.2	160.5	187.7	442	2.9	24	61	32	430
Sri Lanka	25.3	18.7	19.8	21.0	740	1.5	46	73	22	600
East Asia	4,539.9	1,472.6	1,548.3	1,628.0	324	1.0	66	70	35	3,570
China	3,691.5	1,252.2	1,320.2	1,391.9	339	1.1	62	69	28	490
Japan	145.7	125.9	128.1	130.4	864	0.3	277	79	77	31,450
Korea, North	46.5	24.3	26.3	28.5	523	1.8	40	70	61	
Korea, South	38.0	45.7	47.7	49.7	1,203	1.0	72	72	74	7,670
Macau	0.0	0.4	0.5	0.5	42,820	1.2	57		97	
Mongolia	604.3	2.3	2.7	3.0	4	1.4	52	64	55	400
Taiwan	13.9	21.7	22.8	24.0	1,559	1.0	67	74	75	
Southeast Asia	1,734.4	503.6	550.0	600.7	290	1.9	37	64	31	1,070
Brunei	2.2	0.3	0.3	0.4	141	2.4	29	74	67	
Cambodia	69.9	11.2	13.2	15.7	160	2.8	25	50	13	
Indonesia	741.1	204.9	222.0	240.6	276	1.6	43	63	31	730
Laos	91.4	5.1	6.1	7.2	56	2.8	25	52	19	290
Malaysia	127.3	20.9	24.0	27.5	164	2.4	29	71	51	3,160
Myanmar (Burma)	261.2	46.5	51.6	57.3	178	1.9	36	60	25	
Philippines	115.8	71.3	78.9	87.2	616	2.1	33	65	49	830
Singapore	0.2	3.1	3.3	3.6	15,311	1.2	56	74	100	19,310
Thailand	198.1	61.9	65.2	68.7	313	1.4	48	70	19	2,040
Vietnam	127.2	78.5	85.2	92.5	617	2.3	30	65	21	170
Australia	3,069.9	21.9	23.4	24.9	7	0.8	89	78	85	16,760
Australia	2,966.2	18.3	19.5	20.8	6	0.8	91	78	85	17,510
New Zealand	103.7	3.6	3.9	4.1	35	0.9	81	76	85	12,900
Pacific	212.6	6.7	7.6	.87	31	2.4	29	58	18	1,138
Federated States of Micronesia	0.3	0.1	0.1	0.1	435	3.0	23	68	26	
Fiji	7.1	0.8	0.9	0.9	113	2.0	35	63	39	2,140
French Polynesia	1.5	0.2	0.3	0.3	153	2.1	34	70	57	
Guam	0.2	0.2	0.2	0.2	805	2.6	27	74	38	
Marshall Islands	0.1	0.1	0.1	0.1	605	4.0	17	63	65	
New Caledonia	7.4	0.2	0.2	0.2	26	2.0	34	74	70	
Papua New Guinea	178.3	4.3	4.9	5.7	24	2.3	30	57	15	1,120
Solomon Islands	11.0	0.4	0.5	0.6	39	3.7	19	61	13	750
Vanuatu	5.6	0.2	0.2	0.2	33	2.9	24	63	18	1,230
Western Samoa	1.1	0.2	0.2	0.2	171	2.6	27	65	21	980

Map Reading and Interpretation

As can be seen throughout this book, maps are tools that are very useful in gaining an understanding of patterns in geographic space. In fact, they constitute an important visual or *graphic communication* medium whereby encoded spatial messages are transmitted from the cartographer (mapmaker) to the map reader. Of course, this shorthand is necessary because the real world is so complex that a great deal of geographic information must be compressed into the small confines of maps that can fit onto the pages of this book. At the same time, cartographers must carefully choose which information to include; these decisions force them to omit many things in order to prevent cluttering a map with less relevant information. For instance, Figure B in this appendix shows several city blocks in central London but avoids mapping individual buildings because they would interfere with the main information being presented—the spatial distribution of cholera deaths.

MAP READING

Deciphering the coded messages contained in the maps of this book—map reading—is not difficult, and becomes quite easy with a little experience in using this "language" of geography. The need to miniaturize portions of the world on small maps is discussed in the section on map scale (pp. 7–8), and two additional contrasting examples

FIGURE A

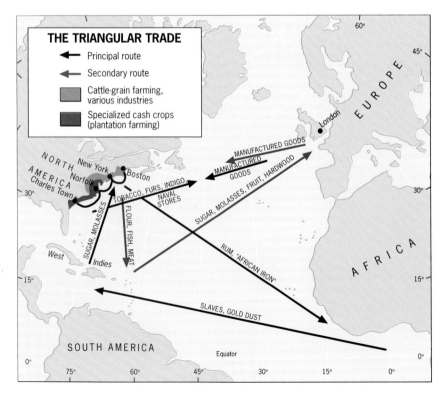

THE TRIANGULAR TRADE

← Principal route

← Secondary route

▪ Cattle-grain farming, various industries

▪ Specialized cash crops (plantation farming)

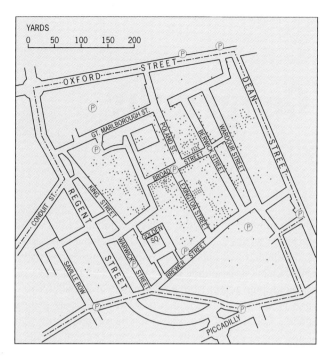

YARDS
0   50   100   150   200

FIGURE B

are provided in Figures A and B. Orientation, or direction, on maps can usually be discerned by reference to the geographic grid of latitude and longitude. *Latitude* is measured from 0° to 90° north and south of the equator (parallels of latitude are always drawn in an east-west direction), with the equator being 0° and the North and South Poles being 90°N and 90°S, respectively. Meridians of *longitude* (always drawn north-south) are measured 180° east and west of the *prime meridian* (0°), which passes through the Greenwich Observatory in London; the 180th meridian, for the most part, serves as the *international date line* that lies in the middle of the Pacific Ocean. Inspection of Figure A shows that north is not automatically at the top of a map; instead, the direction of north curves along every meridian, with all such lines of longitude converging at the North Pole. The many minor directional distortions in this map are unavoidable: it is geometrically impossible to transfer the grid of a three-dimensional sphere (globe) onto a two-dimensional flat map. Therefore, compromises in the form of *map projections* must be devised in which, for example, properties such as areal size and distance are preserved but directional constancy is sacrificed.

MAP SYMBOLS

Once the background mechanics of scale and orientation are understood, the main task of decoding the map's content can proceed. The content of most maps in this book is organized within the framework of point, line, and area symbols, which are made especially clear through the use of color. These symbols are usually identified in the map's *legend*, as in Figure A. Occasionally, the map designer omits the legend but must tell the reader verbally in a caption or within the text what the map is about. Figure B, for instance, is a map of cholera deaths in the London neighborhood of Soho during the outbreak of 1854, with each Ⓟ symbol representing a municipal water pump and each red dot the location of a cholera fatality. *Point symbols* are shown as dots on the map and can tell us two things: the location of each phenomenon and, sometimes, its quantity. The cities of New York and London in Figure A and the dot pattern of cholera fatalities in Figure B (with each red dot symbolizing one death) are examples.

Line symbols connect places between which some sort of movement or flow is occurring. The "triangular trade" among Britain and its seventeenth-century Atlantic colonies (Fig. A) is a good example: each leg of these trading routes is clearly mapped, the goods moving along are identified, and the more heavily traveled principal routes are differentiated from the secondary ones. *Area symbols* are used to classify two-dimensional spaces and thus provide the cartographic basis for regionalization schemes (as can be seen in Fig. I-7 on pp. 16–17). Such classifications can be developed at many levels of generalization. In Figure A, blue and beige areas broadly offset land from ocean surfaces; more specifically, red and green area symbols along the eastern North American coast delimit a pair of regions that specialize in different types of commercial agriculture. Area symbols may also be used to communicate quantitative information: for example, the light-tan-colored zones in Figure I-6 (pp. 14–15) delimit semiarid areas within which annual precipitation averages from 12 to 20 inches (30 to 50 cm).

MAP INTERPRETATION

The explanation of cartographic patterns is one of the geographer's most important tasks. Although that task is performed for you throughout this book, readers should be aware that today's practitioners use many sophisticated techniques and machines to analyze vast quantities of areal data. These modern methods notwithstanding, geographic inquiry still focuses on the search for meaningful *spatial relationships*. This longstanding concern of the discipline is classically demonstrated in Figure B. By showing on his map that cholera fatalities clustered around municipal water pumps, Dr. John Snow was able to persuade city authorities to shut them off; almost immediately, the number of new disease victims dwindled to zero, thereby confirming Snow's theory that contaminated drinking water was crucial in the spread of cholera.

Opportunities in Geography

The chapters of this book give an idea of the wide range of topics and interests that geographers pursue, particularly in the many discussions of concepts (regional and otherwise) and the Focus Essays that survey geography's systematic fields. There are specializations within each of those topics as well as in regional studies—but an introductory book such as this lacks enough space to discuss all of them. This appendix, therefore, is designed to help you, should you decide to major or minor in geography and/or to consider it as a career option.

AREAS OF SPECIALIZATION

As in all disciplines, areas of concentration or specialization change over time. In North American geography early in the twentieth century, there was a period when most geographers were physical geographers, and the physical landscape was the main object of geographic analysis. Then the pendulum swung toward human (cultural) geography, and students everywhere focused on the imprints of human activity on the surface of the Earth. Still later the analysis of spatial organization became a major concern. In the meantime, geography's attraction for some students lay in technical areas: in cartography, in remote sensing, in computer-assisted spatial data analysis, and, most recently, in geographic information systems.

All this meant that geography posed (and continues to pose) a challenge to its professionals. New developments require that we keep up to date, but we must also continue to build on established foundations.

Regional Geography

One of these established foundations, of course, is regional geography, which encompasses a large group of specializations. Some geographers specialize in the theory of regions: how they should be defined, how they are structured, and how their internal components work. This leads in the direction of *regional science*, and some geographers

have preferred to call themselves regional scientists. But make no mistake: regional science is regional geography.

Another, older approach to regional geography involves specialization in an area of the world ranging in size from a geographic realm to a single region or even a State or part of a State. There was a time when regional geographers, because of the interdisciplinary nature of their knowledge, were sought after by government agencies. Courses in regional geography abounded in universities' geography departments; regional geographers played key roles in international studies and research programs. But then the drive to make geography a more rigorous science and to search for universal (rather than regional) truths contributed to a decline in regional geography. The results were not long in coming, and in recent years you have probably seen the issue of ''geographic illiteracy'' discussed in newspapers and magazines. Now the pendulum is swinging back again, and regional geography and regional specialization are reviving. This is a propitious time to consider regional geography as a professional field.

Your Personal Interests

Geography, as we have pointed out, is united by several bonds, of which regional geography is but one. Regional geography exemplifies the spatial view that all geographers hold; the spatial approach to study and research binds physical and human geographers, regionalists, and topical specialists. Another unifying theme is an abiding interest in the relationships between human societies and natural environments. We have referred to that topic frequently in this book; as an area of specialization it has gone through difficult times. Perhaps more than anything, geography remains a field of *synthesis*, of understanding interrelationships.

Geography also is a field science, using ''field'' in another context. In the past, almost all major geography departments required a student's participation in a ''field camp'' as part of a master's degree program; thus many undergraduate programs included field experience. It was one of those bonding practices in which students and faculty with diverse interests met, worked together, and learned from one another. Today, few field camps of this sort are offered, but that does not change what geography is all about. If you see an opportunity for field experience with professional geographers—even just a one-day reconnaissance—take it. But realize this: a few days in the field with geographic instruction may hook you for life.

Some geographers, in fact, are far better field-data gatherers than analysts or writers. In this respect they are not alone: this also happens in archeology, geology, and biology (among other field disciplines). This does not mean that these field workers do not contribute significantly to knowledge. Often on a research team some of the members are better in the field, and others excel in subsequent analysis. From all points of view, however, fieldwork is important.

Geography, then, is practiced in the field and in the office, in physical and human contexts, in generality and detail. Small wonder that so many areas of specialization have developed! If you check the undergraduate catalogue of your college or university, you will see some of these specializations listed as semester-length courses. But no geography department, no matter how large, could offer them all.

How does an area of specialization develop, and how can one become a part of it? The way in which geographic specializations have developed tells us much about the entire discipline. Some major areas, now old and established, began as research and theory building by one scholar and his or her students. These graduate students dispersed to the faculties of other universities and began teaching what they had learned. Thus, for example, did Carl Sauer's cultural geography spread from the University of California, Berkeley.

It is one of the joys of geography that the basics and methods, once learned, are applicable to so many features of the human and physical world. Geographers have specialized in areas as disparate as shopping-center location and glacier movement, tourism and coastal erosion, real estate and wildlife, retirement communities and sports. Many of these specializations began with the interests and energies of one scholar. Some thrived and grew into major geographic pursuits; others remained one-person shows, but with potential. When you discuss your own interests with a faculty adviser, you may refer to a university where you would like to do graduate work. ''Oh yes,'' the answer may be, ''Professor X is in their geography department, working on just that.'' Or perhaps your adviser will suggest another university where a member of the faculty is known to be working on the topic in which you are interested. That is the time to write a letter of inquiry. What is the professor working on now? Are graduate students involved? Are research funds available? What are the career prospects after graduation?

The Association of American Geographers, or AAG (1710 16th Street, N.W., Washington, D.C. 20009-3198 [(202) 234-1450]), recognizes more than 40 so-called Specialty Groups. In academic year 1996–1997, the Specialty Group roster included the following branches of geography:

Africa
Aging and the Aged
American Ethnic Geography
American Indians
Applied Geography
Asian Geography
Bible
Biogeography
Canadian Studies
Cartography
China
Climate
Coastal and Marine Geography
Contemporary Agriculture and Rural Land Use
Cultural Ecology
Cultural Geography
Energy and Environment
Environmental Perception and Behavioral Geography
European
Geographic Information Systems
Geographic Perspectives on Women
Geography Education
Geography of Religions and Belief Systems
Geomorphology
Hazards
Historical Geography
Human Dimensions of Global Change
Human Rights
Industrial Geography
Latin American
Mathematical Models and Quantitative Methods
Medical Geography
Microcomputers
Political Geography
Population
Recreation, Tourism, and Sport
Regional Development and Planning
Remote Sensing
Rural Development
Russian, Central Eurasian, and East European
Socialist Geography
Transportation Geography
Urban Geography
Water Resources

If you write or call the AAG at the address or telephone number given above, they will be glad to send you the newest listing of these Specialty Groups, which also includes the address of the current chairperson of each group. We encourage you to contact the professional geographer who chairs the group(s) you are interested in. These elected leaders are ready, enthusiastic, and willing to provide you with the information you are looking for.

All this may seem far in the future. Still, the time to start planning for graduate school is now. Applications for admission and financial assistance must be made shortly after the *beginning* of your senior year! That makes your junior year a year of decision.

AN UNDERGRADUATE PROGRAM

The most important concern for any geography major or minor is basic education and training in the field. An undergraduate curriculum contains all or several of the following courses (titles may vary):

1. Introduction to Physical Geography I (climatology, elementary oceanography)
2. Introduction to Physical Geography II (natural landscapes, landforms, soils, elementary biogeography)
3. Introduction to Human Geography (principles and topics of cultural and economic geography)
4. World Regional Geography (major world realms)

These beginning courses are followed by more specialized courses, including both substantive and methodological ones:

5. Introduction to Quantitative Methods of Analysis
6. Introductory Cartography
7. Analysis of Remotely Sensed Data
8. Cultural Geography
9. Political Geography
10. Urban Geography
11. Economic Geography
12. Historical Geography
13. Geomorphology
14. Geography of United States-Canada, Europe, and/or other major world realms

You can see how Physical Geography II would be followed by Geomorphology, and how Human Geography now divides into such areas as intermediate and/or advanced cultural and economic geography. As you progress, the focus

becomes even more specialized. Thus Economic Geography may be followed by:

15. Industrial Geography
16. Transportation Geography
17. Agricultural Geography

At the same time, regional concentrations may come into sharper focus:

18. Geography of Western Europe (or other regions)

In these more advanced courses, you will use the technical know-how from courses numbered **5**, **6**, and **7** (and perhaps others). Now you can avail yourself of the opportunity to develop these skills further. Many departments offer such courses as these:

19. Advanced Quantitative Methods (involving numerous computer applications)
20. Advanced Cartography (involving geographic information systems)
21. Advanced Satellite Imagery Interpretation

From this list (which represents only part of a comprehensive curriculum), it is evident that you cannot, even in four years of undergraduate study, register for all courses. The geography major in many universities requires a minimum of only 30 (semester-hour) credits—just 10 courses, fewer than half those listed here. This is another reason to begin thinking about specialization at an early stage.

Because of the number and variety of possible geography courses, most departments require that their majors complete a core program that includes courses in substantive areas as well as theory and methods. That core program is important, and you should not be tempted to put off these courses until your last semesters. What you learn in the core program will make what follows (or should follow) much more meaningful.

You should also be aware of the flexibility of many undergraduate programs, something that can be especially important to geographers. Imagine that you are majoring in geography and develop an interest in Southeast Asia. But the regional specialization of the geographers in your department may be focused somewhere else—say, Africa. However, courses on Southeast Asia are indeed offered by other departments, such as anthropology, history, or political science. If you are going to be a regional specialist, those courses will be very useful and should be part of your curriculum—but you are not able to receive geography credit for them. After successfully completing those courses, however, you may be able to register for an independent study or reading course in the geography of

Southeast Asia if a faculty member is willing to guide you. Always discuss such matters with the undergraduate adviser or chairperson of your geography department.

LOOKING AHEAD

By now, as a geography major, you will be thinking of the future—either in terms of graduate school or a salaried job. In this connection, if there is one important lesson to keep in mind, it is to *plan ahead* (a redundancy for emphasis!). The choice of graduate school is one of the most important you will make in your life. The professional preparation you acquire as a graduate student will affect your competitiveness in the job market for years to come.

Choosing a Graduate School

Your choice of a graduate school hinges on several factors, and the geography program it offers is one of them. Possibly, you are constrained by residency factors, and your choice involves the schools of only one State. Your undergraduate record and grade point average affect the options. As a geographer, you may have strong feelings in favor of or against particular parts of the country. And although some schools may offer you financial support, others may not.

Certainly the programs and specializations of the prospective graduate department are extremely important. If you have settled on your own area of interest, it is wise to find a department that offers opportunities in that direction. If you have yet to decide, it is best to select a large department with several options. Some students are so impressed by the work and writings of a particular geographer that they go to his or her university solely to learn from and work with that scholar.

In every case, information and preparation are crucial. Many a prospective graduate student has arrived on campus eager to begin work with a favorite professor, only to find that the professor is away on a sabbatical leave!

Fortunately, information can be acquired with little difficulty. One of the most useful publications of the AAG is the *Guide to Programs of Geography in the United States and Canada*, published each fall. A copy of this annual directory should be available in the office of your geography department, but if you plan to enter graduate school, a personal copy would be an asset. Not only does the *Guide* describe the programs, requirements, financial aid, and other aspects of geography departments in North America, but it also lists all faculty members and their current research and teaching specializations. Moreover, it also contains a complete listing of the Association's more than 7,000 members and their specializations. You will find the *Guide* indispensable in your decision-making.

You may discover one particular department that stands out as the most interesting and most appropriate for you. But do not limit yourself to one school; after careful investigation, it is best to rank a half-dozen schools (or more), write to all of them for admission application forms, and apply to several. Multiple applications are a bit costly, but the investment is worth it.

Assistantships and Scholarships

One reason to apply to several universities has to do with financial (and other) support for which you may be eligible. If you have a reasonably well-rounded undergraduate program behind you and a good record of achievement, you are eligible to become a teaching assistant (TA) in a graduate department. Such assistantships usually offer full or partial tuition plus a monthly stipend (during the nine-month academic year). Conditions vary, but this position may make it possible for you to attend a university that would otherwise be out of reach. A tuition waiver alone can be worth more than $10,000 annually. Application for an assistantship is made directly to the department; you should write to the department contact listed in the latest *AAG Guide*, who will either respond directly to you or forward your inquiry to the departmental committee that evaluates applications.

What does a TA do? The responsibilities vary, but often teaching assistants are expected to lead discussion sections (of a larger class taught by a professor), laboratories, or other classes. They prepare and grade examinations and help undergraduates tackle problems arising from their courses. This is an excellent way to determine your own ability and interest in a teaching career.

In some instances, especially in larger geography departments, research assistantships are available. When a member of the faculty is awarded a large grant (e.g., by the National Science Foundation) for a research project, that grant may make possible the appointment of one or more research assistants (RAs). These individuals perform tasks generated by the project and are rewarded by a modest salary (usually comparable to that of a TA). Normally, RAs do not receive tuition waivers, but sometimes the department and the graduate school can arrange a waiver to make the research assistantship more attractive to better students. Usually, RAs are chosen from among graduate students already on campus who have proven their interest and ability. Sometimes, however, an incoming student is appointed. Always ask about opportunities.

Geography students also are among those eligible for

many scholarships and fellowships offered by universities and off-campus organizations. When you write your introductory letter, be sure to inquire about other forms of financial aid.

JOBS FOR GEOGRAPHERS

Upon completing your bachelor's degree, you may decide to take a job rather than going on to graduate school. Again, this is a decision best made early in your junior year, for two main reasons: (1) so that you can tailor your curriculum for a vocational objective, and (2) so that you can start searching for a job well before graduation.

Internships

A very good way to enter the job market—and to become familiar with the working environment—is by taking an internship in an agency, office, or firm. Many organizations find it useful to have interns. Internships help organizations train beginning professionals and give companies an opportunity to observe the performance of trainees. Many an intern has ultimately been employed by his or her organization. Some employers have even suggested what courses the intern should take in the next academic year to improve future performance. For example, an urban or regional planning agency that employs an intern might suggest that the intern add a relevant advanced cartography course or urban planning course to his or her program of study.

Some organizations will appoint interns on a continuing basis, say, two afternoons a week around the year; others make full-time, summer-only appointments available. Occasionally, an internship can be linked to a departmental curriculum, yielding academic credit as well as vocational experience. Your department undergraduate adviser or the chairperson will be the best source of assistance.

One of the most interesting internship programs is offered by the National Geographic Society in Washington, D.C. Every year, the Society invites three groups of about eight interns each to work with the permanent staff of its various departments. Application forms are available in every geography department in the United States, and competition is strong. The application itself is a useful exercise, as it tells you what the Society (and other organizations) look for in your qualifications.

Professional Opportunities

A term you will sometimes see in connection with jobs is *applied geography*. This, one supposes, distinguishes geography teaching from practical geography. In fact, however, all professional geography in education, business, government, and elsewhere is "applied." In the past a large majority of geography graduates became teachers—in elementary and high schools, colleges, and universities. More recently, geographers have entered other arenas in increasing numbers. In part, this is related to the decline in geographic education in schools, but it also reflects the growing recognition of geographic skills by employers in business and government.

Nevertheless, what you as a geographer can contribute to a business is not yet as clear to the managers of many companies as it should be. The anybody-can-do-geography attitude is a form of ignorance you will undoubtedly confront. (This is so even in precollegiate education, where geography was submerged in social studies—and often taught by teachers who had never taken a course in the discipline!)

So, once employed, you may have to prove not only yourself but also the usefulness of the skills and capacities you bring to your job. There is a positive side to this. Many employers, once dubious about hiring a geographer, learn how geography can contribute—and become enthusiastic users of geographic talent. Where one geographer is employed, whether in a travel firm, publishing house, or planning office, you will soon find more.

Geographers are employed in business, government, and education. Planning is a profession that employs many geographers. Employment in business has grown in recent years (see the following section). Governments at national, State, and local levels have always been major employers of geographers. And in education, where geography was long in decline, the demand for geography teachers will grow again.

For more detailed information about employment, you should write to the AAG (at the address given earlier) for the inexpensive booklet entitled *Careers in Geography*; check with your geography department, whose office is very likely to have a copy.

Business and Geography

With their education in global and international affairs, their knowledge of specialized areas of interest to business, and their training in cartography, methods of quantitative analysis, and writing, geographers with a bachelor's degree are attractive business-employment prospects. Some undergraduate students have already chosen the type of business they will enter; for them, there are departments that offer concentrations in their areas of interest. Several geography departments, for instance, offer a curriculum that concentrates on tourism and travel—a business in which geographic skills are especially useful.

These days, many companies want graduates with a strong knowledge of international affairs and fluency in at least one foreign language in addition to their skills in other areas. The business world is quite different from academia, and the transition is not always easy. Your employer will want to use your abilities to enhance profits. You may at first be placed in a job where your geographic skills are not immediately applicable, and it will be up to you to look for opportunities to do so.

One of our students was in such a situation some years ago. She was one of more than a dozen new employees doing what was essentially clerical work. (Some companies will use this type of work to determine a new employee's punctuality, work habits, adaptability, and productivity.) One day she heard that the company was considering establishing a branch to sell its product in East Africa. The student had a regional interest in East Africa as an undergraduate and had even taken a year of Swahili-language training. On her own time she wrote a carefully documented memorandum to the company's president and vice presidents, describing factors that should be taken into consideration in the projected expansion. That report evinced her regional skills, locational insights, and knowledge of the local market and transport problems, along with the probable cultural reaction to the company's product, the country's political circumstances, and (last but significantly) this employee's ability to present such issues effectively. She supported her report with good maps and several illustrations. Soon she received a special assignment to participate in the planning process, and her rise in the company's ranks had begun. She had seized the opportunity and demonstrated the utility of her geographic skills.

Geography graduates have established themselves in businesses of all kinds: banking, international trade, manufacturing, retailing, and many more. Should you join a large firm, you may be pleasantly surprised by the number of other geographers who hold jobs there—not under the title of geographer but under countless other titles ranging from analyst and cartographer to market researcher and program manager. These are positions for which the appointees have competed with other graduates, including business graduates. As we noted earlier, once an employer sees the assets a geographer can bring, the role of geographers in the company is assured.

Government and Geography

Government has long been a major employer of geographers at the national and State as well as local level. *Careers in Geography* (the AAG booklet) estimates that at least 2,500 geographers are working for governments, about half of them for the federal government. In the U.S. State Department, for instance, there is an Office of the Geographer staffed by professional geographers. Other agencies where geographers are employed include the De-

fense Mapping Agency, the Bureau of the Census, the U.S. Geological Survey, the Central Intelligence Agency, and the Army Corps of Engineers. Still other employers are the Library of Congress, the National Science Foundation, and the Smithsonian Institution. Many other branches of government also have positions for which geographers are eligible.

Opportunities also exist at State and local levels. Several States now have their own Office of the State Geographer; all States have agencies engaged in planning, resource analysis, environmental protection, and transportation policy-making. All these agencies need geographers who have skills in cartography, remote sensing, database analysis, and the operation of geographic information systems.

Securing a position in government requires early action. If you want a job with the federal government, start at the beginning of your senior year. Every State capital and many other large cities have a Federal Job Information Center (FJIC). (The Washington office is at 1900 E Street, N.W., Washington, D.C. 20415.) You may request information about a particular agency and its job opportunities; it is appropriate to write directly to the personnel office of the agency or agencies in which you are interested. You may also write to the Office of Personnel Management (OPM), Washington, D.C. 20415, which has offices in most large cities.

Planning and Geography

Planning has become one of geography's allied professions. The planning process is a complex one comprising people trained in many fields. Geographers, with their cartographic, locational, regional, and analytical skills, are sought after by planning agencies. Many an undergraduate student gets that first professional opportunity as an intern in a planning office.

Planning is done by many agencies and offices at levels ranging from the federal to the municipal. Cities have planning offices, as do regional authorities. Working in a planning office can be a very rewarding experience because it involves the solving of social and economic problems, the conservation and protection of the environment, the weighing of diverse and often conflicting arguments and viewpoints, and much interaction with workers trained in other fields. Planning is a superb learning experience.

A career in planning can be much enhanced by a background in geography, but you will have to adjust your undergraduate curriculum to include courses in such areas as public administration, public finance, and other related fields. Thus a career in planning itself requires early planning on your part. At many universities, the geography department is closely associated with the planning department, and your faculty adviser can inform you about course requirements. But if you have your eye on a particular office or agency, you should also request information from its director about desired and required skills.

Planning is by no means a monopoly of government. Government-related organizations such as the Agency for International Development (AID), the World Bank, and the International Monetary Fund (IMF) have planning offices, as do nongovernmental organizations such as banks, airline companies, industrial firms, and multinational corporations. In the private sector the opportunities for planners have been expanding, and you may wish to explore them. An important organization is the planners' equivalent of the geographers' AAG: the American Planning Association, 1776 Massachusetts Avenue, N.W., Washington, D.C. 20036. The AAG's *Careers in Geography* provides additional information on this expanding field and where you might pursue its study.

Teaching

If you are presently a freshman or sophomore, your graduation may coincide with the end of a long decline in the geography-teaching profession. Just 30 years ago, teaching geography in elementary or high school was the goal of thousands of undergraduate geography majors. But then began the merging of geography into the hybrid field called "social studies," and prospective teachers no longer needed to have any training in geography in many States. In Florida, for instance, teachers were formerly required to take courses in regional geography and conservation (as taught in geography departments), but in the 1970s those requirements were dropped. Education planners fell victim to the myth that the teaching of geography does not require any training. What was left of geography often was taught by teachers whose own fields were history, civics, or even basketball coaching!

If you read the daily newspapers, you have seen reports of the predictable results. As mentioned earlier, "geographic illiteracy" has become a common complaint (often made by the same education planners who pushed geography into the social studies program and eliminated teacher-education requirements). Now States are returning to the education requirement so that teachers will learn some geography. And geography is returning to elementary and high school curricula—assisted in particular by the National Geographic Society, which supports State-level Geographic Alliances of educators and academic geographers. The need for geography teachers will soon be on the upswing again.

So this may be a good time to consider teaching geography as a career. You should do some research, however, because States vary in their progressiveness in this area. Opportunities will become more widespread after the turn of the century. You should visit the School of Education in your college or university and ask questions about this. Also, write not only the AAG but also the National Council for Geographic Education (NCGE) at Indiana University of Pennsylvania, Indiana, PA 15705. Your State geographic society or Geographic Alliance also may be helpful; ask your department adviser or chairperson for details.

The opportunities in geography are many, but often they are not as obvious as those in other fields. You will find that good and timely preparation produces results and, frequently, unexpected rewards.

The discussion in this appendix has provided a comprehensive answer to the oft-asked question, "What can one do with geography?" If you wish to explore this question further, along with the AAG's *Careers in Geography* we recommend *On Becoming a Professional Geographer*, a book edited by Martin S. Kenzer (Merrill/Macmillan, 1989).

We wish you every success in all your future endeavors. And, speaking for the entire community of professional geographers, we would be delighted to have you join our ranks, should you choose a geography-related career.

APPENDIX D

Video Link Resources

Good Morning America ◼ Videotapes

These videotapes are selected from H. J. de Blij's appearances as Geography Editor on the popular ABC morning news show. Each segment discusses geographically significant issues and world events. A black video icon with the corresponding letter label appears in the margin of the text indicating the appropriate place to introduce the pertinent video segment.

On Location with H. J. de Blij ◼ Videotapes

These 31 five-minute segments narrated by H.J. de Blij contain exclusive, up-to-date footage of geographically significant locations throughout the world. They delve into the physical, economic, historical, social, and political issues relevant to these places and provide insights from a uniquely geographic point of view. A blue video icon with the corresponding number label appears in the margin of the text indicating the appropriate place to introduce the specific segment.

References and Further Readings

(Bullets [•] denote basic introductory works)

INTRODUCTION

Abler, Ronald F., Marcus, Melvin G., & Olson, Judy M., eds. *Geography's Inner Worlds: Pervasive Themes in Contemporary American Geography* (New Brunswick, N.J.: Rutgers University Press, 1992).

Chisholm, Michael. *Modern World Development* (Totowa, N.J.: Barnes & Noble, 1982).

• de Blij, H. J. *Human Geography: Culture, Society, and Space* (New York: John Wiley & Sons, 5th rev. ed., 1996).

• de Blij, H. J. *The Earth: An Introduction to Its Physical and Human Geography* (New York: John Wiley & Sons, 4th rev. ed., 1995).

• de Blij, H. J., & Muller, Peter O. *Physical Geography of the Global Environment* (New York: John Wiley & Sons, 2nd rev. ed., 1996).

Dickinson, Robert E. *The Regional Concept* (London: Routledge & Kegan Paul, 1976).

Fenneman, Nevin M. "The Circumference of Geography," *Annals of the Association of American Geographers*, 9 (1919): 3–11.

• Glassner, Martin I., & de Blij, H. J. *Systematic Political Geography* (New York: John Wiley & Sons, 4th rev. ed., 1989).

• *Goode's World Atlas* (Skokie, Ill.: Rand McNally, 19th rev. ed., 1995).

• Harris, Chauncy D., chief ed. *A Geographical Bibliography for American Libraries* (Washington, D.C.: Association of American Geographers/National Geographic Society, 1985).

Jackson, John Brinckerhoff. *Discovering the Vernacular Landscape* (New Haven: Yale University Press, 1984).

James, Preston E., & Martin, Geoffrey J. *All Possible Worlds: A History of Geographical Ideas* (New York: John Wiley & Sons, 3rd rev. ed., 1993).

Johnston, R. J., et al., eds. *The Dictionary of Human Geography* (Cambridge, Mass.: Blackwell, 3rd rev. ed., 1993).

Livingstone, David. *The Geographical Tradition* (Cambridge, Mass.: Blackwell, 1992).

Mabogunje, Akin L., ed. *The State of the Earth* (Cambridge, Mass.: Blackwell, 1997).

Muehrcke, Phillip C. & Muehrcke, Juliana C. *Map Use: Reading-Analysis-Interpretation* (Madison, Wis.: JP Publications, 3rd rev. ed., 1992).

• Pattison, William D. "The Four Traditions of Geography," *Journal of Geography*, 63 (1964): 211–216.

Rediscovering Geography: New Relevance for the New Century (Washington, D.C.: National Academy of Sciences, 1996).

Sauer, Carl Ortwin. "Cultural Geography," *Encyclopedia of the Social Sciences*, Vol. 6 (New York: Macmillan, 1931), pp. 621–623.

Small, John & Witherick, Michael. *A Modern Dictionary of Geography* (New York: John Wiley & Sons, 3rd rev. ed., 1995).

Waters, Malcolm. *Globalization* (London & New York: Routledge, 1995).

Wegener, Alfred. *The Origin of Continents and Oceans* (New York: Dover, reprint of the 1915 original, trans. John Biram, 1966).

• Wheeler, James O., Muller, Peter O., Thrall, Grant I., & Fik, Timothy J. *Economic Geography* (New York: John Wiley & Sons, 3rd rev. ed., 1998).

Whittlesey, Derwent S., et al. "The Regional Concept and the Regional Method," in James, Preston E. and Jones, Clarence F., eds., *American Geography: Inventory and Prospect* (Syracuse, N.Y.: Syracuse University Press, 1954), pp. 19–68.

CHAPTER I

Applebaum, Anne. *Between East and West: Across the Borderlands of Europe* (New York: Pantheon, 1994)

Archer, Clive. *Organizing Europe: The Institutions of Integration* (London: Edward Arnold, 2nd rev. ed., 1994).

Batalden, Stephen K., & Batalden, Sandra L. *The Newly Independent States of Eurasia: Handbook of Former Soviet Republics* (Phoenix: Oryx Press, 1993).

Blacksell, Mark, & Williams, Allan M., eds. *The European Challenge: Geography and Development in the European Community* (New York: Oxford University Press, 1994).

Brawer, Moshe. *Atlas of Russia and the Independent Republics* (New York: Simon & Schuster, 1994).

Brown, Archie, et al., eds. *The Cambridge Encyclopedia of Russia and the Former Soviet Union* (New York: Cambridge University Press, 1994).

Burtenshaw, David, et al. *The European City: A Western Perspective* (New York: John Wiley & Sons, 1991).

● Butzer, Karl W. *Geomorphology from the Earth* (New York: Harper & Row, 1976).

Castells, Manuel, & Hall, Peter. *Technopoles of the World: The Making of Twenty-First-Century Industrial Complexes* (London & New York: Routledge, 1994).

Chinn, Jeff, & Kaiser, Robert. *Russians as the New Minority: Ethnicity and Nationalism in the Soviet Successor States* (Boulder, Colo.: Westview Press, 1996).

Chisholm, Michael. *Britain on the Edge of Europe* (London & New York: Routledge, 1995).

Chisholm, Michael. *Rural Settlement and Land Use: An Essay in Location* (London: Hutchinson University Library, 3rd rev. ed., 1979).

● Chorley, Richard J., et al. *Geomorphology* (London & New York: Methuen, 1985).

· Clout, Hugh D., ed. *Regional Development in Western Europe* (New York: John Wiley & Sons, 3rd rev. ed., 1987).

Clout, Hugh D., et al. *Western Europe: Geographical Perspectives* (New York: Wiley/Longman, 3rd rev. ed., 1994).

Cohen, Leonard J. *Broken Bonds: The Rise and Fall of Yugoslavia* (Boulder, Colo.: Westview Press, 1993).

Cole, John P., & Cole, Francis J. *The Geography of the European Community* (London & New York: Routledge, 1993).

Crampton, Richard & Crampton, Benjamin. *Atlas of Eastern Europe in the Twentieth Century* (London & New York: Routledge, 1996).

Dawisha, Karen, & Parrott, Bruce. *Russia and the New States of Eurasia: The Politics of Upheaval* (New York: Cambridge University Press, 1994).

Dawson, Andrew H. *The Geography of European Integration: A Common European Home?* (New York: Wiley/Belhaven, 1993).

● de Blij, H. J., & Muller, Peter O. *Physical Geography of the Global Environment* (New York: John Wiley & Sons, 2nd rev. ed., 1996).

● Diem, Aubrey. *Western Europe: A Geographical Analysis* (New York: John Wiley & Sons, 1979).

Diuk, Nadia, & Karatnycky, Adrian. *New Nations Rising: The Fall of the Soviets and the Challenge of Independence* (New York: John Wiley & Sons, 1993).

Drost, Harry. *What's What & Who's Who in Europe* (New York: Simon & Schuster, 1994).

Dunford, M., & Kafkalas, G., eds. *Cities and Regions in the New Europe: The Global-Local Interplay and Spatial Development Strategies* (London: Belhaven Press/Wiley, 1992).

Embleton, Clifford, ed. *Geomorphology of Europe* (New York: Wiley-Interscience, 1984).

Fernandez-Armesto, Felipe, ed. *The Times Guide to the Peoples of Europe* (Boulder, Colo.: Westview Press, 1995).

Fitzmaurice, John. *Damming the Danube: Gabcikovo and Post-Communist Politics in Europe* (Boulder, Colo.: Westview, 1995).

Glebe, Günther, & O'Loughlin, John, eds. *Foreign Minorities in Continental European Cities* (Wiesbaden, West Germany: Franz Steiner Verlag, 1987).

● Gottmann, Jean. *A Geography of Europe* (New York: Holt, Rinehart & Winston, 4th rev. ed., 1969).

Grundy-Warr, Carl, ed. *Eurasia: World Boundaries, Volume Three* (London & New York: Routledge, 1994).

Haggett, Peter. *Locational Analysis in Human Geography* (London: Edward Arnold, 1965). Definition from p. 19.

Hall, Ray, & Ogden, Philip. *Europe's Population in the 1970s and 1980s* (Cambridge, U.K.: Cambridge University Press, 1985).

Hall, Ray, & White, Paul, eds. *Europe's Population: Towards the Next Century* (London: UCL Press/Taylor & Francis, 1995).

Harris, Chauncy D. "Unification of Germany in 1990," *Geographical Review*, 81 (1991): 170–182.

Hoffman, Eva. *Exit into History: A Journey Through the New Eastern Europe* (New York: Viking Press, 1993).

● Hoffman, George W., ed. *Europe in the 1990's: A Geographic Analysis* (New York: John Wiley & Sons [6th rev. ed. of *A Geography of Europe: Problems and Prospects*], 1989).

Hoffmann, Stanley. *The European Sisyphus: Essays on Europe, 1964–1994* (Boulder, Colo.: Westview Press, 1995).

Ignatieff, Michael. *Blood and Belonging: Journeys into the New Nationalism* (New York: Farrar, Straus & Giroux, 1994).

Ilbery, Brian W. *Western Europe: A Systematic Human Geography* (New York: Oxford University Press, 2nd rev. ed., 1986).

Jefferson, Mark. "The Law of the Primate City," *Geographical Review*, 29 (1939): 226–232. Quotation from p. 226.

Johnston, Ronald J., & Gardiner, Vince, eds. *The Changing Geography of the United Kingdom* (London & New York: Routledge, 2nd rev. ed., 1991).

Jones, Alun. *The New Germany: A Human Geography* (New York: Wiley/Longman, 1994).

● Jordan, Terry G. *The European Culture Area: A Systematic Geography* (New York: Harper Collins, 3rd rev. ed., 1996).

Khazanov, Anatoly M. *After the U.S.S.R.: Ethnicity, Nationalism, and Politics in the Commonwealth of Independent States* (Madison, Wis.: University of Wisconsin Press, 1996).

King, Russell, ed. *The New Geography of European Migrations* (New York: John Wiley & Sons, 1993).

Kraus, Michael, & Liebowitz, Ronald D., eds. *Russia and Eastern Europe After Communism: The Search for New Political, Economic, and Security Systems* (Boulder, Colo.: Westview, 1995).

Krause, Axel. *Inside the New Europe* (New York: HarperCollins, 1991).

Lewis, Flora. *Europe: Road to Unity* (New York: Touchstone/Simon & Schuster, 1991).

● McDonald, James R. *The European Scene: A Geographic Perspective* (Englewood Cliffs, N.J.: Prentice-Hall, 1992).

● Mellor, Roy E. H., & Smith, E. Alistair. *Europe: A Geographical Survey of the Continent* (New York: St. Martin's Press, 1979).

Murphy, Alexander B. "The Emerging Europe of the 1990s," *Geographical Review*, 81 (1991): 1–17.

Noin, Daniel, & Woods, Robert, eds. *The Changing Population of Europe* (Cambridge, Mass.: Blackwell, 2nd rev. ed., 1993).

Pinder, David, ed. *Western Europe: Challenge and Change* (New York: Guilford, 1991).

Pounds, Norman J. G. *An Historical Geography of Europe, 1800–1914* (New York: Cambridge University Press, 1985).

Pryde, Philip R., ed. *Environmental Resources and Constraints in the Former Soviet Republics* (Boulder, Colo.: Westview Press, 1995).

Shaw, Denis J. B., ed. *The Post-Soviet Republics: A Systematic Geography* (New York: Wiley/Longman, 1995).

Silber, Laura, & Little, Allan. *Yugoslavia: Death of a Nation* (Harmondsworth, U.K.: Penguin, 1995).

Sommers, Lawrence M. "Cities of Western Europe," in Brunn, Stanley D., & Williams, Jack F., eds., *Cities of the World: World Regional Urban Development* (New York: Harper & Row, 1983), pp. 84–121. Quotation from p. 97.

Turnock, David. *Eastern Europe: An Economic and Political Geography* (London & New York: Routledge, 1989).

Turnock, David. *The Human Geography of Eastern Europe* (London & New York: Routledge, 1989).

Wheeler, James O., & Muller, Peter O. *Economic Geography* (New York: John Wiley & Sons, 2nd rev. ed., 1986), Chapter 13.

Williams, Allan M. *The European Community* (Cambridge, Mass.: Blackwell, 2nd rev. ed., 1994).

Williams, Allan M. *The West European Economy: A Geography of Post-War Development* (Savage, Md.: Rowman & Littlefield, 1988).

CHAPTER 2

Akiner, Shirin. *Islamic Peoples of the Soviet Union* (London: Kegan Paul International, 2nd rev. ed., 1983).

Batalden, Stephen K., & Batalden, Sandra L. *The Newly Independent States of Eurasia: Handbook of Former Soviet Republics* (Phoenix: Oryx Press, 1993).

• Bater, James H. *The Soviet Scene: A Geographical Perspective* (London & New York: Routledge, 1989).

Bater, James H., & French, Richard A., eds. *Studies in Russian Historical Geography* (New York: Academic Press, 2 vols., 1983).

Bobrick, Benson. *East of the Sun: The Epic Conquest and Tragic History of Siberia* (New York: Poseidon Press, 1992).

Brawer, Moshe. *Atlas of Russia and the Independent Republics* (New York: Simon & Schuster, 1994).

• Brown, Archie, et al., eds. *The Cambridge Encyclopedia of Russia and the Former Soviet Union* (New York: Cambridge University Press, 1994).

Chew, Allen F. *Atlas of Russian History: Eleven Centuries of Changing Borders* (New Haven: Yale University Press, 1967).

Chinn, Jeff, & Kaiser, Robert. *Russians as the New Minority: Ethnicity and Nationalism in the Soviet Successor States* (Boulder, Colo.: Westview Press, 1996).

• Critchfield, Howard J. *General Climatology* (Englewood Cliffs, N.J.: Prentice-Hall, 4th rev. ed., 1983).

Dawisha, Karen, & Parrott, Bruce. *Russia and the New States of Eurasia: The Politics of Upheaval* (New York: Cambridge University Press, 1994).

• de Blij, H. J., & Muller, Peter O. *Physical Geography of the Global Environment* (New York: John Wiley & Sons, 2nd rev. ed., 1996).

Demko, George J. et al., eds. *Population under Duress: The Geodemography of Post-Soviet Russia* (Boulder, Colo.: Westview Press, 1996).

Demko, George J., & Fuchs, Roland J., eds. *Geographical Studies on the Soviet Union: Essays in Honor of Chauncy D. Harris* (Chicago: University of Chicago, Department of Geography, Research Paper No. 211, 1984).

Dienes, Leslie. *Soviet Asia: Economic Development and National Policy Choices* (Boulder, Colo.: Westview Press, 1987).

Diuk, Nadia, and Karatnycky, Adrian. *New Nations Rising: The Fall of the Soviets and the Challenge of Independence* (New York: John Wiley & Sons, 1993).

Duncan, W. Raymond, & Holman, G. Paul, Jr., eds. *Ethnic Nationalism and Regional Conflict: The Former Soviet Union and Yugoslavia* (Boulder, Colo.: Westview, 1994).

Dunlop, John B. *The Rise of Russia and the Fall of the Soviet Empire* (Princeton, N.J.: Princeton University Press, 1993).

Fisher, Lois. *Survival in Russia: Chaos and Hope in Everyday Life* (Boulder, Colo.: Westview Press, 1993).

Glenny, Misha. "The Bear in the Caucasus: From Georgian Chaos, Russian Order," *Harper's Magazine*, March 1994, 45–53.

Grundy-Warr, Carl, ed. *Eurasia: World Boundaries, Volume Three* (London & New York: Routledge, 1994).

Hauner, Milan. *What Is Asia to Us? Russia's Asian Heartland Yesterday and Today* (London & New York: Routledge, 1992).

Horensma, Pier. *The Soviet Arctic* (London & New York: Routledge, 1991).

• Howe, G. Melvyn. *The Soviet Union: A Geographical Study* (London & New York: Longman, 2nd rev. ed., 1986).

Hunter, Shireen T. *Transcaucasia in Transition: Nation-Building and Conflict* (Boulder, Colo.: Westview, 1994).

Huttenbach, Henry. *The Caucasus: A Region in Crisis* (Boulder, Colo.: Westview Press, 1996).

Jensen, Robert G., et al., eds. *Soviet Natural Resources in the World Economy* (Chicago: University of Chicago Press, 1983).

Kaiser, Robert J. *The Geography of Nationalism in Russia and the U.S.S.R.* (Princeton, N.J.: Princeton University Press, 1994).

Karklins, Rasma. *Ethnic Relations in the U.S.S.R.: The Perspective from Below* (Winchester, Mass.: Allen & Unwin, 1986).

Khazanov, Anatoly M. *After the U.S.S.R.: Ethnicity, Nationalism, and Politics in the Commonwealth of Independent States* (Madison, Wis.: University of Wisconsin Press, 1996).

Kraus, Michael, & Liebowitz, Ronald D., eds. *Russia and Eastern Europe After Communism: The Search for New Political, Economic, and Security Systems* (Boulder, Colo.: Westview, 1995).

Lapidus, Gail W., ed. *The New Russia* (Boulder, Colo.: Westview, 1994).

Lincoln, W. Bruce. *The Conquest of a Continent: Siberia and the Russians* (New York: Random House, 1993).

• Lydolph, Paul E. *Geography of the U.S.S.R.* (Elkhart Lake, Wis.: Misty Valley Publishing, 1990).

Lydolph, Paul E. *Climates of the Soviet Union* (Amsterdam: Elsevier Scientific Publishing Co., 1977).

Mackinder, Halford J. *Democratic Ideals and Reality* (New York: Holt, 1919).

• Mather, John R. *Climatology: Fundamentals and Applications* (New York: McGraw-Hill, 1974).

Nahaylo, Bohdan, & Swoboda, Victor. *Soviet Disunion: A History of the Nationalities Problem in the U.S.S.R.* (New York: Free Press, 1990).

"The New Russia: Special Report" (cover story) *Time*, December 7, 1992, pp. 32–69.

Nijman, Jan. *The Geopolitics of Power and Conflict: Superpowers in the International System, 1945–1992* (London & New York: Belhaven Press, 1993).

Peterson, D. J. *Troubled Lands: The Legacy of Soviet Environmental Destruction* (Boulder, Colo.: Westview Press, 1993).

Pryde, Philip R., ed. *Environmental Resources and Constraints in the Former Soviet Republics* (Boulder, Colo.: Westview Press, 1995).

"The Rape of Siberia: The Tortured Land" (cover story), *Time*, September, 4, 1995, pp. 42–53.

Remnick, David. *Lenin's Tomb: The Last Days of the Soviet Empire* (New York: Random House, 1993).

Rodgers, Allan, ed. *The Soviet Far East: Geographical Perspectives on Development* (London & New York: Routledge, 1990).

"Russia Reborn: A Survey of Russia," *The Economist*, December 5, 1992, Special Insert, pp. 3–26.

Shaw, Denis J. B., ed. *The Post-Soviet Republics: A Systematic Geography* (New York: Wiley/Longman, 1995).

Smith, Hedrick. *The New Russians* (New York: Avon Trade Paperbacks, updated version, 1991).

Steele, Jonathan. *Eternal Russia: Yeltsin, Gorbachev, and the Mirage of Democracy* (Cambridge, Mass.: Harvard University Press, 1994).

Stewart, John Massey, ed. *The Soviet Environment: Problems, Policies and Politics* (New York: Cambridge University Press, 1992).

● Symons, Leslie, ed. *The Soviet Union: A Systematic Geography* (London & New York: Routledge, 2nd rev. ed., 1990).

Thompson, John M. *Russia and the Soviet Union: An Historical Introduction from the Kievan State to the Present* (Boulder, Colo.: Westview, 3rd rev. ed., 1994).

Valencia, Mark J., ed. *The Russian Far East in Transition: Opportunities for Regional Economic Cooperation* (Boulder, Colo.: Westview Press, 1995).

Wixman, Ronald. *The Peoples of the U.S.S.R.: An Ethnographic Handbook* (Armonk, N.Y.: M. E. Sharpe, 1984).

Wood, Alan, & French, Richard A., eds. *The Development of Siberia: People and Resources* (Basingstoke, U.K.: Macmillan, 1989).

CHAPTER 3

Adams, John S. "Residential Structure of Midwestern Cities," *Annals of the Association of American Geographers*, 60 (1970): 37–62. Model diagram adapted from p. 56.

Allen, James P., & Turner, Eugene J. *We the People: An Atlas of America's Ethnic Diversity* (New York: Macmillan, 1987).

Atlas of North America: Space Age Portrait of a Continent (Washington, D.C.: National Geographic Society, 1985).

● Atwood, Wallace W. *The Physiographic Provinces of North America* (New York: Ginn, 1940).

Bell, Daniel. *The Coming of Postindustrial Society* (New York: Basic Books, 1973). Quotation taken from p. xvi of the 1976 [paperback] Foreword.

Berry, Brian J. L. "The Decline of the Aging Metropolis: Cultural Bases and Social Process," in Sternlieb, George, & Hughes, James W., eds., *Post-Industrial America: Metropolitan Decline and Inter-Regional Job Shifts* (New Brunswick, N.J.: Center for Urban Policy Research, Rutgers University, 1975), pp. 175–185.

● Birdsall, Stephen S., & Florin, John W. *Regional Landscapes of the United States and Canada* (New York: John Wiley & Sons, 4th rev. ed., 1992).

Bone, Robert M. *The Geography of the Canadian North: Issues and Challenges* (Toronto: Oxford University Press, 1992).

Borchert, John R. "American Metropolitan Evolution," *Geographical Review*, 57 (1967): 301–332.

Borchert, John R. "Futures of American Cities," in Hart, John Fraser, ed., *Our Changing Cities* (Baltimore, Md.: Johns Hopkins University Press, 1991), pp. 218–250.

Castells, Manuel, & Hall, Peter. *Technopoles of the World: The Making of Twenty-First-Century Industrial Complexes* (London & New York: Routledge, 1994).

Castles, Stephen, & Miller, Mark J. *The Age of Migration: International Population Movements in the Modern World* (New York: Guilford, 1993).

● Christian, Charles M., & Harper, Robert A., eds. *Modern Metropolitan Systems* (Columbus: Charles E. Merrill, 1982).

Clark, David. *Post-Industrial America: A Geographical Perspective* (London & New York: Methuen, 1985).

Cohen, Robin, ed. *The Cambridge Survey of World Migration* (New York: Cambridge Univesity Press, 1995).

Demko, George J., & Jackson, Michael C., eds. *Populations at Risk: Vulnerable American Groups at the End of the Twentieth Century* (Boulder, Colo.: Westview, 1995).

De Vita, Carol J. "The United States at Mid-Decade," *Population Bulletin*, 50 (March 1996): 1–48.

"For Want of Glue: A Survey of Canada," *The Economist*, June 29, 1991, 20-page supplement.

Garreau, Joel. *Edge City: Life on the New Frontier* (New York: Doubleday, 1991).

Garreau, Joel. *The Nine Nations of North America* (Boston: Houghton Mifflin, 1981).

Gastil, Raymond D. *Cultural Regions of the United States* (Seattle: University of Washington Press, 1975).

● Getis, Arthur, & Getis, Judith, eds. *The United States and Canada: The Land and the People* (Dubuque, Iowa: Wm. C. Brown, 1995).

Girot, Pascal, ed. *The Americas: World Boundaries, Volume Four* (London & New York: Routledge, 1994).

Gober, Patricia. "Americans on the Move," *Population Bulletin*, 48 (November 1993): 1–40.

Gottmann, Jean. *Megalopolis: The Urbanized Northeastern Seaboard of the United States* (New York: Twentieth Century Fund, 1961).

Harris, Chauncy D., & Ullman, Edward L. "The Nature of Cities," *Annals of the American Academy of Political and Social Science*, 242 (1945): 7–17.

Harris, R. Cole. "Regionalism and the Canadian Archipelago," in McCann, Lawrence D., ed., *Heartland and Hinterland: A Geography of Canada* (Scarborough, Ont.: Prentice-Hall Canada, 1982), pp. 458–484.

Hart, John Fraser. *The Land That Feeds Us* (New York: W. W. Norton, 1991).

● Hartshorn, Truman A. *Interpreting the City: An Urban Geography* (New York: John Wiley & Sons, 2nd rev. ed., 1992).

Herzog, Lawrence A. *Where North Meets South: Cities, Space, and Politics on the U.S.-Mexico Border* (Austin, Tex.: University of Texas Press, 1990).

Heyck, Denis L. D., ed. *Barrios and Borderlands: Cultures of Latinos and Latinas in the United States* (London & New York: Routledge, 1994).

Historical Atlas of the United States (Washington, D.C.: National Geographic Society, centennial ed., 1988).

Hudson, John C. *Making the Corn Belt: A Geographical History of Middle-Western Agriculture* (Bloomington, Ind.: Indiana University Press, 1994).

● Hunt, Charles B. *Natural Regions of the United States and Canada* (San Francisco: W. H. Freeman, 2nd rev. ed., 1974).

Janelle, Donald G., ed. *Geographical Snapshots of North America* (New York: Guilford Press, 1992).

Knox, Paul L. *Urbanization: An Introduction to Urban Geography* (Englewood Cliffs, N.J.: Prentice-Hall, 1994).

Lamont, Lansing. *Breakup: The Coming End of Canada and the Stakes for America* (New York: W. W. Norton, 1994).

Lemco, Jonathan. *Turmoil in the Peaceable Kingdom: The Quebec Sovereignty Movement and Its Implications for Canada and the United States* (Toronto: University of Toronto Press, 1994).

Lewis, George J. *Human Migration: A Geographical Perspective* (New York: St. Martin's Press, 1982).

Malcolm, Andrew H. *The Canadians* (New York: Times Books, 1985).

● McCann, Lawrence D., ed. *Heartland and Hinterland: A Geography of Canada* (Scarborough, Ont.: Prentice-Hall Canada, 1982).

● McKnight, Tom L. *Regional Geography of the United States and Canada* (Englewood Cliffs, N.J.: Prentice-Hall, 1992).

Meinig, Donald W. *The Shaping of America: A Geographical Perspective on 500 Years of History; Vol. 1, Atlantic America, 1492–1800* (New Haven: Yale University Press, 1986).

Meinig, Donald W. *The Shaping of America: A Geographical Perspective on 500 Years of History; Vol. 2, Continental America, 1800–1867* (New Haven: Yale University Press, 1993).

Mitchell, Robert D., & Groves, Paul A., eds. *North America: The Historical Geography of a Changing Continent* (Totowa, N.J.: Rowman & Littlefield, 1987).

Muller, Peter O. *Contemporary Suburban America* (Englewood Cliffs, N.J.: Prentice-Hall, 1981).

● Paterson, John H. *North America: A Geography of the United States and Canada* (New York: Oxford University Press, 9th rev. ed., 1994).

● Putnam, Donald F., & Putnam, Robert G. *Canada: A Regional Analysis* (Toronto: Dent, 2nd rev. ed., 1979).

Roberts, Sam. *Who We Are: A Portrait of America Based on the Latest U.S. Census* (New York: Times Books, 1993).

Rooney, John F., Jr., et al., eds. *This Remarkable Continent: An Atlas of United States and Canadian Society and Cultures* (College Station, Tex.: Texas A&M University Press, 1982).

Short, John R. *The Urban Order: An Introduction to Urban Geography* (Cambridge, Mass.: Blackwell, 1996).

Thompson, John Herd, & Randall, Stephen J. *Canada and the United States: Ambivalent Allies* (Athens, Ga.: University of Georgia Press, 1994).

Vance, James E., Jr. *The North American Railroad: Its Origin, Evolution, and Geography* (Baltimore, Md.: Johns Hopkins University Press, 1996).

Vance, James E., Jr. *The Continuing City: Urban Morphology in Western Civilization* (Baltimore, Md.: Johns Hopkins University Press, 1990).

Vance, James E., Jr. *This Scene of Man: The Role and Structure of the City in the Geography of Western Civilization* (New York: Harper's College Press, 1977). Urban realms model discussed on pp. 411–416.

Ward, David. *Cities and Immigrants: A Geography of Change in Nineteenth-Century America* (New York: Oxford University Press, 1971).

Wheeler, James O., & Muller, Peter O. *Economic Geography* (New York: John Wiley & Sons, 2nd rev. ed., 1986). Map adapted from p. 330.

Woodward, C. Vann. "The South Tomorrow," *Time*, "The South Today: A Special Issue," September 27, 1976. Quotation from p. 99.

Yeates, Maurice H. *The North American City* (New York: Harper & Row, 4th rev. ed., 1990). Canadian urban evolution model discussed on pp. 60–67.

Yeates, Maurice H. *Main Street: Windsor to Quebec City* (Toronto: Macmillan of Canada, 1975).

Zelinsky, Wilbur. *Exploring the Beloved Country: Geographic Forays into American Society and Culture* (Iowa City, Iowa: University of Iowa Press, 1995).

● Zelinsky, Wilbur. *The Cultural Geography of the United States: A Revised Edition* (Englewood Cliffs, N.J.: Prentice-Hall, 2nd rev. ed, 1992).

CHAPTER 4

● Augelli, John P. "The Rimland-Mainland Concept of Culture Areas in Middle America," *Annals of the Association of American Geographers*, 52 (1962): 119–129.

● Blakemore, Harold, & Smith, Clifford T., eds. *Latin America: Geographical Perspectives* (London & New York: Methuen, 2nd rev. ed., 1983).

● Blouet, Brian W., & Blouet, Olwyn M., eds. *Latin America and the Caribbean: A Systematic and Regional Survey* (New York: John Wiley & Sons, 2nd rev. ed., 1993).

Booth, John A., & Walker, Thomas W. *Understanding Central America* (Boulder, Colo.: Westview Press, 2nd rev. ed., 1993).

Bryan, Anthony T., ed. *The Caribbean: New Dynamics in Trade and Political Economy* (New Brunswick, N.J.: Transaction Publishers, 1995).

● Butlin, Robin. *Historical Geography: Through the Gates of Space and Time* (London & New York: Routledge, 1993).

Caufield, Catherine. *In the Rainforest: Report from a Strange, Beautiful, Imperiled World* (Chicago: University of Chicago Press, rev. ed., 1991).

Clark, Andrew H. "Historical Geography," in James, Preston E., & Jones, Clarence F., eds., *American Geography: Inventory and Prospect* (Syracuse, N.Y.: Syracuse University Press, 1954), pp. 70–105.

Collier, Simon, et al., eds. *The Cambridge Encyclopedia of Latin*

America and the Caribbean (New York: Cambridge University Press, 2nd rev. ed., 1992).

Cubitt, Tessa. *Latin American Society* (New York: Wiley/Longman, 2nd rev. ed., 1995).

Davidson, William V., & Parsons, James, eds. *Historical Geography of Latin America* (Baton Rouge: Louisiana State University Press, 1980).

Dunkerly, James. *The Pacification of Central America: Political Change in the Isthmus, 1987–1993* (London & New York: Verso, 1994).

● Gilbert, Alan. *Latin America* (London & New York: Routledge, 1990).

Girot, Pascal, ed. *The Americas: World Boundaries, Volume Four* (London & New York: Routledge, 1994).

Gourou, Pierre. *The Tropical World: Its Social and Economic Conditions and Its Future Status* (London & New York: Longman, 5th rev. ed., trans. Stanley H. Beaver, 1980).

Greenfield, Gerald M., ed. *Latin American Urbanization: Historical Profiles of Major Cities* (Westport, Conn.: Greenwood Press, 1994).

Griffin, Ernst C., & Ford, Larry R. "Cities of Latin America," in Brunn, Stanley D., & Williams, Jack F., eds., *Cities of the World: World Regional Urban Development* (New York: HarperCollins, 2nd rev. ed., 1993), pp. 224–265.

Guillermoprieto, Alma. *The Heart that Bleeds: Latin America Now* (New York: Alfred A. Knopf, 1994).

Helms, Mary W. *Middle America: A Culture History of Heartland and Frontiers* (Englewood Cliffs, N.J.: Prentice-Hall, 1975).

"Heroic Illusions: A Survey of Cuba," *The Economist*, April 6, 1996, Special Insert, pp. 1–16.

Herzog, Lawrence A. *Where North Meets South: Cities, Space, and Politics on the U.S.-Mexico Border* (Austin, Tex.: University of Texas Press, 1990).

Huntington, Ellsworth. *Mainsprings of Civilization* (New York: John Wiley & Sons, 1945).

● James, Preston E., & Minkel, Clarence W. *Latin America* (New York: John Wiley & Sons, 5th rev. ed., 1986), Chaps. 2–16.

Knight, Franklin W., & Palmer, Colin A., eds. *The Modern Caribbean* (Chapel Hill, N.C.: University of North Carolina Press, 1989).

Kopinak, Kathryn. *Desert Capitalism: Maquiladoras in North America's Western Industrial Corridor* (Tuscon: University of Arizona Press, 1996).

Krauss, Clifford. *Inside Central America: Its People, Politics, and History* (New York: Touchstone/Simon & Schuster, 1991).

Kurlansky, Mark. *A Continent of Islands: Searching for the Caribbean Destiny* (Reading, Mass.: Addison-Wesley, 1992).

Lockhart, Douglas G., et al., eds. *The Development Process in Small Island States* (London & New York: Routledge, 1993).

Lovell, W. George, & Lutz, Christopher H. *Demography and Empire: A Guide to the Population History of Spanish Central America, 1500–1821* (Boulder, Colo.: Westview Press, 1995).

● Lowenthal, David. *West Indian Societies* (New York: Oxford University Press, 1972).

MacPherson, John. *Caribbean Lands* (London & New York: Longman, 4th rev. ed., 1980).

McCullough, David G. *The Path Between the Seas: The Creation of the Panama Canal, 1870–1914* (New York: Simon & Schuster, 1977).

Momsen, Janet, ed. *Women and Change in the Caribbean* (Bloomington, Ind.: Indiana University Press, 1994).

Myers, Norman. *The Primary Source: Tropical Forests and Our Future* (New York: W. W. Norton, 1984).

Pick, James B. & Butler, Edgar W. *Mexico Megacity* (Boulder, Colo.: Westview Press, 1996).

Pick, James B., & Butler, Edgar W. *The Mexico Handbook: Economic and Demographic Maps and Statistics* (Boulder, Colo.: Westview Press, 1993).

● Richardson, Bonham C. *The Caribbean in the Wider World, 1492–1992: A Regional Geography* (New York: Cambridge University Press, 1992).

Riding, Alan. *Distant Neighbors: A Portrait of the Mexicans* (New York: Alfred A. Knopf, 1985).

Sargent, Charles S., Jr. "The Latin American City," in Blouet, Brian W., & Blouet, Olwyn M., eds., *Latin America and the Caribbean: A Systematic and Regional Survey* (New York: John Wiley & Sons, 2nd rev. ed., 1993), pp. 172–216. Quotation from p. 191; diagram adapted from p. 192.

Sauer, Carl Ortwin. "Foreword to Historical Geography," *Annals of the Association of American Geographers*, 31 (1941): 1–24.

Sealey, Neil. *Caribbean World: A Complete Geography* (New York: Cambridge University Press, 1992).

Sklair, Leslie. *Assembling for Development: Maquila Industry in Mexico and the United States* (Boulder, Colo.: Westview Press, 1989).

Stanislawski, Dan. "Early Spanish Town Planning in the New World," *Geographical Review*, 37 (1947): 94–105.

Uchitelle, Louis. "America's Newest Industrial Belt: Northern Mexico," *New York Times*, March 21, 1993, pp. F1, F4.

Ward, Peter M. *Mexico City* (Boston: G. K. Hall, 1990).

Watts, David. *The West Indies: Patterns of Development, Culture and Environmental Change Since 1492* (New York: Cambridge University Press, 1987).

West, Robert C. *Sonora: Its Geographical Personality* (Austin, Tex.: University of Texas Press, 1993).

● West, Robert C., Augelli, John P., et al. *Middle America: Its Lands and Peoples* (Englewood Cliffs, N.J.: Prentice-Hall, 3rd rev. ed., 1989).

Wilken, Gene C. *Good Farmers: Traditional Agricultural Resource Management in Mexico and Central America* (Berkeley, Calif.: University of California Press, 1987).

Wilson, Patricia A. *Exports and Local Development: Mexico's New Maquiladoras* (Austin, Tex.: University of Texas Press, 1992).

Winn, Peter. *Americas: The Changing Face of Latin America and the Caribbean* (Berkeley, Calif.: University of California Press, 1992).

CHAPTER 5

Augelli, John P. "The Controversial Image of Latin America: A Geographer's View," *Journal of Geography*, 62 (1963): 103–112. Quotation from p. 111.

Barman, R. *Brazil: The Forging of a Nation* (Palo Alto, Calif.: Stanford University Press, 1988).

Becker, Bertha K., & Egler, Claudio A. G. *Brazil: A New Regional Power in the World-Economy: A Regional Geography* (New York: Cambridge University Press, 1992).

Berry, Brian J. L., et al. *The Global Economy: Resource Use, Locational Choice, and International Trade* (Englewood Cliffs, N.J.: Prentice-Hall, 1993).

● Blakemore, Harold, & Smith, Clifford T., eds. *Latin America: Geographical Perspectives* (London & New York: Methuen, 2nd rev. ed., 1983).

● Blouet, Brian W., & Blouet, Olwyn M., eds. *Latin America and the Caribbean: A Systematic and Regional Survey* (New York: John Wiley & Sons, 2nd rev. ed., 1993).

Brawer, Moshe. *Atlas of South America* (New York: Simon & Schuster, 1991).

● Bromley, Rosemary D. F., & Bromley, Ray. *South American Development: A Geographical Introduction* (New York: Cambridge University Press, 2nd rev. ed., 1988).

Brooke, James. "The New South Americans: Friends and Partners," *New York Times*, April 8, 1994, A–3.

Brunn, Stanley D., & Williams, Jack F., eds. *Cities of the World: World Regional Urban Development* (New York: HarperCollins, 2nd rev. ed., 1993).

Caufield, Catherine. *In the Rainforest: Report from a Strange, Beautiful, Imperiled World* (Chicago: University of Chicago Press, rev. ed., 1991).

Caviedes, César N. *The Southern Cone: Realities of the Authoritarian State* (Totowa, N.J.: Rowman & Allanheld, 1984).

● Caviedes, César N., & Knapp, Gregory. *South America* (Englewood Cliffs, N.J.: Prentice Hall, 1995).

Child, Jack. *Geopolitics and Conflict in South America: Quarrels Among Neighbors* (Westport, Conn.: Praeger, 1985).

"Cocaine Wars: South America's Bloody Business," *Time*, February 25, 1985, pp. 26–35.

Collier, Simon, et al., eds. *The Cambridge Encyclopedia of Latin America and the Caribbean* (New York: Cambridge University Press, 2nd rev. ed., 1992).

Crow, John A. *The Epic of Latin America* (Berkeley, Calif.: University of California Press, 4th rev. ed., 1992).

Cubitt, Tessa. *Latin American Society* (New York: Wiley/Longman, 2nd rev. ed., 1995).

Denevan, William M., ed. *Hispanic Lands and Peoples: Selected Writings of James J. Parsons* (Boulder, Colo.: Westview Press, 1988).

Dickenson, John P. *Brazil* (London & New York: Longman, 1983).

Drakakis-Smith, David. *The Third World City* (Boston: Routledge & Kegan Paul, 1987).

Freeman, Michael. *Atlas of World Economy* (New York: Simon & Schuster, 1991).

● Gilbert, Alan. *Latin America* (London & New York: Routledge, 1990).

Gilbert, Alan, & Gugler, Josef. *Cities, Poverty and Development: Urbanization in the Third World* (New York: Oxford University Press, 2nd rev. ed., 1992).

Girot, Pascal, ed. *The Americas: World Boundaries, Volume Four* (London & New York: Routledge, 1994).

Goulding, Michael, et al. *Floods of Fortune: Ecology and Economy Along the Amazon* (New York: Columbia University Press, 1996).

Greenfield, Gerald M., ed. *Latin American Urbanization: Historical Profiles of Major Cities* (Westport, Conn.: Greenwood Press, 1994).

Griffin, Ernst C., & Ford, Larry R. "Cities of Latin America," in Brunn, Stanley D., & Williams, Jack F., eds. *Cities of the World: World Regional Urban Development* (New York: HarperCollins, 2nd rev. ed., 1993), pp. 224–265.

● Griffin, Ernst C., & Ford, Larry R. "A Model of Latin American City Structure," *Geographical Review*, 70 (1980): 397–422. Model diagram adapted from p. 406.

Grigg, David B. *An Introduction to Agricultural Geography* (London & New York: Routledge, 2nd rev. ed., 1995).

Grigg, David B. *The Agricultural Systems of the World: An Evolutionary Approach* (London: Cambridge University Press, 1974).

Guillermoprieto, Alma. *The Heart That Bleeds: Latin America Now* (New York: Alfred A. Knopf, 1994).

Gwynne, Robert N. *Industrialization and Urbanization in Latin America* (Baltimore, Md.: Johns Hopkins University Press, 1986).

● Hartshorn, Truman A. (& Alexander, John W.). *Economic Geography* (Englewood Cliffs, N.J.: Prentice-Hall, 3rd rev. ed., 1988).

● James, Preston E., & Minkel, Clarence W. *Latin America* (New York: John Wiley & Sons, 5th rev. ed., 1986), Chaps. 17–37.

Keeling, David J. *Contemporary Argentina: A Geographical Perspective* (Boulder, Colo.: Westview Press, 1996).

Kent, Robert B. "Geographical Dimensions of the Shining Path Insurgency in Peru," *Geographical Review*, 83 (October 1993): 441–454.

Knox, Paul L., & Agnew, John A. *The Geography of the World Economy* (London & New York: Edward Arnold, 2nd rev. ed., 1994).

Lisansky, Judith. *Migrants to Amazonia: Spontaneous Colonization in the Brazilian Frontier* (Boulder, Colo.: Westview Press, 1989).

Lowder, Stella. *The Geography of Third World Cities* (Totowa, N.J.: Barnes & Noble, 1986).

Maos, Jacob O. *The Spatial Organization of New Land Settlement in Latin America* (Boulder, Colo.: Westview Press, 1984).

Margolis, Mac. *The Last New World: The Conquest of the Amazon Frontier* (New York: W. W. Norton, 1992).

Mercer, Pamela. "Colombia and Coffee Part Ways," *New York Times*, January 9, 1996, A-5.

● Morris, Arthur S. *South America* (Totowa, N.J.: Barnes & Noble, 3rd rev. ed., 1987).

Myers, Norman. *The Primary Source: Tropical Forests and Our Future* (New York: W. W. Norton, 1984).

Nash, Nathaniel C. "Chile: Japan's Backdoor to the West," *New York Times*, April 15, 1993, pp. C1, C6.

Nash, Nathaniel C. "Squalid Slums Grow as People Flood Latin America's Cities," *New York Times*, October 11, 1992, pp. 1, 10.

Page, Joseph A. *The Brazilians* (Reading, Mass.: Addison-Wesley, 1995).

Potter, Robert B. *Third World Urbanization: Contemporary Issues in Geography* (New York: Oxford University Press, 1991).

● Preston, David A., ed. *Latin American Development: Geographical Perspectives* (London & New York: Longman, 1987).

Rengert, George F. *The Geography of Illegal Drugs* (Boulder, Colo.: Westview Press, 1996).

Sanchez-Albornoz, Nicolas. *The Population of Latin America: A History* (Berkeley, Calif: University of California Press, 1974).

Schumann, Debra A., & Partridge, William L., eds. *The Human*

Ecology of Tropical Settlement (Boulder, Colo.: Westview Press, 1987).

Smith, Nigel J. H., et al. *Amazonia: Resiliency and Dynamism of the Land and Its People* (New York & Tokyo: United Nations University Press, 1995).

Steward, Julian H., & Faron, Louis C., eds. *Native Peoples of South America* (New York: McGraw-Hill, 1959).

Stewart, Douglas Ian. *After the Trees: Living on the Transamazon Highway* (Austin, Tex.: University of Texas Press, 1994).

"Survey of Brazil: The Blessed and the Cursed," *The Economist,* December 7, 1991, Special Insert, pp. 3–22.

● Wheeler, James O., Muller, Peter O., Thrall, Grant I., & Fik, Timothy J. *Economic Geography* (New York: John Wiley & Sons, 3rd rev. ed., 1998).

Wilkie, Richard W. *Latin American Population and Urbanization Analysis: Maps and Statistics, 1950–1982* (Westwood, Calif.: UCLA Latin American Center, 1984).

Winn, Peter. *Americas: The Changing Face of Latin America and the Caribbean* (Berkeley, Calif.: University of California Press, 1992).

CHAPTER 6

Ahmed, Akbar S. *Discovering Islam: Making Sense of Muslim History and Society* (Boston: Routledge & Kegan Paul, 1987).

Akiner, Shirin, ed. *Cultural Change and Continuity in Central Asia* (London & New York: Kegan Paul International, 1992).

Amirahmadi, Hooshang, & El-Shakhs, Salah S., eds. *Urban Development in the Muslim World* (New Brunswick, N.J.: Rutgers University, Center for Urban Policy Research, 1993).

Barakat, Halim. *The Arab World: Society, Culture, and State* (Berkeley, Calif.: University of California Press, 1993).

Batalden, Stephen K., & Batalden, Sandra L. *The Newly Independent States of Eurasia: Handbook of Former Soviet Republics* (Phoenix: Oryx Press, 1993).

Beaumont, Peter. *Drylands: Environmental Management and Development* (London & New York: Routledge, 1993).

● Beaumont, Peter, et al. *The Middle East: A Geographical Study* (New York: Wiley/Halsted, 2nd rev. ed., 1988).

Blake, Gerald H., et al. *The Cambridge Atlas of the Middle East and North Africa* (Cambridge, U.K.: Cambridge University Press, 1988).

Brawer, Moshe. *Atlas of Russia and the Independent Republics* (New York: Simon & Schuster, 1994).

Chapman, Graham P., & Baker, Kathleen M., eds. *The Changing Geography of Africa and the Middle East* (London & New York: Routledge, 1992).

Chapman, Graham P., & Baker, Kathleen M., eds. *The Changing Geography of Asia* (London & New York: Routledge, 1992).

Chinn, Jeff & Kaiser, Robert. *Russians as the New Minority: Ethnicity and Nationalism in the Soviet Successor States* (Boulder, Colo.: Westview Press, 1996).

Cloudsley-Thompson, J. L., ed. *Sahara Desert* (Oxford, U.K.: Pergamon Press, 1984).

Cohen, Saul B. "Middle East Geopolitical Transformation: The Disappearance of a Shatter Belt," *Journal of Geography,* 91 (January/February 1992): 2–10.

● Cressey, George B. *Crossroads: Land and Life in Southwest Asia* (Philadelphia: J. B. Lippincott, 1960).

Dawisha, Karen, & Parrott, Bruce. *Russia and the New States of Eurasia: The Politics of Upheaval* (New York: Cambridge University Press, 1994).

Diuk, Nadia, and Karatnycky, Adrian. *New Nations Rising: The Fall of the Soviets and the Challenge of Independence* (New York: John Wiley & Sons, 1993).

Dohrs, Fred E., & Sommers, Lawrence M., eds. *Cultural Geography: Selected Readings* (New York: Thomas Y. Crowell, 1967).

● Drysdale, Alasdair, & Blake, Gerald H. *The Middle East and North Africa: A Political Geography* (New York: Oxford University Press, 1985).

Dutt, Ashok K., et al., eds. *The Asian City: Processes of Development, Characteristics and Planning* (Norwell, Mass.: Kluwer Academic, 1994).

Field, Michael. *Inside the Arab World* (Cambridge, Mass.: Harvard University Press, 1995).

Findlay, Allan M. *The Arab World* (London & New York: Routledge, 1994).

● Fisher, William B. *The Middle East: A Physical, Social and Regional Geography* (London & New York: Methuen, 7th rev. ed., 1978).

Foote, Kenneth E., et al., eds. *Re-Reading Cultural Geography* (Austin, Tex.: University of Texas Press, 1994).

Freeman-Grenville, G. S. P. *The Historical Atlas of the Middle East* (New York: Simon & Schuster, 1993).

Fromkin, David. *A Peace to End All Peace: Creating the Modern Middle East, 1914–1922* (New York: Henry Holt, 1989).

Fukui, Katsuyoshi, & Markakis, John, eds. *Ethnicity and Conflict in the Horn of Africa* (Athens, Ohio: Ohio University Press, 1994).

Fuller, Graham, et al. *Turkey's New Geopolitics: From the Balkans to Western China* (Boulder, Colo.: Westview Press, 1993).

Gilbert, Martin. *Atlas of the Arab-Israeli Conflict* (New York: Oxford University Press, 1994).

Gilsenan, Michael. *Recognizing Islam: Religion and Society in the Modern Middle East* (London: I.B. Tauris, 1992).

Gleason, Gregory. *The Central Asian States: Discovering Independence* (Boulder, Colo.: Westview Press, 1996).

Goldscheider, Calvin. *Israel's Changing Society: Population, Ethnicity, and Development* (Boulder, Colo.: Westview Press, 1996).

Goldschmidt, Arthur, Jr. *A Concise History of the Middle East* (Boulder, Colo.: Westview Press, 5 rev. ed., 1996).

● Gould, Peter R. *Spatial Diffusion* (Washington, D.C.: Association of American Geographers, Commission on College Geography, Resource Paper No. 4, 1969).

Grundy-Warr, Carl, ed. *Eurasia: World Boundaries, Vol. 3* (London & New York: Routledge, 1994).

Gurdon, Charles G., ed. *The Horn of Africa* (New York: St. Martin's Press, 1994).

Hauner, Milan. *What Is Asia to Us? Russia's Asian Heartland Yesterday and Today* (London & New York: Routledge, 1992).

Heathcote, Ronald L. *The Arid Lands: Their Use and Abuse* (London & New York: Longman, 1983).

● Held, Colbert C. *Middle East Patterns: Places, Peoples, and Politics* (Boulder, Colo.: Westview Press, 2nd rev. ed., 1994).

Hiro, Dilip. *Between Marx and Muhammad: The Changing Face of Central Asia* (New York: HarperCollins, 1994).

Hourani, Albert H. *A History of the Arab Peoples* (Cambridge, Mass.: Belknap/Harvard University Press, 1991).

Hourani, Albert, et al., eds. *The Modern Middle East: A Reader* (Berkeley, Calif.: University of California Press, 1994).

Jabbur, Jibrail S. *The Bedouins and the Desert: Aspects of Nomadic Life in the Arab East* (Albany, N.Y.: State University of New York Press, 1995).

Jackson, Peter. *Maps of Meaning: An Introduction to Cultural Geography* (London & New York: Routledge, 1989).

Joffé, George, ed. *North Africa: Nation, State and Region* (London & New York: Routledge, 1993).

Kaiser, Robert J. *The Geography of Nationalism in Russia and the U.S.S.R.* (Princeton, N.J.: Princeton University Press, 1994).

Khazanov, Anatoly M. *After the U.S.S.R.: Ethnicity, Nationalism, and Politics in the Commonwealth of Independent States* (Madison, Wis.: University of Wisconsin Press, 1996).

Kliot, Nurit. *Water Resources and Conflict in the Middle East* (London & New York: Routledge, 1994).

Kotlyakov, V. M. ''The Aral Sea Basin: A Critical Environmental Zone,'' *Environment*, (January–February 1991): 4–9, 36–38.

Kreyenbroek, Philip G., & Sperl, Stefan, eds. *The Kurds: A Contemporary Overview* (London & New York: Routledge, 1992).

Lamb, David. *The Arabs: Journeys Beyond the Mirage* (New York: Random House, 1987).

Lemarchand, Philippe, ed. *The Arab World, The Gulf, and the Middle East: An Atlas* (Boulder, Colo.: Westview Press, 1992).

Lewis, Bernard. *The Middle East: A Brief History of the Last 2,000 Years* (New York: Scribner, 1996).

Lewis, Robert A., ed. *Geographic Perspectives on Soviet Central Asia* (London & New York: Routledge, 1992).

• Longrigg, Stephen H. *The Middle East: A Social Geography* (Chicago: Aldine, 2nd rev. ed., 1970).

Manz, Beatrice F., ed. *Central Asia in Historical Perspective* (Boulder, Colo.: Westview Press, 1994).

Mikesell, Marvin W. ''Tradition and Innovation in Cultural Geography,'' *Annals of the Association of American Geographers*, 68 (1978): 1–16.

• Mostyn, Trevor, ed. *The Cambridge Encyclopedia of the Middle East and North Africa* (New York: Cambridge University Press, 1988).

Newman, David. *Population, Settlement, and Conflict: Israel and the West Bank* (New York: Cambridge University Press, 1991).

Omran, Abdel R., & Roudi, Farzaneh. ''The Middle East Population Puzzle,'' *Population Bulletin*, 48 (July 1993): 1–40.

Park, Chris. *Sacred Worlds: An Introduction to Geography and Religion* (London & New York: Routledge, 1994).

Peters, F. E. *The Hajj: The Muslim Pilgrimage to Mecca and the Holy Places* (Princeton, N.J.: Princeton University Press, 1994).

Peters, Joan. *From Time Immemorial: The Origins of the Arab-Israeli Conflict over Palestine* (New York: Harper & Row, 1984).

Peterson, D. J. *Troubled Lands: The Legacy of Soviet Environmental Destruction* (Boulder, Colo.: Westview Press, 1993).

Pinault, David. *The Shi'ites: Ritual and Popular Piety in a Muslim Community* (New York: St. Martin's Press, 1993).

Prescott, J. R. V. *Political Frontiers and Boundaries* (Winchester, Mass.: Allen & Unwin, 1987).

Pryde, Philip R., ed. *Environmental Resources and Constraints in the Former Soviet Republics* (Boulder, Colo.: Westview Press, 1995).

Rahman, Mushtaqur, ed. *Muslim World: Geography and Development* (Lanham, Md.: University Press of America, 1987).

Rashid, Ahmed. *The Resurgence of Central Asia: Islam or Nationalism?* (Atlantic Highlands, N.J.: Zed Books, 1994).

Robinson, Francis, ed. *Atlas of the Islamic World Since 1500* (New York: Facts on File, 1982).

Sayigh, Yusif A. *Elusive Development: From Dependence to Self-Reliance in the Arab Region* (London & New York: Routledge, 1991).

Schofield, Clive H., & Schofield, Richard, eds. *The Middle East and North Africa: World Boundaries, Vol. 2* (London & New York: Routledge, 1994).

Shaw, Denis J. B., ed. *The Post-Soviet Republics: A Systematic Geography* (New York: Wiley/Longman, 1995).

• Spencer, Joseph E. ''The Growth of Cultural Geography,'' *American Behavioral Scientist*, 22 (1978): 79–92.

Starr, Joyce R., & Stoll, Daniel C., eds. *The Politics of Scarcity: Water in the Middle East* (Boulder, Colo.: Westview Press, 1988).

Stauffer, T. R. ''Water Flows in (Gadhafi's) Pharaonic Project,'' *Christian Science Monitor*, January 8, 1996, pp. 1, 8.

Swearingen, Will D., & Bencherifa, Abdellatif, eds. *The North African Environment at Risk* (Boulder, Colo.: Westview Press, 1996).

Tessler, Mark. *A History of the Israeli-Palestinian Conflict* (Bloomington, Ind.: Indiana University Press, 1994).

Wagner, Philip L., & Mikesell, Marvin W., eds. *Readings in Cultural Geography* (Chicago: University of Chicago Press, 1962).

Wixman, Ronald. *The Peoples of the U.S.S.R.: An Ethnographic Handbook* (Armonk, N.Y.: M. E. Sharpe, 1984).

Wright, Robin B. *Sacred Rage: The Crusade of Modern Islam* (New York: Linden Press/Simon & Schuster, 1985).

CHAPTER 7

''Africa: The Scramble for Existence,'' *Time*, Cover Story, September 7, 1992, pp. 40–46.

Baker, Jonathan, & Pederson, Paul Ove, eds. *The Rural-Urban Interface in Africa: Expansion and Adaptation* (Uppsala, Sweden: Scandinavian Institute of African Studies, 1992).

Bell, Morag. *Contemporary Africa: Development, Culture and the State* (White Plains, N.Y.: Longman, 1986).

• Best, Alan C. G. & de Blij, H. J. *African Survey* (New York: John Wiley & Sons, 1977).

Binns, Tony. *Tropical Africa* (London & New York: Routledge, 1994).

Black, Richard, & Robinson, Vaughan, eds. *Geography and Refugees: Patterns and Processes of Change* (New York: Wiley/Belhaven, 1993).

Bohannan, Paul, & Curtin, Philip D. *Africa and Africans* (Prospect Heights, Ill.: Waveland Press, 4th rev. ed., 1996).

Chapman, Graham P., & Baker, Kathleen M., eds. *The Changing*

Geography of Africa and the Middle East (London & New York: Routledge, 1992).

Christopher, Anthony J. *Colonial Africa: An Historical Geography* (Totowa, N.J.: Barnes & Noble, 1984).

Christopher, Anthony J. *South Africa: The Impact of Past Geographies* (Cape Town: Juta & Co., 1984).

Curtin, Philip D. *The Atlantic Slave Trade* (Madison, Wis.: University of Wisconsin Press, 1969).

Davidson, Basil. *The Black Man's Burden: Africa and the Curse of the Nation-State* (New York: Times Books/Random House, 1992).

• de Blij, H. J. "Africa's Geomosaic Under Stress," *Journal of Geography*, 90 (January/February, 1991): 2–9.

Drakakis-Smith, David. *Urban and Regional Change in Southern Africa* (London & New York: Routledge, 1992).

Fage, J. D. *A History of Africa* (London & New York: Routledge, 3rd rev. ed., 1995).

Fardon, Richard, & Furniss, Graham, eds. *African Languages, Development, and the State* (London & New York: Routledge, 1994).

"The Final Lap: A Survey of South Africa," *The Economist*, March 20, 1993, 26 pp.

Foster, Harold D. *Health, Disease and the Environment* (New York: John Wiley & Sons, 1992).

Freeman-Grenville, G. S. P. *The New Atlas of African History* (New York: Simon & Schuster, 1991).

Gesler, Wilbert M. *The Cultural Geography of Health Care* (Pittsburgh: University of Pittsburgh Press, 1991).

Gleave, Michael B., ed. *Tropical African Development: Geographical Perspectives* (New York: Wiley/Longman, 1992).

Goliber, Thomas J. "Africa's Expanding Population: Old Problems, New Policies," *Population Bulletin*, 44 (November 1989): 1–49.

Goodwin, June, & Schiff, Ben. *Heart of Whiteness: Afrikaners Face Black Rule in the New South Africa* (New York: Scribners, 1995).

• Gould, Peter R. *The Slow Plague: A Geography of the AIDS Pandemic* (Cambridge, Mass.: Blackwell, 1993).

Gourou, Pierre. *The Tropical World: Its Social and Economic Conditions and Its Future Status* (London & New York: Longman, 5th rev. ed., trans. Stanley H. Beaver, 1980).

Griffiths, Ieuan L. L. *The African Inheritance* (London & New York: Routledge, 1995).

Griffiths, Ieuan L. L. *The Atlas of African Affairs* (London & New York: Routledge, 2nd rev. ed., 1994).

• Grove, Alfred T. *The Changing Geography of Africa* (New York: Oxford University Press, 2nd rev. ed., 1994).

Harrison Church, Ronald J. *West Africa: A Study of the Environment and Man's Use of It* (London: Longman, 8th rev. ed., 1980).

Hodd, Michael. *Economies of Africa: Geography, Population, History, Stability, Structure, Performance, Forecasts* (Boston: G. K. Hall, 1991).

Howe, G. Melvin, ed. *Global Geocancerology: World Geography of Human Cancers* (Edinburgh, U.K.: Churchill Livingstone, 1986).

Huke, Robert E. "The Green Revolution," *Journal of Geography*, 84 (1985): 248–254.

Knight, C. Gregory, & Newman, James L., eds. *Contemporary*

Africa: Geography and Change (Englewood Cliffs, N.J.: Prentice-Hall, 1976).

Lamb, David. *The Africans* (New York: Random House, 1983).

Learmonth, Andrew T.A. *Disease Ecology: An Introduction* (New York: Blackwell, 1988).

Lemon, Anthony, ed. *Geography of Change in South Africa* (New York: Wiley/Longman, 1995).

Lewis, Laurence A., & Berry, Leonard. *African Environments and Resources* (Winchester, Mass.: Unwin Hyman, 1988).

Martin, Esmond Bradley, & de Blij, H. J., eds. *African Perspectives: An Exchange of Essays on the Economic Geography of Nine African States* (London & New York: Methuen, 1981).

• Meade, Melinda S., et al. *Medical Geography* (New York: Guilford Press, 1987).

Mehretu, Assefa. *Regional Disparity in Sub-Saharan Africa: Structural Readjustment of Uneven Development* (Boulder, Colo.: Westview Press, 1989).

Meredith, Martin. *In the Name of Apartheid: South Africa in the Postwar Period* (New York: Harper & Row, 1988).

Middleton, John, ed. *Encyclopedia of Sub-Saharan Africa* (New York: Simon & Schuster, 4 vols., 1994).

• Moon, Graham, & Jones, Kelvyn. *Health, Disease and Society: An Introduction to Medical Geography* (New York: Routledge & Kegan Paul, 1988).

Mortimore, Michael. *Adapting to Drought: Farmers, Famines and Desertification in West Africa* (New York: Cambridge University Press, 1989).

• Mountjoy, Alan, & Hilling, David. *Africa: Geography and Development* (Totowa, N.J.: Barnes & Noble, 1987).

Murdock, George P. *Africa: Its Peoples and Their Culture History* (New York: McGraw-Hill, 1959).

Murray, Martin. *The Revolution Deferred: The Painful Birth of Post-Apartheid South Africa* (London & New York: Verso, 1994).

Newman, James L. *The Peopling of Africa: A Geographic Interpretation* (New Haven: Yale University Press, 1995).

O'Connor, Anthony M. *Poverty in Africa: A Geographical Approach* (New York: Columbia University Press, 1991).

Oliver, Roland, & Crowder, Michael, eds. *The Cambridge Encyclopedia of Africa* (Cambridge, U.K.: Cambridge University Press, 1981).

Pakenham, Thomas. *The Scramble for Africa: The White Man's Conquest of the Dark Continent from 1876 to 1912* (New York: Random House, 1991).

Pritchard, J. M. *Landform and Landscape in Africa* (London: Edward Arnold, 1979).

Reading, Alison, J., et al. *Humid Tropical Environments* (Cambridge, Mass.: Blackwell, 1995).

• Senior, Michael, & Okunrotifa, P. *A Regional Geography of Africa* (London & New York: Longman, 1983).

Shannon, Gary W., Pyle, Gerald F., & Bashshur, Rashid L. *The Geography of AIDS: Origins and Course of an Epidemic* (New York: Guilford Press, 1991).

Shillington, Kevin. *History of Africa* (New York: St. Martin's Press, 2 rev. ed., 1995).

Siddle, David, & Swindell, Ken. *Rural Change in Tropical Africa* (Cambridge, Mass.: Blackwell, 1994).

Smith, David M. *Apartheid in South Africa* (Cambridge, U.K.: Cambridge University Press, 3rd rev. ed., 1990).

Smith, David M., ed. *The Apartheid City and Beyond: Urbanization and Social Change in South Africa* (London & New York: Routledge, 1992).

Sparks, Allister. *Tomorrow is Another Country: The Inside Story of South Africa's Road to Change* (New York: Hill & Wang, 1996).

● Stock, Robert. *Africa South of the Sahara: A Geographical Interpretation* (New York: Guilford Press, 1995).

Stren, Richard, & White, Rodney, eds. *African Cities in Crisis: Managing Rapid Urban Growth* (Boulder, Colo.: Westview Press, 1989).

Turner, Bill L., II, Hyden, Goran, & Kates, Robert W., eds. *Population Growth and Agricultural Change in Africa* (Gainesville, Fla.: University Press of Florida, 1993).

Udo, Reuben K. *The Human Geography of Tropical Africa* (Exeter, N.H.: Heinemann Educational Books, 1982).

United Nations High Commission for Refugees. *The State of the World's Refugees: The Search for Solutions* (New York: Oxford University Press, 1995).

CHAPTER 8

Bhardwaj, Surinder M. *Hindu Places of Pilgrimage in India: A Study in Cultural Geography* (Berkeley, Calif.: University of California Press, 1973).

Burki, Shahid J. *Pakistan: A Nation in the Making* (Boulder, Colo.: Westview Press, 1986).

Chapman, Graham P., & Baker, Kathleen M., eds. *The Changing Geography of Asia* (London & New York: Routledge, 1992).

Cohen, Joel E. *How Many People Can the Earth Support?* (New York: W. W. Norton, 1996).

Costa, Frank J., et al., eds. *Urbanization in Asia: Spatial Dimensions and Policy Issues* (Honolulu: University of Hawaii Press, 1989).

Costa, Frank J., et al., eds. *Asian Urbanization: Problems and Processes* (Forestburgh, N.Y.: Lubrecht & Cramer, 1988).

Crossette, Barbara. *So Close to Heaven: The Vanishing Buddhist Kingdoms of the Himalayas* (New York: Alfred A. Knopf, 1995).

Crossette, Barbara. *India: Facing the Twenty-First Century* (Bloomington, Ind.: Indiana University Press, 1993).

Deshpande, C. D. *India: A Regional Interpretation* (New Delhi: Northern Book Centre/Indian Council of Social Science Research, 1992).

Dumont, Louis. *Homo Hierarchus: The Caste System and Its Implications* (Chicago: University of Chicago Press, 1970).

Dutt, Ashok K. "Cities of South Asia," in Brunn, Stanley D., & Williams, Jack F., eds., *Cities of the World: World Regional Urban Development* (New York: HarperCollins, 2nd rev. ed., 1993), pp. 351–387.

● Dutt, Ashok K., & Geib, Margaret. *An Atlas of South Asia* (Boulder, Colo.: Westview Press, 1987).

Dutt, Ashok K., et al., eds. *The Asian City: Processes of Development, Characteristics and Planning* (Norwell, Mass.: Kluwer Academic, 1994).

Er-Rashid, Haroun. *Geography of Bangladesh* (Boulder, Colo.: Westview Press, 1977).

● Farmer, Bertram H. *An Introduction to South Asia* (London & New York: Routledge, 2nd rev. ed, 1993).

Frater, Alexander. *Chasing the Monsoon: A Modern Pilgrimage Through India* (New York: Alfred A. Knopf, 1991).

Gadgil, Madhav, & Guha, Ramachandra. *Ecology and Equity: The Use and Abuse of Nature in Contemporary India* (London & New York: Routledge, 1996).

Gargan, Edward A. "Rising Tide of Hindu Hostility Is Worrying India's Muslims," *New York Times*, September 17, 1993, A1, A5.

Grundy-Warr, Carl, ed. *Eurasia: World Boundaries, Vol. 3* (London & New York: Routledge, 1994).

Isaac, Kalpana. "Sri Lanka's Ethnic Divide," *Current History*, 95 (April 1996): 177–181.

Ives, Jack D., & Messerli, Bruno. *Himalayan Dilemma: Reconciling Development and Conservation* (London & New York: Routledge, 1989).

● Johnson, Basil L. C. *Bangladesh* (Totowa, N.J.: Barnes & Noble, 2nd rev. ed., 1982).

Johnson, Basil L. C. *Development in South Asia* (New York: Viking Penguin, 1983).

Johnson, Basil L. C. *India: Resources and Development* (Totowa, N.J.: Barnes & Noble, 1979).

● Jones, Huw R. *A Population Geography* (New York: Guilford Press, 2nd rev. ed., 1991).

Karan, Pradyumna P. *Bhutan* (Lexington, Ky.: University Press of Kentucky, 1967).

Karan, Pradyumna P., & Ishii, Hiroshi. *Nepal: Development and Change in a Landlocked Himalayan Kingdom* (Tokyo: Institute for the Studies of Languages and Cultures of Asia and Africa, Tokyo University of Foreign Studies, 1994).

Kosinski, Leszek A., & Elahi, K. Maudood, eds. *Population Redistribution and Development in South Asia* (Hingham, Mass.: D. Reidel, 1985).

Lall, Arthur S. *The Emergence of Modern India* (New York: Columbia University Press, 1981).

Lipner, Julius J. *Hinduism* (London & New York: Routledge, 1994).

Lukacs, John R., ed. *The People of South Asia: The Biological Anthropology of India, Pakistan, and Nepal* (New York: Plenum Press, 1984).

McColl, Robert W. "The Insurgent State: Territorial Bases of Revolution," *Annals of the Association of American Geographers*, 59 (1969): 613–631.

McGowan, William. *Only Man Is Vile: The Tragedy of Sri Lanka* (New York: Farrar, Straus & Giroux, 1992).

Malik, Yogendra K., & Singh, V. B. *Hindu Nationalists in India: The Rise of the Bharatiya Janata Party* (Boulder, Colo.: Westview Press, 1994).

Manogaran, Chelvadurai. *Ethnic Conflict and Reconciliation in Sri Lanka* (Honolulu: University of Hawaii Press, 1987).

Mitra, Subrata K., ed. *Subnational Movements in South Asia* (Boulder, Colo.: Westview Press, 1994).

● Muthiah, S., ed. *An Atlas of India* (New York: Oxford University Press, 1992).

Naipaul, V. S. *India: A Million Mutinies Now* (New York: Viking Press, 1990).

● Newman, James L., & Matzke, Gordon E. *Population: Patterns, Dynamics, and Prospects* (Englewood Cliffs, N.J.: Prentice-Hall, 1984).

Noble, Allen G., & Dutt, Ashok K., eds. *India: Cultural Patterns and Processes* (Boulder, Colo.: Westview Press, 1982).

Novak, James J. *Bangladesh: Reflections on the Water* (Bloomington, Ind.: Indiana University Press, 1993).

Oberoi, Harjot. *The Construction of Religious Boundaries: Culture, Identity, and Diversity in the Sikh Tradition* (Chicago: University of Chicago Press, 1995).

• Robinson, Francis, ed. *The Cambridge Encyclopedia of India, Pakistan, Bangladesh, Sri Lanka, Nepal, Bhutan, and the Maldives* (New York: Cambridge University Press, 1989).

Samad, Yunas. *A Nation in Turmoil: Nationalism and Ethnicity in Pakistan* (Thousand Oaks, Calif.: Sage, 1995).

Schwartzberg, Joseph E. *A Historical Atlas of South Asia: 2nd Impression, With Additional Material* (New York: Oxford University Press, 1992).

Siddiqi, A. *Pakistan: Its Resources and Development* (Hong Kong: Asian Research Service, 1985).

Sopher, David E., ed. *An Exploration of India: Geographical Perspectives on Society and Culture* (Ithaca, N.Y.: Cornell University Press, 1980).

• Spate, Oskar H. K., & Learmonth, Andrew T. A. *India and Pakistan: A General and Regional Geography* (London: Methuen, 2 vols., 1971).

• Spencer, Joseph E., & Thomas, William L., Jr. *Asia, East by South: A Cultural Geography* (New York: John Wiley & Sons, 2nd rev. ed., 1971).

Sukhwal, B. L. *India: Economic Resource Base and Contemporary Political Patterns* (New York: Envoy Press, 1987).

van der Veer, Peter, ed. *Nation and Migration: The Politics of Space in the South Asian Diaspora* (Philadelphia: University of Pennsylvania Press, 1995).

Visaria, Leela, & Visaria, Pravin. "India's Population in Transition," *Population Bulletin*, 50 (October 1995): 1–51.

• Weeks, John R. *Population: An Introduction to Concepts and Issues* (Belmont, Calif.: Wadsworth, updated 5th rev. ed., 1994).

Weisman, Steven R. "India: Always Inventing Itself," *New York Times Magazine*, December 11, 1988, pp. 50–51, 62, 64, 110–111.

Wiring, Robert G. *India, Pakistan, and the Kashmir Dispute: On Regional Conflict and Its Resolution* (New York: St. Martin's Press, 1994).

Wolpert, Stanley. *India* (Berkeley, Calif.: University of California Press, 1991).

CHAPTER 9

Amsden, Alice. *Asia's Next Giant: Late Industrialization in South Korea* (New York: Oxford University Press, 1989).

Barnett, A. Doak. *China's Far West: Four Decades of Change* (Boulder, Colo.: Westview Press, 1993).

Bowring, Richard, & Kornicki, Peter, eds. *The Cambridge Encyclopedia of Japan* (New York: Cambridge University Press, 1993).

Buchanan, Keith, et al. *China: The Land and People* (New York: Crown, 1981).

• Burks, Ardath W. *Japan: A Postindustrial Power* (Boulder, Colo.: Westview Press, 3rd rev. ed., 1991).

• Cannon, Terry, & Jenkins, Alan, eds. *The Geography of Contemporary China: The Impact of Deng Xiaoping's Decade* (London & New York: Routledge, 1990).

Castells, Manuel, & Hall, Peter. *Technopoles of the World: The Making of Twenty-First-Century Industrial Complexes* (London & New York: Routledge, 1994).

Chapman, Graham P., & Baker, Kathleen M., eds. *The Changing Geography of Asia* (London & New York: Routledge, 1992).

Cheng, Chu-Yuan. *Behind the Tiananmen Square Massacre: Social, Political, and Economic Ferment in China* (Boulder, Colo.: Westview Press, 1990).

• Chisholm, Michael. *Modern World Development* (Totowa, N.J.: Barnes & Noble, 1982).

Chiu, T. N., & So, C. L., eds. *A Geography of Hong Kong* (New York: Oxford University Press, 2nd rev. ed., 1987).

Chowdhury, Anis, & Islam, Iyanatul. *The Newly Industrializing Economies of East Asia* (London & New York: Routledge, 1993).

Cressey, George B. *Asia's Lands and Peoples: A Geography of One-Third of the Earth and Two-Thirds of Its People* (New York: McGraw-Hill, 3rd rev. ed., 1963).

Cybriwsky, Roman A. *Tokyo: The Changing Profile of an Urban Giant* (Boston: G. K. Hall, 1991).

Dickenson, John P., ed. *Geography of the Third World* (London & New York: Routledge, 2 rev. ed., 1996).

• Drakakis-Smith, David. *Pacific Asia* (London & New York: Routledge, 1992).

Drakakis-Smith, David, & Dixon, Chris, eds. *Economic and Social Development in Pacific Asia* (London & New York: Routledge, 1993).

Dwyer, Denis J., ed. *China: The Next Decades* (New York: Wiley/Longman, 1993).

Fairbank, John King. *China: A New History* (Cambridge, Mass.: Belknap/Harvard University Press, 1992).

Fallows, James. *Looking at the Sun: The Rise of the New East Asian Economic and Political System* (New York: Pantheon, 1994).

Fuchs, Roland J., et al., eds. *Urbanization and Urban Policies in Pacific Asia* (Boulder, Colo.: Westview Press, 1987).

Geelan, P. J. M., & Twitchett, D. C., eds. *The Times Atlas of China* (New York: Van Nostrand, 2nd rev. ed., 1984).

Ginsburg, Norton S., ed. *The Pattern of Asia* (Englewood Cliffs, N.J.: Prentice-Hall, 1958).

Ginsburg, Norton S., & Lalor, Bernard A., eds. *China: The 80s Era* (Boulder, Colo.: Westview Press, 1984).

Goodman, David, ed. *China's Regional Development* (London & New York: Routledge, 1989).

Grundy-Warr, Carl, ed. *Eurasia: World Boundaries, Vol. 3* (London & New York: Routledge, 1994).

Hall, Colin M. *Tourism in the Pacific Rim* (New York: John Wiley & Sons, 1994).

Harrell, Stevan, ed. *Cultural Encounters on China's Ethnic Frontiers* (Seattle: University of Washington Press, 1996).

Harris, Chauncy D. "The Urban and Industrial Transformation of Japan," *Geographical Review*, 72 (January 1982): 50–89.

Hillel, Daniel. "Lash of the Dragon: China's Yellow River Remains Untamed," *Natural History*, August 1991, pp. 28–37.

Ho, Samuel P. S. *Economic Development of Taiwan* (New Haven: Yale University Press, 1978).

Hoare, James, & Pares, Susan. *Korea: An Introduction* (London & New York: Routledge, 1988).

Hodder, Rupert. *Impact Assessment and the West Pacific Rim: An Introduction* (New York: John Wiley & Sons, 1993).

● Hodder, Rupert. *The West Pacific Rim: An Introduction* (New York: John Wiley & Sons, 1992).

Hook, Brian, ed. *The Cambridge Encyclopedia of China* (New York: Cambridge University Press, 2nd rev. ed., 1994).

Hsieh, Chiao-min, & Hsieh, Jean Kan. *China: A Provincial Atlas* (New York: Macmillan, 1992).

Jao, Y. C., & Leung, Chi-Keung, eds. *China's Special Economic Zones: Policies, Problems and Prospects* (New York: Oxford University Press, 1986).

Jingzhi Sun, ed. *The Economic Geography of China* (New York: Oxford University Press, 1988).

Kirkby, Richard J. R. *Urbanization in China: Town and Country in a Developing Economy, 1949–2000 A.D.* (New York: Columbia University Press, 1985).

Knapp, Ronald G., ed. *Chinese Landscapes: The Village as Place* (Honolulu: University of Hawaii Press, 1992).

● Kornhauser, David H. *Japan: Geographical Background to Urban-Industrial Development* (London & New York: Longman, 2nd rev. ed., 1982).

Kristof, Nicholas D., & WuDunn, Sheryl. *China Wakes: The Struggle for the Soul of a Rising Power* (New York: Times Books, 1994).

Lattimore, Owen. *Inner Asian Frontiers of China* (New York: American Geographical Society, 1940).

● Leeming, Frank. *The Changing Geography of China* (Cambridge, Mass.: Blackwell, 1993).

Lo, Chor-Pang. *Hong Kong* (New York: Wiley/Belhaven, 1992).

MacDonald, Donald. *A Geography of Modern Japan* (Ashford, U.K.: Paul Norbury, 1985).

Mackerras, Colin. *China's Minorities: Integration and Modernization in the Twentieth Century* (New York: Oxford University Press, 1994).

Mackerras, Colin, & Yorke, Amanda. *The Cambridge Handbook of Contemporary China* (New York: Cambridge University Press, 1991).

Murata, Kiyogi, & Ota, Isamu, eds. *An Industrial Geography of Japan* (New York: St. Martin's Press, 1980).

Murphey, Rhoads, et al., eds. *The Chinese: Adapting the Past, Building the Future* (Ann Arbor, Mich.: University of Michigan, Center for Chinese Studies, 1986).

Murphy, Alexander B. "Economic Regionalization and Pacific Asia," *Geographical Review*, 85 (April 1995): 127–140.

● Noh, Toshio, & Kimura, John C., eds. *Japan: A Regional Geography of an Island Nation* (Tokyo: Teikoku-Shoin, 1985).

Pannell, Clifton W. "China's Urban Transition," *Journal of Geography*, 94 (May/June 1995): 394–403.

Pannell, Clifton W. "Regional Shifts in China's Industrial Output," *The Professional Geographer*, 40 (February 1988): 19–31.

Pannell, Clifton W., ed. *East Asia: Geographical and Historical Approaches to Foreign Area Studies* (Dubuque, Iowa: Kendall/Hunt, 1983).

● Pannell, Clifton W., & Ma, Laurence J. C. *China: The Geography of Development and Modernization* (New York: Halsted Press/V. H. Winston, 1983).

Pannell, Clifton W., & Torguson, Jeffrey S. "Interpreting Spatial Patterns from the 1990 China Census," *Geographical Review*, 81 (1991): 304–317.

Reischauer, Edwin O. *The Japanese Today: Change and Continuity* (Cambridge, Mass.: Belknap/Harvard University Press, 1988).

Rostow, Walt W. *The Stages of Economic Growth* (New York: Cambridge University Press, 2nd rev. ed., 1971).

Rumley, Dennis, et al., eds. *Global Geopolitical Change and the Asia-Pacific: A Regional Perspective* (Adlershot, U.K.: Avebury, 1996).

Salisbury, Harrison E. *The Long March: The Untold Story* (New York: Harper & Row, 1985).

Schell, Orville. *Mandate of Heaven: A New Generation of Entrepreneurs, Dissidents, Bohemians, and Technocrats Lays Claim to China's Future* (New York: Simon & Schuster, 1994).

● Selya, Roger Mark, ed. *The Geography of China, 1975–1991: An Annotated Bibliography* (East Lansing, Mich.: Asian Studies Center, Michigan State University, 1993).

"Shanghai: City of Glitter and Ghosts," *The Economist*, December 24, 1994, pp. 40–42.

● Simpson, E. S. *The Developing World: An Introduction* (New York: Wiley/Longman, 2nd rev. ed., 1994).

Sit, Victor F. S. *Beijing: The Nature and Planning of a Chinese Capital City* (Chichester, U.K. & New York: Wiley, 1995).

Sit, Victor F. S., ed. *Chinese Cities: The Growth of the Metropolis Since 1949* (New York: Oxford University Press, 1985).

Skidmore, Max J., guest ed. "The Future of Hong Kong," *Annals of the American Academy of Political and Social Science*, 547 (September 1996): special issue.

Smil, Vaclav. "China's Food," *Scientific American*, December 1985, pp. 116–124.

Smil, Vaclav. *The Bad Earth: Environmental Degradation in China* (Armonk, N.Y.: M. E. Sharpe, 1984).

● Smith, Christopher J. *China: People and Places in the Land of One Billion* (Boulder, Colo.: Westview Press, 1991).

Spence, Jonathan D. *The Search for Modern China* (New York: W.W. Norton, 1990).

● Spencer, Joseph E., & Thomas, William L., Jr. *Asia, East by South: A Cultural Geography* (New York: John Wiley & Sons, 2nd rev. ed., 1971).

Terrill, Ross. *China in Our Time: The Epic Saga of the People's Republic from the Communist Victory to Tiananmen Square and Beyond* (New York: Simon & Schuster, 1992).

Theroux, Paul. "Going to See the Dragon: A Journey Through South China, Land of Capitalist Miracles, Where Yesterday's Rice Paddy Becomes Tomorrow's Metropolis, and a Thousand Factories Bloom," *Harper's Magazine*, October 1993, 33–56.

● Tregear, Thomas R. *China: A Geographical Survey* (New York: John Wiley & Sons/Halsted Press, 1980).

Trewartha, Glenn T. *Japan: A Geography* (Madison, Wis.: University of Wisconsin Press, 2nd rev. ed., 1965).

Unwin, Tim, ed. *Atlas of World Development* (New York: Wiley/Longman, 1994).

Veeck, Gregory, ed. *The Uneven Landscape: Geographic Studies in Post-Reform China* (Baton Rouge, La.: Geoscience Publications, 1991).

Vogel, Ezra F. *The Four Little Dragons: The Spread of Industrialization in East Asia* (Cambridge, Mass.: Harvard University Press, 1991).

Wigen, Kären. *The Making of a Japanese Periphery, 1750–1920* (Berkeley, Calif.: University of California Press, 1994).

Wing-Kai Chiu, Stephen, et al. *City States in the Global Economy: Industrial Restructuring in Hong Kong and Singapore* (Boulder, Colo.: Westview, 1996).

Wittwer, Sylvan, et al. *Feeding a Billion: Frontiers of Chinese Agriculture* (East Lansing, Mich.: Michigan State University Press, 1987).

Yabuki, Susumu. *China's New Political Economy: The Giant Awakes* (Boulder, Colo.: Westview Press, trans. Stephen M. Harner, 1994).

Yeung, Yue-man, & Chu, David K. Y., eds. *Guangdong: Survey of a Province Undergoing Rapid Change* (Hong Kong: Chinese University Press, 1994).

● Yeung, Yue-man, & Hu, Xu-Wei, eds. *China's Coastal Cities: Catalysts for Modernization* (Honolulu: University of Hawaii Press, 1992).

Zhao Songqiao. *Physical Geography of China* (New York & Beijing: John Wiley & Sons/Science Press, 1986).

CHAPTER 10

Broad, Robin, & Cavanagh, John. *Plundering Paradise: The Struggle for the Environment in the Philippines* (Berkeley, Calif.: University of California Press, 1994).

Broek, Jan O. M. "Diversity and Unity in Southeast Asia," *Geographical Review*, 34 (1944): 175–195.

Burling, Robbins. *Hill Farms and Padi Fields: Life in Mainland Southeast Asia* (Englewood Cliffs, N.J.: Prentice-Hall, 1965).

Chapman, Graham P., & Baker, Kathleen M., eds. *The Changing Geography of Asia* (London & New York: Routledge, 1992).

Cho, George. *The Malaysian Economy: Spatial Perspectives* (London & New York: Routledge, 1990).

Costa, Frank J., et al., eds. *Urbanization in Asia: Spatial Dimensions and Policy Issues* (Honolulu: University of Hawaii Press, 1989).

Demko, George J., & Wood, William B., eds. *Reordering the World: Geopolitical Perspectives on the Twenty-First Century* (Boulder, Colo.: Westview Press, 1994).

● Dixon, Chris. *South East Asia in the World Economy* (New York: Cambridge University Press, 1991).

Drakakis-Smith, David. *Pacific Asia* (London & New York: Routledge, 1992).

● Dutt, Ashok K., ed. *Southeast Asia: Realm of Contrasts* (Boulder, Colo.: Westview Press, 3rd rev. ed., 1985).

● Dwyer, Denis J., ed. *South East Asian Development* (New York: Wiley/Longman, 1990).

● Fisher, Charles A. *Southeast Asia: A Social, Economic and Political Geography* (New York: E. P. Dutton, 2nd rev. ed., 1966).

Ford, Larry R. "A Model of Indonesian City Structure," *Geographical Review*, 83 (October 1993): 374–396.

● Fryer, Donald W. *Emerging South-East Asia: A Study in Growth and Stagnation* (New York: John Wiley & Sons, 2nd rev. ed., 1979).

Ginsburg, Norton S., et al., eds. *The Extended Metropolis: Settlement Transition in Asia* (Honolulu: University of Hawaii Press, 1991).

● Glassner, Martin I., & de Blij, H. J. *Systematic Political Geography* (New York: John Wiley & Sons, 4th rev. ed., 1989).

Grundy-Warr, Carl, ed. *Eurasia: World Boundaries, Vol. 3* (London & New York: Routledge, 1994).

Hardjono, Joan, ed. *Indonesia: Resources, Ecology, and Environment* (Singapore: Oxford University Press, 1991).

Hartshorne, Richard. "Suggestions on the Terminology of Political Boundaries," *Annals of the Association of American Geographers*, 26 (1936): 56–57.

● Hill, Ronald D., ed. *South-East Asia: A Systematic Geography* (New York: Oxford University Press, 1979).

Hussey, Antonia. "Rapid Industrialization in Thailand, 1986–1991," *Geographical Review*, 83 (1993): 14–28.

Karnow, Stanley. *In Our Image: America's Empire in the Philippines* (New York: Random House, 1989).

Leinbach, Thomas R., & Sien, Chia Lin, eds. *South-East Asian Transport* (New York: Oxford University Press, 1989).

Leinbach, Thomas R., & Ulack, Richard. "Cities of Southeast Asia," in Brunn, Stanley D., & Williams, Jack F., eds., *Cities of the World: World Regional Urban Development* (New York: HarperCollins, 2 rev. ed., 1993), pp. 389–429.

McCloud, Donald G. *Southeast Asia: Tradition and Modernity in the Contemporary World* (Boulder, Colo.: Westview Press, 2nd rev. ed., 1995).

McGee, Terence G. *The Southeast Asian City: A Social Geography* (New York: Praeger, 1967).

Mellor, Roy E. H. *Nation, State, and Territory: A Political Geography* (London & New York: Routledge, 1989).

Murray, Geoffrey, & Perera, Audrey. *Singapore: The Global City-State* (New York: St. Martin's Press, 1996).

Neher, Clark D., & Marlay, Ross. *Democracy and Development in Southeast Asia: The Winds of Change* (Boulder, Colo.: Westview Press, 1996).

Parnwell, Michael, & Bryant, Raymond. *Environmental Change in South-East Asia* (London & New York: Routledge, 1996).

"Powerhouse Penang [Pinang]," *The Economist*, May 22, 1993, p. 39.

● Prescott, J. R. V. *Political Frontiers and Boundaries* (Winchester, Mass.: Allen & Unwin, 1987).

Reading, Alison, J., et al. *Humid Tropical Environments* (Cambridge, Mass.: Blackwell, 1995).

● Rigg, Jonathan. *Southeast Asia—A Region in Transition: A Thematic Human Geography of the ASEAN Region* (New York: HarperCollins Academic, 1991).

Rumley, Dennis, & Minghi, Julian V., eds. *The Geography of Border Landscapes* (London & New York: Routledge, 1991).

Sandhu, Kernial S., & Wheatley, Paul, eds. *Management of Success: The Moulding of Modern Singapore* (Singapore: Institute of Southeast Asian Studies, 1990).

SarDesai, D. R. *Southeast Asia: Past and Present* (Boulder, Colo.: Westview Press, 3rd rev. ed., 1994).

Schofield, Clive H., ed. *Global Boundaries: World Boundaries, Volume One* (London & New York: Routledge, 1994).

"South-East Asian Economies: Dreams of Gold," *The Economist*, March 20, 1993, pp. 21–24.

● Spencer, Joseph E., & Thomas, William L., Jr. *Asia, East by South: A Cultural Geography* (New York: John Wiley & Sons, 2nd rev. ed., 1971).

Ulack, Richard, & Pauer, Gyula. *Atlas of Southeast Asia* (New York: Macmillan, 1988).

Wing-Kai Chiu, Stephen, et al. *City States in the Global Econ-*

omy: *Industrial Restructuring in Hong Kong and Singapore* (Boulder, Colo.: Westview Press, 1996).

Wurfel, David, & Burton, Bruce, eds. *Southeast Asia in the New World Order* (New York: St. Martin's Press, 1996).

Yeung, Yue-man. *Changing Cities of Pacific Asia: A Scholarly Interpretation* (Hong Kong: Chinese University Press, 1990).

CHAPTER II

• Bambrick, Susan, ed. *The Cambridge Encyclopedia of Australia* (New York: Cambridge University Press, 1994).

• Barrett, Rees D., & Ford, Roslyn A. *Patterns in the Human Geography of Australia* (South Melbourne: Macmillan of Australia, 1987).

Bolton, Geoffrey C. *Spoils and Spoilers: Australians Make Their Environment, 1788–1980* (Winchester, Mass.: Allen & Unwin, 1981).

Britton, Steve, et al., eds. *Changing Places in New Zealand: A Geography of Restructuring* (Dunedin, N.Z.: Allied Press, 1992).

Burnley, Ian H. *Population, Society and Environment in Australia* (Melbourne: Shillington House, 1982).

Burnley, Ian H. *The Australian Urban System: Growth, Change, and Differentiation* (Melbourne: Longman Cheshire, 1980).

Camm, J. C. R., & McQuilton, J., eds. *Australia: An Historical Atlas* (Broadway, N.S.W.: Fairfax, Syme & Weldon Associates, 1987).

Connell, J., & Howitt, R., eds. *Mining and Indigenous Peoples in Australasia* (South Melbourne: Sydney University Press, 1991).

Courtenay, Percy P. *Northern Australia: Patterns and Problems of Tropical Development in an Advanced Country* (London & New York: Longman, 1983).

• Cox, C. Barry, & Moore, Peter D. *Biogeography: An Ecological and Evolutionary Approach* (Cambridge, Mass.: Blackwell, 5th ed., 1993).

Cumberland, Kenneth B., & Whitelaw, James S. *New Zealand* (Chicago: Aldine, 1970).

Darlington, Philip J., Jr. *Biogeography of the Southern End of the World* (New York: McGraw-Hill, 1975).

• de Blij, H. J., & Muller, Peter O. *Physical Geography of the Global Environment* (New York: John Wiley & Sons, 2nd rev. ed., 1996).

"Environment and Development in Australia," *Australian Geographer*, 19 (May 1988): 3–220.

Forster, Cliv A. *Australian Cities: Continuity and Change* (Melbourne: Oxford University Press Australia, 1995).

Fuchs, Roland J., et al., eds. *Urbanization and Urban Policies in Pacific Asia* (Boulder, Colo.: Westview Press, 1987).

• Furley, Peter A., & Newey, W. W. *Geography of the Biosphere: An Introduction to the Nature, Distribution, and Evolution of the World's Life Zones* (Stoneham, Mass.: Butterworth, 1982).

Hall, Colin M. *Tourism in the Pacific Rim* (New York: John Wiley & Sons, 1994).

Hamnett, Stephen, & Bunker, Raymond, eds. *Urban Australia: Planning Issues and Policies* (Melbourne: Nelson Wadsworth, 1987).

• Heathcote, Ronald L. *Australia* (New York: Wiley/Longman, 2nd rev. ed., 1994).

Heathcote, Ronald L., ed. *The Australian Experience: Essays in Australian Land Settlement and Resource Management* (Melbourne: Longman Cheshire, 1988).

Heathcote, Ronald L., & Mabbutt, J. A. *Land, Water and People: Essays in Australian Resource Management* (Winchester, Mass.: Allen & Unwin, 1988).

Hodder, Rupert. *The West Pacific Rim: An Introduction* (New York: John Wiley & Sons, 1992).

Hofmeister, Burkhard. *Australia and Its Urban Centers* (Berlin: Gebrüder Borntraeger, 1988).

Hughes, Robert. *The Fatal Shore* (New York: Alfred A. Knopf, 1987).

Huston, M. A. *Biological Diversity: The Coexistence of Species on Changing Landscapes* (New York: Cambridge University Press, 1994).

Illies, J. *Introduction to Zoogeography* (New York: Macmillan, 1974).

Inglis, C., et al., eds. *Asians in Australia: The Dynamics of Migration and Settlement* (Sydney: Allen & Unwin, 1992).

• Jeans, Dennis N., ed. *Australia: A Geography. Volume 1: The Natural Environment* (Sydney: Sydney University Press, 1986); *Volume 2: Space and Society* (Sydney: Sydney University Press, 1987).

Johnston, R. J., ed. *Society and Environment in New Zealand* (Christchurch, N.Z.: Whitcombe & Tombs, 1974).

Lines, William J. *Taming the Great South Land: A History of the Conquest of Nature in Australia* (Berkeley, Calif.: University of California Press, 1992).

Logan, Mal, et al. *Urbanization: The Australian Experience* (Melbourne: Shillington House, 1981).

• McKnight, Tom L. *Oceania: The Geography of Australia, New Zealand, and the Pacific Islands* (Englewood Cliffs, N.J.: Prentice Hall, 1995).

McKnight, Tom L. *Australia's Corner of the World* (Englewood Cliffs, N.J.: Prentice-Hall, 1970).

Maurer, B. *Geographical Analysis of Biodiversity* (Cambridge, Mass.: Blackwell, 1994).

Meinig, Donald W. *On the Margins of the Good Earth: The South Australian Wheat Frontier, 1869–1884* (Chicago: Rand McNally, 1962).

Mielke, H. W. *Patterns of Life: Biogeography of a Changing World* (Winchester, Mass.: Unwin Hyman, 1989).

• Morton, Harry, & Johnston, Carol M. *The Farthest Corner: New Zealand, A Twice Discovered Land* (Honolulu: University of Hawaii Press, 1989).

Ogawa, Naohiro, et al., eds. *Human Resources in Development Along the Asia-Pacific Rim* (Singapore: Oxford University Press, 1993).

Powell, Joseph M. *An Historical Geography of Modern Australia: The Restive Fringe* (New York: Cambridge University Press, 1988).

Powell, Joseph M., & Williams, Michael, eds. *Australian Space, Australian Time: Geographical Perspectives* (New York: Oxford University Press, 1975).

Rich, David C. *The Industrial Geography of Australia* (Sydney: Methuen, 1987).

Serventy, J. *Landforms of Australia* (New York: American Elsevier, 1968).

• Simmons, I. G. *Biogeographical Processes* (Winchester, Mass.: Allen & Unwin, 1983).

Smith, J. M. B., ed. *A History of Australasian Vegetation* (Sydney: McGraw-Hill, 1982).

Spate, Oskar H. K. *Australia* (New York: Praeger, 1968).

Statham, Pamela, ed. *The Origins of Australia's Capital Cities* (Sydney: Cambridge University Press, 1989).

Terrill, Ross. *The Australians* (New York: Simon & Schuster, 1987).

Tivy, Joy. *Biogeography: A Study of Plants in the Ecosphere* (London/New York: Longman, 2nd ed., 1982).

Walmsley, D. J., & Sorenson, A. D. *Contemporary Australia: Explorations in Economy, Society, and Geography* (Melbourne: Longman Cheshire, 1988).

CHAPTER 12

"Antarctica: Is Any Place Safe From Mankind?" *Time*, Cover Story, January 15, 1990, pp. 56–62.

● Bier, James A., cartographer. *Reference Map of Oceania: The Pacific Islands of Micronesia, Polynesia, Melanesia* (Honolulu: University of Hawaii Press, 1995).

Blake, Gerald H., ed. *Maritime Boundaries: World Boundaries, Vol. 5* (London & New York: Routledge, 1994).

Blake, Gerald H., ed. *Maritime Boundaries and Ocean Resources* (Totowa, N.J.: Rowman & Littlefield, 1988).

Borgese, Elizabeth M., Ginsburg, Norton, & Morgan, Joseph R., eds. *Ocean Yearbook, Volume 12* (Chicago: University of Chicago Press, 1996).

Brookfield, Harold C., ed. *The Pacific in Transition: Geographical Perspectives on Adaptation and Change* (New York: St. Martin's Press, 1973).

Brookfield, Harold C., & Hart, Doreen. *Melanesia: A Geographical Interpretation of an Island World* (New York: Barnes & Noble, 1971).

Bunge, Frederica M., & Cooke, Melinda W., eds. *Oceania: A Regional Study* (Washington, D.C.: U.S. Government Printing Office, 1984).

Campbell, Ian C. *A History of the Pacific Islands* (Berkeley, Calif.: University of California Press, 1990).

Carter, John, ed. *Pacific Islands Yearbook* (Sydney, Australia: Pacific Publications, annual).

Couper, Alastair D., ed. *Atlas and Encyclopedia of the Sea* (New York: Harper & Row, 1989).

Couper, Alastair D., ed. *Development and Social Change in the Pacific Islands* (London & New York: Routledge, 1989).

Couper, Alastair D., ed. *The Times Atlas of the Oceans* (New York: Van Nostrand Reinhold, 1983).

Crossley, L. *Explore Antarctica* (New York: Cambridge University Press, 1995).

Damas, David. *Bountiful Island: A Study of Land Tenure on a Micronesian Atoll* (Waterloo, Ont., Canada: Wilfrid Laurier University Press, 1996).

de Blij, H. J. "A Regional Geography of Antarctica and the Southern Ocean," *University of Miami Law Review*, 33 (1978): 299–314.

Freeman, Otis W., ed. *Geography of the Pacific* (New York: John Wiley & Sons, 1951).

Friis, Herman R., ed. *The Pacific Basin: A History of Its Geographical Exploration* (New York: American Geographical Society, Special Publication No. 38, 1967).

● Glassner, Martin I. *Neptune's Domain: A Political Geography of the Sea* (Winchester, Mass.: Unwin Hyman, 1990).

Grossman, Lawrence S. *Peasants, Subsistence Ecology, and Development in the Highlands of Papua New Guinea* (Princeton, N.J.: Princeton University Press, 1984).

Howard, A., ed. *Polynesia: Readings on a Culture Area* (Scranton, Pa.: Chandler, 1971).

Howlett, Diana. *Papua New Guinea: Geography and Change* (Melbourne, Australia: Thomas Nelson, 1973).

Johnston, Douglas M., & Saunders, Philip M. *Ocean Boundary-Making: Regional Issues and Developments* (London & New York: Routledge, 1988).

Karolle, Bruce G. *Atlas of Micronesia* (Honolulu: Bess Press, 2nd rev. ed., 1995).

Kirch, Patrick V. *The Wet and the Dry: Irrigation and Agricultural Intensification in Polynesia* (Chicago: University of Chicago Press, 1994).

Kissling, Christopher C., ed. *Transport and Communications for Pacific Microstates: Issues in Organization and Management* (Suva, Fiji: University of the South Pacific, Institute of Pacific Studies, 1984).

Kluge, P. F. *The Edge of Paradise: America in Micronesia* (New York: Random House, 1991).

Krieger, Michael. *Conversations With the Cannibals: The End of the Old South Pacific* (Hopewell, N.J.: Ecco Press, 1994).

Levison, Michael, et al. *The Settlement of Polynesia: A Computer Simulation* (Minneapolis: University of Minnesota Press, 1973).

Lockhart, Douglas G., et al., eds. *The Development Process in Small Island States* (London & New York: Routledge, 1993).

● McKnight, Tom L. *Oceania: The Geography of Australia, New Zealand, and the Pacific Islands* (Englewood Cliffs, N.J.: Prentice Hall, 1995).

Mitchell, Andrew. *The Fragile South Pacific: An Ecological Odyssey* (Austin: University of Texas Press, 1991).

"Mobility and Identity in the Island Pacific," Special Issue, *Pacific Viewpoint*, 26, No. 1 (1985).

Morgan, Joseph R., ed. *Hawaii* (Boulder, Colo.: Westview Press, 1983).

Nunn, Patrick. *Oceanic Islands* (Cambridge, Mass.: Blackwell, 1994).

● Oliver, Douglas L. *The Pacific Islands* (Honolulu: University of Hawaii Press, 3rd rev. ed., 1989).

Peake, Martin. *Pacific People and Society* (New York: Cambridge University Press, 1992).

Prescott, J. R. V. *Political Frontiers and Boundaries* (Winchester, Mass.: Allen & Unwin, 1987).

Prescott, J. R. V. *The Maritime Political Boundaries of the World* (London & New York: Methuen, 1986).

Quanchi, Max. *Pacific People and Change* (New York: Cambridge University Press, 1992).

Robillard, Albert B., ed. *Social Change in the Pacific Islands* (London & New York: Kegan Paul International, 1991).

Sager, Robert J. "The Pacific Islands: A New Geography," *Focus*, Summer 1988, pp. 10–14.

Segal, Gerald. *Rethinking the Pacific* (New York: Oxford University Press, 1990).

Shenon, Philip. "Isolated Papua New Guineans Fall Prey to Foreign Bulldozers," *New York Times*, June 5, 1994, pp. 1, 6.

Spate, Oskar H. K. *The Pacific Since Magellan, Vol. II: Monopolists and Freebooters* (Minneapolis: University of Minnesota Press, 1983).

Spate, Oskar H. K. *The Pacific Since Magellan, Vol. III: Paradise Found and Lost* (Minneapolis: University of Minnesota Press, 1989).

Spate, Oskar H. K. *The Spanish Lake: A History of the Pacific Since Magellan*, Vol. I (Beckenham, U.K.: Croom Helm, 1979).

● Sugden, David E. *Arctic and Antarctic: A Modern Geographical Synthesis* (Totowa, N.J.: Barnes & Noble, 1982).

Theroux, Paul. *The Happy Isles of Oceania: Paddling the Pacific* (New York: G. P. Putnam's Sons, 1992).

Triggs, Gillian D., ed. *The Antarctic Treaty Regime: Law, Environment and Resources* (New York: Cambridge University Press, 1987).

Vayda, Andrew P., ed. *Peoples and Cultures of the Pacific* (New York: Natural History Press, 1968).

Waddell, Eric, & Nunn, Patrick D., eds. *The Margin Fades: Geographical Itineraries in a World of Islands* (Suva, Fiji: University of the South Pacific, Institute of Pacific Studies, 1994).

Ward, R. Gerard, ed. *Man in the Pacific Islands: Essays on Geographical Change in the Pacific* (New York: Oxford University Press, 1972).

Zurick, David N. "Preserving Paradise," *Geographical Review*, 85 (April 1995): 157–172.

Glossary

Absolute location The position or place of a certain item on the surface of the Earth as expressed in degrees, minutes, and seconds of **latitude**,* 0° to 90° north or south of the equator, and **longitude**, 0° to 180° east or west of the *prime meridian* passing through Greenwich, England (a suburb of London).

Accessibility The degree of ease with which it is possible to reach a certain location from other locations. Accessibility varies from place to place and can be measured.

Acculturation Cultural modification resulting from intercultural borrowing. In cultural geography, the term is used to designate the change that occurs in the culture of indigenous peoples when contact is made with a society that is technologically more advanced.

Agglomerated (nucleated) settlement A compact, closely packed settlement (usually a hamlet or larger village) sharply demarcated from adjoining farmlands.

Agglomeration Process involving the clustering or concentrating of people or activities. Often refers to manufacturing plants and businesses that benefit from close proximity because they share skilled-labor pools and technological and financial amenities.

Agrarian Relating to the use of land in rural communities, or to agricultural societies in general.

Agriculture The purposeful tending of crops and livestock in order to produce food and fiber.

Alluvial Refers to the mud, silt, and sand (collectively *alluvium*) deposited by rivers and streams. *Alluvial plains* adjoin many larger rivers; they consist of such renewable deposits that are laid down during floods, creating fertile and productive soils. Alluvial **deltas** mark the mouths of rivers such as the Mississippi and the Nile.

Altiplano High-elevation plateau, basin, or valley between even higher mountain ranges. In the Andes Mountains of South America, altiplanos lie at 10,000 feet (3000 m) and even higher.

Altitudinal zonation Vertical regions defined by physical-environmental zones at various elevations, particularly in the highlands of South and Middle America. See *puna, tierra caliente, tierra fría, tierra helada,* and *tierra templada.*

*Words in boldface type are defined elsewhere in this Glossary.

Antecedent boundary A political boundary that existed before the **cultural landscape** emerged and stayed in place while people moved in to occupy the surrounding area. An example is the 49th parallel boundary, dividing the United States and Canada between the Pacific Ocean and Lake of the Woods in northernmost Minnesota.

Anthracite coal Hardest and highest carbon content coal (therefore of the highest quality), that was formed under conditions of high pressure and temperature that eliminated most impurities. Anthracite burns almost without smoke and produces high heat.

Apartheid Literally, "apartness." The Afrikaans term given to South Africa's policies of racial separation, and the highly segregated socio-geographical patterns they have produced; a system now being dismantled.

Aquaculture The use of a river segment or an artificial body of water such as a pond for the raising and harvesting of food products, including fish, shellfish, and even seaweed. Japan is among the world's leaders in aquaculture.

Aquifer An underground reservoir of water contained within a porous, water-bearing rock layer.

Arable Literally, cultivable. Land fit for cultivation by one farming method or another.

Archipelago A set of islands grouped closely together, usually elongated into a chain.

Area A term that refers to a part of the Earth's surface with less specificity than **region**. For example, *urban area* alludes very generally to a place where urban development has taken place, whereas *urban region* requires certain specific criteria upon which a delimitation is based (e.g., the spatial extent of commuting or the built townscape).

Areal interdependence A term related to **functional specialization**. When one area produces certain goods or has certain raw materials or resources and another area has a different set of resources and produces different goods, their needs may be *complementary*; by exchanging raw materials and products, they can satisfy each other's requirements. The concepts of areal interdependence and **complementarity** are related: both have to do with exchange opportunities between regions.

Arithmetic density A country's population, expressed as an average per unit area (square mile or square kilometer), without

G-I

regard for its distribution or the limits of **arable** land—see also **physiologic density**.

Aryan From the Sanskrit Arya (''noble''), a name applied to an ancient people who spoke an Indo-European language and who moved into northern India from the northwest. Although properly a language-related term, Aryan has assumed additional meanings, especially racial ones.

Atmosphere The Earth's envelope of gases that rests on the oceans and land surface and penetrates the open spaces within soils. This layer of nitrogen (78 percent), oxygen (21 percent), and traces of other gases is densest at the Earth's surface and thins with altitude. It is held against the planet by the force of gravity.

Austrasia New name for the western Pacific Rim, where a significant regional realignment is now taking place. Includes rapidly-developing countries and parts of countries lining the Pacific from Japan's Hokkaido in the north to New Zealand in the southeast.

Autocratic An autocratic government holds absolute power; rule is often by one person or a small group of persons who control the country by despotic means.

Balkanization The fragmentation of a region into smaller, often hostile political units.

Barrio Term meaning ''neighborhood'' in Spanish. Usually refers to an urban community in a Middle or South American city; also applied to low-income, inner-city concentrations of Hispanics in such western U.S. cities as Los Angeles.

Bauxite Aluminum ore; an earthy, reddish-colored material that usually contains some iron as well. Soil-forming processes such as leaching and redeposition of aluminum and iron compounds contribute to bauxite formation, and many deposits exist at shallow depths in the wet tropics.

Birth rate The *crude birth rate* is expressed as the annual number of births per 1000 individuals within a given population.

Bituminous coal Softer coal of lesser quality than **anthracite** (more impurities remain) but of higher grade than **lignite**. Usually found in relatively undisturbed, extensive horizontal layers, often close enough to the surface to permit strip-mining. When heated and converted to coking coal or *coke*, it is used to make iron and steel.

Break-of-bulk point A location along a transport route where goods must be transferred from one carrier to another. In a port, the cargoes of oceangoing ships are unloaded and put on trains, trucks, or perhaps smaller river boats for inland distribution.

Buffer state See buffer zone.

Buffer zone A set of countries separating ideological or political adversaries. In southern Asia, Afghanistan, Nepal, and Bhutan were parts of a buffer zone between British and Russian-Chinese imperial spheres. Thailand was a *buffer state* between British and French colonial domains in mainland Southeast Asia.

Caliente See *tierra caliente*.

Cartel An international syndicate formed to promote common interests in some economic sphere through the formulation of joint pricing policies and the limitation of market options for consumers. **OPEC** (the Organization of Petroleum Exporting Countries) is a classic example.

Cartogram A specially transformed map not based on traditional representations of scale or area.

Cartography The art and science of making maps, including data compilation, layout, and design. Also concerned with the interpretation of mapped patterns.

Caste system The strict social segregation of people—specifically in India's Hindu society—on the basis of ancestry and occupation.

Cay Pronounced *kee* (and often spelled ''key''). A low-lying small island usually composed of coral and sand. Often part of an island chain such as the Florida Keys or the Bahamas **archipelago**.

Central business district (CBD) The downtown heart of a central city, the CBD is marked by high land values, a concentration of business and commerce, and the clustering of the tallest buildings.

Centrality The strength of an urban center in its capacity to attract producers and consumers to its facilities; a city's ''reach'' into the surrounding region.

Centrifugal forces A term employed to designate forces that tend to divide a country—such as internal religious, linguistic, ethnic, or ideological differences.

Centripetal forces Forces that unite and bind a country together—such as a strong national culture, shared ideological objectives, and a common faith.

Charismatic Personal qualities of certain leaders that enable them to capture and hold the popular imagination, to secure the allegiance and even the devotion of the masses. Gandhi, Mao Zedong, and Franklin D. Roosevelt are good examples in the 20th century.

China Proper The eastern and northeastern portions of China that contain most of the country's huge population, stretching from the Amur River in the far Northeast to the southern border with Vietnam.

Choke point A narrowing of an international waterway to a distance of less than 24 miles, necessitating the drawing of a **median line (maritime) boundary**. These are almost always strategic locations where, presumably, a blockading naval force could ''choke'' off the waterway. Examples are the English Channel between France and the United Kingdom, and the Straits of Malacca between Malaysia and Indonesia.

City-state An independent political entity consisting of a single city with (and sometimes without) an immediate **hinterland**. The ancient city-states of Greece have their modern equivalent in Singapore.

Climate The long-term conditions (over at least 30 years) of aggregate weather over a region, summarized by averages and measures of variability; a synthesis of the succession of weather events we have learned to expect at any given location.

Climatology The geographic study of climates. Includes not only the classification of climates and the analysis of their regional distribution, but also broader environmental questions that concern climate change, interrelationships with soil and vegetation, and human-climate interaction.

Coal See **anthracite** and **bituminous** coal.

Collectivization The reorganization of a country's agriculture that involves the expropriation of private holdings and their incorporation into relatively large-scale units, which are farmed and administered cooperatively by those who live there. This system transformed agriculture in the former Soviet Union, and went beyond the Soviet model in China's program of communization.

Colonialism See **imperialism**.

Common Market Name sometimes given to the 15 countries of the **European Union** that belong to a **supranational** association to promote their economic interests. Six countries originally formed this organization in 1957 under the name European Economic Community (EEC).

Compact state A politico-geographical term to describe a **state** that possesses a roughly circular, oval, or rectangular territory in which the distance from the geometric center to any point on the boundary exhibits little variance. Cambodia, Uruguay, and Poland are examples of this shape category.

Complementarity Regional complementarity exists when two regions, through an exchange of raw materials and/or finished products, can specifically satisfy each other's demands.

Condominium In political geography, this denotes the shared administration of a territory by two governments.

Coniferous forest A forest of cone-bearing, needleleaf evergreen trees with straight trunks and short branches, including spruce, fir, and pine.

Contagious diffusion The distance-controlled spreading of an idea, innovation, or some other item through a local population by contact from person to person—analogous to the communication of a contagious illness.

Contiguous A word of some importance to geographers that means, literally, to be in contact with, adjoining, or adjacent. Sometimes we hear the continental (*conterminous*) United States minus Alaska referred to as contiguous. Alaska is not contiguous to these ''lower 48'' states because Canada lies in between; neither is Hawaii, separated by over 2000 miles of ocean.

Continental drift The slow movement of continents controlled by the processes associated with **plate tectonics**.

Continental shelf Beyond the coastlines of the continents the surface beneath the water, in many offshore areas, declines very gently until the depth of about 660 feet (200 m). Beyond the 660-foot line the sea bottom usually drops off sharply, along the *continental slope*, toward the much deeper mid-oceanic basin. The submerged continental margin is called the continental shelf, and it extends from the shoreline to the upper edge of the continental slope.

Continentality The variation of the continental effect on air temperatures in the interior portions of the world's landmasses. The greater the distance from the moderating influence of an ocean, the greater the extreme in summer and winter temperatures. Continental interiors, of course, tend to be dry as well when the distance from oceanic moisture sources becomes considerable.

Conurbation General term used to identify large multi-metropolitan complexes formed by the coalescence of two or more major urban areas. The Boston-Washington **Megalopolis** along the U.S northeastern seaboard is an outstanding example.

Copra Meat of the coconut; fruit of the coconut palm.

Cordillera Mountain chain consisting of sets of parallel ranges, especially the Andes in northwestern South America.

Core area In geography, a term with several connotations. Core refers to the center, heart, or focus. The core area of a **nation-state** is constituted by the national heartland, the largest population cluster, the most productive region, the area with the greatest **centrality** and **accessibility**, probably containing the capital city as well.

Core-periphery relationships The contrasting spatial characteristics of, and linkages between, the *have* (core) and *have-not* (periphery) components of a national or regional system.

Corridor In general, refers to a spatial entity in which human activity is organized in a linear manner, as along a major transport route or in a valley confined by highlands. Specific meaning in politico-geographical context is a land extension that connects an otherwise **landlocked** state to the sea. History has seen several such corridors come and go. Poland once had a corridor (it now has a lengthy coastline); Bolivia lost a corridor to the Pacific Ocean between Peru and Chile.

Cultural diffusion The process of spreading and adoption of a cultural element, from its place of origin across a wider area.

Cultural ecology The multiple interactions and relationships between a **culture** and its natural environment.

Cultural landscape The forms and artifacts sequentially placed on the physical landscape by the activities of various human occupants. By this progressive imprinting of the human presence, the physical landscape is modified into the cultural landscape, forming an interacting unity between the two.

Cultural pluralism A society in which two or more population groups, each practicing its own **culture**, live adjacent to one another without mixing inside a single **state**.

Culture The sum total of the knowledge, attitudes, and habitual behavior patterns shared and transmitted by the members of a society. This is anthropologist Ralph Linton's definition; hundreds of others exist.

Culture area A distinct, culturally discrete spatial unit; a region within which certain cultural norms prevail.

Culture hearth Heartland, source area, innovation center; place of origin of a major **culture**.

Culture region See **culture area**.

Customs union A **free-trade area** in which members set common tariff rates on imports from outside the trade zone.

Cyclical movement Movement—for example, **nomadic migration**—that has a closed route repeated annually or seasonally.

Death rate The *crude death rate* is expressed as the annual number of deaths per 1000 individuals within a given population.

Deciduous A deciduous tree loses its leaves at the beginning of winter or the start of the dry season.

Definition In political geography, the written legal description (in a treaty-like document) of a boundary between two countries or territories—see **delimitation**.

Deforestation The clearing and destruction of forests (especially tropical rainforests) to make way for expanding settlement frontiers and the exploitation of new economic opportunities.

Deglomeration Deconcentration.

Delimitation In political geography, the translation of the written terms of a boundary treaty (the **definition**) into an official cartographic representation (map).

Delta Alluvial lowland at the mouth of a river, formed when the river deposits its alluvial load on reaching the sea. Often triangular in shape hence the use of the Greek letter whose symbol is △.

Demarcation In political geography, the actual placing of a political boundary on the landscape by means of barriers, fences, walls, or other markers.

Demographic transition model Multi-stage model, based on Western Europe's experience, of changes in population growth exhibited by countries undergoing industrialization. High **birth rates** and **death rates** are followed by plunging death rates, producing a huge net population gain; this is followed by the convergence of birth and death rates at a low overall level. See Figure 8-9.

Demography The interdisciplinary study of population—especially **birth rates** and **death rates**, growth patterns, longevity, **migration**, and related characteristics.

Density of population The number of people per unit area. Also see **arithmetic density** and **physiologic density** measures.

Desert An arid area supporting sparse vegetation, receiving less than 10 inches (25 cm) of precipitation per year. Usually exhibits extremes of heat and cold because the moderating influence of moisture is absent.

Desertification The process of **desert** expansion into neighboring **steppelands** as a result of human degradation of fragile semiarid environments.

Determinism See **environmental determinism**.

Development The economic, social, and institutional growth of national **states**.

Devolution The process whereby regions within a **state** demand and gain political strength and growing autonomy at the expense of the central government.

Dhows Wooden boats with characteristic triangular sail, plying the seas between Arabian and East African coasts.

Diffusion The spatial spreading or dissemination of a **culture** element (such as a technological innovation) or some other phenomenon (e.g., a disease outbreak). See also **contagious**, **expansion**, **hierarchical**, and **relocation diffusion**.

Dispersed settlement In contrast to **agglomerated** or **nucleated** settlement, dispersed settlement is characterized by a much lower **density of population** and the wide spacing of individual homesteads (especially in rural North America).

Distance decay The various degenerative effects of distance on human spatial structures and interactions.

Diurnal Daily.

Divided capital In political geography, a country whose administrative functions are carried on in more than one city is said to have divided capitals.

Domestication The transformation of a wild animal or wild plant into a domesticated animal or a cultivated crop to gain control over food production. A necessary evolutionary step in the development of humankind: the invention of **agriculture**.

Double cropping The planting, cultivation, and harvesting of two crops successively within a single year on the same plot of farmland.

Doubling time The time required for a population to double in size.

Ecology Strictly speaking, this refers to the study of the many interrelationships between all forms of life and the natural environments in which they have evolved and continue to develop. The study of *ecosystems* focuses on the interactions between specific organisms and their environments. See also **cultural ecology**.

Economic tiger One of the burgeoning beehive countries of the western Pacific Rim. Using postwar Japan as a model, these countries have experienced significant modernization, industrialization, and Western-style economic growth since 1980. The three leading economic tigers are South Korea, Taiwan, and Singapore. Hong Kong (Xianggang) was a leading economic tiger before its 1997 reunification with China.

Economies of scale The savings that accrue from large-scale production whereby the unit cost of manufacturing decreases as the level of operation enlarges. Supermarkets operate on this principle and are able to charge lower prices than small grocery stores.

Ecumene The habitable portions of the Earth's surface where permanent human settlements have arisen.

Elite A small but influential upper-echelon social class whose power and privilege give it control over a country's political, economic, and cultural life.

El Niño-Southern Oscillation (ENSO) A periodic, large-scale, abnormal warming of the sea surface in the low latitudes of the eastern Pacific Ocean that produces a (temporary) reversal of surface ocean currents and airflows throughout the equatorial Pacific; these regional events have global implications, disturbing normal weather patterns in many parts of the world.

Elongated state A **state** whose territory is decidedly long and narrow in that its length is at least six times greater than its average width. Chile and Vietnam are two classic examples on the world political map.

Emigrant A person **migrating** away from a country or area; an out-migrant.

Empirical Relating to the real world, as opposed to theoretical abstraction.

Enclave A piece of territory that is surrounded by another political unit of which it is not a part.

Entrepôt A place, usually a port city, where goods are imported, stored, and transshipped; a **break-of-bulk point**.

Environmental determinism The view that the natural environment has a controlling influence over various aspects of human life, including cultural development. Also referred to as *environmentalism*.

Epidemic A local or regional outbreak of a disease.

Erosion A combination of gradational forces that shape the Earth's surface landforms. Running water, wind action, and the force of moving ice combine to wear away soil and rock. Human activities often speed erosional processes, such as through the destruction of natural vegetation, careless farming practices, and overgrazing by livestock.

Escarpment A cliff or steep slope; frequently marks the edge of a plateau.

Estuary The widening mouth of a river as it reaches the sea. An estuary forms when the margin of the land has subsided somewhat (or as ocean levels rise following glaciation periods) and seawater has invaded the river's lowest portion.

Ethnic cleansing The forcible ouster of entire populations from their homelands by a stronger power bent on taking their territory. A process with a long history in Eastern Europe; in the 1990s it resurfaced strongly in former Yugoslavia, particularly in and around Bosnia where it disrupted the lives of millions of Muslims, Serbs, and Croats.

European Union Supranational organization constituted by 15 European countries to further their common economic interests. The ''4–5–6'' method is the best way to learn the names of the members: the 4 giants (Germany, France, Italy, the United Kingdom); the 5 inner countries, all neighbors of Germany (Denmark, the Netherlands, Belgium, Luxembourg, Austria); and the 6 outer countries (Finland, Sweden, Ireland, Portugal, Spain, and Greece).

European state model A **state** consisting of a legally defined territory inhabited by a population governed from a capital city by a representative government.

Evapotranspiration The loss of moisture to the **atmosphere** through the combined processes of evaporation from the soil and transpiration by plants.

Exclave A bounded (non-island) piece of territory that is part of a particular **state** but lies separated from it by the territory of another state.

Exclusive Economic Zone (EEZ) An oceanic zone extending up to 200 nautical miles from a shoreline, within which the coastal **state** can control fishing, mineral exploration, and additional activities by all other countries.

Expansion diffusion The spreading of an innovation or an idea through a fixed population in such a way that the number of those adopting grows continuously larger, resulting in an expanding area of dissemination.

Extraterritoriality Politico-geographical concept suggesting that the property of one **state** lying within the boundaries of another actually forms an extension of the first state.

Favela Shantytown on the outskirts or even well within an urban area in Brazil.

Fazenda Coffee plantation in Brazil.

Federal state A political framework wherein a central government represents the various entities within a **nation-state** where they have common interests—defense, foreign affairs, and the like—yet allows these various entities to retain their own identities and to have their own laws, policies, and customs in certain spheres.

Federation See **federal state**.

Ferroalloy A metallic mineral smelted with iron to produce steel of a particular quality. Manganese, for example, provides steel with tensile strength and the ability to withstand abrasives. Other ferroalloys are nickel, chromium, cobalt, molybdenum, and tungsten.

Fertile Crescent Crescent-shaped zone of productive lands extending from near the southeastern Mediterranean coast through Lebanon and Syria to the **alluvial** lowlands of Mesopotamia (in Iraq). Once more fertile than today, this is one of the world's great source areas of agricultural and other innovations.

Feudalism Prevailing politico-geographical system in Europe during the Middle Ages when land was owned by the nobility and was worked by **peasants** and serfs. Feudalism also existed in other parts of the world, and the system persisted into this century in Ethiopia and Iran, among other places.

Fjord Narrow, steep-sided, elongated, and inundated coastal valley deepened by glacier ice that has since melted away, leaving the sea to penetrate.

Floodplain Low-lying area adjacent to a mature river, often covered by **alluvial** deposits and subject to the river's floods.

Forced migration Human **migration** flows in which the movers have no choice but to relocate.

Formal region A type of **region** marked by a certain degree of homogeneity in one or more phenomena; also called *uniform* region or *homogeneous* region.

Forward capital Capital city positioned in actually or potentially contested territory, usually near an international border; it confirms the **state's** determination to maintain its presence in the region in contention.

Four Motors of Europe Rhône-Alpes (France), Baden-Württemberg (Germany), Catalonia (Spain), and Lombardy (Italy). Each is a high-technology-driven region marked by exceptional industrial vitality and economic success not only within Europe but on the global scene as well.

Fragmented state A **state** whose territory consists of several separated parts, not a **contiguous** whole. The individual parts may be isolated from each other by the land area of other states or by international waters.

Francophone Describes a country or region where other languages are also spoken, but where French is the *lingua franca* or the language of the **elite**. Quebec is Francophone Canada.

Free-trade area A form of economic integration, usually consisting of two or more **states**, in which partners agree to remove tariffs on trade among themselves. Usually accompanied by a **customs union** that establishes common tariffs on imports from outside the trade zone.

Fría See *tierra fría*.

Frontier Zone of advance penetration, usually of contention; an area not yet fully integrated into a national **state**.

Functional region A **region** marked less by its sameness than its dynamic internal structure; because it usually focuses on a central node, also called *nodal* or *focal* region.

Functional specialization The production of particular goods or services as a dominant activity in a particular location.

Geographic realm The basic **spatial** unit in our world regionalization scheme. Each realm is defined in terms of a synthesis of its total human geography—a composite of its leading cultural, economic, historical, political, and appropriate environmental features.

Geometric boundaries Political boundaries **defined** and **delimited** (and occasionally **demarcated**) as straight lines or arcs.

Geomorphology The geographic study of the configuration of the Earth's solid surface—the world's landscapes and their constituent landforms.

Ghetto An intraurban region marked by a particular ethnic character. Often an inner-city poverty zone, such as the black ghetto in the large central city of the United States. Ghetto residents are involuntarily segregated from other income and racial groups.

Glaciation See **Pleistocene Epoch**.

Globalization The gradual reduction of regional contrasts at the world scale, resulting from increasing international cultural, economic, and political exchanges.

Green Revolution The successful recent development of higher yield, fast-growing varieties of rice and other cereals in certain developing countries. This led to increased production per unit area and a temporary narrowing of the gap between population growth and food needs; today, unfortunately, that gap is widening again.

Greenhouse effect The widely used analogy describing the blanket-like effect of the **atmosphere** in the heating of the Earth's surface. The sun's high-energy, incoming radiation passes through the "glass" of the atmospheric "greenhouse" and heats the surface. The heated surface subsequently emits its own, lower-energy radiation, but it is not strong enough to penetrate the "glass" and escape back into outer space; this heat radiated by the warmed surface thus becomes trapped, raising the temperature inside the "greenhouse."

Gross national product (GNP) The total value of all goods and services produced in a country during a given year.

Growing season The number of days between the last frost in the spring and the first frost of the fall.

Growth pole An urban center with certain attributes that, if augmented by a measure of investment support, will stimulate regional economic development in its **hinterland.**

Hacienda Literally, a large estate in a Spanish-speaking country. Sometimes equated with the **plantation**, but there are important differences between these two types of agricultural enterprise (see pp. 208–210).

Heartland theory The hypothesis, proposed by British geographer Halford Mackinder during the first two decades of this century, that any political power based in the heart of Eurasia could gain sufficient strength to eventually dominate the world. Further, since Eastern Europe controlled access to the Eurasian interior, its ruler would command the vast "heartland" to the east.

Hegemony The political dominance of a country (or even a region) by another country. The former Soviet Union's postwar grip on Eastern Europe, which lasted from 1945 to 1990, was a classic example.

Helada See *tierra helada*.

Hierarchical diffusion A form of **diffusion** in which an idea or innovation spreads by trickling down from larger to smaller adoption units. An urban **hierarchy** is usually involved, encouraging the leapfrogging of innovations over wide areas, with geographic distance a less important influence.

Hierarchy An order or gradation of phenomena, with each level or rank subordinate to the one above it and superior to the one below. The levels in a national urban hierarchy are constituted by hamlets, villages, towns, cities, and (frequently) the **primate city**.

High seas Areas of the oceans away from land, beyond national jurisdiction, open and free for all to use.

Highveld A term used in Southern Africa to identify the high, grass-covered plateau that dominates much of the region. Veld means "grassland" in Dutch and Afrikaans. The lowest-lying areas in South Africa are called *lowveld*; areas that lie at intermediate elevations are the *middleveld*.

Hinterland Literally, "country behind," a term that applies to a surrounding area served by an urban center. That center is the focus of goods and services produced for its hinterland and is its dominant urban influence as well. In the case of a port city, the hinterland also includes the inland area whose trade flows through that port.

Humus Dark-colored upper layer of a soil that consists of decomposed and decaying organic matter such as leaves and branches, nutrient-rich and giving the soil a high fertility.

Hydraulic civilization theory The theory that cities able to control irrigated farming over large hinterlands held political power over other cities. Particularly applies to early Asian civilizations based in such river valleys as the Chang, the Ganges, and Indus, and those of Mesopotamia.

Hydrologic cycle The system of exchange involving water in its various forms as it continually circulates among the **atmosphere**, the oceans, and above and below the land surface.

Iconography The identity of a region as expressed through its cherished symbols; its particular **cultural landscape** and personality.

Immigrant A person **migrating** into a particular country or area; an in-migrant.

Imperialism The drive toward the creation and expansion of a colonial empire and, once established, its perpetuation.

Indentured workers Contract laborers who sell their services for a stipulated period of time.

Indochina The eastern component of mainland Southeast Asia, ruled by France during the colonial era; constituted by the countries of Vietnam, Laos, and Cambodia.

Infrastructure The foundations of a society: urban centers, transport networks, communications, energy distribution systems, farms, factories, mines, and such facilities as schools, hospitals, postal services, and police and armed forces.

Insular Having the qualities and properties of an island. Real islands are not alone in possessing such properties of **isolation**: an oasis in the middle of a **desert** also has qualities of insularity.

Insurgent state Territorial embodiment of a successful guerrilla movement. The establishment by anti-government insurgents of a territorial base in which they exercise full control; thus a state within a **state**.

Intermontane Literally, between mountains. The location can bestow certain qualities of natural protection or **isolation** to a community.

Internal migration Migration flow within a **nation-state**, such as ongoing westward and southward movements in the United States.

International migration Migration flow involving movement across international boundaries.

Intervening opportunity The presence of a nearer opportunity that greatly diminishes the attractiveness of sites farther away.

Irredentism A policy of cultural extension and potential political expansion aimed at a national group living in a neighboring country.

Irrigation The artificial watering of croplands. In Egypt's Nile valley, *basin irrigation* is an ancient method that involved the use of floodwaters that were trapped in basins on the floodplain and released in stages to augment rainfall. Today's *perennial irrigation* requires the construction of dams and irrigation canals for year-round water supply.

Isobar A line connecting points of equal atmospheric pressure.

Isohyet A line connecting points of equal rainfall total.

Isolation The condition of being geographically cut off or far removed from mainstreams of thought and action. It also denotes a lack of receptivity to outside influences, caused at least partially by inaccessibility.

Isoline A line connecting points of equal value.

Isotherm A line connecting points of equal temperature.

Isthmus A **land bridge**; a comparatively narrow link between larger bodies of land. Central America forms such a link between North and South America.

Jakota Triangle The easternmost region of the East Asian realm, consisting of *Ja*pan, South *Ko*rea, and *Tai*wan. Anchored by Japan, this is the first full-scale region generated by **Pacific Rim** economic development that has transformed community and society.

Juxtaposition Contrasting places in close proximity to one another.

Land alienation One society or culture group taking land from another. In Subsaharan Africa, for example, European colonialists took land from indigenous Africans and put it to new uses, fencing it off and restricting settlement.

Land bridge A narrow **isthmian** link between two large landmasses. They are temporary features—at least in terms of geologic time—subject to appearance and disappearance as the land or sea-level rises and falls.

Landlocked An interior country or **state** that is surrounded by land. Without coasts, a landlocked state is at a disadvantage in a number of ways—in terms of **accessibility** to international trade routes, and in the scramble for possession of areas of the **continental shelf** and control of the **exclusive economic zone** beyond.

Land reform The spatial reorganization of **agriculture** through the allocation of farmland (often expropriated from landlords) to **peasants** and tenants who never owned land. Also, the consolidation of excessively fragmented farmland into more productive, perhaps cooperatively-run farm units.

Late Cenozoic Ice Age The latest in a series of ice ages that mark the Earth's environmental history, lasting from ca. 3.5 million to 10,000 years ago. As many as 24 separate advances and retreats of continental icesheets took place, the height of activity occurring during the **Pleistocene Epoch** (ca. 2 million to 10,000 years ago).

Latitude Lines of latitude are **parallels** that are aligned east-west across the globe, from 0° latitude at the equator to 90° north and south latitude at the poles. Areas of low latitude, therefore, lie near the equator in the tropics; high latitudes are those in the north polar (Arctic) and south polar (Antarctic) regions.

Leached soil Infertile, reddish-appearing, tropical soil whose surface consists of oxides of iron and aluminum; all other soil nutrients have been dissolved and transported downward by percolating water associated with heavy rainfall.

Leeward The protected or downwind side of a topographic barrier with respect to the winds that flow across it.

Lignite Also called brown coal, a low-grade variety of coal somewhat higher in fuel content than peat but not nearly as good as the next higher grade, **bituminous coal**. Lignite cannot be used for most industrial processes, but it is important as a residential fuel in certain parts of the world.

Lingua franca The term derives from "Frankish language," and applied to a tongue spoken in ancient Mediterranean ports that consisted of a mixture of Italian, French, Greek, Spanish, and even some Arabic. Today it refers to a "common language," a second language that can be spoken and understood by many peoples, although they speak other languages at home.

Littoral Coastal; along the shore.

Llanos Name given to the **savanna**-like grasslands of the Orinoco River's wide basin in the interior parts of Colombia and Venezuela.

Location theory A logical attempt to explain the locational pattern of an economic activity and the manner in which its producing areas are interrelated. The agricultural location theory contained in the **von Thünen model** is a leading example.

Loess Deposit of very fine silt or dust that is laid down after having been windborne for a considerable distance. Loess is notable for its fertility under **irrigation** and its ability to stand in steep vertical walls when **eroded** by a river or (as in China's Loess Plateau) excavated for cave-type human dwellings.

Longitude Angular distance (0° to 180°) east or west as measured from the prime **meridian** (0°) that passes through the Greenwich Observatory in suburban London, England. For much of its length in the mid-Pacific Ocean, the 180th meridian functions as the *international date line.*

Maghreb The region lying in the northwestern corner of Africa, consisting of the countries of Morocco, Algeria, and Tunisia.

Main Street Canada's dominant **conurbation** that is home to more than 60 percent of the country's inhabitants; stretches southwestward from Quebec City in the middle of the St. Lawrence valley to Windsor on the Detroit River.

Mainland-Rimland framework Twofold regionalization of the Middle American realm based on its modern cultural history. The Euro-Amerindian *Mainland*, stretching from Mexico to Panama (except for the Caribbean coastal strip), was a self-sufficient zone dominated by **hacienda** land tenure. The Euro-African *Rimland*, consisting of that Caribbean coastal zone plus all of the Caribbean islands to the east, was the zone of the **plantation** that heavily relied on trade with Europe and North America.

Maquiladora The term given to modern industrial plants in Mexico's northern (U.S.) border zone. These foreign-owned factories assemble imported components and/or raw materials, and then export finished manufactures, mainly to the United States. Most import duties are minimized, bringing jobs to Mexico and the advantages of low wage rates to the foreign entrepreneurs.

Marchland A **frontier** or area of uncertain boundaries that is subject to various national claims and an unstable political history. The term refers to the movement of various national armies across such zones.

Maritime boundary An international boundary that lies in the ocean. Like all boundaries, it is a vertical plane, extending from the seafloor to the upper limit of the air space in the atmosphere above the water.

Median line boundary An international **maritime boundary** drawn where the width of a sea is less than 400 nautical miles.

Because the **states** on either side of that sea claim **exclusive economic zones** of 200 miles, it is necessary to reduce those claims to a (median) distance equidistant from each shoreline. **Delimitation** on the map almost always appears as a set of straight-line segments that reflect the configurations of the coastlines involved.

Megacity A term increasingly used by urban geographers and planners to refer to the world's most heavily populated cities. The term is still an informal one, and there is no generally accepted minimal size; in this book, the term refers to a **metropolis** containing a population of more than 5 million.

Megalopolis Term used to designate large coalescing supercities that are forming in diverse parts of the world; used specifically to refer to the Boston-Washington multi-metropolitan corridor on the northeastern seaboard of the United States, but the term is now used generically with a lower-case *m* as a synonym for **conurbation**.

Mercantilism Protectionist policy of European **states** during the sixteenth to the eighteenth centuries that promoted a state's economic position in the contest with other countries. The acquisition of gold and silver and the maintenance of a favorable trade balance (more exports than imports) were central to the policy.

Meridian Line of **longitude**, aligned north-south across the globe, that together with **parallels** of **latitude** forms the global grid system. All meridians converge at both poles and are at their maximum distances from each other at the equator.

Mestizo The root of this word is the Latin for *mixed*; it means a person of mixed white and Amerindian ancestry.

Metropolis Urban agglomeration consisting of a (central) city and its suburban ring. See **urban (metropolitan) area**.

Metropolitan area See **urban (metropolitan) area**.

Migration A change in residence intended to be permanent. See also **forced**, **internal**, **international**, and **voluntary migration**.

Migratory movement Human relocation movement from a source to a destination without a return journey, as opposed to **cyclical movement**.

Model An idealized representation of reality built to demonstrate its most important properties. A **spatial** model focuses on a geographical dimension of the real world, such as the **von Thünen model** that explains agricultural location patterns in a commercial, profit-making economy.

Monotheism The belief in, and worship of, a single god.

Monsoon Refers to the seasonal reversal of wind (and moisture) flows in certain parts of the subtropics and lower-middle latitudes. The *dry monsoon* occurs during the cool season when dry offshore winds prevail. The *wet monsoon* occurs in the hot summer months, which produce onshore winds that bring large amounts of rainfall. The air-pressure differential over land and sea is the triggering mechanism, with windflows always moving from areas of relatively higher pressure toward areas of relatively lower pressure. Monsoons make their greatest regional impact in certain coastal and near-coastal zones of South Asia, Southeast Asia, and East Asia.

Mulatto A person of mixed African (black) and European (white) ancestry.

Multinationals Internationally active corporations that can strongly influence the economic and political affairs of many countries they operate in.

Nation Legally a term encompassing all the citizens of a **state**, it also has other connotations. Most definitions now tend to refer to a group of tightly-knit people possessing bonds of language, ethnicity, religion, and other shared cultural attributes. Such homogeneity actually prevails within very few states.

Nation-state A country whose population possesses a substantial degree of cultural homogeneity and unity. The ideal form to which most **nations** and **states** aspire—a political unit wherein the territorial state coincides with the area settled by a certain national group or people.

Natural hazard A natural event that endangers human life and/ or the contents of a **cultural landscape**.

Natural increase rate Population growth measured as the excess of live births over deaths per 1000 individuals per year. Natural increase of a population does not reflect either **emigrant** or **immigrant** movements.

Natural resource Any valued element of (or means to an end using) the environment; includes minerals, water, vegetation, and soil.

Nautical mile By international agreement, the nautical mile—the standard measure at sea—is 6076.12 feet in length, equivalent to approximately 1.15 statute miles (1.85 km).

Near Abroad Term used by Russians to refer to the 14 former Soviet Socialist Republics (now independent) that opted to detach themselves from Russia following the 1991 disintegration of the Soviet Union. The current names of these former S.S.R.s are: Estonia, Latvia, Lithuania, Belarus, Ukraine, Moldova, Georgia, Azerbaijan, Armenia, Kazakstan, Uzbekistan, Turkmenistan, Kyrgyzstan, and Tajikistan.

Network (transport) The entire regional system of transportation connections and nodes through which movement can occur.

Nomadism Cyclical movement among a definite set of places. Nomadic peoples mostly are **pastoralists**.

Norden A regional appellation for the Northern European countries of Denmark, Norway, Sweden, Finland, Estonia, and Iceland.

Nucleated settlement See **agglomerated settlement**.

Nucleation Cluster; **agglomeration**.

Oasis An area, small or large, where the supply of water permits the transformation of the surrounding desert into a green cropland; the most important focus of human activity for miles around. It may also encompass a densely populated **corridor** along a major river where **irrigation** projects stabilize the water supply—as along the Nile in Egypt, which can be viewed as an elongated chain of oases.

Occidental Western. See **Oriental**.

Oligarchy Political system involving rule by a small minority, an often corrupt elite.

OPEC The Organization of Petroleum Exporting Countries, the international oil **cartel**. Twelve member-states (as of 1996) are: Algeria, Gabon, Indonesia, Iran, Iraq, Kuwait, Libya, Nigeria, Qatar, Saudi Arabia, United Arab Emirates (UAE), and Venezuela.

Oriental The root of the word oriental is from the Latin for *rise*. Thus it has to do with the direction in which one sees the sun ''rise''—the east; oriental therefore means Eastern. **Occidental** originates from the Latin for *fall*, or the ''setting'' of the sun in the west; occidental therefore means Western.

Orographic precipitation Mountain-induced precipitation, especially where air masses are forced over topographic barriers. Downwind areas beyond such a mountain range experience the relative dryness known as the **rain shadow effect**.

Pacific Rim A far-flung group of countries and parts of countries (extending clockwise on the map from New Zealand to Chile) sharing the following criteria: they face the Pacific Ocean; they evince relatively high levels of economic development, industrialization, and urbanization; their imports and exports mainly move across Pacific waters.

Pacific Ring of Fire Zone of crustal instability along tectonic **plate** boundaries, marked by earthquakes and volcanic activity, that rings the Pacific Ocean basin.

Paddies (paddyfields) Ricefields.

Pandemic An outbreak of a disease that spreads worldwide.

Pangaea A vast, singular landmass consisting of most of the areas of the present continents (including the Americas, Eurasia, Africa, Australia, and Antarctica), which existed until near the end of the Mesozoic era when **plate** divergence and **continental drift** broke it apart. The ''northern'' segment of this Pangaean supercontinent is called *Laurasia*, the ''southern'' part *Gondwana* (see Fig. 7-2).

Parallel An east-west line of **latitude** that is intersected at right angles by **meridians** of **longitude**.

Páramos See *puna*.

Pastoralism A form of **agricultural** activity that involves the raising of livestock. Many peoples described as herders actually pursue mixed agriculture, in that they may also fish, hunt, or even grow a few crops. But pastoral peoples' lives do revolve around their animals.

Peasants In a **stratified** society, peasants are the lowest class of people who depend on **agriculture** for a living. But they often own no land at all and must survive as tenants or day workers.

Peninsula A comparatively narrow, finger-like stretch of land extending from the main landmass into the sea. Florida and Korea are examples.

Peon (*peone*) Term used in Middle and South America to identify people who often live in serfdom to a wealthy landowner; landless **peasants** in continuous indebtedness.

Per capita Capita means *individual*. Income, production, or some other measure is often given per individual.

Perforated state A **state** whose territory completely surrounds that of another state. South Africa, which encloses Lesotho and is perforated by it, is a classic example.

Periodic market Village market that opens every third or fourth day or at some other regular interval. Part of a regional network of similar markets in a preindustrial, rural setting where goods are brought to market on foot (or perhaps by bicycle) and barter remains a major mode of exchange.

Permafrost Permanently frozen water in the soil and bedrock (of cold environments), as much as 1000 feet (300 m) in depth, producing the effect of completely frozen ground. Can thaw near surface during brief warm season.

Physiographic political boundaries Political boundaries that coincide with prominent physical features in the natural landscape—such as rivers or the crest ridges of mountain ranges.

Physiographic region (province) A **region** within which there prevails substantial natural-landscape homogeneity, expressed by a certain degree of uniformity in surface **relief**, **climate**, vegetation, and soils.

Physiography Literally means *landscape description*, but commonly refers to the total physical geography of a place; includes all of the natural features on the Earth's surface, including landforms, **climate**, soils, vegetation, and hydrography.

Physiologic density The number of people per unit area of **arable** land.

Pidgin A language that consists of words borrowed and adapted from other languages. Originally developed from commerce among peoples speaking different languages.

Pilgrimage A journey to a place of great religious significance by an individual or by a group of people (such as a pilgrimage to Mecca for Muslims).

Plantation A large estate owned by an individual, family, or corporation and organized to produce a cash crop. Almost all plantations were established within the tropics; in recent decades, many have been divided into smaller holdings or reorganized as cooperatives.

Plate tectonics Plates are bonded portions of the Earth's mantle and crust, averaging 60 miles (100 km) in thickness. More than a dozen such plates exist (see Fig. I-3), most of continental proportions, and they are in motion. Where they meet one slides under the other, crumpling the surface crust and producing significant volcanic and earthquake activity. A major mountain-building force.

Pleistocene Epoch Recent period of geologic time that spans the rise of humankind, beginning about 2 million years ago. Marked by glaciations (repeated advances of continental icesheets) and milder interglaciations (icesheet retreats). Although the last 10,000 years are known as the *Recent Epoch*, Pleistocene-like conditions seem to be continuing and the present is probably another Pleistocene interglaciation; the glaciers likely will return. See also **Late Cenozoic Ice Age**.

Plural society See **cultural pluralism**.

Polder Land reclaimed from the sea adjacent to shore by constructing dikes and pumping out the water. Technique used widely along the coast of the Netherlands since the Middle Ages to add badly needed additional space for settlements and farms.

Pollution The release of a substance, through human activity, which chemically, physically, or biologically alters the air or water it is discharged into. Such a discharge negatively impacts the environment, with possible harmful effects on living organisms—including humans.

Population (age-sex) structure Graphic representation (profile) of a population according to age and sex.

Population density See **density of population**.

Population distribution The way people have arranged themselves in geographic space. One of human geography's most essential expressions because it represents the sum total of the adjustments that a population has made to its natural, cultural, and economic environments.

Population explosion The rapid growth of the world's human population during the past century, attended by ever-shorter **doubling times** and accelerating *rates* of increase.

Population movement See **migration**; **migratory movement**.

Postindustrial economy Emerging economy, in the United States and a handful of other highly advanced countries, as traditional industry is overshadowed by a higher-technology productive complex dominated by services, information-related, and managerial activities.

Primary economic activity Activities engaged in the direct extraction of **natural resources** from the environment such as mining, fishing, lumbering, and especially **agriculture**.

Primate city A country's largest city—ranking atop the urban **hierarchy**—most expressive of the national culture and usually (but not always) the capital city as well.

Process Causal force that shapes a **spatial** pattern as it unfolds over time.

Protectorate In Britain's system of colonial administration, the protectorate was a designation that involved the guarantee of certain rights (such as the restriction of European settlement and **land alienation**) to peoples who had been placed under the control of the Crown.

Protruded state Territorial shape of a **state** that exhibits a narrow, elongated land extension (or *protrusion*) leading away from the main body of territory. Thailand is a leading example.

Puna In Andean South America, the highest-lying habitable **altitudinal zone**—ca. 12,000 to 15,000 feet (3,600 to 4,500 m)—between the tree line (upper limit of the *tierra fría*) and the snow line (lower limit of the *tierra helada*). Too cold and barren to support anything but the grazing of sheep and other hardy livestock. Also known as the *páramos*.

Push-pull concept The idea that **migration** flows are simultaneously stimulated by conditions in the source area, which tend to drive people away, and by the perceived attractiveness of the destination.

Qanat In **desert** zones, particularly in Iran and western China, an underground tunnel built to carry **irrigation** water by gravity flow from nearby mountains (where **orographic precipitation** occurs) to the arid flatlands below.

Quaternary economic activity Activities engaged in the collection, processing, and manipulation of *information*.

Quinary economic activity Managerial or control-function activity associated with decision-making in a large organizations.

Rain shadow effect The relative dryness in areas downwind of mountain ranges caused by **orographic precipitation**, wherein moist air masses are forced to deposit most of their water content in the highlands.

Realm See **geographic realm**.

Region A commonly used term and a geographic concept of central importance. An **area** on the Earth's surface marked by certain properties.

Regional complementarity See **complementarity**.

Relative location The regional position or **situation** of a place relative to the position of other places. Distance, **accessibility**, and connectivity affect relative location.

Relict boundary A political boundary that has ceased to function, but the imprint of which can still be detected on the **cultural landscape**.

Relief Vertical difference between the highest and lowest elevations within a particular area.

Relocation diffusion Sequential **diffusion process** in which the items being diffused are transmitted by their carrier agents as they evacuate the old areas and relocate to new ones. The most common form of relocation diffusion involves the spreading of innovations by a **migrating** population.

Rural density A measure that indicates the number of persons per unit area living in the rural areas of a country, outside of the urban concentrations.

Sahel Semiarid zone extending across most of Africa between the southern margins of the arid Sahara and the moister tropical savanna and forest zone to the south. Chronic drought, **desertification**, and overgrazing have contributed to severe famines in this area since 1970.

Savanna Tropical grassland containing widely spaced trees; also the name given to the tropical wet-and-dry climate (*Aw*).

Scale Representation of a real0world phenomenon at a certain level of reduction or generalization. In **cartography**, the ratio of map distance to ground distance; indicated on a map as bar graph, representative fraction, and/or verbal statement.

Scale economies See **economies of scale**.

Secondary economic activity Activities that process raw materials and transform them into finished industrial products. The *manufacturing* sector.

Sedentary Permanently attached to a particular area; a population fixed in its location. The opposite of **nomadic**.

Selva Tropical rainforest.

Separate development The spatial expression of South Africa's ''grand'' **apartheid** scheme, wherein nonwhite groups were required to settle in segregated ''homelands.'' During the 1980s, the implementation of this plan collapsed, and its dismantling—together with the system of apartheid—is now well underway.

Sequent occupance The notion that successive societies leave their cultural imprints on a place, each contributing to the cumulative **cultural landscape**.

Shantytown Unplanned slum development on the margins of cities in the developing realms, dominated by crude dwellings and shelters mostly made of scrap wood, iron, and even pieces of cardboard.

Sharecropping Relationship between a large landowner and farmers on the land wherein the farmers pay rent for the land they farm by giving the landlord a share of the annual harvest.

Shatter belt Region caught between stronger, colliding external cultural-political forces, under persistent stress and often fragmented by aggressive rivals. Eastern Europe and Southeast Asia are classic examples.

Shield A large, stable, relatively flat expanse of very old rocks that forms the geologic core of a continental landmass; North America's Canadian Shield is an example.

Shifting agriculture Cultivation of crops in recently cut and burned tropical forest clearings, soon to be abandoned in favor of newly cleared nearby forest land. Also known as *slash-and-burn agriculture*.

Sinicization Giving a Chinese cultural imprint; Chinese **acculturation**.

Site The internal locational attributes of an urban center, including its local spatial organization and physical setting.

Situation The external locational attributes of an urban center; its **relative location** or regional position with reference to other nonlocal places.

Slash-and-burn agriculture See **shifting agriculture**.

Southern Cone The southern, mid-latitude portion of South America constituted by the countries of Chile, Argentina, and Uruguay; often included as well is the southernmost part of Brazil, south of the Tropic of Capricorn (23½° S).

Spatial Pertaining to space on the Earth's surface. Synonym for *geographic(al)*.

Spatial diffusion See **diffusion**.

Spatial interaction See **complementarity**, **transferability**, and **intervening opportunity**.

Spatial model See **model**.

Spatial morphology See **territorial morphology**.

Spatial process See **process**.

Special Economic Zone (SEZ) Manufacturing and export center within China, created in the 1980s to attract foreign investment and technology transfers. Despite special tax benefits and other economic incentives, the response from abroad (except from Hong Kong) was initially limited. Five SEZs—all located on southern China's Pacific coast—currently operate: Shenzhen (by far the most successful), Zhuhai, Shantou, Xiamen, and Hainan Island in the far south.

Squatter settlement See **shantytown**.

State A politically organized territory that is administered by a sovereign government and is recognized by a significant portion of the international community. A state must also contain a permanent resident population, an organized economy, and a functioning internal circulation system.

State capitalism Government-controlled corporations competing under free market conditions, usually in a tightly regimented society. South Korea is a leading example.

Steppe Semiarid grassland; short-grass prairie. Also, the name given to the semiarid climate type (*BS*).

Stratification (social) In a layered or stratified society, the population is divided into a **hierarchy** of social classes. In an industrialized society, the working class is at the lower end; **elites** that possess capital and control the means of production are at the upper level. In the traditional **caste system** of Hindu India, the ''untouchables'' form the lowest class or caste, whereas the still-wealthy remnants of the princely class are at the top.

Subduction In **plate tectonics**, the **process** that occurs when an oceanic plate converges head-on with a plate carrying a continental landmass at its leading edge. The lighter continental plate overrides the denser oceanic plate and pushes it downward.

Subsequent boundary A political boundary that developed contemporaneously with the evolution of the major elements of the **cultural landscape** through which it passes.

Subsistence Existing on the minimum necessities to sustain life; spending most of one's time in pursuit of survival.

Suburban downtown In the United States (and increasingly in other highly advantaged economies), a significant concentration of diversified economic activities around a highly **accessible** suburban location, including retailing, light industry, and a variety of major corporate and commercial operations. Late-twentieth-century coequal to the American central city's **central business district (CBD)**.

Superimposed boundary A political boundary emplaced by powerful outsiders on a developed human landscape. Usually ignores pre-existing cultural-spatial patterns, such as the border that now divides North and South Korea.

Supranational A venture involving three or more national **states**—political, economic, and/or cultural cooperation to promote shared objectives. The **European Union** is one such organization.

System Any group of objects or institutions and their mutual interactions. Geography treats systems that are expressed **spatially**, such as **regions**.

Systematic geography Topical geography: cultural, political, economic geography, and the like.

Takeoff Economic concept to identify a stage in a country's **development** when conditions are set for a domestic Industrial Revolution, which occurred in Britain in the late eighteenth century

and in Japan in the late nineteenth following the Meiji Restoration.

Tectonics See **plate tectonics**.

Tell The lower slopes and narrow coastal plains along the Atlas Mountains in northwesternmost Africa, where the majority of the **Maghreb** region's population is clustered.

Templada See *tierra templada*.

Terracing The transformation of a hillside or mountain slope into a step-like sequence of horizontal fields for intensive cultivation.

Territoriality A country's or more local community's sense of property and attachment toward its territory, as expressed by its determination to keep it inviolable and strongly defended.

Territorial morphology A **state's** geographical shape, which can have a decisive impact on its spatial cohesion and political viability. A **compact** shape is most desirable; among the less efficient shapes are those exhibited by **elongated, fragmented, perforated**, and **protruded** states.

Territorial sea Zone of seawater adjacent to a country's coast, held to be part of the national territory and treated as a segment of the sovereign **state**.

Tertiary economic activity Activities that engage in *services*—such as transportation, banking, retailing, education, and routine office-based jobs.

Theocracy A **state** whose government is under the control of a ruler who is deemed to be divinely guided or under the control of a group of religious leaders, as in post-Khomeini Iran. The opposite of the theocratic state is the secular state.

Tierra caliente The lowest of the **altitudinal zones** into which the human settlement of Middle and South America is classified according to elevation. The *caliente* is the hot humid coastal plain and adjacent slopes up to 2,500 feet (750 m) above sea level. The natural vegetation is the dense and luxuriant tropical rainforest; the crops are sugar, bananas, cacao, and rice in the lower areas, and coffee, tobacco, and corn along the higher slopes.

Tierra fría The cold, high-lying **altitudinal zone** of settlement in Andean South America, extending from about 6,000 feet (1850 m) in elevation up to nearly 12,000 feet (3600 m). **Coniferous** trees stand here; upward they change into scrub and grassland. There are also important pastures within the *fría*, and wheat can be cultivated. Several major population clusters in western South America lie at these altitudes.

Tierra helada The highest and coldest **altitudinal zone** in Andean South America (lying above 15,000 feet [4,500 m]), an uninhabitable environment of permanent snow and ice that extends upward to the Andes' highest peaks of more than 20,000 feet (6,000 m).

Tierra templada The intermediate **altitudinal zone** of settlement in Middle and South America, lying between 2,500 feet (750 m) and 6,000 feet (1850 m) in elevation. This is the ''temperate'' zone, with moderate temperatures compared to the *tierra caliente* below. Crops include coffee, tobacco, corn, and some wheat.

Time-space convergence The increasing nearness of places that occurs as modern transportation breakthroughs progressively reduce the time-distance between them. The trip by boat from New York to San Francisco before the Civil War took months. After 1870, the transcontinental railroad cut the travel time to less than two weeks; by 1930, trains made the journey in three days. After 1945, propeller planes made the trip in about 12 hours; and by 1960, non-stop jet planes achieved today's travel time of five hours.

Toponym Place name.

Totalitarian A government whose leaders rule by absolute control, tolerating no differences of political opinion.

Transculturation Cultural borrowing that occurs when different **cultures** of approximately equal complexity and technological level come into close contact. In **acculturation**, by contrast, an indigenous society's culture is modified by contact with a technologically superior society.

Transferability The capacity to move a good from one place to another at a bearable cost; the ease with which a commodity may be transported.

Transition zone An area of **spatial** change where the peripheries of two adjacent realms or regions join; marked by a gradual shift (rather than a sharp break) in the characteristics that distinguish these neighboring geographic entities from one another.

Tropical savanna See **savanna**.

Tsunami A seismic (earthquake-generated) sea wave that can attain gigantic proportions and cause coastal devastation.

Turkestan Northeasternmost region of the North Africa/Southwest Asia realm. Known as Soviet Central Asia before 1991, its five (dominantly Islamic) former S.S.R.s have become the independent countries of Kazakstan, Uzbekistan, Turkmenistan, Kyrgyzstan, and Tajikistan.

Unitary state A **nation-state** that has a centralized government and administration that exercises power equally over all parts of the state.

Urbanization A term with several connotations. The proportion of a country's population living in urban places is its level of urbanization. The **process** of urbanization involves the movement to, and the clustering of, people in towns and cities—a major force in every geographic realm today. Another kind of urbanization occurs when an expanding city absorbs rural countryside and transforms it into suburbs; in the case of cities in disadvantaged countries, this also generates peripheral **shantytowns**.

Urban (metropolitan) area The entire built-up, non-rural area and its population, including the most recently constructed suburban appendages. Provides a better picture of the dimensions and population of such an area than the delimited municipality (central city) that forms its heart.

Urban realms model A spatial generalization of the large, late-twentieth-century city in the United States. It is shown to be a widely dispersed, multi-centered metropolis consisting of increasingly independent zones or *realms*, each focused on its own **suburban downtown**; the only exception is the shrunken central realm, which is focused on the **central business district** (see Figs. 3-9 and 3-10).

Veld Open grassland on the South African plateau, becoming mixed with scrub at lower elevations where it is called *bushveld*. As in Middle and South America, there is an **altitudinal zonation** into **highveld**, *middleveld*, and *lowveld*.

Voluntary migration Population movement in which people relocate in response to perceived opportunity, not because they are forced to move.

Von Thünen model Explains the location of agricultural activities in a commercial, profit-making economy. A **process** of spatial competition allocates various farming activities into concentric rings around a central market city, with profit-earning capability the determining force in how far a crop locates from the market. The original (1826) Isolated State model now applies to the continental scale (see Fig. 1-4).

Watershed Drainage basin occupied by a complete river system, constituted by the main river and its entire set of tributary streams and rivers.

Water table When precipitation falls on the soil, some of the water is drawn downward through the pores in the soil and rock under the force of gravity. Below the surface it reaches a level where it can go no further; there it joins water that already saturates the rock completely. This water that "stands" underground is *groundwater*, and the upper level of the zone of saturation is the *water table*.

Weather The immediate and short-term conditions of the **atmosphere** that impinge on daily human activities.

Windward The exposed, upwind side of a topographic barrier that faces the winds that flow across it.

World geographic realm See **geographic realm**.

List of Maps and Figures

List of Photo Credits

Introduction

Page 3: © Alex Quesada/Matrix. **Page 6:** © Craig Aurness/West Light. **Pages 12–37:** H. J. de Blij.

Chapter 1

Page 48: © Tibor Bognar/The Stock Market. **Page 49:** H. J. de Blij. **Page 51:** © Leo de Wys, Inc. **Page 59:** © DeRichemond/The Image Works. **Page 61:** © Stephane Compoint/Sygma. **Pages 67–84:** H. J. de Blij. **Page 88 top:** H. J. de Blij. **Page 88 bottom:** © Peter Carmichael/Aspect Picture Library. **Pages 91–98:** H. J. de Blij. **Page 102:** © Tomislav Peternek/Sygma. **Page 105:** © Amanda Merullo/Stock, Boston.

Chapter 2

Page 113: H. J. de Blij. **Page 124:** © Sovfoto/Eastfoto. **Page 137:** © Jeremy Nicholl/Matrix. **Page 138:** © Sygma. **Page 143:** H. J. de Blij. **Page 144:** © Lada Cais/Sovfoto/Eastfoto. **Page 147:** © EPIX/Sygma. **Page 148:** © Peter Blakely/SABA. **Page 150:** H. J. de Blij.

Chapter 3

Page 158: © Dembinsky Photo Associates. **Page 161:** H. J. de Blij. **Page 162:** © Kathleen Campbell/Gamma Liaison. **Page 171:** © Uniphoto, Inc. **Page 177:** © Arthur C. Smith/Grant Heilman Photography. **Page 179:** Courtesy The North Star Steel Company. **Pages 184 & 190:** H. J. de Blij. **Page 196:** © John Edwards/Tony Stone Images/New York, Inc.

Chapter 4

Page 203: H. J. de Blij. **Page 204:** © Michael Fogden/Oxford Scientific Films. **Page 210 top:** © J. P. Courau/DDB Stock Photos. **Page 210 bottom:** © Jean-Gerard Sidaner/Photo Researchers. **Page 212:** H. J. de Blij. **Page 213:** © Jim Forbes. **Pages 214 & 219:** H. J. de Blij. **Page 220:** © Alex McLean/Landslides. **Page 222:** © Byron Augustin/Tom Stack & Associates. **Page 228:** © Peter Poulides. **Page 229:** H. J. de Blij.

Chapter 5

Pages 239–248: H. J. de Blij. **Page 253:** © James Brooke/New York Times Pictures. **Page 255:** © Stephanie Maze/Woodfin Camp & Associates. **Pages 258–269:** H. J. de Blij.

Chapter 6

Page 280: Office of National Du Tourisme Tunisien. **Page 283:** © Steve Vidler/Leo de Wys, Inc. **Pages 290–310:** H. J. de Blij. **Page 317:** Comstock, Inc. **Page 320:** © Marcus Rose/Panos Pictures.

Chapter 7

Page 336: © Jose Azel/Aurora. **Page 341:** H. J. de Blij. **Page 345:** © M. & E. Bernheim/Woodfin Camp & Associates. **Pages 350–367:** H. J. de Blij.

Chapter 8

Page 380: © Bettmann Archive/Corbis. **Page 385:** © Dilip Mehta/The Stock Market. **Page 389:** © Porterfield/Chickering/Photo Researchers. **Pages 393 & 398:** H. J. de Blij. **Page 399:** © Jacques Jangoux/Tony Stone Images/New York, Inc. **Page 400:** © Steve McCurry/Magnum Photos, Inc. **Pages 403 & 408:** H. J. de Blij.

Chapter 9

All photos by H. J. de Blij.

Chapter 10

Pages 480–498: H. J. de Blij. **Page 500:** J. Apicella/Cesar Pelli & Associates. **Pages 502 & 504:** H. J. de Blij. **Page 507:** © Bill Cardoni/Bruce Coleman, Inc.

Chapter 11

All photos by H. J. de Blij.

Chapter 12

All photos by H. J. de Blij.

Geographical Index (Gazetteer)

Index